Petrography to Petrogenesis

Petrography to Petrogenesis

M. J. Hibbard

Department of Geological Sciences
Mackay School of Mines
University of Nevada, Reno

Prentice Hall, Englewood Cliffs, New Jersey 07632

Library of Congress Cataloging-in-Publication Data

Hibbard, M. J.
    Petrography to petrogenesis / M. J. Hibbard.
      p.   cm.
    Includes index.
    ISBN 0-02-354145-8
    1. Petrology.   2. Petrogenesis.   I. Title.
  QE431.2.H53   1995
  552'.03—dc20   93-23101
                                                 CIP

Editor: *Robert McConnin*
Production Supervisor: *Helen Wallace*
Production Manager: *Francesca Drago*
Text Designer: *Eileen Burke*
Cover Designer: *Cathleen Carbery*
Cover Illustration: *M. J. Hibbard*

The cover photo shows a tapered thin section of sodian augite of a limburgite from Thetford, Vermont. Interference colors ranging from first-order yellow (upper left) to third-order green (lower right) are maintained in epitaxic non-sodian augite formed by incongruent direct melting along fractures.

© 1995 by Prentice-Hall, Inc.
A Division of Simon & Schuster, Inc.
Englewood Cliffs, New Jersey 07632

All rights reserved. No part of this book may be
reproduced, in any form or by any means,
without permission in writing from the publisher.

Printed in the United States of America

10 9 8 7 6 5 4 3 2 1

ISBN 0-02-354145-8

Prentice-Hall International (UK) Limited, *London*
Prentice-Hall of Australia Pty. Limited, *Sydney*
Prentice-Hall Canada Inc., *Toronto*
Prentice-Hall Hispanoamericana, S.A., *Mexico*
Prentice-Hall of India Private Limited, *New Delhi*
Prentice-Hall of Japan, Inc., *Tokyo*
Simon & Schuster Asia Pte. Ltd., *Singapore*
Editora Prentice-Hall do Brasil, Ltda., *Rio de Janeiro*

Preface

Hand specimens and thin sections of rocks contain many clues to the nature of processes responsible for their formation. Optical examination of the mineralogical and physical features of rocks in thin section is **petrography**, and interpretation of these attributes is to model a picture of **petrogenesis**. Whenever other types of data are available, such as tectonic setting, rock and mineral chemistry, and submicrostructural images, they logically become a part of the genetic modeling.

Petrography is complimented nicely by study of hand specimens from which the thin sections are derived, in part because the choice of section direction through the specimen controls what appears in thin section. Outcrops are observed and described in the "field," but only hand specimens end up on the laboratory table, where there is more opportunity for thorough investigation.

The descriptive, data collecting exercise is the "work" part of a scientific investigation. The interpretive stage is the "fun" part because creation of a genetic model generates a sense of accomplishment and allows for a measure of subjectivity. The investigator becomes personally involved, which, of course, is the basis of much healthy professional give-and-take.

The question arises as to how much petrogenesis should be presented in a textbook in view of its inherently controversial nature. A philosophical view is that it is not important to focus on any particular view, popular or unpopular, but rather on the method of genetic modeling that should be a primary concern in education. Thus the process of interpretation can be presented in a textbook in the spirit of example rather than fact. If the method of genetic interpretation is demonstrated, the student has the tools to make his or her own judgments. Nevertheless, instructors are likely to make it more comfortable for students, as they learn to reason geologically, by providing them with tidy rock classifications and genetic models of general acceptance, since, after all, there must be some common ground as a basis of communication. This textbook is full of interpretations of rock-forming processes that will please some and irritate others. With that in mind, there is an attempt here to show the line of thought connecting the data and the interpretive model so that the reader can evaluate better the worth of the interpretation.

At what educational level can the techniques of optical mineralogy and petrography be introduced? If optical mineralogy is treated as a means-to-an-end, as it is in this text, any sophomore in college can be using the petrographic microscope for purposes of mineral identification in eight to ten lectures and five 3-hour laboratories. Optical phenomena such as relative refractive index, extinction positions, and birefringence are quickly grasped by students if the phenomena are imaged with a video camera mounted on a petrographic microscope and projected onto a screen, where the optical phenomena can be manipulated for all to see. Familiarization with the textural attributes of rocks takes about six lectures and three 3-hour laboratory sessions if there is liberal use of line drawings and photomicrographs (or video). This approach leaves approximately a quarter or one half of a semester for a systematic journey through the various rock groups.

For geoscientists in fields such as geological engineering, hydrogeology, and geophysics, this second half of an introductory petrology course can be tailored to be a terminal exercise. For geology majors, the second half of an introductory course constitutes a menu of things to come.

The reference list at the end of the book is extensive, yet many important works necessarily have been omitted. Some older references have been included for classical value and historical perspective. In many cases, a more recent publication on a particular subject has been chosen over more original works on the same subject simply because the newer work cites the earlier works.

I wish to extend my thanks to the following individuals who have read critically parts of the manuscript: Richard Beane, Harold Bonham (Nevada Bureau of Mines & Geology), James Carr (University of Nevada), John Caskey (University of Nevada), Dhanesh Chandra (University of Nevada), Eric H. Christiansen (Brigham Young University), Bevan M. French (NASA), Karen H. Fryer (Ohio Wesley University), Patrick Goldstrand (Oak Ridge National Laboratory), Anthony J. Gow (CRREL Department of the Army), David Green (University of Nevada), Lawrence J. Herber (California State Polytechnic University), Raymond Joesten (University of Connecticut), Jack Kepper, Albert M. Kudo (University of New Mexico), Mary Lahren (University of Nevada), Maureen Leshendok (University of Nevada), John Longshore (Humboldt State University), Tim K. Lowenstein (State University of New York, Binghamton), Thomas Lugaski (University of Nevada), Berry W. Lyons (University of Nevada), Stephen A. Nelson (Tulane University), Stephen A. Norton (University of Maine), Carlos C. Plummer (California State University, Sacramento), Paul H. Reitan (State University of New York, Buffalo), Charles P. Sabine (University of Nevada), Richard A. Schultz (University of Nevada), Miles L. Silberman (U.S. Geological Survey), Joseph Tingley (Nevada Bureau of Mines & Geology), Steven G. Wesnousky (University of Nevada, Reno), and Robert A. Wiebe (Franklin & Marshall College).

My thanks are also extended to those individuals who supplied photos, thin sections, or rocks that have been incorporated in one way or another into the work: Charles R. Bacon, Neely H. Bostick, John Caskey, Henry S. Chafetz, Mario Coniglio, Edward F. Duke, Cynthia Dusel-Bacon, James R. Firby, Robert O. Fournier, Bevan French, Patrick Goldstrand, Anthony J. Gow, Cordell T. Gray, Ray Gutschick, Michael W. Grutzeck, Ezat Heydari, Marty Houhoulis, Martin Jackson, Jack Kepper, Barry Kues, Mary Lahren, Gilbert Lafreniere, Maureen P. Leshendok, Tim K. Lowenstein, John Lyons, Howard MacCarthy, Jr., Robert Meyer, Douglas McLelland, Daniel B. Nahon, Chris Osterberg, Mike Owens, James J. Papike, Tom H. Pearce, Fred Peterson, Jack Quade, Keith Rigby, William I. Rose, Jr., Thea Robbins,

J.M. Rouchy, Charles P. Sabine, Charles A. Sandberg, Richard A. Schweickert, Richard A. Schultz, Norman J. Silberling, Miles L. Silberman, Burt Slemmons, Richard E. Stoiber, Andy Stroud, Robert Symonds, Chris J. Talbot, Joseph A. Vance, Andreas Wetzel, Robert A. Wiebe, Richard Yund, Donald H. Zenger, Mike Zientek, and Don Zigament.

The following individuals are highly skilled photo technicians, sympathetic to the need for preparation of quality photographs: Gordon Allen, Carol Bakken, Ted Cook, Theresa Danner, Keith Davis, Ronald R. Evenson, Lindl Hinkson.

Very personal thanks go to John B. Lyons, Peter Misch (deceased), and Joseph A. Vance for their generous sharing of information and encouragement through the course of my career and the preparation of this work, and gratitude goes to my students, especially Geoffrey Gardner and Maureen Lechendok, for tolerating courses designed to test potential textbook material.

Finally, my thanks go to my understanding wife, Terrie Nault, and daughters Hillary and Anita, and early on, Monique and Dorothea, who each were so tolerant of papers too often spread all over the kitchen table and of screeching halts along the highway to take yet another photo of that lens cap perched on the rock face.

M. J. Hibbard

Brief Contents

1 Introduction   1

Part I Optical Mineralogy   4

2 Orthoscopic Optical Mineralogy   6
3 Conoscopic Optical Mineralogy   37
4 Rock-Forming Minerals   43
5 Plagioclase Composition   76
6 Petrographic Measurements   95

Part II The Anatomy of Rocks   105

7 Physical Attributes of Rocks   107
8 Textural Interpretation   130

Part III Classification of Rocks   141

9 Traditional Rock Classifications   143
10 A Unified Classification of Rocks   156

Part IV Fundamental Processes of Rock Formation   159

11 Classification of Rock-Forming Processes   161
12 Chemical Activity in Aqueous Solutions   164
13 Phase Relations in Crystallizing Magma and Melting Rock   174
14 Crystal Growth and Dissolution in Fluids   190
15 Dynamics of Solids with Gases and Liquids   202
16 Deformation and Recrystallization of Solids   209

Part V Rocks Formed by Magmatic Processes   221

17 Homogeneous Magmatic Rocks   221
18 Cumulate Magmatic Rocks   235
19 Mixed Magmatic Rocks   242
20 Assimilation-Hybrid Magmatic Rocks   261

Part VI Rocks Formed by Solid-State Processes   273

21 Metamorphic Rocks   275
22 Low-Temperature Recrystallized Rocks   312
23 Cataclasites   319
24 Impactites   324
25 Low-Temperature Deformed Rocks   332

x   Brief Contents

Part VII Rocks Formed by Magmatic and Solid-State Processes   343

    26   Magmatic Breccia   346
    27   Dynamomagmatic Gneisses   351
    28   Anatexites   357
    29   Restitites   365
    30   Injection Migmatites   368

Part VIII Rocks Formed by the Mechanical Interaction of High-Temperature Fluids or Gases with Rocks or Magma   371

    31   Pyroclastic Rocks   373
    32   Hydrothermal Breccias   383

Part IX Rocks Formed by Precipitation and Reactive Crystallization Involving High-Temperature Fluids   387

    33   Magmagenic Pegmatites   389
    34   Hydrothermal Rocks   396
    35   Fumarolites   424

Part X Rocks Formed by Mechanical and Chemical Processes Involving Low-Temperature Aqueous Fluids   429

    36   Clastic Sedimentary Rocks   431
    37   Biosedimentary Rocks   454
    38   Evaporites   481
    39   Low-Temperature Precipitative Rocks   498
    40   Weathered Rocks   503

Color Interference Chart   16

Color Plates 1 to 8   208

Appendixes   525

    Conversion Factors   525
    Table of Atomic Properties   527
    Geological Time Scale   530

References   533

Contents

1 **Introduction** 1

Part I Optical Mineralogy 4

2 **Orthoscopic Optical Mineralogy** 6

   The Optics of Minerals 6
   Refractive Index 7
   Optical Classification of Minerals 11
   Color and Pleochroism 18
   Extinction and Illumination—
     The Analyzing Polarizer 21
   Birefringence and Interference Colors 26

3 **Conoscopic Optical Mineralogy** 37

   Conoscopic Light and Interference Figures 37
   Conoscopic Optical Information 38

4 **Rock-Forming Minerals** 43

   Eighty-Five Rock-Forming Minerals 43
   Opaque Minerals in Thin Section 72
   Submicroscopic Aggregates of Minerals and
     Amorphous Materials 73
   Accessory Minerals 74

5 **Plagioclase Composition** 76

   Plagioclase as a Solid Solution Series 76
   Determinative Methods 76
   The a-Normal Extinction-
     Angle Method 77
   Presentation of Compositional Data
     of Zoned Plagioclase 94

6 **Petrographic Measurements** 95

   Phase Analysis 95
   Size Analysis 97
   Shape Analysis 97
   Orientation Analysis 97

Part II · The Anatomy of Rocks 105

7 Physical Attributes of Rocks 107

Fundamental Constituents 107
Rock Texture 108

8 Textural Interpretation 130

The Pleasure and Anguish of Textural Interpretation 130
Method of Textural Analysis 130
Textural Interpretation Based on Mineral Assemblages 131
Two-Dimensional Views of Three-Dimensional Objects 131
First Impression, Comparative Anatomy, and the Illusions of Nature 134

Part III · Classification of Rocks 141

9 Traditional Rock Classifications 143

Classifiable Rock Attributes 143
Traditional Classifications 144
Status of Traditional Rock Classifications 155

10 A Unified Classification of Rocks 156

Components of the Unified Rock Classification 156

Part IV · Fundamental Processes of Rock Formation 159

11 Classification of Rock-Forming Processes 161

Fundamental Rock-Forming Processes 161
General Rock-Forming Processes 163

12 Chemical Activity in Aqueous Solutions 164

Properties of Water and Aqueous Phase 164
Aqueous Solutions 165
Geologically Important Aqueous Systems 166
Aqueous Phase in the Geological Environment 170

13 Phase Relations in Crystallizing Magma and Melting Rock 174

Magma and Lava 174
Phase Equilibria 175
Some Fundamental Phase Equilibria Relationships 175
Effect of Elevated Pressure on Equilibrium 181
Effect of H_2O on Silicate Phase Relations 181
Rock-Forming Magmatic Systems 184

14 Crystal Growth and Dissolution in Fluids 190

Nucleation in Fluids 190
Crystal Growth in Liquids 192
Crystal Dissolution 198
Direct Crystal Melting 201

15 Dynamics of Solids with Gases and Liquids 202

Dynamics of Fluids in Massive or Framework Rock 202
Dynamic Behavior of Mineral and Rock Grains in Fluids 203

16 Deformation and Recrystallization of Solids 209

The Effect of Heat 209
The Effect of Confining Pressure 209
The Effect of Deviatoric Stress 210
The Effect of Time 211
Temperature and Pressure
 Within the Earth 211
Interrelation of Temperature, Confining
 Pressure, Deviatoric Stress, and
 Time 212
The Effect of Fluid 213
Physical Expression of
 Rock Deformation 213

Part V Rocks Formed by Magmatic Processes 219

17 Homogeneous Magmatic Rocks 221

Textural Attributes of Homogeneous
 Magmatic Rocks 221
Examples of Magmatic Solidification 221
Distinction Between Crystallization
 of Supercooled Magma and
 Devitrification of Glass 230

18 Cumulate Magmatic Rocks 235

Processes Leading to Mineral
 Concentration Not Representing
 Magma Composition 235
Crystallization of Cumulates 239

19 Mixed Magmatic Rocks 242

Textural and Mineralogical
 Predictions 242
Textures Compatible with
 Magma Mixing 244

20 Assimilation-Hybrid Magmatic Rocks 261

What Is Assimilation? 261
The Processes of Assimilation 262
Textural Expression of Assimilation:
 Some Examples 264
The Scale of Assimilation 271

Part VI Rocks Formed by Solid-State Processes 273

21 Metamorphic Rocks 275

The Metamorphic Regime 275
Chemical Limits of Metamorphism 275
Protoliths of Metamorphism 276
Metamorphic Mineral Reactions 278
Texture of Metamorphic Rocks 285
Recrystallization and
 Physical Environment 289
Reconstruction of Metamorphic History 298

22 Low-Temperature Recrystallized Rocks 312

Glacial Ice 312
Salt Diapirs and Evaporites in
 Thrust Belts 316

23 Cataclasites 319

The Regime of Brittle Faulting 319
Textural Characteristics of Cataclasites 320

24 Impactites 324

Impacts and Natural Shock Waves 324
Shock Waves and Shock Metamorphism 326
Extraterrestrial Impactites 329

25 Low-Temperature Deformed Rocks 332

Low-temperature Low-Strain-Rate Deformation 332
Deformation Mechanisms 332
Deformation of Soft Sediments 335
Deformation of Sedimentary Rock 337
Brittle Deformation or Ductile Deformation? 340

Part VII Rocks Formed by Magmatic and Solid-State Processes 343

26 Magmatic Breccia 346

Brittle Fragmentation 346
Incorporation into Magma 347
Examples of Magmatic Breccia 347

27 Dynamomagmatic Gneisses 351

Granitic Gneisses 351
The Late-Magmatic Dynamomagmatic Environment 351
Dynamomagmatic Textures 353

28 Anatexites 357

Melting of Rock 357
Mineralogical Composition of Anatexites 359
Can Anatectic Migmatites Be Identified? 360
An Example of Partial Anatexis 360
Migmatization 361
Magma Bodies of Suspected Anatectic Origin 363

29 Restitites 365

Identification of Restitite 365
Genesis of Restitite Minerals 365
Suspect Restitites 366

30 Injection Migmatites 368

Source of Magma 368
Classic Injection Migmatites 370

Part VIII Rocks Formed by the Mechanical Interaction of High-Temperature Fluids or Gases with Rocks or Magma 371

31 Pyroclastic Rocks 373

The Eruptive Stage 373
The Transportive and Depositional Stages 377
Epigenetic Deposition and the Transition to Clastic Sedimentary Rocks 380

32 Hydrothermal Breccias 383

Explosive Rock Brecciation 383
Hydrothermal Breccia Bodies 384
Examples of Hydrothermal Breccia 384

Part IX Rocks Formed by Precipitation and Reactive Crystallization Involving High-Temperature Fluids 387

33 Magmagenic Pegmatites 389

Pegmatite Texture 389
Crystallization of H_2O-Bearing Granitic Magma 389

34 Hydrothermal Rocks   396

Hydrothermal Solutions   396
Hydrothermal Interaction with Rock   398
Hydrothermal Systems   409

35 Fumarolites   424

Fumarole Exhalations   424
Fumarole Incrustations   425

Part X Rocks Formed by Mechanical and Chemical Processes Involving Low-Temperature Aqueous Fluids   429

36 Clastic Sedimentary Rocks   431

Source, Transportation, and Accumulation of Sediments   431
Compaction and Diagenesis   433
Physical Characteristics of Siliciclastic Sedimentary Rocks   438
Compositional Characteristics of Siliciclastic Sedimentary Rocks   443
Siliciclastic Depositional Environments   444

37 Biosedimentary Rocks   454

Biota and Rock-Forming Process   454
The Remains of Organisms in Biosedimentary Rocks   455
Submicroscopic Biogenic Material   471
Recycled Biogenic Material   472
Nonbiogenic Material in Biosedimentary Rocks   473
Diagenesis   473
Rock Type and Depositional Environment   477

38 Evaporites   481

Precipitation of Evaporite Minerals   481
Dissolution of Evaporite Minerals   482
Modification of Evaporite Mineral Assemblages by Reaction   484
Carbonates as Evaporites   485
Physical Configuration of Evaporite Systems   485
Textural Interpretation of Evaporite Genesis   486

39 Low-Temperature Precipitative Rocks   498

The Thinolite Story   498
Precipitation Resulting from Mixing of Solutions   499
Precipitation Resulting from External Factors Without Mixing   502

40 Weathered Rocks   503

Weathering As a Rock-Forming Process   503
Rock–Fluid Reaction in the Weathering Zone   504
Weathering As a Function of Rock Type   505
Weathered Rocks   509

Color Interference Chart   16

Color Plates 1 to 8   208

Appendixes   525

Conversion Factors   525

Table of Atomic Properties   527

Geological Time Scale   530

References   533

Index   575

1

Introduction

The focus of this text is on the data of **petrographic analysis** and its relation to **rock-forming processes**. Parts I and II present the methods and terminology of **petrography**, which in this work includes the study of **hand specimens** as well as **thin sections**. A microscope utilizing polarized transmitted light (Figure I.1) is the instrument required to examine thin sections of rocks. Quantitative determinations of volume percent of minerals and other rock-forming grains in rocks are made with point-counting equipment attached to the petrographic microscope, Figure I.1. Automatic imaging equipment is now available for the same purpose and for rapid characterization of shapes and sizes of rock-forming grains (Figure I.2).

Rocks are made from pre-existing rocks and from magmas, aqueous solutions, and gases. Interaction of these **fundamental materials** at various temperatures, pressures, and kinetic conditions, including the participation of biological processes, occur in many different ways, generating a seemingly limitless variety of rocks.

Every effort has been made to keep **description** separated from **interpretation**. However, since descriptive information is the basis of interpretation of rock-forming process, there must be a bridge between them. Part III contains a discussion of the strengths and weaknesses of linking description to interpretation by way of **rock classification**. Part IV makes the tie between description and interpretation by constraining chemical and physical processes capable of forming rocks.

Fundamental processes of rock formation described in Part IV include (1) chemical behavior of aqueous solutions, (2) phase relations in melt solutions, (3) nucleation and crystallization of minerals in melts, aqueous solutions, and gases, (4) dynamics of physical interaction of solids in fluids and gases, and (5) deformation and recrystallization of solids. These phenomena are neither purely chemical nor purely physical, but they are "fundamental" from a geological point of view.

Parts V to X describe and interpret the origin of rocks of all types. Formation of a rock typically involves several fundamental processes. Therefore, it is useful to speak of **general processes** of rock formation, such as "magmatic," "metamorphic," "sedimentary," "weathering," and "hydrothermal." A generalized rock-forming process may be dominated by a particular fundamental process. For example, some magmatic rocks result mainly from crystallization of magma, whereas in a dynamic environment fundamental mechanical behavior between crystals and liquid becomes important as well in determining the final chemical and physical attributes of the rock.

It is fashionable and expedient to characterize the formation of rocks in the context of **systems** larger

Part 1 Optical Mineralogy

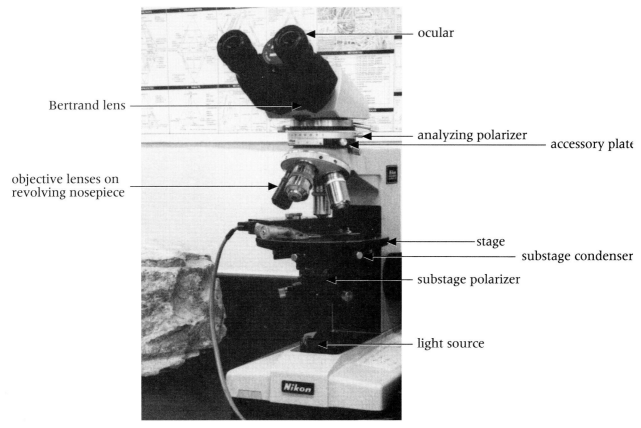

Figure I.1
Nikon petrographic microscope with the Swift Automatic Point Counting System consisting of an electrically driven mechanical stage mounted on the microscope stage attached to a point counter (not shown). The background is a portion of *Mineral and Rock Table,* compiled by P. Lof (Elsevier, 1982).

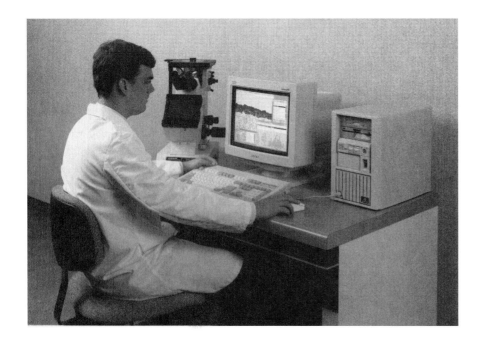

Figure I.2
Omnimet 3 Image Analyzer for determining grain size and area percent of phases present. (Courtesy Buehler Ltd., Lake Bluff, Illinois.)

than the scale of hand specimens and thin sections. There are convecting magmatic systems, fluvial systems, and hydrothermal systems, each of which produces a variety of rock types and involves many fundamental processes and, typically, several generalized processes. A convecting magmatic system cannot be defined on the basis of a single hand specimen, but characterization of rock-forming processes at the hand-specimen and thin-section scale is likely to be very useful in characterization of a system of which the hand specimen and thin section are integral parts.

Note to the Student:

In the photo captions that accompany the text, you will see two abbreviations. They are as follows:

PPL = plane polarized light
X-polar = crossed polars.

Part *I*

Optical Mineralogy

Identification of minerals in hand specimens is a tentative exercise at best. Microscopic confirmation or revision of preliminary mineral identities is a routine exercise in petrographic investigations. Some minerals, such as those in the clay group, do not normally occur as crystals large enough for optical examination. X-ray diffraction is the technique of choice in that case and is essential for augmentation of petrographic studies involving fine-grained rocks.

Identification of most rock-forming minerals with the petrographic microscope can be made by using light that is transmitted through the mineral grains parallel to the optic train of the microscope, which results in an orthoscopic view (Chapter 2). Supplemental or confirmatory identification can be made in which incident light enters the crystalline structure of a mineral oblique to the optic train direction, resulting in an conoscopic view (Chapter 3). Once a petrographer is familiar with the common rock-forming minerals, there is only casual need to obtain more optical information beyond that obtainable by orthoscopic techniques. In fact, observations of relative refractive index, extinction angle, twinning, and cleavage are generally more diagnostic of the varieties of some minerals than the optical data obtained conoscopically. This is true for the feldspars, and these are the most common rock-forming minerals. Nevertheless, the occasion is likely to arise when the conoscopic information is critical for identification of unfamiliar minerals and for positive distinction between well-known minerals, such as quartz and untwinned oligoclase plagioclase that are orthoscopically very similar in some metamorphic rocks.

Access to a listing of optical properties of rock-forming minerals is essential for mineral identification and verification (Chapter 4). All minerals have values of refractive index and birefringence that can be easily estimated, but some minerals can be identified on the basis of pleochroism, cleavage, or crystal morphology without major concern for other optical properties. For example, cleavage and crystal habit are major identifying properties of mica, and the pleochroism of biotite easily distinguishes it from muscovite. As another example, it is well known that the distinction between amphiboles and pyroxene in hand specimens based on cleavage intersection angles is subject to error. Pyroxenes are commonly twinned on {100}. Parting along this plane makes an angle with the right angle pyroxene {110} cleavage that simulates the 56° and 124° cleavage intersection angles made by {110} amphibole cleavage. Distinction between twin and cleavage direction is readily determined in thin section.

Plagioclase is the most abundant mineral in the Earth's crust, and as a result it is a major parameter in the classification of igneous rocks. Plagioclase is a solid solution series ranging from pure albite to pure anorthite. The series is temperature dependent, meaning that compositional zoning of crystals reflects the thermal record of the rock in which the plagioclase forms. These two factors, (1) widespread occurrence and (2) the importance of plagioclase zoning in petrogenesis, justify detailing of a technique used to determine plagioclase composition (Chapter 5). Emphasis is on the a-normal technique for the simple reason that this writer has found it to be easy to use and at an accuracy level appropriate for most petrographic studies.

The fundamentals of identification of rock-forming minerals in thin section is logically followed by discussion of techniques used to (1) quantify the amounts of minerals and other rock-forming materials in rocks, (2) quantify the size and shape of grains, and (3) characterize the orientation of grains in rocks (Chapter 6).

The mineralogical mode of rocks is a quantitative expression of mineral content. The importance of the mode is underscored by the fact that it is a particularly useful parameter in the classification of plutonic magmatic rocks. The traditional manual method of point counting to obtain the mode, using a mechanical stage mounted on the petrographic microscope to obtain volume percent of phases (mineralogical modes and non-mineral modes), is being replaced by automated image analysis. These techniques involve phase discrimination (equivalent to the mode), as well as providing a means for characterizing size and shape of rock-forming grains. The forecast is that with automated imagery, there may be very significant empirically obtained petrographic information that would never be obtained by the tedious traditional point-counting method.

Petrofabric analysis is a procedure for determining the dimensional and crystallographic orientation of rock-forming grains in rocks, primarily as a basis for a kinematic reconstruction of deformed rocks. Automatic imaging techniques, using multiple thin sections, is likely to let the petrofabric analysis part of petrographic investigation become a powerful tool in structural analysis of geologic terrains.

2

Orthoscopic Optical Mineralogy

The Optics of Minerals

The behavior of ordinary light as it passes through a crystal is governed by the electric fields established by the particular elements present and their arrangement in the crystal lattice. Crystal fragments and very thin sections of most minerals will transmit light and are observed with the **petrographic microscope** (Stoiber and Morse, 1981; Kerr, 1977; Wahlstrom, 1979; Jones and Bloss, 1980; Shelley, 1985; Ehlers, 1987; Nesse, 1991; Gribble and Hall, 1992). Sections of minerals 30 microns thick are convenient since (1) preparation of **thin sections** (Humphries, 1992) of minerals and rocks any thinner is technically difficult, and (2) optical effects useful for mineral identification are optimal at about 30 microns. Minerals that do not transmit light at this thickness can be examined by using a microscope that reflects light from polished cut surfaces of these opaque minerals.

The petrographic microscope is essential for **mineral identification** and characterization of **rock texture**. Each mineral has its own **optical properties** and can be grouped with other minerals with similar properties into an optical classification. Identification of unknown minerals can then be made by finding their position in the optical classification. Optical properties relate closely to physical properties, such as **cleavage** and **twinning**, and to chemical properties that are expressed as compositional zoning. The size and shape of mineral grains and the interrelations of grains in rocks are the fundamental attributes of rock texture.

If light is transmitted through crystals parallel to the microscope axis, the view is **orthoscopic**. Orthoscopic techniques are routine for the study of rock-forming materials and their textural relationships in thin sections. With a little experience, the more common rock-forming minerals, of which there are about 40, depending on one's point of view, can be routinely identified by orthoscopic techniques.

The anatomy of a thin section is shown in Figure 2.1. The orientation of mineral grains in thin sections is fixed in two dimensions. In some cases, there is a need to observe the mineral grains of rocks in three dimensions, especially if there is preferred lattice or shape orientation of minerals, acquired during dynamic rock-forming processes. One way to achieve this is to cut three sections at right angles to each other. If that additional expense cannot be justified, the **universal stage** is an alternative (Chapter 6). This instrument attaches to the flat stage of the petrographic microscope and permits limited rotation of a thin section out of the plane of the microscope stage. This effectively rotates a crystal structure so that optical effects can be observed along the third dimension.

2 Orthoscopic Optical Mineralogy

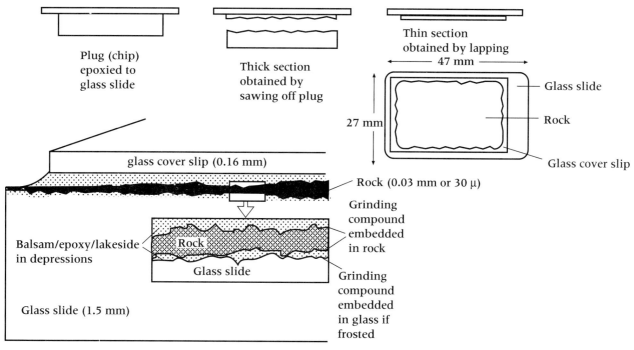

Figure 2.1
Anatomy of a thin section.

Refractive Index

Refractive index (r.i.) is an optical characteristic of minerals that is extremely useful in distinguishing certain minerals from others that are optically similar in other respects. It is defined as

refractive index =
light wave velocity in vacuum / light wave velocity in solid (or fluid)

If the refraction of light in a vacuum is arbitrarily taken as 1, the r.i. of air is calculated to be 1.00029 at 15°C and 760 mm mercury (1 atmosphere); that of water at 20°C is 1.333, and that of most glass slides used in the preparation of thin sections is about 1.525.

The refractive index of minerals that transmit light in thin section range between 1.434 for fluorite and 3.22 for hematite, the latter being nearly opaque. Immersion liquids can be prepared with known refractive indices (Hurlbut, 1984) ranging over those indices of most minerals. The refractive index of unknown minerals can then be compared with the selected liquids (Stoiber and Morse, 1981; Jones and Bloss, 1980). Grains of crushed minerals are sequentially immersed in liquids of different refractive index until there is a match, at which point the refractive index of the mineral has the same value as the liquid. Some minerals require consideration of the specific orientation of a grain as it lies on the glass slide in the liquid; that is, the refractive index varies with grain orientation. Refractive index of minerals in uncovered thin sections can be determined by using a special oil-immersion technique (Laskowski et al., 1979) on minerals or mineral grains.

In thin section relative refractive index can be determined by comparing with adjacent minerals or mounting medium of known index. There are four fundamental and very useful interface relationships between materials as observed in thin sections, wherein the refractive index of one material is compared with that of another (Figure 2.2): (1) two crystals of the same mineral or different minerals, with lattice orientations yielding similar refractive indices; (2) two crystals of the same mineral or different minerals, with lattice orientations yielding different refractive indices; (3) a crystal in contact with natural glass or amorphous material, and (4) a crystal in contact with mounting material (typically, epoxy).

The useful optical effects that occur along the interface between these pairs of materials depend on the nature of the materials, their orientation if crystalline, and the geometry of the interface. Since the interface between two substances is not likely to be exactly planar, and since the orientation of the interface with respect to the plane of the thin section can be expected to be highly variable, light from the substage condensing optical system will have angular relationships with the interface. An angular relation is responsible for the phenomenon known as refraction. No refraction can

8   Part 1   Optical Mineralogy

← – movement of Becke line as focal plane is raised
h – higher-presenting r.i.
l – lower-presenting r.i.

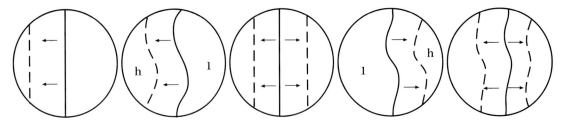

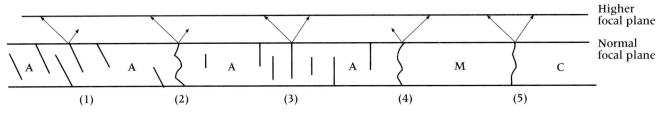

Figure 2.2
Situations of relative refractive index resolved by Becke line method and determination of cleavage orientation.

(1) Inclined cleavage in crystal A. (2) Interface between two crystals of A (anisotropic) in different orientations. (3) Vertical cleavage in crystal A, or identical orientation of two crystals of A. (4) Interface between a given orientation of crystal A and a higher index material M (isotropic or anisotropic crystal, glass, or epoxy). (5) Interface between material M and crystal C with overlapping r.i. and specific orientations providing near match of r.i.

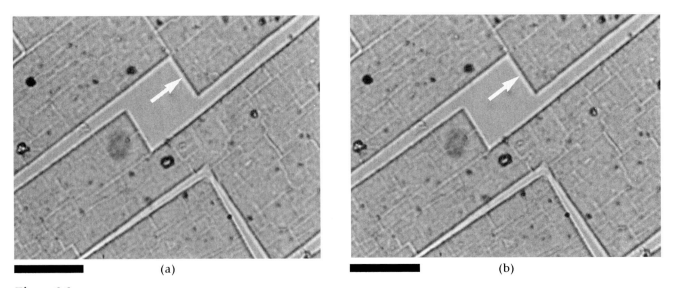

Figure 2.3
Determination of relative refractive index between mounting epoxy (1.540) and cleaved sanidine using Becke line method. PPL. Bar scales = 0.08 mm.

(a) Becke line (arrow) moves into epoxy as focal plane distance is increased, indicating epoxy has higher r.i. than sanidine. (b) Becke line (arrow) moves into sanidine if focal plane distance is decreased, giving opposite, but not standard practice, results.

occur if light travels at right angles to the plane separating two materials, even if they have different refractive indices. The glass slide of the thin section underlies a film of epoxy, and the glass cover slip overlies a layer of some bonding material such as balsam (Figure 2.1). Neither surface presents an interface with these adhesives that has an angular relation with respect to the microscope stage, and there are no special optical effects.

Refraction (and reflections) occur along the interface between materials, resulting in a deflection of light into the substance of higher refractive index. This phenomenon is related to a reduction of light velocity in the substance of higher refractive index and is ultimately related to differences in element content and arrangement in the crystal lattice.

For the petrographer, the useful optical phenomena are (1) movement of a concentration of light parallel to the interface between materials and (2) a lightening or darkening along the interface. These effects are best observed using either a 10× or 20× objective lens. Partial closure of the substage iris diaphragm also enhances the effects.

The first effect utilizes light transmitted directly from below the stage (**central illumination**) into the interface zone between two materials. The blur of light is known as the **Becke line**, in honor of F. Becke, a pioneering mineralogist and petrographer. The line moves laterally, parallel to the trace of the interface, into the substance of higher refractive index as the distance between the materials and the microscope objective lens is increased (Figures 2.2 and 2.3).

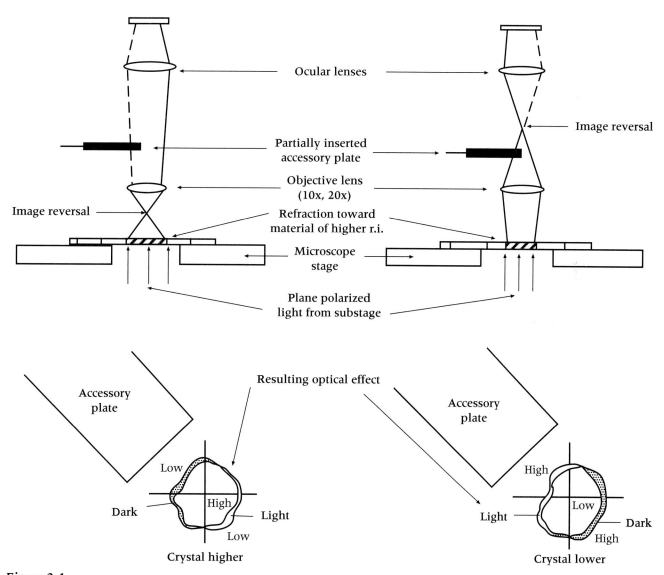

Figure 2.4
Optic train in polarizing microscope in which relative refractive index is being determined by oblique illumination method, and resulting optical effects. Modified from Stoiber and Morse (1972).

Light traveling up along a cleavage in a mineral will also "spread" as the focal distance is increased (Figure 2.2). The behavior of the light traveling along a cleavage is very useful in determining the orientation of the cleavage surface relative to the plane of the thin section. When the distance between the crystal surface and the microscope objective lens is increased, the light "line" will separate equally into the two parts of the crystal if the cleavage plane is perpendicular to the thin section (and thus to the microscope stage). If the cleavage is inclined, the light will move as a line parallel to the cleavage trace but sideways in a direction opposite to the direction of dip of the cleavage (Figure 2.2).

The second way of determining relative refractive index is by observing optical phenomena obtained by creating **oblique illumination**. This procedure consists of inserting an obstacle (commonly the leading edge of an accessory plate frame) partially blocking the train of light in the microscope above the objective lens. With the shadow encroached close to the interface zone between the two materials being compared, the interface zone will be lighter, darker, or have yellow–orange or light blue **color fringes**. The exact position of these effects with respect to the materials involved varies according to the geometry of the interface, the types of materials being compared, and the amount of difference in refractive index. The position is not important; it is the nature of the effect (lightening or darkening) that is useful in determining relative refractive index.

If the interface zone becomes lighter, the material on the side of this zone from which the shadow has encroached has a higher refractive index than the adjacent material (Figure 2.4). If the interface zone becomes darker, the material on the same side as the shadow has a lower refractive index (Figure 2.4). If a yellow or orange color fringe appears, the material on the shadow side will have a very slight higher presenting refractive index; if a light blue color fringe appears, the material on the shadow side has a very slight lower presenting refractive index. Some relations of relative refractive index as generated by oblique illumination are shown in the photomicrographs of Figure 2.5.

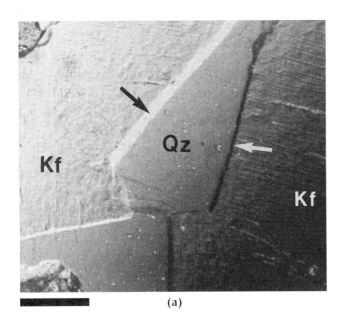
(a)

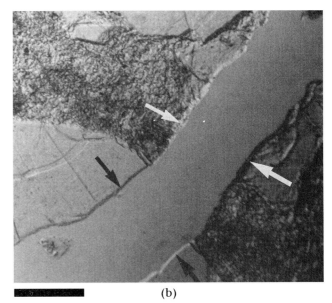
(b)

Figure 2.5
Determination of relative refractive index using oblique illumination (shadow encroaches from lower right). PPL. Bar scales = 0.12 mm.

(a) Wedge of quartz in K-feldspar, producing light at qz → Kf interface (black arrow), and dark at Kf → qz interface (white arrow), either indicating quartz has greater r.i. than K-feldspar. Quartz may stand up or be in depression for observer.
(b) Epoxy-filled fracture in porphyritic glassy volcanic rock. Epoxy → plagioclase interface is dark (smaller black arrow), and plagioclase → epoxy interface is light (larger black arrow), either indicating plagioclase has higher r.i. than epoxy. Epoxy → glass interface is light (smaller white arrow) and glass → epoxy interface is dark (larger white arrow), either indicating epoxy has a higher r.i. than glass. Epoxy in plagioclase phenocryst may stand up in relief for observer, whereas epoxy in glass may be in depression for observer.

If the presenting refractive index of one material is much different than another, a **relief** difference can be observed. The greater the difference in relative refractive index, the greater the sense of relief, and it can be enhanced by establishing the oblique illumination shadow. Relief is a three-dimensional optical effect. Unfortunately, the sense of relief is not reliable; what stands up in relief and what is in depression varies from one viewer to the next, as it may with stereo pairs of air photographs. The design of the microscope, the viewing orientation, and even the momentary mental state of the viewer seem to effect the sense of relief. The reader may be able to reverse the sense of relief in certain portions of the photomicrographs of Figure 2.5 by turning them upside down.

Materials that have a large difference in refractive index will appear very distinct from each other. For example, garnet "stands out" in comparison to quartz and mica with which it is commonly associated because (1) garnet has a much higher refractive index in comparison with these other minerals with which it is in contact, and (2) refractions also occur between the uneven surface (unpolished) of the garnet and the mounting epoxy, producing a relatively dark mottled effect. **Cleavage** also produces local darkening where it is developed and filled with epoxy (Figure 2.6a). If a mineral stands out in relief, it does not necessarily mean that it has a high refractive index. Fluorite appears to stand out in relief in relation to the mounting epoxy in a fracture (Figure 2.6b), but actually the fluorite has a very low refractive index in comparison with the epoxy—that is, strong negative relief.

Difference in relief may be very minor or very obvious. If it is obvious, a scan of the thin section can be made by moving the thin section with oblique illumination set in place, observing the location of grains that contrast with others. The contrast between the refractive index of K-feldspar (in any orientation) and that of plagioclase (Figure 2.6c) or quartz is significantly marked and is amenable to this procedure. Such a **relief scan** is useful regardless of which mineral appears to be in depression or elevation. Which mineral has the actual lower or higher relative refractive index can be determined by observing the lightening or darkening along interfaces as generated by oblique illumination or by using the Becke line method. Once the mineral identities are established, the scan can be used to approximate the volume percentage of each mineral phase.

Optical Classification of Minerals

Opaque Minerals

Minerals in thin section that do not transmit light are classified as **opaque**. The form of the opaque mineral may be useful information, but no optical effects can be generated. If a portable illuminator (even a flashlight) is available, distinction may be made between minerals such as pyrite (brassy) and magnetite (silvery); however, distinction between magnetite and ilmenite requires use of the reflecting microscope. The

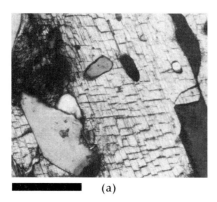
(a)

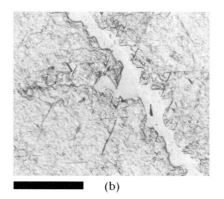
(b)

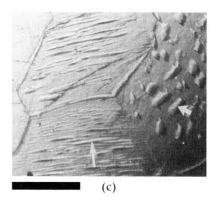
(c)

Figure 2.6
Optical sense of refractive index.

(a) Epoxy in cleavage of kyanite crystal gives a sense of high relief (positive in this case). PPL. Bar scale = 0.2 mm. (b) High negative relief of fluorite falsely mimics high positive relief. PPL. Bar scale = 0.2 mm. (c) Perthitic albite has marked positive relief in comparison to three host K-feldspar crystals as viewed in oblique illumination. Albite lamella (thin arrow) appear as patches (short thick arrow) if intersected transversely. PPL. Bar scale = 0.2 mm.

difference in color between pyrite (brassy) and chalcopyrite (yellow–brassy) can be observed, especially if crystals can be observed side by side.

If the substage light source is left on when the makeshift incident light source is applied, some useful optical effects can be observed for certain materials. For example, a mat of clay mineral crystals, limonite, or leucoxene may appear more or less opaque with only substage light and may be mistaken for metallic oxides or sulfides. The non-metallic character of these materials becomes readily apparent if some incident light is applied. Additionally, the fine granular character of such materials commonly becomes apparent if both substage and incident light are applied simultaneously.

Isotropic and Anisotropic Minerals

Non-opaque minerals include all minerals that will transmit light when cut to the standard 0.03 mm (30 microns) thickness of thin sections. The minerals are divided into two groups based on fundamental behavior of the transmitted light.

Isotropic minerals, such as fluorite and garnet, have refractive index values that are the same for any direction of light travel through the crystal. This is the same as saying that the lattice and chemical composition of the crystals are optically equivalent in all directions. Crystallographically, these are the isometric minerals; optically they are isotropic.

Light from the substage optical system in the petrographic microscope is polarized in an east–west (E–W) direction in newer microscopes and (north–south) N–S in older models. E–W polarized light entering isotropic crystals is refracted to a degree that is determined by the crystallographic–compositional nature of the particular mineral species. The refractive index is the same for any position of stage rotation and for any cut through the crystal that was determined in the preparation of the thin section. The refraction of light occurs laterally away from the polarization plane, but since the crystal is isotropic the same refraction occurs through any direction in the crystal that is presented to the plane of polarized light. Consequently, the relative refractive index effects would be the same in all directions if the lower polarizer were removed, since an infinite number of refractions, all of the same magnitude, would be contributing to the refraction effect.

Anisotropic minerals are hexagonal, tetragonal, orthorhombic, monoclinic, and triclinic. Consider light travel through hexagonal and tetragonal crystals first. The unit cell of any mineral in either of these crystal systems is elongated or shortened in one direction, unlike isometric lattices that are "equi-dimensional." If there is an E–W plane of polarized light traveling upward into a hexagonal or tetragonal crystal whose c-crystallographic axis lies E–W on the microscope stage, refraction will take place to a degree determined by the mineral examined. So far this is the same as refraction in isotropic minerals.

What happens if the crystal is rotated 90° and light is observed as it travels along the plane in the crystal that is perpendicular to the c-axis direction? The degree of refraction will be different since the presenting crystal structure is different. Hexagonal and tetragonal minerals have two principal directions that yield a maximum and minimum refractive index, and any intermediate value can be obtained by sectioning these crystals in different directions.

Since orthorhombic, monoclinic, and triclinic crystals are defined by three crystallographic axes of different lengths, the crystal structures have three principal planes that are structurally distinct from each other. This fact suggests that there will be a maximum, a minimum, and some unique intermediate value of refractive index for orthorhombic, monoclinic, and triclinic crystals, and intermediate values are obtainable by different section directions as well.

Polarized light must be available for the examination of refractive index along specific planes in anisotropic crystals. If non-polarized light were used, the relative refractive index of an anisotropic mineral could still be compared with adjacent materials, but the particular orientation of the crystal would be of no consequence. In such a case, the presenting refractive index would be an average of all the directions presenting themselves to light vibrating in an infinite number of directions. This information would be useful, but there is the opportunity to go one step further. By using light polarized into a plane, there can be determination of a specific planar direction in the crystal having its own unique value of refractive index. This provides more detailed optical information which can be used for more elaborate optical classification and for more detailed characterization of specific minerals.

Light passing through crystal lattices can be visualized as passing between ions or clusters of ions known as "motifs," the building units of Bravais lattices. For the plane polarized light to travel through a crystal without change in direction, except for lateral diffraction parallel with the polarization direction previously described for isotropic crystals, it must travel along a plane in the lattice that has equivalent arrangement of ions on either side. In this way the electrical component of light interacts with the electron clouds of the ions without change in direction of the light propagation. This equivalence is realized in all isometric crystals, in hexagonal and tetragonal crystals that are oriented with the c crystallographic axis oriented perpendicular to the microscope stage, and "optic axial" directions (defined in a later section) in orthorhombic, monoclinic, and triclinic crystals, oriented vertically.

For all other orientations of anisotropic crystals there are two possibilities: (1) when the crystal lattice is oriented symmetrically with E–W polarization the light passes through undeviated (except for lateral shift previously described), and (2) when the crystal lattice is in any intermediate position (as obtained by stage rotation) the light is deviated and split into two mutually perpendicular planes of light. These new orientations are governed by the electrical fields within the crystal lattice, which effectively deviate light to new directions. Any of the "deviated" vibration directions, when placed E–W, generates its own value of refractive index as a function of lateral light deviation (refraction) from the trace of those directions.

Double Refraction

When substage polarized light is split into two mutually perpendicular planes in certain orientations of anisotropic crystals, there is said to be **double refraction**. Observation of this phenomenon can be directly observed with a non-polarized light source as follows. If a small circle (or other image) is viewed through a very clear calcite crystal, the light from the circle is non-polarized as it passes up to the calcite crystal. As the light enters the calcite crystal, refraction will occur in certain directions, depending on the orientation of the lattice and the geometry of the calcite–air interface. Since the light initially is vibrating in an infinite number of directions, all possible positions for refraction are activated. Two such positions of refraction can be found for any orientation of the calcite crystal except the one in which the c crystallographic axis is vertical. Thus most positions of calcite will generate "double refraction." By inspection it is seen that a double image of the circle lines up along the c-crystallographic-axis direction (Figure 2.7). If the crystal is rotated, there is no change in the double image relative to the c axis—as the axis moves, so does the double image. Evidently, the anisotropic nature of the calcite lattice has resulted in light traveling along two different paths; thus, we see two images. Since the propagation directions are different, the light must be traveling through different lattice configurations, resulting in two different velocities. Refraction along these directions must be different since light is traveling along two different planes in the anisotropic crystal. If the calcite were cut to thin section thickness, the double image would not be noticeable; there must be a long divergent travel distance in the crystal for a separation of image to become visible.

Both dots represent light that is now polarized by the crystal. Is the light we see as one dot vibrating parallel to, at right angles to, or at some intermediate angle to that of the other dot? If a sheet of polarizing film is placed between the circle and the calcite crystal, the following phenomena can be observed (Figure 2.7). If the polarization direction of the film is perpendicular to crystallographic c, only one image is seen and it is in the position of the circle that appeared to be "floating" above and beyond the other image seen without the film. The light forming the image that has disappeared must have been vibrating N–S (parallel to crystallographic c) and has been canceled by the E–W polarization of the polarizing film.

If the crystal (or film) is turned 90°, again only one circle is seen, but this time it is the image that appeared to be the real one on the paper. The "floating" image is not seen because its light must have been vibrating E–W (normal to crystallographic c); now, that direction is normal to the polarization in the polarizing film and is canceled.

It can be seen that light traveling in a plane that is normal to crystallographic c is strongly refracted (displaced image), whereas polarized light traveling along a plane in which crystallographic c lies is not refracted at all.

Now place the polarizing film so that the polarization direction is at 45° position to crystallographic c. The double image reappears because the light polarized by the film is repolarized into two mutually perpendicular planes of wave vibration moving through the crystal along different paths, creating the double image (Figure 2.7).

Now observe a thin section containing two orientations of calcite with crystallographic c lying in the plane of the section and the microscope stage (Figure 2.7). With the c-normal direction placed E–W, the calcite crystal acquires a mottled darkening, indicating a large contrast in refractive index between its uneven surface and mounting materials. A relative high refractive index compared with adjacent minerals or epoxy would be indicated if checked by the Becke line or the oblique illumination method. The surface of the crystal in this orientation is mottled dark and light.

With crystallographic c lying parallel to E–W polarization, there is much less refraction and the crystal "washes out" (Figure 2.7). In this case, the mottling or topography is not seen, since in this position there is less of a difference in refractive index between the calcite and the mounting materials. The refractive index compared with surrounding minerals would be low.

Suppose the calcite crystal is rotated so that there is a view looking down the c axis. In this orientation, calcite behaves like an isotropic crystal. There is no double image, and the relative refractive index in thin section does not vary as the microscope stage is rotated.

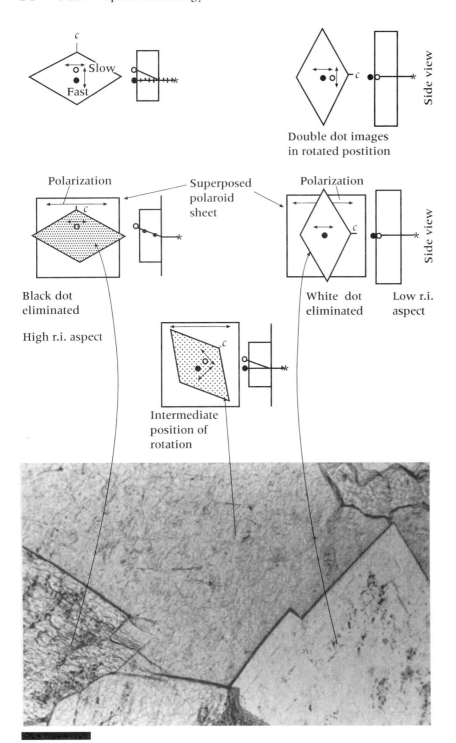

Figure 2.7
Double refraction demonstrated by double image produced in clear calcite. Non-polarized light vibrates in both *c* and *c*-normal directions. Polaroid sheet placed beneath crystal and above dot acts as an E–W polarizer, illuminating dot image produced by N–S vibrating light in crystal. Appearance of calcite crystals in thin section (PPL; bar scale = 0.17 mm) using polarizing microscope is shown correlated with the different orientations of the calcite crystal.

Uniaxial and Biaxial Anisotropic Crystals

Anisotropic crystals are further divided into two subgroups. **Uniaxial** crystals include all minerals in the hexagonal and tetragonal systems. They have one **optic axis**, and it coincides with the *c* crystallographic axis. **Biaxial** minerals are those in the orthorhombic, monoclinic, and triclinic crystal systems. They have two optic axes, and neither axis coincides with a crystallographic axis.

The orientation of uniaxial minerals in thin section is shown as three fundamental cases (Figure 2.8). In one situation, the *c* axis is perpendicular to the microscope stage (Figure 2.8a). The refractive index of

2 Orthoscopic Optical Mineralogy   15

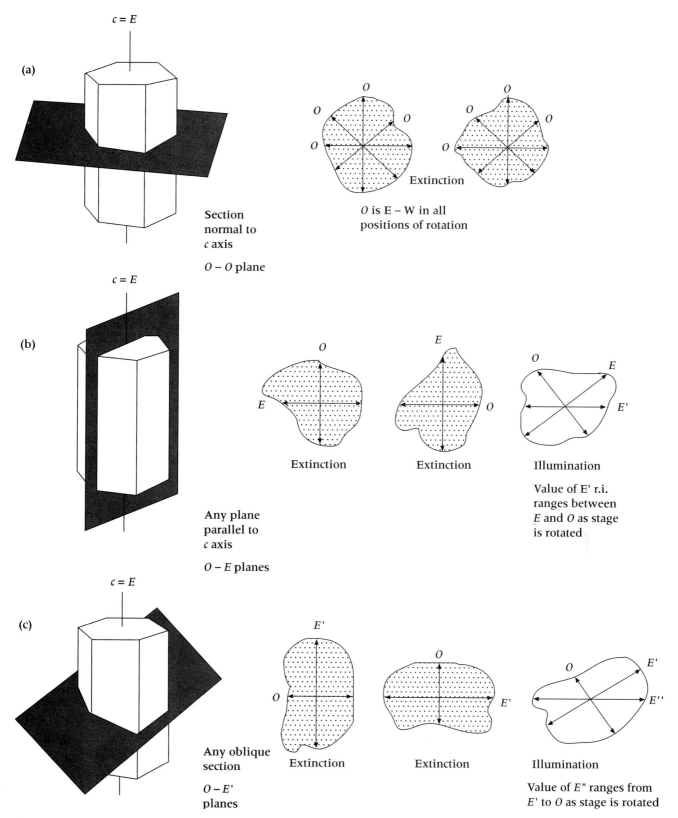

Figure 2.8
The figure presents principal vibration directions for various sections through a uniaxial anisotropic crystal such as quartz.

the mineral in this position is the same for all positions of stage rotation. The refractive index is designated as n_ω and is the value for the infinite number of vibration directions lying parallel to the stage and therefore perpendicular to the c axis. These directions can be defined as the **O vibration directions**, and they are **principal vibration directions**.

In another situation the uniaxial crystal is cut so that the c axis lies in the plane of the thin section and the microscope stage (Figure 2.8b). The c-axis direction can be defined as the **E vibration direction** and is another **principal vibration direction**. The refractive index in this direction is different and is designated as n_ϵ. The direction perpendicular to the c axis, in the same section of the crystal, is the **O vibration direction** that was observed before with the c axis vertical. The value of n_ϵ may be greater or less than n_ω depending on the particular mineral. If n_ϵ is greater than n_ω the mineral has a **positive optic sign**; if it is less, it has a **negative optic sign**.

The third situation includes any slice through the crystal that is neither parallel nor perpendicular to crystallographic c (Figure 2.8c). Of course, this would be the common situation in the thin sectioning of a rock containing a more-or-less random orientation of uniaxial crystals, such as an aggregate of quartz crystals in a quartzite or calcite crystals in a marble. In this case, the O principal vibration direction will always be presented, but only partial values of n_ϵ will occur. This is the **E' vibration direction** along which the value of $n_{\epsilon'}$ depends on the angular relationship between crystallographic c and the plane of the thin sectioning and, of course, the mineral species. The values of n_ω; n_ϵ; and $n_{\epsilon'}$ can be determined, or compared relatively with adjacent materials, only when the corresponding vibration direction is placed parallel to the direction of polarized light generated in the substage polarizer, since that is the direction along which the light is transmitted through the crystal and is the position from which deviation by refraction occurs.

Biaxial anisotropic crystals are uniquely described as having the three principal vibration directions X, Y, and Z. These directions have refractive indices associated with them designated as n_α; n_β; and n_γ respectively. These three principal vibration directions are mutually perpendicular. The actual refractive index values for these principal directions vary according to the mineral species, but $n_\alpha < n_\beta < n_\gamma$ always applies. Conversely, it can be recalled that uniaxial minerals may have n_ω greater than or less than n_ϵ.

Values of refractive indices can be compared with adjacent materials in thin sections by placing any principal vibration direction (X, Y, or Z) or any partial principal direction (X' or Z') parallel to the substage polarization. There is no Y' since Z' and X' converge on Y itself. Since thin sections are two dimensional, not all vibration directions can be viewed in single sections. The various fundamental presenting vibration directions for biaxial minerals are illustrated with one mineral sectioned in various directions (Figure 2.9). Although the vibration directions and associated refractive indices will be different for other biaxial minerals, the various planes that can be viewed will be the same.

The orientation of the vibration directions with respect to the crystallographic axes varies for each crystal system. In the orthorhombic system, the principal vibration directions are parallel to the crystallographic axes a, b, and c. The vibration directions correspond to different crystallographic axes according to the mineral species.

In monoclinic crystals, the principal vibration direction Y is commonly parallel to crystallographic b (amphiboles and pyroxenes), but X or Z is parallel to b in some minerals. Neither X, Y, nor Z is parallel to crystallographic a or c. In triclinic crystals, none of the principal vibration directions parallel crystallographic axes.

In routine petrographic studies, there is no need (nor the opportunity in thin sections) to determine the actual values of refractive index for the various vibration directions in anisotropic crystals. Relative refractive index can be determined and is a very useful determinative parameter. If a crystal changes in appearance as the stage is rotated, we know immediately that the mineral is anisotropic. If relief differences are marked, as they are for calcite, it can further be assumed that the mineral has a wide range of refractive index, whether it be uniaxial or biaxial.

The range of refractive indices for some common rock-forming minerals is shown in Figure 2.10. The maximum and minimum values reflect the particular mineral species, but since some minerals have variable composition the range is much larger to include all compositions, as it has been for cordierite. The plagioclase feldspars have a wide range in values as well, but here they have been grouped into individual ranges for albite, oligoclase, andesine, and labradorite. The presenting relative refractive index is determined by whatever vibration direction happens to be aligned parallel to the E–W polarization. Some very useful relations occur, even though the particular vibration directions cannot be identified.

For example, K-feldspar has a lower refractive index than plagioclase feldspar except for some orientations of albite. K-feldspar has a lower refractive index than quartz, no matter what the orientations of the K-feldspar and quartz are presenting. Conversely, plagioclases in the oligoclase–andesine range will be very close to quartz. Differences in orientation of the quartz and plagioclase will generate only minor differences in relative refractive index. If orientation combinations

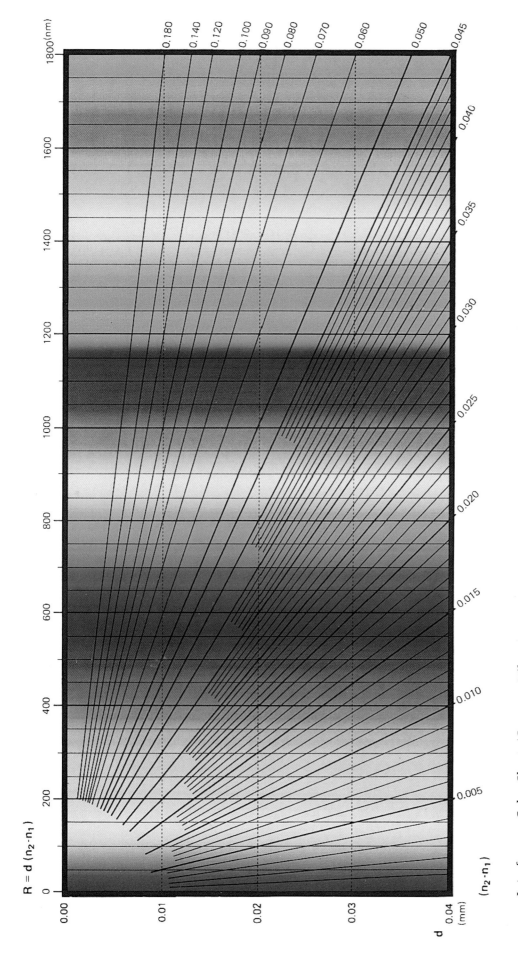

Interference Color Chart. (Courtesy Nikon.)

2 Orthoscopic Optical Mineralogy   17

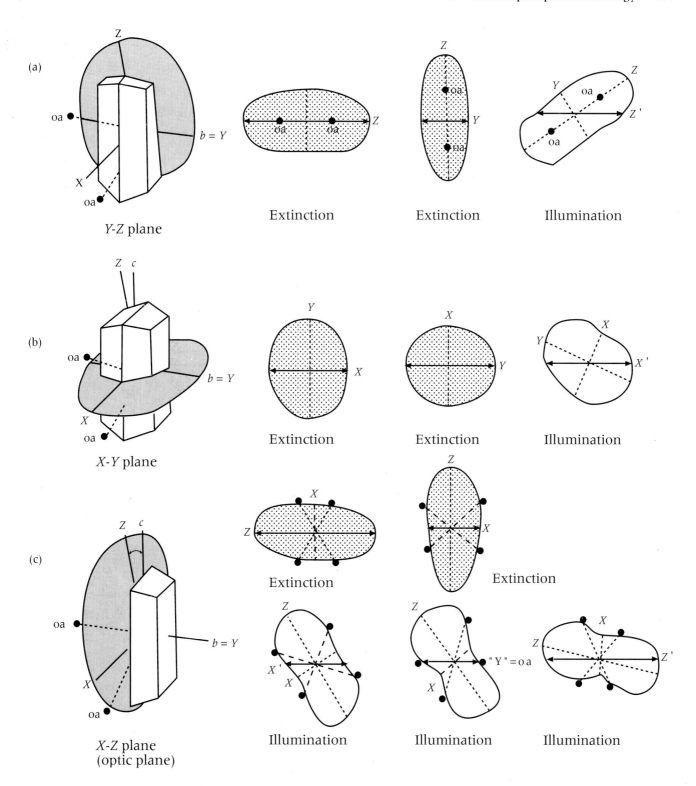

Figure 2.9
Principal vibration directions for various sections through a biaxial anisotropic crystal such as hornblende. Sections containing X′ and/or Z′ not shown.

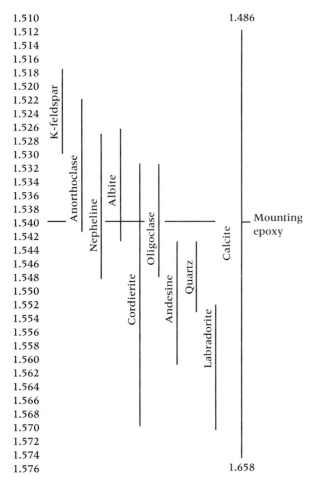

Figure 2.10
Range of refractive index for some common rock-forming minerals indicating importance of crystal orientation (presenting r.i.) in relative refractive index determinations. Relation to mounting epoxy r.i. (1.540) shown. Extreme range of calcite r.i. illustrates that only minerals with narrow range of r.i. are useful for relative r.i. determinations.

produce a near match, color fringes appear. Color fringes consist of light blues on one side of the interface between two materials being compared and orange to yellow colors on the other side.

A mineral with a wide range in refractive index, such as calcite, is not useful in relative refractive index determinations (Figure 2.10). In one orientation, calcite will be typically less than feldspars and quartz; in a position 90° from the first position, calcite will be relatively much higher than these minerals.

The crystallographic axes in minerals are not visible features, although their positions commonly can be located if the identity of the mineral is already known or suspected and if the mineral has some visible feature, such as twinning, cleavage, or crystal faces, to which the crystallographic axes have some

known relation. For relative refractive index work with anisotropic minerals, it is not necessary to know which vibration direction is yielding the higher or lower index value.

The refractive index of mounting epoxy is commonly 1.540. The relations of common rock-forming minerals to this epoxy may be very useful (Figure 2.10). Examination of interfaces between crystals and epoxy can be made along cracks, "plucked" areas, and at the margins of the section if the epoxy has not been scraped off the glass slide.

Color and Pleochroism

The color of a mineral results from the absorption of some wavelengths of visible light and the transmission of others. For minerals, chemical composition and crystal structure are the primary controlling factors. Some rock-forming minerals have **natural color** in thin section, but minerals such as feldspars and quartz are colorless. Color is a very handy diagnostic property of minerals in thin section, since no special optical manipulation is required to observe it. Although there is some variation in color perception, and particularly in its verbalization, relative differences in color from one mineral to the next are readily observable.

Minerals such as calcite and gypsum commonly range from white to colorless in hand specimens and, therefore, can be predicted to be colorless in thin section. Darker-colored minerals commonly contain iron and are more likely to appear colored in thin section. For example, iron-poor diopside [$CaMgSi_2O_6$] is very light green in hand specimen but colorless in thin section. The hedenbergite member of the diopside–clinoferrosilite solid solution series is iron bearing [$Ca(Mg,Fe)Si_2O_6$] and is green even in thin section. Iron-poor pink grossularite garnet [$Ca_3Al_2Si_3O_{12}$] is colorless in thin section but becomes pink, if there is a compositional shift to the iron-rich andradite member [$Ca_3Fe_2Si_3O_{12}$] of the grossularite–andradite series. Other iron-bearing or ferromagnesium minerals include biotite, hornblende, and aegirine–augite. These minerals are commonly colored in thin section. Exceptions include dark red–brown almandine garnet [$Fe_3Al_2Si_3O_{12}$] and green olivine [$(Mg,Fe)_2SiO_4$] which are both commonly colorless in thin section.

If an anisotropic mineral is colored as seen in thin section, there will be color variation. In this case, color depends on the path that the polarized light from the microscope takes through the crystal, just as refractive index varies in that respect. Variability of color, as related to orientation of anisotropic crystals with respect to substage polarized light, is known as **pleochroism**.

The color of isotropic minerals does not depend on crystal orientation, and minerals such as some garnets and spinels are the same color no matter how they are oriented in the thin section—they are non-pleochroic. It is important to remember that pleochroism is observed without the upper polarizer in place. Colors observed with crossed polars are "interference colors" (next section), and they are quite unlike the colors obtained with one direction of plane polarized light.

Biotite is a good example of a pleochroic mineral (Figure 2.11). A section of a biotite crystal cut perpendicular to the excellent basal cleavage generally will be elongated parallel to the trace of the cleavage. If the cleavage trace and elongation direction are positioned parallel to the direction of substage polarization, the biotite crystal will be dark brown, green, or red–brown, depending on the specific composition of the biotite. The cleavage itself has no effect on the color produced. If the crystal is rotated 90°, the polarized light passing through the crystal at right angles to the cleavage trace appears only as a light tan, confirming that biotite is a pleochroic mineral.

The following experiment is useful in the understanding of pleochroism. The color of a certain biotite in thin section varies from dark brown to tan as described above. If the same biotite is observed with unpolarized transmitted light, it will be a light brown, more or less the average of the dark brown and tan obtained with polarized light. This effect can be seen by holding the thin section containing the biotite in front of a lamp and observing the biotite crystal with a hand lens. If the thin section is rotated in front of such a light source, the biotite exhibits no change in color and is therefore non-pleochroic under these conditions.

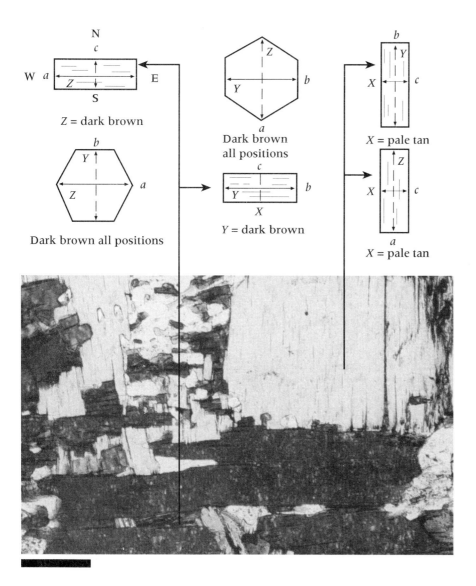

Figure 2.11
Biotite pleochroism shown with two intergrown crystals in opposing orientations. E–W substage polarization. PPL. Bar scale = 0.2 mm.

A basal section of this biotite—that is, a section cut parallel to cleavage and perpendicular to crystallographic c—is a dark brown viewed with polarized light and remains that color even if the crystal is rotated. The lack of color change or apparent pleochroism is due to the fact that the vibration directions in all planes perpendicular to the basal plane all have nearly identical refractive indices and optical properties. Sections of biotite crystals intermediate between a position parallel to and one perpendicular to the basal plane yield intermediate color characteristics and will be pleochroic if viewed with polarized light.

Uniaxial pleochroic minerals have two extremes of color or color intensity, namely those observed with the O vibration direction and E vibration direction alternately placed parallel to substage polarization. Biaxial minerals have the three principal directions (X,Y,Z) and, therefore, three principal color directions if the mineral is pleochroic. Since the full color effect can only be observed with the principal vibration directions placed in the plane of the microscope stage and parallel to substage polarization, more than one crystal orientation (a minimum of two) must be examined in thin sections of biaxial minerals to obtain

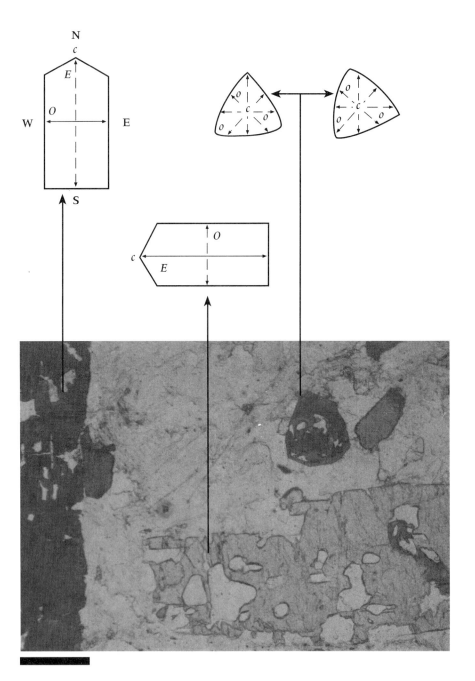

Figure 2.12
Tourmaline pleochroism shown with three crystals in different orientations. E–W substage polarization. PPL. Bar scale = 0.6 mm.

a three-dimensional **pleochroic formula**. For biaxial minerals, a pleochroic formula is the indication of color for *X, Y,* and *Z* as they are viewed parallel to substage polarization. For uniaxial minerals the pleochroic formula is the color for *O* and the color for *E*.

The relationship among pleochroic colors, vibration directions, and crystallographic axes for the uniaxial mineral tourmaline is shown in Figure 2.12. In this case, the prism elongation direction (*E* vibration direction) placed parallel to substage polarization generates nearly no color at all, whereas the short direction across the prism form (*O* vibration direction) generates a dark orange–brown. The direction *O* is always parallel to polarization in the prism cross section (Figure 2.12) and maintains the strong color even if the stage is rotated. The pleochroic formula for this tourmaline is *O* = deep orange–brown, *E* = colorless.

In practice, the optical directions of minerals in thin section using only orthoscopic views can be related to crystal structure only if the crystals have some visible reference feature such as cleavage, crystal face, crystal elongation, or twin plane. However, the fact that a mineral is colored or pleochroic is very diagnostic in itself, even if vibration directions associated with the colors are not identified.

Extinction and Illumination— The Analyzing Polarizer

Light that has passed through crystals in thin section is either (1) vibrating parallel to the original polarization as fixed by the substage polarizer or (2) repolarized into two mutually perpendicular vibration planes, neither of which are parallel to the lower polarization direction. Insertion of another polarizer, above the objective lenses of the microscope, can distinguish these two possibilities, and it is therefore known as the **analyzing polarizer**.

The polarization direction in the analyzing polarizer is at right angles to that of the lower polarizer (Figure 2.13). If the analyzing polarizer is in, all light is canceled. A similar cancellation of light can be achieved by superposing two biotite crystals in two thin sections. Since non-polarized light passing through an anisotropic crystal such as biotite is polarized by the biotite itself, one of the biotites acts as the primary polarizer and the other becomes the analyzer. Hold the thin sections in front of a lamp producing non-polarized light. Use two biotite crystals sectioned at a high angle to the basal cleavage in separate thin sections. If the cleavage of one biotite crystal is superposed at right angles to the cleavage of the other, extinction occurs. There are principal vibration directions nearly parallel to cleavage, making the cleavage a good reference for orientation in this experiment.

Polarized light passing up through glass, mounting epoxy, and all orientations of isotropic minerals is undeviated and canceled if the upper polarizer is in place (Figure 2.13). This phenomenon is known as **extinction** and results from **crossed polars** (X-polars). Extinction is maintained for these materials even if the section is rotated. Location of isotropic mineral grains such as fluorite and garnet may be facilitated if they occur with other minerals that do not remain at extinction when the stage is rotated.

Anisotropic minerals also allow polarized light from the lower polarizer to pass through undeviated if the principal vibration directions are oriented parallel to the lower polarization (Figure 2.14a). A principal vibration direction is always parallel to lower polarization in optic-axis sections, resulting in extinction at all positions of stage rotation (Figure 2.14c).

For sections that are not normal to an optic axis, there are an infinite number of intermediate positions of stage rotation where principal vibration directions are not in a plane parallel to the lower polarization. In these cases, repolarization occurs, and insertion of the upper polarizer does not result in extinction (Figure 2.14b). The two mutually perpendicular planes of repolarized light have an angular relation to the plane of polarization of the upper polarizer and are partially resolved to that plane resulting in light transmission to the viewer.

Two fundamental effects occur for all anisotropic minerals observed between crossed polars: (1) starting at an extinction position (i.e., with a principal vibration direction or its projection parallel to lower polarization), extinction occurs every 90° of rotation, alternating with maximum illumination at the 45° position, or (2) extinction occurs at all positions of stage rotation, meaning that a principal vibration direction is continuously parallel to lower polarization as the crystal is rotated.

In the first case, extinction occurs if any of the principal vibration directions *O, E, X, Y,* or *Z* lies in the plane of the microscope stage and parallel to lower polarization. Extinction also occurs if the vibration direction is not actually parallel to the stage but lies in a vertical plane that is parallel to lower polarization. In this case *E'*, *X'*, or *Z'* projected to the stage surface lies parallel to polarization. For oblique cuts through crystals, the vibration directions will also project to the stage plane from nonvertical planes, and *E'*, *X'*, or *Z'* may be brought parallel to lower polarization. In any

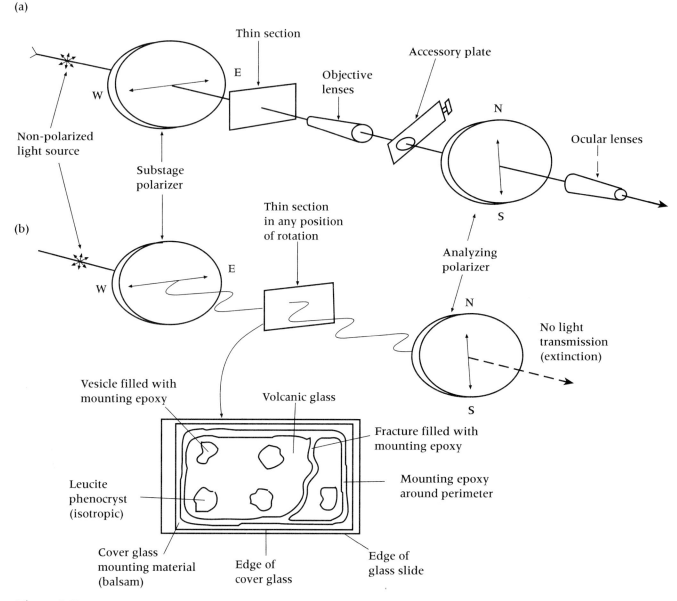

Figure 2.13

(a) Optic train of polarizing microscope showing position of lower and upper polarizer and other microscope components. (b) Illustration of how crossed polars results in extinction and how this phenomena occurs when isotropic materials are viewed between crossed polars.

other position of stage rotation, extinction does not occur, and there is repolarization into two mutually perpendicular planes with light being resolved in the upper polarizer.

The intensity of this resolved light varies from none at all at extinction to maximum brightness in the 45° position with respect to the polarization directions. In practice, maximum brightness appears to persist even when the position of rotation is as little as 6° (Figure 2.15).

In the second case of fundamental effects with crossed polars, extinction occurs in all positions of rotation if the crystal has been sectioned perpendicular to an optic axis. There is one possible section for uniaxial crystals, two for biaxial crystals.

An example of these two fundamental cases of extinction is shown in Figure 2.16. Three crystals of quartz are shown in two different positions of stage rotation. One crystal remains extinct in both positions of stage rotation, indicating that it is a section normal

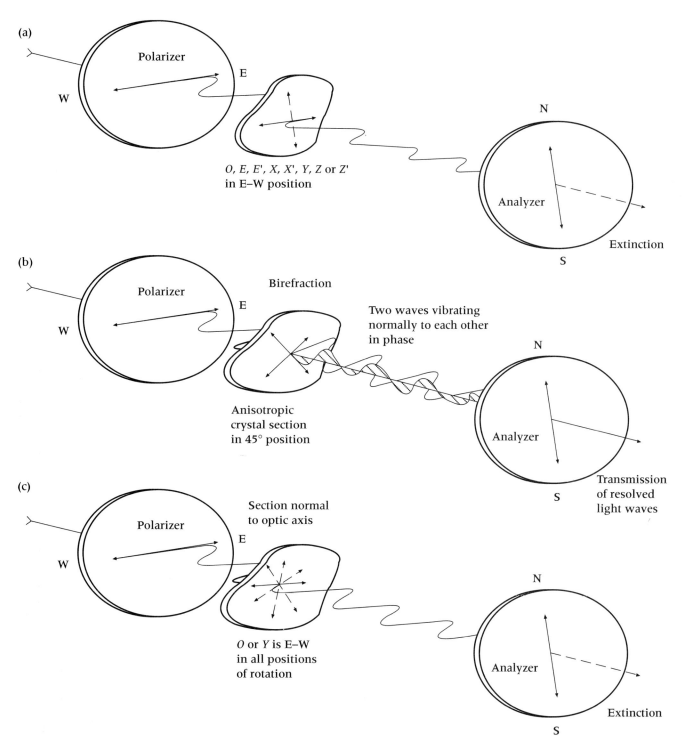

Figure 2.14
The effect of orientation of anisotropic crystal sections on light with crossed polars.

(a) Light vibrating in any principal vibration plane of crystal in E–W position is canceled by upper polarizer. (b) In any position not E–W, light entering crystal is split into two mutually perpendicular vibration directions with partial resolution into N–S plane of upper polarizer. (c) Section cut perpendicular to optic axis always presents a principal vibration direction E–W, with undeviated light being canceled in upper polarizer.

Figure 2.15
Transition from extinction to apparent maximum illumination has taken place within 16° as shown by angle described (black lines) by position of cleavage in extinction portion of crystal (bottom) and cleavage in illuminated portion (top). Medium gray is achieved with as little as 6° deviation from extinction position (middle portion). Enstatite crystal with deformation bands. Deformed enstatite crystal. X-polars. Bar scale = 0.5 mm.

to an optic axis. Two crystals go in and out of extinction, indicating that principal vibration directions are sequentially in line with substage polarization, resulting in no double refraction and cancellation of all light by the analyzing polarizer.

A fibrous variety of quartz, known as chalcedony, may occur with the fibers radiating from a center. The fibers are elongate either in the c-crystallographic direction or along an a axis. Each fiber is an individual crystal. A fiber should be extinct (1) if its O vibration direction is parallel to substage polarization), (2) if its E vibration direction is parallel to substage polarization), and (3) if its optic axis (c axis) is in a vertical position. These conditions are satisfied where fibers are N–S, E–W, and vertical. The corresponding extinction forms an extinction cross (Figure 2.17), the center of which is the position of fibers that are normal to the microscope stage. As the stage is ro-

(a)

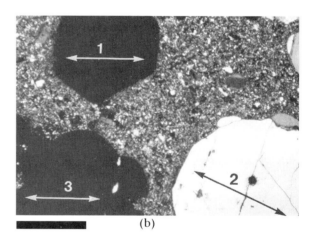
(b)

Figure 2.16
Extinction positions in three quartz phenocrysts.

(a) Crystal (1) with hexagonal shape is optic-axis view (white double arrow is O parallel to E–W polarization); crystal (2) has either E' or O parallel to E–W polarization (white double arrow); and crystal (3) has E' or O (black double arrow) not parallel to E–W. (b). Rotation of thin section results in maintained extinction for crystal (1) (O is E–W); illumination of crystal (2) (no principal vibration direction E–W); and extinction of crystal (3), in which a principal vibration direction has been brought parallel to E–W. X-polars. Bar scale = 1.6 mm.

Figure 2.17
Extinction position in K-feldspar "fibers" that have a principal vibration direction oriented parallel and perpendicular to E–W substage polarization, defining an "extinction cross." Transition from complete extinction to apparent maximum illumination occurs in fibers slightly out of alignment, analogous to the gray "subgrain" in the enstatite Figure 2.15. Devitrification spherulites in silicic volcanic, with sanidine crystal at one center (lower right). X-polars. Bar scale = 0.5 mm.

tated, non-vertical crystal fibers systematically move in and out of these extinction positions, and, although the fibrous cluster turns, the extinction cross remains in a fixed position as orientations proper for extinction are eliminated and others move in to take their place.

Another example of extinction is illustrated by the three crystals of andalusite shown in Figure 2.18. The extinction of the elongate prismatic section occurs when the elongate direction is placed E–W (as shown) or N–S. In either position, a principal vibration direction is parallel to E–W substage polarization. Since the extinction position is parallel to crystal elongation, it is termed **parallel extinction**. The other two crystals shown are prism cross sections. In one, the extinction position bisects the angle defined by the intersection of the prism faces, and this is termed **symmetrical extinction**. The other cross section is illuminated, and no principal vibration direction is parallel to substage polarization in its shown position.

If extinction occurs at some angle to crystal elongation, such as a non-paired crystal face, twin plane, or cleavage it is known as **inclined extinction**. The degree of inclination with respect to cleavage in amphiboles and clinopyroxenes (viewed on 010) is

Figure 2.18
Extinction in relation to crystal morphology. Extinction parallel ("parallel extinction") to long prism section (bottom) and symmetrical ("symmetrical extinction") with respect to angle of intersection of prism faces (white arrow) of andalusite. X-polars. Bar scale = 1 mm.

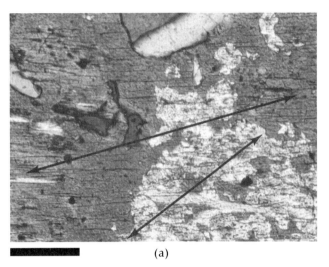 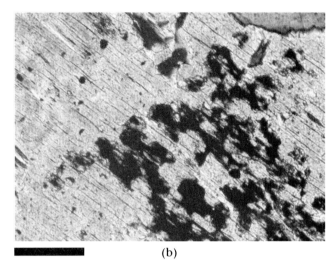

Figure 2.19
Extinction angle between Z and c for clinoamphibole (hornblende) that has replaced all but some interior patches of clinopyroxene. X-polars. Bar scale = 0.15 mm.

(a) Extinction position (Z) for clinoamphibole (longer arrow) defines a significantly smaller angle with respect to cleavage than extinction position (Z) for clinopyroxene (shorter arrow). (b) Extinction position for the clinopyroxene obtained by rotating section so that Z in clinopyroxene is E–W.

very important in (1) distinguishing clinoamphiboles and clinopyroxenes, and, in some cases, (2) identifying the variety of amphibole or pyroxene. Clinoamphiboles typically have a much smaller inclined extinction angle in comparison to most clinopyroxenes (Figure 2.19). The **extinction angle** is obtained by orienting cleavage or the trace of a {110} crystal face parallel to the E–W (or N–S) cross hair in the microscope ocular, reading the microscope stage position (in degrees), and then turning the stage until the mineral grain is extinct. The angular difference between the new position and the starting position is the extinction angle.

Birefringence and Interference Colors

As previously discussed, when light waves pass through anisotropic crystals in most positions of stage rotation, there is birefraction generating light vibrating in two planes perpendicular to each other and traveling in different directions. It was indicated that since these new directions of light travel encounter different configurations of ions in the crystal lattice, they travel at different velocities.

Any two vibration directions in anisotropic crystals have their own value of refractive index associated with it and have their own travel velocity. Except for optic-axis views, there is always a larger value (n_h) and a lower value (n_l) of refractive index. The difference between the high and low values for any given crystal in any given orientation can be expressed as the **presenting birefringence**:

$$\delta_p = n_h - n_l$$

Except for thin sections cut through crystals perpendicular to an optic axis, all sections of anisotropic crystals will present birefringence. **Maximum birefringence** (δ), however, is presented only with a particular orientation of the crystal. Maximum birefringence for **uniaxial crystals** is either $n_\omega - n_\epsilon$ or $n_\epsilon - n_\omega$ depending on the relative values of refractive index along the principal vibration directions O and E for the mineral. Maximum birefringence for biaxial minerals is always $n_\gamma - n_\alpha$. For both uniaxial and biaxial crystals, the principal vibration directions, E–O or X–Z, must lie in the plane of the microscope stage for observation of the optical effects relating to maximum birefringence. All anisotropic crystals can present any birefringence between the maximum for the specific mineral being examined to no birefringence in sections in which the difference between the high- and low-value index becomes zero; these are the optic-axis sections.

The two birefracted waves passing through anisotropic crystals are referred to as the **fast wave** (or ray) and the **slow wave** (or ray), as an expression of their different velocities. Observe the effects of birefraction using a monochromatic light source, such as the yellow–orange produced by a sodium vapor lamp with a wavelength of 589 nanometers (nm). The slow direction for sodium light traveling through an anisotropic crystal will result in a lag with respect to sodium light traveling along the fast direction. This lag is known as **retardation** or **path difference** Δ. The longer the two waves are traveling at different velocities through the crystal, the more retardation there will be. Thickness (t) of the crystal, then, becomes a major factor. The relationship is

$$\Delta = t(n_h - n_l)$$

If thickness is held constant at the standard section thickness (30 microns), the amount of retardation will depend on the mineral being observed and its orientation in the thin section, since both of these factors relate to $n_h - n_l$ (Figure 2.20). Retardation of one wave of the sodium light with respect to the

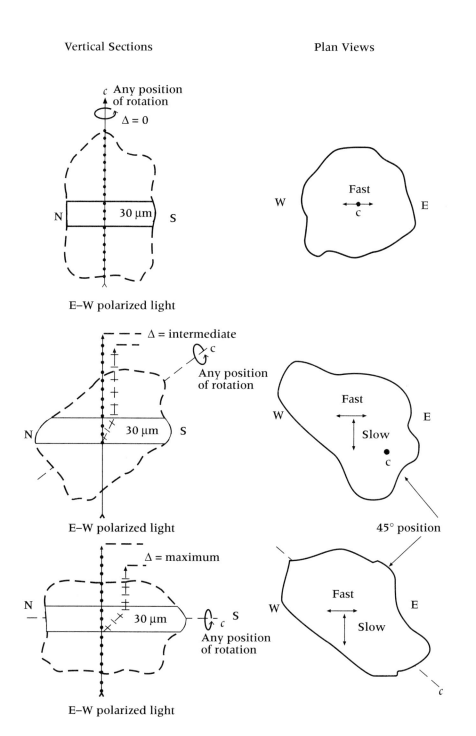

Figure 2.20
Relation of crystal orientation to path difference and retardation shown for a uniaxial crystal using a monochromatic light source.

other is expressed as a **phase difference** or **phasal difference** (P).

Light enters an anisotropic crystal vibrating in a plane governed by the lower polarizer. When the light exits the anisotropic crystal along two perpendicular planes, one wave is retarded relative to the other. Two waves in phase and vibrating in a common plane result in constructive interference (yielding maximum brightness), whereas if they are completely out of phase there is destructive interference (darkness). Now examine two cases of retardation for a single wavelength of light. A retardation of $\frac{1}{2}$ and a retardation of 1 are shown in Figure 2.21. In both cases substage polarized light is converted to two planes of light

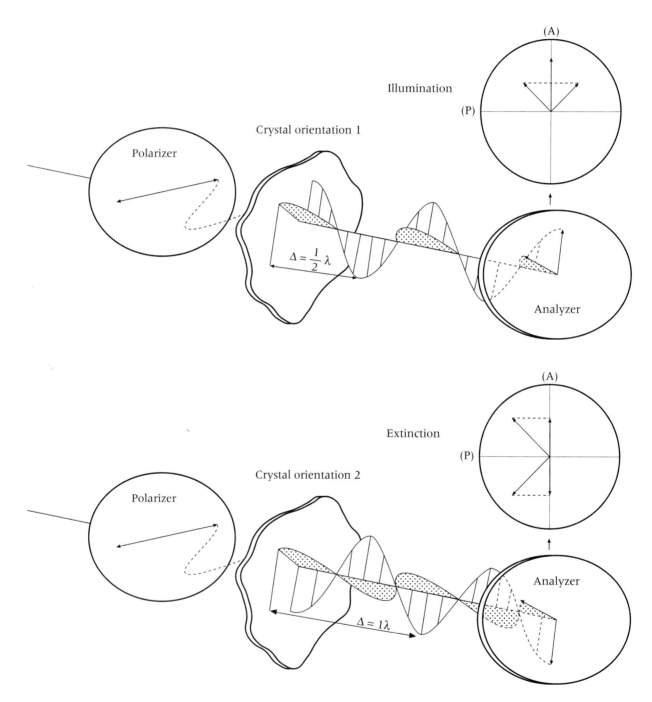

Figure 2.21
Resolution of monochromatic light in analyzing polarizer that has been birefracted and retarded a half-wavelength (crystal orientation 1), yielding illumination, and one full wavelength (crystal orientation 2), yielding extinction.

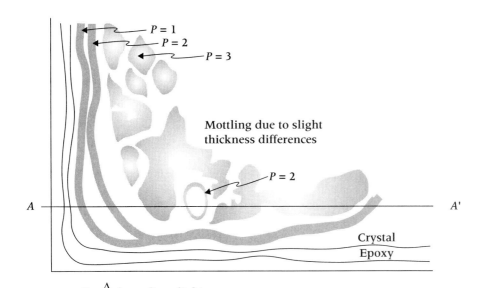
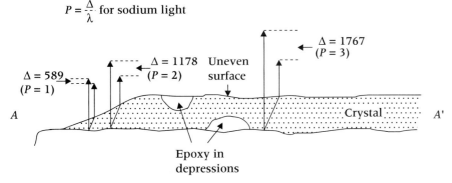

Figure 2.22
Extinction bands and extinction areas generated in variable-thickness anisotropic crystal using monochromatic sodium light. Principal vibration directions are in the 45° position. Extinction occurs where phase difference is 1, 2, and 3.

if the principal vibration directions are not parallel to substage polarization—for example, if they are in the 45° position. If the two waves are in phase, there should be constructive interference and intense light. If the two waves are out of phase, there should be destructive interference and darkness.

A crystal of variable thickness (Figure 2.22), using polarized sodium light, for example, should be black at those places where the thickness is just the value to make $\Delta = \frac{1}{2}\lambda$ of sodium light. We look at the crystal and see no black areas or bands. What is wrong?

Remember that the slow and fast sodium light waves are vibrating in planes at right angles to each other, not in a common plane. Since the effects of wave interference can only be seen if the waves are vibrating in a common plane, how can this condition be created? The analyzing polarizer only accepts light vibrating in a single plane. If our crystal is viewed with the analyzing polarizer in place, components of both waves as they vibrate in the single plane of the analyzer can be observed. Resolution of waves retarded half of a wavelength generates apparent constructive interference (with the analyzer in place), and resolution of waves retarded a full wavelength appears as destructive interference as shown in Figure 2.21.

Now re-examine our crystal viewed with polarized sodium light with the analyzing polarizer in place. **Extinction bands** and extinction surfaces appear (Figure 2.22), indicating that destructive interference is occurring in these places. Since destructive interference occurs at whole wavelengths of retardation, there should be extinction when the retardation is $n\lambda$ for sodium light. The wavelength of sodium light is 589 nm, so extinction occurs when the retardation is $589n$, $1178n$, and $1767n$. The relation is expressed as the phase difference:

$$P = \frac{\Delta}{\lambda}, \quad \text{where } \Delta = t(n_h - n_l)$$

Suppose $n_h - n_l$ were 0.06 for our mineral with a given orientation in the thin section. Since 0.03 mm = $3 \times 10^4 n$, $\Delta = 3 \times 10^4 \times 0.06 = 1800n$. Since λ for sodium light is $589n$, destructive interference will take place according to the calculation

$$P = 1800/589 = 3$$

A birefringence has been selected that will result in extinction of the entire crystal, but suppose the crystal tapers from 0.03 mm to zero thickness (Figure 2.22). In thin section this could be observed along the

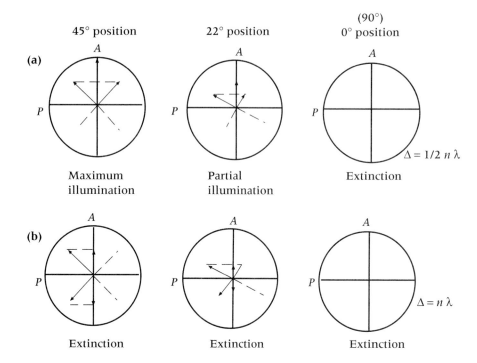

Figure 2.23
Effect of retardation and orientation of two principal vibration directions on the appearance of an anisotropic crystal viewed between crossed polars using monochromatic light.

(a) Retardation is $\frac{1}{2}n$ wavelength, causing constructive interference and illumination. (b) Retardation is $1n$ wavelength, resulting in destructive interference and cancellation of light in all positions of stage rotation in upper polarizer.

perimeter where lapping may have reduced mineral grains to less than the maximum 0.03 mm standard thickness. Such a reduction may also occur where there are depressions in the rock (and mineral) surface or where tapering is related to overlap with other grains. Consider the tapered crystal in which epoxy used to mount the rock plug (chip) onto the glass slide or to mount the cover glass onto the rock section after lapping has filled in below or above the tapered crystal, effectively bringing the thickness up to 0.03 mm (Figure 2.22). Since epoxy has no birefracting properties, t in the phase difference equation becomes a variable:

$$P = \frac{t(0.06)}{589n}$$

Two extinction bands are generated (Figure 2.22), one at $P = 1$ and another at $P = 2$. The extinction surface is at $P = 3$, although it can be expected to be somewhat mottled since the surfaces of minerals are not perfectly planar.

What happens if the crystal is rotated out of the 45° position? When a zero or the 90° position is reached, the entire crystal is extinct regardless of thickness differences (Figure 2.23a). In these positions there is no birefraction, and all sodium light is canceled by the analyzing polarizer. Of course, if $\Delta = n\lambda$ (Figure 2.22), there would be extinction in all positions if the mineral section were evenly $P = 1$, $P = 2$, or $P = 3$.

The sodium light visible between the extinction bands and areas is of maximum intensity with the vibration directions set at the 45° position. Intensity drops undetectably slowly and then very rapidly as the plane of substage polarization is approached (Figure 2.15). As the angular relation between the vibration directions in the crystal and the analyzing polarizer changes, so does the resolution. In effect the amplitudes of the waves resolved are reduced, resulting in a reduction of light transmitted to the observer (Figure 2.23a). Since the positions of extinction bands and areas do not change with respect to the crystal as rotation occurs, it can be concluded that there is no change in retardation as the stage is turned.

The optical principles attendant to monochromatic light passing through anisotropic crystals apply to all the wavelengths of light making up **white light**. Since white light is used in routine optical studies, the optical phenomena become more complex. White light consists of the combination of all the eye-sensitive colors ranging from violet to red, corresponding to wavelengths from $400n$ to $700n$. With white light, wavelength becomes a variable in the equation

$$P = \frac{\Delta}{\lambda}$$

This means that if P is a whole number for a given wavelength, it will be some other number for the other wavelengths present.

Extinction bands cannot occur for white light as a unit, since at a retardation providing total destructive

2 Orthoscopic Optical Mineralogy   31

Average wavelengths in nanometers for the various colors of white light									
410 violet	440 indigo	470 blue	515 green	560 yellow 589 sodium	620 orange	700 red	retardation $t(n_h - n_l)$	Resulting interference color observed	Order
							100	Gray	1st
							200	White	
							300	Yellow	
410							400	Orange	
	440	470					500	Red 550	
			515	560			600	Violet	2nd
					620	700	700	Blue	
820	880						800	Green	
		940					900	Yellow	
			1030				1000	Red	
							1080	Violet	
				1120			1200	1100 Blue	
1230					1240		1300	Green	3rd
	1320						1400	1400 Yellow	
		1410					1500	Orange	
			1545				1600	Red 1650	
1640				1680			1700	Blue	
	1760						1800	Green	
		1880		1860			1900	Yellow	4th
							2000	Orange	
2050		2060					2100	2100 Orange	
							2200	Pink 2240	

Figure 2.24
Component wavelengths of "white light" and their destructive interference at whole multiples of wavelength retardation. Observed interference colors result from dominance of certain colors governed by retardation provided by the crystal. Modified from Kerr (1977).

interference for a given wavelength of color other colors will be in constructive interference. Some part of the visible spectrum will always be transmitted. The colors transmitted, resulting from retardations obtained in a crystal, are known as interference colors (Figure 2.24). Colors for which a whole-number value of its wavelength is close in value to the prevailing retardation will be reduced or absent; those farthest away will be transmitted as some color mix. For example, red (actually magenta or purplish red) will be transmitted when the prevailing retardation of 550, 1100, or 1650 is realized, even though monochromatic red light has a wavelength of 700n, with repetition at 1400n and 2100n (Figure 2.24). Notice that most colors repeat in groups of colors known as orders.

If the analyzing polarizer is not in place, no interference colors are observed, just as no extinction bands were observed using monochromatic light. How are interference colors generated in the analyzing polarizer? If thickness and difference in presenting refractive index of a crystal are just right to provide a phasal difference for yellow light at $\frac{1}{2}\lambda$, there is resolution of yellow light in the analyzer. At the same time there is destructive interference for the blue light, which will be retarded one full wavelength. Other colors will be proportionally resolved or canceled, leaving an observed interference color that is the **presenting interference color**, which has a direct relation to **presenting birefringence**.

Determination of Birefringence

Every anisotropic mineral has its own maximum birefringence as governed by its maximum range of refractive index. If the maximum birefringence of a mineral can be determined, it constitutes a characteristic that is very useful for discriminating one mineral from another. How then can the maximum birefringence be determined if the high and low values of refractive index are not known for a mineral that has not been identified? If we had that information, a simple subtraction of the low index value from the high value yields the birefringence. If the numerical value of birefringence were known, we could inspect an interference color chart (Figure 2.25 and interference color chart insert, p. 16) and find what would be the highest interference color that should be observed. Worked in reverse, by searching for a crystal orientation that presents the highest interference color, the charts can be used to find the numerical value of birefringence without knowledge of the actual values of refractive index.

There is a need to find the highest interference color for a given mineral type. Since it is the maximum birefringence that is so important, several crystals of the same mineral are examined in thin section to ensure that an orientation is presenting that yields close to maximum birefringence. Scanning a thin section containing an aggregate of crystals typically results in a reasonably close estimate of maximum birefringence as indicated by interference color. If there is a single crystal in thin section, and unless there is additional information, there is no way to determine whether the

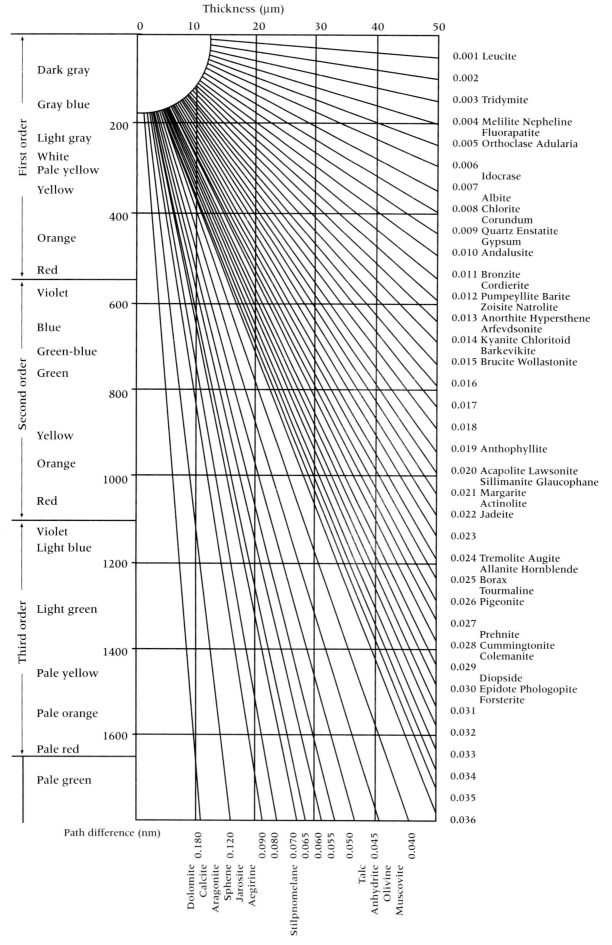

presenting interference color is maximum or some "lower" color.

Since colors repeat in several orders, it becomes imperative to identify not only the color but also the order to which it belongs. In practice, the birefringence of most of the rock-forming minerals is in the first order and the lower part of the second order. This fact is particularly convenient because these lower-order colors are more distinct than higher-order colors that become progressively pastel in character (thus the utility of thin section being 30 microns thick). However, a few minerals such as calcite and dolomite are common rock-forming minerals but have very high interference colors. Since there are few other minerals that have such high colors, the extreme birefringence is very revealing in itself, even if the precise color and order cannot be determined. Interference colors are very indistinct above order 3.

Identification of a specific interference color in the first, second, and third orders can be accomplished by (1) looking at tapered edges of crystals where colors descend in sequence as thickness decreases and (2) using what is known as the **red-1 retardation plate**. In both procedures, the sequence of color changes is identical, and the maximum interference color for a given crystal orientation can usually be identified.

Tapered crystal edges are very common in thin section. Several examples are shown in photomicrographs of Plate 1. Tapering is likely to occur at the margins of thin sections and toward open fractures or plucked areas. If a white interference "color" appears, other colors can be traced very easily since white only occurs in the first order. For example the sillimanite crystal shown in Plate 1A is white at one end of the prism. Colors rise along the prism to yellow, red, and finally blue. The maximum presenting birefringence is blue, about 0.021 on the interference color chart (Figure 2.25). This must be blue of the second order because there is no blue in the first order and there was no sequence of colors leading to third-order blue.

Minerals with higher birefringence may have color bands on the tapered margins. The very edge of the olivine crystal shown in Plate 1B is white. Colors can be traced all the way to fourth-order green. One way to do this is to look for the occurrence of red bands (or another selected color), much as marker beds are used in stratigraphic correlations. Three red bands along the thin edges of the olivine crystal shown in Plate 1B indicate the color of the main surface of the grain (pale green) must be above third-order red, namely green of the fourth order.

Since crystal orientation affects presenting birefringence, in addition to thickness, the interplay between orientation and thickness in a single thin section becomes important. The quartzite shown in Plate 1C thins toward the margin of the thin section. Some orientations of the quartz crystals in the interior of the section present an orange interference color. This orange must be in the first order because thin edges of some of these crystals are white, and white is only a first-order color. Other quartz crystals in the interior zone range from yellow to white to gray to black. Except for the likelihood that some of the crystals are in an extinction position (even some orange and yellow crystals may be extinct), the range in colors is an expression of crystal orientation, progressing toward black for optic-axis sections.

Notice that there are no orange nor yellow crystals in the marginal zone of the thin section (Plate 1C). This indicates that orientations presenting these colors in the interior are now too thin and present white instead. Yellows become white, grays become dark grays, and dark grays become black in the tapered zone.

What is the maximum birefringence of this mineral? Since there are many crystal orientations to observe, the maximum is likely to be presenting. The maximum presenting birefringence is orange, but is this the birefringence (δ) of the mineral? Based on other information, it is known that this rock is a quartzite. The birefringence of quartz is 0.009, or a cream color (Figure 2.25). The orange and yellow presenting in our thin section is higher than cream color and the section is too thick. The pale yellow or near-cream-colored crystals in the middle zone are an indication that the portion of the section there is about 0.03 mm (30 microns) thick. The margin is thinner and the interior is thicker than a standard this section. Not all thin sections are prepared properly, and the investigator should be on the alert for sections, or portions of sections, that are too thin or too thick. If quartz is yellow, and certainly if feldspar ($\delta = 0.005 - 0.007 =$ white) is yellow, the section is too thick.

The second way of determining presenting birefringence and birefringence (δ) is by using the red-1 accessory plate. The plate is a crystal (formerly gypsum, now quartz) section cut to yield a retardation of $550n$ (a magenta red). Insertion of the plate between a crystal and the upper polarizer raises or lowers any presenting interference color by an amount equivalent to $550n$. Being able to raise or lower interference colors by a whole order has the following advantage. If a presenting interference color is second order, but you are not sure about that, lowering to first order presents some unique first-order color, perhaps even white. Raising to third-order generates pastel colors, characteristic of the third and fourth orders.

Figure 2.25
Michel–Lèvy chart giving relation of interference colors to thickness and birefringence of common rock-forming minerals.

The red-1 plate is anisotropic and has two positions of mutually perpendicular vibration directions lying parallel to the plane of the thin section. The plate is inserted either from the NW or SE position, and the vibration directions are always in the 45° position. The position of the slow direction is always SW–NE, and the fast direction is, therefore, NW–SE.

Except for optic-axis views, all anisotropic crystals also have slow and fast directions presenting in the section of a crystal. When the stage is rotated so that the slow direction in the crystal coincides with the NE–SW oriented slow direction in the red-1 plate (and fast will then coincide with fast), retardation is increased and interference colors "rise." If the fast direction in the crystal is superposed with the slow direction in the plate, retardation is reduced and interference colors "fall" (Figures 2.26, 2.27).

The presenting interference color of the anhydrite crystal shown in the photomicrograph of Plate 2A is shades of blue with local mottling to red (and white). The mottled coloration results from slight differences in thickness. Plucking during thin sectioning has produced a hole in the mineral, its shape being guided by the right-angle cleavage characteristic of anhydrite. The hole is black with crossed polars since glass and epoxy are isotropic. If the crystal is in a position for constructive interference with the red-1 plate inserted, colors rise (Plate 2B). Black becomes the magenta of the red-1 plate, the thin white margin of the crystal facing the hole becomes blue (white + magenta = blue), and the major surface of the crystal becomes red (red + red = second-order red) and green (blue + red = third-order green).

If the anhydrite crystal is in a position that results in destructive interference with the red-1 plate inserted, the black hole again becomes magenta of the red-1 plate, the white rim remains white (first-order red – white = white), and the major surface of the crystal becomes mottled white (blue – red = white) and black (red – red = black). Since the birefringence of anhydrite is about 0.04 (third-order blue) and since only second-order blue is presenting (Plate 2A), the crystal must be in an orientation that does not present maximum birefringence (assuming 0.03-mm thickness).

The red-1 retardation plate is also used to determine whether the fast or slow vibration direction lies parallel to prism elongation. This determination is useful in identifying some minerals that may not have many other diagnostic properties. The test is to observe a rise or fall in interference colors when the red-1 plate is inserted over the mineral. If colors "rise," the slow direction in the red-1 plate is parallel to the slow direction in the mineral, and, correspondingly, fast parallels fast. If the slow direction in the crystal is parallel to elongation, the crystal is **length slow**, and if the fast direction is parallel it is **length fast**. The long section of andalusite oriented NW–SE in Plate 3A has a pale yellow presenting birefringent color. Insertion of the red-1 plate raises the color to second-order greenish yellow (Plate 3B). This must be second-order greenish yellow because a blue shows along margins and blue is already in the second order. This rise in colors indicates that the slow direction in the mineral is across the length of the prism. Thus, the andalusite must be length fast.

There are two prism cross sections in Plate 3. One is at an extinction position (Plate 3A) and presents only the magenta of the red-1 plate (Plate 3B). The other section changes from white to yellow with insertion of the red-1 plate. This indicates that a diagonal of the square prism section positioned NE–SW is a fast direction, since colors have fallen (magenta – white = first-order yellow). This means that the slow of the red-1 plate is superposed over a fast direction in the min-

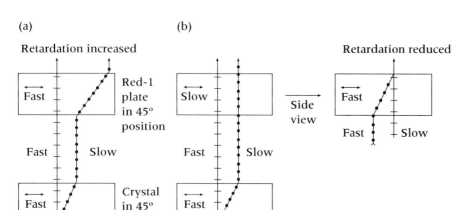

Figure 2.26
Effect on path of fast and slow waves passing through anisotropic crystal and then through red-1 retardation plate. With crystal in 45° position, retardation will either be increased (a) or decreased (b). Modified from Kerr (1977).

2 Orthoscopic Optical Mineralogy   35

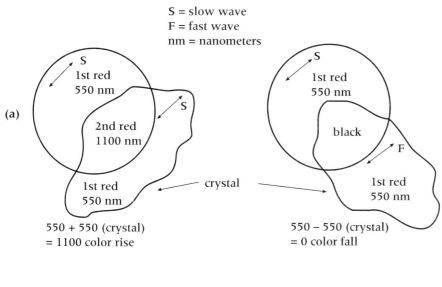

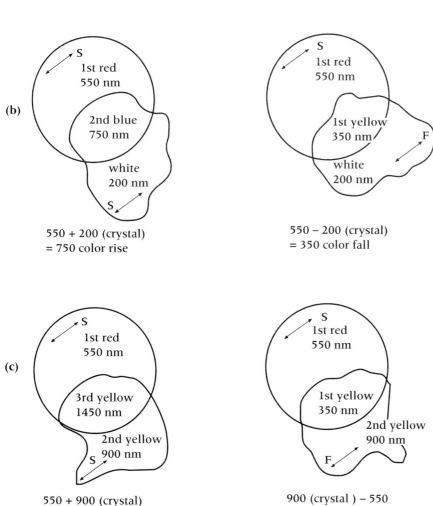

Figure 2.27
Diagrammatic examples of rise or fall of interference colors obtained by insertion of red-1 retardation plate. Since fast and slow directions in plate are known, fast and slow directions in crystal can be determined.

(a) Mineral such as lawsonite.
(b) Mineral such as feldspar.
(c) Mineral such as diopside.

of the red-1 plate is superposed over a fast direction in the mineral.

There is another accessory plate useful in determining the slow and fast directions in minerals. This is the **mica (glimmer) plate**, and its retardation is half that of the red-1 plate (ca. 137n), and the rise or fall of interference colors will be correspondingly less.

Anomalous interference colors typify some anisotropic minerals with low to very low birefringence. Such a property is very useful for identification since only a few minerals have this property, namely, clinozoisite, idocrase, melilite, some chlorite, and iron-poor zoisite. The anomalous colors commonly are deep blue, indigo, purple, or brown, but they are not equivalent to any "normal" interference color. The abnormal color may appear only for certain orientations of the mineral, but the appropriate orientation is likely to occur if there are several crystals in the thin section. The abnormal colors are related to an optical characteristic known as **dispersion**. Dispersion is related to difference in refractive index for the various wavelengths of light making up white light and varies according to the mineral species and its orientation. Dispersion is also related to differing absorption power of crystals. Refractive index values given for minerals actually refer to values obtained using sodium light; otherwise a range of indices would be reported corresponding to refraction at other wavelengths.

Interference colors may be modified for another reason. Interference colors of pleochroic minerals are anomalous in a sense, since the interference color is superposed with the "natural" color of the mineral (in thin section) in its various orientations. Minerals with relatively strong pleochroism, such as biotite and hornblende, present these modified interference colors, and somewhat more care must be exercised in determining just what the color and order are for these minerals. This type of modification is much less obvious than the anomalous interference colors related to dispersion and is not very useful in mineral identification.

3

Conoscopic Optical Mineralogy

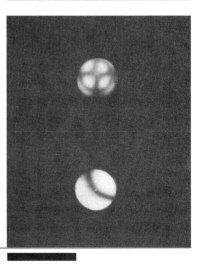

(top) Calcite (uniaxial) optic axis interference figure that a has a few degrees 2V of isogyre separation (slightly biaxial) due to lattice strain. Crystal is in marble that orthoscopically has calcite crystals with bent twin lamellae and serrated grain interfaces. (bottom) Anhydrite (biaxial) optic axis interference figure with curvature indicating about 40° 2V.

The conoscopic technique requires a crystal large enough so that optical effects from adjacent grains do not interfere. A "high"-power objective (45×–60×) must be used to obtain the desired optical effects; thus a crystal as small as 0.1 mm in diameter is quite satisfactory. Clay minerals and any other mineral that occur as exceedingly fine-sized crystals are not suitable for optical work of any kind and must be identified by x-ray diffraction. Rocks that are very coarse grained, or those that contain only a few grains of the mineral being identified, may not have a grain in proper orientation for the critical conoscopic observations to be made. Most granular rocks will have some suitable grain orientations.

Conoscopic Light and Interference Figures

Petrographic microscopes are designed with a condensing lens that can be swung into the optic train just below the thin section (Figure 3.1). The lens should be as close to the glass of the thin section as possible; this can be adjusted by racking the substage unit up. In this position the lens focuses E–W polarized light within the crystal being observed. The crossover light vectors are received in the high-power objective. If the mineral is in proper focus, the objective lens will be very close to the top of the thin section (be careful!). Most high-power objectives have a spring-loaded lens that depresses on contact when the operator inadvertently racks the microscope stage with the thin section up onto the objective lens (the microscope tube is racked down in older models).

Since light is emitted along different paths from the interior of a crystal, a partial three-dimensional view of the crystal structure is obtained (Wahlstrom, 1979; Nesse, 1991). Since some of the light will be traveling along principal vibration planes in the lattice, there will be areas that are in extinction with the upper polarizer in place. Other areas will yield double refraction and will be light with variable intensity and with variable presenting interference colors, just as would be expected of different sections and positions of extinction and near extinction in orthoscopic views.

The extinction areas form **isogyres**. The configuration of isogyres and an interisogyre area make up an **interference figure**. The nature of this figure is a function of the mineral, its orientation, and the position of stage rotation; thus, it is an optical expression of crystal structure and a clue to mineral identity.

One more manipulation must be made before interference figures can be actually observed. If an ocular is removed, small figures can be seen "far" down the microscope tube. With the ocular inserted, light is

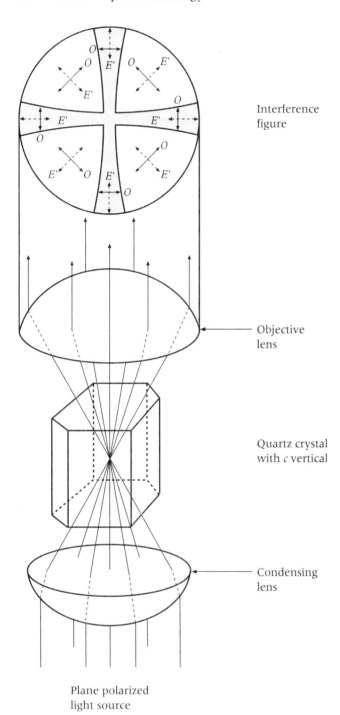

Figure 3.1
Conversion of light within crystal by means of condensing lens. Light emitted from crystal is seen as an interference figure. Modified from Nesse (1991).

so dispersed that figures are not seen at all. The best view of an interference figure can be obtained by inserting the **Bertrand lens** (installed in most microscopes) located between the retardation plate position and the ocular(s). With the lens in place, interference figures can be viewed in a much broader field of view.

Conoscopic Optical Information

Since an interference figure consists of extinct and non-extinct portions, it follows that only anisotropic crystals can generate figures. The character of diagnostic interference figures relates to (1) **uniaxial** and **biaxial** character, (2) the location of the high and low positions of refractive index relative to the optic axes (**optic sign**), and (3) the angle of intersection of the two axes in the biaxial group (**optic angle**). Any or all of this information may be useful in mineral identification.

Not all crystal orientations yield diagnostic interference figures. Concentrate on the optimal crystal orientations because they are totally reliable regardless of the experience of the petrographer. Try to select a crystal that has the lowest presenting birefringence. If the mineral occurs as a crystal aggregate, the lowest presenting birefringence can be seen in relation to higher presenting birefringence. If only one or two crystals occur in the entire thin section, conoscopic work may have to be abandoned, except for the investigator that is experienced in interpreting figures generated by sections that are not perpendicular to optic axes. The lowest presenting birefringence is a position normal or near normal to the optic axis of uniaxial crystals or to either of the optic axes of biaxial crystals.

For low-birefringent minerals, such as quartz and feldspars, birefringence normal to an optic axis produces extinction, a decidedly black extinction (Figure 2.16). For higher-birefringent minerals there is a **dispersion** effect that prevents the crystal from being

Figure 3.2
Interference figures generated by light transmission through various directions in (a) uniaxial minerals and (b) biaxial minerals. Views are shown with a principal vibration direction E–W followed by the optical effects appearing in the 45° position.

3 Conoscopic Optical Mineralogy

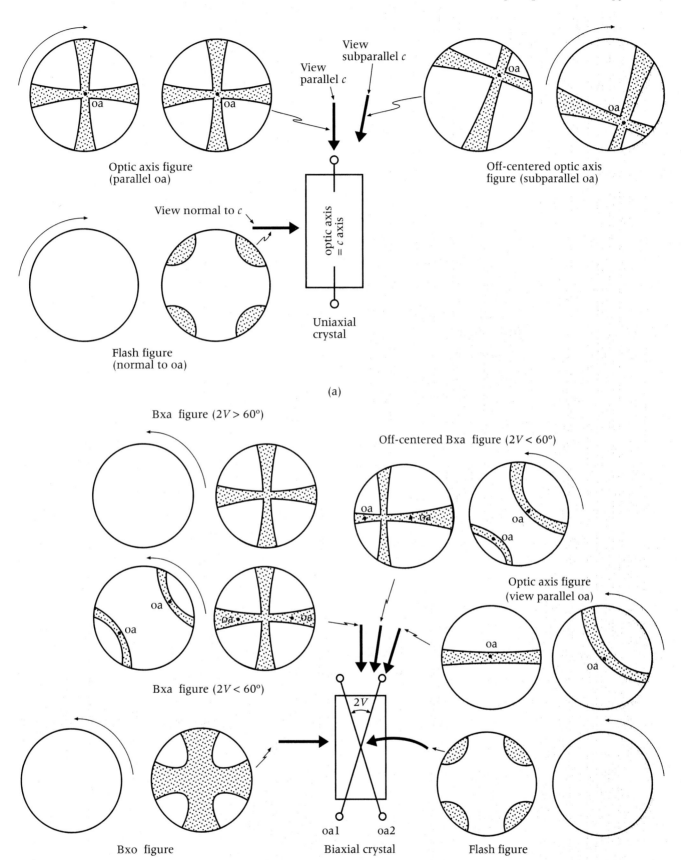

(a)

(b)

totally extinct; that is, the extinction position varies a little for each component of white light. As a result, even a section perfectly normal to an optic axis of a mineral with "normal" interference colors will present some shade of gray, bluish gray, or even bluish white if it has relatively high birefringence (Plate 4A).

Interference Figures

If an interference figure contains a well-defined, relatively thin **extinction cross** that is **centered** in the field of view, rotation of the crystal on the stage will result in one of two phenomena. If the cross remains stationary, the mineral is uniaxial and the view is down the optic axis and, thus, down the c-crystallographic axis. If the cross splits into two arcuate isogyres, which either remain or move out of the field of view (depending on the mineral species) as the crystal is turned to a 45° position, the mineral is biaxial. If the center of the cross in the starting position is not at the center of the field of view, the figure is **off-centered**. As the stage is rotated, a uniaxial off-centered cross will migrate around the field of view, and the biaxial off-centered cross will break into two isogyres of different size. These phenomena are sketched in Figure 3.2.

If an interference figure contains a single isogyre that does not leave the field of view on stage rotation, the mineral is biaxial and the view is down or close to one of the optic axes. If the section is perfectly normal to the optic axis, the isogyre will be straight in positions parallel to polarization. In the 45° position, it may be straight or arcuate, depending on the mineral species. If it is arcuate, the direction of curvature swings about a pivot point (optic-axis intersection point) every 90° of rotation. If the section is not exactly perpendicular to the optic axis, the pivot point of the isogyre will appear to wander as the stage is rotated. These relations are sketched in Figure 3.2b.

Determination of Biaxial Optic Angle (2V)

An imaginary line lying in a plane containing both optic axes of biaxial minerals and bisecting the acute angle made by the two optic axes is known as the **acute bisectrix** or **Bxa**. If the mineral section is perpendicular to the Bxa, there will be a well-defined extinction cross that separates into two isogyres with stage rotation, as previously described. A similar cross is formed in an **obtuse bisectrix (Bxo)** view, but this figure is less distinct than a Bxa (Figure 3.2b).

The acute angle defined by the optic axes is known as the optic angle and is designated $2V$. The magnitude (degrees) of $2V$ may be a critical diagnostic

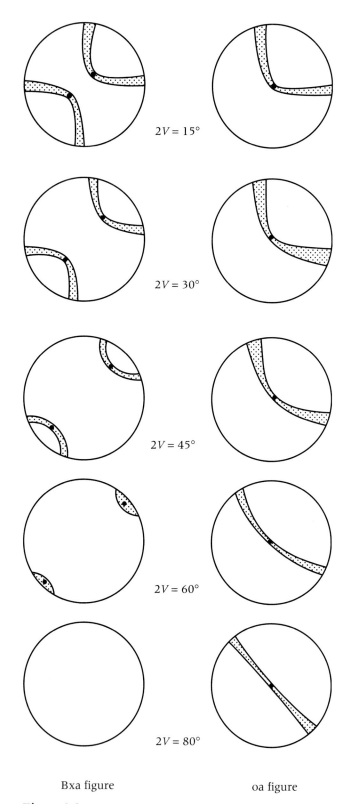

Bxa figure   oa figure

Figure 3.3
Estimation of 2V from separation of isogyres at 45° position (Bxa figures) and estimation of 2V from curvature of isogyres (biaxial optic-axis figures).

optical property for some minerals (e.g., pigeonite, sanidine); it is always a property that can be used in a confirmatory sense.

An estimate of the value of 2V can be made as follows (Tobi, 1956; Kamb, 1958; Wilcox, 1966; Bloss, 1981). If the section is a Bxa view, and if the isogyres do not leave the field of view on stage rotation, the distance between the pivot points (optic axis intersection points) is a function of the angle $2V$ (Figure 3.3). If the section is normal to an optic axis, $2V$ is a function of the curvature of the isogyre in the 45° position (Figure 3.3).

Determination of Optic Sign

If the high-refractive direction in uniaxial crystals is the E direction, the mineral is **uniaxial positive**. If the O direction has the higher refractive index, it is **uniaxial negative**. At first glance, it might seem that such a determination of optic sign could be made by observing relative refractive index with respect to some physical feature such as cleavage, twinning, or crystal form. There are two problems. First, the difference between the high- and low-index directions may be undetectable for low-birefringent minerals by cursory examination: second, crystal grains in rocks commonly have no expression of reference physical features. In rare cases it can be done. For example, the high- and low-index directions in a calcite rhomb are obvious (Figure 2.7) and the low-index direction is in the c-axis direction. Thus, calcite has been determined to be uniaxial negative without the use of interference figures. But calcite occurs most commonly as grains without clear-cut morphological expression, and in that case the relation of refractive index to crystal form is lacking.

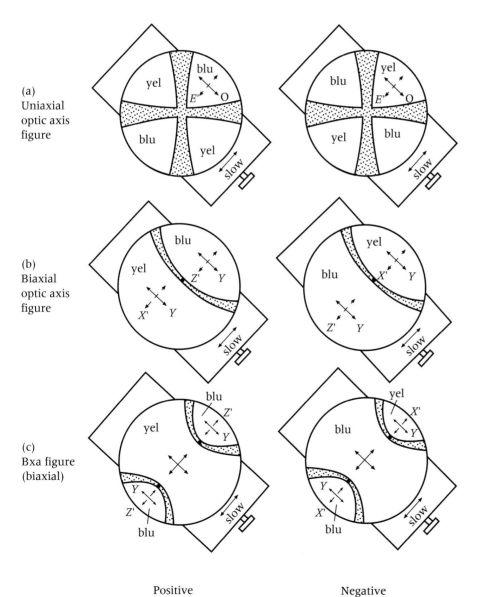

Figure 3.4
Optic sign determinations by generating interference colors with red-1 plate for uniaxial optic-axis figure (a), biaxial optic-axis figure (b), and Bxa figure (c). Example for mineral with first-order white birefringence such as quartz and feldspar.

A determination of optic sign of uniaxial minerals can be made very confidently if a centered or off-centered optic-axis figure is obtained. Insertion of the red-1 retardation plate causes changes in the interference colors presenting between the isogyres. The orientation of the slow and fast directions in the plate relate to a rise or fall in color or color bands in the various sectors of the figure (Figure 3.4a), similar to the determination of length slow and length fast. In practice, the change in color(s) in a particular sector is selected by the observer consistently. There are only a few uniaxial rock-forming minerals, but optic sign may be important in verification of identity, such as a positive sign would be for quartz.

Determination of optic sign for biaxial minerals is also a simple retardation manipulation, but only if there is obvious curvature of the isogyre in an optic-axis figure (Figure 3.4b), or if there is a confirmed Bxa figure (Figure 3.4c). If $2V$ is between about 60° and 90°, the isogyres in a Bxa figure leave the field of view and there is possible confusion with a Bxo figure (Figure 3.2b). If $2V$ is less than about 60°, the isogyres do not leave the field of view and a Bxa is ensured. The rise or fall in interference color or color bands can now be generated in the fields of the Bxa in the 45° position by inserting the red-1 plate (Figure 3.4c), and a specific sector is consistently watched by the observer.

If the curvature of a single isogyre in an optic-axis view of a biaxial mineral is very slight or nonexistent (45° position), the $2V$ is near 90°, and the distinction between positive and negative is very difficult to determine. If the mineral has a lower $2V$, curvature can be seen and the rise or fall in colors generated by insertion of the red-1 plate, on the concave and convex sides of the isogyre, is an indication of optic sign (Figure 3.4b). It is common practice to watch for a rise or fall of color or color bands in the sector on the concave side of the isogyre.

Rock-Forming Minerals

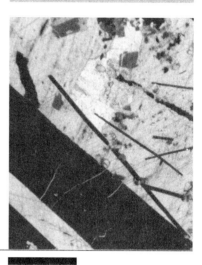

Specular hematite (black) in ankerite. Accordion-like bundles consist of chlorite. X-polars. Bar scale = 1.4 mm.

Most rocks contain some combination of minerals within the group: quartz, alkali feldspar, plagioclase, mica, chlorite, clay mineral, amphibole, pyroxene, dolomite, and calcite. Expansion of this list to about 85 minerals covers most of the rock-forming minerals encountered in examination of common rocks.

Eighty-five rock-forming minerals, not including clay minerals, are arranged according to **birefringence** and **refractive index** in Figure 4.1. These two optical properties roughly separate these common rock-forming minerals into groups when plotted against each other. This provides an initial means of narrowing the possibilities during mineral identification.

Table 4.1 lists the properties of 184 minerals, including the 85 in Figure 4.1. The other minerals are either more specific varieties of minerals or minerals that have restricted occurrence in rocks. Comprehensive works on the optical properties of rock-forming minerals are Deer et al. (1962a,b, 1963a,b, 1966, 1978, 1982, 1986), Troger (1979), MacKenzie and Guilford (1980), Phillips and Griffen (1981), Ehlers (1987), and Smith and Brown (1988).

Eighty-Five Rock-Forming Minerals

The most diagnostic properties of the 85 minerals in Figure 4.1 are listed alphabetically. In cases where the optical relations to crystal morphology are particularly useful in mineral identification, a sketch of these relations is provided (but text references are not). Minerals appearing in photomicrographs throughout the text are indicated.

Actinolite $Ca_2(Mg,Fe^{2+})_5Si_8O_{22}(OH)_2$ Monoclinic. Figure 4.2 An amphibole (124°/56° cleavage, diamond-shaped sections, small $Z \wedge c$) that is identical to tremolite in section unless it has clear-to-green pleochroism with increasing Fe content, especially **ferroactinolite**. Green actinolite in hand specimens may be clear in section. Actinolite occurs in Al-free rocks, such as some metamorphosed carbonates and some metamorphosed ultramafic rocks. Pleochroic varieties very similar to hornblende (Al bearing) and, in the absence of chemical data, terms such as **actinolitic hornblende** and **calcic clinoamphibole** can be used.

continued on p. 59

44 Part 1 Optical Mineralogy

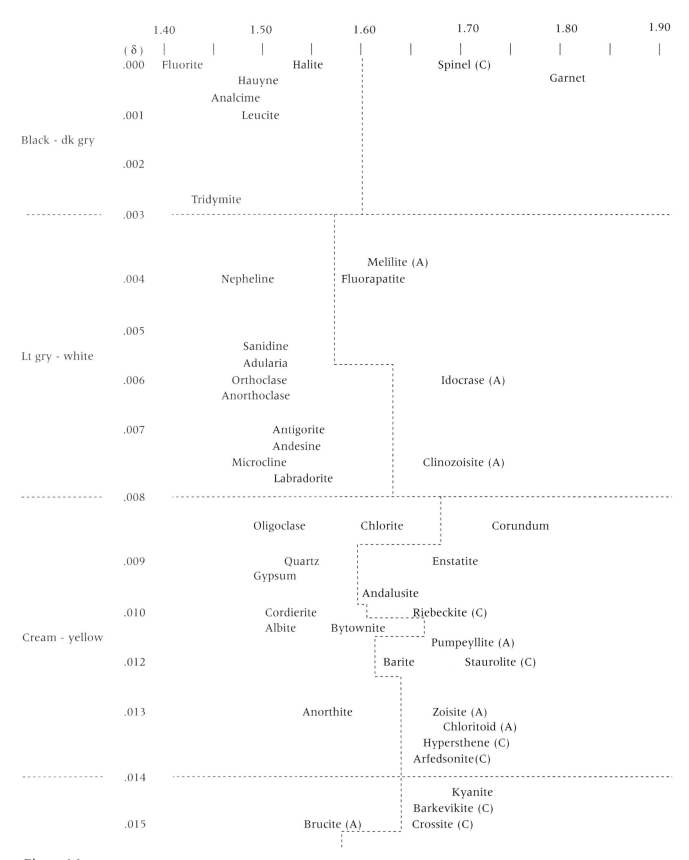

Figure 4.1

Distribution of common rock-forming minerals with respect to birefringence and refractive index. C = colored or pleochroic. A = anomalous interference color.

4 Rock-Forming Minerals

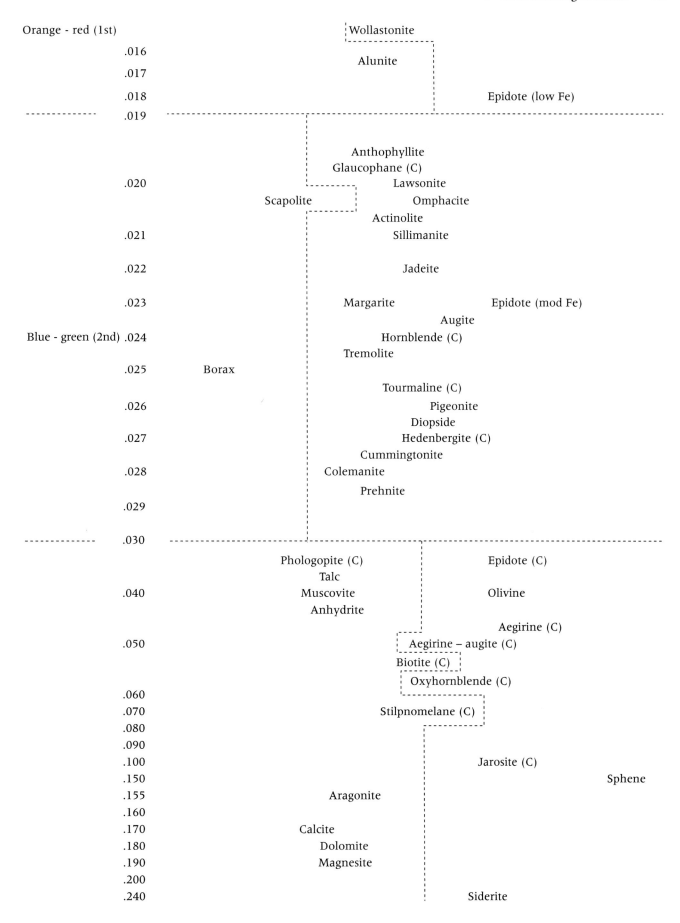

Table 4.1
Physical and Optical Properties of Rock-Forming Minerals

Mineral (abbreviation) Formula	Hardness Specific Gravity	Cleavage/ Parting (P)	Twinning Color (ts)	Habit Color (hs)	Range of Biref. (δ) Low–High R.I.	Crystal System Sign 2V (deg)
Acmite (acm) [see aegirine]						
Actinolite (act) $Ca_2(Mg,Fe^{2+})_5Si_8O_{22}(OH)_2$	5–6 3.1–3.4	{110} {100}P	{100} clear/pale yel/grn	prismatic dk grn	0.019–0.024 1.620–1.690	monoclinic (−)75–84
Adularia (adu) $KAlSi_3O_8$	6 2.5	{001}{010}	Carlsbad clr	wedge-shaped wht, salmon	0.005 1.518–1.525	monoclinic (−)35–50
Aegirine (aeg) $NaFe^{3+}(SiO_3)_2$	6 3.4–3.6	{110} {100}	{100} grn,yel	prismatic blk	0.036–0.060 1.720–1.840	monoclinic (−)58–90
Aegirine–augite (ae–au) $(Na,Ca)(Fe^{3+},Fe^{2+},Mg,Al)(SiO_3)_2$	6 3.4–3.6	{110} {100}	{100} grn,yel	prismatic blk, dk grn	0.040–0.066 1.70–1.76	monoclinic (+)70–90
Albite (ab) $NaAlSi_3O_8$ [also plagioclase An_{0-10}]	6 2.6	{001}{010}	albite, pericline clr	tabular wht	0.01 1.528–1.542	triclinic (+)78–84
Albite (high)(ab) [disordered structure]					0.007–0.008 1.526–1.541	triclinic (−)52–54
Alkali feldspar (alk) $(Na,K)AlSi_3O_8$						
Allanite (all) $(Ca,Ce,La)_2(Al,Fe^{3+},Fe^{2+})_3 O(SiO_4)Si_2O_7)(OH)$	5–6.5 3.4–4.2			tabular blk, dk brn	0.013–0.036 1.690–1.828	monoclinic (−)40–90, (+)57–90
Almandine (alm) $Fe_3Al_2(SiO_4)_3$	7–7.5 4.3		brn,yel,grn	dodecahedral red-brn	1.830	isometric
Alunite (alu) $KAl_3(SO_4)_2(OH)_6$	3.5–4 2.6–2.9	{0001}	clr	rhomb-tab wht,gry,pnk	0.010–0.023 1.568–1.601	trigonal (+)
Amesite (ams) $(Mg_4Al_2)(Si_2Al_2)O_{10}(OH)_8$	2.5–3 2.7	{001}	clr	platy pale grn	0.010–0.017 1.590–1.620	monoclinic (+)10–30
Amphibole (amp) $NaCa_2(Mg,Fe)_5(Si,Al)_8O_{22}(OH)_2$						
Analcime (anl) $Na(AlSi_2)O_6 \cdot H_2O$	5–5.5 2.2–2.3	{001}{111}	{001}{110} clr	trapezohedral clr, wht	0.000–0.001 1.479–1.493	isometric
Anatase (ana) TiO_2	5.5 3.9		red-brn, yel-brn	dipyramidal brn	0.073 2.488–2.561	tetragonal (−)
Andalusite (and) Al_2SiO_5	6.5–7.5 3.1	{110}	clr, pnk	prismatic wht, pnk, tan	0.009–0.011 1.629–1.650	orthorhombic (−)71–88

Paramount Publishing International

(PRENTICE HALL/HARVESTER WHEATSHEAF/ELLIS HORWOOD/ALLYN &
BACON/WOODHEAD FAULKNER/APPLETON & LANGE)
Campus 400, Maylands Ave, Hemel Hempstead, Herts., HP2 7EZ, England
Tel: Hemel Hempstead (0442) 881900   Fax:(0442) 257115   Telex: 82445
Registration: London 714516   VAT Registration No. GB 490 5885 08

DR R F CHEENEY
GRANT INSTITUTE GEOLOGY
UNIVERSITY OF EDINBURGH
WEST MAINS ROAD
EDINBURGH EH9 3JW

REFERENCE
IU160984

Date and Tax Point
2/12/94

INSPECTION COPY

W/H Order No. P384974

SBN	TITLE	AUTHOR	PRICE
0023541458	PETROGRAPHY PETROGENESIS	HIBBARD	£ 19.95

This book is sent to you for 60 days examination following: YOUR REQUEST BY LETTER.

When you have had the opportunity to examine the book and its suitability to your reading requirements and considered the options outlined below, please complete the details below and return this form.

☐ This book will be recommended for purchase as a main course text on the following course: _____
_____
Beginning on: _____   Student Numbers: _____   I am therefore retaining the book as a desk copy.

NB Books may only be retained, without purchase, if they are the main recommended or dual recommended text.
If you wish to retain the book as a supplementary recommended text or for your own interest, please refer to the new boxed section below.
Which book shop will you inform of your students' requirements? _____

Comments on the book (Academic quality, course applicability, etc): _____
_____

We would be grateful for the names of colleagues whom you think may be interested in this book:
_____
_____

Book(s) currently recommended as main text: _____

☐ I return the book herewith.

☐ I am retaining the book and enclose a cheque for £/$ _____ made payable to **INTERNATIONAL BOOK DISTRIBUTORS LTD**

☐ I am retaining the book and wish you to debit my credit card account with the amount of £ _____
My Access/American Express/Diners Club No. is: ☐☐☐☐☐☐☐☐☐☐☐☐☐☐
Name: _____   Expiry Date: _____
Home Address: _____
(Required by Credit Card Companies)

☐ I am retaining the book and require an invoice.

---

* UK AND EIRE ACADEMICS ONLY *

As a major textbook publisher we understand that academics need to keep up to date with new publications and developments in your subject area. We are therefore offering you the opportunity to purchase this book at 40% discount. If you would like to take advantage of this offer please fill out the relevant section below.

I am retaining the book and enclose a cheque for £ _____ (retail price - 40%) made payable to:
INTERNATIONAL BOOK DISTRIBUTORS LTD
I am retaining the book and wish you to debit my credit card account with the amount of £ _____ (retail price -40%) (DETAILS AS ABOVE)

Signed _____

Telephone No: _____ Ext: _____ Date: _____

PLEASE QUOTE REFERENCE WHEN REPLYING

INTERNATIONAL BOOK DISTRIBUTORS LTD
TERMS AND CONDITIONS OF SALE

1. **GENERAL**

   (A) In these Conditions "The Company" means International Book Distributors Limited and, where applicable, any other company which is part of the Simon & Schuster Group or any third party for whom International Book Distributors provides a contracted service. "The Customer" means the person, firm or company placing an order with The Company.

   (B) All orders are accepted and goods supplied subject to the following Terms and Conditions which shall govern the contract and cannot be altered by The Customers' Terms of Purchase. No addition to or variation from these Terms and Conditions shall be binding on The Company unless it is in writing and signed by a duly authorised representative of The Company.

   (C) Goods must not be sold to the general public before their publication date, namely that date in respect of any of the goods which is shown on the invoice or despatch documentation or which The Company otherwise indicates as the first day they may be sold to the general public.

2. **ORDERS**

   (A) The Company reserves the immediate right, at any time, (without prejudice to any other remedy) to terminate the agreement constituted by these Conditions or to cancel any uncompleted order or to suspend delivery in the event that any amounts payable by The Customer are overdue or there is any breach by The Customer of any of its obligations under these Conditions or for any other reason which at the discretion of The Company justifies such action.

   (B) Prices are subject to change without prior notification (before or after goods are invoiced, in the latter case only as a consequence of pricing or invoicing error).

   We reserve the right to charge any extra costs incurred by The Company in meeting The Customer's order requirements.

   (C) Any orders outstanding with The Company after the termination or expiry of a distribution agreement between The Company and the publisher concerned will be transferred to the publisher's new distributor together with any money paid in advance of despatch of the order for cash sales. (Alternatively The Company will refund such monies paid in advance to The Customer depending on the distribution agreement).

3. **DELIVERY AND RISK**

   (A) Goods will be delivered to the delivery address shown on The Company's invoice or to the Customer's designated Shipper or Agent and, if none is shown, to the person to whom the invoice is addressed/despatched. Any delivery dates are given as estimates only and in no circumstances shall The Company be liable for any loss whatsoever suffered or caused through late delivery or non-delivery. Neither the Company nor its carriers are obliged to provide loading or unloading facilities on delivery.

   (B) The risk of loss and/or damage (but not title) to goods supplied by The Company shall pass to The Customer when they are delivered to The Customer or other person to whom The Company has been authorised by The Customer to deliver the goods, whether expressly or by implication, and (Accordingly The Customer should insure the goods thereafter against such risks as may be commercially prudent.

   (C) Any damage to the goods in transit, or shortages in the goods delivered, must be notified to the relevant carrier within 10 days of receipt (packing and contents to be held for inspection). On no account will claims be considered if notified outside this period.

4. **RETURNS**

   (A) Prior written authorisation for returned goods must be obtained. Returns will only be considered at its discretion by The Company within 12 months of supply. Returns permission must be requested from the publisher of the goods in question. Authorisation will be subject to the returns conditions and policies imposed by the publishers concerned which are in addition to the terms and conditions set out herein. Authorisation by a publisher's representative does not confer automatic credit for returns if the returns conditions and policies are not adhered to.

   (B) ISBN and full details of the books requested to be returned must be provided. The relevant invoice number must be quoted. If this is not possible, the minimum information required is the month and year of supply. Failure by The Customer to provide correct information will delay the processing procedure. The Company reserves the right to refuse to credit any goods returned by The Customer where no evidence of purchase is provided.

   (C) Subject to the provisions of this paragraph 4, old editions may be returned within three months of the publication of a new edition provided such returns are further made within 12 months of original supply.

   (D) Unauthorised returns will, at The Company's discretion, be sent back to The Customer at The Customer's risk and expense, or be credited at a reduced rate or subject to the imposition of some other penalty.

   (E) Goods returned from exhibitions must be listed and packed separately with the complete number of parcels stated. Invoice numbers must always be quoted for these returns.

   (F) Defective and incorrect supplies should be returned immediately quoting the relevant invoice number and the reason for return.

   (G) All parcels returned by The Customer should be clearly marked as returned goods and should be enclosed with the full details showing the reason for the return. Only complete books may be returned and not title pages unless otherwise authorised in advance by The Company, in writing.

   (H) The Company will not accept books back for credit unless they are in mint condition, and the titles in question are not out of print.

   (I) The Company advise that all returns should be delivered by a carrier who can provide proof of delivery. The Company is not liable for any returns lost in transit. All returns should be securely packed to ensure safe arrival. Returns remain the responsibility and property of The Customer until receipt in The Company's Warehouse. The Customer is liable for any shortages or damages during transit. All returns are made at The Customer's expense and accordingly The Company will not accept any charges levied by shipping/transport agents.

   (J) The Company will not give returns permission for software or books specially ordered from our USA offices and special price deals.

   (K) A copy of the Authorised Returns Note must be returned with the books.

5. **TITLE**

   (A) Notwithstanding any other provision of these Conditions, the ownership of all goods supplied pursuant to these Conditions shall remain vested in The Company (which reserves the right to dispose of them) until The Company has received payment in full of all debts owing by The Customer to The Company.

   (B) In the event that payment is overdue in whole or in part or upon the commencement of any act or proceedings in which The Customer's solvency is involved. The Company may without prejudice to any of its other rights recover or resell the goods or any of them and may enter upon The Customer's premises by its servants or agents for that purpose. The Customer agrees to pay The Company all costs of repossession.

   (C) Where The Customer sells the goods, prior to acquiring the ownership of them, all money received from such sale shall be held by The Customer as trustee for The Company until all sums due to The Company from The Customer have been duly paid.

6. **PAYMENT**

   (A) Payment terms are as agreed by The Customer and The Credit Manager in writing. Time is of the essence with respect to The Customer's obligations hereunder. Payment may not be withheld, or delayed by The Customer for unauthorised returns or otherwise without the prior written agreement of The Company.

   (B) The Company reserves the right to charge interest on overdue amounts accruing on a daily basis from the date payment is due until the date of actual payment both before and after judgement. The rate of interest charged will be equal to 3% above Standard Chartered Bank plc base rate from time to time in force.

7. **LIABILITY**

   (A) The Company shall be liable for death or personal injury resulting from negligence of The Company, its servants or agents (but not independent contractors) while acting in the course of their employment by The Company.

   (B) The Company does not make or give any warranty, representation or undertaking as to the quality of the goods, their correspondence with description or fitness for purpose, that the goods are not defamatory, injurious, obscene, unlawful or in breach of copyright or in any other manner whatsoever.

   (C) Save as and to the extent provided by these Conditions The Company shall not in any circumstances be liable to The Customer or any successor or assignee of The Customer in respect of any loss of whatsoever nature occurring to The Customer arising from the supply of goods or from non-delivery, delayed delivery, damage to or loss of the goods owing to any act or omission by The Company (including negligence) or any cause not within The Company's control including (without limitation) fire, flood, accident, strike, riot, lock-out, trade dispute, industrial action, terrorism, nuclear accident, war, insurrection, act or restraint of Government.

8. **TERMINATION AND GENERAL**

   (A) The agreement constituted by these Conditions shall terminate forthwith if any order is made for the bankruptcy of or an effective resolution is passed for the winding-up of The Customer or if The Customer being a company is unable to pay its debts within the meaning of Section 123 of the Insolvency Act 1986, or any statutory re-enactment or modification thereof, or makes a composition with creditors or if a supervisor, receiver, administrator, administrative receiver or other encumbrancer takes possession of or is appointed over the whole or any part of the assets of The Customer.

   (B) If the agreement between The Company and a publisher expires or is terminated for any reason The Company may terminate the agreement constituted by these Conditions forthwith or at any time thereafter in relation to the goods supplied by that publisher.

   (C) The termination of the agreement constituted by these Conditions shall not affect any rights or obligations of the parties arising prior to such termination.

   (D) All contracts under these Conditions shall be governed by and construed in accordance with the laws of England and all disputes shall be submitted to the exclusive jurisdiction of the English courts.

4 Rock-Forming Minerals

Mineral	H	SG	Cleavage	Twinning	Color	Habit	Birefringence / RI	Crystal system / 2V
Andesine (ads) [plagioclase An_{30-50}]	6–6.5	2.66–2.68	{001}{010}	{010} peric, Carls	clr	tabular wht, gry	0.008 / 1.544–1.562	triclinic (+)78–90
Andradite (ard) $Ca_3Fe_2(SiO_4)_3$	6.5–7	3.8			lt brn	dodecahedral dk brn, blk	1.887	isometric
Anglesite (ang) $PbSO_4$	2.5–3	6.3	{001}{210}		clr	tabular wht	0.017 / 1.878–1.895	orthorhombic (+)60–75
Anhydrite (anh) $CaSO_4$	3.5 3		{010}{100} {001}	{011}	clr	tabular wht, gry	0.044 / 1.570–1.614	orthorhombic (+)44
Anorthite (an) $CaAl_2Si_2O_8$ (also An_{90-100})	3.5–4 2.9–3.1		{001}{010}	{010}, peric, Carls	clr	tabular wht, lt brn	0.013 / 1.573–1.590	triclinic (−)77–79
Anorthoclase (anc) $(Na,K)AlSi_3O_8$ (Ab_{63-90})	6–6.5 2.7		{001}{010}	{010}, peric, Carls	clr	tabular wht, gry	0.006–0.007 / 1.524–1.534	triclinic (−)42–52
Anthophyllite (amp) $(Mg,Fe)_7Si_8O_{22}(OH)_2$	6 2.85–3.39		{110}		clr	bladed wht, gry, brn	0.013–0.025 / 1.580–1.722	orthorhombic (+)75–90, (−)70–90
Antigorite (atg) $Mg_6Si_4O_{10}(OH)_8$	5.5–6 2.8–3.3		{110}		clr–pal brn	prismatic gry, brn	0.004–0.009 / 1.550–1.574	monoclinic (−)20–60
Apatite (apa) $Ca_5(PO_4CO_3)_3(F,OH,Cl)$	2–3 2.5–2.6		{0001}		clr, pal grn	platy grn	0.001–0.017 / 1.603–1.667	hexagonal (−)
Aragonite (ara) $CaCO_3$	5 2.9–3.5		{010}	{110}, cyclic clr		prismatic clr,wht	0.155 / 1.530–1.685	orthorhombic (−)18
Arfevdsonite (arf) $Na_3Fe_4Fe^{3+}Si_8O_{22}(OH)_2$	3.5–4 2.95		{110} {100}P	{100}	yel,grn-blu	prismatic blk	0.012–0.015 / 1.670–1.720	monoclinic (−)0–70
Augite (aug) $(Ca,Mg,Fe^{2+},Na)(Mg,Fe^{2+},Fe^{3+},Al,Ti)(Si,Al)_2O_6$	5–6 3.3–3.5		{110} {100}P,{001}P	{100}{001}	clr	prismatic blk	0.018–0.030 / 1.670–1.760	monoclinic (+)25–60
Azurite (azu) $Cu_3(OH)_2(CO_3)_2$	5–6 3.2–3.6		{100}{011}		blu	tabular prismatic azure blu	0.108 / 1.730–1.830	monoclinic (+)68
Barite (bar) $BaSO_4$	3–3.5 2.8–3.3		{001}{210}		clr	tabular wht	0.012 / 1.646–1.649	orthorhombic (+)36–40
Barkevikite (bak) $NaCa_2(Fe^{2+},Mg,Mn)_5(Si_7Al)O_{22}(OH)_2$	5–6 3.3–3.4		{110} {100}P	{100}	yel brn-red, brn–dk brn	prismatic brn blk	0.012–0.018 / 1.680–1.710	monoclinic (−)40–50
Biotite (bt) $K_2(Mg,Fe^{2+})_{5-6}Al_{0-1}(Si_{5-6}Al_{2-3})O_{20}(OH,F)_4$	2.5–3 2.7–3.3		{001}		tan-red brn,brn,grn	platy blk	0.040–0.070 / 1.560–1.700	monoclinic (−)0–25
Boehmite (boe) $AlO(OH)$	3.5–4 3		{010}		clr	platy wht	0.015 / 1.640–1.670	orthorhombic (+)80

continued

48   Part 1   Optical Mineralogy

Table 4.1 continued

Mineral (abbreviation) Formula	Hardness Specific Gravity	Cleavage/ Parting (P)	Twinning Color (ts)	Habit Color (hs)	Range of Biref. (δ) Low-High R.I.	Crystal System Sign 2V (deg)
Borax (brx) $Na_2B_4O_7 \cdot 10H_2O$	2–2.5 1.7	{100}	clr	stubby prismatic wht, gry	0.025 1.447–1.472	monoclinic (−)40
Bowlingite (bow) [goethite + clays + others]						
Bronzite (bz) $(Mg,Fe)_2(SiO_3)_2$	5–6 3.3–3.4	{110} {100}P	{100}exsolution pal yel–pal grn	stubby prismatic bronze brn	0.010–0.012 1.677–1.700	orthorhombic (−)57–90
Brookite (bro) TiO_2	5.5–6 4.1		yel brn–red brn	tabular yel brn, red brn	0.117 2.583–2.700	orthorhombic (+)0–30
Brucite (bru) $Mg(OH)_2$	2.5 2.4	{0001}	clr	platy wht, pal grn, brn	0.010–0.020 1.559–1.600	hexagonal (+)
Bytownite (by) [plagioclase An_{70-90}]	6–6.5 2.7	{001}{010}	{010},peric, Carls clr	tabular wht, gry, brn, grn brn	0.010 1.564–1.584	triclinic (−)77–90, (+)87–90
Calcite (cc) $CaCO_3$	3 2.7	{10$\bar{1}$1} {01$\bar{1}$2}P	{01$\bar{1}$2}{0001} clr	rhomb, scalenodedral clr,wht,blu,pnk	0.172 1.486–1.658	trigonal (−)
Cancrinite (can) $(Ca,Na)_{7-8}(AlSiO_4)_6 (CO_3,SO_4,Cl)_{1-2} \cdot 1-5H_2O$	5–6 2.3–2.5	{10$\bar{1}$0}	clr	prismatic wht	0.012–0.028 1.495–1.530	hexagonal (−)
Carbon (crb) C						
Carnallite (car) $KMgCl_3 \cdot 6H_2O$	2.5 1.6		clr	pseudo prismatic wht	0.030 1.465–1.497	orthorhombic (+)70
Cassiterite (cas) SnO_2	6–7 6.9–7	{100}	{101} yel, red-brn	bipyram yel, brn	0.090–0.010 1.990–2.100	tetragonal (+)
Celadonite (cel) [see glauconite]						
Cerrusite (cer) $PbCO_3$	3–3.5 6.5	{110}{021}	{110} clr	tabular clr,wht,gry	0.273 1.803–2.076	orthorhombic (−)8
Chabazite (chb) $Ca_2(Al_4Si_8)O_{24} \cdot 13H_2O$	4.5 2.1	{10$\bar{1}$1}	{0001}{10$\bar{1}$1} clr	rhombohedral wht	0.002–0.010 1.470–1.494	trigonal (−)
Chalcedony (chc) [fibrous variety of SiO_2]	6.5–7 2.5–2.6		clr, pale yel, brn	colloform clr, many colors		
Chalcopyrite (cpy) $CuFeS_2$	3.5–4 4.1–4.3		opaque	tetrahedral brass yel		tetragonal
Chamosite (chm) $(Fe_5^{2+}Al)(Si_3Al)O_{10}(OH)_8$	2–3 3.1–3.4	{001}	yel, grn	platy grn	0.002–0.010 1.640–1.670	monoclinic (−) small

4 Rock-Forming Minerals

Mineral	H	G	Cleavage/Form	Color	Habit	Birefringence / n	System / Optic sign / 2V
Chert (cht) [variety of quartz]							
Chlorite (chl) $(Mg,Fe,Al)_{12}(Si,Al)_8 O_{20}(OH)_{16}$	2–3	2.6–3.3	{001}	clr–pal grn	platy grn	0.001–0.015 / 1.560–1.680	monoclinic (–))0–40, (+) 0–60
Chloritoid (ctd) $(Fe^{2+},Mg,Mn)_2(Al,Fe^{3+})Al_3O_2(SiO_4)_2(OH)_4$	6.5	3.2–3.8	{001}	pal yel–pal grn	platy dk grn	0.006–0.022 / 1.715–1.740	monoclinic (–))55–90, (+))36–90
Chlorophaeite (cph) [limonite + chlorite + serpentine]							
Chondrodite (chn) $2(Mg_2SiO_4)\cdot Mg(OH,F)_2$	6–6.5	3.1–3.2	{001}	clr–yel brn	tabular yel brn, brn red	0.025–0.037 / 1.592–1.675	monoclinic (+))64–90
Chromite (chr) $FeCr_2O_4$	5.5	5		opaque	octahedral blk	1.900–2.120	isometric
Chrysocolla (cry) $Cu_2H_2Si_2O_5(OH)_4$	2–4	2–2.4		clr–pal grn	fibrous blu grn	0.023 / 1.460–1.54	orthorhombic (–) small
Chrysotile (chy) $Mg_3Si_2O_5(OH)_4$	2.5	1.5		clr–pal yel grn	fibrous gry, yel, grn	0.008 / 1.529–1.567	monoclinic (–))30–50
Clinoamphibole (cam) [see hornblende, tremolite, etc.]							
Clinochlore (clc) $(Mg_5Al)(Si_3Al)O_{10}(OH)_8$	2–3	4.3	{001}	clr–pale grn	platy grn	0.002–0.009 / 1.570–1.60	monoclinic (+))0–40
Clinoptilolite (clp) $(Na,K,Ca)_{2-3}Al_3(Al,Si)_2Si_{13}O_{36}\cdot 12H_2O$	3.5–4	2.1–2.2	{010}	clr	platy wht	0.001–0.010 ? / 1.487–1.515 ?	monoclinic (+))0–70
Clinopyroxene (cpx) [see diopside, augite, etc.]							
Clinozoisite (clz) $Ca_2Al_3O(Si_2O_7)(SiO_4)(OH)$	6–7	3.2–3.4	{001}	clr	prismatic gry	0.004–0.012 / 1.703–1.734	monoclinic (+))14–90
Coesite (coe) [high pressure form of SiO_2]	7.5	2.93	{100}{021}	clr	fine granular clr		
Colemanite (col) $Ca_2B_6O_{11}\cdot 5H_2O$	4–4.5	2.4	{010}{101}	clr	prismatic clr, wht	0.028 / 1.586–1.614	monoclinic (+))55
Collophane (cop) [cryptocrystalline apatite]							
Cordierite (cdt) $Mg_2Al_3(Si_5Al)O_{18}$	7–7.5	2.5–2.7	{110}{130} wedges	clr	prismatic gry, gry blu	0.005–0.018 / 1.522–1.578	orthorhombic (–))40–90, (+))75–90
Corundum (cor) Al_2O_3	9	3.9–4.1	{0001}P {10$\bar{1}$1}P	clr	prismatic gry blu, brn, yel, red	0.008–0.009 / 1.759–1.772	hexagonal (–)

continued

Table 4.1 continued

Mineral (abbreviation) Formula	Hardness Specific Gravity	Cleavage/ Parting (P)	Twinning Color (ts)	Habit Color (hs)	Range of Biref. (δ) Low–High R.I.	Crystal System Sign 2V (deg)
Cristobalite (crs) SiO_2	6.5 2.3		clr	pseudocubic clr, wht	0.003 1.484–1.487	tetragonal (−)
Crossite (cro) $Na_2Mg_3Fe_2(Si_4O_{11})_2(OH)_2$	6 3.1–3.3	{110}	{100} lt yel-blu-vio	prismatic blu blk	0.012–0.016 1.650–1.700	monoclinic (+)(−)0–90
Cummingtonite (cum) $Mg_7(Si_4O_{11})_2(OH)_2$	5–6 3.1–3.4	{110}	{100} cir–pale brn	prismatic dk grn brn	0.022–0.035 1.630–1.700	monoclinic (+)60–90
Cuprite (cup) Cu_2O	3.5–4 5.8–6.1		red	octahedral red	2.849	isometric
Diaspore (dia) AlO(OH)	6.5–7 3.3–3.5	{010}		platy wht	0.040–0.048 1.682–1.752	orthorhombic (+)84–86
Diopside (di) $Ca(Mg,Fe)Si_2O_6$	6.5 3.2	{110} {100}P	{100}{001} clr	stubby prismatic lt grn	0.029–0.030 1.66–1.70	monoclinic (+)57–58
Dolomite (dol) $CaMg(CO_3)_2$	3.5–4 2.8	{1011}	{0221}{0001} clr	rhombohedral wht, brn, pnk	0.179 1.500–1.679	trigonal (−)
Dumortierite (dum) $(Al,Fe^{3+})_7O_3(BO_3)(SiO_4)_3$	7–8.5 3.2–3.4	{100}	{110}cyclic clr-blu	bladed prismatic dk blu, violet	0.011–0.027 1.655–1.723	orthorhombic (−)15–52
Enstatite (en) $Mg_2(SiO_3)_2$	5–6 3.2–3.3	{110}	clr	prismatic gry, brn	0.008–0.010 1.650–1.682	orthorhombic (+)35–90
Epidote (ep) $Ca_2Al_2Fe^{3+}O(SiO_4)(Si_2O_7)(OH)$	6–7 3.2–3.4	{001}{100}	{100} clr-pal yel	prismatic grn	0.012–0.049 1.715–1.797	monoclinic (−)64–90
Fayalite (fay) Fe_2SiO_4	6.5 4.2–4.3	{010}	clr-pal yel	stubby prismatic lt grn	0.052 1.827–1.879	orthorhombic (−)46
Feldspar (fld) [see plagioclase and alkali feldspar]						
Fibrolite (fib) [fibrous variety of sillimanite]						
Fluorapatite (fap) $Ca_5(PO_4)_3F$	5 2.9–3.5		clr	prismatic yel, grn	0.003–0.005 1.629–1.650	hexagonal (−)
Fluorite (flu) CaF_2	4 3.18	{111}	{111} clr/pale vio	cubic vio/grn/yel	1.433–1.435	isometric
Forsterite (fo) $(Mg,Fe)_2SiO_4$	7 3.2		clr	equant pale grn	0.035 1.635–1.670	orthorhombic (+)82
Galena (gal) PbS	2.5 7.5	{100}	opaque	cubic lead gry	4.015	isometric
Garnet (gar) [see almandine, grossularite, etc]					1.134–1.887	isometric

4 Rock-Forming Minerals 51

Mineral	Hardness	Cleavage	Color	Habit	Birefringence / Indices	Optical / System
Garnierite (gnr) [Ni-bearing serpentine]	2.5–3.5					
Gibbsite (gib) Al(OH)$_3$	2.3–2.4	{001}	clr	tabular wht	0.019 1.568–1.600	monoclinic (+)0–40
Glauconite (glt) (K,H$_3$O)$_2$(Fe^{3+},Al,Fe^{2+},Mg)$_4$ (Si$_{7-7.5}$Al$_{1-0.5}$)O$_{20}$(OH)$_4$	2 2.4–3	{001}	pale yel-grn	platy grn	0.014–0.032 1.560–1.650	monoclinic (−)0–20
Glass (gls) [obsidian, tachylite, pseudotachylite, lechatelierite]	5–6 2.3–2.9		clr, lt brn	mass, perlitic brn, blk	1.480–1.650	amorphous
Glaucophane (gla) Na$_2$Mg$_3$Al$_2$(Si$_4$O$_{11}$)$_2$(OH)$_2$	6 3–3.3	{110}	lt yel-vio-blu	prismatic blu blk	0.016–0.024 1.59–1.66	monoclinic (−)0–50
Goethite (goe) FeO(OH)	5–5.5 4.3	{010}	yel-brn	fib, platy yel brn	0.138–0.140 2.260–2.515	orthorhombic (−)0–27
Graphite (gra) C	1–2 2.2	{0001}	opaque	platy metallic blk		hexagonal
Grossularite (gro) Ca$_3$Al$_2$Si$_3$O$_{12}$	6.5–7 3.5		clr	dodecahedral yel, red, red-brn	0.000–0.001 1.734	isometric
Gypsum (gyp) CaSO$_4$·2H$_2$O	2 2.3	{010}{100} {$\bar{1}$11}	{100} clr	tabular, rhombic sec clr, wht	0.009–0.010 1.519–1.531	monoclinic (+)58
Halite (hal) NaCl	2–2.5 2.1	{100}	clr	cubic clr	1.544	isometric
Hastingsite (htg) NaCa(Mg,Fe^{2+})$_4$(Al,Fe) [(Si$_3$Al)O$_{11}$]$_2$(OH)$_2$	5–6 3.3	{110} {100}P	lt yel-grn-dk grn	prismatic dk grn	0.020–0.021 1.650–1.700	monoclinic (−)45–90
Hauynite (huy) (Na,Ca)$_{4-8}$Al$_6$Si$_6$O$_{24}$(SO$_4$)$_{1-2}$	5.6–6 2.4–2.5	{111}	clr, pal blu	dodecahedral dk blu, gry, grn	0.000–0.001 1.496–1.505	isometric
Hedenbergite (hed) CaFe(SiO$_3$)$_2$	5.5 3.5	{110} {100}P{001}P	yel grn-grn	stubby prismatic dk grn, blk	0.025–0.026 1.720–1.760	monoclinic (+)62–63
Hematite (hem) Fe$_2$O$_3$	5–6 4.7	{0001}P {10$\bar{1}$1}P	red thin edges	platy, earthy blk	0.028 3.150–3.220	hexagonal (−)
Hemimorphite (hmp) Zn$_4$(OH)$_2$Si$_2$O$_7$·H$_2$O	5 3.4	{110}	clr	sheaflike clusters wht, gry, pal grn	0.022 1.614–1.636	orthorhombic (+)46
Heulandite (heu) (Ca,Na$_2$)(Al$_2$Si$_7$)O$_{10}$·6H$_2$O	3.5–4 2.2	{010}	clr	platy wht	0.001–0.007 1.487–1.512	monoclinic (+)30
Hornblende (ho) Ca$_2$(Mg,Fe^{2+})$_4$(Al,Fe^{3+}) (Si$_7$Al)O$_{22}$(OH)$_2$	5–6 3–3.4	{110}	grn–olive grn	prismatic blk	0.015–0.034 1.610–1.710	monoclinic (−)35–90
Hydrogrossular (hyg) Ca$_3$Al$_2$[(SiO$_4$)$_x$(OH)$_4$]$_3$	6–6.5 3–3.6		clr	octahedral wht, lt brn, pnk	1.670–1.680	isometric

continued

52   Part 1   Optical Mineralogy

Table 4.1   continued

Mineral (abbreviation) Formula	Hardness Specific Gravity	Cleavage/ Parting (P)	Twinning Color (ts)	Habit Color (hs)	Range of Biref. (δ) Low-High R.I.	Crystal System Sign 2V (deg)
Hypersthene (hy) $(Mg,Fe)_2(SiO_3)_2$	5–6 3.4–3.5	{110} {100}P	pnk–pal grn	stubby prismatic dk grn brn	0.012–0.015 1.690–1.730	orthorhombic (−)50–57
Ice (ice) H_2O	1.5 0.9167		{0001} clr	dendritic, granular clr, wht	0.0014 1.306–1.308	hexagonal (+)
Iddingsite (ids) [goethite + clays]						
Idocrase (ido) $Ca_{10}(Mg,Fe^{2+})_2Al(Si_2O_7)_2(SiO_4)_5(OH,F)_4$	6–7 3.3–3.4		clr	prismatic yel-grn, yel-brn	0.001–0.012 1.690–1.746	tetragonal (+)
Illite (ill) $K_{1-1.5}Al_4(Si_{6.5-7}Al_{1-1.5})O_{20}(OH)_4$	1–2 2.6–2.9	{001}	clr	platy wht	0.030–0.035 1.540–1.610	monoclinic (−)0–10
Ilmenite (ilm) $FeTiO_3$	5–6 4.79	{10$\bar{1}$1}P	{0001}{10$\bar{1}$1} blk	tabular, skeletal blk		hexagonal (+)
Jadeite (jad) $NaAl(SiO_3)_2$	6 3.2–3.4	{110}	{100} clr	stubby prismatic wht, pal grn	0.012–0.032 1.650–1.690	monoclinic (+)70–75
Jarosite (jar) $KFe_3(SO_4)_2(OH)_6$	3 3.25	{0001}	pale yel–yel brn	platy orange brn	0.101–0.105 1.713–1.820	hexagonal (−)
Kaersutite (kar) $NaCa_2(Mg,Fe^{2+})_4TiSi_6Al_2O_{22}(OH)_2$	5–6 3.2	{110} {100}	{100} yel brn to red brn	blk	0.019–0.083 1.670–1.772	monoclinic (−)66–84
Kaolinite (kao) $Al_2Si_2O_5(OH)_4$	2–3 2.5–2.7	{001}	clr	platy wht	0.005–0.008 1.560–1.570	monoclinic (−)23–60
Kernite (ker) $Na_2B_4O_7 \cdot 4H_2O$	2.5–3 1.9	{100}{001} {$\bar{2}$01}	{110} clr	stubby prismatic clr, wht	0.034 1.454–1.488	monoclinic (−)80
K-feldspar (Kf) $KAlSi_3O_8$	[see sanidine, orthoclase, microcline, adularia]					
Kyanite (ky) Al_2SiO_5	4–7.5 3.6	{100}{010} {001}P	clr	bladed gry, lt blu	0.012–0.016 1.712–1.734	triclinic (−)78–84
Labradorite (lab) [plagioclase An_{50-70}]	6–6.5 2.69–2.71	{001}{010}	{010},peri, Carls clr	tabular gry, blu	0.007–0.008 1.555–1.573	triclinic (+)77–86
Larnite (lar) Ca_2SiO_4	6 3.28–3.31	{100}	{100} clr	tabular clr, wht, gry	0.023–0.025 1.707–1.740	monoclinic (+)63–73
Laumontite (lau) $Ca(Al_2Si_4)O_{12} \cdot 4H_2O$	3–4 2.23–2.41	{010}{110}	{100} clr	prismatic wht	0.008–0.016 1.502–1.525	monoclinic (−)25–47
Lawsonite (law) $CaAl_2Si_2O_7(OH)_2 \cdot H_2O$	7–8 3.1	{001}{010}	{110} clr	tabular clr, gry	0.019–0.021 1.665–1.686	orthorhombic (+)76–87

4 Rock-Forming Minerals

Mineral	Hardness	Cleavage	Color	Other	Birefringence	RI	Habit	Crystal system / 2V
Lepidochrosite (lpc) FeO(OH)	5–5.5 4.1	{010}	yel-org-yel brn	platy brn, red	0.570 1.940–2.510			orthorhombic (−)83
Lepidolite (lep) $K_2(Li_{3-4},Al_{2-3})_{5-6}(Si_{8-6}Al_{0-2}O_{20}(OH,F)_4$	2.5–4 2.1–3.3	{001}	clr	platy clr, pnk, pal violet	0.018–0.038 1.520–1.590			monoclinic (−)0–58
Leucite (leu) $KAlSi_2O_6$	5.5–6 2.47–2.50			trapezohedral clr	0.001 1.508–1.511			isometric
Leucoxene (lux) [Ti oxides such as rutile, anatase, brookite]								
Limonite (lim) $FeO(OH) \cdot nH_2O$								
Magnesite (mgs) $MgCO_3$	3.5–4.5 3–3.4	$\{10\bar{1}1\}$	clr	crystals very rare wht, gry, pnk	0.191 1.509–1.700			trigonal (−)
Magnetite (mag) Fe_3O_4	5 4.7		opaque	octahedral blk	opaque			isometric
Malachite (mch) $Cu_2(OH)_2CO_3$	3.5–4 3.9–4.1	$\{\bar{2}01\}\{010\}$	clr-pal grn-dk grn	acicular green	0.254 1.655–1.909			monoclinic (−)43
Manganite (mng) MnO(OH)	4 4.3	{010}	opaque	prismatic blk	0.28 2.250–2.530			monoclinic (+)small
Margarite (mar) $CaAl_2(Si_2Al_2)O_{10}(OH)_2$	3.5–4.5 3–3.1	{001}	clr	platy gry pnk, lt yel	0.010–0.032 1.590–1.650			monoclinic (−)26–67
Melilite (mel) $(Ca,Na)_2(Mg,Al)(Si,Al)_2O_7$	5–6 2.9–3	{001}	clr	tabular yel, grn, brn	0.000–0.008 1.632–1.670			tetragonal (+)(−)
Merwinite (mer) $Ca_3MgSi_2O_8$	6 3.1–3.3	{100}{611}	clr	equant clr, pal grn	0.018 1.706–1.724			monoclinic (+)70
Microcline (mic) $(K,Na)AlSi_3O_8 (Kf_{100-92})$	6–6.5 2.5–2.6	{001}{010}	grid pattern	tabular wht, salmon, grn	0.007–0.008 1.518–1.522			triclinic (−)66–68
Microcline cryptoperthite (mcp) $[Kf_{92-20}]$			clr		0.008–0.009 1.515–1.536			triclinic (−)66–90, (+)87–90
Monazite (maz) (Ce,La,Th)PO_4	5–5.5 4.6–5.4	{100}	clr-yel brn	tabular yel brn, red brn	0.045–0.075 1.770–1.851			monoclinic (+)6–19
Monticellite (mtc) $Ca(Mg,Fe)SiO_4$	5.5 3.1–3.3	{031} cyclic	clr	prismatic clr, wht, gry	0.014–0.020 1.639–1.680			orthorhombic (−)70–90
Montmorillonite (mtm) $(\frac{1}{2}Ca,Na)_{0.67}(Al_{3.33}Mg_{0.67})Si_8O_{20}(OH)_4 \cdot nH_2O$	1–2 2–2.7	{001}	clr	platy wht	0.020–0.030 1.480–1.600			monoclinic (−)0–30
Muscovite (ms) $KAl_2(Si_3Al)O_{10}(OH)_2$	2.5–4 2.7–2.8	{001}	clr	platy clr	0.036–0.049 1.550–1.610			monoclinic (−)30–47

continued

Table 4.1 continued

Mineral (abbreviation) Formula	Hardness Specific Gravity	Cleavage/ Parting (P)	Twinning (ts) Color (ts)	Habit (hs) Color	Range of Biref. (δ) Low-High R.I.	Crystal System Sign 2V (deg)
Natrolite (nat) $Na_2(Al_2Si_3)O_{10} \cdot 2H_2O$	5 2.2	{110}	clr	prismatic, fibrous wht, gry	0.012–0.013 1.473–1.496	orthorhombic (+)58–64
Nepheline (ne) $Na_3KAl_4Si_4O_{16}$	5.6–6 2.5–2.6		clr	prismatic wht	0.003–0.005 1.526–1.549	hexagonal (−)
Nontronite (ntr) $(Ca,Na)Fe_4^{3+}(SiAl)O_{20}(OH)_4 \cdot nH_2O$	1–2 2–2.7	{001}	yel gr-grn	platy grn	0.030–0.040 1.530–1.640	monoclinic (−)25–70
Nosean (nos) $Na_8(Al_6Si_6O_{24})SO_4$	5.5 2.2–2.4		{111}not in t.s. clr, pal blu	dodecahedral wht, gry, blu	0.000–0.001 1.496–1.505	isometric
Oligoclase (olg) [plagioclase An_{10-30}]	6–6.5 2.6	{001} {010}	{010}, peric, Carls clr	tabular wht, pnk	0.007–0.008 1.534–1.552	triclinic (−)86–90, (+)84–90
Olivine (ol) $(Mg,Fe)SiO_4$	6.5–7 3.2–4.3		clr	stubby prismatic grn	0.035–0.052 1.635–1.879	orthorhombic (−)46–87, (+)82–90
Omphacite (omp) $(Ca,Na)(Mg,Fe^{2+},Fe^{3+},Al)(SiO_3)_2$	5–6 3.2–3.3	{110} {100}P	{100} clr	stubby prismatic dk grn	0.018–0.027 1.660–1.72	monoclinic (+)58–83
Opal (opl) $SiO_2 \cdot nH_2O$	5.6–6.5 1.9–2.2		clr, lt brn	colliform wht, play of colors	isotropic 1.435–1.460	amorphous
Orthite [see allanite]						
Orthoamphibole (oam) [see Anthophyllite]						
Orthoclase (or) $(K,Na)AlSi_3O_8$ [also Kf_{100-85}]	6 2.5	{001} {010}	Carlsbad clr	tabular wht, salmon, red	0.005 1.518–1.525	monoclinic (−)35–50
Orthoclase cryptoperthite (orp) [Or_{85-20}]					0.006–0.008 1.520–1.536	monoclinic (−)50–90
Orthopyroxene (opx) [enstatite, bronzite, hypersthene, etc.]						
Oxyhornblende (oxh) $Ca_2Na(MgFe^{3+},Al,Ti)_5[(Si_3Al)O_{11}]_2(OH)_2$	5–6 3.2–3.3	{110}	{100} brn-red brn	prismatic blk	0.018–0.083 1.650–1.800	monoclinic (−)56–88
Palagonite (pal) [altered basaltic glass]						

4 Rock-Forming Minerals   55

Mineral	H	Density	Cleavage	Habit	Color	δ / n	System / 2V
Paragonite (par) $NaAl_2(Si_3Al)O_{10}(OH)_2$	2.5	2.7–2.9	{001}	platy	clr	0.028–0.038 / 1.560–1.610	monoclinic (−)0–40
Pargasite (pag) $NaCa_2Mg_4(Al,Fe^{3+})[(Si_3Al)O_{11}]_2(OH)_2$	5–6	3–3.1	{110} {100}P	prismatic	blk, brn	0.020 / 1.610–1.670	monoclinic (+)70–90
Penninite (pen) $Mg_5(Al,Fe)(OH)_8(Al,Si)_4O_{10}$	2–2.5	2.6–2.8	{001}	platy	clr–pale grn	0.000–0.006 / 1.560–1.600	monoclinic (+)0–20
Periclase (pcl) MgO	5.5–6	3.5–3.6	{100}	cubic	clr	— / 1.736–1.745	isometric
Perovskite (prv) $CaTiO_3$	5.5	5.1–5.2	{111}	cubic, octahedral	wht, gry, yel, brn clr, pale yel, brn	0.000–0.002 / 2.27–2.40	pseudoisometric
Phengite (phg) $K_2Al_3(Mg,Fe^{2+})Si_7AlO_{20}(OH,F)_4$	2.5–3	2.7–2.9	{001}	platy	wht, pale grn clr, pale grn	0.028–0.049 / 1.530–1.640	monoclinic (−)30–47
Phlogopite (phl) $KMg_3(Si_3Al)O_{10}(OH)_2$	2–2.5	2.7–2.9	{001}	platy	brn clr–pale brn	0.030 / 1.535–1.565	monoclinic (−)0–15
Piedmontite $Ca_2(Al,Fe^{3+},Mn^{3+})_3O\cdot SiO_4\cdot Si_2O_7(OH)$	6–6.5	3.4–3.5	{001}	prismatic	red brn, blk pal yel–red brn	0.025–0.082 / 1.725–1.832	monoclinic (+)50–86
Pigeonite (pgn) $(Mg,Fe^{2+},Ca)(Mg,Fe^{2+})(SiO_3)_2$	6	3.3–3.4	{110} {001}	stubby prismatic	grn brn, blk clr	0.023–0.029 / 1.680–1.750	monoclinic (+)0–32
Plagioclase (pl) $(NaSi,CaAl)AlSi_2O_8$							
Polyhalite (pol) $K_2MgCa_2(SO_4)_4\cdot 2H_2O$	3.5	2.7	{10$\bar{1}$} {010}P	platy, fibrous	clr, wht clr	0.019 / 1.548–1.567	triclinic (−)62–70
Powellite (pow) $Ca(Mo,W)O_4$	3.5–4	4.2	{112}	pyram	yel brn clr, yel, brn, grn	0.011 / 1.967–1.978	tetragonal (+)
Prehnite (prh) $Ca_2Al(AlSi_3)O_{10}(OH)_2$	6–6.5	2.8–2.9	{001}	radiating platy	pale yel grn, wht clr	0.020–0.035 / 1.610–1.673	orthorhombic (+)60–70
Prochlorite [ripidolite] $(Mg,Fe^{2+},Al)_6(Si,Al)_4O_{10}(OH)_8$	2–3	2.8–3.0	{001}	platy	grn pale grn–grn	0.004–0.010 / 1.590–1.620	monoclinic (+)20–50
Pseudolucite (psl) [nepheline + K-feldspar]							
Pumpellyite (pum) $Ca_2Al_2(Mg,Fe^{2+},Fe^{3+},Al)(SiO_4)(Si_2O_7)(OH)_2(H_2O,OH)$	6	3.2–3.3	{001}	tabular, bladed	grn brn pal yel–grn	0.002–0.022 / 1.674–1.764	monoclinic (+)10–85
Pyrite (pyt) FeS_2	6–6.5	5		cubic brass	opaque opaque		isometric

Table 4.1 *continued*

Mineral (abbreviation) Formula	Hardness Specific Gravity	Cleavage/ Parting (P)	Twinning (ts) Color (ts)	Habit Color (hs)	Range of Biref. (δ) Low–High R.I.	Crystal System Sign 2V (deg)
Pyrolusite (pyl) MnO_2	2–6 4.5–5	{110}	opaque	dendrites blk	opaque	tetragonal
Pyrope (pyr) $Mg_3Al_2(SiO_4)_3$	7.5 3.5		clr, pal red	dodecahedral red	1.714	isometric
Pyrophyllite (pyp) $Al_2Si_4O_{10}(OH)_2$	1–2 2.8–2.9	{001}	clr	platy wht	0.050 1.534–1.601	monoclinic (−)53–62
Pyroxene (pyx) $(Ca,Mg,Fe,Ti,Na)_2(Al,Si)_2O_6$						
Quartz (qz) SiO_2	7 2.65		clr	prismatic clr, wht, blk, vio, yel	0.009 1.544–1.553	trigonal (+)
Richterite (ric) $NaCaNa(Mg,Fe)_5Si_8O_{22}(OH)_2$	6 2.9–3.2	{110}	{100} clr–pal yel/red	prismatic brn, red brn, grn brn	0.017 1.606–1.623	monoclinic (−) mod
Riebeckite (reb) $Na_2Fe_3^{2+}Fe_2^{3+}(Si_4O_{11})_2(OH)_2$	6 3.1–3.4	{110}	{100} yel brn–dk blu	prismatic blk	0.007–0.012 1.690–1.720	monoclinic (−)50–90, (+)70–90
Rutile (rut) TiO_2	6–6.5 4.2–5.6	{110}	{011} knee yel brn–red brn	prismatic red, blk	0.285–0.296 2.605–2.903	tetragonal (+)
Salite (sal) $Ca(Mg,Fe^{2+})(SiO_3)_2$	6 3.3	{110} {100}P	{100}{001} clr–pal grn	stubby prismatic grn, grn brn	0.028–0.29 1.670–1.730	monoclinic (+)57–58
Sanidine (san) $(K,Na)AlSi_3O_8 (Kf_{100-37})$	6 2.5	{001}{010}	Carlsbad clr	tabular clr, wht	0.005–0.006 1.518–1.530	monoclinic (−)18–42
Sanidine (high) [more disordered structure]						
Saponite (sap) $(½Ca,Na)_{0.67}Mg_6(Si_{7.33}Al_{0.67} O_{20}(OH)_4 \cdot nH_2O$	1–2 2–2.7	{001}	clr	platy wht	0.010–0.040 1.480–1.590	monoclinic (−)moderate
Sapphirine (sap) $(Mg,Fe,Al)_8O_2(Al,Si)_6O_{18}$	7.5 3.4–3.5		clr	tabular gry, pal blu	0.004–0.007 1.701–1.734	mono/tri (+)66–90, (−)50–90
Scapolite (sca) $(Ca,Na)_4[(Al,Si)_3Al_3Si_6O_{24}] (Cl,CO_3)$	4.5–5 6.1	{101}	{110} clr	dipyram wht, fluor blu-wht	0.004–0.037 1.534–1.600	tetragonal (−)
Scheelite (she) $CaWO_4$	5–6 2.5–2.8		clr	prismatic wht, gry, lt grn	0.016 1.918–1.934	tetragonal (+)
Schorl (shr) [see tourmaline]	7–7.5 3–3.2		lt blu–blk	prismatic blk		

Mineral (abbr) / Formula	H	G	Cleavage	Twin	Habit	Color (t.s.)	Color	δ	n	System	sign/2V
Sericite (ser) [fine-grained white mica, commonly muscovite]											
Serpentine (srp) [see antigorite] Mg₃Si₂O₅(OH)₄											
Siderite (sid) FeCO₃	4–4.5	3.9	{101̄1}		rhombohedral	clr–pale brn	brn	0.242	1.633–1.875	trigonal	(−)
Sillimanite (sil) Al₂SiO₅	6.5–7.5	3.2	{100}		prismatic	clr	wht	0.020–0.023	1.653–1.684	orthorhombic	(+)20–30
Smithsonite (sms) ZnCO₃	3.5–4.5	4.4	{101̄1}		rhombohedral	clr	wht, pal brn, pal grn	0.228	1.621–1.849	trigonal	(−)
Sodalite (sod) Na₈Al₆Si₆O₂₄Cl₂	5.5–6	2.1–2.4		{111}not in t.s.	dodecahedral	clr, pal blu	blu	0.000–0.001	1.483–1.487	isometric	
Soda-niter (son) NaNO₃	1.5–2	2.2	{101̄1}		rhombohedral	many	clr, wht	0.248	1.337–1.585	trigonal	(−)
Spessartite (sps) Mn₃Al₂(SiO₄)₃	7–7.5	4.2			dodecahedral	clr	brn-red, blk		1.800	isometric	
Sphalerite (spa) ZnS	3.5–4	3.9–4.1	{110}		tetrahedral	clr, pale yel-brn	yel-brn, blk	opaque	2.370–2.500	isometric	
Sphene (sph) CaTiSiO₅	5–5.5	3.4–3.5	{110}	{100}	diamond shapes	pal brn	dk brn	0.100–0.192	1.800–2.110	monoclinic	(+)17–56
Spinel (spl) MgAl₂O₄	7.5–8	3.5–4.0		{111}P	dodecahedral	{111}	grn, red, brn		1.719	isometric	
Spodumene (spd) LiAlSi₂O₆	6.5–7	3–3.2	{110}		prismatic	clr	wht	0.014–0.027	1.650–1.680	monoclinic	(+)54–69
Spurrite (spu) Ca₅(SiO₄)₂CO₃	5	3	{001}		granular	clr	gry	0.039	1.640–1.679	monoclinic	(−)40
Staurolite (stt) Fe₂Al₉O₆(SiO₄)₄(O,OH)₂	7–7.5	3.7–3.8	{010}	{023}{232}	prismatic	clr-org yel	brn	0.009–0.015	1.736–1.762	monoclinic	(+)80–90
Stilbite (stb) (Ca,Na₂,K₂)(Al₂Si₇)O₁₈·7H₂O	3.5–4	2.1–2.2	{010}	cruciform	platy	clr	wht	0.006–0.013	1.482–1.513	monoclinic	(−)30–49
Stilpnomelane (stl) (Fe³⁺,Fe²⁺,Mg,Mn,Al)₅₋₆(Si₄O₁₀)₂(OH)₄	3–4	2.5–2.9	{001}		platy	clr-dk brn	blk	0.030–0.110	1.543–1.745	triclinic	(−)≃0
Stishovite (stv) [high pressure form of SiO₂]					massive		clr				
Sulfur (sul) S	1.5–2.5	2			dipyramidal	clr	pal yel	0.287	1.958–2.245	orthorhombic	(+)69
Sylvite (slv) KCl	2	1.9	{100}		cubic, octahedral	clr	clr, red		1.490	isometric	
Talc (tal) Mg₃Si₄O₁₀(OH)₂	1	2.6–2.8	{001}		platy	clr	wht, lt grn	0.040–0.050	1.538–1.600	monoclinic	(−)0–30

continued

Table 4.1 continued

Mineral (abbreviation) Formula	Hardness Specific Gravity	Cleavage/ Parting (P)	Twinning Color (ts)	Habit Color (hs)	Range of Biref. (δ) Low-High R.I.	Crystal System Sign 2V (deg)
Thenardite (thn) Na_2SO_4	2.5–3 2.6	{010}	{110} clr	tabular, prismatic wht	0.021 1.464–1.485	orthorhombic (+)83
Thompsonite (tps) $NaCa_2(Al_5Si_5)O_{20} \cdot 6H_2O$	5–5.5 2.1–2.3	{010}{100}	clr	blades, acicular clr, gry	0.006–0.021 1.497–1.544	orthorhombic (+)42–75
Titanite (tit) [see sphene]						
Topaz (tpz) $Al_2(SiO_4)(F,OH)_2$	8 3.4–3.5	{001}	clr	prismatic clr, yel, red, blu	0.008–0.011 1.606–1.638	orthorhombic (+)48–68
Tourmaline (tor) $Na(Mg,Fe,Li,Al)_3Al_6(Si_6O_{18})(BO_3)_3(OH,F)_4$	7 3–3.2		clr-blu,grn,brn	prismatic blk, brn, grn, blu,yel	0.015–0.035 1.610–1.698	trigonal (–)
Tremolite (trm) $Ca_2Mg_5Si_8O_{22}(OH)_2$	5–6 3–3.1	{110}	{100} clr	prismatic wht	0.022–0.026 1.600–1.650	monoclinic (–)84–88
Tridymite (trd) SiO_2	7 2.2		{10$\bar{1}$6} clr	wedges clr, wht	0.002–0.004 1.469–1.483	orthorhombic (+)35–90
Trona (trn) $Na_3H(CO_3)_2 \cdot 2H_2O$	2.5–3 2.1	{100}	clr	fibrous, bladed clr, wht, gry	0.128 1.412–1.540	monoclinic (–)72
Ulexite (ulx) $NaCaB_5O_9 \cdot 8H_2O$	2.5 1.9	{010}{1$\bar{1}$0}	{010}{100} clr	blades, fibrous clr, wht	0.023 1.491–1.520	triclinic (+)73–78
Uvarovite (uvr) $Ca_3Cr_2(SiO_4)_3$	7.5 3.8		pale grn	dodecahedral dk grn	1.86	isometric
Vermiculite (vmc) (Mg,Ca)(Mg,Fe^{2+})$_5$(Fe^{3+},Al)(Si$_5$Al$_3$)O$_{20}$(OH)$_4$	1.5 2.4	{001}	clr-grn brn	platy grn brn	0.020–0.30 1.520–1.580	monoclinic (–)0–8
Wairakite (wai) $Ca(AlSi_2O_6)_2 \cdot 2H_2O$	5.5–6 2.2	{100}	{110} clr	pseudo-octahedral wht	0.004 1.498–1.502	monoclinic (+)70–90
White mica (whm) [muscovite, paragonite, phengite]						
Wollastonite (wo) $CaSiO_3$	4.5–5 2.9–3.1	{100}{001} {$\bar{1}$02}	{100} clr	bladed wht	0.013–0.017 1.615–1.662	triclinic (–)36–60
Xenotime (xtm) YPO_4	4–5 4.3–5.1	{110}	clr-pale yel	prismatic yel brn, red brn	0.095–0.107 1.719–1.826	tetragonal (+)
Zircon (zir) $ZrSiO_4$	7.5 4.6		{111} clr	prism with dipyram yel brn	0.042–0.065 1.920–2.015	tetragonal (+)
Zoisite (zo) $Ca_2Al_3O(SiO_4)(Si_2O_7)(OH)$	6 3.1–3.3	{010}	clr	prismatic gry	0.005–0.020 1.685–1.725	orthorhombic (+)0–60

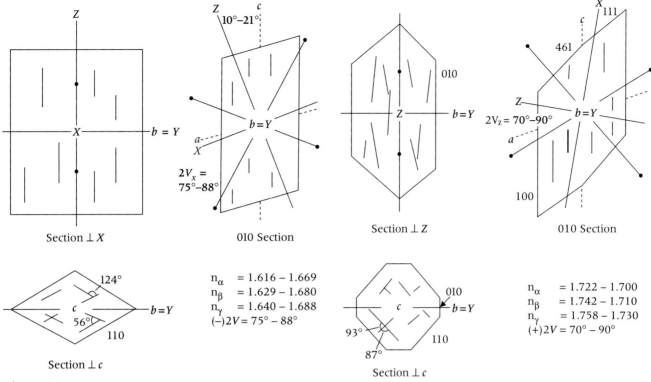

Figure 4.2
Actinolite.

Figure 4.4
Aegirine–augite.

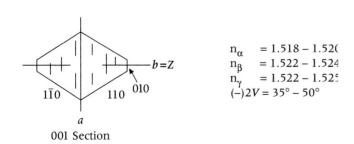

Figure 4.3
Adularia.

Adularia (K,Na)AlSi$_3$O$_8$(Kf$_{90-100}$) Triclinic or Monoclinic Figures 7.4c, 34.25. Authigenic in sedimentary rocks and low-temperature hydrothermal. Domination of {110} prism results in diamond-shaped crystals. Variable symmetry: if triclinic it is a **microcline**; if monoclinic an **orthoclase**, and crystals may have domains of both. Disordered structures due to relatively rapid crystallization.

Aegirine NaFe^{3+}(SiO$_3$)$_2$ Monoclinic Figure 7.14c. A clinopyroxene with 90° cleavage but very small $Z \wedge c$, even smaller than clinoamphiboles. Very dark green to yellow pleochroism is diagnostic alone.

Aegirine–Augite (Na,Ca)(Fe^{3+},Fe^{2+},Mg,Al)(SiO$_3$)$_2$ Monoclinic Figure 7.14c. This clinopyroxene has 90° cleavage but $Z \wedge c$ similar to clinoamphiboles. It has green to yellow pleochroism, without the green–blues characterizing arfedsonite, some hornblende, and blue amphiboles.

Albite NaAlSi$_3$O$_8$ Triclinic Figures 7.11c, 7.14e, 37.13c. The only plagioclase (albite–pericline twins, nearly right-angle cleavage) that has refractive indices below quartz and mounting epoxy in most orientations. Pure albite is also an alkali feldspar. See Figure 5.9 for optical orientation.

Alunite KAl$_3$(SO$_4$)$_2$(OH)$_6$ Trigonal Figures 34.5b, 40.16, 40.24a. Similar to brucite but lacking the brownish anomalous interference colors. Occurs typically in

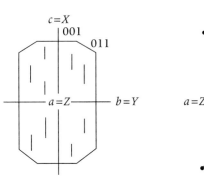
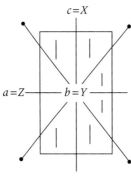
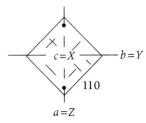

$n_\alpha = 1.629 - 1.640$
$n_\beta = 1.633 - 1.644$
$n_\gamma = 1.638 - 1.650$
$(-)2V = 71° - 88°$

Figure 4.5
Andalusite.

weathered rocks that originally contained sulfides. Submicroscopic crystal aggregates common.

Analcime $Na(AlSi_2)O_6 \cdot H_2O$ Isometric. Isotropic colorless mineral in matrix of alkalic igneous rocks such as teschenites and analcime basalts. Dodecahedrons occur in vesicles.

Andalusite Al_2SiO_5 Orthorhombic Figures 2.18, 34.5c; Plate 3a,b. Square or diamond-shaped sections have symmetrical extinction and present 90° cleavage. Sections parallel to prism are length fast (unlike enstatite) and have parallel extinction. Graphite (± quartz) cross is **chiastolite** (Figure 21.33a). Occurs in aluminous schists, unlike barite, which is superficially similar but optically positive.

Andesine $(NaSi,CaAl)AlSi_2O_8$ (An_{30-50}) Triclinic Figure 5.16. A plagioclase (multiple twinning common, near 90° cleavage) of "intermediate" composition along with **oligoclase**. See Figure 5.9 for optical orientation.

Anhydrite $CaSO_4$ Orthorhombic Figure 38.10; Plate 2. Right-angle cleavage, twinning at 45° to cleavage, and rectangular sections are characteristic. Parallel extinction, unlike some orientations of colemanite, which does not have twinning. Occurs in evaporites.

Anorthite $(CaAl_2Si_2O_8)$ (also An_{90-100}) Triclinic. Calcic plagioclase (albite–pericline twins, nearly right-

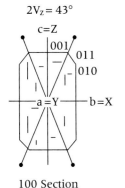
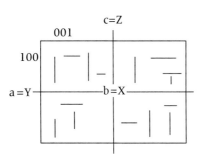
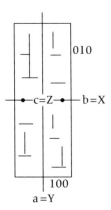

$n_\alpha = 1.570$
$n_\beta = 1.576$
$n_\gamma = 1.614$
$(+)2V = 43°$

Figure 4.6
Anhydrite.

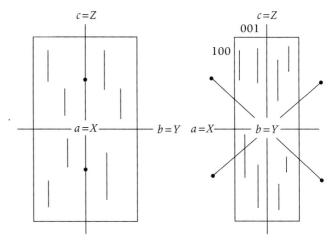
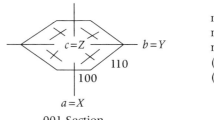

$n_\alpha = 1.588 - 1.694$
$n_\beta = 1.602 - 1.710$
$n_\gamma = 1.613 - 1.722$
$(+)2V = 70° - 90°$
$(-)2V = 75° - 90°$

Figure 4.7
Anthophyllite.

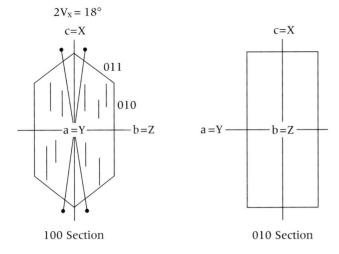
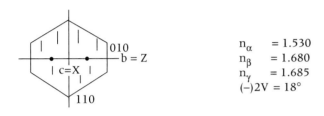

Figure 4.8
Aragonite.

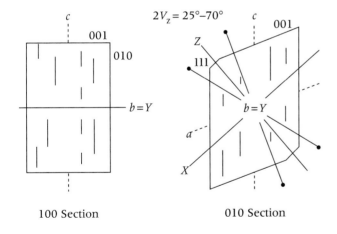
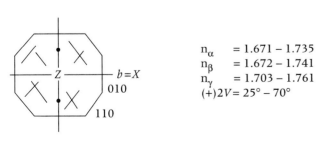

Figure 4.9
Augite.

4 Rock-Forming Minerals

angle cleavage) occurring in sodium-poor environments, such as in metamorphosed carbonates. See Figure 5.9 for optical orientation.

Anorthoclase $(Na,K)AlSi_3O_8$ (Ab_{63-90}) Triclinic. An **alkali feldspar** (low r.i., feldspar cleavage) occurring in volcanic and shallow intrusive rocks. Twinning similar to that of microcline, but lamellae are thinner and less distinct. Generally not visibly perthitic.

Anthophyllite $(Mg,Fe)_7Si_8O_{22}(OH)_2$ Orthorhombic. An amphibole (124°/56° cleavage) having parallel extinction. An aluminous variety is known as **gedrite**.

Antigorite $Mg_3Si_2O_5(OH)_4$ Monoclinic Plate 6C. Nonfibrous **serpentine**. Crinkled micaceous habit, lack of pleochroism and association with olivine are characteristic. Optically indistinguishable from the variety **lizardite**. High-Mg chlorites have higher birefringence; other chlorites are pleochroic. Brucite has higher birefringence (first-order yellow–orange) and is uniaxial. Fibrous serpentine is **chrysotile** (asbestos), having somewhat higher birefringence (first-order yellow) than antigorite–lizardite.

Aragonite $CaCO_3$ Orthorhombic Figure 39.2. Similar to calcite and dolomite but lacks the rhombohedral cleavage and uniaxial optics. May have multiple parallel twins or cyclic twins. Typically radial fibrous in fossils (Figure 37.16b).

Arfedsonite $Na_3Fe_4Fe^{3+}Si_8O_{22}(OH)_2$ Monoclinic. Prismatic habit, 124°/56° cleavage, diamond-shaped sections, brownish yellow, yellow–green, and blue–green pleochroism, optically positive, and occurrence in alkalic igneous rocks is characteristic. Arfedsonite and iron-rich variety, **eckermanite**, are similar to common hornblende and pargasite; however, these amphiboles generally do not occur in alkalic rocks.

Augite $(Ca,Mg,Fe^{2+},Na)(Mg,Fe^{3+},Al,Ti)(Si,Al)_2O_6$ Monoclinic Figures 2.19, 7.8f, 7.11f, 7.11h. A **clinopyroxene** (90° cleavage, large $Z \wedge c$) (cpx) typical in mafic igneous rocks, less colorless than olivine, nonpleochroic. Very similar to diopside (Al-free clinopyroxene that occurs in Al-deficient rocks such as some metamorphosed carbonates and metamorphosed ultramafic rocks). Variety **titanaugite** has green–yellow–violet pleochroism and is commonly zoned.

Barite $BaSO_4$ Orthorhombic. Parallel and symmetrical extinction. Multiple (4) directions of cleavage, some of which yield 90° angles. Length slow (like enstatite), but unlike enstatite it occurs in hydrothermal veins and in sedimentary rocks. Very similar to **celestite**.

Barkevikite $NaCa_2(Fe^{2+},Mg,Mn)_5(Si_7Al)O_{22}(OH)_2$ Monoclinic. A clinoamphibole (124°/56°, diamond-shaped sections) with red–brown to yellow–brown pleochroism and occurrence in alkalic igneous rocks is likely to be barkevikite. Oxyhornblende is similar but occurs in mafic volcanic rocks.

Biotite $K(Mg,Fe^{2+})_{5-6}Al_{0-1}(Si_{5-6}Al_{2-3})O_{20}(OH,F)_4$ Monoclinic. Figures 2.11, 7.14c, 8.9c; Plate 6a,b. Characterized by micaceous habit, tan to brown pleochroism. Clear to light green biotite (high ferric iron) is similar to chlorite, but biotite is without anomalous interference colors. Red–brown biotite is high in Ti (Figures 7.14c, 14.8b). The deeper colors are parallel to cleavage, unlike tourmaline. Pleochroic halos around zircon (and other accessory minerals containing U and Th) inclusions are common.

Borax $Na_2B_4O_7 \cdot 10H_2O$ Monoclinic. Water soluble. Specially prepared sections are required. A {110} cleavage intersects at 90°. These are cut by a {100} cleavage. Anomalous blue and brown interference colors in some orientations.

Brucite $Mg(OH)_2$ Hexagonal. The crinkled micaceous habit, brownish anomalous interference colors are characteristic of brucite. Similar to some high-Mg chlorites. Associated with periclase or olivine in metamorphosed dolomitic rocks.

Bytownite $(CaAl,NaSi)AlSi_2O_8$ (An_{70-90}) Triclinic. A calcic plagioclase (albite/pericline twins, nearly right-angle cleavage). See Figure 5.9 for optical orientation. All calcic plagioclases are susceptible to alteration to sericite, epidote, zoisite, calcite, and zeolites.

Calcite $CaCO_3$ Hexagonal Trigonal Figures 2.7, 21.13a. Characterized by rhombohedral cleavage and twinning and by birefringence that is so high that it is not very noticeable. Optically indistinguishable from dolomite, ankerite, and most siderite, except that these minerals typically have some associated alteration to iron oxides. Can be radial fibrous in fossils and ooids (Figure 37.25a).

Chlorite $(Mg,Fe,Al)_{12}(Si,Al)_8O_{20}(OH)_{16}$ Monoclinic Figure 21.7; Plate 6b. Micaceous habit, light green to colorless pleochroism and anomalous blue or brown birefringence are characteristic. Chloritoid is similar but has a higher r.i. and may have sector zoning (Figure 21.33b). Mg-chlorite in serpentine-bearing rocks is nonpleochroic. **Penninite** is a common alteration of biotite in granitic rocks, having Berlin blue anomalous interference colors. **Chamosite** and **berthierine** are common in iron-rich sediments. Varieties such as **amesite**, **prochlorite**, and **clinochlore** are not optically distinguishable.

Chloritoid $(Fe^{2+},Mg,Mn)_2(Al,Fe^{3+})Al_3O_2(SiO_4)_2(OH)_4$ Monoclinic Figures 21.33b, 21.37e,f. A chlorite-like mineral with unusually high refractive index is likely to be chloritoid. Pleochroic in light greens and blue–greens, commonly with hourglass sector zoning. Interference color tends to be somewhat anomalous yellow to brown.

Clinozoisite $Ca_2Al_3O(Si_2O_7)(OH)$ Monoclinic. Anomalous blue and yellowish green interference colors are indicative. Extinction parallel to elongation and cleavage.

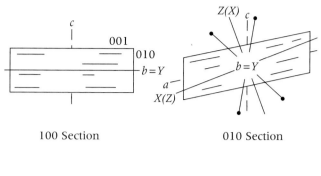

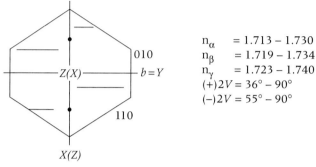

$n_\alpha = 1.713 - 1.730$
$n_\beta = 1.719 - 1.734$
$n_\gamma = 1.723 - 1.740$
$(+)2V = 36° - 90°$
$(-)2V = 55° - 90°$

Figure 4.10
Chloritoid.

Colemanite $Ca_2B_6O_{11} \cdot 5H_2O$ Monoclinic Figure 38.12. Similar to anhydrite, including 90° cleavage. Sections near (010) have inclined extinction, unlike anhydrite. Occurs in evaporites.

Cordierite $Mg_2Al_3(Si_5Al)O_{18}$ Orthorhombic Figures 7.11i, 7.14b, 21.8, 21.9a, 21.31b, 28.2. Cyclic wedge-shaped twins and chloritic alteration along margins and in microveins distinguish from quartz and nontwinned oligoclase.

Corundum Al_2O_3 Hexagonal. Tapered prismatic sections with parallel extinction. **Sapphirine** is biaxial and may be pleochroic from clear to pale blue. Neither associated with quartz.

Crossite $Na_2Mg_3Fe_2(Si_4O_{11})_2(OH)_2$ Monoclinic Figure 21.19c, Plate 7a. Blue amphibole (124°/56° cleavage, diamond-shaped sections) with blue to violet to tan pleochroism. These colors are darker than those of glaucophane, but the position of the optic plane across the direction of elongation is diagnostic. Nevertheless, $(-)2V$ ranges through 0 (uniaxial) to biaxial (+) with the optic plane parallel to elongation.

Cummingtonite $Mg_7(Si_4O_{11})_2(OH)_2$ Monoclinic. An amphibole (124°/56° cleavage, diamond-shaped sections, small $Z \wedge c$) similar to tremolite, except it may be pleochroic to pale brown and have multiple

4 Rock-Forming Minerals 63

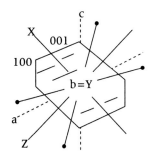

Figure 4.11
Clinozoisite.

$n_\alpha = 1.703 - 1.715$
$n_\beta = 1.707 - 1.725$
$n_\gamma = 1.709 - 1.734$
$(+)2V = 14° - 90°$

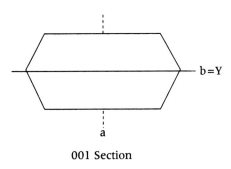

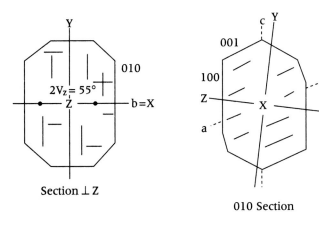

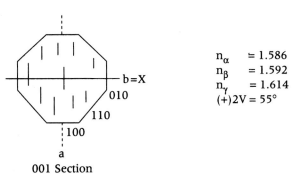

$n_\alpha = 1.586$
$n_\beta = 1.592$
$n_\gamma = 1.614$
$(+)2V = 55°$

Figure 4.12
Colemanite.

twinning. Most commonly has (+) sign, unlike tremolite. Expect to find it in high-Mg rocks, and the iron end member, **grunerite**, in high-iron rocks.

Diopside $CaMgSi_2O_6$ Monoclinic. A calcic clinopyroxene (90° cleavage, large $Z \wedge c$) common in metamorphosed carbonate rocks. Less common in igneous rocks than is augite. Optically indistinguishable from augite and the iron-bearing variety **salite**.

Dolomite $CaMg(CO_3)_2$ Hexagonal Trigonal Plate 8b,c. Very similar to calcite, requiring staining for distinction. Characterized by rhombohedral cleavage, twinning, and extreme birefringence. Occurs as rhombohedral crystals in sedimentary rocks more commonly than calcite.

Enstatite $Mg_2(SiO_3)_2$ Orthorhombic Figures 2.15, 17.7c. Pyroxene cleavage (90°), parallel extinction in all sections parallel to prism. Length slow (unlike andalusite) and may have CPX exsolution lamella (Figure 7.16d). Enstatite has 0–10% $FeSiO_3$, and **bronzite** (Figures 18.6, 18.7b, c) has 70–90% $FeSiO_3$.

Epidote $Ca_2Al_2Fe^{3+}O(Si_2O_7)(SiO_4)(OH)$ Monoclinic Figure 21.34a. Elongate sections show one direction of cleavage and have parallel extinction. Low-iron epidote has no color; high-iron epidote is **pistacite** with clear to yellow pleochroism. Very similar to **lawsonite**, the latter having two cleavages at 90° and associating typically with blue amphiboles.

Fluorapatite $Ca_5(PO_4)_3(F,OH,Cl)$ Tetragonal. Hexagonal prism cross sections and rectangular sections of prisms with parallel extinction along with its clarity

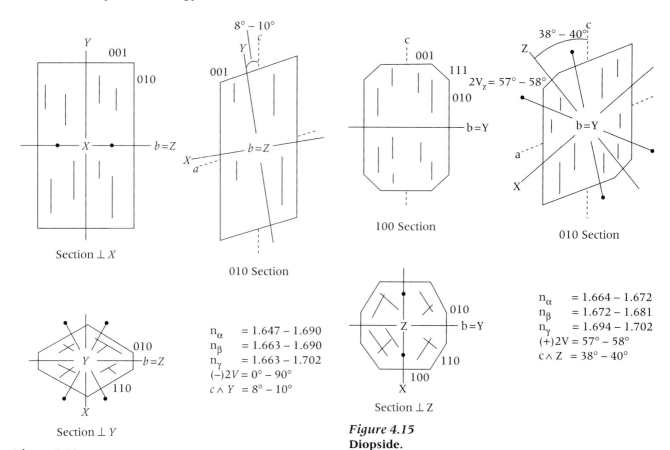

Figure 4.13
Crossite.

Figure 4.15
Diopside.

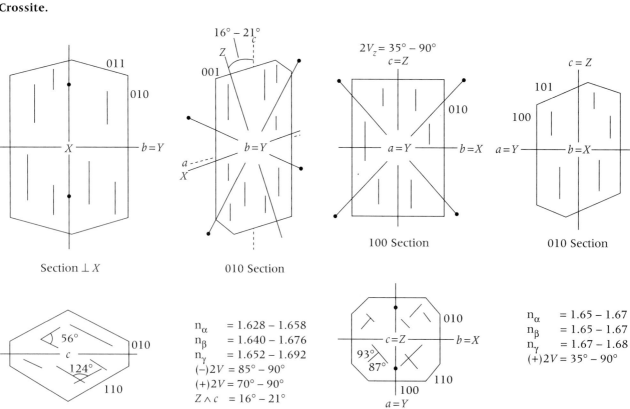

Figure 4.14
Cummingtonite.

Figure 4.16
Enstatite.

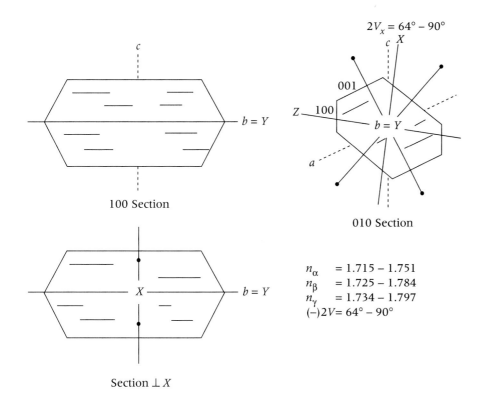

Figure 4.17 Epidote.

n_α = 1.715 − 1.751
n_β = 1.725 − 1.784
n_γ = 1.734 − 1.797
$(-)2V$ = 64° − 90°

are diagnostic. Typically fibrous habit in fossils (Figure 37.21).

Fluorite CaF_2 Isometric Figures 2.6b, 34.28. Isotropic, colorless, and very low refractive index. Octahedral cleavage typically results in intersections defining triangular fragments.

Garnet $X_3Y_2(SiO_4)_3$ Isometric. Dodecahedral colorless crystals yield equant polygonal sections. **Grossularite** (Figure 34.20a) occurs in calc–silicate rocks. The OH-bearing variety **hydrogrossular** (Figure 7.11d) has birefracting growth zones, typically occurring in skarns where it has an **andradite** component (Figure 34.20a). **Pyrope** occurs in metamorphosed mafic and ultramafic rocks such as eclogite. **Almandine** is the typical garnet of pelitic schists (Figures 21.17, 21.32a, 21.37a,b). **Spessartine** occurs in aplites, pegmatites, blueschists (Figure 21.7), and rarely in vapor-phase cavities in rhyolite (Figure 17.12d). **Uvarovite** is the rare Cr-bearing garnet of some peridotites and serpentinites.

Glaucophane $Na_2Mg_3Al_2Si_8O_{22}(OH)_2$ Monoclinic Figure 21.25. A blue amphibole (124°/56° cleavage, diamond-shaped sections) with blue–violet–tan pleochroism. Similar to crossite but weaker pleochroism. Optic plane is (010).

Gypsum $CaSO_4·2H_2O$ Monoclinic Figures 38.5c, 38.9, 38.12. Prominent {010} cleavage parallel to elongation has parallel extinction in this orientation. A single {100} twin is common. Speckled extinction similar to muscovite.

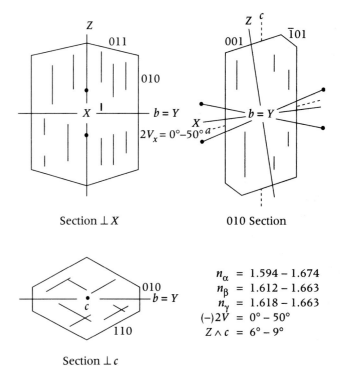

n_α = 1.594 − 1.674
n_β = 1.612 − 1.663
n_γ = 1.618 − 1.663
$(-)2V$ = 0° − 50°
$Z \wedge c$ = 6° − 9°

Figure 4.18 Glaucophane.

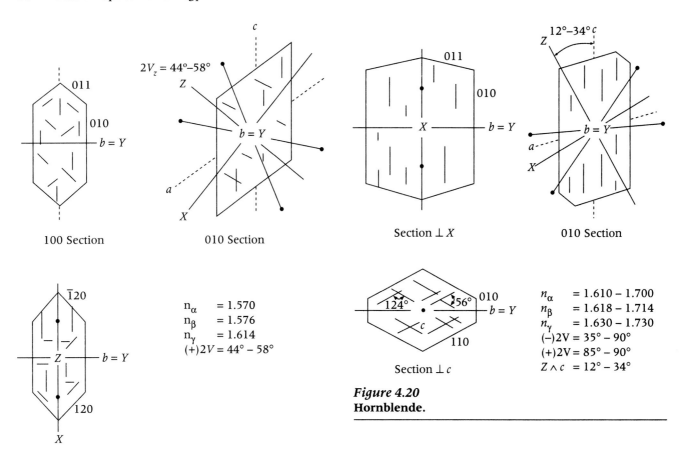

Figure 4.19
Gypsum.

Figure 4.20
Hornblende.

Halite NaCl Isometric Figures 35.4, 38.6c. Rarely prepared in thin section since it is soluble in water. Isotropic, clarity, and cubic cleavage are indicative.

Hauyne $(Na,Ca)_{4-8}(Al_6Si_6O_{24})(SO_4)_{1-2}$. Equant polygon sections as phenocrysts in silica-undersaturated volcanic rocks. May be very pale green or blue and may show several directions of cleavage (dodecahedral). Leucite has diagnostic twinning. Optically indistinguishable from **sodalite** and **nosean** (Figure 7.8h). Analcime does not form phenocrysts.

Hedenbergite $CaFe(SiO_3)_2$ Monoclinic. A calcic clinopyroxene (90° cleavage, large $Z \wedge c$) with iron more abundant than magnesium (**ferrosalite** has somewhat lower Fe content), resulting in pleochroism in greens. **Aegirine–augite** is similar but optically negative.

Hornblende $Ca_2(Mg,Fe^{2+})_4(Al,Fe^{3+})(Si_7,Al)O_{22}(OH)_2$ Monoclinic Figures 2.19, 7.11g,h, 20.9. **Common hornblende** is an amphibole (124°/56° cleavage, diamond-shaped sections, small $Z \wedge c$) with tan–olive green–bluish green pleochroism, stronger colors than most actinolite. Variety **pargasite** is optically positive, and variety **hastingsite** has a small 2V.

Hypersthene $(Mg,Fe)_2(SiO_3)_2$ Orthorhombic Figures 21.9a, 28.2b. A prismatic crystal in this δ and r.i. range, having parallel extinction, 90° cleavage, and pale pink to pale green pleochroism.

Idocrase (Vesuvianite) $Ca_{10}(Mg,Fe^{2+})_2Al_4(Si_2O_7)_2(SiO_4)_5(OH,F)_4$ Tetragonal. Anomalous indigo and brown interference colors, and association with diopside and grossularite are typical. Lack of cleavage distinguishes it from low-Fe zoisite.

Jadeite $NaAl(SiO_3)_2$ Monoclinic Figure 21.19a. A pyroxene (90° cleavage, large $Z \wedge c$) typically associated with blue amphiboles. Iron-bearing varieties can have anomalous blue and brown interference colors, similar to Fe-bearing zoisite, which, however, has parallel extinction. Similar to omphacite (in eclogite) and the calcic clinopyroxenes (not in blueschists).

Jarosite $KFe_3(SO_4)_2(OH)_6$ Trigonal Figure 40.23. A nonpleochroic orange–yellow is typical. Occurrence in altered rocks is useful. There is a solid solution series to alunite, but alunite is clear in section. Uniaxial optics is confirming.

Kyanite Al_2SiO_5 Triclinic Figure 16.5c. Two directions of cleavage parallel to elongation are nearly at 90° in prism cross sections similar to andalusite. However, kyanite has inclined extinction. A basal parting perpendicular to cleavage generates a "fragmented" aspect (Figure 2.6a).

Labradorite $(NaSi,CaAl)AlSi_2O_8$ (An_{50-70}) Triclinic Figures 5.7, 21.6b. The relatively calcic plagio-

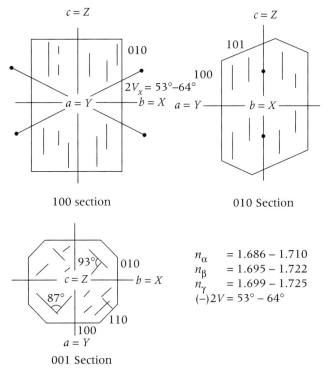

Figure 4.21
Hypersthene.

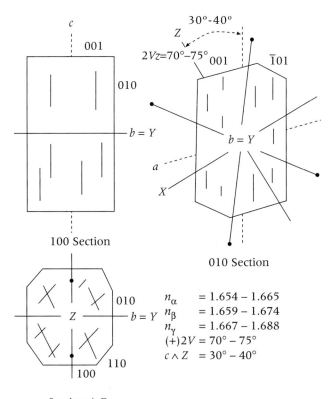

Figure 4.22
Jadeite.

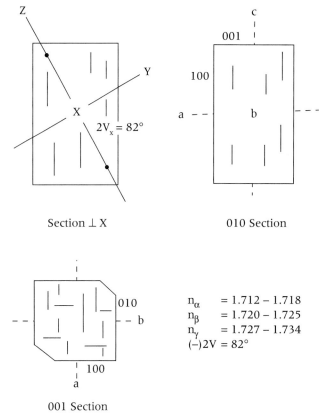

Figure 4.23
Kyanite.

clase (characteristic albite–pericline twinning and right-angle cleavage) of mafic igneous rocks. See Figure 5.9 for optical orientation.

Lawsonite $CaAl_2Si_2O_7(OH)_2 \cdot H_2O$ Orthorhombic Figure 21.10a. Similar to nonpleochroic epidote (low Fe) except for two cleavages at 90°. Parallel extinction. Association with blue amphiboles is characteristic.

Leucite $KAlSi_2O_6$ Isometric Figure 7.11j. Equant polygon sections as phenocrysts in silica-undersaturated (quartz-free) volcanic rocks. Multiple twinning similar to microcline but in more than two directions.

Pseudoleucite is leucite pseudomorphically replaced by fine-granular nepheline and K-feldspar.

Magnesite $MgCO_3$ Hexagonal Trigonal. This carbonate has rhombohedral cleavage but not twinning that is so characteristic of calcite and dolomite. Occurrence in altered ultramafic rocks or metamorphosed dolostone, along with brucite, is useful but not confirming.

Margarite $CaAl_2(Si_2Al_2)O_{10}(OH)_2$ Monoclinic. A "white mica" with high indices of refraction and low birefringence. Difficult to confirm optically.

Melelite $(Ca,Na)_2(Mg,Al)(Si,Al)_2O_7$ Tetragonal. Most commonly as phenocrysts in quartz-free volcanic

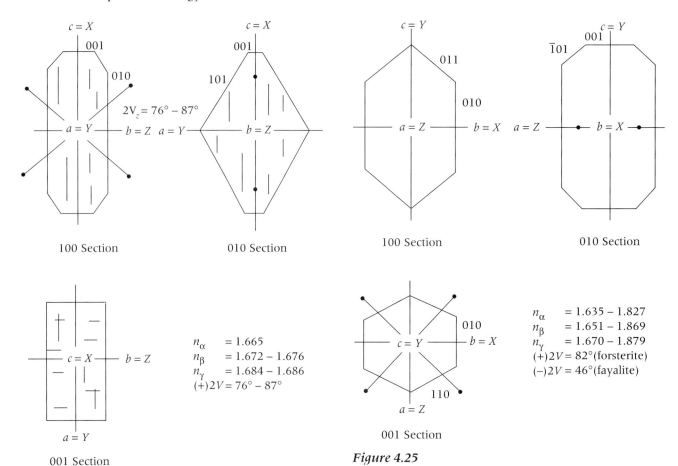

Figure 4.24
Lawsonite.

Figure 4.25
Olivine.

rocks, appearing as square, rectangular, or octagonal sections. Anomalous blue interference colors distinguish from nepheline.

Microcline $(K,Na)AlSi_3O_8$ (Kf_{92-100}) Triclinic Figures 7.11e, 7.14a, 17.9b. A K-rich alkali feldspar of plutonic magmatic rocks and metamorphic rocks. Characterized by crosshatched twins that are diffuse and discontinuous compared to equivalent albite and pericline twinning in plagioclase. Twinned microline inverts from a monoclinic predecessor (probably in an aqueous environment), and local twin development in crystals may indicate coexistence of microcline and orthoclase. Perthitic intergrowths with albite are common (Figures 7.8c, 7.16b, 14.3b), ranging from cryptoperthites to coarse perthites of granitic pegmatites (Figure 7.16a).

Muscovite $KAl_2(Si_3Al)O_{10}(OH)_2$ Monoclinic Figures 8.9a, 8.9c, 17.10c; Plates 4a, 6a. The most common "white mica." Optically indistinguishable from paragonite and nearly so from margarite, talc, and pyrophyllite. Variety **phengite** contains some Mg and Fe; **hydromuscovite** contains more water. Most **sericite** (Figures 34.16a, 40.20b) is fine-grained muscovite.

Nepheline $Na_3KAl_4Si_4O_{16}$ Hexagonal Figures 7.4b, 7.8h. Hexagonal or rectangular sections with parallel extinction, zoning common, intergrowths with albite common, poor cleavage unlike feldspar, does not occur with quartz.

Oligoclase $(NaSi,CaAl)AlSi_2O_8$ (An_{10-30}) Triclinic Figures 5.19, 7.15c. Plagioclase (albite–pericline twins and near right-angle cleavage) of "intermediate" composition. Nontwinned crystals in metamorphic rocks are very similar to quartz and cordierite. "Sericitic" alteration common (unlike quartz), lack of progressive extinction (unlike quartz), and lack of chloritic alteration (unlike cordierite) are indicative. See Figure 5.9 for optical orientation.

Olivine $(Mg,Fe)SiO_4$ Orthorhombic Figures 7.4e, 14.6b, 14.8d; Plates 1b, 6c. Subequant crystals with intersecting prism faces, lack of cleavage, parallel extinction, clarity, alteration to minerals such as **iddingsite** along fractures are all characteristic of olivine. Epidote and pyroxenes have cleavage.

Omphacite $(Ca,Na)(Mg,Fe^{2+},Fe^{3+},Al)(SiO_3)_2$ Monoclinic. A pyroxene (90° cleavage, large $Z \wedge c$) that has clear to pale green pleochroism, weaker than aegirine–augite. Jadeite is not pleochroic. Occurrence with garnet in eclogite and other high-pressure sodium-

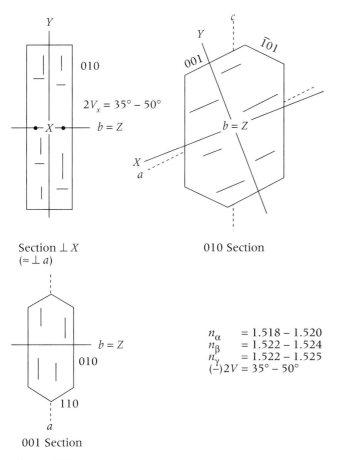

Figure 4.26
Orthoclase.

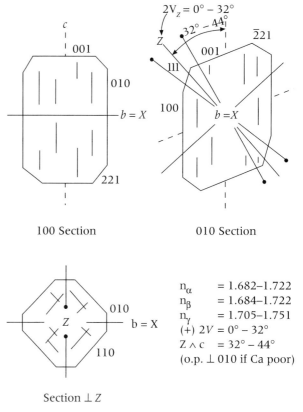

Figure 4.27
Pigeonite (low Ca).

bearing rocks is typical, unlike diopside, salite, and hedenbergite.

Orthoclase (K,Na)AlSi$_3$O$_8$ (Kf$_{20-100}$) Monoclinic Figures 14.4, 7.11b, 17.4e, 17.9a, 17.10a. An **alkali feldspar** (low r.i. and two directions of cleavage at right angles) occurring most commonly in shallow intrusive magmatic rocks. Optic angle between 40° and 70° contrasts with small angles of sanidine. Optically indistinguishable from submicroscopically twinned microcline. Without specific knowledge of lattice dimensions, a generalization such as "alkali feldspar" or "K-feldspar" is preferred usage. May be perthitic (Figures 2.6c, 17.11c) and have a Carlsbad twin. "Dusty" kaolinitic alteration common.

Oxyhornblende Ca$_2$Na(Mg,Fe^{3+},Al,Ti)$_5$[(Si$_3$Al)O$_{11}$]$_2$ (OH)$_2$ Monoclinic. An amphibole (124°/56° cleavage, triangular sections, and small $Z \wedge c$) that has greenish yellow to red–brown pleochroism. Very similar to barkevikite and **kaersutite**, although these minerals have somewhat larger (−)2V.

Phologopite KMg$_3$(Si$_3$Al)O$_{10}$(OH)$_2$ Monoclinic. A mica with clear to pale brown pleochroism. Occurs in metamorphosed impure carbonates and ultramafic rocks such as kimberlite.

Pigeonite (Mg,Fe^{2+},Ca)(Mg,Fe^{2+})(SiO$_3$)$_2$ Monoclinic. Very similar to augite except for small (+)2V (0°–32°). Occurs in volcanic rocks. Inverts to orthopyroxene and augite in slow-cooling environments.

Prehnite Ca$_2$Al(AlSi$_3$)O$_{10}$(OH)$_2$ Orthorhombic. Sheaflike clusters of crystals with parallel extinction are common. Basal cleavage across elongation. Similar to pumpellyite.

Pumpeyllite Ca$_2$Al$_2$(Mg,Fe^{2+},Fe^{3+},Al)(SiO$_4$)(Si$_2$O$_7$) (OH)$_2$(H$_2$O,OH) Monoclinic. Radial subparallel clusters common. Light yellow to green pleochroism, anomalous blue or yellow–brown interference colors common. May not be easily distinguishable from clinozoisite and zoisite if crystals are small and poorly formed.

Quartz SiO$_2$ Hexagonal Figures 2.16, 14.9b, 14.10, 16.3a, 17.4c, 25.2, 31.5a. Clarity of crystals, common presence of at least some progressive extinction, lack of alteration, lack of twinning and cleavage are diagnostic. Rhombohedral cleavage may develop in impactites and in very thin grains in tapered thin sections. Phenocrysts in volcanics and crystals in epithermal veins typically are well formed and yield hexagonal cross sections. Excellent reference mineral in thin section because of common occurrence and ease of identification. Cream-colored maximum

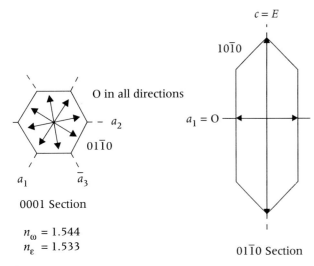

Figure 4.28 Quartz.

birefringence indicates appropriate thickness of thin section. Uniaxial interference figure distinguishes from untwinned oligoclase and untwinned cordierite (see oligoclase).

Riebeckite $Na_2Fe_3Fe_2(Si_4O_{11})_2(OH)_2$ Monoclinic. A nearly blue–black amphibole (124°/56° cleavage) occurring in alkaline igneous rocks. Aegirine is green–black.

Sanidine $(K,Na)AlSi_3O_8$ (Ab_{20-60}) Monoclinic Figures 17.2a, 17.13a, 19.6. The "glassy" alkali feldspar of volcanic rocks. Clarity similar to quartz but low r.i., Carlsbad twins, and biaxial optics make the distinction. Optic angle small in comparison with orthoclase. Not visibly perthitic. Intergrowths with cristobalite and tridymite in spherulites of rhyolites. **High sanidine** has optic plane parallel to (010), whereas **low sanidine** has the optic plane normal to (010).

Scapolite $(Na,Ca)_4[(Al,Si)_3Al_3Si_6O_{24}](Cl,CO_3)$ Tetragonal. Several directions of cleavage parallel to elongation, intersecting at 45° in cross sections. Parallel

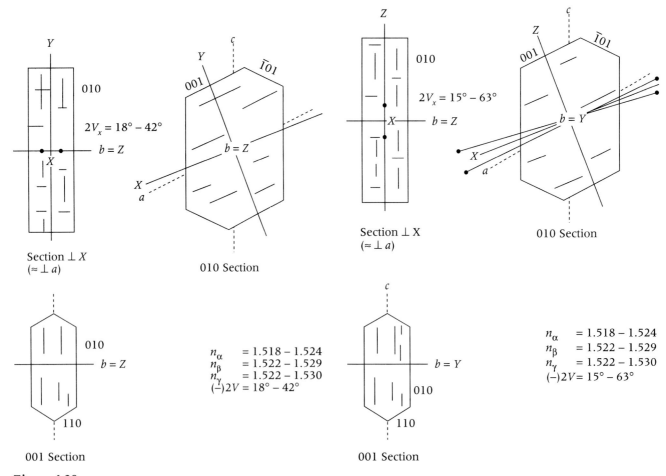

Figure 4.29

(a) Sanidine (low). (b) Sanidine (high).

extinction. **Marialite** (Na end member))has lower δ and **meionite** (Ca end member) has higher δ. Uniaxial (−) is a good confirming character.

Siderite $FeCO_3$ Hexagonal Trigonal. Rhombodedral cleavage and twinning similar to calcite and dolomite. May be pale brown, and commonly is altered to iron oxides (Figure 36.3b).

Sillimanite Al_2SiO_5 Orthorhombic Figure 8.9a; Plate 1a. One cleavage parallels elongation, and this has parallel extinction. One cleavage symmetrical with diamond-shaped prism sections is characteristic. Length slow, unlike andalusite. Kyanite has inclined extinction. Anthophyllite has amphibole cleavage and occurs in high-Mg rocks, not high-Al rocks.

Spinel XAl_2O_4 (X = Mg,Fe,Zn, or Mn) Isometric. Nonpleochroic colors in greens or browns. Rhombic shapes due to sectioning of octahedrons.

Sphene $CaTiSiO_5$ Monoclinic Figures 19.13, 20.5b; Plate 6b. Nonpleochroic pale brown and diamond-shaped sections are characteristic. Similar to **allanite** (pleochroic), **monazite** (pale yellow), **xenotime** (uniaxial), **rutile** (uniaxial), **cassiterite** (uniaxial).

Staurolite $Fe_2Al_9O_6(SiO_4)_4(O,OH)_2$ Monoclinic (nearly orthorhombic) Figures 21.26a, 21.35a, 21.37a. Golden to colorless pleochroism very diagnostic. Diamond-shaped sections with symmetrical extinction. Elongate sections are rectangular and have parallel extinction, unlike kyanite, with which it may be associated in aluminous schists.

Stilpnomelane $(Fe^{3+},Fe^{2+},Mg,Mn,Al)_{5-6}(Si_4O_{10})(OH)_4$ Monoclinic. Pleochroic from tan to brown or green, and micaceous habit make it similar to biotite. However, stilpnomalane does not have the "speckled extinction" so characteristic of biotite and phologopite.

Talc $Mg_3Si_4O_{10}(OH)_2$ Monoclinic. Optically indistinguishable from white mica. Association with Mg minerals such as dolomite, serpentine, olivine, anthophyllite, enstatite is quite diagnostic.

Tourmaline $Na(Mg,Fe,Li,Al)_3Al_6(Si_6O_{18})(BO_3)_3(OH,F)_4$ Hexagonal Trigonal Figures 2.12, 34.6b. Pleochroic

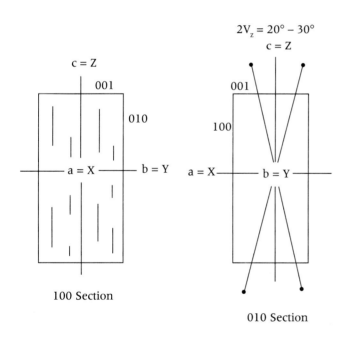
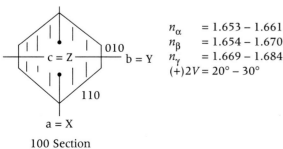

Figure 4.30
Sillimanite.

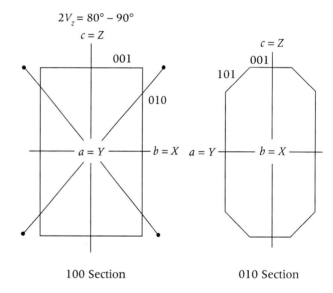
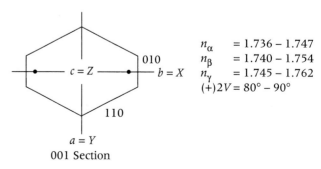

Figure 4.31
Staurolite.

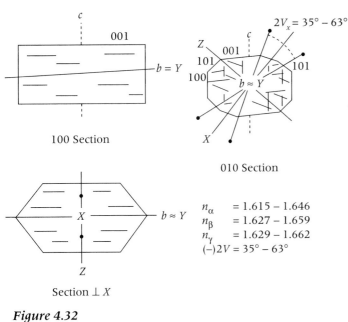

Figure 4.32
Wollastonite.

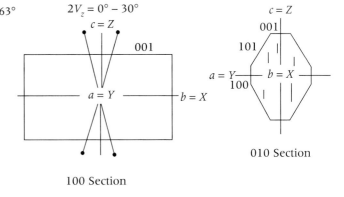

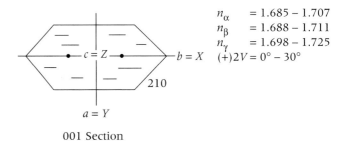

Figure 4.33
Zoisite (low Fe) (α).

in blue, pink, brown, green, gray, depending on composition; typically defining zoning. Darkest colors are generated with polarization aligned perpendicular to elongation. Parallel extinction and triangular cross sections are confirming characteristics. **Schorl, dravite,** and **elbait** are varieties.

Tremolite $Ca_2Mg_5Si_8O_{22}(OH)_2$ Monoclinic Figure 21.10b. A nonpleochroic amphibole (124°/56° cleavage, diamond-shaped sections, small $Z \wedge c$) very similar to **cummingtonite**. The latter commonly has multiple twinning, may have some pale brown pleochroic color, and is mostly optically positive. Wollastonite has near 90° cleavage in sections across length. Very similar to **richterite**, but this mineral commonly has pale yellow to pale blue–green pleochroism.

Tridymite SiO_2 Orthorhombic (pseudohexagonal) Figure 31.7. Wedge-shaped twins, very low birefringence, and biaxial optics distinguish from quartz. Occurs in siliceous volcanics, especially in the matrix or in cavities. **Cristobalite** is another form of silica occurring in volcanics, but its presence can rarely be confirmed optically.

Wollastonite $CaSiO_3$ Triclinic Figure 7.11k Cross-sections of elongate crystals show three directions of cleavage, two of which are nearly at 90°. This orientation also presents higher birefringence (orange–red) than sections parallel to elongation (white–gray), unlike tremolite. Indistinguishable from tremolite in hand specimens.

Zoisite $Ca_2Al_3O(Si_2O_7)(SiO_4)(OH)$ Orthorhombic. Difficult to confirm optically. Elongate sections have parallel extinction and a single cleavage like clinozoisite and epidote. Low-Fe variety (α) has Berlin blue anomalous interference colors similar to idocrase, but the latter is uniaxial and without cleavage. Zoisite with blue to greenish yellow anomalous colors is very similar to clinozoisite. High-Fe zoisite (ß) is very similar to medium-Fe epidote. Melilite has similar anomalous interference colors but is uniaxial and is restricted to a primary occurrence in mafic volcanic rocks. The major mineral in "saussuritized" plagioclase is zoisite.

Opaque Minerals in Thin Section

Some indication of the identity of opaque minerals occurring in thin section can be made by observing grain morphology and appearance in reflected light. A flashlight or penlight provides enough light for this procedure. (See also Pasteris, 1983). Even more information can be obtained if thin sections are polished (uncovered).

Chalcopyrite $CuFeS_2$. Lack of crystal form is typical, yellow brass (not just brass) in reflected light.

Chromite $FeCr_2O_4$ Figure 18.7b. Gray metallic aspect in reflected light. Similar to magnetite. Occurrence in olivine-bearing ultramafic rocks.

Cinnabar HgS Figures 34.29c,d. Reddish in reflected light. Earthy masses and as impurity in opal and chalcedony.

Galena PbS. Cubic cleavage and bright metallic aspect in reflected light. Cubic form may occur.

Graphite C. Platy habit is characteristic, occurrence in metamorphosed rocks is most common. Molybdenite is physically similar but occurs in quartz veins and has bluish hues in reflected light.

Ilmenite $FeTiO_3$ Figures 19.11a, 28.2c. Platy and/or skeletal habit common. Gray metallic aspect in reflected light. Rims of leucoxene alteration are more likely on ilmenite than magnetite.

Magnetite Fe_3O_4 Figures 34.18, 36.8a. Triangular and rhombic sections of octahedra are typical. Gray metallic aspect (not brassy) in reflected light.

Pyrite FeS_2 Figure 34.4. Cube form common (Figure 34.21e), brassy in reflected light. Pyrrhotite is very similar but does not occur as cubes and may present a basal parting.

Specular Hematite Fe_2O_3. Platy habit (Figure 36.8a and Chapter 4 frontispiece photo) and deep red color along thin edges are diagnostic.

Submicroscopic Aggregates of Minerals and Amorphous Materials

Some minerals are abundant in rocks but rarely develop into large enough crystals for microscopic examination of physical features and optical properties. For example, if the diameter of a crystal is less than the standard thickness of a thin section (0.03 mm), the presenting birefringence will be too low. Identification may be impossible even if individual crystals can be seen with high-power magnification. If crystals cannot be seen, the material may still be crystalline or it may be amorphous or glass. **Amorphous** materials such as collophane and limonite consist of crystals, whereas natural glass commonly contains both crystalline and noncrystalline components.

Alunite Commonly as submicroscopic crystalline aggregates in hydrothermally altered and weathered rock that contained sulfides (Figure 40.24a). Association with quartz results in hard pinkish white rock.

Bauxite Contains hydrous aluminum oxide minerals such as **boehmite**, **diaspore** and particularly **gibbsite** (Figure 40.9), commonly admixed with limonites and clay minerals.

Bentonite Altered volcanic ash containing mostly the clay mineral montmorillonite.

Bowlingite Green alteration product of olivine, consisting of goethite and chlorite.

Carbon and Hydrocarbons Occur as submicroscopic particles in dark calcareous rocks, as **kerogen** in oil shale (Figure 37.2), as bitumens associated with petroleum, and as coal (Figure 37.3).

Celadonite A variety of glauconite.

Chalcedony Fibrous variety of quartz that typically has submicroscopic pores. Colloform or spherulitic (radial fibrous) form typical (Figure 34.2). Fibers that elongate parallel to *a* axis (hexagonal) are length slow. These fibers are twisted, bringing crystallographic *c* vertical here and there along the length of the fiber, resulting in a patchy extinction. Fibers that elongate parallel to crystallographic *c* are length fast. Colored colloform varieties are known as agate (Figure 40.17).

Chert Massive "granular" aggregate of fibrous quartz containing submicroscopic pores. Generally white or gray (Figures 36.3a, 37.24b,d), ranging to black carbon-bearing **flint** (Chapter 37 frontispiece photo) and red hematite-bearing **jasper** (Figure 34.27, 36.8a).

Chlorites Much chlorite in rocks occurs as very small crystals mixed with clay minerals in the clay fraction of sedimentary and epiclastic volcanic rocks, and with clay minerals and zeolites in weathered and hydrothermally altered mafic rocks in which there is a source of iron and magnesium. Common in amygdules (Figure 34.17).

Chlorophaeite Orange to green alteration mineral occurring in volcanic rocks, containing mostly **limonite**, **chlorite**, and **serpentine**.

Chrysocolla Blue–green in reflected light. Resembles fibrous and colloform habit of chalcedony (Figure 40.19). Similar to turquois.

Clay Minerals Common constituents of mudrocks, hydrothermally altered rocks and weathered rocks. The principal minerals are **kaolinite** (Figure 36.6b), **montmorillonite** (Figure 34.5a), illite, and **vermiculite** (Figure 40.2). Similar to very-fine-grained **white micas** and pyrophyllite. White matts in reflected light, more-or-less opaque in transmitted light. Admixtures with **zeolites**, **leucoxene**, **alunite**, and **gypsum** occur.

Collophane Mostly **apatite**. Yellowish brown to gray in reflected light. Oolitic or colloform habit common in rock known as **phosphorite**.

Garnierite An apple-green nickel serpentine, resulting from weathering of Ni-bearing ultramafic rocks.

Glass Volcanic glass most common, but also can be generated along faults as **pseudotachylite** and in impactites as **shock glass** (Figures 24.3, 24.5, 24.6). Colorless (Figures 17.1a, 17.2b) to browns and black (**sideromelane**) (Figures 17.2c, 17.3a) and **obsidian** (Figure 17.1b), depending on submicroscopic crystal content. Gas cavities, perlitic cracks (Figure 17.12b), or shard (Figure 31.5) form may occur. Refractive index varies from 1.48 to 1.60, depending on composition, but is generally greater than opal.

Glauconite Typically occurs as green pellets consisting of very fine flakes. Occurs in arenites and carbonates (Figure 36.3b), and is an indicator of marine environment. **Celadonite** variety occurs in vesicles.

Hematite Earthy variety imparts red coloration to sandstones and cherts. Admixtures with limonite common. Opaque submicroscopic masses in thin section (Figures 36.6a, 36.8e, 40.6).

Hydrobiotite Biotite interlayered with **vermiculite** (Figure 40.2).

Jarosite Amber yellow microscopic and submicroscopic crystalline aggregates (Figure 40.23). Association with limonite very common.

Iddingsite Red–brown to orange–brown (transmitted light) alteration product of olivine, consisting of mainly of **goethite** and **hematite**.

Leucoxene Typically contains **rutile**, **brookite**, and **hematite**, and is commonly associated with ilmenite. White to light gray nonmetallic in reflected light.

Limonite Hydrous iron oxides (Figures 40.22, 40.24b) consisting of minerals such as **goethite** and **lepidochrosite**. Color in reflected light is brown, but varies to red–brown with hematite present, to yellow–brown with jarosite, and to black–brown with contained manganese minerals.

Opal Colorless (Figures 32.2, 34.2b, 34.29c) to pale brown or gray. Very low r.i. = 1.40 to 1.46, isotropic, and colloform habit typical. **Opal-A** changes to opal-CT, a disorded cristobalite with tridymite stacking occurring as lepispheres (Figure 40.16) consisting of clusters of blade-shaped crystals. Packing of spheres generates a diffracting gradient resulting in a variety of play of colors.

Palagonite Yellowish brown to greenish brown alteration product of volcanic glass; r.i. ranges from 1.47 to 1.63.

Pseudoleucite Form the same as leucite but more birefringence, evidently due to the presence of fine-granular **nepheline** and **K-feldspar** pseudomorphic after **leucite**.

Pyrophyllite Resembles very fine aggregates of talc, gibbsite, clay minerals, and sericite. See clay minerals.

Sericite Very fine-grained white mica (Figure 34.5e, 34.16a, 40.20b), most commonly muscovite. See clay minerals.

Sulfur Figure 35.3a May occur in very fine-grained aggregates. Pale yellow and weakly pleochroic in thin section. Extreme birefringence seen as white. Very high r.i. Association with fumarolites, hot springs, and evaporites. In evaporites, forms by bacterial reduction of sulfates.

Wad Black mixtures of manganese oxides or hydroxides (**pyrolusite**, **manganite**, **psilomelane**) that may be admixed with limonite.

Zeolites Large volumes of small crystals occur in very-fine-grained sedimentary rocks and as alteration fractions in rocks capable of generating sodium, calcium, and aluminum. Larger crystals occur in vesicles and veinlets but as such are not major rock-forming minerals. Minerals include **analcime**, **natrolite**, **thompsonite**, **stilbite**, **chabazite**, **heulandite**, and the varieties **clinoptilolite** and **laumontite**.

Accessory Minerals

Some minerals typically occur as small grains in small amounts in rocks. Such **accessory minerals** may have special petrogenetic importance, such as indicators of sediment provenance and as a source of rare earths that are detected when the rocks in which they occur are chemically analyzed. Some of the minerals listed are accessory in some rocks yet major constituents in others.

Allanite(orthite) $(Ca,Ce,La)_2(Al,Fe^{3+},Fe^{2+})_3O(Si_2O_7)(SiO_4)(OH)$. Brown to dark brown pleochroism, zoning common. Contains rare earths cerium and lanthanium and concentrates other rare earths, especially the LREE. Also contains uranium and thorium.

Anatase TiO_2. Very similar to **rutile** and **brookite**. Diamond-shaped sections. Occurs in igneous rocks and in clastic sedimentary rocks.

Apatite $Ca_5(PO_4,CO_3)_3(F,OH,Cl)$ Figures 19.11a, 19.13, and Chapter 14 frontispiece photo. Occurs as clear stubby or slender prisms in magmatic rocks. High r.i. but low birefringence (middle first order). Concentrates rare earths, especially the middle REE. Also may concentrate scandium, yttrium, uranium, thorium.

Cassiterite SnO_2. Tin-bearing granitic rocks contain small grains of cassiterite, as well as sediments derived from such rocks. Brownish, reddish, or yellowish prisms with dipyramidal terminations (same as zircon). Very similar to rutile and anatase, but has higher birefringence than zircon. Concentrates zirconium. Typically contains tantalum and may contain scandium.

Corundum Al_2O_3. Can occur as accessory in low-silica magmatic rocks, high-Al metamorphic rocks, and in clastic sedimentary rocks. Clear prismatic crystals with very high r.i. and low birefringence (first order).

Epidote $Ca_2Al_2Fe^{3+}O(SiO_4)(Si_2O_7)(OH)$. May occur as a primary magmatic mineral. Clear or clear to yellow pleochroic prismatic. Concentrates uranium and thorium. The manganese variety **piedmontite** may occur in blueschists and greenschists.

Ilmenite $FeTiO_3$. Common opaque accessory mineral in igneous and metamorphic rocks. Also occurs as a detrital sediment.

Magnetite $Fe^{2+},Fe_2^{3+}O_4$. Typical opaque accessory in igneous rocks, metamorphic rocks, and in clastic sedimentary rocks.

Monazite $(Ce,La,Th)PO_4$. Similar to zircon, sphene, zenotime, and epidote. High r.i. and birefringence.

More-or-less equant, clear or pale green–yellow crystals. Contains the rare earths cerium and lanthanum and also thorium. Concentrates rare earth elements, especially the LREE. Also contains uranium and thorium. Pleochroic halos are typical if in biotite.

Perovskite $(Ca,Na,Fe^{2+},Ce)(Ti,Nb)O_3$. Colorless to dark brown, equant (pseudocubic) crystals. Occurs as accessory mineral typically in mafic and alkaline igneous rocks. Confirmation of identity is difficult. Contains the rare earth cerium and may concentrate others.

Rutile TiO_2. Occurs as minute red–brown stubby prismatic crystals with very high r.i. and birefringence in igneous and metamorphic rocks, and as a detrital mineral in sedimentary rocks. Long hairlike needles in quartz are common.

Sphene $CaTiSiO_5$ Figures 19.13, 20.5b; Plate 6b. Brownish mineral with very high r.i. and birefringence. Diamond-shaped sections are typical. Concentrates rare earth elements, especially the middle REE. Also contains uranium, thorium, scandium, yttrium.

Xenotime YPO_4. Colorless to yellow–brown, prismatic crystals with very high r.i. and birefringence. Similar to zircon, monazite, sphene, rutile, and cassiterite. May produce pleochroic halos in biotite along with zircon. Occurs in granitic rocks, schists, gneisses, and clastic sedimentary rocks. Some concentration of rare earth elements and uranium.

Zircon $ZrSiO_4$. Well-formed prisms with dipyramidal terminations. Common occurrence in granitic rocks (Plate 3c and Chapter 14 frontispiece photo), but also in metamorphic and clastic sedimentary rocks. Clear to pale brown. Very high r.i. and high birefringence (second order). Crystals less than 30 microns present lower birefringence. Very similar to xenotime. Most important concentrator of uranium and thorium. Also concentrates hafnium.

5

Plagioclase Composition

Pericline twin plane in calcic core (An_{80}) has distinctly different orientation in less calcic rim (An_{50}) (arrows) as viewed on nearly (010) (rhombic section). Large arrow points to faint trace of albite twins. X-polars. Bar scale = 0.66 mm.

Plagioclase as a Solid Solution Series

Plagioclase composition can vary between pure albite ($NaAlSi_3O_8$) and pure anorthite ($CaAl_2Si_2O_8$). Albite is the lower-temperature phase, anorthite the higher. Intermediate compositions can be generalized as $(NaSi,CaAl)AlSi_2O_8$, but specific intermediate compositions are conveniently indicated by reference to the anorthite component. For example, An_5 indicates nearly pure albite, whereas An_{40} is a plagioclase with 40% anorthite and 60% albite. The composition of a plagioclase reflects the temperature conditions existing during its crystallization. However, this statement is true only to the extent that the chemical environment can supply the appropriate growth components. For example, even though pure albite forms at a lower temperature than does pure anorthite, its field of stability is extended to higher temperatures in the absence of the calcium needed to form the anorthite component. If calcium is available, a more calcic plagioclase will form instead.

Determinative Methods

Petrographic examination of plagioclase-bearing rock includes an estimation of the anorthite content of the plagioclase. The position of principal vibration directions with respect to cleavage traces or certain twin planes varies with composition and allows for correlation of extinction angle with composition (Jones and Bloss, 1980). A single extinction-angle measurement relates to the composition of a nonzoned crystal. Multiple angles are recorded for zoned crystals.

For most investigations, an approximation of composition to within about 2–10% An is possible with exinction-angle measurements, accurate enough for most petrographic work. Some investigations are focused on detailed examination of plagioclase zoning, and the electron microprobe then becomes the technique of choice. Although accuracy can be improved with this microchemical technique, the main advantage is its ability to analyze areas only a few microns in diameter in thin crystals, which might be important in delicately zoned crystals, including those in which zoning is defined by trace-element distribution, for which optical expression is lacking.

The use of the microprobe can hardly be justified in the general appraisal of plagioclase composition in a hand specimen sampled from a large body of rock. The rock body is not likely to be homogeneous with respect to plagioclase distribution, nor can the composition of the crystals be expected to be consistent throughout the body. There could be variation of plagioclase composition within a given thin section as well as misleading compositional variations related to the sectioning of zoned crystals in which the core of a crystal might not be intersected (see Chapter 8). This statistical nightmare precludes, in most circumstances, the need for more compositional accuracy beyond that obtainable with extinction-angle measurements.

Other optical methods of determining plagioclase composition exist. Refractive index can be determined for cleaved grains broken from crystals and emersed in oils of known refractive index (Chayes, 1952; Laskowski et al., 1979; Hurlbut, 1984). Optic axial angles (J.V. Smith, 1958; R.L. Smith 1960; Tobi, 1956) can be determined for crystals in thin sections; for crystals not in thin section, a spindle stage can be used to determine optic axial angles (Bloss, 1981). The relations of optical directions to more than one twin type is a variation of the extinction-angle method (Tunell, 1952, Slemmons, 1962, 1963; Tobi and Kroll, 1975). All of these techniques are more involved than the extinction-angle method and therefore are not adaptable to routine petrographic analysis.

The a-Normal Extinction-Angle Method

The extinction-angle method presented here is informally referred to as the **a-normal extinction-angle method**. The a-normal method is a refinement of the well-known Michel-Lèvy extinction-angle method and of the Rittman zone method.

In the a-normal method, a crystal is found that has been sectioned close to a plane that is perpendicular to crystallographic a. This requirement at first might seem to be statistically restrictive, but in practice the following has been realized through many years of usage: (1) even in coarse-grained rocks, such as granites, it is rare that a crystal cannot be found in a single thin section that is not at least close to the a-normal position, and (2) if the **universal stage** is employed (Chapter 6), there are many more crystals in a thin section that can be used for the determination, since they can be rotated into the a-normal position. The practicality of setting up the universal stage to measure only a few plagioclase compositions can understandably be questioned. If plagioclase in numerous thin sections is to be determined, the use of the U-stage is warranted, since the time saved in locating a-normal sections greatly exceeds the time required to mount and adjust the U-stage. Not all laboratories are equipped with a U-stage, but that certainly does not preclude use of the a-normal method.

The a-normal method consists of four steps:

1. Locate a crystal with crystallographic a perpendicular to the microscope stage or up to about 30° from perpendicular if the crystal is to be rotated on the U-stage.
2. Identify the direction that is the trace of the {010} cleavage or (010) albite twin law composition plane.
3. Measure the angle from the trace of the (010) plane to the nearest extinction position (which will be X') and repeat for areas of different composition in zoned crystals.
4. Convert these angular values to percent An by using the data, for instance, of Tobi and Kroll (1975).

Each step of the a-normal method is discussed as an ideal case and a nonideal case. The reader may be thinking that the a-normal method is complicated and saves no time over other methods. With a little practice, however, it is an efficient way to estimate plagioclase composition.

Step 1 Location of a-Normal Position

The Ideal Case. Either twinning or cleavage can be used to locate a position normal to crystallographic a, but since most plagioclase is twinned, and since twin lamella are more apparent than cleavage, twinning provides the ideal case for locating a.

Make a rapid scan of a thin section for plagioclase containing two sets of sharp (perpendicular to the twin planes) twin lamellae oriented at nearly right angles to each other. Repeated twinning of the **albite twin law** and the **pericline twin law** is very common in plagioclase, especially in magmatic rocks (Table 5.1). The composition plane of the albite law is parallel to (010) cleavage (Figure 5.1), but the composition plane of pericline twins is variable with respect to (001) cleavage (Figure 5.22; p. 80). The composition plane of pericline twins is subparallel to (001) for plagioclase in the An_{30-50} range, being exactly parallel at $An_{40'}$ in which case the twin is a special kind of pericline twin known as an **acline-A twin** (Figure 5.). Unlike the composition plane of albite twinning, which shows no change in orientation as it crosses zones of varying composi-

Table 5.1
Feldspar Twin Laws

Twin Law	Comp. Plane	Twin Axis	Type	Occurrence
Albite	(010)	nor (010)	Repeated	Very common
Pericline	(h01)	b axis	Repeated	Common
Acline A	(001)	b axis	Variety of pericline	
Carlsbad	Mostly (010)	c axis	Simple, penetration	Common
Albite–Carlsbad	Mostly (010)	Normal c, ∥ (010)	Simple, penetration	Common
Acline B	(100)	b axis	Repeated	Rare
X–Acline B	(100)	Normal b ∥ (100)	Repeated	Rare
Ala B	(010)	a axis	Repeated	Rare
Albite–ala B	(010)	Normal b ∥ (010)	Repeated	Rare
Ala A	(001)	a axis	Repeated	Rare
Esterel	(0k1)	a axis	Repeated	Rare
X–Carlsbad	(100)	Normal c	Repeated	Rare
Baveno (right)	(021)	Normal (021)	Simple	Rare
Baveno (left)	(021)	Normal (021)	Simple	Rare
Manebach	(001)	Normal (001)	Simple	Rare
Manebach–ala A	(001)	Normal a, ∥ (001)	Repeated	Rare
Manebach–Acline A	(001)	Normal b, ∥ (001)	Repeated	Rare

tion (Figure 7.8a), the orientation of the trace of the pericline twin, as viewed on (010), varies with compositional zoning (Figure 5.3; p. 81).

The intersection of (010) and (001) defines a line that is crystallographic a. Thus, for plagioclases of intermediate composition two sets of reasonably sharp twins at right angles to each other define a position that is close to a-normal.

The Non-Ideal Cases. If a plagioclase crystal is untwinned, location of a crystal with two sharp cleavages intersecting at nearly right angles (plagioclase is triclinic) is the a-normal position. A check on this can be made by using a 10× or 20× objective lens and uncrossed polars. Move from a focus position on a given cleavage trace to a defocused position just above the crystal. If light moves equally away from the trace of the cleavage as this focal plane is changed, the cleavage is perpendicular to the microscope stage and it will be sharp. Traces of the cleavages in larger crystals are commonly developed in thin sections of plagioclase, but it may be necessary with fine-grained rocks to find crystals at the margin of the thin section, where grinding may have facilitated cleavage development (Figure 5.4a; p. 81).

If the plagioclase has only one set of sharp twins, look for a sharp cleavage trace nearly perpendicular to this twin direction (Figure 5.4b; p. 81). If the twins are albite, the cleavage is (001), and if the twins are pericline the cleavage is (010), but in either case, for intermediate plagioclase compositions, an approximate a-normal position has been located.

If the plagioclase is suspected to be very albitic or very anorthitic, perhaps based on rock type, it must be

5 Plagioclase Composition

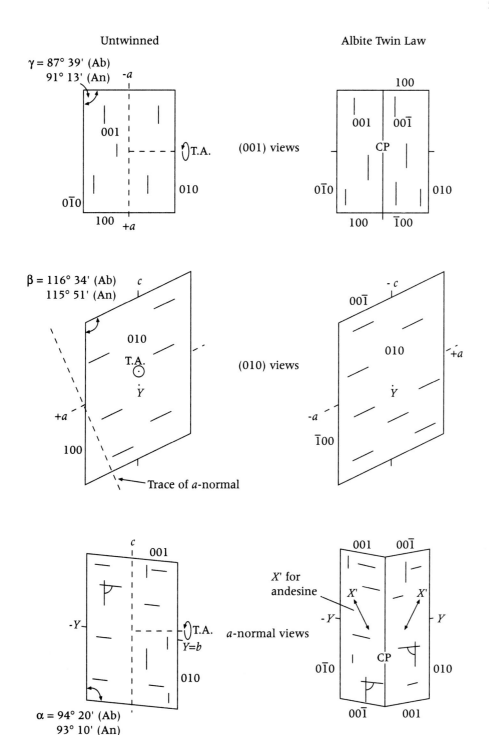

Figure 5.1; text p. 77
Twin axis and composition plane of albite twin law in relation to plagioclase morphology, *a*-normal plane, and cleavage.

⊢ = *a*-normal position showing acute angle

T.A. = twin axis

— = vertical cleavage

remembered that pericline twins will be "fuzzy" in the *a*-normal position (Figure 5.2), although two sets of twin lamellae will still be at nearly right angles to each other. In this case, location of a position normal to (001) is made by observing cleavage. (1) Location of two sharp cleavages that are nearly at right angles to each other fixes the *a*-normal position. (2) Location of a sharp set of twin lamellae that has parallel sharp cleavage (must be albite or acline-A twin law) and another cleavage (without associated twin lamellae) at about right angles to the cleavage-parallel lamellae fixes the *a*-normal position. (3) If there is one set of

continued on p. 84

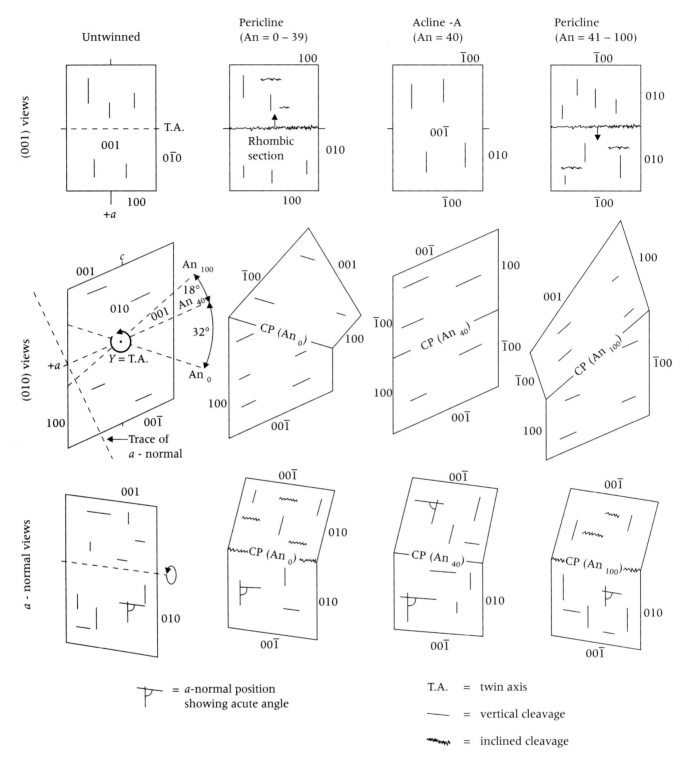

Figure 5.2; text p. 77
Twin axis and composition plane of pericline and acline-A twin laws in relation to plagioclase morphology, *a*-normal plane, and cleavage.

5 Plagioclase Composition   81

Figure 5.3; text p. 78
Angle between pericline twin lamellae and (001) cleavage (two arrows and white line) as viewed on (010) varies from about 10° in interior more calcic plagioclase (left) to about 2° in marginal zone (partly in extinction, upper right) reflecting systematic change in orientation of rhombic section with composition. Twin separating upper and lower portions (illuminated) of crystal is also a pericline twin. X-polars. Bar scale = 0.12 mm.

(a)

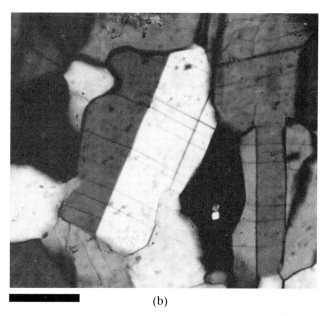
(b)

Figure 5.4; text p. 78

(a) *a*-Normal section of small plagioclase in which (010) [NE–SW] and (001) [NW–SE] cleavages (arrow) have developed near margin of thin section where thinning has occurred. X-polars. Bar scale = 0.05 mm. (b) Plagioclase crystals (center and lower right) have sharp albite twin lamellae and sharp (001) cleavage. Example of usefulness of twinning and one cleavage direction in locating *a*-normal position in small crystals. X-polars. Bar scale = 0.05 mm.

Figure 5.5; text p. 84
Twin axis and composition plane of Carlsbad twin law in relation to plagioclase morphology, *a*-normal position, and cleavage.

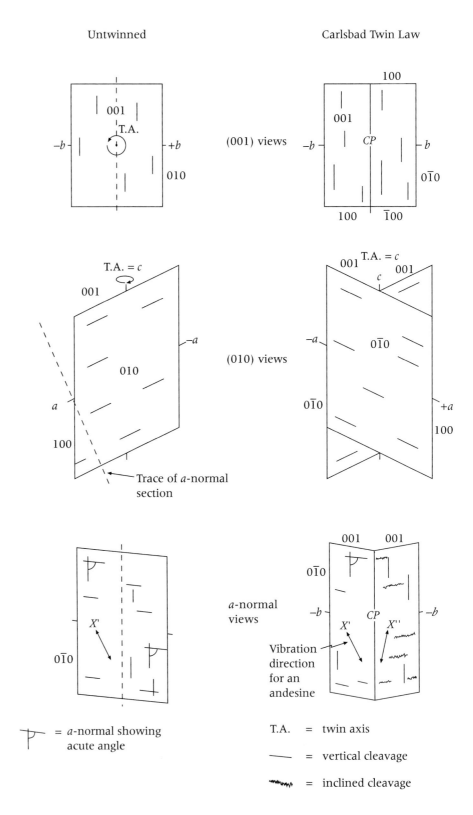

Untwinned | Albite–Carlsbad Twin Law

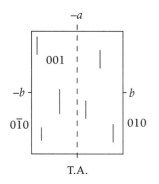
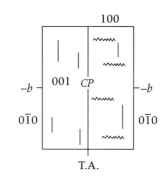

(001) views

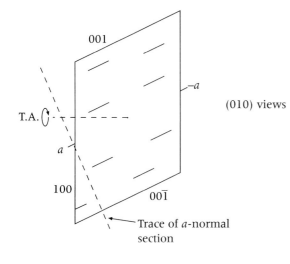
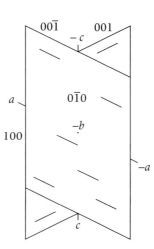

(010) views

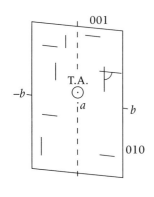
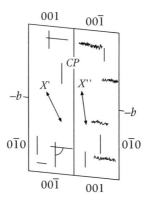

a-normal views

⊢ = *a*-normal showing acute angle

T.A. = twin axis
— = vertical cleavage
⁓⁓⁓ = inclined cleavage

Figure 5.6; text p. 84
Twin axis and composition plane of albite–Carlsbad twin law in relation to plagioclase morphology, *a*-normal position, and cleavage.

5 Plagioclase Composition   83

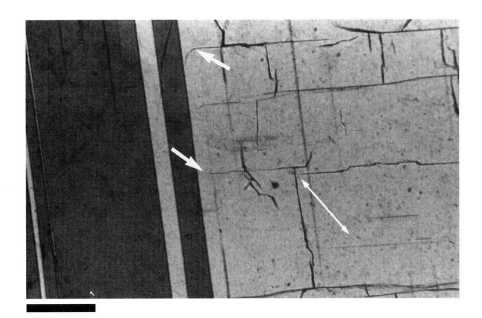

Figure 5.7
Trace of (010) portion of Carlsbad twin located by termination of (001) cleavage (middle thick arrow) and curvature of (001) cleavage (upper thick arrow). Note lack of (010) cleavage in the several albite twin lamellae on the entire left side of the Carlsbad twin and the well-developed (001) cleavage on the right side. Well-developed (010) cleavage forms an acute angle with (001) cleavage in which X' lies (thin double arrow). Angle between X' and (010) is 35° (An_{67}). X-polars. Bar scale = 0.15 mm.

sharp twin lamellae and another fuzzy set of lamellae that is parallel to sharp cleavage, the fuzzy twins must be pericline and the section is *a*-normal.

Carlsbad (Figure 5.5; p. 82) and albite–Carlsbad (Figure 5.6; p. 83) twins are common in igneous rocks (Table 5.1), and location of *a*-normal positions within them may be required. These twins are known as "penetration twins," and since there is typically one twin for the Carlsbad law the relation is known as a "simple twin." The albite–Carlsbad twin relation may repeat. Albite and pericline twins typically occur within the lamellae of these twins. The trace of the composition surface of the penetration twins is generally parallel to the trace of (010), but there may be a "jog" in the trace since the composition surface consists of several planar surfaces in different positions, thereby affecting the location of the twin interface in sections that intersect the planar surfaces.

The trace of the penetration twin surface is mostly coincident with the trace of (010) and is therefore useful in quickly locating crystals with at least one direction that has *a*-normal potential. Only one half of these penetration twins can present *a*-normal in a given section. The apparent absence (or obscurity) of (001) cleavage in the half not presenting *a*-normal is diagnostic of either the Carlsbad or albite–Carlsbad twin relation (Figure 5.7).

The trace of a Manebach twin plane is parallel to that of pericline twinning in the *a*-normal position and may be mistaken for the Carlsbad direction. Distinction can be made in step 2, but rarely does the Manebach relation occur (Table 5.1).

Step 2   Identification of the (010) Direction

The Ideal Case.   Once an *a*-normal section is located for a crystal, or for a particular portion of a crystal having a penetration twin, the (010) direction can be identified for most common compositions of plagioclase as follows. Place a set of twin lamellae or a

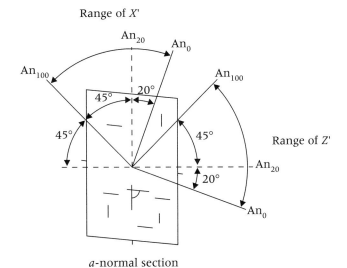

Figure 5.8; text p. 86
Position of X' (and Z') with respect to plagioclase morphology and cleavage for all compositions, as viewed in *a*-normal section.

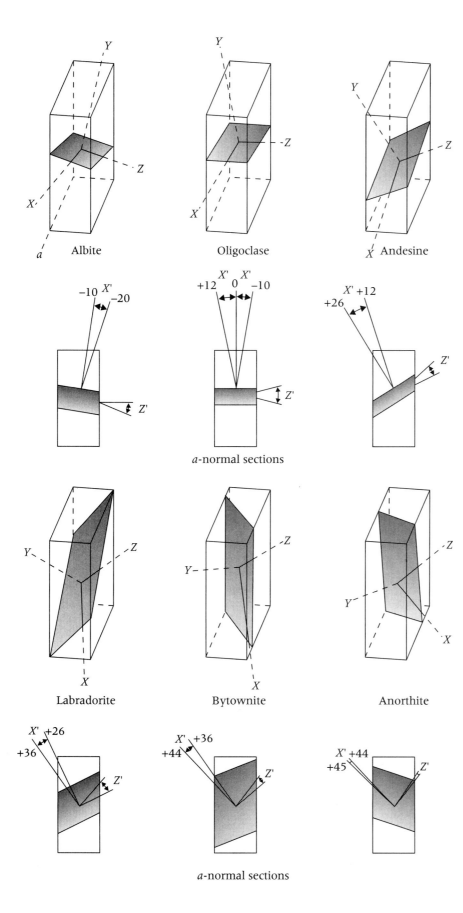

Figure 5.9; text p. 86
Position of optic plane, X, Z, and Y for the major plagioclase compositional groups. Position of X' shown in a-normal sections, respectively.

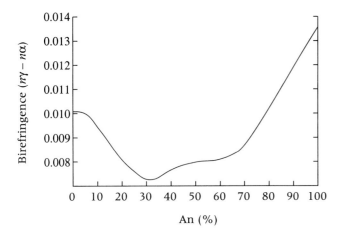

Figure 5.10
Variation of birefringence with change in plagioclase composition.

cleavage direction NW–SE. Insert the red-1 accessory plate. If a blue interference color is obtained, (010) is NW–SE. If a yellow interference color is obtained, (010) is NE–SW. The explanation of these phenomena is as follows.

In the *a*-normal position, there will be X' and Z' vibration directions lying in the plane of the microscope at right angles to each other. X' is closer than Z' to the trace of (010) in all plagioclases except anorthite (Figures 5.8, p. 84; 5.9, p. 85). Since the birefringence in *a*-normal sections of plagioclase will be in the general range of white to slightly off-white toward cream color (Figure 5.10), placement of a red-1 retardation plate between the crystal (not at extinction) and the analyzing polarizer results in either a blue (red-1 + white → blue) or a yellow (red-1 − white → yellow) interference color (Plate 4b). Retardation

Figure 5.11; text p. 88
Determination of trace of (010) for plagioclase using red-1 plate for various combinations of cleavage, twinning, and cleavage + twinning. At least some portion of each crystal is *a*-normal (except crystal 6). Only crystals 1, 7, and 8 can be seen to be *a*-normal without additional twin or cleavage information.

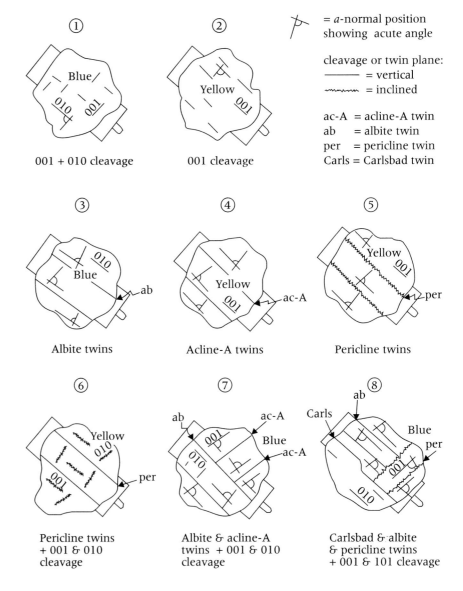

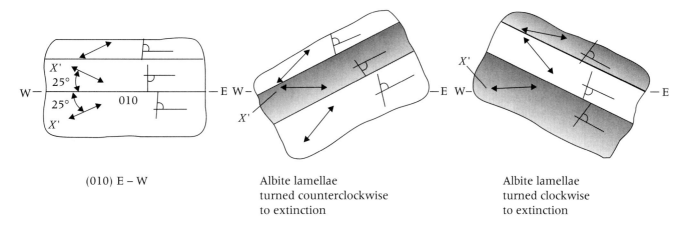

Figure 5.12; text p. 89
Position of X' and extinction position for alternating portions of albite twins. Angle between X' and trace of (010) is 25°.

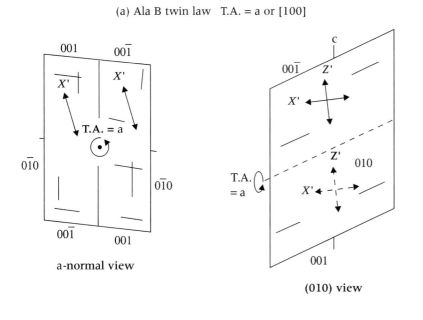

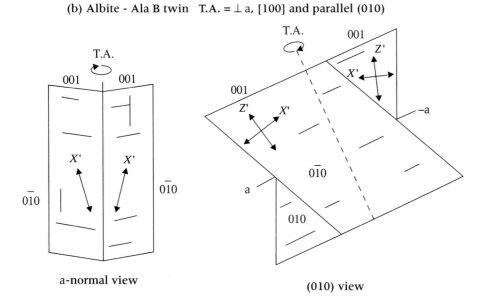

Figure 5.13; text p. 90
Twin axis and composition plane of ala B twin law (a) and albite–ala B twin law (b) in relation to crystal morphology, *a*-normal section, and cleavage.

addition (blue) will occur when the trace of (010) cleavage, the trace of the albite composition plane, or the trace of a Carlsbad or albite–Carlsbad twin plane are placed NW–SE (Figure 5.11; p. 86), since X' is generally in that direction and it roughly corresponds to the fast vibration direction in the retardation plate. Retardation subtraction (yellow) occurs when the trace of the (001) cleavage or the trace of the pericline (and acline-A) composition plane is placed NW–SE because Z' of the crystal now corresponds to the fast direction of the red-1 plate.

The Non-Ideal Cases. For plagioclase more calcic than about An_{90}, ambiguity arises when locating X'. For these very calcic compositions, the extinction angle between X' and the trace of (010) is 40°–45°. In this situation, a very similar angular relation exists between X' and the trace of the (001) cleavage and between Z' and the trace of the (001) cleavage (Figure 5.8). If the cleavages cannot be identified by some other means, the angular relation between X' and the trace of (010) as determined in step 3 can be misidentified and the extinction angle measured will be in error by several degrees. Therefore, use of the red-1 plate in distinguishing (010) from (001) does not apply to very calcic plagioclase, since the interference color obtained is magenta red instead of blue or yellow.

How then can the specific composition of very calcic plagioclase be determined? If a very calcic plagioclase crystal is zoned to less calcic compositions, the (010) direction can easily be verified. The identification of X' relative to Z' is first determined in the less calcic portion by using the red-1 plate. Location of X' in the less calcic portion would automatically locate it in the most calcic portion of the crystal, since the zones are likely to be crystallographically continuous.

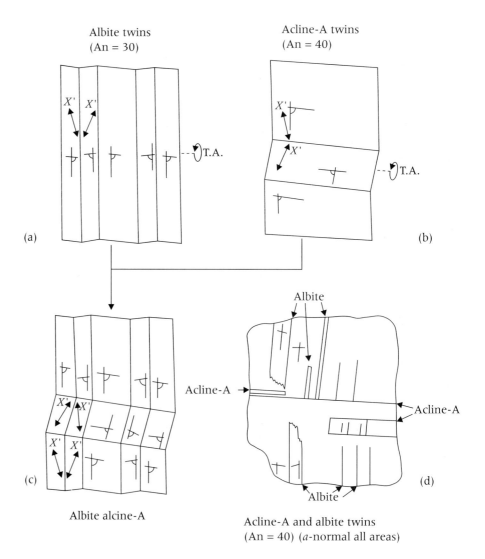

Figure 5.14; text p. 90 **Albite twins (a), acline-A twins (b), and combination albite and acline-A twins (c) shown in *a*-normal sections. Actual occurrence of combined twins shown in (d).**

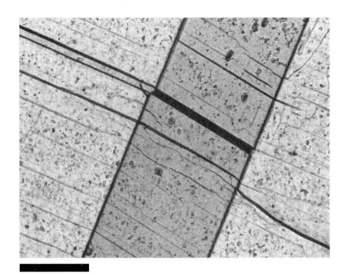

Figure 5.15; text p. 90
Large albite twin lamellae (center) cut by small pericline twin (in extinction). Pericline twin also occurs in portion of albite lamella on left side. Note jog in orientation of (001) cleavage as it crosses the central albite twin. X-polars. Bar scale = 0.12 mm.

Step 3 Measurement of the Angle Between X' and (010)

The Ideal Case. The angle between X' and the trace of (010) is obtained by placing either the (010) cleavage trace or sharp albite twin lamellae (both identified in step 2) parallel to substage polarization and then turning to the nearest extinction position by trying a clockwise and a counterclockwise stage rotation. The angle is obtained by recording the positions on the perimeter scale of the microscope stage and making the subtraction.

The Non-Ideal Cases. In a nonzoned crystal or within a specific zone of a zoned crystal, the angle between X' and (010) should have the same value, but opposing directions, in adjacent albite twin lamella (Figures 5.1, 5.12; p. 87). A Carlsbad twin may superficially resemble an albite twin, especially if no repeated albite twin lamellae are present. In this case, in the a-normal position, the extinction angles will still have the opposite sense of direction with respect to (010), but they will

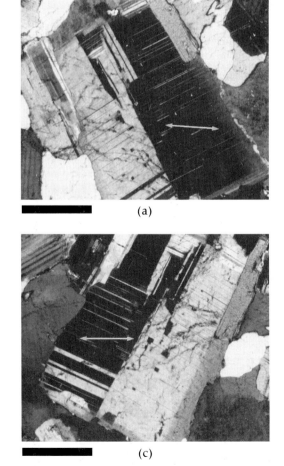

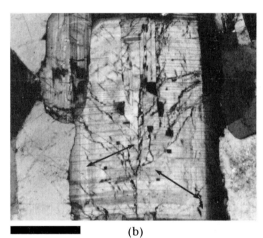

Figure 5.16; text p. 90
Major pericline twin with subsidary pericline and albite twin lamellae.

(a) Position of X' (white arrow) shown at extinction position for right half of twin. (b) Position of X' (black arrows) in both halves of twin in position providing equal illumination. Note small extinct squares in crystal interior (patchy zoning). (c) Position of X' (white arrow) shown at extinction for left half of twin. X-polars. Bar scale = 0.6 mm.

have different values (Figure 5.5) because one portion of the twin is not in the *a*-normal position) (Figure 5.7). For an albite–Carlsbad twin the extinction angles will have the same sense of direction with respect to (010), but then they must have different values (Figure 5.6).

Ala B twins, although rare (Table 5.1), are repeated like albite and pericline twins. Extinction positions are parallel in each pair lamella (Figure 5.13; p. 87), and there can be no confusion with albite twins. However, another rare repeated twin, known as the albite–ala B law (Table 5.1), is optically identical to albite twinning in an *a*-normal section (Figure 5.13). In this case, the extinction angle measured is applicable to determination of composition even if the distinction between albite and albite–ala B has not been made (or discovered). If desired, the presence of the rare albite–ala B twin can be verified by locating sections perpendicular to (010) that are not also *a*-normal. Extinction will not be equal in opposing members of an albite–ala B twin as they will always be for albite twins sectioned anywhere normal to (010).

Location of *a*-normal, (010), and X' in crystals with several types of twins can be facilitated if the twin relations are identified. The relations in crystals with both albite and pericline twinning are shown in Figure 5.14; p. 88 and photomicrograph Figures 5.15; p. 89 and 5.16; p. 89.

It has been shown that only certain parts of penetration twins can be at *a*-normal. A typical relation of albite, pericline, and Carlsbad (or albite–Carlsbad) twinning in a single crystal is shown in Figure 5.17. Some lamellae are too thin for accurate observation of extinction, and only the thicker lamellae are useful in obtaining extinction-angle data (Figure 5.17). In this example the pericline twins are not sharp even though they are in an *a*-normal position.

Step 4   Conversion of Extinction Angle to Percent An

The Ideal Case.   Since plagioclase composition varies systematically with crystal structure (Figure 5.9), it may at first seem simple enough to plot the angle between X' and (010) against percent An, thereby obtaining a diagram that can thereafter be used to convert extinction-angle data to plagioclase composition (Figure 5.18). Table 5.2 lists the values of An corresponding to extinction angles taken in the *a*-normal position. Since the optics of high-temperature plagioclase are distinctly different than low-temperature plagioclase, especially for intermediate and sodic plagioclase, the extinction-angle values are different (Figure 5.18 and Table 5.2). Selection of the appropriate values must be based on an initial evaluation of the environment in which the plagioclase formed. Volcanic plagioclases are high temperature; plutonic, metamorphic, and hydrothermal plagioclases are low temperature.

Inspection of Figure 5.18 and Table 5.2 reveals that there is an overlap of extinction values for low- and high-temperature optics from 0° to 15°. If the plagioclase is known to be albitic or intermediate in composition, there is no problem in selecting the appropriate curve (Figure 5.18) and reading An content as a function of extinction angle.

The Non-Ideal Cases.   If plagioclase is in the range An_{15}–An_{25}, choice of curve (Figure 5.18) may not be possible without eliminating the ambiguity. The position of X' for all compositions of plagioclase less than An_{20} lie on the opposite side of (010) with respect to plagioclase greater than An_{20} (Figure 5.19). For the composition An_{20}, X' lies parallel to (010).

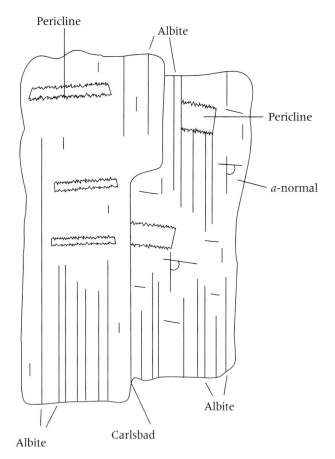

Figure 5.17
Sketch of crystal showing relation of Carlsbad, albite, and pericline twins. *a*-Normal position only in right half of Carlsbad twin in which pericline twin planes are not vertical (vertical only for acline-A).

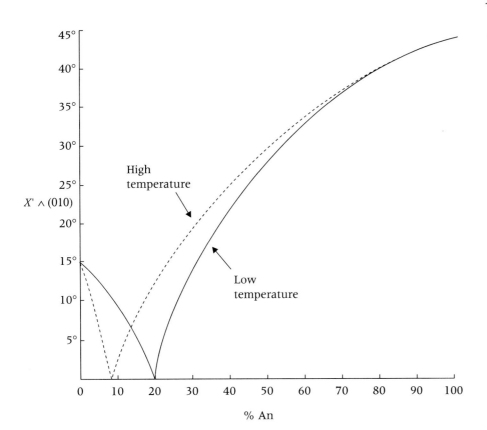

Figure 5.18
Angle between X' and (010) shown for low- and high-temperature plagioclase optics as a function of composition.

Table 5.2
Conversion of Angles Between X' and (010) of Plagioclase to Percent An

Ext. Angle	Weight Percent An		Ext. Angle	Weight Percent An		Ext. Angle	Weight Percent An		Ext. Angle	Weight Percent An	
	Low T	High T		Low T	High T		Low T	High T		Low T	High T
−15	0	0	+02	22	11	+19	36	27	+36	65	65
−14	3	1	+03	22	11	+20	37	28	+37	68	67
−13	6	1	+04	23	12	+21	39	30	+38	71	70
−12	8	1	+05	23	13	+22	41	31	+39	73	73
−11	9	2	+06	24	14	+23	42	33	+40	75	75
−10	10	2	+07	25	15	+24	45	35	+41	77	80
−09	11	2	+08	25	16	+25	46	37	+42	80	85
−08	12	3	+09	26	17	+26	48	40	+43	90	90
−07	14	3	+10	26	18	+27	51	42	+44	100	100
−06	15	4	+11	27	19	+28	53	43			
−05	16	5	+12	27	20	+29	54	44			
−04	17	6	+13	28	21	+30	55	46			
−03	18	7	+14	29	22	+31	56	49			
−02	19	8	+15	30	23	+32	57	52			
−01	20	8	+16	31	24	+33	58	56			
00	20	9	+17	33	25	+34	60	60			
+01	21	10	+18	34	26	+35	63	63			

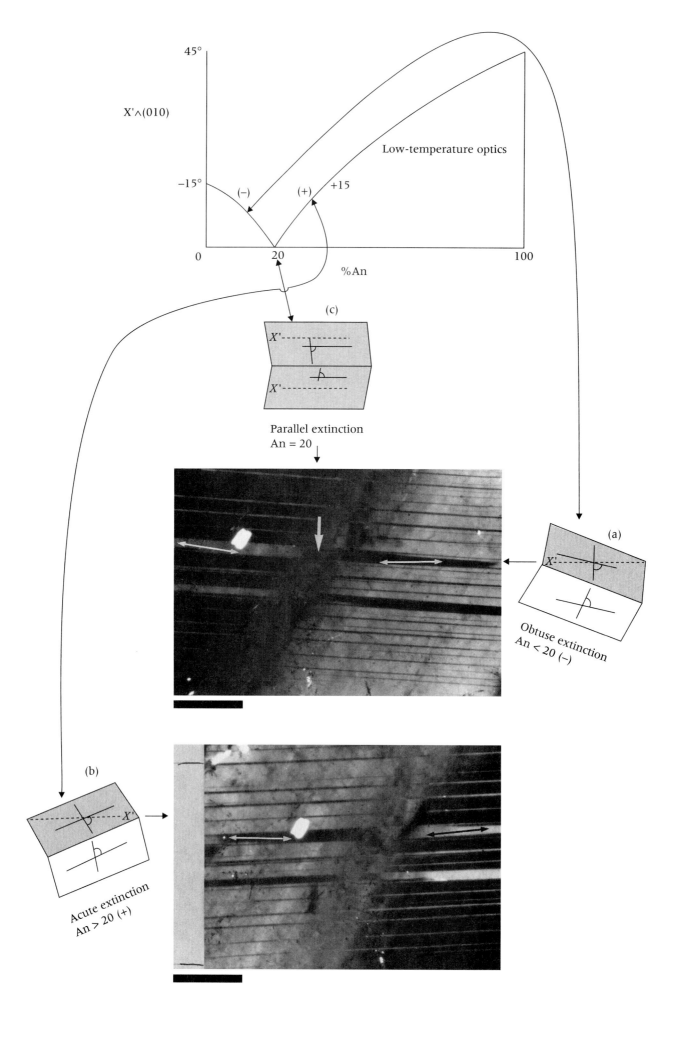

Figure 5.19
Position of X′ (double-headed arrows) relative to a thin albite twin lamella in zoned plagioclase (center of crystal is located to the left) (X-polars, bar scale = 0.16 mm). Position of X′ relative to the lamellae of an albite twin and angle produced by intersection of (010) and (001) cleavage shown exaggerated in small sketches.

(a) X′ in the obtuse angle is negative extinction and An is less than 20, -7° (An$_{14}$) as determined in the photographed crystal. (b) X′ in the acute angle is positive extinction and An is greater than 20, 6° (An$_{25}$). (c) X′ and extinction parallel to (010) occur only for An$_{20}$. Twin disappearance zone (An$_{20}$) shown with short white arrow is maintained with any position of stage rotation.

If both cleavages are very well developed in a perfectly oriented *a*-normal section, the position of X′ with respect to the angle described by their intersection is useful in resolving the problem. Since all plagioclase is triclinic, the angle of intersection of the (010) and (001) cleavage is not 90°; there is an acute angle and a complementary obtuse angle described by the intersection. If X′ (nearest position of extinction with respect to (010)) lies somewhere in the acute angle, the extinction is **acute (positive) extinction** (Figure 5.7). If it lies in the obtuse angle, it is **obtuse (negative) extinction**. Obtuse extinction occurs for compositions less than An$_{20}$, whereas compositions greater than An$_{20}$ have acute extinction. An extinction angle of zero corresponds to An$_{20}$. The relation of extinction position to the intersection of two cleavages

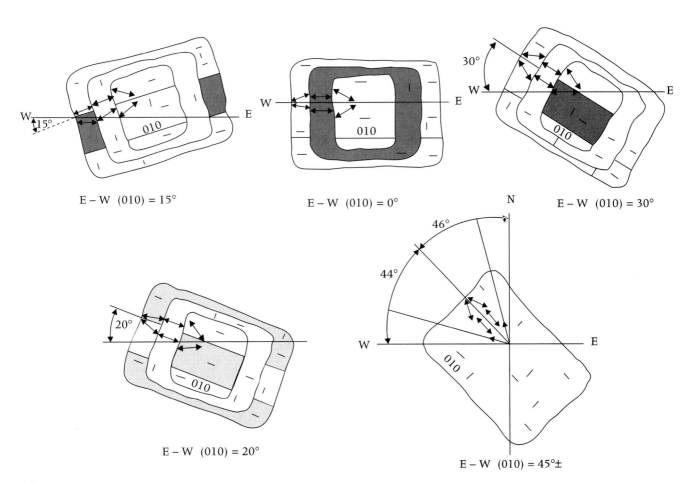

Figure 5.20; text p. 94
Extinction positions (X′) for various portions of a plagioclase crystal having both zoning and albite twinning. Incomplete extinction at 20°.

that is near a right angle is subject to error and cannot be expected to be a routine procedure. Fortunately, there are other clues that can be evaluated in resolving the "overlap" region of extinction angles.

A zone in a crystal that is of An_{20} composition shows only faint or no expression of albite twinning (Figure 5.19; p. 93). The twinning is very apparent in the zoning region on either side of this "twin disappearance zone" where the composition is greater than An_{20} in one direction and less than An_{20} in the other. The disappearance-zone effect maintains in any position of stage of rotation except, of course, at extinction positions. For this reason, twinning in nonzoned plagioclase close to An_{20} is weak or might even seem to be absent.

All visible twinning will also "disappear" in those positions of stage rotation in which the vibration directions in each twin are symmetrical with respect to polarization of the microscope. For a-normal sections this disappearance is in the 45° position as well as in the N–S or E–W position (Figure 5.20; p. 93). The "disappearance" can be made perfect in the 45° position by fine adjustment of rotation, but it is not perfect in the other positions because perfect normality to (010) is not likely to be obtained unless the crystal is rotated on the universal stage.

In a zoned crystal, twin "disappearance" that is not dependent on position of stage rotation (i.e., at composition An_{20}) can be a very useful reference point in resolving the extinction-angle data in the ambiguous region. For example, progressive zoning in magmatic plagioclase is typically from higher An cores to lower An rims. If there is a "disappearance" zone in these magmatic crystals, the composition of the plagioclase An can be expected to increase (acute extinction) inward from this An_{20} zone and decrease (obtuse extinction) outward. Thus, an extinction angle of 6° in an adjacent zone inboard relates to An_{25} on one curve, and an angle of 5° outboard relates to An_{15} on the other curve (Figure 5.18 and Table 5.2). Of course, if this "normal" zoning trend is in doubt, as it may well be in metamorphic rocks, verification by observation of extinction with respect to the acute or obtuse angle of cleavage intersection may be required.

There are other clues to the location of more calcic portions of zoned crystals, and thus to the sense of zoning. More calcic portions tend to alter to minerals such as sericite, sericite + epidote, or calcite. An altered core and/or zone can then indicate the position of more calcic plagioclase and therefore the direction of An change from an An_{20} zone elsewhere in the same crystal. As another example, birefringence of plagioclase varies with composition, and very calcic plagioclase generates a cream-colored interference color rather than the white characterizing most other composition.

If plagioclase is nonzoned or weakly zoned, as it is in most metamorphic rocks, the extinction-angle ambiguity for plagioclase near An_{20} must be resolved by observing either acute or obtuse extinction. If the plagioclase is near pure albite or approaching An_{30}, the extinction-angle ambiguity may be resolved by comparing the presenting refractive indices of the plagioclase with adjacent quartz or mounting epoxy.

Presentation of Compositional Data of Zoned Plagioclase

Most plagioclase in rocks has at least a hint of compositional zoning. In the a-normal method, what should be measured for a given zoned plagioclase crystal and how should the information be presented? A report on the composition of zoned plagioclase should include the following genetically important information: (1) maximum and minimum An values, (2) a diagrammatic presentation of the distribution of the compositional variations, and (3) an estimate of the average (bulk) composition of the plagioclase crystal, with due consideration of the smaller volume of interior zones in comparison with the larger volume of outer zones.

Since plagioclase is most conveniently expressed as percent An, a plot of An against location along a traverse from the core to the rim of the crystal generates a graphic picture of plagioclase zoning (Figure 7.9). Several crystals may have to be measured to ensure that core composition has been intersected. The resulting **zoning curve** should include the highest and lowest An value for the crystal and the configuration of the various zones. For zones that are difficult to measure, such as delicate oscillatory zoning, a diagrammatic simulation can be shown as part of the zoning curve.

The average composition of a given crystal can be generalized. Subjective estimation of an "average extinction angle" for the crystal, with due consideration of the fact that a zone of given thickness has more volume near the rim compared with one of the same thickness near the core, is likely to be more accurate than a simple arithmetic averaging of the high and low compositional values.

6

Petrographic Measurements

Phase Analysis

Minerals and other rock-forming materials are **phases**. If a rock consists of 100% mineral grains, the volume percent of each of the mineral phases constitutes the **mineralogical mode**. If grains are rock fragments, either crystalline, glassy, or amorphous, they become part of the mode as well. In addition, if grains are too small for identification they are collectively considered **matrix** (which may be 100% if there are no other grains present). Matrix may also consist of identifiable grains (for which a grain mode could be determined) in which there exist grains of a larger size fraction. In both cases, the matrix constitutes some volume percent of the rock.

Phase analysis is a matter of discriminating the area fraction of different phases in rocks that are visually distinct from each other. For randomly oriented grains of phases, the area fraction is known to be close to the volume fraction (Galwey and Jones, 1966; Cashman and Ferry, 1988). This is convenient, since point counting and image analysis are done on images derived from thin sections that are inherently two dimensional.

Manual Methods (Point Counting)

Point counting is done either on thin sections or on slabbed surfaces of very coarse-grained rocks. The mechanical aspect of point counting is relatively simple. A thin section is mounted on a **mechanical stage** attached to the microscope stage. Two gear trains oriented at right angles to each other allow movement of the thin section in specified increments in two directions. As the section is moved along traverses, the cross-hair intersection sequentially locates at a point over rock-forming materials. Tabulation of point occurrences over similar material during the course of the traverses can be made on any counter module where the counter keys are assigned to the various rock-forming materials. Semi-automatic point counters have an electrical connection between the mechanical stage and the counter, automatically moving the thin section one space with each count recorded.

Thin sections of very coarse-grained rocks do not generate meaningful point-count information. As an alternative, a transparent film having a grid pattern can be placed on a slabbed rock surface and point counting can proceed manually (Jackson and Ross, 1956; Fitch, 1959; Smithson, 1963).

Grains may be too small or too indistinct from other grains of different composition to be properly point-counted. In some cases, staining of minerals in uncovered thin sections (or rock slabs) aids in identification. K-feldspar may be stained yellow, and plagioclase red (Bailey and Stevens, 1960; Ruperto et al., 1964; Houghton, 1980). Calcite may be stained red, leaving associated dolomite unstained. Stains may be selected that differentiate between nonferroan and ferroan calcite or dolomite and between calcite and high-Mg calcite (Friedman, 1959; Dickson, 1965; see also Adams et al., 1984; Dravis and Yurewicz, 1985). There is also a stain test for fluorite (Sharp et al., 1977), cordierite (Boone and Wheeler, 1968), paragonite (Laduron, 1971), and brucite (Haines, 1968). Voids in almost any rock may be filled with a stain (including fluorescent stains) introduced into an impregnating epoxy (Yanguas and Dravis, 1985) before the thin section is made or by staining epoxy resin exposed in the voids of uncovered thin sections, polished thick sections, or slabbed surfaces (Ruzyla and Jezek, 1987).

Elucidation of subtle differences in composition may be very important in petrogenetic analysis. Photographs of minerals as they are bombarded with an electron beam that causes a discriminating luminescence can be very revealing (Marshall, 1988); these photos may be qualitatively examined or quantitatively read by automatic imaging techniques. This **cathode luminescence** technique is particularly useful in the definition of secondary quartz overgrowths on clastic quartz grains and in identifying different generations of plagioclase that is not optically possible (Hopson and Ramseyer, 1990).

Statistical Analysis of Point-Counting Data

Statistical analysis of point-counting data depends on a number of variables. Meaningful petrogenetic interpretation depends on (1) field sampling (does the sample represent the rock body or not), (2) direction of hand-specimen sectioning (important for rocks with directional fabric), and (3) choice of traverse and point spacing (dependent on the grain size of the rock) (Chayes, 1956; Jackson and Ross, 1956; Plafker, 1956; Bayly, 1960; Solomon, 1963; van der Plas and Tobi, 1965). A general rule of thumb for obtaining semiquantitative data from thin sections of medium-grained rocks (1–5 mm) is that the traverse and point spacing should be selected so as to generate about 1500 points spaced over the entire rock surface presenting in a standard thin section (Chayes, 1956; Hutchison, 1974). More than one thin section or a slabbed surface may be required for determining the mode of coarser-grained rocks. Of course, there are problems in counting finer-grained rocks in which the grains may be difficult to identify.

Automated Image Analysis

Image analysis systems consist of a scanner (video camera) mounted on a petrographic microscope, an electron microscope, or a macrostand. Images of phases (including voids) are converted to numeric values based on a gray scale in a computer-driven processor (analyzing module) and displayed as pixel points on a monitor. Phases appear as different shades of gray and may be displayed in pseudocolor, as chosen by the user, or converted to a binary (black-and-white) image. If there is little or no contrast between phases as viewed with a petrographic microscope, photographs may be taken in any manner that produces contrast between the phases of interest, including the scanning electron microscope and the Nomarski interference contrast device, both described in this chapter. These photos may then be "read" by the video camera mounted on a macrostand.

The **scanning electron microscope (SEM)** can produce excellent images of various phases in the backscatter electron mode, which shows differences in average atomic weight and therefore differences in composition (Goldstein et al., 1981). X-ray maps of the elements can be generated that delineate the various phases present. These maps may be automatically processed to generate a mineralogical mode or a phase mode in general.

Nomarski interference contrast microscopy enhances microtopography generated by differential etching of polished sections of rocks containing grains. Hydrochloric acid is good for carbonates and some silicates; fluoboric acid etches the more calcic zones in plagioclase, and hydrofluoric acid may be required for other silicates. Etching shows the location of grains of similar composition and of internal features such as compositional zoning (see Figure 7.8b). Modal analysis of volcanic rocks may be done with an accuracy unattainable with polarized light (Pearce and Clark, 1989).

X-Ray Diffraction Phase Analysis

Quantitative x-ray diffractometry is a technique for determining the abundance of various carbonate phases in very fine-grained carbonate rocks (Fang and Zevin, 1985; Martinez and Plana, 1987).

Size Analysis

Grain sizes in rocks have traditionally been estimated by visual inspection of thin sections and rock slabs for purposes of application to rock classifications. Grains of clastic sediments and some clastic sedimentary rocks can be separated from their matrix and sieved to determine the quantities of size fractions. Grain-size measurements can also be made on acetate peels (Gutteridge, 1985) (see the section on shape analysis).

More accurate means of grain-size measurement in thin sections have generated data that have promise of being petrogenetically significant (Marsh, 1988; Cashman and Ferry, 1988). The automated image analyzer can be used to measure grain dimensions on a quantitative basis. Images of grains of different sizes are generated with the optical microscope, the SEM, the Nomarski technique, or by simply photographing rock slabs. The digitized images of grains generated by these techniques can be processed for size information, including measurement of the longest horizontal cord for each grain, the average number of particle intersections per unit length of transect line, or the greatest width and length.

Several considerations must be addressed so that **crystal size distributions (CSD)**—plots of cumulative numbers of grains against size—can be properly evaluated. Thin sections present two-dimensional views of three-dimensional solids. There are important stereological considerations in the conversion from the area measurements to true grain-size distribution per unit volume (Gray, 1970; Marsh, 1988b; Cashman and Ferry, 1988).

One thin section of a rock that has dimensional fabric is an obvious problem, for which multiple thin sections is an obvious step in its solution. Another problem is how a grain is intersected by the plane of the thin section. Intersection of a grain corner will give the false impression of small size. Although the maximum frequency can be expected to be displaced to a smaller size, the departure from true distribution may be statistically insignificant (Galwey and Jones, 1966). It would seem that this problem is compounded if the grains had extreme dimensional morphology, such as being platy or rodlike, and of course this would be a major consideration if there were a preferred orientation.

Shape Analysis

Characterization of grain shape is a matter of establishing some easily observed criteria such as the presence or absence of crystal faces, the configuration of grain surfaces (curvature, flatness, irregularity), and the distribution and geometric configuration of these surfaces with respect to the whole crystal (external only or interior of cellular grains).

Qualitative characterization of grain morphologies has traditionally been made in thin section with the optical microscope: however, quantification with **automated imagery** is under development.

Characterization of grain shape of exceedingly small grains can be done with the scanning electron microscope. Qualitative characterization of grain shape by SEM-generated **secondary electron images** is now routine (Bertin, 1975; Goldstein et al., 1981). The SEM has the advantage of being able to look at very small grains and of producing images with considerable depth of field. The study of very fine-grained rocks such as volcanic rocks, mud rocks, carbonates, chalk (Figures 37.8a, 37.8b), diatomaceous rocks (Figure 37.7b), weathered and altered rocks, and sublimates in fumarolites (Figures 35.4a, 35.4b) has particularly benefited from this technique.

Acetate peels record an impression of an etched rock surface on a thin sheet of acetate film; the technique is particularly useful with carbonate rocks (Adams et al., 1984; Gutteridge, 1985) and sulfate rocks (Mandado and Tena, 1986). Examination of crystal morphologies and textural relations replicated by acetate peels is done by direct qualitative observation, optical microscope photography, and SEM photomicrography (Brown, 1986).

Orientation Analysis

Grains in rocks may have **dimensional** and/or **crystallographic** preferred orientation. Image analysis systems that measure grain shape can also record the orientation of the long and short axes of grains that are not equant. A three-dimensional picture requires measurements of the grains on three mutually perpendicular surfaces, either three slabbed surfaces of a hand specimen or three thin sections cut from those slabs.

Single-crystal grains may or may not have crystallographic preferred orientation. Nondimensional single-crystal grains such as quartz sand clasts in quartz arenites are not likely to have preferred orientation of crystal structure. Quartz grains in some metamorphic rocks have marked dimensional orientation as well as crystallographic orientation. This orientation suggests that the forces responsible for grain deformation are those that produced the crystallographic orientation (Chapter 16). An example of such dual preferred orientation and its determination is described in the next section.

Microscopic Petrofabric (Orientation) Analysis

The **universal stage** has an E–W, N–S, and vertical axis of rotation (three axes, but some stages have five axes), allowing rotation of grains in thin sections into the third dimension (Figure 6.1). The procedure for mounting and adjusting the universal stage and for making corrections for high angle of tilt is given in Emmons (1943) and Turner and Weiss (1963).

In microscopic **petrofabric analysis**, many measurements of crystallographic or dimensional orientation are made in thin section. Since the orientation of one grain with respect to another and to the host rock must be known, data must be collected while the thin section is oriented in three dimensions. This orientation is done by means of a rectangular guide attached to the upper hemisphere plate, movable in one direction only (Figure 6.2). Each time a new grain is located for measurement, the thin section must be parallel to the sides of the guide. The orientation of the thin sections with respect to the hand specimen (Figure 6.3) must also be known, requiring good communication between the petrographer and the thin-section preparator. Orientation of the hand specimen to field structures is a matter of taking oriented samples (Prior et al., 1987).

Consider the following example of **crystallographic orientation analysis** using the universal stage. The procedure for determining the c-axis orientation of two grains of quartz occurring in a mylonitic quartzite is traced in Figure 6.3. The quartz grains have dimensional as well as crystallographic preferred orientation. The triaxial ellipsoidal (or tabular) dimensional aspect of the quartz grains in the mylonite are shown in the block diagram of Figure 6.3. Since **crystallographic preferred orientation** (**CPO**) of the quartz is not observable in hand specimens, the procedure for determining the CPO is necessarily microscopic.

Thin Section Normal to Kinematic X (Z-Y Section).
An approximation of the orientations of crystallographic c of the quartz crystals in the Z-Y thin section (Figure 6.3) is made by using the red-1 plate to determine the fast and slow directions in the crystals. In the 45° position, either rotated with the inner stage of the U-stage or with the microscope stage (with or without the U-stage), a population of quartz grains will be blue, whereas others will be yellow (Plate 7b). This indicates that the c axis of the blue group presents as E' (quartz is optically positive) subparallel to the

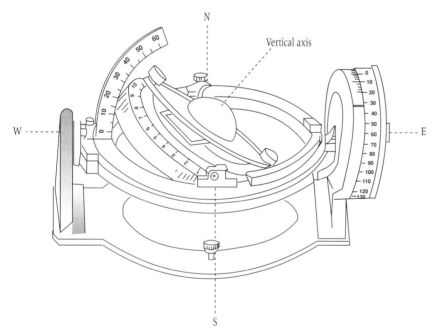

Figure 6.1
Sketch of universal stage showing E–W, N–S, and vertical axes. Microscope stage to which stage is screwed serves as an additional vertical axis.

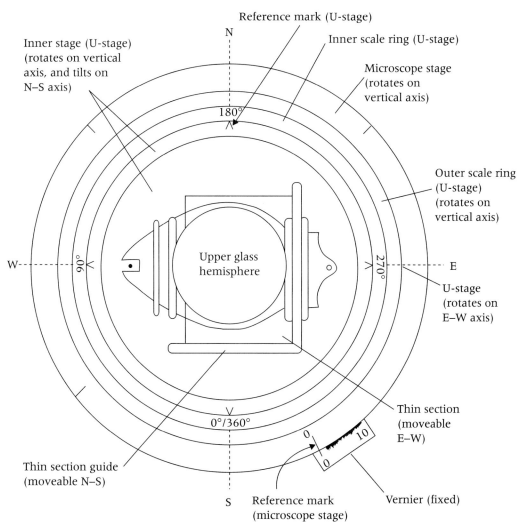

Figure 6.2
Plan view of the universal stage showing how a thin section is held in fixed orientation with a metal guide. Also shown are the location of scales and reference marks used in recording bearings of rotation on inner vertical axis of U-stage and vertical axis of microscope stage.

slow direction in the plate, and that the c axis of the yellow group presents as E' (close to O) subparallel to the fast direction in the plate.

The procedures described in the following section could be used to determine the precise orientation of quartz crystals within the two population groups in the Z-Y section. Crystallographic c axes of the crystals would be plotted on an equal-area stereonet.

Thin Section Normal to Kinematic Y (Z-X Section). Now consider thin section Z-X (Figure 6.3). Crystals (1) and (2) shown in the Z-Y thin section (greatly enlarged) are also shown in the Z-X thin section so that the relationships can be directly compared between the two sections.

Step 1 consists of placement of the two crystals in the 45° position, which in this example is a position of illumination for both crystals. Insertion of the red-1 plate generates a magenta red color for crystal (1), indicating a position (E'') nearly normal to the optic axis, and a blue for crystal (2), indicating E' is generally NE–SW (E'' indicates a position very close to crystallographic c (E), whereas E' are all other positions between E and O).

Step 2A brings crystal (1) to an extinction position with a 14° inner stage rotation, placing E'' in an

continued on p. 103

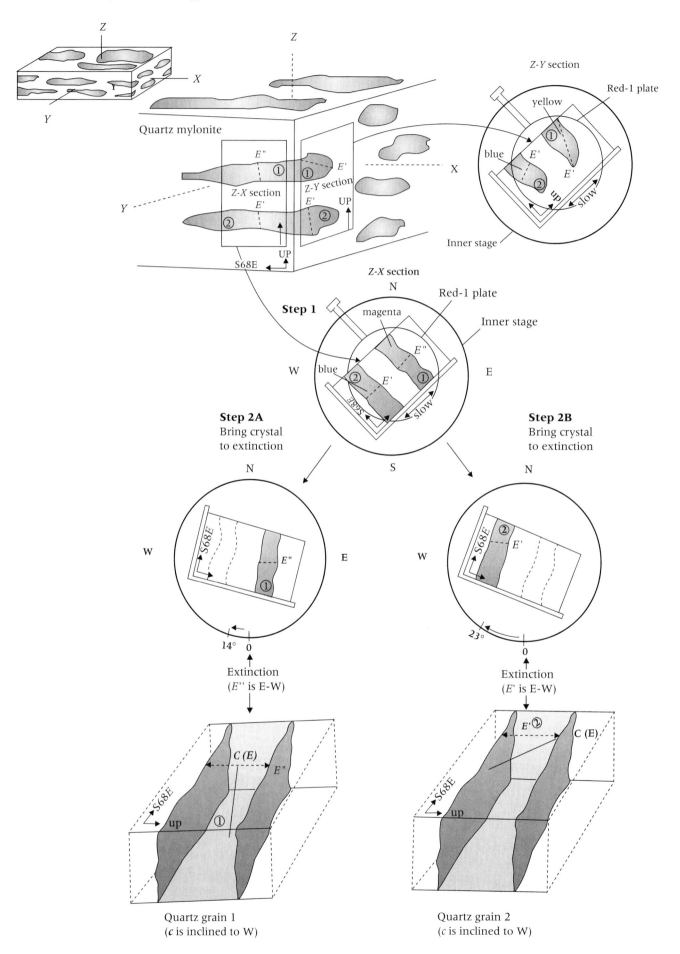

Step 3A
Destroy extinction by rotating microscope stage (clockwise chosen)

Step 3B
Destroy extinction by tilting on E–W axis (northerly chosen)

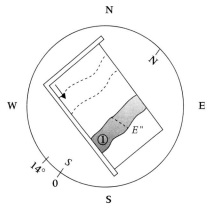
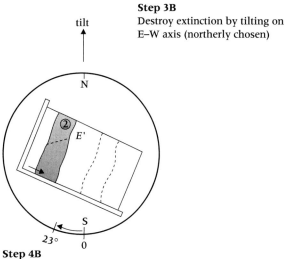

Step 4A
Tilt on N–S axis of inner stage to extinction (10° northwesterly). E (*c*) is now vertical and extinction maintains with microscope rotation. Adjust inner stage rotation and tilt if necessary (none shown). (Note: if *c* initially had been inclined easterly, tilt on N–S axis would have been found to be southeasterly to bring E (*c*) vertical.)

Step 4B
Rotate inner stage (10° clockwise shown) and tilt on N–S 29° (easterly shown) to bring E (*c*) parallel to microscope stage, indicated by maintenance of extinction with tilt on E–W axis. Repeat rotation and tilt by trial and error until this position is reached (10° additional rotation shown). (Note: if E (*c*) initially had been inclined easterly, inner stage rotation would have been counterclockwise and tilt westerly).

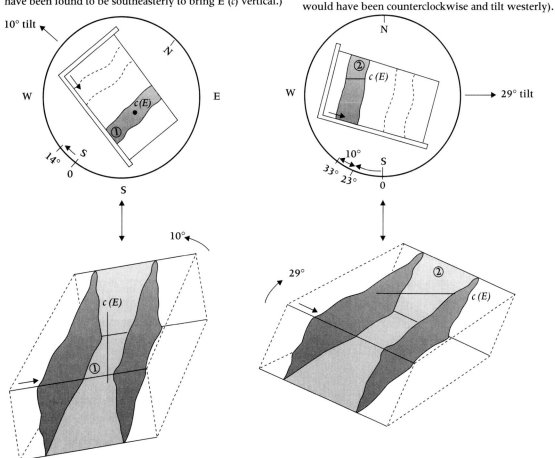

Figure 6.3
Sketch of field-oriented mylonitic quartzite from which two oriented thin sections have been derived. Section Z-X is normal to Y kinematic axis, presenting thin elongate quartz grains. Section Z-Y is normal to X kinematic axis, presenting relatively small slightly flattened quartz grains. There are two populations of quartz grains with different *c*-axis orientation, qualitatively seen expressed by interference colors in Plate 7b. See text for discussion of steps leading to plot of crystallographic *c* for the two crystals. From Gaudemer and Tapponnier (1987) for contoured plots of multiple *c* axes.

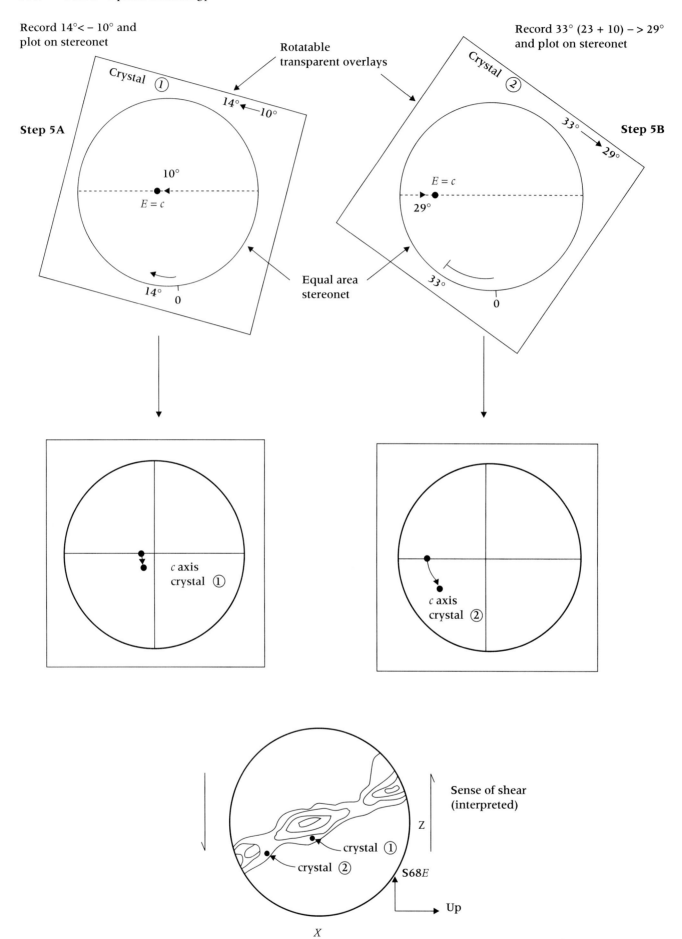

E–W position. $E(c)$ is now inclined either easterly or westerly in the vertical plane containing $E(c)$. If the extinction position was the result of O being placed E–W, rotation of c to vertical in step 3A would not be possible and a return to step 2A would be required.

Step 3A consists of rotation of the microscope stage (clockwise chosen) so that crystal (1) is in the illuminated 45° position as shown.

Step 4A brings $E(c)$ to a vertical position. A northwesterly 10° tilt on the N–S inner axis of the U-stage brings crystal (1) to an extinction position that maintains with any position of microscope stage rotation, meaning that E'' was presenting E–W in step 2A and that it was inclined westerly (if it had been O, no extinction position would be found and a restart in step 2A with a 90° inner stage rotation would be required). Some adjustment of inner stage vertical axis rotation may be required, as well as the tilt on N–S (trial and error).

Step 5A consists of plotting the c-axis on an equal-area stereonet, using the 14° rotation and 10° tilt data. In this example, the stereonet is set up such that its plane is parallel to the X-Z plane, with X parallel to the N–S axis of the stereonet and Z parallel to the E–W axis. To plot the orientation of a c axis, one simply reverses, on the stereonet (lower hemisphere), the rotations of the thin section that were used to locate the c axis. For crystal (1), the c axis was located by bringing it to vertical. On the stereonet, this corresponds to a point in the center of the net. To plot the original orientation of the c axis on the stereonet, first rotate (clockwise) the transparent overlay to a 14° position, then plot a point on the E–W line of the stereonet, 10° to the west of center. Counterclockwise rotation of the overlay 14° brings the c axis point to its actual position in the lower hemisphere of the stereonet.

Returning to crystal (2), step 2B (Figure 6.3) brings crystal (2) to an extinction position in a 23° inner stage rotation, placing E' in an E–W position. E ($=c$) is now either inclined easterly or westerly (as shown by the block diagram for this example) in the vertical plane containing E.

Step 3B consists of destroying extinction by tilting on E–W axis (northerly chosen in Figure 6.3).

Step 4B is a trial-and-error procedure bringing E ($=c$) into the plane of the microscope stage in an E–W orientation. Rotate the inner stage on its vertical axis and tilt on the inner stage N–S axis (trial and error) until extinction maintains with any tilt on E–W axis of the universal stage. E ($=c$) is now parallel to the E–W axis of the U-stage and the microscope stage.

Step 5B is the plot of the c axis on the stereonet using the data of rotation and tilt. In this case, rotation during the trial-and-error adjustments added 10° to the initial 23° rotation (= 33°) whereas the tilt is 29° easterly (needing no adjustment in this case). The c axis was located by bringing it to a horizontal E–W orientation in 33° (clockwise). To plot the original orientation of the c axis, rotate the overlay clockwise 33° and restore the vertical axis rotation by plotting a point along the stereonet E–W line 29° from the west margin of the stereonet. Counterclockwise rotation of the overlay brings the c-axis point to its actual position in the lower hemisphere of the equal area stereonet.

In practice, if E is inclined less than 45°, E is brought to parallelism with the microscope stage. If greater than 45°, E is brought to a vertical position. If E is steep but not close to vertical, E' can be located with the red-1 plate and rotated to vertical. The magenta color and uncertainty as to the identity of E'' and O obtains only if E is near vertical.

Part II

The Anatomy of Rocks

Physical attributes of rock-forming mineral grains and nonmineral grains, including their physical relation to each other, constitute rock texture (Chapter 7). This textbook focuses on the visual characterization of rock texture by means of line drawings and photographs. Of course, verbal characterization is necessary in order to communicate with script and to enable the development of numerical codings of both objective observations and subjective interpretations of texture to enable computerized and quantitative comparisons and presentations to be performed.

Existing textural nomenclature has evolved from the beginnings of petrographic observation in the latter part of the nineteenth century and the early part of the twentieth century. Knowledge of rock-forming processes has increased dramatically in the last 30 years or so, and it is seems safe to say that the data of petrography and particularly its nomenclature have not kept pace. It seems to be a case of too little and too much terminology. A deficiency of well-defined terms arises as new textural knowledge is accumulated or as there becomes a need to be more specific, and a plethora of historically generated, vaguely defined terms is characteristic of any evolving science.

Since "familiarity breeds content" much of the traditional textural nomenclature is essentially retained in this text, even though the attitude "call it what you like, but let's look at the photograph" is promoted. Nonetheless, some liberties have been taken here, and a few new terms and some reapplied terms are presented as alternatives in the spirit of providing a more objective language and one more compatible with the enormous increase of petrographic and petrogenetic information acquired in recent times.

Textural interpretation is the fun part of petrologic investigation and discussions of texture genesis are threaded throughout the text. It is the intent in Chapter 8 to make the student aware of the pitfalls inherent in the study of three-dimensional objects (rocks) in two dimensions (thin sections).

7

Physical Attributes of Rocks

Augite oscillatory zoning.
X-polars. Bar scale = 0.6 mm.

Fundamental Constituents

The **fundamental constituents** of rocks are minerals, submicroscopic crystalline and noncrystalline material, glass, and voids. Minerals are naturally occurring crystals (C) and are by far the most common rock-forming material. Submicroscopic crystalline material, such as the clay-sized crystals in mud rocks, micritic carbonates, chert, and devitrified glass, and amorphous or cryptocrystalline material, such as opal and allophane, are designated here as **submicroscopic material (SM)**.

Some rocks consist of **glass (G)**. Obsidian, pumice, sideromelane, pseudotachylite (fault derived), and lechatelierite (mostly meteor impact origin) are the natural glasses. Although the isotropic nature of glass is the key to its identification, some uncertainty may arise if amorphous materials are present, since they too are isotropic. Very small crystals in glass are microlites and crystallites; their presence may be detected, but optical identification is not feasible, and they are considered a part of the glass.

Voids (V) in rocks are fundamental constituents because they have size, shape, and volume just like solid constituents. Voids include vesicles in lavas, miarolitic cavities in granites, primary pore spaces in clastic rocks, open spaces in breccias, and solution cavities in carbonate rocks.

Distribution of Fundamental Constituents

Crystals, submicroscopic material, glass, and voids may be evenly distributed in rocks or occur in visually definable portions of rocks, where they are defined here as **units**. Rocks may be dominated by units, such as grainstones, pyroclastic flow tuffs, and breccias, but more commonly there is a **matrix** that encloses the units. Here are some examples:

Crystalline Units

1. Small microdiorite inclusion in granodioritic matrix
2. Orbicules in orbicular diorite
3. Quartzite pebble in sandy matrix
4. Schist clast in litharenite
5. Fossil in sparry or micritic calcite matrix

Submicroscopic Material Units

1. Kf–tridymite spherulite in rhyolitic glass matrix

2. Micritic pelloids in sparry calcite matrix
3. Cherty clasts in claystone matrix
4. Jasper breccia fragments in crystalline quartz matrix
5. Mud rock intraclasts in mud rock matrix

Glass Units

1. Glass shards in glass matrix
2. Pumice fragments in glass matrix

Voids in Units

1. Vesicular lapilli in scoriaceous basalt
2. Vesicular pumice fragment in rhyolitic tuff
3. Solution cavities in fragments of calcrete

Grains as Constituents of Rocks

Rock material occurring as any visible physical entity within the scale of a hand specimen or thin section is a **grain**. Therefore, grains have size and shape. A grain may be (1) a single crystal (crystal grain), (2) a cluster of crystals (polycrystalline grain unit), (3) a unit consisting of submicroscopic material (submicroscopic unit), or (4) a unit consisting of glass (glass unit).

A grain may have had a significant mechanical history, such as those in siliciclastic sedimentary rocks and those in cataclasites and in pyroclastic rocks. In contrast, crystal grains in rocks that have crystallized in magmas and from aqueous solutions have had a relatively nonmechanical history.

Rock Texture

Size, shape, and interrelations of grains constitute **rock texture** (Cross et al., 1906). The interrelations of grains relates closely to their orientation and distribution in rocks. Consequently, even though orientation and location are traditional fabric elements of particular interest to structural geologists, they may also be considered as textural attributes at the thin section and hand-specimen scales. Furthermore, internal features of crystals such as zoning, twinning, and dislocations are also presented here in the context of texture because their development typically is related to the same dynamic and kinetic phenomena that generate the traditional rock textures.

Grain Size

Grain size (GS) can be expressed as its mean diameter or maximum diameter in millimeters. Measurement is made by laying a scale over the grain or by placing the grain under the micrometer scale of the microscope eyepiece and reading the dimension(s). If the grain is equant, only one measurement is needed. If the grain is nonequant, such as an elongate or tabular grain, several measurements can be made and a mean diameter calculated. Generalizations are necessary, not only because there is rarely justification to carefully measure each and every grain, but also because a two-dimensional section through a rock produces many artificial sizes and shapes of grains resulting from the intersection of corners and edges of grains. A statement of grain size is typically made on the basis of longest dimension.

How are generalized grain-size data defined? Grains too small for microscopic mineral identification is our lower limit, and the largest crystal or boulder you can imagine is our upper limit. Divisions have been made to accommodate natural occurrences, not by simple arithmetic partitioning. For example, volcanic magmatism typically produces grains ranging in size from glass (no crystals) to grains much less than a millimeter in diameter. In this case a few divisions are drawn within this size range to accommodate variations in volcanic rocks that seem to be useful in classification and characterization of rock-forming process.

Plutonic magmatism and metamorphic recrystallization typically yield crystals ranging in size from about a millimeter to several centimeters. This size range is much larger than quenched magma in the volcanic environment, but establishment of a proportionally larger number of subdivisions is not warranted.

Size of Individual Grains. The millimeter grain-size divisions in the following list have been adopted on the basis of three assumptions: (1) generalization of grain-size data is necessary to facilitate petrographic observations, even though automated measuring devices promise to refine these observations; (2) establishment of size-range groups is necessary so that the generalized data can be compared from one rock to another (classification), and (3) the boundaries defining these groups can be established to accommodate certain visual considerations and natural occurrences of rocks:

Very coarse grained	>50		
Coarse grained	>5	to	<50
Medium grained	>1	to	<5
Fine grained	>0.1	to	<1
Very fine grained	>0.01	to	<0.1 (10–100 μ)
Glass/SM	glass	to	<0.01 (<10 μ)

What are the justifications for adoption of these specific limits? The 5-mm break was chosen because (1) one-half of a centimeter is an easy dimension to visualize (it is about 1/4 inch), (2) many plutonic magmatic rocks have grain sizes from 5 mm to about 1 cm, (3) it is the break between medium and coarse in many igneous rock classifications, and (4) the traditional break between granules and pebbles in sedimentary rocks is close, at 4 mm.

The 1-mm break was chosen because (1) 1 mm is easy to visualize (the mark of a medium-size pen is close to 1 mm), (2) it is the break between coarse and very coarse sand in sedimentary rock classifications, (3) it is the break between fine- and medium-size crystals in most igneous rock classifications, and (4) it is the size below which the use of a hand lens or microscope becomes necessary.

The 0.1-mm break was chosen because (1) it is the approximate size above which granularity in a hand specimen is sensed and below which the sense of granularity is lost even with a hand lens, (2) it is near the break between very fine sand and fine sand in most classifications of sedimentary rocks, (3) it is the size above which, in thin section, mineral grains are easily identified with "low" and "medium" power objectives, and (4) it is easier to read on a micrometer scale than 0.125 mm.

Crystals between 0.01 mm and 0.1 mm commonly can be identified with "high" power magnification. Crystallinity below 0.01 mm (10 microns) may be apparent, but mineral identification is uncertain at best. Crystals shown as SEM images are typically a few microns in diameter. A "feel" for sizes expressed in microns is shown by the 40-μ- (0.04 mm) wide galena crystal in relation to the tip of a push pen (Figure 7.1).

Since the distinction between glass, amorphous material, and exceedingly small crystals may not be obvious, particularly if there is a mixture of these materials, no subdivisions are proposed below 0.01 mm for petrographic investigations unaided by SEM images.

Size Within Grain Aggregates. **Equigranular** implies that the generalized longest diameter of grains in a rock is within the limits defining one of the established size groups. For example, a fine-grained rock contains grains whose long diameters lie mostly between the limits of 0.1 mm and 1 mm. If the range is from just above 0.1 mm to close to 5 mm, the rock would be characterized as **seriate inequigranular** because two size ranges are involved. But suppose the range of diameters is from 0.5 to 1.5 mm, bridging the 1-mm boundary? Such a rock would have been equigranular if the size limits originally had been placed differently, but it is inequigranular with respect to the established limits since sizes range across the 1-mm boundary. Does this significantly undermine genetic relevance of grain-size partitioning? From a genetic point of view, the most important aspect of grain size is the actual size of the crystals and the deviation from the mean, not whether the rock is called equigranular or inequigranular.

Inequigranular rocks may have a bimodal or even a trimodal grain-size distribution. **Bimodal inequigranular** rocks are those containing phenocrysts (magmatic), porphyroblasts (metamorphic, hydrothermal), or other such grains that are obviously larger than a finer-grained fraction of grains occurring in the same rock. In some cases, what is "obvious" bimodality may prove to be a continuous range of grain size in the light of quantitative grain-size measurements (Marsh, 1988b). **Trimodal inequigranular** are typically rocks with a matrix of glass or submicroscopic material (SM), in which there are two size groups of crystals.

The larger grains in either bimodal or trimodal inequigranular rocks commonly are called "megacrysts," in avoidance of the genetic terms "phenocrysts," "porphyroblast," and "porphyroclast" when only a descriptive term is justified. However, a term such as **porphyrograin** is a descriptive term that does not present an inconsistency with the "micro," "meso," and "macro" scales of geological observation, nor is it limited to crystals. Porphyrograins include single crystals, polycrystalline grains, and grains consisting of submicroscopic material or glass. **Microporphyrograins** is an appropriate term for many very fine-

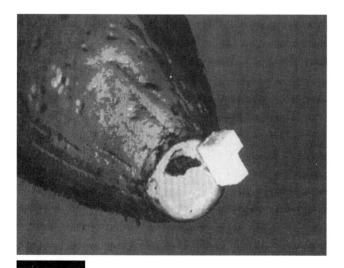

Figure 7.1
Sense of dimension relative to tip of push pen. Long diameter of galena crystal is 40 microns.

110 Part 2 The Anatomy of Rocks

GRAIN SIZE

Equigranular

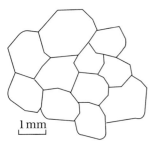
Equant crystal grains
Medium–grained

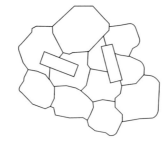
Equant + inequant crystal grains
Medium–grained

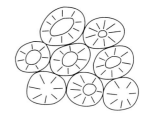
Equant – spherical unit grains
Medium–grained

Seriate Inequigranular

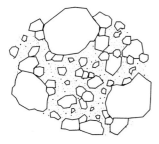
Equant crystal grains
Very fine–grained → medium–grained

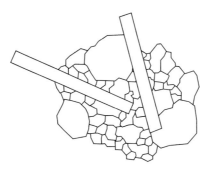
Inequant crystal grains
Fine–grained → coarse–grained

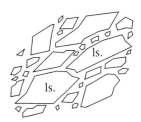
Equant + inequant unit grains
Fine–grained → medium–grained

Bimodal Inequigranular

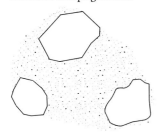
Equant crystal grains
Very fine–grained + medium–grained

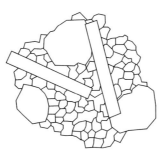
Inequant crystal grains
Fine–grained + medium–grained

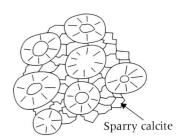
Equant unit grains +
equant crystal grains
Fine–grained + medium–grained

Trimodal Inequigranular

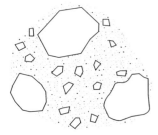
Equant crystal grains
Very fine–grained +
Fine–grained + medium–grained

Inequant crystal grains
Glass + fine–grained +
Medium–grained

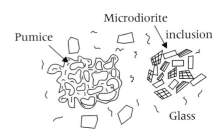
Equant unit grains +
equant crystal grains
Glass + fine–grained (crystals) +
medium–grained (units)

grained and glassy rocks that contain microscopic, yet relatively large, grains. A rock containing a size range of porphyrograins could appropriately be termed **seriate porphyrograined**, one genetic variety of which is "seriate porphyritic."

How many crystals of much larger relative size must a rock have to be considered "porphyrograined" or "microporphyrograined"? The presence of a single porphyrograin (or microporphyrograin) could conceivably have just as much important genetic significance as a rock containing 90% porphyrograins. It seems wise, therefore, not to ignore a very low percentage of porphyrograins, nor a very low percentage of matrix, to accommodate semantic simplicity.

Examples of grain size and grain-size variations are sketched in Figure 7.2.

Grain Morphology

The **shape** of grains (i.e., grain morphology, GM) can be expressed by reference to three-dimensional axes describing either equant, tabular, platy, or prismatic solids. These solids may be single crystals or "units" of any rock-forming material. For crystals, shapes reflecting crystal structure are referred to in this text as **eustructural**, whereas shapes not reflecting crystal structure are **astructural**, and intermediate shapes are **substructural**. Crystal surfaces may be crystal faces, planar surfaces that are irrational with respect to crystal structure (such as polygonal mosaics), or nonplanar surfaces that are curving or irregular.

Some crystals are incomplete. Terms such as "skeletal," "dendritic," "spherulitic," and "cellular" are expressions of this incompleteness, but a general term such as **integrity**, introduced here, allows for a nonspecific classification. Complete crystals have high integrity, whereas a completely cellular crystal has low integrity. Such a scaling is not intended to be sensitive to the presence or absence of crystal faces.

A crystal that has inclusions of other grains is not a complete crystal either, and it too has a reduced integrity. A crystal that has a few cells, a few inclusions, or simply an irregular surface represents intermediate integrity. Incomplete noncrystalline grains or grains composed of submicroscopic material can also be scaled according to integrity. For example, a pumice fragment is a glass grain with very low integrity.

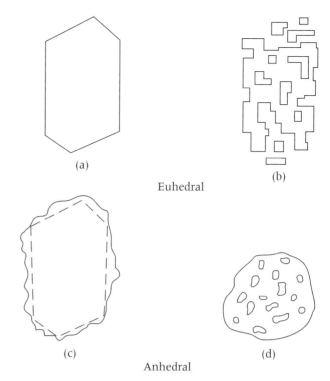

Figure 7.3
Sketches of crystal grains showing limitation of traditional crystal shape classification.

Crystals A and B are very different (A is complete, B is cellular), yet both are totally bounded by crystal faces, thus satisfying the criteria for "euhedral." Crystals C and D are very different (C is nearly "euhedral," D is far from "euhedral"), yet both are without crystal faces, thus satisfying the criteria for "anhedral."

This classification of crystal morphology in terms of crystallographic expression of shape and integrity is compared with the traditional morphological classification of crystals in Figure 7.3. The traditional terms are "euhedral" (crystals bounded by crystal faces), "subhedral" (crystals bounded by some crystal faces), and "anhedral" (crystals without crystal faces). The emphasis here is on the nature of the final surfaces of crystals as to whether they are crystal faces. The crystals in Figure 7.4a,b,c,d are mostly bounded with crystal faces and would be good examples of "euhedral"

Figure 7.2
Sketched examples of grain size as classified into four fundamental groups. Examples include a variety of grain shapes and sizes. Scale shown for equigranular rock containing equant crystal grains applies to all other sketches.

morphology or, as proposed, "eustructural" morphology. A eustructural olivine crystal with medium integrity is shown in Figure 7.4e.

One of the crystals sketched in Figure 7.3 and shown in the photograph in Figure 7.5 is shaped as a perfectly grown crystal of plagioclase feldspar, yet its final surfaces are not crystal faces. Allowances can be made for such dilemmas with traditional classification only if the scale of observation is taken into consideration. If the surface of a crystal approximates crystallographic planes, let that plane be considered a crystal face even though at some higher magnification it can be seen that there are no faces. The plagioclase shown in Figure 7.5 clearly has no crystal faces, even observed with a low-power objective, but its overall shape reflects nearly perfect feldspar crystallographic form.

A shape-integrity classification (Figure 7.6; p. 114) is designed to eliminate these potential discrepancies and to be more sensitive to other aspects of grain morphology, particularly the completeness of crystals.

There must also be a morphological classification of noncrystal grains, and one is included in Figure 7.6. Of course, in this case there is no concern with crystal faces nor with how shape may reflect crystal structure. Nevertheless, classification for noncrystal grains is sensitive to integrity, sphericity, roundness, and angularity just as it is with crystals.

Morphological characterization of a single grain in a rock is of little petrogenetic value by itself. How

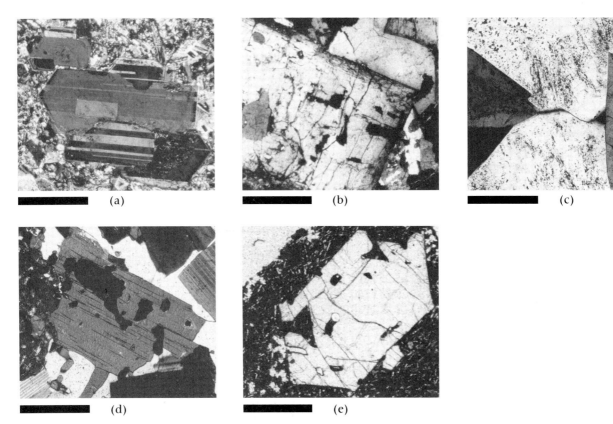

Figure 7.4
Examples of crystal morphology closely reflecting crystal structure ("eustructural"), with variable completeness of crystal ("integrity").

(a) Complete (high "integrity") eustructural ("euhedral") plagioclase phenocrysts in andesite. X-polars. Bar scale = 0.8 mm. (b) Eustructural ("euhedral") nepheline crystal with a few inclusions (medium integrity). X-polars. Bar scale = 0.8 mm. (c) Eustructural ("euhedral") adularia crystals with high integrity. X-polars. Bar scale = 0.18 mm. (d) Eustructural plagioclase (right) and biotite (center) in quartz diorite, both with medium integrity. X-polars. Bar scale = 0.18 mm. (e) Eustructural olivine with medium integrity reflecting local skeletal–cellular morphology. Vesicle (upper left). Picrite basalt, Oahu, Hawaii. PPL. Bar scale = 0.8 mm.

can the morphological characteristics of grains in rocks be expressed as an aggregate for the rock as a whole? For crystals, the traditional usage is as follows. If the majority of crystals in a rock are "euhedral," the aggregate texture with respect to crystal morphology is "panidiomorphic granular." If the majority of crystals are subhedral, the rock is "hypidiomorphic granular." A rock with mostly anhedral crystals is "allotriomorphic granular."

In some cases these terms are a reasonable expression of crystal morphology in rocks. However, in many other cases there are several populations of grains, each with distinctive grain morphology. There is no reason why a subordinate morphological type should be any less important from a genetic point of view. The garnet–mica–quartz schist, sketched in Figure 7.7; p. 115, has garnets with crystal faces ("euhedral" or "eustructural"), mica "plates" with basal crystal faces and ragged edge terminations ("subhedral" or "substructural"), and polygonized quartz with no crystal faces ("anhedral" or "astructural").

Crystal Zoning

Crystal zoning (CZ) chiefly reflects changes in composition and crystal structure as the crystal grows. Minerals that are members of a solid solution series, such as plagioclase feldspar (Figures 7.8a,b; p. 116), olivine, clinopyroxenes (Chapter 7 frontispiece photo), and amphiboles typically develop growth zoning, recording not only the chemical–structural changes but also the shape of the crystal as it enlarges. Thus, crystal zoning is a record of crystal morphology prior to final growth, and this is petrogenetically significant.

For example, a plagioclase crystal may grow in a magma by enlargement parallel to crystal faces, but final growth of the crystal may compete with adjacent crystals for space and finish with an irregular nonrational surface. If classification of grain morphology is based on this final growth morphology alone, the major record of "euhedral" growth has been ignored.

Expression of zoning is seen as differences in (1) extinction position, (2) pleochroism, and/or (3) birefringence. Zoning may also be expressed by former differences in composition that later are the locations of exsolved phases (Figure 7.8c). This expression is typical in alkali feldspars in slow-cooling magmatic systems. Another well-known "secondary" expression of zoning is the sericitic alteration of more calcic zones of plagioclase.

Zoning related chiefly to differences in extinction position includes morphological changes, such as the change from nonfibrous quartz to the fibrous variety chalcedony. Zonal arrangement of cellular and noncel-

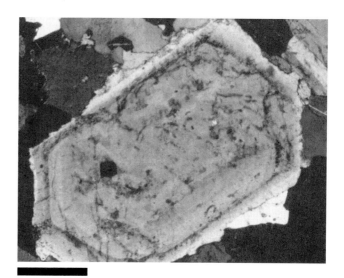

Figure 7.5
Nearly perfect crystal of plagioclase lacking crystal faces. Traditionally classified as anhedral, here classified as "eustructural" (reflecting crystal structure) with "medium integrity" (incomplete, irregular surfaces). X-polars. Bar scale = 0.6 mm.

lular portions of plagioclase crystals (Figures 7.8a, 14.8a, 14.9a, 14.11) is rather common in volcanic rocks. This is a morphological zoning, but it is also typically compositional since the cellular plagioclase is typically different in composition from the noncellular portion. Variations in triclinicity of K-feldspar may be observed as abrupt or gradual differences in extinction position. Zoning may also be expressed by the preferred location of impurities, such as inclusions of other mineral phases (Figures 7.8d, 21.33a), and by the preferred location of fluid inclusions.

Zoning may develop as a progressive change in composition and/or crystal structure, or it may occur abruptly. **Progressive zoning** records a continuous change in composition and accompanying crystal structure. The classic example of progressive zoning is "normal zoning" in magmatic plagioclase, resulting from a gradual decrease in the anorthite content from core to rim of the crystal. Optically this is expressed by a progressive change in extinction position (Figure 7.8e). "Normal" indicates that crystallization takes place in response to falling temperature, as it does in magmatic and hydrothermal systems.

Step zoning applies to abrupt changes in composition that partition crystals into areas of distinct optical character (Figures 7.8f,g,h). **Oscillatory zoning** is a fine-scale repetitive version of step zoning and/or progressive zoning. Each oscillation in an oscillatory

GRAIN MORPHOLOGY

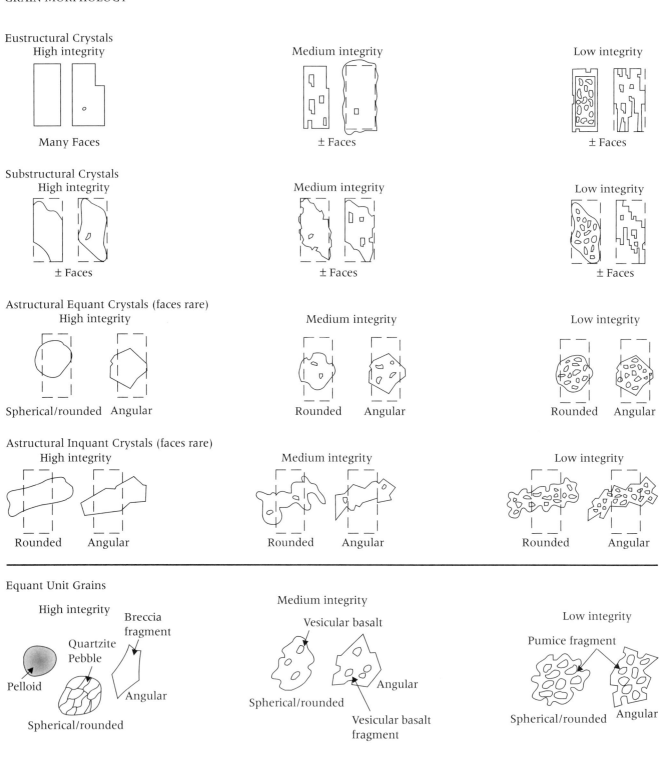

zoned plagioclase (Figures 7.8a,b) typically begins with a step to higher anorthite content followed by a progressive return to the original or a lower anorthite content. Combinations of these zoning types in single crystals are rather common (Figure 7.8a).

If zoning conforms to the general shape of a crystal, it is **continuous zoning** (Figure 7.8g,h). If it is localized at certain crystallographic regions, it is **sector zoning** (Figure 7.8f). If the zoning is localized in several areas in the crystal, with little or no preference for certain structural positions, it is known as **patchy zoning** (Figure 5.16).

The various types of zoning, their combinations and location in crystals are shown in Figure 7.9; p. 117.

Grain Discontinuities

In addition to twinning and zoning, there are several other intracrystalline attributes of crystals that have optical expression. They include optical imperfections relating to the nature and location of dislocations (Chapter 16), including **progressive extinction** ("wavy" or "undulatory" extinction) (Figure 16.5a), deformation lamellae (Figure 16.5b), **kinkbands** (Figure 16.5c), subgrains (Figure 16.5d), and **deformation "bands."** Both crystal grains and unit grains may have **microfractures** and **microfaults**, both being nondistributive grain discontinuities.

Progressive extinction is one expression of a strained crystal lattice. Besides the progressive nature of extinction, deformation of a crystal lattice may be detected simply by the distorted ("bent") shape of the crystal and/or by the presence of arcuate twin lamellae with tapered terminations (Figure 16.6).

Subgrains look like shadowy crystals within crystals. The orientation of subgrains (or subcrystals) deviates only slightly from the orientation of the host crystal. If there is more deviation of lattice, the subgrains are considered to be new crystals (Chapter 16). "Ribbon structure" applies mainly to extremely elongated quartz crystals occurring in some quartzofeldspathic metamorphic rocks (Figure 16.5e).

Some examples of crystal discontinuities are sketched in Figure 7.10; p. 118.

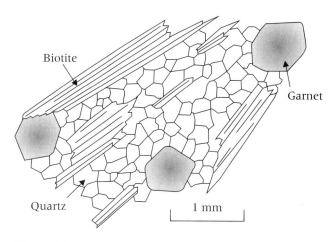

Figure 7.7
Sketch of garnet–mica schist as an example of a rock containing three distinct morphological types of crystals. The quartz is astructural–equant–high integrity–angular–no faces, the biotite is substructural–medium integrity–some faces, and the garnet is eustructural–high integrity–many faces.

Crystal Twinning

The crystallographic properties of twins have an extensive literature, but little is known about the occurrence and petrogenetic significance of twinned crystals. It is known, however, that complexly twinned plagioclase (albite–pericline–Carlsbad) is common in magmatic rocks and rare (absence?) in rocks formed by recrystallization processes. It is known that alkali feldspar **Baveno twins** (Figure 7.11a; p. 120) are common in certain granites, but it is not known what special circumstances are required. The same can be said for **Manebach twins** (Figure 7.11b), chessboard plagioclase (Figure 7.11c), and twinned birefracting garnet (Figure 7.11d).

The petrogenetic significance of microcline twinning (Figures 7.11e, 17.9b, 17.9c) seems to be the best understood. Microcline (triclinic) cannot grow twinned. It has been demonstrated that twinned microcline must have had a monoclinic alkali feldspar

Figure 7.6
Classification of crystal grain and noncrystal ("unit") grain morphology (shape). Crystals (feldspar example) are classified first relative to their expression of crystal structure, including equant or inequant for the astructural group, second according to their integrity (completeness), and third as to the presence or absence of crystal faces, sphericity, roundness, or angularity. "Unit" grains are first classified as equant or inequant, second according to integrity, and third as to sphericity, roundness, or angularity.

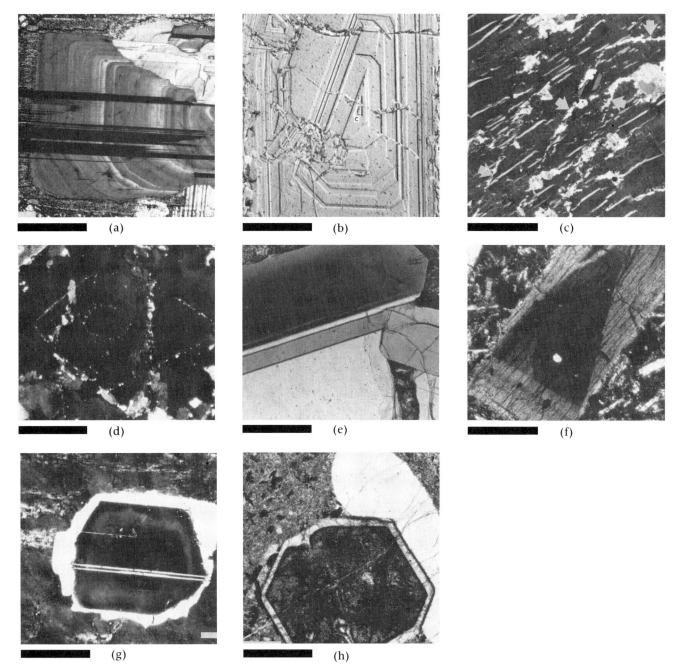

Figure 7.8
Crystal zoning.

(a) Zoned and twinned plagioclase phenocryst in andesite. Oscillatory zoning with calcic "spikes" (white zones) is succeeded by a thin rim zone consisting of spongy cellular plagioclase of about the same composition as the calcic spikes. Note orientation of albite twin lamellae (black lamellae) unaffected by zoning, and Carlsbad twin (upper right). X-polars. Bar scale = 0.8 mm. (b) Normarski interference image of zoned plagioclase from a rhyolite. Bar scale = 79 μm. (Courtesy T. H. Pearce, Queen's University.) (c) Oscillatory zoning in alkali feldspar defined by exsolved albite (arrows). X-polars. Bar scale = 0.8 mm. (d) Zoning in quartz crystal of a granodiorite defined by the position of two "zones" of inclusions. The inner zone defines a round stage of growth; the outer zone outlines a beta quartz form (trigonal dipyramid). X-polars. Bar scale = 0.2 mm. (e) Normal progressive zoning of plagioclase. X-polars. Bar scale = 0.2 mm. (f) Step zoning (sector type) in augite with some oscillatory zoning in outer portion. X-polars. Bar scale = 0.2 mm. (g) Step zoning in plagioclase consisting of an albitic rim and an intermedate composition core. X-polars. Bar scale 0.2 mm. (h) Step zoning in nosean phenocryst defined by clear rim and "dusty" core. Nepheline phenocryst (white). Phonolite. PPL. Bar scale = 0.8 mm.

predecessor, such as orthoclase. The transformation is evidently facilitated by the presence of an aqueous phase and/or by shearing (Eggleton and Buseck, 1980), underscoring the petrogenetic value of this type of twinning.

Repetitive rhombohedral twinning of calcite (Figure 21.13a) is another example of twinning indicating something special about environment. In this case, the twins are secondary, forming mechanically in a dynamic environment. In sharp contrast, cementing sparry calcite in carbonate rocks is typically untwinned.

Some twins are so common that their absence is the indicator of special conditions. For example, the lack of albite-twinned plagioclase is characteristic of some metamorphic rocks in which the plagioclase occurs in polygonal mosaics with quartz.

Not all twins are easily identified by routine optical investigation. Plagioclase twins may be easily misidentified. For example, ala A and ala B twins are very similar to albite twins, and Carlsbad twins are very similar to albite–Carlsbad twins (Chapter 5).

Sketches of some of the more common twin types are presented in Figure 7.12; p. 122.

Grain Relations

Even the casual observer has to be impressed by some of the intricate patterns of grain interrelation generated by rock-forming processes. Grain relations (GR)

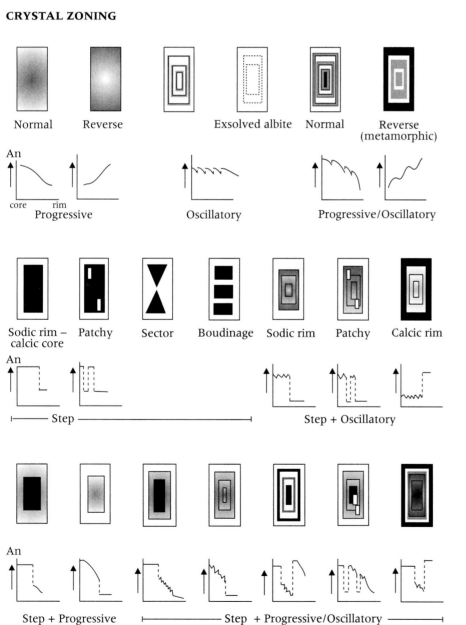

Figure 7.9
Classification of crystal zoning shown with sketches and zoning curves. Zoning curves plot An percent against distance from center of crystal. "Normal" implies compositional change from core to rim in response to falling temperature. "Reverse" implies change in composition in response to increase in temperature (increase in An for plagioclase).

GRAIN DISCONTINUITIES

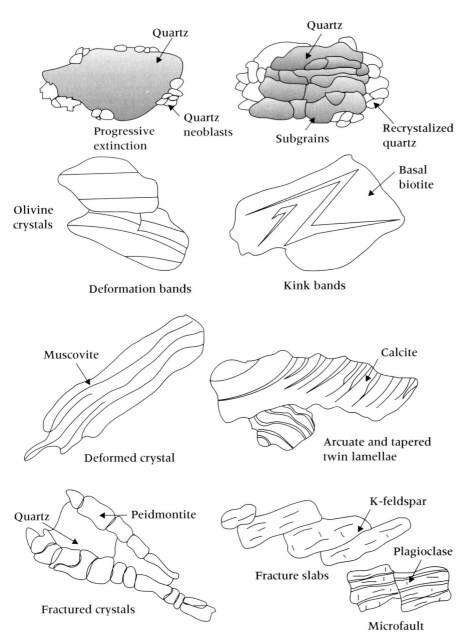

Figure 7.10
Sketches of grain discontinuities.

are grouped as follows: granular, intergranular, transecting, including, and mantling. Some glassy rocks and some rocks with submicroscopic material will have no grain relations, since either there are no grains or grains are isolated in these matrix materials and do not contact each other. The grains of some rocks have only granular (side by side) relationships. However, most rocks with grains have a dominant granular texture and other grain relations as well. An illustrated classification of grain relations is shown in Figure 7.13; p. 122.

Granular Texture. A **granular** relation consists of grains in contact with each other, in a more-or-less side-by-side relationship. The size and shape of the participating grains are variable, and their interface may be simply planar or curved, or complexly sutured or scalloped. Apparent inclusion of one grain in another may result from direction of sectioning of a nonplanar interface (Chapter 8), yet the real relationship is side by side.

Crystalline rocks such as granites, schists, and crystalline limestones are dominated by granular grain relations. Rocks dominated by the presence of clastic, cataclastic, or pyroclastic grains, such as arenites, cataclasites, and ash flow tuffs, are also characterized by granular texture.

Intergranular Texture. An **intergranular** relation can be described as the occurrence of any relatively small volume rock-forming material located between and confined by grains. For example, the sodic plagioclase located between larger grains of alkali feldspar (Figure 8.7a) is an intergranular relation, as is the quartz located between plagioclase crystals (Figure 7.4d). "Intergranular" has also been used for the occurrence of relatively small pyroxene crystals located between larger plagioclase laths, typical of some basaltic and diabasic rocks. This relation is viewed in the present classification as seriate inequigranular or bimodal inequigranular in the context of grain size, as well as granular in the context of grain relations.

Transecting Texture. A **transecting** relationship between grains includes glass, submicroscopic material, or crystals in **microveins** (and "mesoveins" of hand specimens) that have a transecting relationship with grains or nongrained rock material. Typically, the veins result from precipitations and reactive replacements in fractures and faults (originally grain discontinuities). For example, the sets of microfractures shown in Figure 7.14a, p. 126; contain sericitic mica and as such are considered to be a transecting texture. Similarly, chlorite-bearing microveins in the cordierite crystal in Figure 7.14b are considered a transecting relation—in this case a useful texture in distinguishing the cordierite and the adjacent microvein-free quartz.

Vein material may be of the same or similar composition as the host. The transecting microvein shown in Figure 7.14e is albite in a more calcic plagioclase host. The transecting microvein in Figure 33.2 has both quartz and K-feldspar; the latter has an epitaxial relation with the K-feldspar host crystal.

A transecting relation may not be veinlike. The stilbite occurring between biotite cleavage fragments shown in Figure 7.14c has filled "space" provided by extension of the biotite crystal, but it has also filled around the biotite fragments. Another transecting relation is shown by the covellite–idaite replacement veins in bornite (Figure 7.14d).

There are many other transecting possibilities, some of which are sketched in Figure 7.13.

Including Texture. **Including** grain relationships do not necessarily mean that a pre-existing grain was included by a later-forming grain. There are seven subtypes of including relations (Figure 7.13) whose names are given here in a way that avoids genetic connotation.

One including relation applies to a relatively large host grain containing other grains that are orders of magnitude smaller. The composite grain is here termed a **poikigrain**. This relationship is common in metamorphic rocks in which porphyrograins are **poikiloblasts**. Garnet and staurolite porphyroblastic poikiloblasts typically contain many small grains of quartz (Figures 21.17, 21.35a).

Poikigrains of magmatic rocks are commonly phenocrysts. Certain hybrid granitic magmatic rocks have poikigrains consisting of randomly oriented grains of plagioclase and biotite in a much larger K-feldspar host (Figure 19.14). Alkali feldspar phenocrysts in two-feldspar granites and granodiorites commonly contain small plagioclase crystals that commonly have preferred crystallographic orientation with respect to the host crystal (Figure 17.4e, 17.10a). Small, more-or-less round grains of quartz in plagioclase or K-feldspar of some inequigranular granitic magmatic rocks is an including texture. The quartz is commonly restricted to the outermost portion of the host crystal defining a localized zone of inclusions (Figure 17.4c 17.7a). Magmatic poikigrains, including phenocrysts, have **poikilitic** texture.

Another including relation consists of included and partially included grains that are much larger in relation to the host grains than those inclusions of poikigrains. The volume of the host crystal typically is less than the aggregate volume of the inclusions. The composite grain is defined here as an **oikigrain**. **Oikocrysts** with a pyroxene host are common in some cumulate rocks (Figure 18.7a,b,c), and **ophitic texture**, consisting of plagioclase laths in pyroxenes (amphiboles and biotite as well), is considered here to be an oikigrain variety characteristic of some gabbroic and diabasic magmatic rocks.

Graphic texture is primarily represented by the occurrence of "graphic" quartz in either K-feldspar or sodic plagioclase (Figure 7.15a, p. 127), but other mineral pairs have the relationship. A **graphicgrain** has the graphic texture.

Vermicular texture is an including relation characterized by "wormlike" (**vermicular**) forms of a mineral phase in a larger host mineral. The composite is a **vermiculargrain**. The well-known variety of this texture, **myrmekite**, consists of vermicular quartz in a more dominant sodic plagioclase host (Figure 7.15c). A vermicular relation between clinopyroxene and plagioclase is shown in Figure 7.15d, in which the two

phases are about equal volume. A vermicular relation between biotite and quartz is shown in Figure 7.15e, which coexists with myrmekite. Some very fine examples of vermicular texture involve sulfides, among them chalcocite–bornite, krennerite–pyrrhotite, and oxide silicate pairs such as chromite in orthopyroxene (Ramdohr, 1981).

Coexistence and gradation of graphic and vermicular texture characterize the rock **granophyre**. For example, quartz in K-feldspar and plagioclase may have both straight and curved interfaces (Figure 7.15b). **Granophyric texture** is a term that generalizes the coexistence of graphic and vermicular morphologies. A **symplectic** relationship refers to intergrowths of two minerals without making morphological distinctions. The term **eutectoid** refers specifically to pairs that are known to have crystallized simultaneously at a eutectic (Chapter 13).

Cellular texture is an including relation that applies to incomplete crystals resulting from cellular growth (Chapter 14) (both biotic and abiotic) or partial dissolution. Plagioclase may be either boxy cellular (Figure 14.5a) or spongy cellular (Figures 5.1, 14.11, 19.6b). Cellular olivine and pyroxene are common in quenched volcanic rocks. Cells may also occur in noncrystal grains such as the chambers in corals (Figure 37.10b,c) and fusilinids (Figure 37.6b). A cell implies that "space" enclosed by crystalline or noncrystalline wall material is subsequently filled with crystalline or noncrystalline material such as glass, submicroscopic material such as micrite, organic material, epitaxial crystals (K-feldspar in plagioclase cells), and nontaxial crystals (quartz in K-feldspar cells). Petrified wood is a good example of floral cell filling. The definition of cellular-including cannot be purely descriptive because recognition of the cellular form as deriving via growth or resorption is an interpretative exercise. Examples of cellular-including texture are sketched in Figure 7.13.

Lamellar, lensoid, and **rod textures** apply to mineral pairs resulting from exsolution in solids (Figure 7.13), simultaneous crystallization in liquids, and reactive replacement. **Perthite** (Figure 7.16a, p. 128) and **antiperthite** are well-known examples resulting from exsolution.

Superposed texture having an inclusion configuration (Figure 7.13) typically results from pseudomorphic or partial pseudomorphic replacement of one material by another. For example, calcic cores and zones of plagioclase commonly are replaced by epidote, calcite, or sericite (Figure 34.16a). Sericitization of andalusite may be partial or complete. Interpretation is required in cases where the genetic history may not be clear. Transition of superposed-including to transecting texture is one such situation.

Mantle Texture. A **mantle** relationship most commonly is characterized by the occurrence of one mineral

continued on p. 126

Figure 7.11
Crystal twinning.

(a) Combination Baveno right and Baveno left twins in alkali feldspar of two-feldspar granite porphyry. Arrow points to microvein discussed in Chapter 27. X-polars. Bar scale = 1.5 mm. (b) Manebach twin of alkali feldspar with typical "herringbone" pattern of exsolved albite. X-polars. Bar scale = 0.7 mm. (c) Chessboard albite twinning consisting of albite and pericline twins. X-polars. Bar scale = 0.7 mm. (d) Wedged-shaped twinning in birefracting calc-garnet (hydrogrossular). X-polars. Bar scale = 0.7 mm. (e) Microcline-type albite and pericline twinning in K-feldspar. X-polars. Bar scale = 0.18 mm. (f) Penetration twin of augite. Note how one portion of twin (dark individual) is apparently diminished (due to section view), simulating replacement by the other. X-polars. Bar scale = 1.5 mm. (g) (100) twinning of hornblende shown in relation to (1) the two directions of (010) cleavage (124°–56°) and the typical amphibole prism cross section (upper) and (2) a section approximately parallel to (010) (lower). X-polars. Bar scale = 0.6 mm. (h) Multiple (100) twinning in clinopyroxene showing angular relation to near-right-angle pyroxene cleavage. Note mantling hornblende (arrow). X-polars. Bar scale = 0.18 mm. (i) Cyclic zoning of cordierite. X-polars. Bar scale = 0.7 mm. (j) Multiple-direction twinning in leucite. Note similarity of twin lamella shape to microcline twinning. X-polars. Bar scale = 0.7 mm. (k) Repetitive (100) twinning in wollastonite in relation to (1) end view (010) and well-developed cleavage (right) and (2) elongate section (left) (arrows point to two twin planes). X-polars. Bar scale = 0.18 mm.

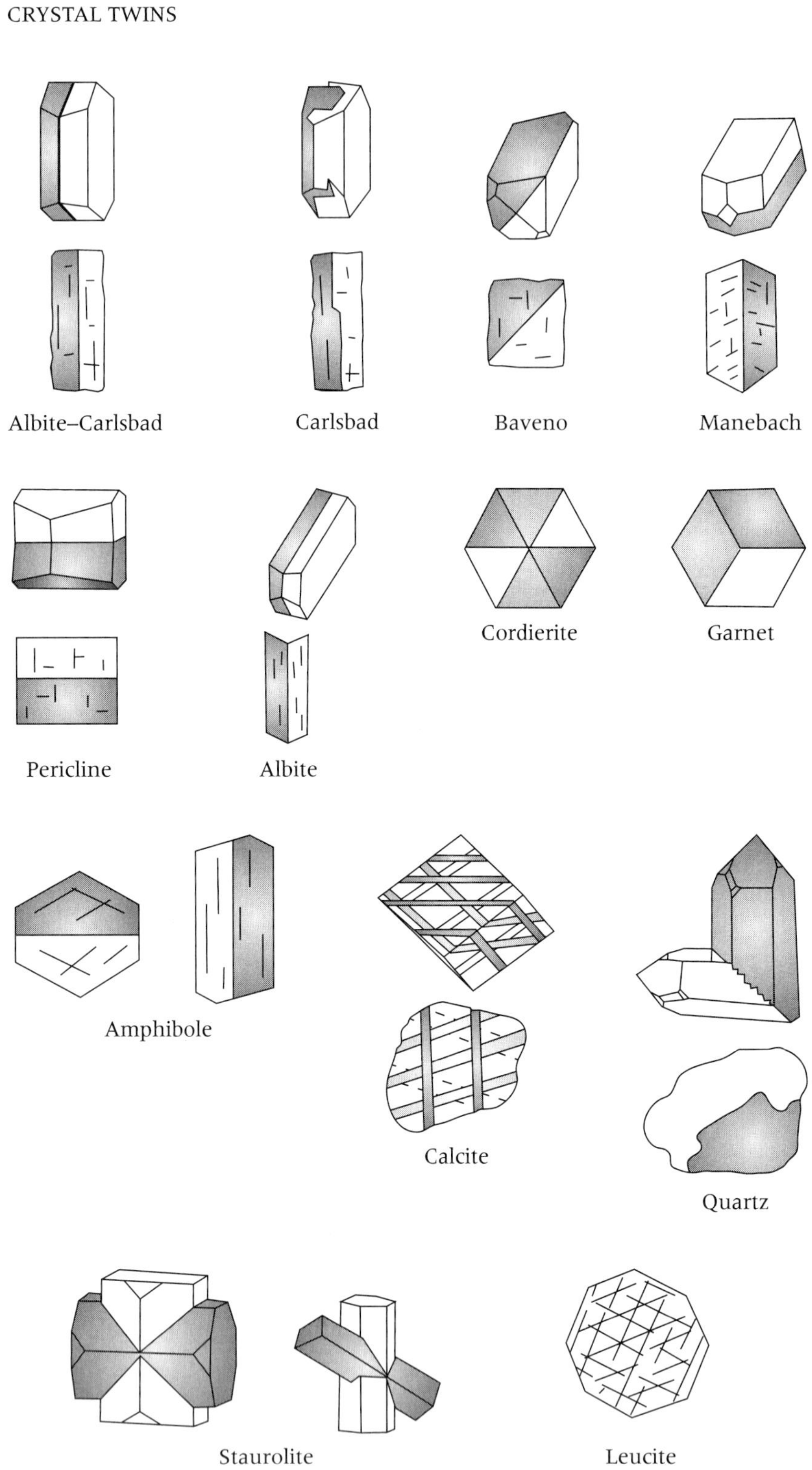

122   Part 2   The Anatomy of Rocks

CRYSTAL TWINS

Figure 7.12
Sketches of some of the more common twinned crystals.

Albite–Carlsbad   Carlsbad   Baveno   Manebach

Pericline   Albite   Cordierite   Garnet

Amphibole   Calcite   Quartz

Staurolite   Leucite

Figure 7.13 →
Sketched examples of grain relations classified as granular, intergranular, transecting, including, and mantled.

7 Physical Attributes of Rocks

GRAIN RELATIONS

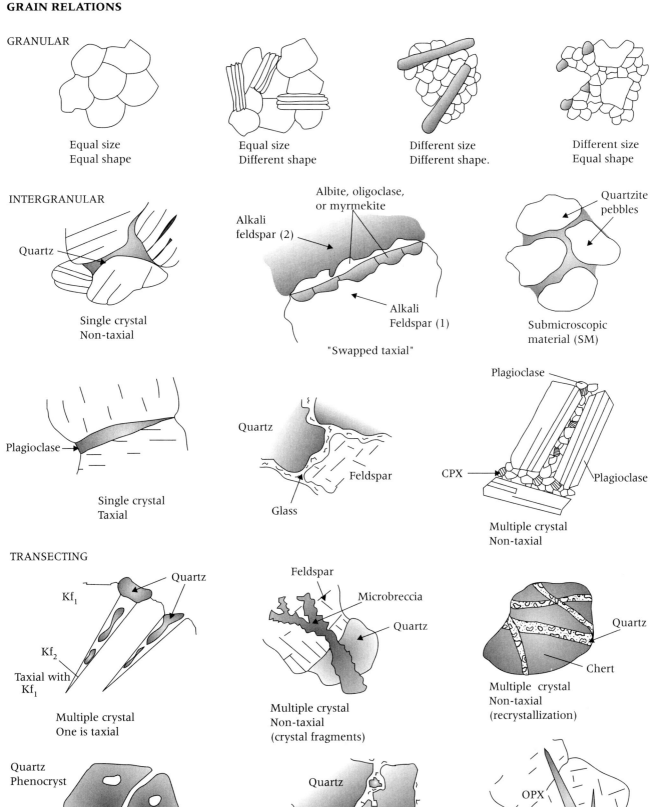

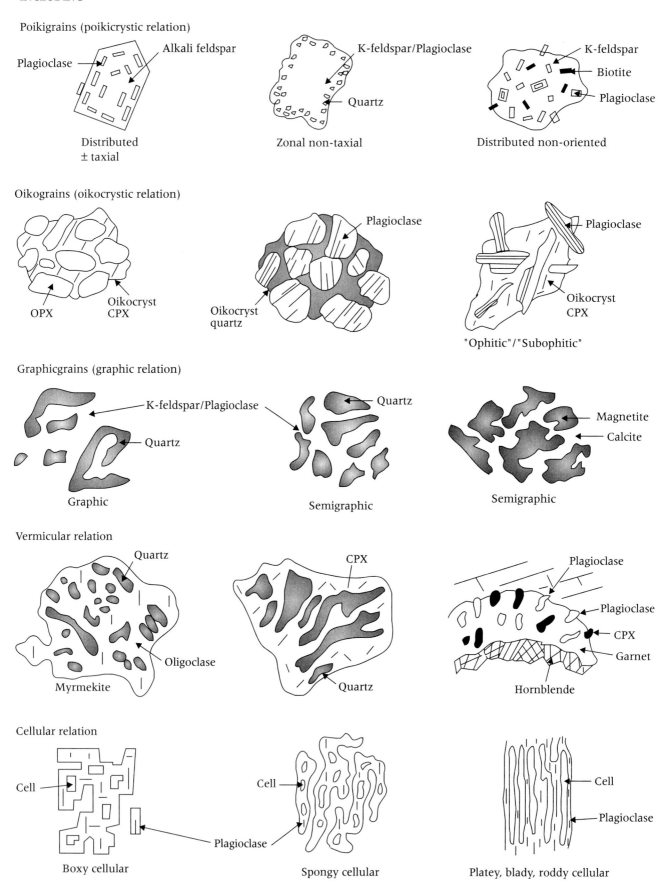

Figure 7.13 continued

Lamellar, lensoid, rod relations

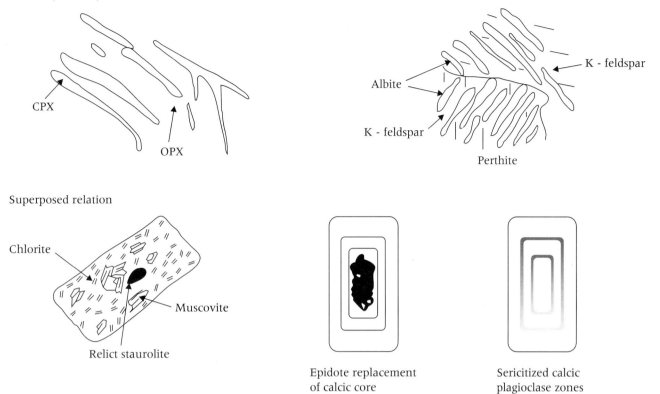

Superposed relation

MANTLING

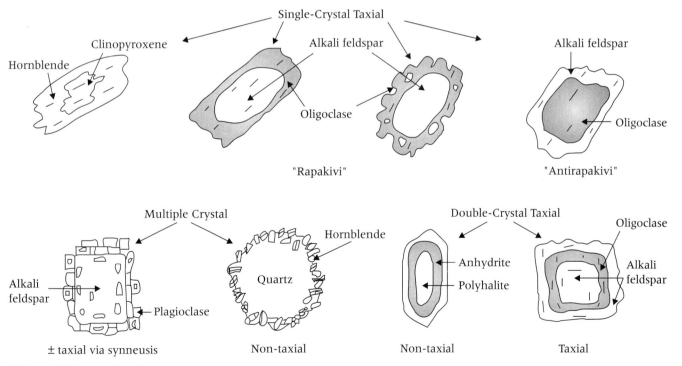

Figure 7.13 continued

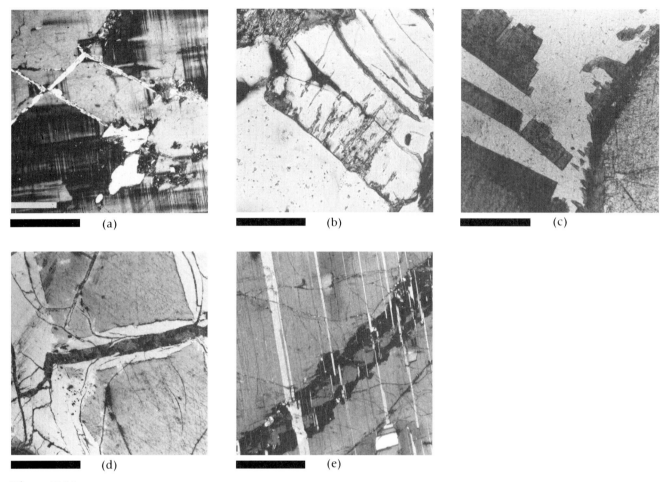

Figure 7.14
Transecting grain relations.

(a) Sericitic white mica in conjugate fracture set in microcline. X-polars. Bar scale = 0.7 mm. (b) Chlorite at rim and in fractures of cordierite (upper right). Lack of alteration in adjacent quartz (lower left) provides useful distinguishing criteria. X-polars. Bar scale = 0.18 mm. (c) Zeolite along cleavage and surrounding Ti-biotite in shonkinite. Note aegirine mantle on aegirine–augite (lower right). PPL. Bar scale = 0.18 mm. (d) Crosscutting and peripheral replacement of bornite by covellite (black) and idaite (white borders). Irian Jaya, Indonesia. Reflected light. Bar scale = 0.7 mm. (e) Albite microvein transecting more calcic plagioclase. Eucrite, Ardnamurchan, Scotland. X-polars. Bar scale = 0.6 mm.

phase rimming (mantling) another (Figure 7.13). The mantling phase may be a single crystal or a cluster of crystals. There may be an epitaxial relation between mantle and host core phase, especially if there is crystallographic similarity (feldspar on feldspar), or there may be no crystallographic continuity if crystal structures are very dissimilar (hornblende on quartz). Epitaxial single-crystal mantles are the most common. The well-known relation between clinopyroxene and mantling clinoamphibole is shown in Figure 7.11h. An aegirine mantle on aegirine–augite is shown in Figure 7.14c. The famous rapakivi mantling, in which oligoclase is mantled on K-feldspar, is shown in Plate 5a, and the antirapakivi relation (K-feldspar on plagioclase) is shown in Plate 5b.

Fabric

Fabric is an expression of **location** and **orientation** of rock-forming materials in rocks (Figure 7.17; p. 129). If there is no preferred location and no preferred orientation, the fabric is **nondirectional**. If there is pre-

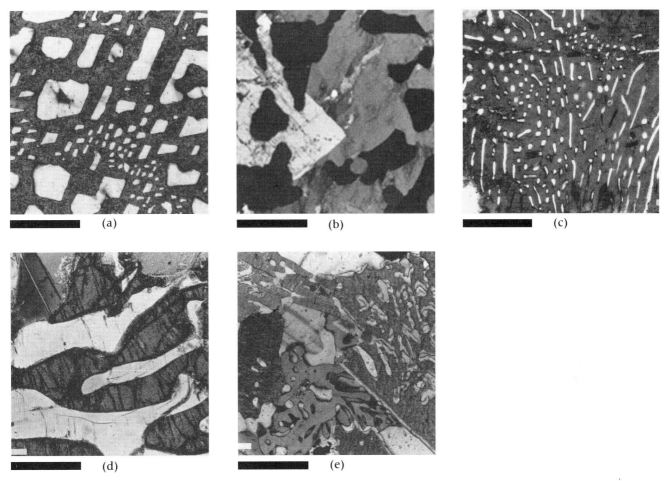

Figure 7.15
Graphic- and vermicular-including grain relations.

(a) Graphic quartz in K-feldspar. X-polars. Bar scale = 0.18 mm. (b) Semigraphic (graphic–vermicular) quartz in K-feldspar (right) and plagioclase (left). X-polars. Bar scale = 0.7 mm. (c) Vermicular quartz in oligoclase (myrmekite). X-polars. Bar scale = 0.18 mm. (d) Vermicular plagioclase in clinopyroxene. X-polars. Bar scale = 0.18 mm. (e) Vermicular quartz in biotite (right and upper left) and vermicular quartz in plagioclase (myrmekite) (lower left). X-polars. Bar scale = 0.18 mm.

ferred orientation, preferred location, or some combination of the two, the rock has a **directional** fabric.

Rock-forming materials contributing to **preferred location** are of two general types; (1) compositional and (2) physical. Preferred location of hornblende with respect to feldspar and quartz in a layered gneiss is a compositional feature. Color layering as controlled by trace amounts of hematite in sandstones or magnetite in obsidian is also a **compositional preferred location.**

Physical preferred location includes bedding and laminations of sedimentary rock in which **grain size** is the dominating control. There may also be differences in **grain shape** that have preferred location. Elongate minerals such as amphiboles and platy minerals such as micas may locate preferentially apart from equidimensional minerals such as quartz and feldspars, but this is also a compositional preferred location. Most examples of preferred location relate to differences in grain composition and grain morphology.

Preferred orientation of rock-forming materials relates to the orientation of inequant grains. Preferred **dimensional orientation** may or may not have **preferred crystallographic orientation**. For example, quartz in some metamorphic tectonites may have preference for crystallographic c to lie normal or

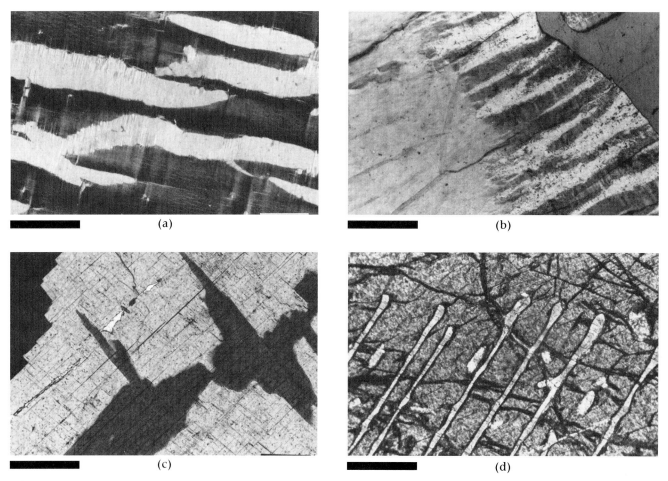

Figure 7.16
Lamellar-including grain relations.

(a) Albite lamellae in microcline (perthite). X-polars. Bar scale = 0.17 mm. (b) Albite lamellae in marginal portion (microcline) of alkali feldspar crystal, indicating preferential exsolution. X-polars. Bar scale = 0.7 mm. (c) Patchy lamellar K-feldspar in sodic plagioclase (antiperthite). X-polars. Bar scale = 0.17 mm. (d) Cpx exsolution lamellae in Opx host. X-polars. Bar scale = 0.17 mm.

subnormal to a plane of flattening (Figure 6.3). Such quartz grains typically also have grain dimension preferred orientation with the shortest dimension of the grain lying in the plane of schistosity and the longest dimension defining a lineation (stretching lineation). Mineral grains such as amphiboles and micas commonly have preferred dimensional orientation in rocks, but since the shape of these grains is a reflection of their crystal structure the preferred orientation is also crystallographic. This orientation was not that of the stretched and flattened quartz previously described. Dimensional orientation may also be defined by noncrystal grains such as rock fragments in clastic sedimentary rocks and pyroclastic rocks.

Preferred orientation and preferred location of rock-forming materials have two fundamental structural expressions. **Planar fabric** derives from preferred locations and preferred orientations of platy or tabular dimensional grains. **Linear fabric** derives from preferred orientation of elongate grains. Tabular grains (or triaxial ellipsoids) can contribute to both planar and linear aspects of rock fabric.

Figure 7.17
Sketches of example rock fabrics classified according to nondirection and directional fabric elements.

7 Physical Attributes of Rocks 129

FABRIC
NONDIRECTIONAL

Single-crystal equant grains
Quartz sand

Polycrystalline equant unit grains (quartzite pebbles)

SM equant unit grains (biopelloids)

SM equant unit grains (spherulitic rhyolite)

Single-crystal inequant (magmatic feldspar)

Polycrystalline or SM inequant grains (cataclasite)

Unit grains inequant (fusilinids)

PREFERRED ORIENTATION

Crystals (mica schist) ± folded

Crystals (phenocrysts)

c axis
Inequant oriented crystal grains (quartz)

Inequant polycrystalline grains (metaconglomerate)

Inequant unit grains (glass shards)

Inequant polycrystalline or SM unit grains (cataclasite)

Inequant unit grains (flattened oolite)

PREFERRED LOCATION

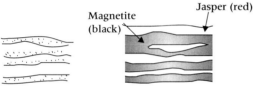

SM grains (water-lain tuff)

Magnetite (black), Jasper (red)
"Banded" ironstone

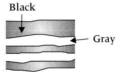
Black, Gray
Layered obsidian

Lisagang rings (limonite in rhyolite)

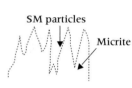
SM particles, Micrite
Stylolite

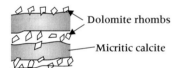
Dolomite rhombs, Micritic calcite
Partly dolomitized micrite

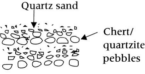
Quartz sand, Chert/quartzite pebbles
Grain size grading

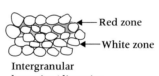
Red zone, White zone
Intergranular hematite / limonite (quartz arenite)

Point aggregates (sparry calcite in micrite)

PREFERRED ORIENTATION AND PARALLEL PREFERRED LOCATION

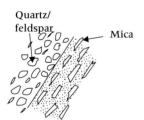
Quartz/feldspar, Mica
Layered schist

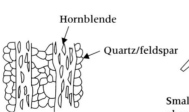
Hornblende, Quartz/feldspar
Layered gneiss

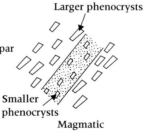
Larger phenocrysts, Smaller phenocrysts
Magmatic layering

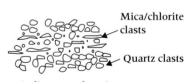
Mica/chlorite clasts, Quartz clasts
Sedimentary layering

PREFERRED ORIENTATION AND PERPENDICULAR PREFERRED LOCATION

Comb layering (magmatic)

Orbicule (magmatic)

Layered chalcedony

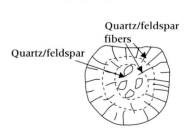
Quartz/feldspar fibers, Quartz/feldspar
Spherulite rhyolite

8

Textural Interpretation

It is testimony to the marvelous adaptability of geological observations that well exposed and thoroughly studied localities can provide competent petrologists with evidence for diametrically opposed conclusions. Whatever the popularly accepted theory of the moment may be, one can usually find evidence to support it.

Alexander R. McBirney (1979)

The Pleasure and Anguish of Textural Interpretation

The most enjoyable part of petrologic investigation is in construction of genetic models. Such models must be made with limited, sometimes outright skimpy data, but this is the challenge and intrigue of research—this is when we have the chance of being right and take the risk of being wrong. Rock-forming processes are varied and complex, yet in any given set of circumstances, for a given rock, there can be only one set of processes by which the rock formed. This is worth reflecting on when interpretation of geologic data seems hopeless and we catch ourselves saying: "the rock could have formed in a number of ways."

The textural features of rocks are a record of rock-forming process "written in the stone" (as Peter Misch would say, quoting his mentor V. M. Goldschmidt). Textural relations are a form of writing, in a sense, but our deciphering of this script is soberly primitive. Witness the history of interpretation of the simple intergrowth of quartz and plagioclase known as **myrmekite**. The paper by Evans (1974) entitled "Myrmekite One Hundred Years Later" attests to our inability to successfully explain what "should" be a relatively simple relationship. In view of our humble position as petrologists, our interpretation of rock-forming process is necessarily tentative and guarded. Even when we develop a genetic model that we are "sure" is correct, we have learned not to use words such as "never," "always," and "proof" when the model is discussed because exceptions have an ugly way of surfacing sooner or later.

Method of Textural Analysis

Any method of data analysis is useful as long as it leads to construction of genetic models that are close to the "truth." We like to construct models that are "reasonable" in the light of what we have learned about geological processes. However, a model that is later shown to be close to the truth may at first seem to be "unreasonable," but as long as the data is consistent with the model it should be acceptable. What is popular and unpopular can have little to do with the truth. Witness Wegeners' bold position with respect to his continental drift model, and Bunsen's (1851) "preposterous" view that two magmas of entirely different composition and viscosity can mix to form an homogeneous hybrid magma.

There are two methods of data analysis: **empirical observation** and **rational thought** (Mackin, 1963). Empirical observation typically consists of a plot of data with respect to two variables, such as frequency of occurrence of a particular texture against rock composition. The results of such a "fishing expedition" may reveal some genetic tie between texture and rock composition, providing a shortcut in the process of genetic modeling. Empirical observation and application to a problem also includes the principle of **uniformitarianism** ("the present is the key to the past") (James Hutton and Charles Lyell, eighteenth and nineteenth centuries). Here it is possible to stumble onto a rock-forming process in progress that nicely applies to rock preserved in the geologic record. Certainly the present is the key to the past in many situations, such as sedimentation and volcanism, but it is not useful, for example, in relating low-oxygen atmospheres to weathering phenomena in the pre-Cambrian nor to the formation of myrmekite in granites deep in the Earth's crust.

Rational thought typically utilizes the **method of multiple working hypotheses** (Chamberlin, 1897; J. G. Johnson, 1990; Railsback, 1990). Rational thought depends on our knowledge, experience, and on how clever we can be. Rational thought is "thinking through" a problem, accepting what is most reasonable and rejecting the less reasonable, more-or-less following the "method of multiple working hypotheses." Thinking through a problem requires knowledge. We either try to make an analogy with experimental results or to active processes occurring in the field (uniformitarianism), or we try to imagine processes based on our knowledge of chemistry and physics.

Textural Interpretation Based on Mineral Assemblages

Reactions between solids or between solids and fluids produce new minerals that coexist in some textural configuration. The types of minerals involved and their distributions can be all that is needed to explain the textural relation.

For example, what is the origin of the texture consisting of small quartz grains included in staurolite (Figure 21.35a)? There needs to an explanation of what originally occupied the volume now occupied by staurolite. It is well known that staurolite can grow at the expense of biotite and muscovite by reactive replacement in response to elevated temperatures. This explains why there are no inclusions of biotite or muscovite in the staurolite since these minerals are the reactants used up where staurolite crystallizes. Quartz remains in the staurolite as inclusions because it is not used in the reaction. But quartz grains outside the staurolite are larger (Figure 21.35a), suggesting that either (1) the interior grains are partially dissolved or (2) the coarseness of matrix quartz is due to secondary recrystallization. In case (2), enlargement of quartz within staurolite is not possible because it became isolated from silica sources. In summary, the nature and distribution of mineral phases are the clue to textural interpretation in many cases.

Two-Dimensional Views of Three-Dimensional Objects

Observation of textural relations in thin section is limited to two dimensions. Intersection of interfaces between grains can yield apparent relationships that are susceptible to misinterpretation (Figure 8.1). Some misleading relationships can be resolved by locating a third-dimensional view of the same grain relation in the same thin section or in sections cut in different directions.

If the interface between grains is not planar, a section may produce a false including relation of one grain in another (Figures 8.1a, 8.2). Similarly, sections through skeletal or cellular grains may yield what appear to be inclusions in the grain (Figures 8.1b–d). Elongate inclusions such as graphic quartz in feldspar appear completely different, depending on section direction (Figures 8.1e, 33.4a). Patch perthite appears as lenticular perthite in cross section (Figures 8.1f, 2.6c). A false sense of replacement can occur as the result of matrix minerals overlapping quartz grains within the thickness of a thin section (McBride, 1987).

Sections through zoned crystals can present misleading zoning configurations (Figure 8.1g). Zoning of magmatic plagioclase is a powerful record of magmatic consolidation "written in the crystal," and apparent zoning configurations generated by section direction must be recognized as such. Reasonable estimates of average composition of zoned plagioclase can only be made if sections cut the core of the plagioclase crystal.

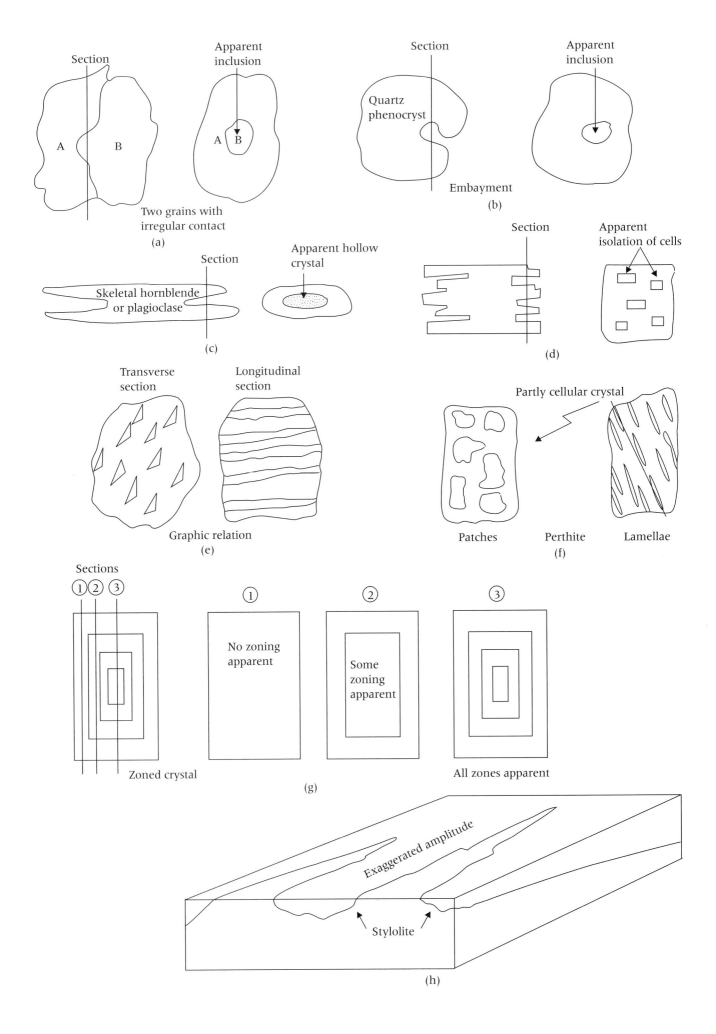

8 Textural Interpretation

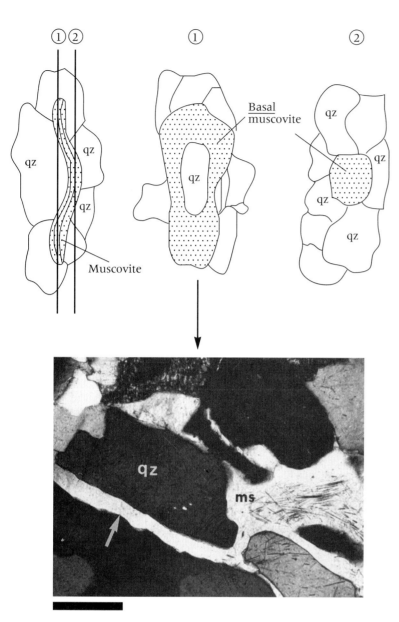

Figure 8.2

(a) Sections through a quartz–muscovite metamorphic rock in which the muscovite crystal has been wrapped around quartz. Section (1) presents an apparent inclusion of quartz in a larger muscovite crystal, whereas section (2) presents a side-by-side grain relation. (b) Example of relation (1) in which an including relation is realized as well as an intergranular relation of muscovite between quartz crystals (arrow). X-polars. Bar scale = 0.15 mm.

The amplitude of stylolites can be enormously exaggerated if the surface on which the reference material (clays, iron oxides, carbonaceous remains) is intersected at a low angle (Figure 8.1f).

A double mantle on a sanidine phenocryst can present a paradoxical sequence of mantling (Figure 8.3). In one section, fine spongy cellular plagioclase is mantled by a coarse boxy cellular plagioclase mantle. In another section, the fine spongy type appears to mantle the coarse boxy type, and in yet another the spongy type is absent. The sequence of this double mantle on the sanidine host is crucial to its interpretation. An interpretation based on an assumption that the sequence reversal is real would be totally different from an interpretation based on the knowledge that the sequence reversal is only apparent. This means that a single thin section might present a textural relation that is entirely misleading.

Figure 8.1

Real and apparent textural relations between and in rock-forming grains as controlled by location and direction of thin sectioning.

134 Part 2 The Anatomy of Rocks

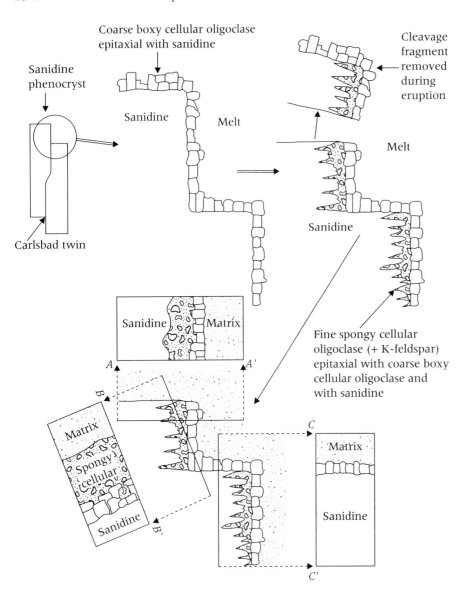

Figure 8.3
Three different apparent mantling relations generated by the sectioning of a single sanidine phenocryst.

Sections A-A' and B-B' have opposing sequences of mantling, and section C-C' has omitted the fine spongy cellular plagioclase.

The presenting intersection angle of more than two cleavage or two twin directions is fundamentally dependent on section direction. The characteristic 90° cleavage of pyroxenes appears more like the 124°/56° cleavage characterizing amphiboles if intersection is at an intermediate angle to the length of the prism. Intersection of albite twin lamellae and coexisting pericline twin lamellae may seem to be at an "impossible" angle in some sections (Figure 8.4). Trace of cleavage planes may not be visible and trace of twin planes are at least somewhat "fuzzy" if sections are at low angle to these planar surfaces. Intersection of a single grain may generate shapes that are not characteristic of the grain as a whole. A triangular-shaped crystal can result from sectioning across the corner of a crystal (Figure 8.4a).

First Impressions, Comparative Anatomy, and the Illusions of Nature

Tree Branches or Shrimp Burrows?

First impressions of rock genesis may be embarrassingly incorrect. Consider certain fossil remains (Figures 37.13a,b) occurring in the Cliff House Sandstone (Upper Cretaceous) of the Mesa Verde and Chaco Canyon regions of the Colorado Plateau. For many years these remains were thought to be fossilized tree

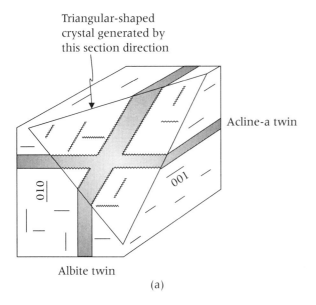

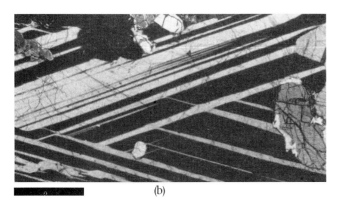

Figure 8.4

(a) Apparent acute angle of intersection of albite and acline-A twins in plagioclase generated by corner section direction. Shape of crystal generated by section direction is triangular (an apparent morphology). (b) Example of apparent twin intersection. X-polars. Bar scale = 0.7 mm.

branches (Barry S. Kues, personal communication). A paper by Weimer and Hoyt (1964) describes shrimp burrows in the coastal sands of Georgia that are "identical" to the "branches" in the Cliff House Sandstone. The burrows bifurcate in branchlike fashion and have a central open tube not unlike the core region of some plants. The knobby exterior of the burrows was thought to be unusual but perhaps characteristic of a certain Cretaceous flora. This situation is a graphic example of misleading first impressions and a fine example of how the principle of uniformitarianism can save the day.

Reactive Replacement or Direct Precipitation?

Albite, oligoclase, and quartz-bearing varieties (myrmekite) commonly occur along the interface between alkali feldspar crystals in granitic rocks. The plagioclase crystals typically have an epitaxial relation to the adjacent host crystals. If there is a single row of crystals presenting, the epitaxial relation is to only one of the adjacent alkali feldspar grains. If there is a double row of crystals, the epitaxial relation involves both alkali feldspar grains (Figure 8.5). Surprisingly, as seen in some section views, in the case of the double row, crystals of one row are epitaxial with respect to the host alkali feldspar grain on the other side of the other row, not epitaxial to the alkali feldspar grain with which they are in direct contact (Plate 4c, Figure 8.6, 8.7a). This "swapped" relationship has been noted by Ramberg (1962), Phillips (1964), and Phillips et al., (1972). The eye-catching relation has a simple explanation if the direction of sectioning is taken into consideration (Hibbard, 1979; Figure 8.6). The "swapped" relation does not appear if the double row of crystals occurs as an intracrystalline microvein in an alkali feldspar crystal (Phillips et al., 1972; Hibbard, 1979). The "swapped" relation can still occur (Hibbard, 1979; Figures 8.5b, 8.7b) but it is not expressed by difference in optical orientation since the two portions of the host grain have the same structural orientation.

Polycrystalline "microveins" can locate either between grains (intercrystal) or within grains (intracrystal) (Hibbard, 1980). They can form by at least five different processes: (1) filling of open space, (2) relegation of fluid to intergranular position as crystals enlarge, (3) filling of potential space (extension with simultaneous fluid infiltration, applying to intragranular fracture), (4) reactive replacement, and (5) melting followed by crystallization of melt. The first option cannot occur at high pressures where myrmekite-bearing rocks form. A model of precipitation of myrmekite from fluids (Hibbard, 1979) between alkali feldspar grains and in a fracture within a single alkali feldspar grains in shown in Figure 8.5b. This model does not require replacement nor melting of alkali feldspar. It only requires that there be intergranular and intragranular silicate melt present.

Nevertheless, there are many examples of reactive replacement that are morphologically similar to precipitations between grains and within fractured or cleaved grains. Reactive replacement along a high-energy zone such as a cleaved crystal proceeds bilaterally away from the initial cleavage plane into the adjacent halves of the host crystal. Such a relation is shown in Figure 8.8, where simultaneous dissolution–crystallization has

continued on p. 138

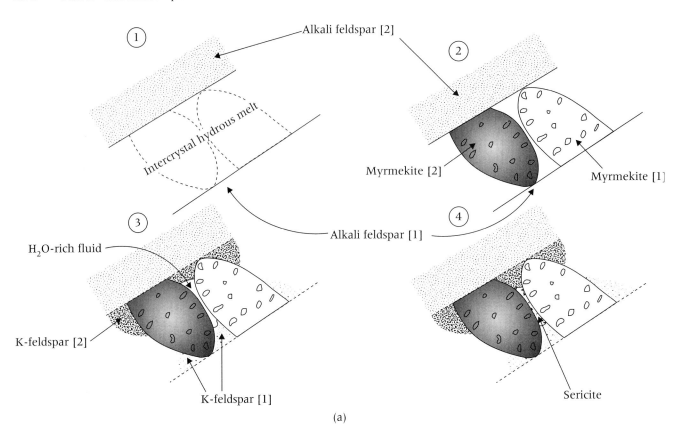

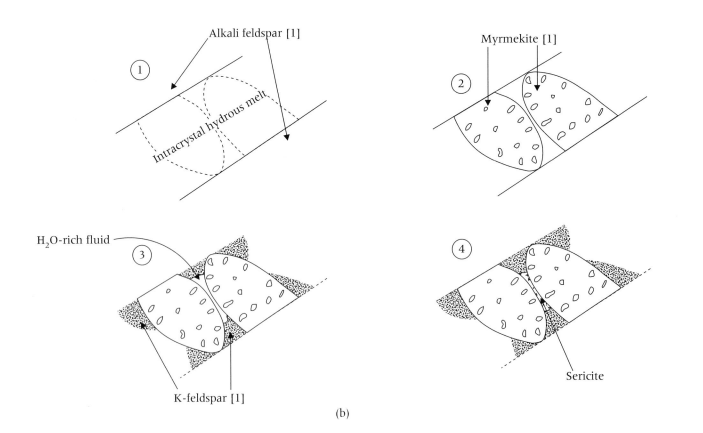

Figure 8.5

Sequential development of epitaxial myrmekite lobes between (a) two alkali feldspar crystals, and (b) two parts of a single alkali feldspar crystal.

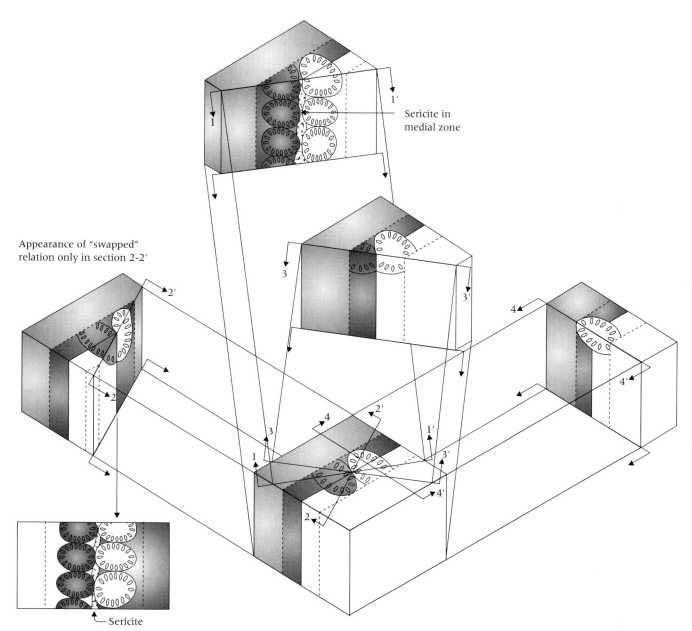

Figure 8.6
Diagrammatic presentation of section views through myrmekite intergranular between two differently oriented alkali feldspar crystals. Section 2–2' generates the "swapped" relation shown in Figure 8.7 and Plate 4c. Section 1–1' and redrawn section 2–2' show the location of additional lobes of myrmekite and the medial position of sericitic white mica.

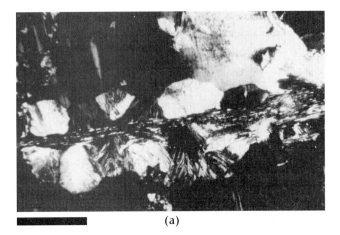
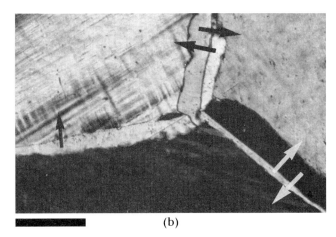

Figure 8.7
(a) Two rows of myrmekite lobes with a medial zone containing sericite in a single host microcline. The lobes are convext toward host microcline. See Figures 8.5b and 8.6 for explanation. X-polars. Bar scale = 0.6 mm. (b) The apparent "swapped" relation of intergranular sodic plagioclase with respect to three microcline crystals. Arrows begin in plagioclase grain and end in host microcline with which it is epitaxial. In one case, there is only one intergranular plagioclase grain between microcline grains (left). X-polars. Bar scale = 0.6 mm.

advanced outward from a medial plane that was the original cleavage surface in a clinopyroxene (Hibbard and Sjoberg, 1994). A similar relation, involving reactive replacement without melting, is shown in Plate 6c. Here serpentine replacement has advanced bilaterally into olivine from what probably were fractures in the olivine.

The medial line (plane intersection) between rows of sodic plagioclase and rows of myrmekite lobes (Figures 8.5b, 8.7a) is similar to the replacement relations in the clinopyroxene and olivine described earlier. However this comparison of textural anatomy does not work. There "appears" to be replacement of alkali feldspar by myrmekite, but the relations in Figures 8.5a, 8.6 indicate that a non-replacive origin is a viable option.

The occurrence of white mica in the medial zone of the double row of myrmekite lobes (Figure 8.7a) indicates that fluids are present at a very late stage, converting feldspar to mica. If sericitization is essentially contemporaneous with the myrmekite, the location of the sericite in the medial zone has a simple geometric explanation (Figures 8.5a, 8.6). If sericite had been involved in a reactive replacement process, instead of the non-replacive model shown, it could be expected to occur at the reaction front, along the curved surfaces of the myrmekite lobes. However, there is no sericite in this position.

Myrmekite facing alkali feldspar crystals in deformed granitic rocks is also controversial (Chapter 27), and consideration of the apparent relations presenting in two-dimensional thin-section views is relevant. One point of view is that the myrmekite crystallizes from late-stage hydrous magma along with K-feldspar in the pressure shadow region associated with alkali feldspar and plagioclase crystals that have crystallized at an earlier magmatic stage (Hibbard,

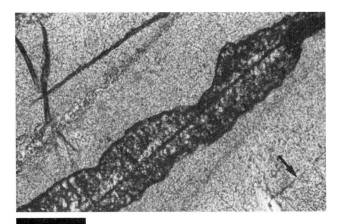

Figure 8.8
Example of dissolution–crystallization proceeding laterally along somewhat wavy front into a crystal from a medial cleavage plane. Parallel cleavage without reaction is shown by the arrow. Spongy-cellular clinopyroxene has been generated from original magmatic clinopyroxene of slightly different composition (see Hibbard and Sjoberg, 1994). PPL. Bar scale = 0.15 mm.

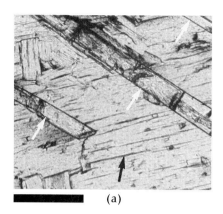
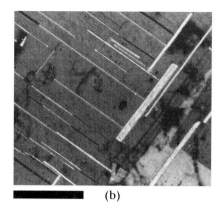
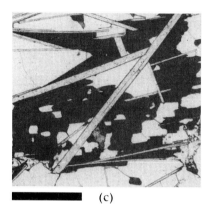

Figure 8.9
Textural relations useful in comparative textural analysis.

(a) Two sillimanite crystals (white arrows) lying at high angle to muscovite cleavage (black arrow). PPL. Bar scale = 0.18 mm. (b) White mica blades in plagioclase cleavage indicating cleavage-controlled fluid penetration and reaction. X-polars. Bar scale = 0.18 mm. (c) Crosscutting relations between muscovite, biotite, and quartz suggestive of simultaneous crystallization. X-polars. Bar scale = 0.7 mm. See discussion in text.

1987). It has been observed that rigid crystals of the earlier alkali feldspar and plagioclase do not occur in some of the myrmekite-lined K-feldspar augen (Simpson and Wintsch, 1989; Vernon, 1991a), suggesting that relocation of fluids in pressure shadows does not generally apply. However, a view of the relations in three dimensions (Figure 27.4d) provides an explanation why rigid crystals do not appear in some views of myrmekite-lined K-feldspar augen, locations that do have pressure-shadow relations to earlier-formed crystals.

Sequential or Simultaneous Crystallization?

Prisms of sillimanite lying across a muscovite crystal at high angle to muscovite cleavage are shown in Figure 8.9a. The first impression might be that the sillimanite has formed at the expense of muscovite in response to increasing grade of metamorphism in accord with the following well-known equation:

$$2 \text{ muscovite} + 2H^+ \rightarrow 3 \text{ sillimanite} + 2K^+ + 3 \text{ quartz} + 3H_2O$$

Such a reaction is presumed to initiate by the presence of at least an intergranular film of aqueous fluid. This fluid may have entered the muscovite along fractures that happened to have developed across the cleavage direction. More aqueous fluid is generated as dehydration reaction proceeds. Potassium ions and silica would be removed in the throughgoing aqueous fluid.

The problem with this reactive replacement interpretation, consisting of sillimanite forming at the expense of muscovite, is that there is no evidence of fractures (or healed fractures) in the muscovite beyond the ends of the sillimanite crystals, and there is no sillimanite produced along muscovite cleavage. In contrast, muscovite has clearly replaced plagioclase along two directions of plagioclase cleavage and twinning (Figure 8.9b). By comparison, the sillimanite–muscovite begs another explanation.

A crosscutting relation between muscovite, biotite, and quartz, comparable to the sillimanite–muscovite crosscutting relation, is shown in Figure 8.9c. Here the crosscutting relations appear to be indiscriminate. In particular, there is a large muscovite crystal that cuts across biotite and several quartz crystals, not unlike the sillimanite that crosscuts the large muscovite (Figure 8.9a). In the case of the biotite and muscovite, there is no compelling reason, such as a change of temperature, for the muscovite to form at the expense of biotite; both minerals can be stable together through a wide range of temperature and pressure conditions (Chapter 21). Furthermore, reaction of biotite to muscovite along biotite cleavage would be expected (cf. Figure 8.9b) if conditions did favor a biotite

→ muscovite reaction, common in retrograde metamorphism and hydrothermal alteration. Although some muscovite is parallel to biotite cleavage, most of it is not (Figure 8.9c).

Lacking a compelling reason for a biotite → muscovite reaction, it is likely that both biotite and quartz, plus muscovite, crystallized simultaneously. The micas asserted their strong crystallographic control over quartz. By this textual comparison, simultaneous growth of sillimanite and muscovite (Figure 8.9a) is indicated, with the sillimanite prisms being the dominant crystal form.

Part **III**

Classification of Rocks

Rock classification is a fundamental scientific exercise that groups similar rocks together in expectation that they have similar origins. Groupings are based on mineralogical and various physical attributes. Since classification is a scientific exercise, it is expected to follow the scientific method in which description precedes interpretation. Are shortcuts allowed? Some think not. ". . . little good can come when a science yields to the social pressure that rewards modeling and theorizing while scorning description without 'theory'" (Mandelbrot, 1983).

Petrographers are never completely satisfied with rock classifications because there is continual acquisition of new information that needs to be classified. How can rock classifications be constructed so that they may "grow with the science"? One way is to refine and expand the traditional classifications of igneous, sedimentary, and metamorphic rocks (Chapter 9). As an example, it was not until the 1960s that the distinction between pyroclastic flows and lava flows was clearly established. A subclassification of the igneous rocks was needed that allowed separation of the mechanical (pyroclastic) process from the lava crystallization process. This distinction has resulted in several versions of pyroclastic rock classification. Another, more recent, example of refinement and expansion is the classification of fault rocks. Such rocks originally were considered to be metamorphic rocks, not deserving of separate classification and not even of special consideration within the metamorphic rock classification. There are now classifications of fault rocks separated from metamorphic rock classifications.

Another way for classification to "grow with the science" is by way of an open-ended global rock classification (Carr and Hibbard, 1991, Chapter 10). This has the advantage of not being locked into traditional classifications in which there must be a general sense of rock genesis so that the appropriate classification can be selected. An igneous-looking rock is likely to be classified within the igneous rock classification even if it is later determined to have had no component of magmatic history. A global classification provides a vehicle for clear separation of description from interpretation because there is no choice of classification required at the outset.

A global classification also is more amenable to revision and expansion because the artificial partitions erected within and between rock classifications are eliminated. "Pigeon-hole classifications . . . are useful as aids to memory and for purposes of comparison, but they suggest clear-cut distinctions where none exist . . . " (Williams, Turner, and Gilbert, 1954).

A global rock classification is sensitive to classification of rocks that have formed by more than one process. There is no traditional classification for dual-process rocks such as a "metabasalt" or a "hydrothermally altered andesite," whereas a global classification can readily accommodate such rocks.

Many examples of rocks must be "forced" into a traditional rock classification, depending on the experience and prejudice of the classifier. Here are several examples in the form of rhetorical questions:

1. Should a fine-grained fully recrystallized carbonate rock occurring in a stratigraphic section with slaty shale and partially recrystallized quartz arenite be classified as a limestone (sedimentary rock implied) or a marble (metamorphic rock implied)?
2. Should granitic gneisses containing both metamorphic and magmatic features be classified as igneous or metamorphic rock? "To group with truly igneous rocks superficially similar products of metamorphism and metasomatism can only confuse an already difficult and complex natural situation by obscuring genetic implications otherwise obvious in common igneous assemblages" (Williams, Turner and Gilbert, 1982, p. 85).
3. Should "metarhyolite" be forced into the metamorphic rock classification because the youngest rock-forming process was partial recrystallization?
4. Is it useful in the genetic interpretation of granites to place a granite with 0% mafic minerals in the same classification pigeonhole with a granite containing 89% mafic minerals (IGU classification), or are these extremes so rare that they can be ignored?
5. A mass of granitic pegmatite occurs in sillimanite schist with no obvious relation to magmatic activity. Should the pegmatite be classified as an igneous rock or a metasomatic metamorphic rock?
6. How should migmatites be classified? Should the classification of migmatites themselves suffice, as a generalized rock-forming process, or should the magmatic and metamorphic processes involved in their formation be distinguished and classified?

The global rock classification presented in Chapter 10 is offered as one possible means of solution to these types of classification problems and others that are sure to arise in the future.

9

Traditional Rock Classifications

Classification, by whatever scheme, should conclude, rather than start, a petrographic inquiry.

(E-an Zen, 1988, Annual Review of Earth
and Planetary Science, *16:* 47)

Classifiable Rock Attributes

Grain Composition

Mineral Grains. Most rocks contain mineral grains. These minerals are indirect indicators of rock chemistry and of pressure–temperature conditions that prevail during rock formation. Such information is genetically important and the reason that mineral content is fundamental to most rock classifications. The **mineralogical mode** is a volume percent expression of mineral content. Proportions of mineral types are the basis for groupings within classifications.

Very fine-grained rocks, including most volcanic rocks and all mud rocks, are not adaptable to mineralogical classification because the mineralogical mode cannot be accurately determined. As a consequence, classification of such rocks is based on bulk chemistry. Nevertheless, chemical analyses of very fine-grained and glassy igneous rocks can be converted into a **mineralogical norm**, consisting of a calculated assemblage of minerals, that approximates but does not equal the mineralogical mode. Norm calculations and their variations have been applied mainly to igneous rocks (Cross et al., 1902; Niggli, 1936; Kelsey, 1965), but there is a calculation for metamorphic rocks as well (Barth, 1962).

Nonmineral Grains. Many sedimentary rocks and pyroclastic rocks, among others, contain nonmineral grains, including crystalline and noncrystalline lithic fragments. Siliciclastic sedimentary rocks typically contain crystalline fragments such as chert and quartzite, but coals contain noncrystalline grains known as "macerals." Pyroclastic volcanic rocks are characterized by glass shards and pumice fragments that can be considered as noncrystalline lithic fragments. Other grains in rocks are not lithic fragments; that is, they are not characterized by derivation from pre-existing rock or magma. For example, oolites and peloids are important constituents of some carbonate sedimentary rocks and are nonmineral grains (not single crystals or crystal fragments), typically consisting of fine crystalline calcite, aragonite, or dolomite.

The volume percent of nonmineral grains is a primary parameter in the classification of sedimentary rocks because the types of grains present are an important indicator of sediment source (provenance) and of postdepositional lithification processes (diagenesis).

Grain Size

By far the most common physical attribute of rocks used in classification is grain size. The reason is that most rocks consist of grain aggregates and their sizes are relatively easy to measure. Grain size has considerable genetic significance. For example, it is an expression of the energy environment in sedimentation and of nucleation density in magmas and solutions.

Magmatic rock classifications have only a fine-grained and coarse-grained category, intended to be more-or-less equivalent to rates of nucleation in volcanic and plutonic environments. Grain size is the basis of distinction between mud rocks, siltstones, sandstones, and conglomerates in clastic sedimentary rock classifications. Metamorphic rock classifications include expression of grain via rock name; hornfelses are relatively finegrained and granofelses are relatively coarse grained. Similarly, phyllite implies finegrained, whereas schist implies relatively coarse grained.

Color

The color of volcanic rocks is useful in preliminary classification of volcanic rocks, since direct observation of mineral content is limited due to fine-grain size. Very felsic volcanic rocks are generally light colored, and mafic igneous rocks are dark. There are so many exceptions to these generalizations, due, for example, to hydrothermal alteration and weathering processes, that color is not a first-order classification attribute. Coloration can be imparted by the presence of very small and very minor amounts of materials that do not qualify as grains on the hand-specimen–thin-section scale. Finely divided iron oxides and hydroxides in the Wingate and Entrada Sandstones and the Redwall Limestone of the Colorado Plateau give these rocks their red coloration. Similarly, small amounts of magnetite are responsible for the blackness of obsidian, and weathered rocks are typically variegated. These colors and color variations may be dramatic, but they are not amenable to rock classification.

Fabric

A principal classification attribute of metamorphic rocks is fabric, indicated as either directional (foliated) and nondirectional (nonfoliated) fabric or indirectly via rock names such as schist (directional implied) and hornfels (nondirectional implied).

Fabric is a major parameter in the classification of sedimentary structures but not of sedimentary rocks. Most sedimentary rocks have planar fabric resulting from the process of sedimentation generating bedding and laminations. Conversely, magmatic rocks may acquire planar and/or linear fabric under special circumstances, but since most magmatic rocks are nondirectional there is little need for an expression of fabric in their classification.

Traditional Classifications

Igneous Rocks

The Streckeisen (1967, 1973, 1976), Streckeisen and Le Maitre (1979), and Le Maitre (1989) classification of igneous rocks, generated through the auspices of the International Union of Geological Sciences (IUGS), is in wide use at the present time. The classification is fundamentally mineralogical, deriving from similar mineralogical classifications established during the early part of this century and enlarged with great detail by Albert Johannsen (Johannsen, 1931–1938).

The IUGS igneous rock classification makes a distinction between volcanic and plutonic rocks, implying that the distinction between such rocks can be made on the basis of occurrence or grain size. There is a classification diagram for plutonic rocks (Figure 9.1) and another for volcanic rocks (Figure 9.2), both of which are easy to use as long as mineral content is known and minerals are the main constituents.

The IUGS classification is not a vehicle by which rocks can be identified as being igneous, nor does it provide direct information indicating either volcanic or plutonic affinity. This means that a rock must first be identified as being igneous and then designated as either volcanic or plutonic before the classification can be used. Effectively, a rock must be classified outside of the classification before it can be classified within it and assigned a rock name.

An igneous origin may be apparent on the basis of chemical and physical attributes observed on the

continued on p. 147

Figure 9.1
Classification of plutonic igneous rocks based on Streckeisen (1976) and LeMaitre (1989).

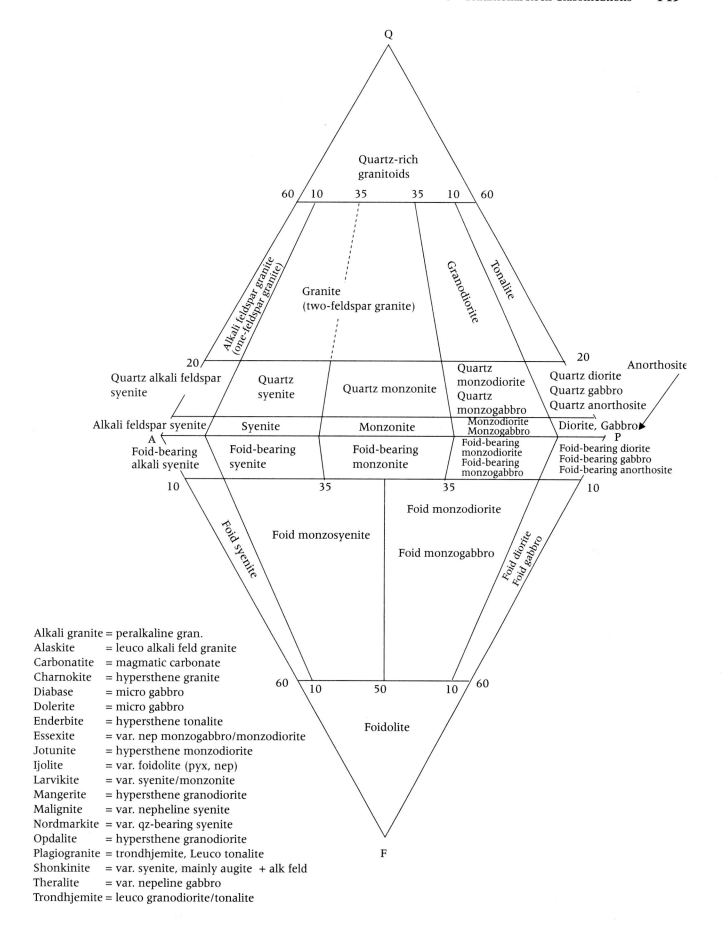

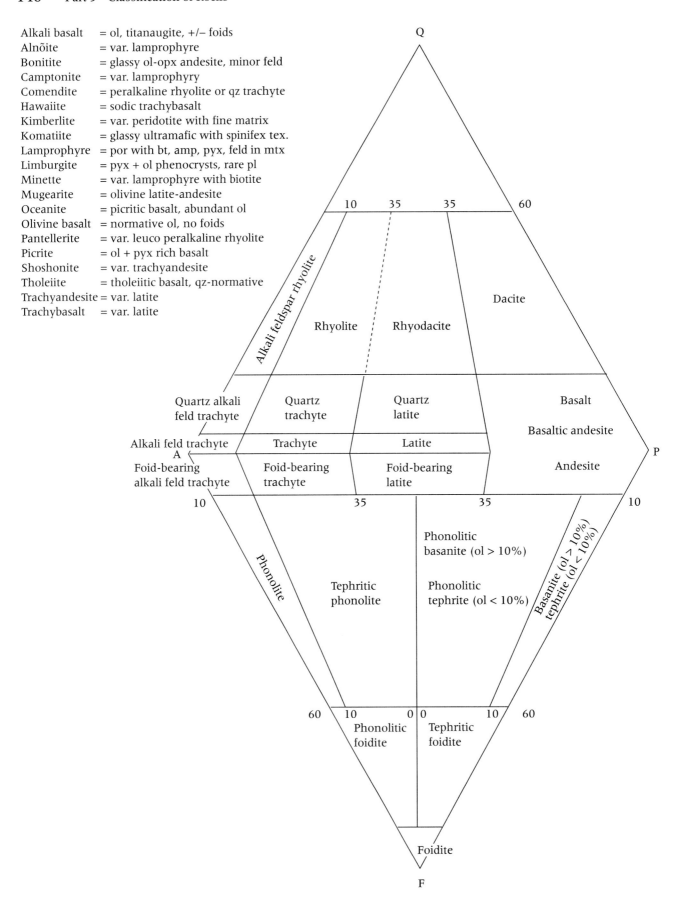

Figure 9.2
Classification of volcanic igneous rocks based on Streckeisen (1967) and LeMaitre (1989).

hand-specimen–thin-section scale or on the basis of field relations. The finer-grained magmatic rocks are assumed to be mostly volcanic and the coarser-grained magmatic rocks are assumed to be mostly plutonic. Rocks transitional between volcanic and plutonic cannot be identified simply by an intermediate grain size. Grain size is not reliable—many basalts are coarser grained than quench phases in plutonic systems (such as aplites and microdiorites). Additionally, some coarse-grained "plutonic" rocks crystallize in high-level magma chambers, not far from subvolcanic and surface volcanic activity.

There are further limitations of the IUGS classification. A triangular diagram is a simple way to present mineral variations involving three mineral phases. In reality, most rocks contain more than three phases

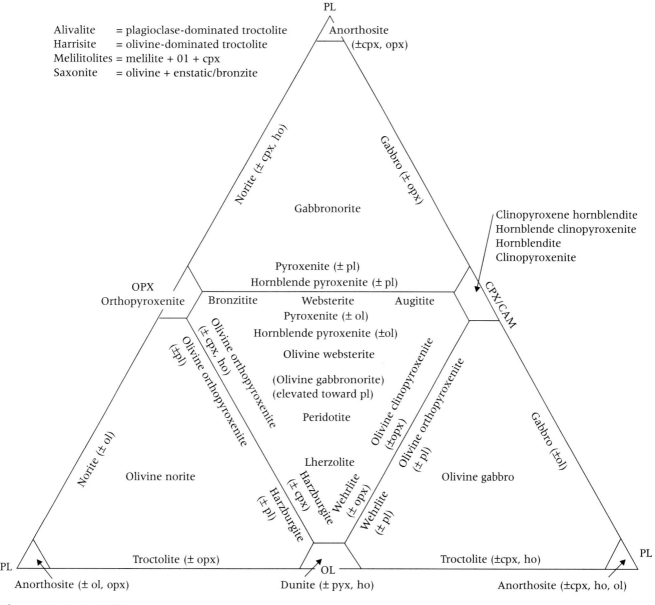

Figure 9.3; text p. 148
Classification of mafic and ultramafic rocks based on Streckeisen (1973) and LeMaitre (1989) presented as a composite diagram.

and certain assumptions must be made if the simplicity of the triangular diagram is to be maintained. One assumption is that the mafic mineral content of rocks containing quartz, feldspar, or feldspathoids is not paramount in genetic interpretation. The IUGS classification does not distinguish a granite that contains a few percent biotite–hornblende and one that is dominated by these minerals.

Additional diagrams showing many rock categories based on mineral percentages have been created for the mafic and ultramafic igneous rocks (Streckeisen, 1973). A composite of these diagrams is presented in Figure 9.3; p. 147. Curiously, no such detail of classification is provided for rocks such as granites, even though granites are much more abundant. "It may seem especially unfortunate that the commonest igneous rocks are usually those most vaguely defined" (Williams, Turner, and Gilbert, 1982, p. 67).

A serious limitation to any mineralogical classification of igneous rocks is the barrier presented by very fine-grained rocks and glassy rocks. The mineral content of very fine-grained rocks, such as micrites and cherts, can be approximated with some confidence, but the mineral content of volcanic rocks is more elusive. Nevertheless, a clue to mineralogical content, or chemical composition of glass, is provided by phenocrysts. For example, consider three volcanic rocks with 5% phenocrysts set in a very fine-grained matrix. Phenocryst assemblages consisting of (1) plagioclase only, (2) plagioclase + quartz, and (3) plagioclase + quartz + alkali feldspar in each case is an indicator of probable matrix mineral content in accord with the principles of phase equilibria. Accordingly, rock (1) could have mainly plagioclase in the matrix, rock (2) would have at least plagioclase and quartz in the matrix, and rock (3) must have all three minerals in the matrix. Very fine-grained or glassy igneous rocks completely devoid of phenocrysts or microphenocrysts are rare.

The IUGS classification does not distinguish **magmatic** and **pyroclastic** igneous rocks. Pyroclastic igneous rocks have been classified on the basis of clast type (Figure 9.4a) and size (Figure 9.4b) rather than mineralogy, since constituent grains are dominantly glassy or very fine crystalline. Some suggestion of composition is indicated by phenocrysts that are erupted along with rapidly solidifying magma.

There are no comprehensive classifications for variants of igneous rocks such as **pegmatites, magmatic breccias, assimilation-modified igneous**

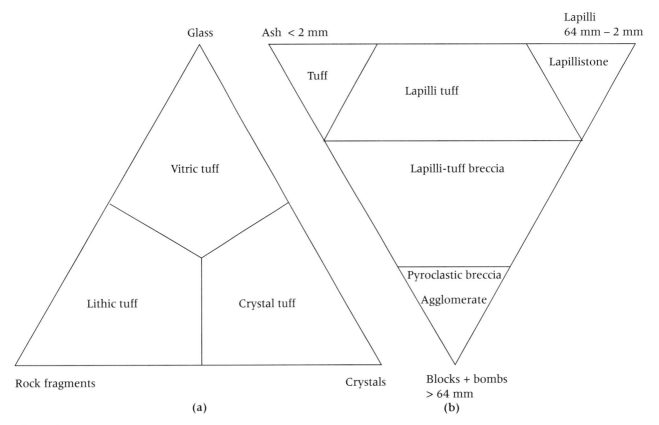

Figure 9.4
Classification of pyroclastic rocks after
(a) Pettijohn (1975) and Schmid (1981), (b) Fisher (1966).

chemical group \ type of metamorphism	Static Thermal Recrystallization (± inherited directional fabric)			
	Finer-grained protoliths		Coarser-grained protoliths	
	lower grade	higher grade	lower grade	higher grade
carbonate	hornfelsic siliceous ls/dol	granofelsic marble	non-directional (±inherited directional)	directional fabric (acquired)
carbonate-siliceous	hornfelsic siliceous ls/dol calc-silicate hornfels	calc-silicate granofels	fine-grained: hornfelsic	phyllitic mylonitic
carbonate-argillaceous	hornfelsic marlstone calc-silicate hornfels	calc-silicate granofels	coarse-grained: granofelsic	schistose gneissose
argillaceous	argillite hornfelsic mudrock-shale	Al-silicate hornfels/granofels		
quartzose	hornfelsic qz arenite hornfelsic orthoquartzite	granofelsic quartzite	hornfelsic conglomerate	granofelsic qz cong/rudite
quartzo-feldspathic	hornfelsic rhyolite	qz-mica hornfels/granofels	hornfelsic conglomerate hornfelsic granitics	qz-feld-mica granofels
mafic-intermediate	hornfelsic basalt/greenstone	ho-pl (pyx) granofels/hornfels	hornfelsic gabbro	ho-pl granofelsic amphibolite pyroxene granonfels (granulite)
ultramafic	n.a.	n.a.	hornfelsic/granofelsic serpentinite	hornfelsic/granofelsic steatite talc-anthophyllite granofels

Note: the box at the upper right of the table (carbonate through argillaceous rows, coarser-grained protoliths columns) indicates generic nomenclature: non-directional (±inherited directional) vs directional fabric (acquired), subdivided into fine-grained (hornfelsic / phyllitic mylonitic) and coarse-grained (granofelsic / schistose gneissose).

chemical group \ type of metamorphism	Dynamothermal Recrystallization (± stage of static thermal recrystallization)			
	Finer-grained protoliths		Coarser-grained protoliths	
	lower grade	higher grade	lower grade	higher grade
carbonate	phyllitic/mylonitic limestone	schistose/gneissose marble	n.a.	n.a.
carbonate-siliceous	phyllitic siliceous ls/dol calc-silicate phyllite	calc-silicate gneissose marble	n.a.	n.a.
carbonate-argillaceous	phyllitic marlstone calc-silicate phyllite	calc-silicate gneissose marble	n.a.	n.a.
argillaceous	slate, phyllite	mica schist	n.a.	n.a.
quartzose	phyllitic mylonitic quartzite	mylonitic quartzite gneissose quartzite	phyllitic/mylonitic conglomerate	quartz schist/gneiss
quartzo-feldspathic	phyllitic/mylonitic rhyolite phyllitic/mylonitic arenite	quartzo-feldspathic gneiss quartz-mica schist	phyllitic/mylonitic granite/grandiorite	quartz-feldspar gneiss granitic gneiss
mafic-intermediate	greenschist, blueschist	schistose/gneissose amphibolite ho-(pyr) schist/gneiss	gneissose gabbro phyllitic wacke	gneissose amphibolite pyr-pl gneiss (granulite)
ultramafic	n.a.	n.a.	schistose serpentinite	talc schist anthophyllite-tremolite schist

Figure 9.5; text p. 150
Classification of metamorphic rocks based on protolith type, grade of metamorphism, and static–dynamic environment. Box at the upper right indicates generic nomenclature.

rocks, magma-mixed magmatic rocks, **cumulate magmatic rocks**, or magmatic rocks acquiring metamorphic attributes during late-stage flow (**dynamomagmatic rocks**). However, this is a classification of lunar igneous rocks (Stöffler et al., 1980).

Magmatic rocks that have been weakly metamorphosed cannot be readily classified. These are the **meta-igneous** rocks, such as metabasalt and metarhyolite that contain both metamorphic and magmatic attributes. They are analogous to rocks that have been weakly weathered or hydrothermally altered.

There are rocks in which crystallization of magma is not the only rock-forming process involved in their formation. These rocks are **migmatites** and may be generated by **magma injection**, **anatectic melting**, or **metasomatism** (Mehnert, 1968).

Extension of magmatism to hydrothermal activity leads to the formation of **hydrothermal rocks**, which consist of alteration assemblages derived from previous rocks and include direct precipitations in spaces, potential spaces, and as wholesale replacements. Characterization of mineral assemblages occurring in hydrothermally altered rock (Burnham, 1962) is analogous to metamorphic facies (Turner, 1968; Winkler, 1979). No classification of hydrothermal rocks includes physical attributes as well as mineral assemblages. Detailed mineralogical–textural classification of hydrothermally altered rocks would have to include a description of relict protolith (Creasey, 1966) in addition to the alteration assemblage. The open chemical system inherent in hydrothermal systems enormously increases the diversity of the resulting altered rocks, not only affecting the alteration assemblage but promoting precipitation of vein minerals. The result is an unmanageable assortment of altered rocks that defy comprehensive classification. The same problem applies to the low-temperature equivalent of hydrothermally altered rocks—the **weathered rocks**.

Metamorphic Rocks

Classification of metamorphic rocks is by fabric, grain size, and mineralogy, the latter being an expression of bulk chemistry. Some classifications plot grain size (fine and coarse) against fabric (foliated or nonfoliated). The fine-grained nonfoliated rocks are **hornfels**, the fine-grained foliated rocks are **phyllites** and **slates**. Coarse-grained nonfoliated rocks are **granofelses** or **granulites**, and the coarser-grained foliated rocks are **schists** and some **gneisses**. The term "granofels" is favored over "granulite" because the latter is also a metamorphic facies, and a "granulite" from a textural point of view need not be in the "granulite facies." **Mylonite** can be considered to be a special variety of schist, forming in ductile fault zones.

Some classifications of metamorphic rocks emphasize mineralogy as a reflection of protolith composition (e.g., Spry, 1969). Other classifications relate mineralogy to conditions of temperature and pressure in which the rock formed. The latter are the facies and grade classifications. In combination with grain size and fabric, a mineralogical metamorphic rock classification consists of three fundamental attributes identifiable in the rock name itself. For example, a garnet–staurolite schist indicates staurolite grade was reached and a relatively coarse-grained rock with planar fabric was formed. Furthermore, it is implied that the rock contains micas and quartz (or the garnet and staurolite would not have formed) and that the chemical composition of the protolith was argillaceous,

(pelitic) since garnet, staurolite, and mica derive fundamentally from clay minerals.

Some metamorphic rock names, entrenched in traditional classification, indicate mineralogy and grain size but not fabric. For example, "marble" implies carbonate mineralogy and relatively coarse grain size, but does not distinguish directional and nondirectional varieties. "Amphibolite" indicates amphibole plus plagioclase mineralogy, relatively coarse grain size, but the name is nonspecific with respect to fabric.

A metamorphic rock classification incorporating compositional and physical parameters is shown in Figure 9.5; p. 149. The classification has both descriptive and genetic components. For example, directional and nondirectional rocks (descriptive) are equated with dynamic and static conditions of metamorphism respectively (interpretative). There is also division based on the grain size of protoliths (Figure 9.5; see also Spry, 1969); this is a genetic consideration that is

		Random — Fabric	Foliated	
Incohesive		Fault breccia	?	
		Fault gouge	?	
Cohesive	Glass	Pseudotachylite	?	
	Tectonic reduction in grain size dominates grain growth by recrystallization + neomineralization	Crush breccia / Fine crush breccia / Crush microbreccia		0–10%
		Protocataclasite	Protomylonite	10–50%
		Cataclasite	Mylonite (Phyllonite)	50%–90%
		Ultracataclasite	Ultramylonite	90–100%
	Grain growth pronounced	?	Blastomylonite	

Figure 9.6
Classification of cataclasites and related rocks after Sibson (1977).

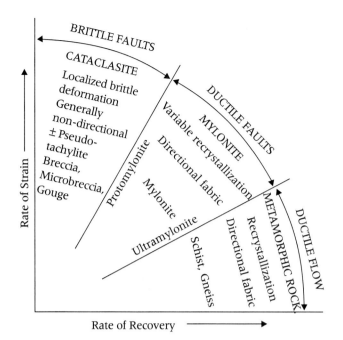

Figure 9.7
Classification of fault rocks modified from Weis et al. (1984).

reflected in the texture of the resulting metamorphic rock, especially those that are weakly metamorphosed.

Cataclasites are formed by cataclastic processes and are considered either a variety of metamorphic rock or a submetamorphic rock formed primarily as a result of deformation. Cohesiveness and fabric (foliated and nonfoliated) are the basis of subdivision within the field of cataclasis (Sibson, 1977; Figure 9.6; p. 150). Strain rate and temperature are the constraints of cataclasites (Wise et al., 1984; Figure 9.7).

The lower limit of metamorphism based on mineral reactions bridges the P–T region between diagenesis and very low-grade metamorphism. Therefore, from a mineralogical point of view, cataclasites are not metamorphic rocks because such reactions have not occurred. From a fabric point of view cataclasites are metamorphic rocks because there has been deformation. Cataclasites are generated in fault zones where the rate of strain is extreme and mineralogical reaction precluded. Brittle deformation is dominant, and heat may be generated well into the metamorphic field, even to the point of melting. Cataclasites viewed as variants of metamorphic rock are comparable to pyroclastic rocks being variants of magmatic rocks.

Rocks deformed at low temperature, low pressure, and slow strain rates, with or without significant recrystallization, such as in mélanges, orogens, and along aseismic faults, are distinct from metamorphic rocks and cataclasites. These rocks are abundant but have no mineralogical–textural classification. In most

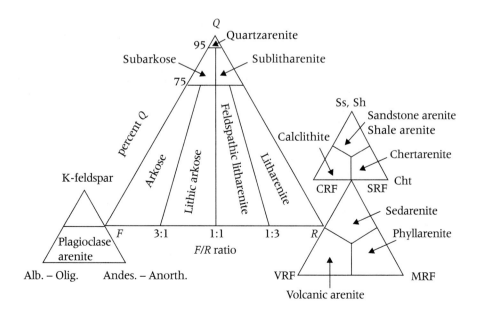

Figure 9.8
Classification of sandstones after Folk (1974).

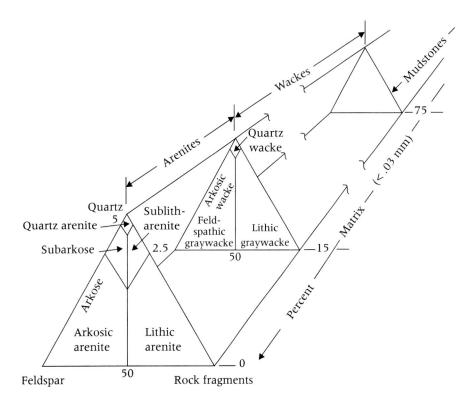

Figure 9.9
Classification of clastic sedimentary rocks after Dott (1964).

cases, designation is made by reference to the style of deformation, retaining some rock name that resides in some other classification, such as folded limestone or folded crystalline limestone. Some rocks in this category are dominated by recrystallization, such as intrusive rock salt and glacial ice.

A very special type of metamorphic rock is formed by meteoric impact. There is no comprehensive classification of **impactites**, but there is a classification of lunar breccias (Stöffler et al., 1980).

Sedimentary Rocks

Classification of sedimentary rocks most commonly distinguishes clastic, biochemical, and inorganic chemical processes. Detailed classifications of siliciclastic and carbonate rocks emphasize clast and rock composition and less so on the association of biotic activity. Clastic sedimentary rock classifications are based on the varying proportions of clasts of different material content and grain size. The most popular traditional classification of sandstones is shown in Figure 9.8; p. 151 in which categories are established on the basis of clast composition and, to a lesser extent, clast morphology.

The relation of sandstones to the finer-grained siliciclastic (dominated by silicate mineral grains) sedimentary rocks, ranging from wackes to mud rocks, is

% clay-size constituents			0 – 32	33 – 65	66 – 100
			Gritty	Loamy	Fat or slick
Nonindurated	Beds	> 10 mm	Bedded silt	Bedded mud	Bedded claymud
	Laminae	< 10 mm	Laminated silt	Laminated mud	Laminated claymud
Indurated	Beds	> 10 mm	Bedded siltstone	Mudstone	Claystone
	Laminae	< 10 mm	Laminated siltstone	Mudshale	Clayshale
Metamorphosed	Degree of metamorphism ↓		Quartz argillite	Argillite	
			Quartz slate	Slate	
			Phyllite and/or mica schist		

Figure 9.10
Classification of mudrocks after Potter et al. (1980).

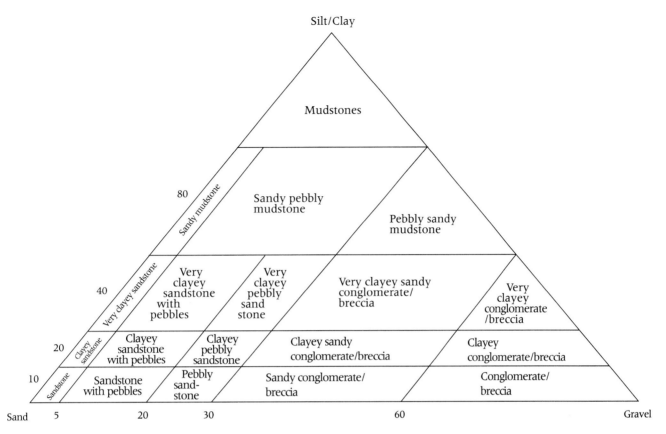

Figure 9.11
Classification of conglomerates and breccias modified from Piper and Rodgers (1980).

shown in another popular diagram (Figure 9.9). There are detailed classifications of shales (Potter et al., 1980; Figure 9.10) in which grain size and fabric (laminae or beds) are the chief parameters. Classification of conglomerates and breccias in which sand and finer particles vary in amounts relative to pebble-sized clasts is shown in Figure 9.11.

Carbonate sedimentary rocks are classified according to the proportions of very fine-grained carbonate grains (carbonate "mud") with respect to larger grains, known as "allochems" (Dunham, 1962; Figure 9.12) and (Folk, 1959, 1962; Figure 9.13). Allochems include intraclasts, oolites, pelloids, and fossil fragments. Larger cementing grains of carbonate,

Original components not organically bound together during deposition					Components organically bound during deposition
Contains carbonate mud			No carbonate mud	
Mud supported		Grain supported		
< 10% allochems	> 10% allochems			
Mudstone	Wackestone	Packstone	Grainstone	Boundstone

Figure 9.12
Classification of limestones after Dunham (1962).

Volumetric allochem composition	> 10% allochems		< 10% allochems			Undisturbed reef and bioherm rocks
	Sparry calcite > Micrite	Micrite > Sparry calcite	1 – 10% allochems	< 1% allochems		
> 25% Intraclasts	Intrasparite	Intramicrite	Intraclasts Intraclast-bearing micrite	Micrite or if sparry patches present, dismicrite		
< 25% Intraclasts / >25% Ooids	Oosparite	Oomicrite	Ooids Ooid-bearing micrite		Most abundant allochems	
< 25% Ooids volume ratio is: bioclasts: peloids / 3:1	Biosparite	Biomicrite	Bioclasts Fossiliferous micrite			
3:1 to 1:3	Biopelsparite	Biopelmicrite				Biolithite
1:3	Pelsparite	Pelmicrite	Peloids Peloid-bearing micrite			

Figure 9.13
Classification of limestones after Folk (1959, 1962).

known as "sparite," are a part of the Folk classification. These classifications are largely descriptive, but they depend somewhat on the identification (interpretation) of materials accumulated as part of the sedimentary process from those that form during lithification (diagenesis).

Classification of siliciclastic and carbonate mixtures is shown in Figure 9.14 based on Mount (1985).

Classification of carbonaceous sediments is given in Pettijohn (1975); see Figure 9.15a,b.

Epiclastic volcanic rocks have both volcanic and clastic sedimentary attributes (Fisher, 1961; Wright et al., 1980; Schmid, 1981). Ash fall into water bodies and fluvial reworking of volcaniclastics are processes bridging volcanism and clastic sedimentation. Sedimentation involving volcanic debris is likely to merge

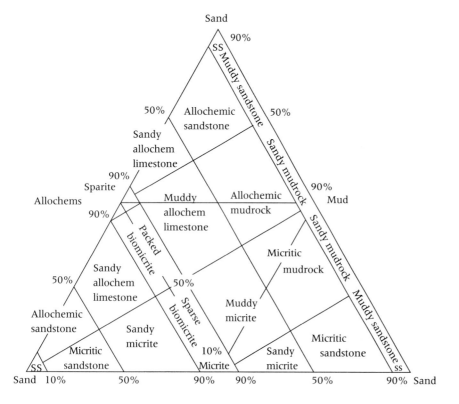

Figure 9.14
Classification of mixed carbonate and siliciclastic sedimentary rocks after Mount (1985).

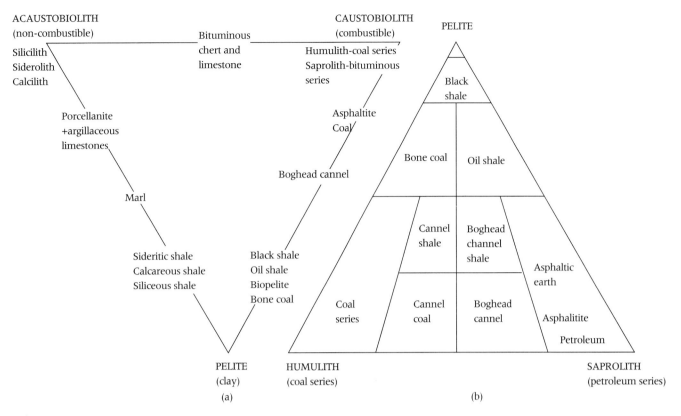

Figure 9.15
Classification of bioliths and biopelites (a) and carbonaceous rocks (b) after Pettijohn (1975).

with sedimentation of nonvolcanic debris, and the variety of rocks so produced is immense, defying classification.

There is no comprehensive classification of **evaporite sedimentary rocks**, but several classifications focus on specific minerals. For example, a textural classification of halite in evaporites is presented by Havorka (1987), and a structural–textural of anhydrite is given by Maiklem et al. (1969).

General classifications at the hand-specimen–thin-section scale exist for **phosphorites** (Slansky, 1986) and **ironstones** (Dimroth, 1976). Although there are well-known petrographic features of coals, classifications are directed more toward source material and percentage of combustible compounds (Francis, 1961; van Krevelen, 1981).

Status of Traditional Rock Classifications

This review of rock classification at the hand-specimen–thin-section scale has shown that (1) there is little or no classification for some rocks and (2) no classification of any rock group is purely descriptive. Each classification has its strengths and weaknesses; in most cases there is a compromise between description and interpretation, motivated by the desire (1) to have a single classification diagram and (2) to keep the classification simple so that it will be used.

10

A Unified Classification of Rocks

"To stimulate advancements in their science, geologistis must be willing to revise their classifications as they are to make new observations and new interpretations."

(Mason Hill, 1963, p. 173)

Toward a Unified Rock Classification

Complete objectivity in rock classification is possible only if the potentially subjective step of selection of traditional classification is eliminated. This means that all rocks must be classified together, even though classification of rocks of very different derivation may at first seem unreasonable and unproductive.

Most traditional rock classifications contain components of descriptive and interpretative information. This historical development underscores the need to have both of these components in a classification—that is, a need to relate raw data to rock genesis. A unified classification can accommodate both objectives by having two parts. The first part consists of a mechanism and a format for objectively obtaining and organizing physical and chemical data. The second part allows for interpretation of rock-forming process. Both parts are amenable to revision as new data are collected and new interpretations are formulated. A unified classification can be open ended, whereas traditional classifications are inherently constrained.

Components of the Unified Rock Classification

Descriptive Part of Unified Classification

The mineralogical, chemical, and textural attributes of any rock can be expressed by the formula

$$RX = FC + MN + CH + TX$$

where FC = fundamental constituents, MN = minerals, CH = chemistry, and TX = texture.

Many rocks are uniform with respect to material content and physical relationships of constituent grains. Other rocks have certain discontinuities within themselves on the scale of hand specimens and thin sections observed with the optical microscope. To provide for classification of both types of rocks, the concept of "units" occurring within rocks was established in Chapter 7.

Rocks without units are defined by

$$RX = FC + MN + CH + TX$$

Rocks containing units are defined by

$$FC + MN + CH + TX \text{ (matrix)} + FC + MN + CH + TX \text{ (units)}$$
$$= FC + MN + CH + TX \text{ (rock)}$$

The Components of the Rock Formula: Fundamental Constituents The fundamental constituents (FC) of rocks (and of units and matrix if applicable) are crystals (C), submicroscopic materials (SM), glass (G), and voids (V), as defined in Chapter 7.

Mineralogical Content. The mineralogical content (MN) of a rock consists of a listing of all mineral species present. MN is a vehicle by which the presence of a mineral can be indicated regardless of its abundance. The importance of this consideration can be illustrated with the following examples. The presence of even one small relict grain of staurolite in a quartz–mica schist is a key indicator of metamorphic grade and, therefore, very important in petrogenetic interpretation. The presence of a trace of glauconite in a sedimentary rock may be a clue to, or confirmation of, marine depositional environment. Even a minor amount of nepheline in a syenite is evidence of silica undersaturation.

Chemistry. The chemistry (CH) of a rock can be estimated from its mineralogical content (e.g., Dietrich and Sheehan, 1964). Of course, for rocks with considerable unidentifiable submicroscopic crystalline or amorphous material, or glass, the results of this calculation will be less meaningful and take a subordinate role in the classification scheme. For crystal-bearing rocks the mineralogical mode is determined first. Using estimated mineral composition in relation to the volume percent of the mineral present, the weight percent of simple compounds (oxides for most rocks) can be calculated for the rock as a whole. If the composition of SM is known, such as it commonly is for materials such as chert, micrite, and limonite, it is included as mineral in the calculation.

Texture. The texture of a rock consists of grain size (GS), grain morphology (GM), crystal zoning (CZ), grain discontinuities (GD), crystal twinning (CT), grain relations (GR), and fabric (FB) (Chapter 7).

Comparison with a Data Base The unified classification proposed (Carr and Hibbard, 1991) consists of a rock data base to which information on unknown rocks can be compared. The data of FC, MN, CH, and TX for each rock of a data base rock group are assembled and digitized. Then the data of FC, MN, CH, and TX are determined for an unknown sample and entered into a computer program that proceeds to look for the closest match in the data base. The data base is accessed by a neural network computer algorithm.

An unknown sample is descriptively classified when the closest match is made in terms of FC, MN, CH, and TX. For example, an unknown rock is found to consist of crystals and glass (FC). The crystals are quartz and sanidine (MN), and bulk rock chemistry is estimated to be silicic and alkaline (CH). The unknown rock contains fine-grain-size glass shards and larger pumice grains all with granular relation to each other and with preferred dimensional orientation (TX). The rock is found to be most similar to a certain rock of the data base that was originally classified by traditional methods as a welded tuff.

Interpretive Part of Unified Classification

Each of the data base rocks has been assigned a rock name and alternative rock names based on traditional methods of classification and interpretation. Each data base rock also is assigned a coded rock-forming process (listed in Carr and Hibbard, 1991) based on current knowledge and understanding (Chapter 11). Assignment of rock-forming process to the data base samples is tentative and subject to revision and expansion within the format of the computer program. The rock-forming process for the welded tuff data base sample involves interaction of silicate melt, high-T solids, and high-T aqueous-phase characteristic of pyroclastic eruption, pyroclastic flow, and welding. Since this is the best match for the unknown rock sample, it is tentatively presumed that the unknown sample is a welded pyroclastic flow tuff.

Part *IV*

Fundamental Processes of Rock Formation

The process of rock formation can be described in general or fundamental terms. For example, weathering can lead to the formation of weathered rocks, the term "weathering" conveying the general sense of process leading to the conversion of pre-existing rock to new rock. In detail, weathering involves the fundamental processes crystallization and dissolution, active in a low-temperature environment in the presence of rock, water, and air. Even more specific characterization of processes is embodied in each fundamental process. For example, "precipitation" involves mechanisms of nucleation, kinetics of attachment to nuclei and crystal surfaces, as well as mechanisms of transport (diffusion and convection) of growth material to the crystal–fluid interface.

Assigning general and fundamental rock-forming processes to rocks that contain evidence of more than one origin requires considerable latitude in concept. For example, a weakly weathered granite containing incipient clay alteration of feldspars and partial conversion of biotite to vermiculite is still mostly a magmatic rock, whereas some bauxites represent complete weathering of a certain magmatic rock. A similar situation exists for hydrothermally altered rocks; some are weakly altered, whereas some have no indication of pre-alteration lithology. The same situation exists for weakly metamorphosed rocks and for rocks that are partly recrystallized or deformed at submetamorphic temperatures. For purposes of rock classification the scheme presented in Chapter 10 does not depend on the degree of weathering, hydrothermal alteration, metamorphism, low-temperature recrystallization, or low-temperature deformation. For purposes of process classification, partial conversions from one rock type to another can be ignored, with the focus on processes that are assumed to be more-or-less complete.

Classification of Rock-Forming Processes

His [Henry Clifton Sorby (1826–1908)] descriptive work is of the highest calibre, but it was all the time governed by his knowledge of, and search for, processes."

J.R.L. Allen (1977)

Fundamental Rock-Forming Processes

Rocks are made from **solids, melts, aqueous solutions**, and **gases. Fundamental processes** are defined as **non-crystalline solidification, crystallization, dissolution** and **sublimation, melting, mechanical fragmentation, inter- and intracrystalline creep**, and **mechanical interaction**.

Non-crystalline solidification (Chapter 14) is the process by which melts (chiefly silicate melts) become glass in response to rapid undercooling. Glasses form in quenched magma (volcanic glass), in brittle faults (pseudotachylite), in meteor-impacted rock (shock glasses), and by lightning strikes (fulgurites). Solidification also includes transformation of gels to amorphous solids, such as silica gel to opal-A, and other processes by which amorphous solids form.

Crystallization (Chapters 12, 13, 14, 16) occurs in melt, water, aqueous solutions, gases, and solids. Crystallization in magmas (melt solutions) generates crystals. Pure phase melts, such as silica melt, are very rare and produce glass (lechatelierite) instead of crystals. Crystallization in aqueous solutions is **precipitation**. In most cases, water or supercritical aqueous phase is the dominant constituent of the solvent, although in carbonate systems dissolved carbon dioxide may constitute a high mole fraction of the fluid. Crystallization in water commonly involves biotic as well as inorganic processes. Formation of crystals directly from gases is one type of **sublimation**. **Recrystallization** in solids occurs either as regrowth of minerals without mineralogical change or generation of new minerals via mineralogical reactions. Most "solid-state" crystallization is assisted by H_2O in the form of at least an intergranular film of fluid, making this type of crystallization a variety of aqueous solution crystallization. Recrystallization without fluids present occurs at great depths in the Earth's crust and in the mantle, but it also occurs in materials that recrystallize readily at low temperature such as ice and halite.

Dissolution (Chapters 12, 14) is the dissolving of solids (chiefly minerals) into melts, aqueous solutions, and gases. This is a rock-forming process in the sense that the dissolution is incomplete, physically and chemically modifying pre-existing solids that are eventually preserved in the final rock. **Reactive**

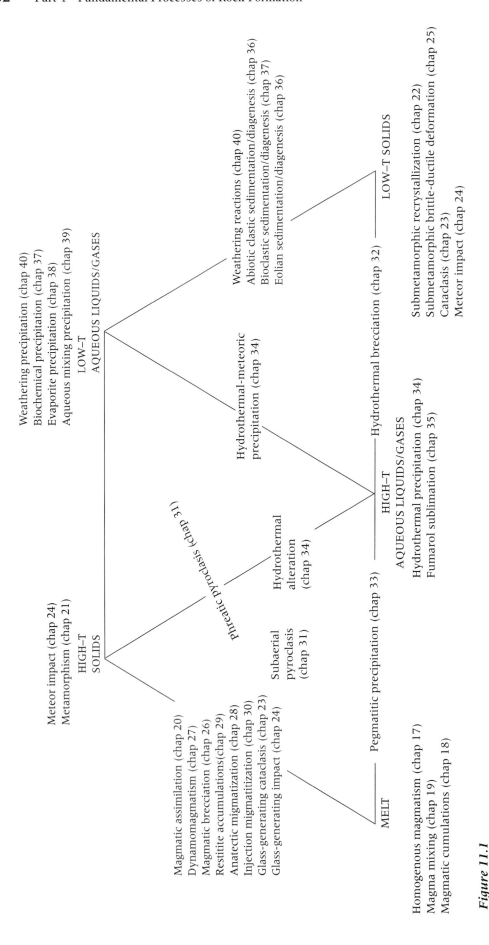

Figure 11.1
Rock-forming processes keyed to text chapters arranged according to five rock-forming materials. Some processes appear in more than one position in the triple triangle diagram as a means of accommodating process variation within a general process catagory. Fifteen single rock-forming processes are indicated at triangle apices. Sixteen processes involve two rock-forming processes, positioned on appropriate tie line. Subaerial pyroclasis involves three rock-forming materials. Phreatic pyroclasis involves four rock-forming materials.

replacement is a very common form of dissolution. Transformation of solids to gases is another type of **sublimation**. Sublimation of solids (Chapter 14) is rare, being confined to morphological reshaping of ice crystals in contact with the atmosphere and sublimates in contact with fumarolitic gases and the atmosphere.

Melting (Chapter 13) is much less common than dissolution of crystals into melts (melt solution) because higher temperatures are required. For example, in a magmatic system, dissolution of quartz at the interface with silicic magma can place in the range of 600–1000°C, whereas direct melting at sites within quartz crystals isolated from ambient melt occurs only if temperatures are at least 1700°C.

Mechanical fragmentation (Chapters 15, 16) includes the formation of autobreccias in moving magma, magma and hydraulic fracturing of pre-existing rock, impacts related to meteor collision, sedimentation grain collisions in air and water, explosive eruptions in solids and magmas resulting from rapid release of fluids, and by shear related to tectonic deformation.

Inter- and intracrystalline creep (Chapter 16) occurs within crystal lattices and as an intergranular process resulting in strain of rocks. It occurs at elevated temperatures in a dynamothermal metamorphic setting, and at low temperatures in salt intrusions, salt glaciers, ice glaciers, and some deformed carbonate rocks.

Mechanical interaction (Chapter 15) as a contributor to the formation of rocks is mainly the relocation of mineral and rock grains resulting from differential of flow of the grains with respect to melt, water, supercritical aqueous phase, air, and other gases. For example, there is (1) gravity and convective accumulation of crystals in magmas and in melt-extracted anatectic systems, (2) transportation, sorting, and deposition of clastic sediments in water and air, and (3) winnowing of particles in volcanic eruption columns.

General Rock-Forming Processes

Generalized rock-forming processes are shown with respect to rock-forming materials in Figure 11.1 and are the basis of the chapter subjects in Parts V–X. Temperature has a major effect on the behavior of solids and of aqueous solutions, and has been incorporated into the generalized classification of rock-forming processes (Figure 11.1). This permits, for example, distinction between (1) brittle and ductile deformation, (2) water transport of clastic sediments (cold water) compared with hydrothermal activity (hot water), and (3) particle transport in air compared with transport in hot volcanic gases.

12

Chemical Activity in Aqueous Solutions

Water, during the 4 billion years it has been on Earth, has drastically affected crustal and upper mantle chemical and tectonic processes, because of its mobility and strong reactivity, forming Earth-sized and local geochemical cycles, facilitating the generation of magma, crustal differentiation, seawater composition, volcanoes, geothermal systems, and hydrothermal fluids that formed ore bodies

(Frank W. Dickson, personal communication)

Properties of Water and Aqueous Phase

The phase relations in the H_2O system are shown in Figure 12.1. The **solid**, **liquid**, and **gas** phases coexist at the triple point (0.0098°C, 0.006 atm (4.58 mm Hg)), well out of range of natural pressure conditions, as is the sublimation-pressure (ice–gas curve) extending to lower temperatures (Figure 12.1). The liquid–ice boundary curve (Figure 12.1) indicates fundamental departure from most other liquid–solid equilibria. The slope of this curve indicates that the melting point of ice decreases with increasing pressure, meaning that the density of the liquid phase is greater than that of the solid phase. Therefore, pressure favors the liquid phase, and within a glacier there can be melting along crystal interfaces where there is maximum compression. Films of water must relocate in places of minimum compression and refreeze as a process of solution redeposition, contributing to change in shape of the glacier that can be observed as glacial flow.

The **boiling point** of pure water is located on the vapor-pressure (liquid-gas) curve at 100°C and one atmosphere pressure. The slope of the liquid-gas boundary curve indicates that as the pressure is raised more heat is required for vaporization.

The **critical point** of pure water is at 374.4°C and 218 atmospheres (221 bars) (Figure 12.1), beyond which there is no physical distinction between liquid and gas (vapor). As the temperature and pressure of water (liquid) and vapor (gas) in equilibrium increases, thermal expansion causes the liquid to become less dense. Simultaneously, the gas becomes denser as the pressure rises, and at the critical point the densities of the two phases become identical and distinction between them disappears. For pure water, this single phase is now a **supercritical aqueous phase** (a fluid). The density of this phase at the critical end point is 0.4 g/cc, compared with 1.000 g/cc at

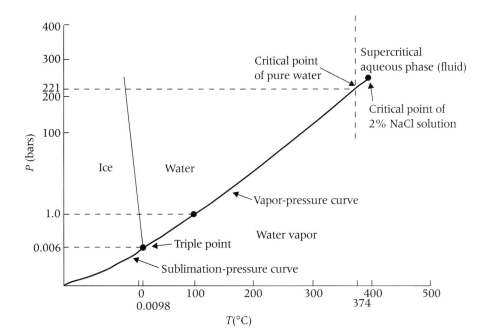

Figure 12.1
Phases of H_2O and critical point of pure H_2O and with 2% HCl. Note that triple point is not realized in geological environments (0.006 bar).

4°C and 1 atm pressure. A supercritical aqueous phase is able to bond with solute, in marked contrast to water vapor below the critical point, which can bond with little or no solute, as shown by the evaporation of water from a salt solution leaving salt as a precipitate.

Therefore, the chemical behavior of H_2O associated with rock-forming materials can be expected to be different above the critical point. Since H_2O associated with magmatism, higher grades of metamorphism, and high-temperature hydrothermal activity is well above the critical temperature and pressure (of pure water) at depth within the crust, it can be expected to have considerable solvent properties. The critical point of water is increased with respect to temperature and pressure if the solute has a low volatility, such as halite (Figure 12.1). If the solute is a gas, such as carbon dioxide, the critical point is lowered instead.

Aqueous Solutions

Water as a Solvent

The dissolving power of water is enhanced by the dipole character of the water molecule. The **solutions** produced have chemical properties that depend on the nature of the solute.

Solution of minerals may be **congruent** or **incongruent**. If congruent, the composition of the solution reflects the composition of the mineral. If the solution is incongruent, the composition of the solution is different because an intermediate solid phase forms. For example, the solution (or dissolution) of polyhalite produces gypsum as a new phase, in equilibrium with a new liquid.

Dissolution of solids may result in ions (Na^+), complex ions (CO_3^{-2}), or molecules (H_4SiO_4) existing in the solution. If a molecule or molecular cluster is large enough, it occurs as a **colloid** (1–5 nm) or a **gel** (5 nm–1 μm) in suspension. Larger particles, such as clays, are less likely to be in suspension and settle out in response to gravity. Colloids are influenced by gravity, unlike ions and simple molecules, but colloidal suspensions can be maintained for long periods of time.

Gases also dissolve in water. Carbon dioxide dissolves in natural waters producing carbonic acid. Such a solution is very important in the precipitation and dissolution of minerals such as calcite, aragonite, and dolomite.

Equilibrium in Aqueous Solutions

The behavior of an aqueous solution in rock-forming environments is governed by fundamental chemical phenomena that are expressed as **equilibrium constants** and **solubility products**, familiar to students that have a course in inorganic chemistry.

Consider the crystallization and dissolution of halite. Evaporation of water from a saturated halite solution shifts the equilibrium, resulting in crystallization of halite. Conversely, dissolution of halite crystals in equilibrium with a saturated solution occurs if the solution is mixed with freshwater, such as can occur by fluvial input into an evaporite basin.

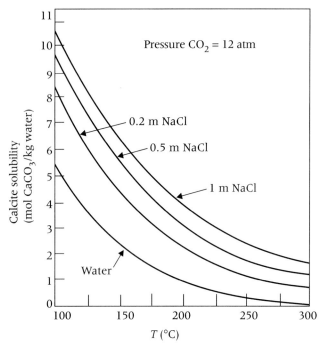

Figure 12.2
The solubility of calcite in H_2O and in NaCl solutions at carbon dioxide partial pressure of 12 atm (Ellis, 1963). Solubility decrease with temperature rise.

the boiling point may be raised, and the freezing temperature may be reduced.

Dilute solutions are closer to ideal behavior than are concentrated solutions. Nonideal behavior in concentrated solutions is quantified as the **activity coefficient**, a measure of the chemical activity reduced from ideal levels assumed by equilibrium constants and solubility products. Activities relate to **ionic strength** that is a function of charge and concentration. As ionic strength of solutions increases, the activity coefficient (i.e., effective concentration) decreases.

Effect of Pressure and Temperature on Equilibrium

Changes in temperature and pressure effect equilibrium as well as addition or subtraction of solute. More halite can be dissolved in water if the water temperature is raised. At 100°C the solubility of halite is 391.2 g/cc, compared with 357 g/cc at 0°C. Since change in temperature changes the equilibrium, the activities must also change. Sluggish reactions are not indicated by the equilibrium constant, the activity coefficient, nor the reaction itself. Very concentrated solutions, such as the brines in evaporitic systems, depart significantly from ideal behavior of dilute solutions.

Less halite can be dissolved in a solution that already contains sodium ions from another source. This is known as the **common ion effect**, namely that solubility is decreased by the presence of one of its own ions in the solution. Conversely, the presence of dissimilar ions may make a salt more soluble. If foreign ions, along with water dipole molecules, cluster around the sodium and chloride ions, these ions may not be able to get together to precipitate halite. As a result, more halite can dissolve than if the solvent were pure water. For example, more calcite can be dissolved in sodium chloride solutions than in pure water (Figure 12.2). This has significance in geological circumstances where calcite in equilibrium with groundwater of meteoric origin is prone to dissolution if there is mixing with ocean water along coastal regions.

Activity of Ions and Molecules in Solution

Dissolution of ionic minerals such as halite does not yield totally "free" ions. Incomplete dissociation of ions from each other may leave ion pairs or clusters in a solution, even though separation from the crystalline state is complete. As a result, there is departure from "ideal behavior." Vapor pressure may be varied,

Geologically Important Aqueous Systems

A few of the more common chemical systems, important in rock formation, involving H_2O are discussed in this chapter. The minerals and textures of rocks formed in the presence of aqueous solutions typically contain a record of both dissolution and precipitation (crystallization), and is the reason why the factors promoting forward and reverse reaction in aqueous solutions are geologically applicable.

NaCl–H_2O System (Halite–Aqueous Phase)

The occurrence of halite in evaporites illustrates the importance of this system at low temperatures. The occurrence of hydrogen chloride in volcanic gases and brine–halite in fluid inclusions of rocks generated by hydrothermal processes are indicators of the importance of this system at elevated temperatures. There are no sodium chloride magmas.

The solubility of halite in water is only slightly sensitive to temperature, depending mainly on concentration (Figure 12.3). Water freezes to ice at 0°C, but this freezing temperature is depressed as the concentration of salt is increased. At temperatures below the freezing curve, ice and brine coexist. At extremely low temperatures (−21°C) everything freezes, producing crystals of hydrohalite (NaCl·2H$_2$O) in ice, or ice crystals in hydrohalite if the salt concentration is very high. Hydrohalite is the stable form of halite at very low temperatures. It is known to form in Siberian salt lakes in the winter when evaporation takes place at very low temperatures. At very high concentrations of salt, hydrohalite coexists with brine above −21°C but below 0°C. Above 0°C halite coexists with saturated brine solutions if the concentration of salt is about 30%.

The equilibrium curve separating liquid water from water vapor shifts to higher temperatures with sodium chloride as solute. This is the same as saying that the vapor pressure of brines is less than that of pure water. Eventually the saturation curve is reached, and crystalline halite coexists with brine and water vapor.

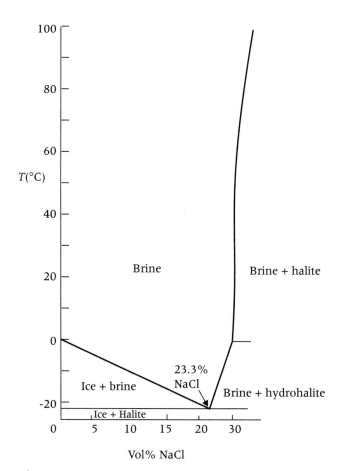

Figure 12.3
The system NaCl–H$_2$O (Braitsch, 1971).

SiO$_2$–H$_2$O System (Silica–Aqueous Phase)

The SiO$_2$–H$_2$O system is extremely important throughout the temperature range of geological activity, ranging from biogenic associations in sedimentary systems to hydrothermal quartz veins and wall rock silicification (Williams et al., 1985; Williams and Crerar, 1985). Pure quartz magmas, even those containing dissolved aqueous phase, do not occur in the geological environment because of the high temperatures required (1700°C at 1 bar). Quartz veins are not the result of crystallization of silica melt, which is why they are not called quartz dikes.

Silica occurs in several polymorphic forms (Figure 13.3). Beta quartz (high quartz) crystallizes in magmas and the higher temperature range of metamorphism. In aqueous environments, silica occurs as alpha quartz (low-T quartz), fibrous quartz (chert–chalcedony), opal-CT (disordered cristobalite–tridymite), and opal-A (amorphous silica with attached water).

Solubility of quartz at 25°C is very low (6–10 ppm) unless the pH (an expression of the concentration of the H$^+$ ion in aqueous solutions) is high (low concentrations of H$^+$). At the same temperature, the solubility of opal-CT is 20–30 ppm, and the solubility of opal-A is 60–130 ppm. These solubility relations are shown in Figure 12.4. The solubility of these silica polymorphs increases with temperature up to about 350°C (Figure 12.4). Most dissolved silica comes from dissolution of silicate minerals, biogenic silica precipitates, and volcanic glass, not quartz itself. Accumulations of biogenic silica, in the form of microfossil tests, spicules, and spines, have an extensive surface area exposed to aqueous solutions. This aids in the dissolution process.

Seawater contains only 2–14 ppm silica, well below the saturation level. Streams and groundwater contain about 10–60 ppm silica. How then do diatoms, radiolarians, siliceous sponges, and sea urchins form in the ocean? The extraction must be **enzymatically catalyzed**, not dependent on chemical saturation. In fact, the very low concentration of silica in seawater is in part attributed to the presence of these organisms.

Silica reacts with water according to the equation

$$SiO_2 + 2H_2O = H_4SiO_4 \text{ or } Si(OH)_4 \text{ or } SiO_2 \cdot nH_2O$$

Silicic acid (H$_4$SiO$_4$) is even weaker than carbonic acid and, as such, is not affected by pH much below 9. Above a pH of 9 the solubility of quartz is significantly increased, but such a high pH is rare in nature.

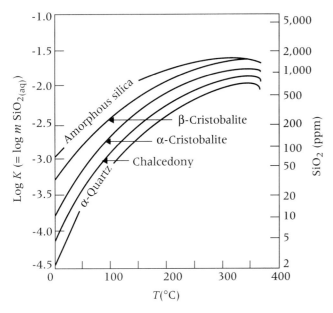

Figure 12.4
Solubility of silica polymorphs increasing with temperature (Walther and Helgeson, 1977).

Silica colloids in the form of hydratic gels (polymers) form in silica supersaturated water solutions. Such gels may form in the hot spring environment where the solubility of silica is significantly increased.

Abiogenic precipitation of silica typically is in the form of opal-A, even though quartz is thermodynamically predicted. If nucleation and growth of a thermodynamically stable phase are kinetically slow (quartz), a phase that is less thermodynamically stable (opal-A) but kinetically more labile (high nucleation rate) forms instead. Sudden cooling, say of hot spring water, tends to induce polymerization and formation of colloidal amorphous silica with attached water molecules. A drop in pH or concentration by evaporation would have the same effect.

With time opal-CT forms at expense of opal-A even though, once again, quartz is the thermodynamically stable phase. Quartz forms from opal-CT through a recrystallization process. Thus, the precipitation sequence of silica begins with opal-A, the most soluble form of silica.

CO_2–H_2O System (Carbon Dioxide–Aqueous Phase)

Carbon dioxide is abundant in the atmosphere and is readily available to dissolve in water. CO_2 also is a major component of volcanic gases along with water vapor.

CO_2 dissolves in water, forming carbonic acid according to the equation

$$CO_2(air) + H_2O = H_2CO_3$$

It can be seen in Figure 12.5 that solubility increases with increasing temperature at elevated pressures, but decreases with increasing temperature at low pressures.

Dissociation of carbonic acid occurs according to the equation

$$H_2CO_3 = 2H^+ + CO_3^{2-}$$

The equilibrium relations in a "simple" solution of CO_2 in water are actually complex, because any of the following ions may be present in the solution: H^+, OH^-, H_2CO_3, HCO_3^-, CO_3^{2-}.

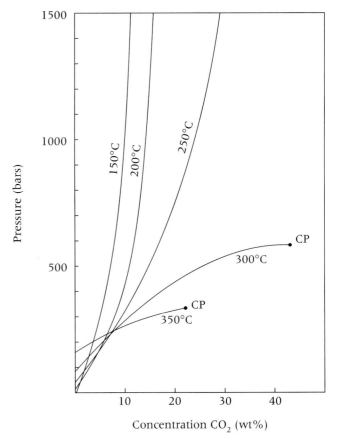

Figure 12.5
Solubility of CO_2 in pure H_2O at various temperatures. Solubility increases with temperature at high pressures, and decreases with temperature at low pressures (Takenouchi and Kennedy, 1965).

CaCO₃–H₂O System (Calcite–Aqueous Phase)

The $CaCO_3$–H_2O system has major ramifications in the understanding of carbonate evaporites and biochemical carbonate sedimentary rocks (Bathurst, 1975). At elevated temperatures, carbonate minerals appear in hydrothermal veins and at still higher temperatures as water-bearing magmatic carbonatites and related intrusive systems such as kimberlites.

The solubility of calcite in pure cold water is only 0.014 g/L, increasing to 0.018 g/L in hot water. This means that, like halite, this mineral is more soluble as temperature increases. However, unlike halite, the effect of CO_2 already dissolved in natural waters has a drastic effect on the solubility of calcite, actually generating a chemical system in which the solubility of calcite decreases with increasing temperature (Figure 12.6).

Water in contact with air has some dissolved CO_2. This means that the water is slightly acid since it must contain some carbonic acid. The fundamental equation relating the solution of calcite in such water is shown as

$$H_2O + CO_2$$
$$\Updownarrow$$
$$CaCO_3 + H_2CO_3 = Ca^{2+} + 2HCO_3$$

Any chemical process that increases the amount of CO_2 favors dissolution of calcium carbonate to maintain equilibrium. Any process that decreases the amount of CO_2 results in the precipitation of calcium carbonate. For example, CO_2-bearing groundwater will tend to dissolve limestone along fractures, producing cavities or caves. If the water table lowers, these cavities contain air. Then if CO_2-bearing groundwater under some pressure, as a result of confinement in the groundwater system, enters these caves along fractures, an escape of CO_2 gas from this water, because of a slight drop in the partial pressure of CO_2, results in precipitation of calcium carbonate, typically in the form of stalactites and stalagmites (speleothems). In fluvial systems, a slight drop in CO_2 pressure may occur as the result of turbulence, promoting local encrustations of carbonate minerals on rock surfaces.

An increase in temperature results in a decrease in the solubility of CO_2 in water, at low pressure, and a lower partial pressure of CO_2 related to this heating favors precipitation of calcium carbonate. Thus, minerals such as aragonite and calcite are less soluble at higher temperatures, quite unlike halite, which is more soluble at higher temperatures.

Organisms remove CO_2 from the environment in the process of photosynthesis, which can cause precipitation of calcium carbonate. Conversely, decay of organic matter produces CO_2, and carbonate minerals can be dissolved.

Much surface ocean water is near saturation with respect to calcium carbonate. This is a favorable environment for precipitation of calcium carbonate especially if assisted by organisms. Deep-ocean water is markedly undersaturated with respect to calcium carbonate. The reason is that with an increase in pressure and a decrease in temperature, more CO_2 can be held in solution. This prevents precipitation of calcium carbonate in deep waters and favors the dissolution of calcareous tests (planktonic origin) settling from above. The rate of dissolution of these calcareous biotic remains compensates for the rate of their supply. Thus the calcite compensation depth (CCD) is the depth separating environments of precipitation and dissolution. The CCD varies according to the ocean involved, but is generally 3.5–5 km for calcite and 2–3 km for aragonite. On the basis of solution chemistry alone, the CCD should be about 500 m. This apparent discrepancy means that the kinetics of dissolution and precipitation of calcium carbonate is a major controlling factor.

Acid waters generated by the weathering of rocks containing sulfides readily dissolve calcium carbonate:

$$CaCO_3 + 2H^+ \rightarrow Ca^{2+} + H_2O + CO_2 \text{ (or } HCO_3^-\text{)}$$

According to solubility product calculations, aragonite is slightly more soluble than calcite. Nevertheless, aragonite forms where, on the basis of solubility, calcite should form. In this case aragonite probably

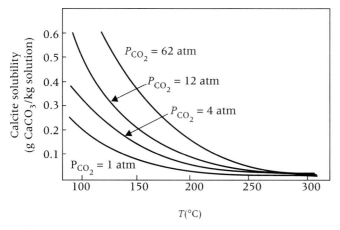

Figure 12.6
Solubility of calcite in H_2O at various partial pressures of CO_2. Solubility decreases with temperature but increases with pressure (Ellis, 1959, 1963).

forms **metastably** and with time reverts to calcite. This is kinetically very slow because nucleation is required to make the transition from orthorhombic structure to hexagonal structure. Shells consist of aragonite or calcite, and in the case of pelecypods, there may be alternating layers of each. High temperature favors aragonite, as attested by the common occurrence of aragonite in travertine.

Dolomite is another carbonate mineral that has great importance in sedimentary rocks. An equilibrium relation can be shown as

$$CaMg(CO_3)_2 = Ca^{2+} + Mg^{2+} + 2CO_3^{2-}$$

This equilibrium does not seem to have much relevance in aqueous solutions of geological interest. In the first place, dolomite cannot be synthesized in the laboratory, and the equilibrium constant can only be approximated. In the second place, even though seawater is approximately saturated in dolomite, there is no evidence for direct dissolution nor precipitation in seawater. Dolomite contains Ca and Mg ions in different planes. The structure is very ordered, and evidently the kinetics of reaction are extremely slow at normal temperatures. When there is competition for Ca ions, calcite (or aragonite) precipitates instead of dolomite.

A geologically more pertinent reaction is "conversion" of calcite to dolomite:

$$2CaCO_3 + Mg^{2+} \rightarrow CaMg(CO_3)_2 + Ca^{2+}$$

Since ion diffusion in crystals is exceedingly slow, **dolomitization** takes place more likely by dissolution of calcite in the presence of Mg ions with simultaneous dolomite precipitation.

Magnesium may also occur in calcite, albeit in disordered fashion. High-Mg calcite and aragonite form in modern carbonate sediments in preference to calcite and dolomite. Evidently the presence of magnesium depresses the precipitation of calcite but not of aragonite. Low-Mg calcite (< 5% $MgCO_3$) also forms, and it is more stable in seawater than pure calcite.

$CaSO_4$–H_2O System (Anhydrite–Aqueous Phase)

Anhydrite and gypsum are important constituents of evaporite sedimentary rocks. They also occur in hydrothermally altered rocks in which calcium and sulfate are available. The solubility of anhydrite decreases with rising temperature (Figure 12.7). The equilibrium relation is

$$CaSO_4 = Ca^{2+} + SO_4^{2-}$$

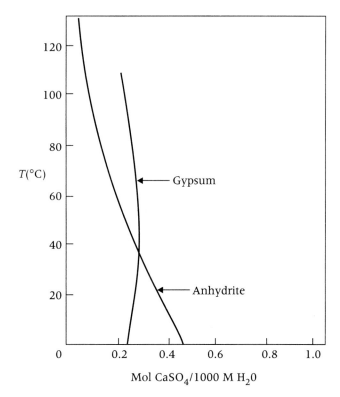

Figure 12.7
Solubility of anhydrite and gypsum in water at various temperatures. Solubility of anhydrite decreases with temperature, whereas solubility of gypsum is little affected (Braitsch, 1971).

Anhydrite has a very close association with gypsum. The chemical relation is

$$\underset{\text{anhydrite}}{CaSO_4} + 2H_2O = \underset{\text{gypsum}}{CaSO_4 \cdot 2H_2O}$$

The solubility of gypsum is nearly independent of temperature (Figure 12.7). In aqueous solutions, gypsum is the stable form below 42°C (Figure 12.7). Above this temperature, anhydrite is the stable form, but since nucleation energies are higher for anhydrite, gypsum may crystallize in its place. Only gypsum forms in laboratory experiments.

Aqueous Phase in the Geological Environment

In very few places are rocks formed that are completely free of H_2O (Fyfe et al., 1978). The principal locations and associations of H_2O are shown in Figure 12.8. Since chemical activity involving aqueous phase

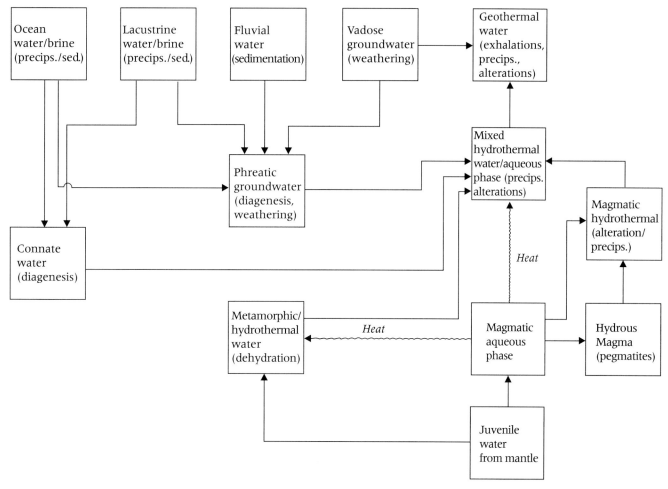

Figure 12.8
Occurrence of H$_2$O in the geological environment.

is typically in some way dependent on temperature, a brief discussion of the participation of aqueous phase in rock-forming processes is made according to environment temperature.

Low-Temperature Environments

Low-temperature aqueous phase occurs on the Earth's surface as (1) **fluvial water**, (2) **lacustrine water**, and (3) **ocean water** (Drever, 1988). Water in open spaces of rock materials above the water table is in the **vadose** zone, and water below the water table is in the **phreatic** zone. The phreatic zone includes water in rock materials beneath lakes and oceans. Water of the vadose and phreatic zones is collectively known as **groundwater**. Surface water and groundwater have been in recent contact with the atmosphere and as such is **meteoric water**.

The chemical environment of surface water and groundwater varies from oxidizing to reducing and from acid to basic, plotting conveniently on an Eh–pH diagram (Figure 12.9). Oxidizing environments are those in which free oxygen is available. Reducing environments are isolated from the atmosphere and therefore from oxygen. **Eh** is the geochemical expression of the tendency for oxidation or for reduction (**redox potential**) of a given element in aqueous solution. The reduction of H$^+$ to H$_2$ is taken as a standard and assigned the value of 0 volts. As an example, the strong tendency for Al (as artificially prepared) to oxidize to Al^{3+} is indicated by a standard oxidation potential of +1.66 V, and the fact that metallic Al does not occur in the natural environment is indicated by a standard reduction potential of −1.66 V. In comparison, the standard reduction potential of Cu^{2+} is +0.34, which is indicative of the well-known occurrence of copper as a native element in a reducing environment.

Mixing of meteoric phreatic water with ocean phreatic water along an interface in coastal regions is likely to shift equilibria. For example, dolomitization of limestone may be favored in the zone of mixing.

Settling of clastic grains and precipitation of minerals in lakes, oceans, and the continental margin basins where seawater and freshwater meet results in sedimentation. Sedimentation associated with fluvial activity is mainly clastic, much of which is dumped into the standing bodies of water. Groundwater is particularly instrumental in the formation of weathered rocks and rocks that result from precipitation such as **calcrete** and **speleothems** (cave deposits).

Connate water occurs in the pores of buried sediments and their indurated equivalents. The water was trapped as meteoric water, but as connate water it has been out of contact with the atmosphere for a long time. Connate water solutions are particularly important in the **diagenesis** of sedimentary rocks.

With continued burial, connate water is squeezed out of sediments, leaving only a film of water molecules rather strongly attached to mineral and rock grains. Sooner or later dehydration reactions occur and water that was tied up in minerals, such as clays and gypsum forming in the surface environment, is now liberated at depth. This water mixes with pore water, and this composite connate water is progressively expelled as sedimentary rocks are converted to metamorphic rocks. Some water is retained at lower grades of metamorphism in the form of new hydrous minerals, such as micas, chlorites, and amphiboles.

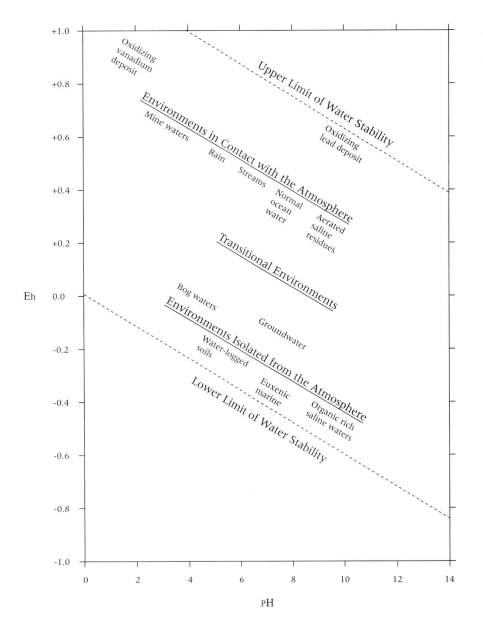

Figure 12.9
Occurrence of water in the natural range of Eh and pH (Garrels and Christ, 1965).

Medium- and High-Temperature Environments

Higher-temperature environments occur at depths in accord with the geothermal gradient and near magmatic heat sources. At high temperatures the water is above its critical point and is **aqueous phase**. Magma may supply heat at any depth and right at the Earth's surface. This means that meteoric water and connate water may be heated and become **hydrothermal water**.

As indurated sedimentary rocks and low-grade metamorphic rocks are subjected to extreme grades of metamorphism, dehydration reactions take place and generate **metamorphic water**, which is at great depth where the temperature and pressure are great, but it can also be generated at shallow depths in zones of contact thermal metamorphism. In any event, metamorphic water can be expected to blend with meteoric water and connate water or, more realistically, with hydrothermal water derived from meteoric and connate water.

Juvenile water is aqueous phase derived from the mantle. Most of this water finds its way into the crust dissolved in basaltic magma that has formed anatectically from mantle rocks and has subsequently intruded into the crust. Crystallization of this water-bearing basaltic magma liberates a hydrothermal phase. This is **magmatic water**. Another source of magmatic water is from siliceous magmas generated by the anatexis of crustal rocks. Although this water is magmatic water, it is not juvenile, since melting of siliceous crustal rocks incorporates water that mostly had a connate and meteoric beginning.

Expelled magmatic water is a major source of **hydrothermal water**. Hydrothermal waters typically consist of magmatic water mixed with meteoric water, particularly where magmas crystallize in the upper crust where the liberated magmatic water interfaces with meteoric groundwater and connate water. It is also reasonable to expect mixing with metamorphic water at greater depths.

Hydrothermal water, whatever the sources, may arrive at the surface (subaerial or subaqueous) in the form of hot springs. This is **geothermal water**. Aqueous phase may also arrive at the surface in the form of water vapor issuing from fumaroles associated with volcanic activity.

13

Phase Relations in Crystallizing Magma and Melting Rock

Experiment is a necessary check on inference from observations on the natural materials and in turn provides a chemical basis for hypotheses on origins, which may be tested in the field and modified to give a nearer approach to the mechanism of rock genesis.

J.F. Schairer (in Bowen, 1956)

Magma and Lava

Silicate melt contains mostly silicon and oxygen along with lesser amounts of aluminum, magnesium, iron, calcium, sodium, and potassium. In the geological environment, the silica content of magma varies from about 55% to 70%, resulting in the formation of magmatic rocks ranging from mafic to felsic and from alkaline to silicic. Magma originates in the mantle and lower crust, forming intrusive magmatic bodies in the upper crust and extrusive magmatic bodies on the Earth's surface. The deeper intrusive bodies are mostly plutons and the shallower intrusive bodies are typically dikes, sills, and laccoliths. Magmatic bodies nearly or partly extrusive include necks, plugs, and domes, whereas lava flows are totally extrusive.

The chemical components of magma do not occur as ions or complex ions floating around in a solvent as they do in aqueous solutions. Instead they form complex molecules that link together (polymerize), giving "structure" to the melt. Loss of heat from magma allows for decidedly more organization of these chemical components into **crystalline silicate solids**, and if heat loss is extremely rapid, as it may be in a volcanic environment, lava solidifies to **glass**.

An increase in **geostatic pressure** favors transformation of silicate liquid to crystalline solids, even if there is no heat loss. In the geological world, this relation is of little consequence since magmas generally are less dense than their rock surroundings and are prone to move up to lower-pressure regions rather than down to higher-pressure regions. Furthermore, there is no tectonic mechanism that can place a batch of magma into a higher-pressure region. Consequently, loss of heat is the fundamental reason for crystallization of magma.

Generation of magma by **anatexis** (melting of pre-existing rock) is more or less the opposite of **magmatic crystallization**. In melting, heated crystal structures of minerals become unstable, breaking down to form melts. Melting occurs in rocks that invariably have several mineral phases. This means that melt solutions are formed, not pure melts equivalent to specific mineral compositions.

Transformation from the crystalline state to melts is favored by reduction of geostatic pressure, except if there is H_2O dissolved in the magma. This is **adiabatic melting**, occurring at sites of extension (pressure reduction) along deep faults where rocks are hot. For example, Hawaiian eruptions are invariably preceded by deep-focus earthquakes, strongly suggesting that there is a genetic relation between changes in rock pressure and generation of magma. Once a magma is formed and is on its way toward the Earth's surface, crystallization begins in response to heat loss. Once again, reduction of pressure may lead to formation of melt solution by **partial dissolution** of early formed crystals without addition of heat.

Phase Equilibria

The object of **phase equilibria** experimentation is to find out what the temperature and pressure conditions are when the various types of magmatic rocks form. Rocks have different melting and freezing temperature ranges. Basalt magmas are much hotter than water-bearing granitic magmas.

Mineral phases forming in experimentation are determined by the chemical composition of the starting materials. The **phase state**, solid or liquid in the case of magmas, is recorded at various temperatures and pressures (Ehlers, 1972; Ernst, 1976; Morse, 1980). Thermodynamic modeling of phase relations in magmatic systems is also possible (Nekvasil, 1988).

Application of experimental or thermodynamic data to real rocks is made with caution for at least three reasons; (1) natural magmas are chemically much more complex than simulated experiments or calculated compositions; (2) not all magmatic rocks represent the composition of magma from which they crystallized (dunite represents mechanical accumulation of olivine crystals, not a magma of olivine composition); and (3) some magmatic rocks result from heterogeneous mixing of one magma with another of different composition.

Some Fundamental Phase Equilibria Relationships

Diopside–Anorthite System

The fundamental relation between the liquid state and the solid state in a system containing two components is shown with the diopside–anorthite system diagram (Figure 13.1; Bowen, 1915). This magmatic system consists of two minerals that have a **binary eutectic** relationship. The chemical components in the system are $CaMgSi_2O_6$ and $CaAl_2Si_2O_8$, and the bulk composition is similar to some mafic magmatic rocks.

A magmatic **phase diagram** establishes boundary conditions between the liquid and solid states as governed by temperature, pressure, and composition (P–T–X). In systems without H_2O as an additional component, pressure does not have as much effect on the phase relations as does temperature and composition. Consequently, pressure is held constant in this

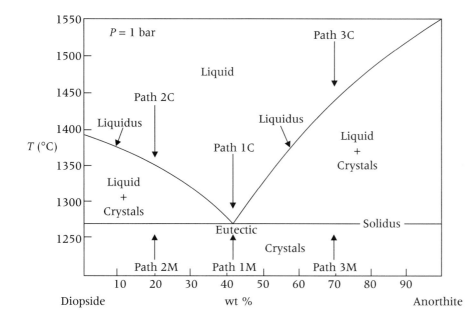

Figure 13.1
The diopside–anorthite system at 1 bar (Bowen, 1915). See text for discussion of paths.

experiment, and the pertinent relationships can be shown with a two-dimensional diagram.

The **liquidus** is a boundary curve representing temperatures above which all crystals are dissolved in a hypothetical anatectic system containing diopside and anorthite, and below which crystallization may begin in a magmatic system. The **solidus** is the boundary between the very first liquid to form anatectically or the very last magmatic liquid to crystallize.

Any point on the liquidus represents an equilibrium between melt and one type of crystal. A point on the solidus means that there is an equilibrium between two mineral phases. The **eutectic** point is a position where the liquidus and solidus meet, where the melt has a specific composition and is in equilibrium with two mineral phases.

Diopside and anorthite are silicate minerals with different crystal structures. Consequently, there are no minerals with compositions intermediate between diopside and anorthite. This lack of a **solid solution** relation dictates the particular configuration of the binary phase diagram. The eutectic is located at about 42% anorthite and 58% diopside, at which point the two solid phases and one liquid phase coexist. For this system, the eutectic is at 1270°C at 1 atm pressure, without the liquidus-lowering effect of water, which was not included in the experiment.

The melting (or freezing) temperature of pure diopside or pure anorthite is systematically lowered by the presence of the other. This is the same as saying that the bonding of ions in their respective crystal structures is inhibited by the presence of foreign ions.

The physical (textural) relations between crystals of diopside and anorthite can be predicted by tracing the **path of crystallization**, also referred to as the **path of changing melt composition**. Crystallization begins when the melt is cooled and the liquidus is reached for some specified starting composition. For example, the cooling path (1C, Figure 13.1) results in nucleation and crystallization of diopside and anorthite simultaneously, since the point of liquidus intersection is the eutectic point. In reality, nucleation may not occur precisely at a eutectic nor on any liquidus (Chapter 14), but for purposes of demonstration of phase equilibria it is assumed that there is no supercooling of melts and the process takes place under equilibrium conditions. At the eutectic, diopside and anorthite may crystallize in a side by side or an intergrowth textural relation (depending on rate of cooling and other factors, Chapter 14), continuing until all the melt is used up.

If the initial composition of a melt were weighted in favor of diopside, intersection of the liquidus with drop in temperature (path 2C) results in the nucleation and growth of diopside. Crystallization of diopside depletes the melt in equivalent components, shifting the composition toward anorthite—that is, down the liquidus as heat is taken from the system. Crystallization along the liquidus can be traced all the way to the eutectic point, at which point anorthite begins to crystallize as well. At the eutectic, simultaneous crystallization continues until the melt is used up. Similarly, if the initial composition of a melt were rich in anorthite growth components (path 3C), anorthite would begin to crystallize first, followed by simultaneous growth at the eutectic.

Reversal of these crystallization relations can be traced in melting experiments, simulating anatexis. If there is heating of a rock containing 58% diopside and 42% anorthite (path 1M, Figure 13.1), beginning of melting occurs at the eutectic temperature and the composition of these first and last melts will be the eutectic composition. For any other proportion of diopside with respect to anorthite, the first melt to form occurs along the interface of any two crystals of diopside and anorthite, and this first melt has the composition of the eutectic. This occurs because the eutectic composition has the lowest melting temperature. However, if the bulk composition of the "rock" is toward diopside (path 2M) or anorthite (path 3M), away from the eutectic, the less abundant mineral will be totally melted before the other, effectively tracing the **path of melting** back up the liquidus, in reverse of the path of crystallization.

Leucite–Silica System

The leucite–silica system (Figure 13.2; Schairer and Bowen, 1947) is a very important subsystem of the more complex granitic and syenitic systems. In systems capable of crystallizing a silica mineral, quartz is generated at low temperatures in systems that have H_2O as an additional component. Tridymite and/or cristobalite occur in high-temperature "dry" systems. The silica polymorphs are shown in relation to temperature and pressure in Figure 13.3.

In systems that have the appropriate chemical components to crystallize leucite and a silica mineral, K-feldspar appears as an additional phase. The stability fields of tridymite and leucite (Figure 13.2) are not contiguous, but are separated by K-feldspar since the reaction leucite + silica mineral → K-feldspar is likely to occur. Therefore, in a melt system, or in a "rock" system being melted, there is the expected formation of K-feldspar as well as leucite and quartz.

If a melt is compositionally weighted toward leucite, **incongruent melting** (anatexis) or **incongruent freezing** (magmatic crystallization) occurs. Beginning of crystallization on the liquidus produces leucite (path 1C) even though there may be enough silica in the system to produce K-feldspar. Leucite is stable at higher temperatures than K-feldspar, so it

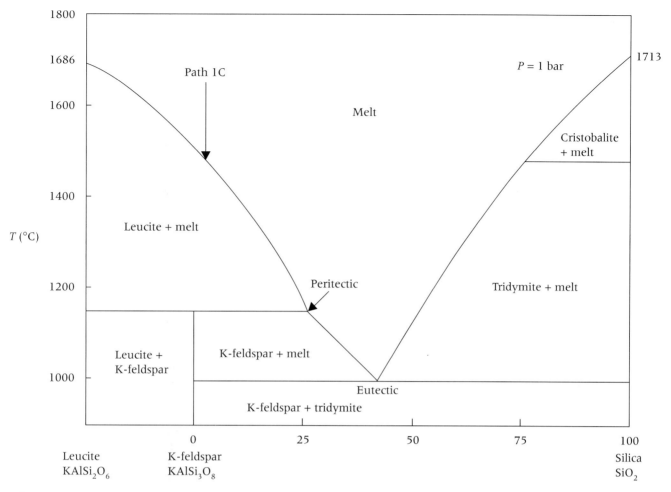

Figure 13.2
The leucite–silica system at 1 bar (Schairer and Bowen, 1947). See text for discussion of paths.

appears first at the higher temperatures. However, as crystallization of leucite proceeds, the melt composition moves down along the liquidus to the **peritectic**. At that point, K-feldspar is stable, and a reaction can occur between leucite crystals and the silica-richer melt to produce K-feldspar. Since the eutectic has not been reached and a new phase (K-feldspar) is appearing, the relation is termed **incongruent**. If the temperature is held at the peritectic (or slightly below it), or if there is slow progression down the liquidus toward the eutectic, the conversion of leucite to K-feldspar could go to completion as long as melt has access to leucite. However, in natural systems the leucite will tend to acquire a protective mantle of K-feldspar, thus limiting reaction. Regardless of the amount of leucite converted to K-feldspar, crystallization of K-feldspar directly from the melt occurs along the liquidus between the peritectic and the eutectic, at which point both K-feldspar and tridymite crystallize simultaneously until melt is depleted.

Albite–Anorthite System (Plagioclase Series)

The albite–anorthite system is a binary system with no eutectic (Figure 13.4; Bowen, 1913). Since there is a continuous solid solution series between albite and anorthite, crystallization produces one crystalline phase rather than two. There probably is no natural magmatic system consisting of plagioclase constituents alone, but the system is a very important part of more complex systems that produce most of the known magmatic rocks.

A continuous series is conducive to the crystallization of zoned crystals. In a sense, zones in minerals of a solid solution series are different minerals even though they occur in a single crystal. Starting with a melt composition equivalent to 40% anorthite and 60% albite, the liquidus is intersected at that composition on cooling (Figure 13.4). The composition of the

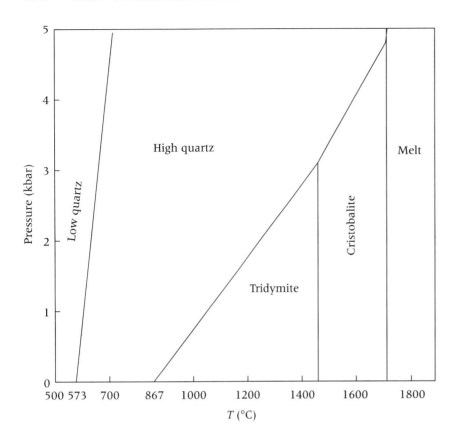

Figure 13.3
Pressure–temperature dependence of silica polymorphs (Tuttle and Bowen, 1958).

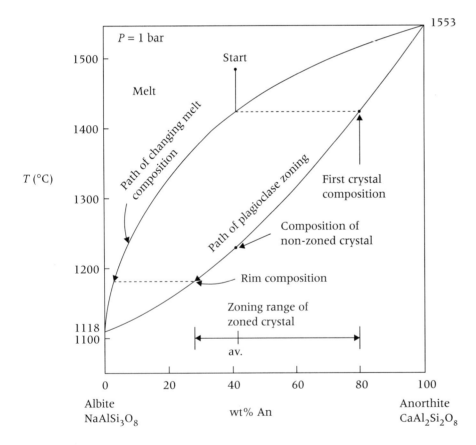

Figure 13.4
The albite–anorthite system at 1 atm, with crystallization path for An_{40} (Bowen, 1913). See text for discussion of crystallization paths.

first crystals at liquidus intersection are much more anorthitic than the composition of the initial melt, as shown by the horizontal line intersecting the solidus at about 80% anorthite. As crystallization proceeds, as a consequence of heat loss, the melt becomes progressively enriched in albitic component, and the path of changing melt composition moves down the liquidus toward albite.

What is happening to the developing crystalline phase as temperature falls? The early forming, more calcic plagioclase may react with melt, generating progressively more sodic plagioclase while the melt progressively changes its composition toward albite. If such reactions continue, maintaining **equilibrium**, the final plagioclase crystals would be nonzoned and of a composition identical to that of the initial melt, namely 40% anorthite.

In natural magmatic systems, plagioclase most commonly is zoned. Evidently, continuous reaction is inhibited by the armoring effect of the outer layers of crystal, eliminating contact of new melt with old crystal and effectively preventing exchange of ions. Compositional zoning is the result. It is a state of overall **disequilibrium** because the interior of the crystal is at a different energy state than the exterior portions with respect to the remaining melt. **Normal progressive zoning** may result, being a continuous (progressive) change (decreasing An) in composition with respect to falling (normal case) temperature. The path of changing melt composition moves down the liquidus to very albitic compositions; the path of crystallization moves from the composition of the very first crystal, approximately An_{80}, to rim plagioclase more albitic than the starting composition (An_{30} for this example). The average composition of the plagioclase must still be An_{40}.

The most common type of zoned plagioclase in magmatic rocks is **normal progressive oscillatory zoning**. Oscillations result from momentary but repeated reversal of composition. Each reversal starts with a more calcic composition and finishes with a more sodic composition in equilibrium with the melt. Each calcic peak is less calcic than in the previous oscillation. The result is an overall normal progressive zoning trend with superposed oscillations. The reason for oscillatory growth is most likely related to disparities in growth rates relative to diffusion rates at the melt–crystal interface. The **diffusion–supersaturation** model of oscillatory zoning is described in Chapter 14.

Nonzoned magmatic plagioclase does occur, but it is not common. If the melt in the vicinity of a crystallizing plagioclase is continuously refreshed, continued growth of a fixed composition is possible (Chapter 18).

Phase relations in the plagioclase series are also important in consideration of anatectic melting of plagioclase-bearing rock. Since plagioclase is a very common mineral in a wide variety of rocks, the chances of it being involved in anatectic melting are very good. Melting of zoned and nonzoned plagioclase is a complex process because of the plagioclase solid solution. There is a propensity for melts formed to be in equilibrium with crystals of different composition (Chapters 14, 28), just as the first crystals forming in a crystallizing magma are much more calcic than the coexisting melt.

Sanidine–Albite System (Alkali Feldspar Series)

The sanidine–albite system (Bowen and Tuttle, 1950) is a pseudobinary system with solid solution. At 1 atm (no dissolved H_2O), a melt equivalent to 35% sanidine and 65% albite crystallizes as a homogeneous unzoned alkali feldspar (anorthoclase) at the **minimum** temperature (1063°C) (Figure 13.5). This is not a eutectic

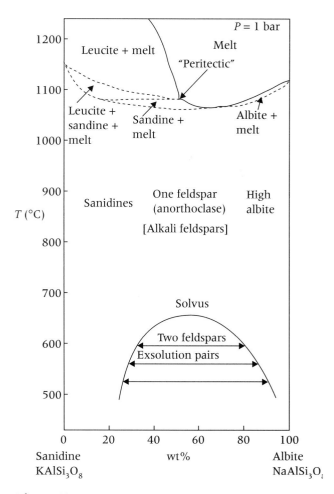

Figure 13.5
The alkali feldspar series at 1 bar (Bowen and Tuttle, 1950).

as it was in the diopside–anorthite system, where two (not one) mineral phases crystallized simultaneously.

Crystallization of melt compositions rich in $KAlSi_3O_8$ yield leucite, just as they did in the leucite–tridymite system (Figure 13.2). This leucite reacts with the melt (if physically permitted) to form a sanidine at a point equivalent to the peritectic in the leucite–silica system. If the original melt composition is not extremely rich in K-feldspar component, crystallization may proceed to the minimum. Zoning to progressively more albitic alkali feldspar is expected under nonequilibrium conditions.

Alkali feldspar containing 25–95% albite component is not stable at temperatures below the solvus (Figure 13.5). The solvus is a boundary curve separating the field of stability of a single homogeneous alkali feldspar (such as anorthoclase) at high temperature, from the lower-temperature field in which two alkali feldspars are generated by **exsolution**. The resulting **perthitic** alkali feldspar is characterized by lamellar alternation of albite and K-feldspar (Figures 2.6c, 7.16a,b). A homogeneous alkali feldspar is unstable below the solvus, but it may survive metastably in the rapid-cooling volcanic environment. Perthite is characteristic of alkali feldspar granites and syenites, where slow cooling in the plutonic environment allows exsolution to take place.

The instability in the alkali feldspar series at subsolvus temperatures arises from the disparity of ionic size of potassium in comparison with the much smaller sodium ion, both of which try to fit into the feldspar structure in the same site. When a homogeneous alkali feldspar is slowly cooled, lattice strain is relieved by ion migration, leading to the appearance of separate mineral phases.

Detection of zoning in sanidines, anorthoclases, and orthoclases is more difficult than in plagioclase in which composition is very sensitive to differences in crystal structure and optical orientation. Nevertheless, alkali feldspar zoning may have optical expression, most commonly defined by exsolved albite (Figures 7.8c, 14.3b).

Effect of Elevated Pressure on Equilibrium

The effect of geostatic pressure on phase relations in silicate systems, though generally not large, can mean the difference between crystal growth and crystal dissolution if a mineral is on or near its liquidus. A partial dissolution texture, for example, may indicate that a crystal-bearing magma has experienced a sudden decrease in confining pressure, as long as the system is relatively "dry" and the effects of H_2O are not relevant.

The liquidus and solidus boundaries are raised with respect to temperature as the confining pressure increases in dry systems (Figure 13.6). In natural systems, in which pressure is reduced as intrusion takes place, the liquidus–solidus shift to lower temperatures. Crystals that formed at higher pressures find themselves well above the liquidus, dissolving back into the magma.

The shift in position of the liquidus and solidus is shown for the simple binary eutectic system albite–silica (Figure 13.6a) and in the plagioclase system (Figure 13.6b).

The effect of pressure on the crystallization of magma or melting of rock to form magma can be shown for a single composition by systematically varying pressure and temperature (Figure 13.6c) If a magma composition happens to be a eutectic or minimum composition, there is no distinction between liquidus and solidus. This combined "liquidus–solidus" rises with respect to temperature as pressure is increased (Figure 13.6c). Most magma compositions are not this simple, and the liquidus and solidus are separated by a crystal-plus-melt region. In that case, both the liquidus and solidus rise with increase in pressure.

Effect of H_2O on Silicate Phase Relations

Aqueous phase can enter (dissolve into) silicate melt as OH^{1-} ions by breaking (depolymerizing) oxygen bridges that bond $(Al,Si)O_4$ tetrahedral "molecules" in the melt. The following reaction is useful in illustrating the depolymerization:

$$(Al,Si)O_4 - O + H_2O = (Al,Si)O_4 + 2(OH)^{1-}$$

In effect, the structural condition of the melt containing dissolved H_2O is much farther from a quasi-crystalline state existing before aqueous phase was introduced, and the magma is less viscous (Hess, 1977; Mysen, 1988). The temperature (or temperature range) at which H_2O-bearing magmas freeze is lower than "dry," more polymerized magmas because more thermal energy has to be abstracted from the system before the effects of depolymerization can be overcome and reorganization can occur all the way to the crystalline state.

Magmas under higher geostatic pressure can dissolve more aqueous phase (Figure 13.7). Therefore, a magma generated at greater depth will dissolve more H_2O, if H_2O is available, than will a magma nearer the surface. It follows that if a magma already is saturated with H_2O, intrusion to a region of lower pressure will result in a separation of H_2O as an aqueous phase

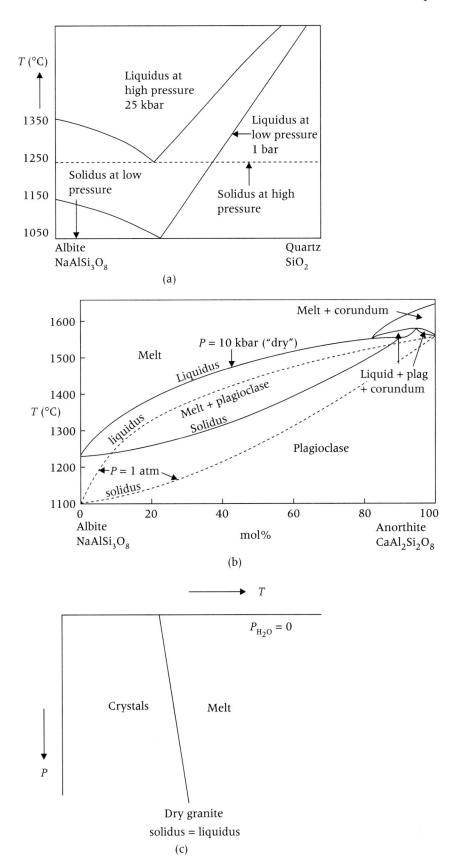

Figure 13.6
The effect of confining pressure on silicate phase equilibria.

(a) The albite–silica system (data from Bell, 1964). (b) The plagioclase series (Lindsley, 1964). (c) The "dry" granite system (Tuttle and Bowen, 1958).

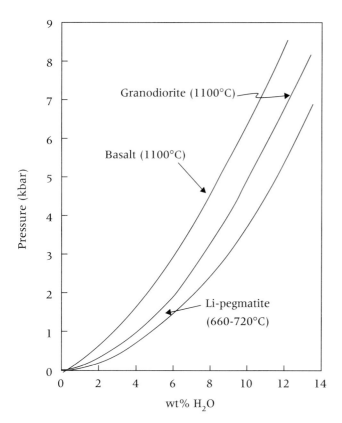

Figure 13.7
Solubility of H_2O in magmas of basaltic, granodioritic, and Li–pegmatite composition (Hamilton et al., 1964).

(Whitney, 1975). Since it has been shown that H_2O depolymerization lowers the liquidus, it follows that the presence of H_2O will result in anatectic melting at lower temperatures.

There are other effects of H_2O on the crystallization of magma, besides lowering of a liquidus and solidus. If H_2O is added to the diopside–anorthite system, the appearance of amphibole can be expected (depending on pressure and temperature) since

diopside + H_2O → amphibole

If diopside is the first mineral phase to appear, its continued crystallization results in a concentration of dissolved H_2O in the melt, since diopside does not use H_2O in its crystal structure. Crystallization of anorthite has the same effect since it, too, is anhydrous. As partial pressure of H_2O increases, amphibole may become the stable phase in preference to diopside. Epitaxial amphibole mantles on augite and partial replacements of augite are very common in dioritic rocks. This is part of Bowen's (1928) "reaction series."

The effect of H_2O dissolved in melts of the K-feldspar–albite system at various pressures is shown in Figure 13.8. The "dry" system at 1 bar (Figure 13.5) has a liquidus–solidus at very high temperatures. At 1 kbar with the melt saturated with H_2O, the liquidus–solidus is lowered (Figure 13.8a), and it is lowered further at 2 kbar (Figure 13.8b). There is also a steady, but small, increase in the solvus temperature with increasing pressure. At 5 kbar and saturation with H_2O, the liquidus and solidus are lowered so much that the solidus and solvus intersect (Figure 13.8c).

Notice that the field of leucite is progressively reduced and finally eliminated. Effectively, the liquidus and solidus being lowered by the effects of dissolved H_2O favors direct crystallization of alkali feldspar rather than leucite.

Intersection of the solidus with the solvus has even more important consequences. Instead of a single alkali feldspar (anorthoclase) crystallizing at the minimum in the K-feldspar–albite system, two feldspars—one richer in potassium and one richer in sodium—may crystallize in equilibrium (Figure 13.8c). In effect, the minimum has become a eutectic; that is, two solid phases are in equilibrium with a melt. Therefore, crystallization of water-bearing magma at depth under higher pressure will yield both a potassium-rich feldspar (orthoclase instead of sanidine at these lower temperatures) and a sodium-rich feldspar, if the initial composition is generally in a middle range between the end members. Although a side-by-side crystallization of a K-feldspar and an albite seems to be indicated by these phase relations, Tuttle and Bowen's "subsolvus granites" (Tuttle and Bowen, 1958), there seem to be few natural occurrences, suggesting that granite systems are not saturated in H_2O during the early stages of crystallization (Robertson and Wyllie, 1971; Wyllie, 1983).

The shift of the liquidus and solidus of the plagioclase series related to water pressure is shown in Figure 13.9. The shift is about 300° lower if the water pressure (equal to confining pressure) is 5 kbar, much greater than the reverse shift of 100° or so, obtained by raising the confining pressure to 10 kbar in a dry system (Figure 13.6b).

Melting of a rock in the presence of an inadequate supply of aqueous phase results in the reverse process (Wyllie, 1977; Thompson, 1982). Initial melting could produce a water-bearing melt at temperatures reflecting lowered liquidus and solidus. As melting proceeds, in response to increasing temperature, the melt may become progressively more undersaturated with respect to aqueous phase if the initial availability of water was limited. Continued melting of the rock may not occur unless the temperature continues to rise, keeping pace with the shift of the liquidus and solidus to higher temperatures as the melt progressively becomes "dryer."

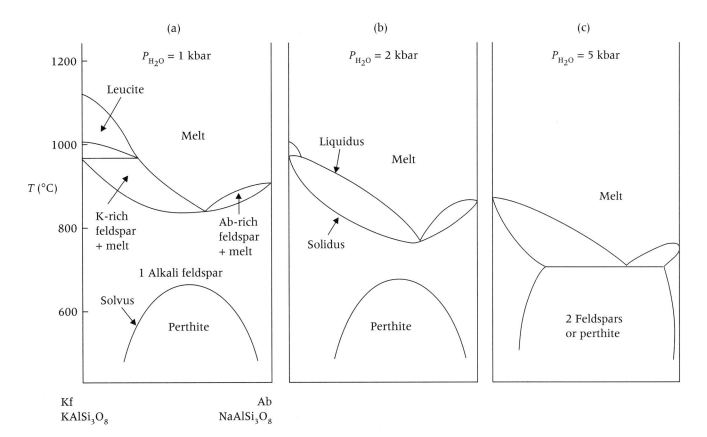

Figure 13.8
The alkali feldspar system at 1 kbar, 2 kbar, and 5 kbar water pressure (Bowen and Tuttle, 1950; Tuttle and Bowen, 1958; Yoder et al., 1957; and Morse, 1969).

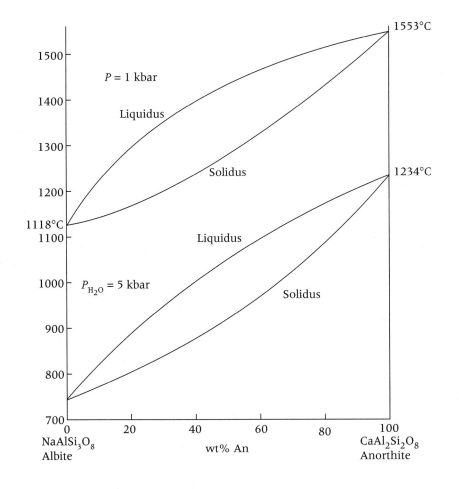

Figure 13.9
The plagioclase system at 5 kb H$_2$O pressure in comparison with the "dry" system at 1 bar (Bowen, 1913; Yoder et al., 1957).

If an H_2O-bearing magma becomes saturated in H_2O because of crystallization of anhydrous mineral phases, and/or reduction in geostatic pressure attendant to intrusion (less H_2O soluble at lower pressure, Figure 13.7), there may be a buildup of aqueous-phase pressure, leading to rupture of the confining magma chamber walls and rapid escape of the aqueous phase. This is a **pressure quench**, resulting in rapid crystallization of remaining melt.

Rock-Forming Magmatic Systems

The System Plagioclase–Clinopyroxene (Gabbro–Basalt)

If $NaAlSi_3O_8$ is added to the diopside–anorthite system, a **ternary phase diagram** is generated (Figure 13.10). The albite–anorthite–diopside system approximates magma from which gabbro and basalt (without orthopyroxene and olivine) crystallize (Hamilton and Anderson, 1967). However, most natural systems such as this contain augite instead of pure diopside and at least some dissolved H_2O.

Ternary diagrams have a vertical temperature axis. They can be shown as either a two-dimensional projection of liquidus surfaces (Figure 13.10a) or as a three-dimensional form (Figure 13.10b). A ternary system contains three binary subsystems; in this example they are the albite–anorthite, anorthite–diopside, and albite–diopside systems (Figure 13.10c–e).

Various paths of crystallization can be traced showing the sequence of crystallization and predicted textural development. One such path (path 1, Figure 13.10) begins in the melt region, proceeding vertically down to an intersection of the plagioclase liquidus surface where relatively calcic plagioclase begins to crystallize. The composition of the melt descends down the liquidus surface, across the isotherms, toward the **boundary curve** between diopside and plagioclase. If plagioclase is zoning (disequilibrium crystallization), it becomes progressively more sodic, reflected in the curved path convex toward Ab apex. Intersection of the boundary curve occurs and diopside begins to crystallize along with plagioclase. Two solid phases are now in equilibrium with a melt and the boundary curve is a **cotectic line**. Crystallization of two phases continues down the cotectic line, toward albite, until all the melt is used up.

If rates of nucleation and crystal growth are appropriate, there may be a textural indication that plagioclase began to crystallization prior to pyroxene. Well-formed laths of plagioclase included in pyroxene ("ophitic" or **oikicryst** texture) are common in mafic magmatic rocks, suggesting a somewhat earlier start of plagioclase crystallization (Figure 17.5d). If the nucleation rate of diopside is relatively high, many smaller grains of diopside will concentrate between tabular plagioclase in a granular ("intergranular") texture.

The System Plagioclase–Clinopyroxene–Olivine (Olivine Gabbro–Basalt)

The system plagioclase–clinopyroxene–olivine contains olivine as a third mineral phase (ignoring spinel), approximating olivine basalt and olivine gabbro. Phase relations are shown with a three-dimensional diagram (Figure 13.11). The front face of this tetrahedron is the albite–anorthite–diopside ternary system with its cotectic line curving across its face (Figure 13.10). Within the tetrahedron there is a **ternary cotectic line** representing the intersection of three boundary surfaces (pl–ol, di–ol, and pl–di). Isotherms are not conveniently shown within the tetrahedron, so in their place liquidus temperatures are shown at strategic points and the direction of falling temperature on the ternary cotectic line is indicated with an arrow.

A path of crystallization (path 1) beginning in plagioclase volume moves to the forsterite–plagioclase boundary surface, where plagioclase crystallization is joined by forsterite. The path then moves across the fo–pl surface to the ternary cotectic line, where diopside becomes the third mineral phase crystallizing simultaneously with plagioclase and olivine.

Olivine gabbros and olivine basalts commonly contain an orthopyroxene. Phase relations in the quinary system plagioclase–olivine–clinopyroxene–enstatite–quartz cannot be shown in a single diagram. Since forsteritic olivine is unstable in the presence of quartz, the phase relations predict that enstatite will form in place of forsterite or in addition to forsterite, depending on the amount of additional silica in the system (Bowen, 1914, 1928; Schairer, 1957; Schairer and Yoder, 1962).

It is generally agreed that basalt is generated in the Earth's mantle. This means that the mantle must contain rocks capable of generating melts of basaltic composition. First melts in such rocks can be expected to be the "lowest melting fraction." The rocks being melted may never have been through a magmatic cycle themselves. Rocks such as eclogite can melt totally to form basaltic magma. Partial melting of peridotites can also generate basaltic magma.

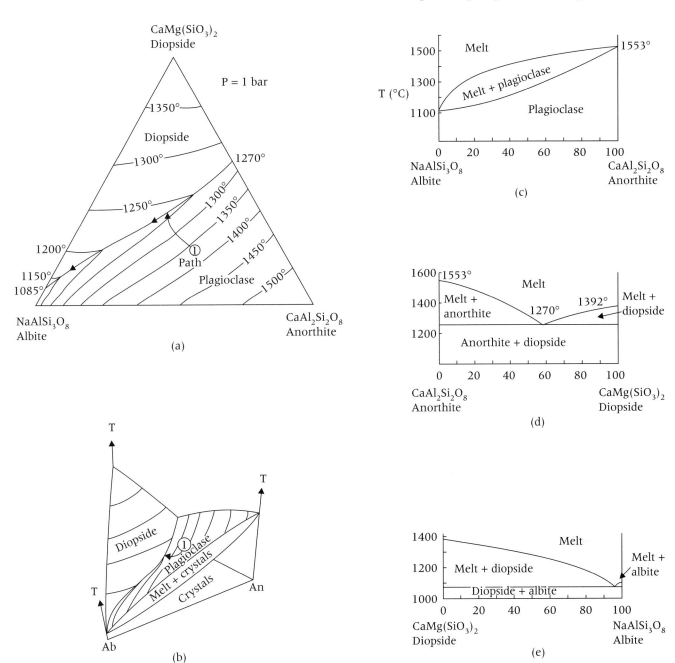

Figure 13.10
The albite–anorthite–diopside system at 1 bar (Bowen, 1915). See text for discussion of crystallization path.

The System Quartz–Alkali Feldspar (One-Feldspar Granites–Rhyolites)

Siliceous magmas capable of generating granites and rhyolites without plagioclase crystallize an alkali feldspar in the form of sanidine, anorthoclase, or orthoclase (perthite or submicroscopic "cryptoperthite" is common). Since albite and K-feldspar are in solid solution or in a perthitic relation (rarely in a side-by-side relation), there is one feldspar (alkali feldspar) and the rocks are one-feldspar granites and one-feldspar rhyolites.

Phase relations in the alkali feldspar–granite system are shown with a ternary projection and a three-

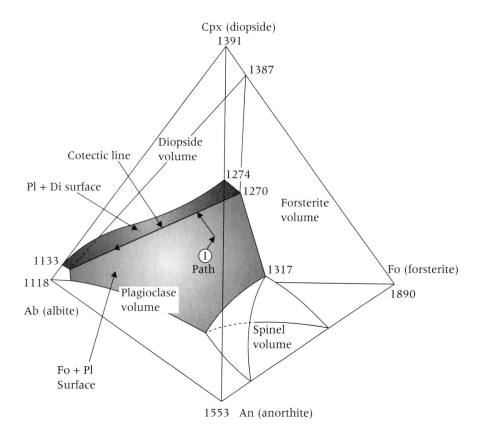

Figure 13.11
The albite–anorthite–diopside–olivine system at 1 bar (Yoder and Tilley, 1962). See text for discussion of crystallization path.

dimensional sketch (Figure 13.12). The participating binary systems are albite–quartz, K-feldspar–albite, and K-feldspar–quartz.

The dry system at 1 bar pressure (Figure 13.12) has a phase boundary between the quartz and alkali feldspar fields. This boundary is a line toward which isotherms decrease in value, generating a "thermal trough." A low-temperature point in the trough lies near the center of the ternary diagram. This point is the **minimum** and is the site of Bowen's "minimum melt," having a composition of approximately 1/3qz + 1/3Kf + 1/3ab. At the minimum, the liquidus coincides with the solidus. Since this is the "dry" and, therefore, high-temperature system, there is a leucite field toward the $KAlSi_3O_6$ apex.

A common path of crystallization (path 1) begins on the alkali feldspar boundary surface, reflecting an initial composition skewed toward K-feldspar. Crystallization of a K-rich alkali feldspar drives the melt composition toward the boundary curve, at which point quartz and an alkali feldspar crystallize together. The melt composition then moves down the boundary curve toward the minimum where the melt is eventually exhausted.

Phase relations in the "wet" ternary granite system are shown in Figure 13.13 (Tuttle and Bowen, 1958; Steiner et al., 1975). The pressure is 5 kbar and the melt is saturated in H_2O. There is a slight shift of the pseudoternary eutectic composition away from the quartz apex in comparison with the "dry" system (Figure 13.12). In addition, there is a separation of the alkali feldspar field into albites and orthoclases separated by a cotectic line. This means that two feldspars theoretically can form side-by-side (or as a nonperthitic intergrowth) in the H_2O-bearing system.

Points on the cotectic line have two solid phases in equilibrium with one melt. The point of intersection of the liquidus and solvus surfaces now becomes a **ternary eutectic**. The values of the isotherms on the thermal surfaces are considerably lower than those in the dry system (Figure 13.12), reflecting the depolymerization effect of H_2O. In addition, the field of leucite has been eliminated.

Path 1 (Figure 13.13) represents a typical situation. Crystallization starts with an orthoclase (or sanidine) reaching the thermal valley some distance from the eutectic. On the boundary curve (cotectic line), quartz crystallizes simultaneously with orthoclase and the composition of the melt moves to the eutectic. At that point an albitic plagioclase will be the third phase to crystallize. In a slow-cooling environment this late-crystallizing albite is relegated to an intergranular position, possibly represented by the textural relations shown in Figure 8.7a.

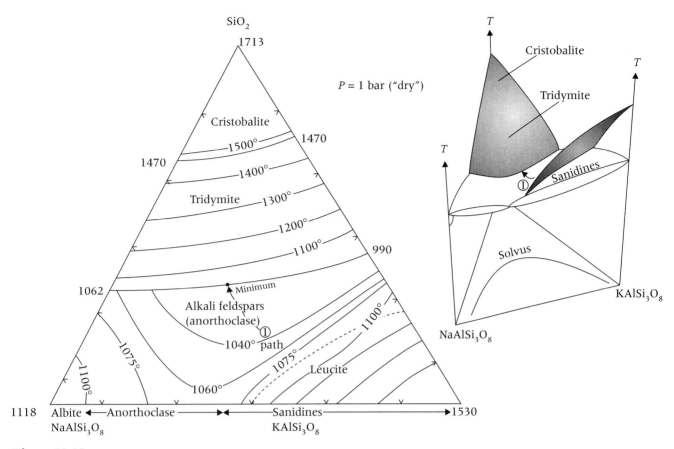

Figure 13.12
The ternary "granite" system at 1 bar (Schairer, 1957; Tuttle and Bowen, 1958). See text for discussion of crystallization path.

The System Quartz–Alkali Feldspar–Plagioclase (Two-Feldspar Granites and Rhyolites, Granodiorites, and Rhyodacites)

Anorthite added to the ternary granitic–rhyolitic system generates a calc–alkaline quaternary system (quinary, with H_2O) corresponding to a much wider variety of naturally occurring magmatic rocks. Adding even small amounts of anorthite component to the ternary system has the effect of ensuring crystallization of two different feldspars rather than a single alkali feldspar, regardless of the partial pressure of H_2O (Franco and Schairer, 1951; Yoder et al., 1957; Carmichael, 1963; von Platen, 1965; James and Hamilton, 1969; Brown and Fyfe, 1970; Piwinskii and Wyllie, 1970; Robertson and Wyllie, 1971; Burnham, 1979).

The presence of anorthite component effectively limits the partitioning of albite component into alkali feldspar by making plagioclase more calcic than albite. There is no solid solution series between anorthite and K-feldspar because of the ionic charge and ionic size disparities between calcium and potassium. Two-feldspar granites with an "intermediate" plagioclase and an alkali feldspar such as perthitic or crypto-perthitic orthoclase or microcline are very common.

The general form of the quaternary granitic system at a H_2O pressure of 5 kbar is shown in a tetrahedron (Figure 13.14). The surfaces shown within the tetrahedron are **cotectic surfaces** (not liquidus surfaces) separating volumes in which quartz, alkali feldspar, and plagioclase are stable. Intersection of the quartz–plagioclase surface with the alkali feldspar–plagioclase surface is the **quaternary cotectic line** ("univariant curve"). The line represents the location of melts in equilibrium with three crystal phases.

Path 1 (Figure 13.14) starts in plagioclase space. Crystallization of plagioclase zoning to more albitic compositions can be expected. As a result, the path of changing melt composition moves away from the plagioclase sideline along an arc, reflecting zoning to more albitic compositions. At intersection of the two-feldspar cotectic surface, alkali feldspar (an orthoclase) begins

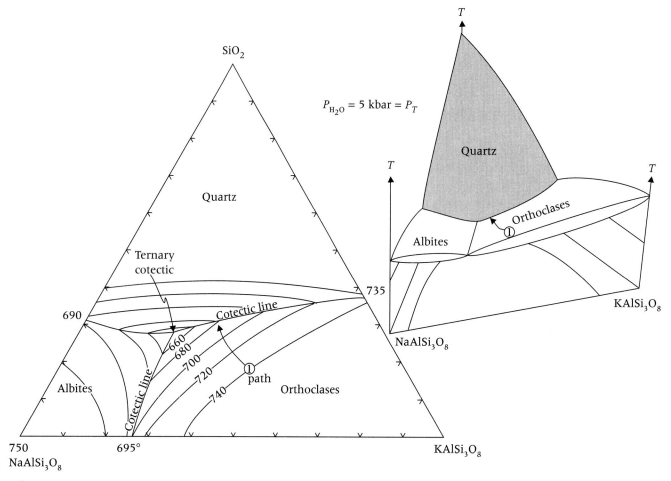

Figure 13.13
The ternary "granite" system at 5 kb H$_2$O pressure (Luth et al., 1964). See text for discussion of crystallization path.

to crystallize. Simultaneous crystallization of plagioclase and alkali feldspar moves the melt composition to the cotectic line at which point quartz is an additional crystallizing phase. Then the melt composition moves down the cotectic toward the ternary eutectic. The system finishes crystallizing at a point somewhat above the base of the ternary system since orthoclase can accommodate a little anorthite component.

Path 2 (Figure 13.14) starts in plagioclase space, moving to the plagioclase–orthoclase feldspar cotectic surface and then down to the cotectic line to the ternary eutectic. At the eutectic, quartz begins to crystallize. Path 3 begins in plagioclase space as well, but the path leads directly to the eutectic where quartz, alkali feldspar, and albitic plagioclase crystallize simultaneously. Path 4 leads from plagioclase space to the quartz phase boundary surface. Such an initial composition is deficient in K-feldspar component, characterizing some granodiorites and quartz diorites. The path of changing melt composition (path 4) down the quartz–plagioclase surface leads to the quartz–plagioclase cotectic and then to the ternary eutectic.

All of these paths of crystallization begin with crystallization of plagioclase, followed by simultaneous crystallization of two or more phases. Textural development is extremely variable, depending chiefly on nucleation and crystallization rates. In general, later-crystallizing phases can be expected to be relegated to intergranular positions where they contact other phases that started to crystallize early. Simultaneous crystallization of quartz, orthoclase, and oligoclase commonly generates a granular relation (Figure 17.4b), but an intergranular micrographic intergrowth of quartz and K-feldspar is well known (Figures 17.4a, 17.6b, 17.7b), with sodic plagioclase probably being relegated to rim growth on larger adjacent plagioclase crystals.

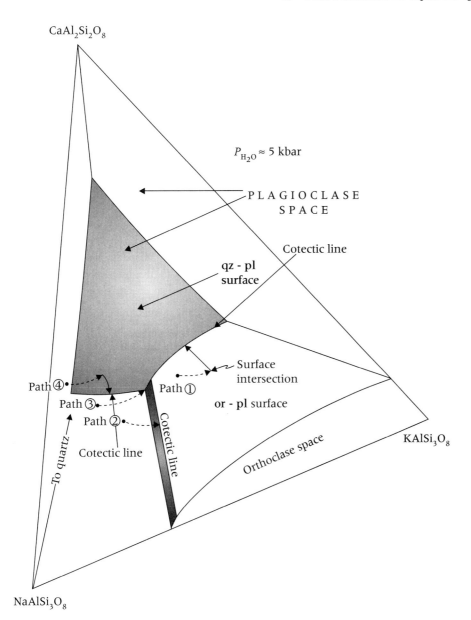

Figure 13.14
The quaternary two-feldspar "granite" system (Yoder et al., 1957; Carmichael, 1963; von Platen, 1965; James and Hamilton, 1969, Presnall and Bateman, 1973, Winkler et al., 1975). See text for discussion of crystallization paths.

Magmatic Systems Producing Feldspathoids with Feldspars

Silica-deficient magmatic rocks containing feldspathoids can be shown as an extension of the ternary granite system to $NaAlSiO_4$ (carnegieite) and $KAlSiO_4$ (kalsilite) (Bowen, 1928; Schairer, 1957). The phase relations apply to nepheline- and leucite-bearing rocks containing alkali feldspars, and minerals such as nosean, sodalite, and hauyne. Rocks such as foyaites and phonolites are produced.

With addition of anorthite, forming a quaternary system (quinary if H_2O is included) plagioclase-bearing feldspathoidal–feldspathic rocks, such as tephrites and theralites, are modeled experimentally (Schairer and Bowen, 1947; Carmichael, 1965; Hamilton and MacKenzie, 1965). Both nepheline and leucite may occur in silica-undersaturated intermediate and mafic magma systems.

Crystal Growth and Dissolution in Liquids

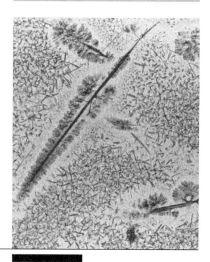

Clinopyroxene dendrites (scopulites) crystallized prior to crystallites, clearing melt in their vicinity of crystallite growth components. Vitrophyry, Scotland. PPL. Bar scale = 0.16 mm.

Nucleation in Liquids

Nucleation Theory

Nucleation is the first step in crystal growth in magmas and aqueous solutions (and gases), and this occurs by either heterogeneous or homogeneous processes (Jackson et al., 1967; Kirkpatrick, 1975; Dowty, 1980).

Heterogeneous nucleation occurs on pre-existing crystalline substrates, such as the surfaces of minute crystals or larger crystal and rock surfaces. Dislocation and impurity sites on these surfaces are particularly effective in initiating nucleation. Heterogeneous nucleation is the general case in natural environments in which liquids are inherently contaminated with solid particles.

Homogeneous nucleation is a process that is energetically much more difficult than heterogeneous nucleation. Nevertheless, it may be geologically important, especially in those situations in which undercooling and supersaturation is indicated, such as at the lower and upper surfaces of lavas and in evaporite systems. Cooling of magmas and aqueous solutions (and gases) results in a decrease of atomic and ionic mobility (diffusive motion). Liquidus and solubility curves define the thermodynamic location of the boundary between the liquid and solid states, but, in reality, crystallization occurs at lower temperatures. If the system is relatively homogeneous, there may be considerable undercooling of magmas and supersaturation of solutions required before crystallization can occur. The reason for this is that a nucleus must form before crystal growth can proceed. Embryo solids are not stable because the surface free energy is largely due to unsatisfied bonds, in comparison to the volume free energy associated with a small cluster of atoms. The energy relation is

$$\Delta G = \text{net free-energy change} = 4r^2 \gamma + 4/3\, r^3 \Delta G_V$$

where G_V = free energy associated with volume of solid
γ = specific surface energy
r = spherical radius

A stable nucleus can only form if the energy of the system is reduced (generally by heat loss) to a level at which a nucleus of critical size (expressed as a radius) is stable so that crystal growth can proceed. The critical size forms at an energy level that is the **free energy of activation for nucleation**, which is an energy level related to the net free-energy change (Figure 14.1). Below the critical size, the surface energy of the embryo solid is large compared with the volume energy, and dissolution of the particle is favored. Stable nuclei are formed during thermal fluctuations (affecting the ΔG_V term) as the natural system is further cooled. Undercooling has the effect of storing energy that is rapidly released as **latent heat of crystallization** after formation of the stable nucleus and ensuing crystal growth.

Framework silicates, such as feldspars, require more undercooling to generate nuclei than do chain structure pyroxenes, and even less undercooling is required for independent tetrahedral minerals, such as olivine. Additionally, iron and magnesium in a siliceous magma may cause a breakdown of the Si–O framework in the melt, in effect promoting more rapid (easier) nucleation of the chain and sheet silicates and suppressing nucleation of feldspar and quartz (Naney and Swanson, 1980).

If undercooling of a melt is extreme, molecular or ionic clusters are unable to diffuse together to form a nucleus, and chemical components solidify into glass. Similarly, it is easier for some materials to form noncrystalline solids even though thermodynamics predicts crystal formation. For example, opal-A forms in preference to quartz in low-temperature aqueous fluids containing silica gels.

Effect of Nucleation Rate

If a magma is cooled to a temperature at which only a few stable nuclei form and is held near that temperature as crystallization on these nuclei proceeds, all the growth materials will have to diffuse to these growth centers. The result is a rock containing a few relatively large crystals. Growth rate is not the controlling factor: the size of the crystals depends on the amount of growth material and the **nucleation density** (number of nuclei per volume). If magma is undercooled to a greater degree, more stable nuclei are able to form, which is expressed as a higher rate of nucleation. Growth on these nuclei results in many crystals and a finer-grained rock.

Phenocrysts and **microphenocrysts** are large compared with the grain size of their **matrix**. This indicates that the phenocrysts must have grown on a smaller number of nuclei in comparison to a high nucleation density in matrix melt. Picture a magma at depth cooling very slowly to environment. Only a few nuclei form since undercooling is minimal in this situation. Now if there is intrusion into the shallow crust or eruption to the surface where heat loss is greatly accelerated, the nucleation rate increases drastically and the small crystals of the matrix are formed. This is **two-stage crystallization** caused by two-stage cooling (slow then fast). If the matrix solidifies to glass rather than crystallizing, porphyritic glass is formed (Chapter 17). Such a rock is a **vitrophyre** and is unquestionably generated by two stages of cooling. If the size of phenocrysts are microscopic and the matrix is submicroscopic crystalline or glassy, the rock is **microporphyritic**.

Three-stage crystallization magmatic rocks are represented by volcanic rocks that have two size fractions of phenocrysts and a crystalline or glassy matrix (Chapter 17). A bimodal phenocryst grain-size distribution can be related to two stages of cooling followed by a quench of the remaining melt attendant to surface or near-surface eruption, but the bimodality may be due to other factors as discussed here.

It is not always possible to demonstrate that two stages of crystallization, or two stages of crystallization in a three-stage sequence, are the result of two- or three-stage cooling. Contrasting crystal-size fractions in magmatic rocks can originate in a way that is not governed by change in environment, such as a second or third pulse of intrusion or extrusion (Dowty et al., 1974; Walker et al., 1976; Swanson, 1977).

Many large plutons of two-feldspar granite contain relatively large (typically 4–6 cm) phenocrysts of alkali feldspar with a medium- or coarse-grained matrix (Figure 17.10b), and there is little or no reason to postulate more than one pulse of emplacement on the basis of other geological information. It does little good to postulate that the alkali feldspar crystals are larger because they began to grow before any of the other minerals. In the first place, it is nucleation density that controls crystal size, not duration of growth or growth rate. In the second place, the bulk composition of two-feldspar granites containing alkali feldspar phenocrysts typically plots in plagioclase space of the quaternary granitic system (Figure 13.14). This means that plagioclase, not alkali feldspar, begins to crystallize first. This is texturally verified by the presence of small plagioclase inclusion throughout alkali feldspar phenocrysts (Figure 17.10a; Hibbard, 1965). Two-stage cooling of these porphyritic granites is precluded on the basis of this textural observation alone; the small fraction of plagioclase crystals precedes growth of the larger alkali feldspar crystals. Evidently, the relatively large size of the alkali feldspar is due to a lower nucleation rate for that phase compared with coexisting plagioclase and quartz.

A sense of this nucleation rate difference can even be detected in a two-stage rock clearly formed by two stages of cooling. The two-feldspar granite porphyry shown in Figure 17.4f contains phenocrysts of alkali feldspar, quartz, plagioclase, hornblende, and biotite in a very fine-grained matrix consisting of quartz, K-feldspar, and less albitic plagioclase. The first stage of nucleation is represented by the phenocryst assemblage and the second stage is indicated by a much higher nucleation density. Some of the alkali feldspar crystals are much larger than the other phenocrysts, suggesting a relatively lower nucleation rate for the alkali feldspar during the first stage of the two-stage event and supporting the view that porphyritic two-feldspar granites represent a single stage of crystallization during continuous cooling.

Nucleation rate is also likely to play a major role in the formation of very large alkali feldspar crystals in granitic pegmatites of hydrous magmatic origin. The large to gigantic size of crystals suggests that nucleation rate is relatively low in aqueous phase that receive alkali feldspar growth materials from co-existing magma (Jahns and Burnham, 1969).

Crystal Growth in Liquids

Crystallization Theory

Once the nucleation process proceeds beyond the critical radius (Figure 14.1), crystal growth commences. The process of attachment of ions and molecules to growth surfaces involves reaction at the crystal surface–fluid interface. The crystalline state represents a lower-energy state than a melt solution (aqueous solution, gas) in which crystals grow. The crystalline state represents a lower-energy state than glass as well.

When undercooling occurs, three factors control the rate of crystallization: (1) the attachment mechanism, (2) diffusion of growth material to the attachment site, and (3) the convection of growth material to the attachment site. The first two processes are inevitable. The third may not occur under certain circumstances, such as the nearly static condition within a stagnate boundary layer within a magma chamber.

The driving force for diffusion to the crystal surface–fluid interface and attachment is dominantly thermal in most geological situations, since heat transfer is much faster than mass transfer. If the thermal gradient is low, crystallization will be slow and crystals will be complete crystals. If the gradient is steep, such as at contacts with "cold" wall rock or between "cold" and "hot" mingling magmas, there will be undercooling and rapid growth with the potential for development of cellular morphology.

The thermal gradient established between crystal and fluid is modified as crystallization proceeds due to the latent heat of crystallization generated during crystallization. **Thermal undercooling** can only be maintained if there is rapid transfer of heat to the ambient fluid. Cellular crystal growth (including dendritic growth) may be a good way to transfer the latent heat to the cooler surroundings. Consider the tip of a dendrite or the tip of a dendrite arm or the corner of a cube. Any such protuberance finds itself in a cooler local environment because what heat is generated is more readily absorbed in the relatively larger volume of fluid located away from the main body of the crystal. Hence, there is a greater level of under-

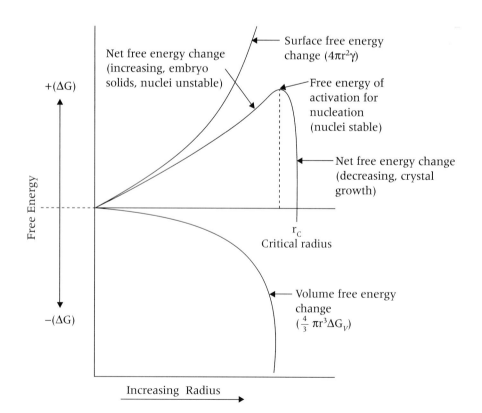

Figure 14.1
Net change in free energy as a function of change in surface and volume free energy as related to the formation of stable nuclei and unstable embryo particles (Flinn and Trojan, 1990).

cooling, and the growth rate should be faster, extending dendrite arms and corners of crystals. Additionally, **constitutional undercooling** can occur as the composition of a fluid changes due to compositional changes occasioned by rejection of some elements by a growing crystal. A protuberance, such as a dendrite arm, grows progressively faster the farther it gets from the main crystal surface because in that direction the liquidus or solubility curve is rising and undercooling is increasing.

Hopper crystals (Figure 14.2) may form as the result of greater supersaturation at the corners and edges of, for example, a cube, compared with the faces of the cube, because there is a reduction of supersaturation at the faces where crystallization is occurring if there is insufficient replenishment of fresh solution. Relatively accelerated growth on the cube edges and especially at the corners, where supersaturation is greatest, generates the hopper morphology.

Interface Attachment Kinetics Crystal growth rates apply in two senses: (1) rate of overall enlargement of crystal and (2) rate of growth in relation to crystallographic direction (Kirkpatrick, 1977). Attachment of ions or molecules on crystals is most difficult on planar surfaces and is facilitated by rough surfaces. However, once a nucleus does form on a planar surface, addition of growth units is facilitated and there is growth by **layer spreading**. If multiple nuclei are able to form on a rough surface, crystal growth may be more or less simultaneous (**continuous growth**) across the surface. If a crystal grows with a screw dislocation, there is a raised edge on the growth plane to which attachment can take place with relative ease with continuing spiral growth. Certain crystallographic planes may accept growth materials more easily than others. This leads to sector zoning (Hollister and Gancarz, 1971; Dowty, 1976).

The rate at which a crystal grows is, with some exceptions, inversely proportional to interplanar spacing within crystal lattices. In effect, incoming ions or ion clusters are more likely to remain attached to a crystal if the position of the new growth layer is closer (stronger bonding) to the one beneath it. Fewer attachments or temporary attachments occur on a growth surface where the bonding is weaker because the growth unit is more likely to be swept away before attachment is secured and growth units become part of the crystal lattice. For example, the $\{111\}$ plane of halite has the closest spacing, $\{110\}$ is intermediate, and $\{100\}$ the widest spacing. As a result, growth rate of halite might be expected to be greatest in the $\{111\}$ direction. Octahedral faces do not develop on halite, and if they were artificially prepared rapid growth in the $\{111\}$ direction would eliminate such faces, leaving only the cube faces. Thus, faces with larger interplanar spacings grow more slowly but dominate the form of the crystal. If the initial halite grain were a sphere, as it might be along the shore of an evaporite basin, growth would take place normal to $\{111\}$, generating a cube.

Dominance of one crystal face over another must depend on factors other than interplanar spacing. A corner (or edge) of a crystal developed during an early stage of crystallization can evolve into a face as crystallization proceeds (Figures 14.3b, 17.4d). What initially appears to be a high-rate direction becomes a relatively slow-rate direction.

Diffusion Rate in Fluids Diffusion is another rate-controlling factor in crystal growth. As crystal growth

Figure 14.2
Artificial hopper bismuth crystals, similar to natural halite hoppers. Bar scale = 7 mm.

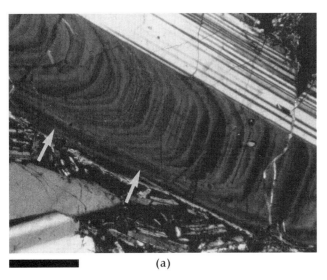
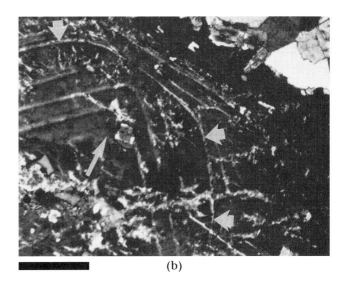

Figure 14.3; text p. 197
Noncellular crystal growth and zoning as indicator of growth morphology and direction velocity.

(a) Plagioclase oscillatory zoning indicating high growth velocity in crystallographic *a* direction compared with *b* direction as indicated by extreme attenuation of growths zones parallel (010) (arrows). Basalt. X-polars. Scale = 0.6 mm. (b) Zoning in alkali feldspar defined by difference in extinction position and exsolved albite. Early edge (corner?) of crystal (long arrow) becomes irregular "face" of final crystal against matrix (upper right), since this became the low-velocity direction. Possible resorption surface (three short arrows). X-polars. Bar scale = 0.6 mm.

occurs, there is a continuous change in composition of the fluid effectively slowing the growth rate unless there is a continual supply of growth components to the crystal–fluid interface. Diffusion of ions in the fluid or convection of solute-bearing fluid can maintain crystallization.

In complex multicomponent solutions, there is a reduction of some chemical components at the fluid–crystal interface as a result of crystallization. This sets up a chemical gradient in the fluid between the interface and the ambient fluid. The steepness of this **compositional gradient** depends on the rate of crystallization and the rate of diffusion. If the rate of growth is slow (as determined by the rate of heat loss to environment, rate of dissipation of latent heat, or the nature of attachment reactions), additional growth materials may have enough time to diffuse in from the ambient melt. If diffusion rates are drastically reduced, as they may be in extreme undercooling, crystallization will cease as growth materials become unavailable.

There is an interplay between diffusion rate, saturation level, and growth rate that may be the reason for some types of oscillatory zoning, such as that of plagioclase (Figure 7.8a). This is the **diffusion–supersaturation** model (Vance, 1962a; Bottinga et al., 1966; Sibley et al., 1976; Muncill and Lasaga, 1988), stating that if the rate of diffusion to the growth surface lags behind crystal growth, there is a temporary cessation of growth until a new supply of growth components can diffuse in. Crystallization may not resume at exactly the composition at which it terminated; that is, supersaturation occurs before renewed crystallization begins. In that case, new growth will be of a different composition (higher anorthite content in the case of plagioclase), and a compositional zone will be produced.

Diffusional Exchange Between Crystal and Fluid
Crystallization in melt solutions in which there are components of a continuous solid solution series, such as the binary plagioclase series (Figure 13.4), or a peritectic relation such as in the K-feldspar–silica system, involving conversion of leucite to K-feldspar (Figure 13.2), is prone to readjustment of solid-phase composition as crystallization proceeds.

In the plagioclase system, there could be a continuous reaction (Bowen, 1928) between melt and plagioclase (equilibrium crystallization), rather than zoning. If reaction occurs, relatively calcic plagioclase is converted to relatively sodic plagioclase by diffusion of cal-

Figure 14.4; text p. 198
Polygonal mosaic resulting from simultaneous growth of K-feldspar crystals in aplite dike in two-feldspar granite pluton. Harrison Pass, Nevada. Note triple point (arrow). X-polars. Bar scale = 0.12 mm.

cium out of a crystal at the same time that there is diffusion of sodium into the crystal. The occurrence of unzoned plagioclase in some rocks suggests that this equilibrium crystallization may apply to some natural silicate systems. However, delicately zoned magmatic plagioclase is testimony to lack of a homogenizing reaction during crystallization. As indicated in Chapters 13 and 18, unzoned plagioclase may form in those situations where fresh melt is supplied to the growing plagioclase rather than by reaction with melt in a closed system. Nevertheless, the process of **coring** is common in metal systems (Moffatt et al., 1964) and may occur in silicate systems. Coring results from partial replacement of a crystal contained in a fluid by ion diffusion in and out of the crystal, leaving an unreplaced portion in the core.

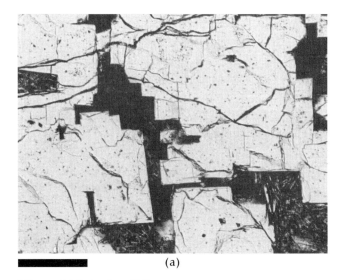
(a)

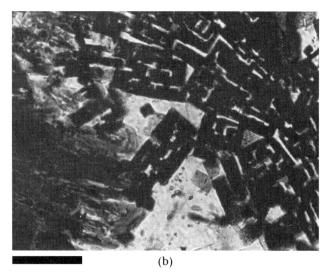
(b)

Figure 14.5; text p. 198
Boxy cellular growth.

(a) Boxy plagioclase viewed along crystallographic *a* (note cleavage) in glassy matrix. Crystal-bearing glass matrix contrasting with crystal-free glass deep within cellular structure may represent repositioning of phenocryst in magma system. Basalt dike, Trinity Range, Nevada. (Sample courtesy John Heggeness.) X-polars. Bar scale = 0.14 mm. (b) Honeycomb boxy cellular orthoclase with quartz formed in supercooled rhyolitic magma. Indian Springs, Nevada. (Sample courtesy Paul Purington.) X-polars. Bar scale = 0.06 mm.

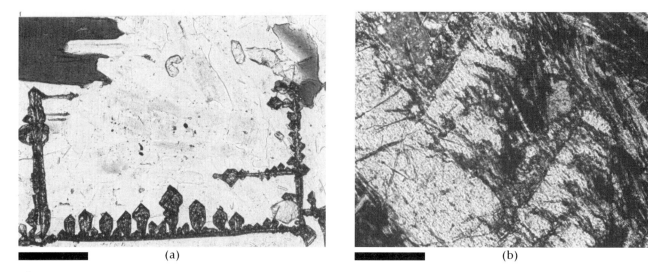

Figure 14.6; text p. 198
Dendritic cellular crystal growth.

(a) Dendritic zircon in hypersthene-clinopyroxene monzonite of Ringing Rocks Magmatic Complex, 25 km southeast of Butte, Montana. PPL. Bar scale = 0.06 mm. (Courtesy Andy Stroud.) (b) Dendritic arms of olivine crystal in komatiite. Ontario, Canada. X-polars. Bar scale = 0.14 mm.

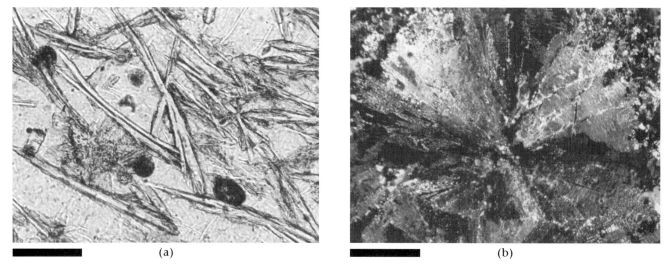

Figure 14.7; text p. 198
Spherulitic cellular crystal growth.

(a) "Horsetail" plagioclase in glass matrix. PPL. Bar scale = 0.06 mm. (b) Spherulite of granophyric K-feldspar–quartz in granular and granophyric matrix, representing crystallization in strongly undercooled rhyolitic melt. Stillwater Range, Nevada. X-polars. Bar scale = 1.2 mm.

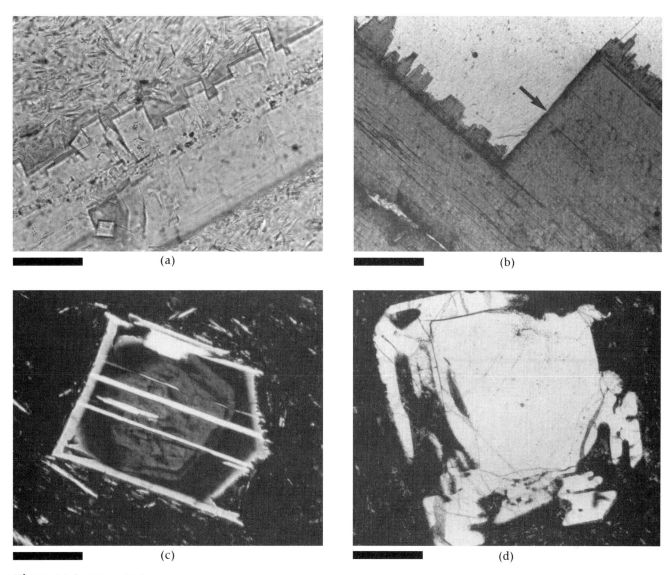

Figure 14.8; text p. 198
Transition from noncellular to boxy cellular crystal growth.

(a) Local final boxy cellular growth of otherwise noncellular plagioclase. Coso Range, California. (Sample courtesy C. Bacon, U.S. Geological Survey.) PPL. Bar scale = 0.06 mm. (b) Incomplete crystal growth in final zone parallel to base in biotite. Note absence of growth on crystal face normal to base (arrow). Shonkinite. Montana. PPL. Bar scale = 0.14 mm. (c) Skeletal–cellular terminal growth of noncellular zoned plagioclase, as a result of undercooling event leading to final quench to matrix glass. X-polars. Bar scale = 0.1 mm. (d) Dendritic extensions of earlier noncellular quartz phenocryst in rhyolite. Rounded arms of dendrites may represent a mild dissolution stage prior to quench of melt to glass. St. Francis Mountains, Missouri. X-polars. Bar scale = 0.5 mm.

Crystal Growth Morphologies

Crystallization in fluids is characterized by enlargement on planes parallel to rational crystallographic directions. The shape of crystals during growth is elegantly recorded by compositional zoning such as plagioclase and alkali feldspar (Figures 14.3, a,b, p. 194; 17.4d). As previously discussed, corners and edges of early crystals may become planes as growth proceeds. Sublimates tend to develop "anisotropic" morphologies, such as elongated halite and magnetite crystals (Figure 35.4a; Symonds, 1993), possibly as a result of directional flow of solutions (Sunagawa, 1981).

Final crystallization commonly results in interference with adjacent crystals, precluding development of crystal faces (Figures 7.5; 14.3b). In some cases, final growth adjustment results in more-or-less polygonal shapes with triple points (Figure 14.4; p. 195), similar to those generated by solid-state recrystallization (Figure 16.8c,d) and adcumulate processes (Figure 18.6).

The dominant factor controlling rate of crystallization is the degree of undercooling of melts and the degree of supersaturation of solutions. Rapid crystallization results in formation of imperfect crystals. At low undercooling the **integrity** (see Chapter 7) of a crystal is high; the crystals are complete, without morphological imperfections except for possible lack of facets due to late-stage interference with adjacent crystals.

At moderate to high undercooling or supersaturation, crystals are not perfectly formed. These are called **cellular** crystals, in a very general sense, in which the cells may be inclosed within a crystal or be partly open to the exterior where they contribute to an irregular exterior of a crystal. Cellular growth morphology (not corresponding strictly to the metallurgical definition) includes hoppers, skeletal forms, boxworks, spongy morphologies, and a variety of dendrites and spherulites (Tiller, 1964; Bryan, 1972; Lofgren, 1974, 1980, 1983).

Boxy cellular crystals resulting from crystal growth are shown in Figure 14.5, p. 195, a,b; and dendritic cellular crystals are shown in Figures 14.6, p. 196, a,b; 14.8, p. 197, d; 14.9a,b. Cellular crystals characterized by radiating morphology range from "horsetails" (Figure 14.7a) to spherulites (Figure 14.7, p. 196; b).

A crystal may be only partly cellular, such as the localized boxy cellular morphology shown in Figure 14.8a,b, the skeletal cellular morphology extending from a noncellular core (Figure 14.8c), and a dendritic–skeletal morphology associated with a noncellular core (Figure 14.8, p. 197; d). There may be variation of cellular morphology within a given crystal. A change from boxy cellular to dendritic cellular morphology is shown in Figures 14.9a,b. The reasons for changes in growth morphology, especially change from noncellular to cellular, in some cases can be correlated with displacement of growing crystals to an environment of greater undercooling.

Crystal Dissolution

Dissolution Theory

Detachment of ions or ion clusters from crystal lattices at interfaces with liquids is the opposite of crystal growth. In aqueous solutions, the process is simply **dissolution**, whereas evaporation of ice crystals, say on the surface of a glacier, is a **volatilization** process. In melts, the process is **direct melting** if the composition of a melt is the same as that of the crystal phase.

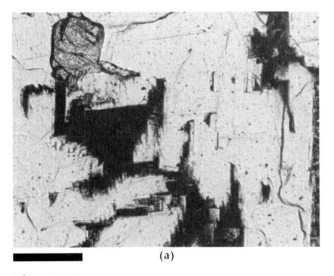
(a)

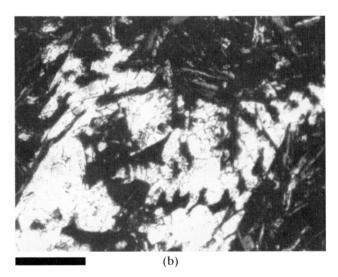
(b)

Figure 14.9
Transition from boxy cellular crystal growth to dendritic growth.

(a) Boxy cells lined with finer, somewhat dendritic cellular plagioclase during late-stage quench. Crystal-bearing matrix (far right) contrasts with crystal-free glass in cells, as in Figure 14.5a. Basalt, Pah Rah Range, Nevada. X-polars. Bar scale = 0.14 mm. (b) Somewhat boxy olivine phenocryst with dendritic extensions. Hawaiian basalt. X-polars. Bar scale = 0.14 mm.

If the composition of a melt is different from that of the crystal, the process is **dissolution melting**.

If a crystal of halite is placed in an undersaturated brine solution, dissolution can occur. Detachment of ions can be expected to take place most rapidly in directions of weakest lattice bonding, generally in a direction perpendicular to planes with greatest lattice spacing, or, stated another way, dissolution is faster on planes containing high packing density of ions. If a crystal were a sphere, as it can be prepared in the laboratory or occurring as a natural spherical clastic grain, this phenomena can be observed (Heimann et al., 1970). The spheres become faceted as dissolution proceeds, rather than just becoming spheres with smaller radii. A spherical halite grain can be expected to dissolve preferentially parallel to {100} forming a cube, since this is the direction of greatest interplanar spacing. In this case, the most rapid dissolution direction is the least rapid growth direction. If the halite grain were a cube to begin with, the most rapid dissolution would be at the corners and edges, irrespective of closer interplanar spacing, because ions in these positions are less bonded to the crystal lattice.

Dissolution of salts such as halite generates **electrolyte solutions**. Dissolution of minerals such as calcium carbonate into water is a process known as **hydrolysis**, in which hydrogen or hydroxyl ions are generated. Both of these examples represent **congruent dissolution** because no new solid phases are formed during the dissolution process. Dissolution of a mineral such as polyhalite in an aqueous solution results in simultaneous precipitation of gypsum and is called **incongruent dissolution**.

Dissolution melting also may be congruent or incongruent. Partial dissolution of quartz phenocrysts back into a silicic magma is congruent dissolution. Dissolution melting of a solid solution mineral, such as plagioclase, presents two possibilities. If the plagioclase is zoned, dissolution into a melt of basaltic composition can be visualized as the reverse of the original nonequilibrium crystallization, although in reality this is not likely to occur for kinetic reasons. However, if the plagioclase is unzoned, partial dissolution generates a melt that is not in equilibrium with the surface of the crystal. Adjustment toward equilibrium can occur by diffusion of sodium out of the crystal and diffusion of calcium into the crystal, generating a more calcic plagioclase in equilibrium with the more sodic melt (Tsuchiyama, 1985). This is a variation of incongruent melting dissolution and is a close parallel to coring, which is common in metal systems, as previously discussed.

Dissolution of K-feldspar in a dry siliceous melt of appropriate composition (Figure 13.2) generates leucite before the entire system is melted. This is incongruent melting in which a new mineral phase is generated rather than compositional modification by lattice diffusion of an original mineral phase.

Dissolution into melts and aqueous solutions is driven by differences in chemical potential. An undersaturated condition represents this potential difference. Three fundamental factors affect dissolution rate: (1) detachment kinetics, (2) diffusion rates and directions, and (3) convection of the liquid (Kuo and Kirkpatrick, 1985).

The interface liquid must be at least slightly undersaturated in the components of the adjacent solid in order for dissolution to occur. Dissolution results in a increased concentration of chemical components of the solid in the adjacent liquid. A concentration profile is established from the interface out into the ambient liquid. For example, dissolution of quartz into a silicate melt of mixed composition results in a higher concentration of silica at the interface. This is a "normal" profile, and silica is expected to diffuse "down" the profile to lower concentrations in the ambient melt.

Dissolution of complex minerals generates a variety of elements. The profiles for each element are different because diffusion rates vary. In some cases, diffusion profiles are inverted, and "uphill" diffusion occurs. For example, a high silica content at the dissolution interface results in more silica polymerization. An element such as potassium tends to diffuse toward such polymerized melts, regardless of the concentration of the element (Zhang et al., 1989).

Dissolution ceases if the influx of elements at the interface cannot be removed because saturation is reached (at constant P and T) and the driving force for dissolution no longer exists. This means the diffusion rate can control the rate of dissolution.

Removal of solute from the dissolution front can also be achieved by convection of the liquid rather than solely by diffusion in the liquid. Turbulence or streamflow can maintain an undersaturated liquid at the interface.

Partial Dissolution Crystal Morphologies

Dissolution in melt solutions and aqueous solutions characteristically results in rounding of corners and edges of crystals since detachment from planar surfaces is energetically much more difficult. Dissolution can also result in embayment of crystals. This type of dissolution probably takes place at relatively high energy sites where there are dislocations, fracture surfaces, twin planes, intracrystal compositional boundaries, or simply a preferred direction through the crystal lattice may control internal dissolution (Vance, 1965). Embayment may be so active that **spongy**

cellular morphology is generated (Tsuchiyama, 1985; O'Brien et al., 1988).

Dissolution into a crystal would seem to be self-arresting, since diffusion or convection away from the dissolution front should be inhibited by partial isolation from the ambient liquid. For example, it is generally agreed that the perfectly spherical embayments of quartz phenocrysts (Figures 14.10a,b) result from dissolution. How can silica undersaturation be maintained within the enlarging spherical cell? Perhaps the quartz phenocryst is in a turbulent environment and enough silica can escape or be flushed out of the cell. Rupturing of quartz phenocrysts is good evidence for such turbulence (Figure 14.10b).

A more appealing mechanism for spherical dissolution in quartz is presented by Donaldson and Henderson (1988), based on the experiments of Heimann (1973). This model requires the presence of a gas bubble at the embayment front. Circular transport of materials around the bubble–melt interface is envisioned as a "coring" mechanism that takes on solutes at the crystal–melt interface, delivering them to the ambient melt at the mouth of the embayment. This type of coring is a removal process, distinct from the coring resulting from ionic diffusion in and out of crystal lattices, as previously described.

The distinction between cellular growth textures and cellular dissolution textures may not be apparent, although boxy cellular morphology can be confidently identified as a growth texture (Figure 14.5a). In some cases, spongy cellular morphology can be demonstrated to be the result of dissolution melting (Tsuchiyama, 1985). The contrast between boxy and spongy cellular morphology can be observed in a single crystal (Figure 14.11), suggesting that there was an episode that included both growth and dissolution, perhaps resulting from a change in total pressure in an erupting magmatic system and/or temperature changes related to magma mixing. Similar morphological expression of growth and dissolution can be expected in evaporite systems, related to diurnal and seasonal temperature changes and to concentration changes related to evaporation and episodic flooding.

Direct Crystal Melting

There are few geological situations in which there is direct melting of a single mineral phase. Other than some impactites (Chapter 24), only monomineralic magmatic rocks, such as dunites and anorthosites, would be candidates for an origin requiring extremely

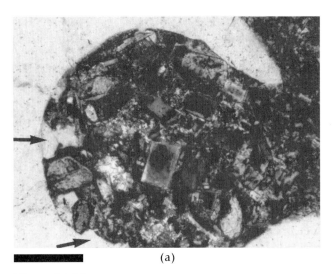
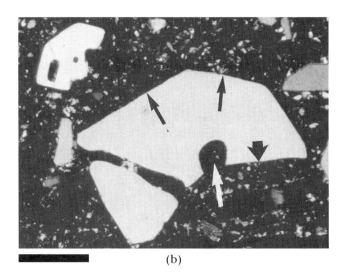

(a) (b)

Figure 14.10
Crystal resorption in magma.

(a) "Bubble"-cored embayment in quartz phenocryst. Note postresorption epitaxial growth quartz (arrows). Reheating in magma-mixing system may have contributed to the resorption. Andesite dike, Carson Range, Nevada. X-polars. Bar scale = 0.14 mm. (b) Quartz phenocryst with crystal faces (thin black arrows) was resorbed by coring (white arrow) and subsequently fractured (thick black arrow = fracture surface). Welded Tuff. Nevada. X-polars. Bar scale = 0.6 mm.

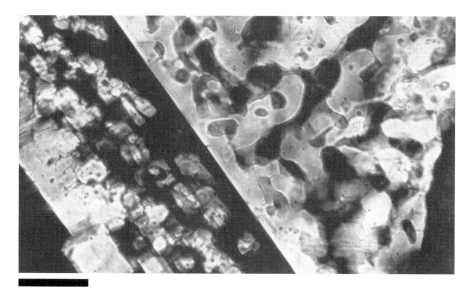

Figure 14.11
Contiguous portions of cellular plagioclase consisting of spongy cellular inner portion (implying partial dissolution) and boxy cellular outer portion (implying growth). X-polars. Bar scale = 0.05 mm. Coso Range, California. (Courtesy C. Bacon, U.S. Geological Survey.)

high magmatic temperatures—in the range of direct melting of respective source rock. The direct melting temperature is greater than that required for dissolution melting; for example, at 1 bar pressure, pure forsterite melts at 1890°C (Figure 13.11), anorthite melts at 1553°C (Figure 13.4), quartz melts at 1710°C (Figure 13.3), and leucite melts at 1686°C (Figure 13.2).

Incongruent direct melting of plagioclase simultaneously generates a more calcic plagioclase by lattice diffusion (Tsuchiyama and Takahashi, 1983), and incongruent melting of a pyroxene simultaneously generates a pyroxene of different composition (Hibbard and Sjoberg, 1994). In both of these examples, a spongy cellular crystal morphology is generated.

15

Dynamics of Solids with Gases and Liquids

The operations of erosion and sedimentation may be seen in progress at the present time, and the part of geology dealing with these processes has thus been built up on strictly Lyellian principles. For reasons which are sufficiently obvious, this method is not applicable to igneous action.

Alfred Harker (1909)

But being exceedingly difficult, if not actually impossible, to observe magma in its various stages, a physical theory of magmatism has only now begun to be needed and developed. Although the chemistry of silicate melts is a mature science, physical igneous petrology is in its infancy.

Bruce D. Marsh (1984)

Dynamics of Fluids in Massive or Framework Rock

Water and Aqueous Phase in Rock

Water and supercritical aqueous phases can (1) fracture rock, (2) migrate to zones of dilation caused by shearing, and (3) permeate rock that has interconnecting fractures or pores. These processes have several common manifestations.

Hydrofracturing occurs if the fluid pressure in a rock crack exceeds the minimum principal stress in the rock by an amount equal to the tensile strength of the rock. This process is significant in the brittle deformation regime where fractures, joints, fracture cleavage, and similar features are formed. Weakening of bonds by way of mineralogical reactions at the tip of a crack promotes slow crack propagation by a process known as **stress corrosion cracking** (Anderson and Grew, 1977).

Dilational pumping of aqueous fluids in permeable rock occurs along shear planes where there is local reduction in pressure, resulting from protrusions of rock on irregular surfaces riding up on protrusions on the opposing fault surface. This is a variation of the **Bagnold effect** (Bagnold, 1966) and is important in the localization of fluids that participate in reactions and deposit minerals in the form of syndeformational low- and high-temperature veins.

Infiltration of aqueous fluids into porous rocks occurs by **capillary action** in the absence of a pressure gradient, and by **permeation**, if driven by a pressure gradient. Permeability affects the rate and location of chemical reorganization occurring during rock-forming processes such as diagenesis, weathering, and hydrothermal alteration.

Volcanic Gases in Rock

Explosive fragmentation of rock in volcanic vents generates **volcanic breccias** and **ashes**. Although magma is disaggregated as well, there is direct fragmentation of solidified magma temporarily choking the throat of volcanoes, and fragmentation of vent wall rock that solidified during prior eruptions. Magmatic gases containing a large fraction of aqueous phase vapor exsolve from magmas as the magmas move from deep high-pressure regions toward the Earth's surface. If the rate of gas exsolution is high in relation to diffusion through the magma and leakage into wall rock, the strength of the confining mass of solidifying magma is exceeded, thus producing an explosive eruption.

Eruptions related to release of gases from magma at depth generates **diatremes** (Chapter 32), pipelike structures containing fragmented rock derived from the mantle, the crust, and rock formations exposed at the surface directly above the diatreme. **Pyroclastic dikes** (Chapter 31) and **hydrothermal explosion breccias** (Chapter 32) are rocks of similar origin.

Magma in Rock

Interaction of magma with rock includes (1) dike propagation, (2) dilational pumping, (3) filtration of magma through its own crystalline framework, (4) extraction of anatectic melt, (5) diapirism, and (6) stoping.

Magmatic pressure can cause fracturing and diking, similar to the process of hydraulic fracturing. **Dike propagation** depends on maintenance of magmatic pressure and suppression of crystallization. A dike swarm has the effect of keeping the environment heated to near liquidus temperature, favoring propagation. **Injection migmatites** contain an assortment of dikelike forms whose distribution is likely to be controlled partly by magmatic pressure.

Infiltration of magma through **crystal mushes** in a magma chamber is similar to percolation of water through porous sandstones. Adcumulate magmatic rocks are generated by this mechanism (Chapter 18). A melt produced by partial anatexis may be **filter-pressed** from its source by tectonic compression, as long as there has been enough melting to establish (1) interconnecting exit paths and (2) enough thermal inertia to keep the melt from crystallizing. At least 25% porosity is considered minimal for extraction to occur (Chapter 28).

Anatectic melting produces silicate melt that is less dense than its crystalline equivalent. This means that a body of magma or crystal-bearing magma will tend to buoyantly rise. Such a density inversion may result in **diapiric flow**. The flow path through the Earth's lithosphere must be "softened up" or the flow will be so slow that the melt will lose heat to environment and crystallize. If there are successive pulses of magma taking the same route through the lithosphere, the softened condition of the wall rock can be maintained. A diapir moves by changing place with its near-solidus or migmatitic roof and wall rock that flows ductilely around the advancing diapir, filling in behind the diapir as it moves by and requiring no regional dilation.

Collapse of roof or wall rock into a magma chamber occurs as blocks of rock are detached along bedding and fracture planes and if the density of the rock is greater than that of the magma. Although **magmatic stoping** was once thought to be a major process of emplacement of plutons, it is now considered to be important only at shallow depths because blocks take up about double the volume of their unbrecciated rock equivalent and a "room problem" exists. Unless there is regional tectonic dilation (Tikoff and Teyssier, 1992), stoping is restricted to near-surface environments where brittle tectonic adjustments can be made and magma can exit to the surface, thereby providing more room for further stoping. Stoping in a magma chamber of fixed volume in the plutonic environment chokes on itself, terminating the stoping process.

Dynamic Behavior of Mineral and Rock Grains in Fluids

Creation of Rock and Mineral Grains

Mineral and rock grains are formed from rock by (1) tectonic fragmentation, (2) fracture propagation, (3) mineralogical reactions that break bonds between interlocking grains, (4) explosive fragmentation, and (5) rock impacts. Grains are formed from fluids by (1) nucleation and growth of crystals and (2) explosive disaggregation of magma.

Tectonic fragmentation typically occurs along faults (fault breccia) and in fold hinges (crush breccia). **Fracture propagation** is an expansion process caused by freezing water and pressurized fluid. **Mineralogical reactions** occur along bedding, foliation, fractures, and crystal boundary surfaces where there has been fluid infiltration. Crystallization of easily cleaved mica and clay minerals has the effect of weakening initially strong interlocking relations between minerals such as feldspar and quartz. Shrink–swell nature of some clay minerals also contributes to grain production.

Grains produced as a result of **explosive fragmentation** of rock is typified by pyroclastic eruptions. When pressurized gas exceeds the strength of its confining rock, fragmentation occurs. Grains are also formed when grains impact bedrock or other grains. Meteor impact is the extreme and rare case, but collision of free-falling grains with rock below a cliff and the downslope tumbling of grains onto rock below are widespread ongoing processes.

Grains are also formed by crystallization of magma and precipitation in aqueous solutions and gases. Dynamic interaction between these crystals and their host fluid is immediate in comparison to grains that must be detached from rock before they dynamically interact with fluids in, for example, a fluvial or eolian setting. Disaggregation of magma accompanies rock fragmentation in an eruptive volcanic environment.

The size of grains produced from rock ranges from gigantic slide blocks to clay-size particles. Blocks of rock at the base of a cliff (talus) and sand-size grains (grus) at the base of a rock surface are two common occurrences of grains whose source is easily identified. The size of grains produced from fluids is limited by nucleation and growth rates. Giant crystals are rare, occurring only in certain pegmatites.

Fundamental Dynamics of Grains in Fluids

A force causes acceleration of a grain, and it may be described as a push or a pull. Flow of fluid against a grain is a push or impact force, whereas gravitational force exerts a pull on grains and fluids. Grains resting on rock surfaces or, more commonly, on other grains have a temporary weak chemical bonding to those rock and grain surfaces, strengthened by gravitational forces that pull these solids together. Grains temporarily suspended in fluid are pulled down to a sedimentation surface by the force of gravity as well.

Fluid Flow Fluid flow can be (1) laminar, (2) streamline, (3) turbulent, or (4) oscillatory. In all cases, movement of grains by fluid flow is part of the rock-forming process.

Directional flow is accompanied by at least some turbulence and in some cases by a large component of turbulence. For example, relatively smooth channels of fluvial systems are characterized by **laminar flow** (somewhat streamline in narrow channels), whereas in rough channels there is considerable turbulence. Obstacles to flow generate vortices and wakes, which are the ingredients of **turbulent flow**. Interaction of fluid with suspended solids at high flow velocities causes turbulence as well. Even fluid impacting another fluid, such as at the water–air interface in streamflow, generates turbulence. Turbulent flow has components of flow that are directed laterally, inclined, upward, and downward.

Nondirectional flow is oscillatory flow associated with wave action and, in the broad sense, tides. **Oscillatory flow** generated by wave motion along lacustrine and marine shorelines is caused by eddies (turbulence). If the depth of the water is less than about half of the wavelength of waves ("wave base"), there is an interaction of water motion and grains on the floor. The orbital motion of water "molecules" in a wave becomes elliptical when the wave "feels" the bottom. An extreme flattening of this elliptical motion generates a to-and-fro or back-and-forth movement of grains (especially sand grains) that can be described as oscillatory, typically with some tendency for a net forward motion.

Movement of Grains by Fluid Flow If a force is applied to a grain from the side, such as provided by the impact of a fluid moving in one direction along a horizontal surface, the bonding between two grains may be strained or broken. There are many geological circumstances where flow is at least locally in a lateral direction, including parts of fluvial, eolian, and tidal systems, and along the roof and floor of a magma chamber.

If there is a component of fluid flow in a downward direction, there is a component of gravity acting in the same direction. There is downward motion associated with turbulent eddies and convective eddies and cells, ebb tides, beach backwash, and water flow down fluvial channels.

There may also be a component of fluid flow in an upward direction acting against gravity. There is an upward component of flow in eddies associated with turbulent flow and the upward part of convection cells. Upwardly directed impact of fluid against grains either initiates or maintains suspension. Upward flow is also realized by dewatering and **fluidization** of sediments soon after deposition. Rapid degassing in volcanic pipes carrying pyroclastic tephra into the atmosphere is another example of upward flow. In both cases, upward flow is in response to a pressure differential.

Regardless of the direction of fluid flow, there is a **shear stress** (force per unit area, or dynes/cm^2) established between a grain and the solid on which it rests when impacted by a fluid. If the shear stress exceeds **frictional strength**, bonds are broken and the grain either rolls or slides.

Fluid impacting against a grain must flow around that grain, resulting in another shear relation established between the grain and the fluid. Shear in this case results in the deformation of the fluid since fluid next to the grain moves at a lower rate (because of **drag**) than fluid farther from the fluid–solid interface. Shear also applies to the grain, but deformation within the grain is only potential since most grains associated with fluids are rigid and do not deform.

The rates of shear depend on the viscosity of the fluid and velocity of flow. Air, water, water slurry (water containing suspended rock and mineral particles), magma, and magma mushes (magma containing crystals) are the common fluids in the geological environment. **Viscosity** is given in units of poise (poise = g/cm-sec), centipoise, or Pascal seconds (Pa-sec = kg/cm-sec), where 10 poise = 1 Pa-sec. The Pascal (Pa) is an expression of pressure, with 10 Pa ≈ 1 bar or 1 atm.

Newtonian fluids such as air, water, and crystal-free magma have no strength, and their viscosity does not change as shear rate increases. The viscosity of magma is orders of magnitude greater than water, just as water is much more viscous than air. **Non-Newtonian fluids** such as water slurries (mudflows) and magma "mushes" have some yield strength, and their viscosities generally increase with higher shear rates. Some **debris flows** behave as **Bingham plastics** (significant yield strength, variable viscosity).

The greater the viscosity of the fluid, the greater the shear force (at a given velocity) applied to the grain during flow. The viscosity of water at 20°C is 1.0×10^{-2} poise, compared with air under the same conditions at 1.8×10^{-4} poise. Thus, water is about 55 times more viscous than air at 20°C. In contrast, the viscosity of H_2O-free basalt magma at 1250°C and 1 atm is 10^7 poise, which means that a grain of given size and density is most easily moved by flowing magma and least easily moved by air, each flowing at the same velocity. Similarly, the impact force on a grain increases as the flow velocity increases; that is, the force of resistance is proportional to the kinetic energy of the fluid as well as to the viscosity of the fluid.

Since most grains consist of silicate minerals of similar density, grain density is rarely a major factor in grain motion. In contrast, it takes more force to move a large grain in comparison to a small grain along a friction surface, because the greater size usually means a greater weight and therefore a stronger "bonding" to the surface. With a specified force, therefore, smaller grains move easier (faster) than larger grains of comparable density.

The shape of a grain also determines the force needed to move it. The force needed to slide a cubic grain lying face down on a surface is about the same force needed to slide a spherical grain of equal weight, since the "bonding" at the sphere contact is much stronger than any one of the distributed "bonds" on the undersurface of the cube grain. Dry contacts between grains are not as strongly bonded as those in which a film of water is present. The water molecule acts as a "gluing agent," and wet sand is more difficult to move than dry sand as long as the water is not in excess and pressurized.

Laminar flow of fluid passing around and over a grain has a higher velocity than the ambient fluid velocity, because the same amount of fluid must pass through a smaller volume of space in order to maintain an overall constant fluid flow velocity in the system. The higher-velocity zone is necessarily a zone of reduced pressure into which the grain tends to move. The result is the **Bernoulli lift force** that reduces the strength of "bonds" between the grain and the underlying surface or grain. Reduction of bonding strength allows for easier lateral transport; elimination of bonding results in suspension.

Settling and Rising of Grains in "Stationary" Fluids Crystals of lower density than their enclosing fluid will rise in response to **buoyancy**. Crystallization of plagioclase may result in buoyant rise of the feldspar. Grains momentarily suspended in fluids, such as the finer debris produced by meteor impact and pyroclastic eruption, by vibrational suspension caused by seismic waves, by grain collisions and turbulent flow in fluvial and eolian systems, and by Bernoulli lift, eventually are returned to a rock or grain surface by gravity.

Grains settling through fluids do so (in general) in accord with **Stokes' law** of settling. Grains settling in fluids have a shear relation established between them and the entraining fluid. Since fluid of higher viscosity and a given flow rate can generate greater force against a grain lying on a horizontal surface than can a lower viscosity fluid, it follows that more resistance to settling is associated with higher-viscosity fluids. Furthermore, a large grain, free-falling through a Newtonian fluid, will reach the bottom before a smaller grain of comparable density because the amount of drag developed as the grain moves through the fluid is proportionally more for the larger-surface-area grain in relation to its volume. Of course, if there were no water and no air, the grains would reach the ground at the same time because the acceleration of gravity is independent of mass.

The Stokes relation applies to settling in air, water, and magma. Since the viscosity of water is about

55 times that of air, settling in air is orders of magnitude faster. Similarly, settling of crystals in magma, which is 10^9 more viscous than water, is considerably slower.

Settling velocities are also reduced in slurries and mushes because of their high viscosity. As a result, for example, large rock fragments can be transported in debris flows in part because they do not immediately settle in the slurry matrix fluid.

Avalanching (Gravity Flow) Grains perched on a sloping rock or grain aggregate surface may move downslope in response to either vibration or oversteepening, without the assistance of downward-directed fluid flow. **Avalanching** occurs on the lee slopes of ripples and dunes and on talus slopes where rockslides occur. Gravity is the dominating force, with the rate of downward movement being governed by interference (collisions, shears) of participating solids. Flow is characteristically turbulent.

Rapid downslope movement is initiated in at least five situations: (1) mechanical oversteepening of slope (such as rotation of a fault block), (2) material oversteepening by way of addition of grains upslope (such as pyroclastic accumulations on the upper regions of a volcano, and grain accumulation at the ridge crest of a dune or ripple), (3) removal of downslope support (such as by dip–slip fault displacement or lateral fluvial erosion, (4) addition of mass by way of grain wetting and pore filling, and (5) loss of strength related to sudden increase in pore fluid pressure that reduces the pressure at grain contacts leading to liquidlike behavior (**liquefaction**).

Grain Deposition in Subaerial Environments

Movement and accumulation of mineral and rock grains where air is the enclosing fluid is the **subaerial environment**. It is dominated by **eolian** processes in the desert environment, but it includes the accumulation of loess and volcanic ash.

A very high ratio of grains to air characterizes subaerial **gravity flows**, such as dry **debris flows** (rockslides), **grain flows**, and **ash flows** (if volcanic gases are substituted for air). Grain-to-grain interactions are more influential than air (gas) to grain dynamics in the movement of these materials.

Sands of the eolian environment are generated by the **sorting** of the coarser and finer grain fractions. Wind cannot move coarse sand, granules, cobbles, and gravel, leaving such grains as lag. Clay and very fine silt-sized grains are removed by winds and eventually dispersed over the land and water surfaces. The fines may be deposited as loess if there is an unusually voluminous source of fine sediments, such as that of periglacial plains.

The net result of this wind sorting is the accumulation of coarse silt and fine sand in the form of sand dunes. Cross stratification is characteristic of dunes, resulting from migration of dunes by way of erosion on the stoss (upwind) side of the dune and deposition on the lee side. The inclined foreset laminations are typically truncated by another set of laminations representing a new phase of accumulation and perhaps a different prevailing wind direction.

Asymmetric ripples form on the stoss surfaces of dunes. Movement of sand forming the larger dunes and the smaller ripples is mainly by saltation (temporary suspension), but includes sliding and rolling along the traction carpet. Movement is slow and progressive (**creep**). Grains are momentarily suspended by upward forces such as the upward components of turbulent flow, by the Bernoulli force, and by impacting of moving grains with stationary grains. Grain collisions generate grain rounding that is diagnostically eolian. **Sand flow** (avalanching) occurs on the lee side of dunes and ripples.

At high wind velocities, ripples flatten out and planar beds form (Allen, 1977; Collinson and Thompson, 1989; Figure 36.14). Evidently, the force of the wind breaks the bonds between grains that are strong enough to build slopes at lower velocities. Such planar bedding is not likely to be preserved, however, since, as the wind abates, ripples immediately form once again.

The fallout of tephra is governed by Stokes' law of settling. Although **normal graded bedding** (coarse at the bottom) occurs, the settling of relatively large pumice fragments may be slow enough, due to their low density, to generate **inverse (reverse) graded bedding**. Inverse grading may also occur as a result of grain avalanching down the steeper lee side of dunes and ripples.

Grain Deposition in Subaqueous Environments

Movement and deposition of grains in the **subaqueous environment** includes (1) gravity-controlled channel flow (fluvial), (2) open-water lacustrine and oceanic unidirectional current flow driven by differences in water density (temperature, salinity), (3) marginal lacustrine and oceanic wave-generated oscillatory flow, (4) tidal oscillatory flow, and (5) grain settling in water bodies where there is sediment input from pyroclastic eruption, wind suspensions, and decelerating threads of sediment-bearing fluvial flow into standing-water bodies, forming deltas. In addition, water-saturated accumulations of grains, with a high ratio of solids to water, are subaqueous in the sense that pore spaces are filled with water, the water playing an important role in the dynamics of flow when the water-saturated grains become suspended and water-saturated **debris flow** occurs. Internally, debris flows are subaqueous, but as a unit they can be subaerial or subaqueous. Fluidization associated with **soft sediment deformation** is a subtle variation of subaqueous flow that may also occur subaerially or subaqueously. **Turbidity flows** are confined to the subaqueous environment because they are characterized by turbulent flow in which the slurry as a unit is momentarily suspended in water. **Turbidites** contain many different bed forms, including planar stratification, graded beds, ripples, and scour features.

Fluvial flow is overall unidirectional, ranging from dominantly lamellar to dominantly turbulent, locally with streamline characteristics. Grain movement and deposition, and the resulting bed forms, vary according to flow velocity and grain size. The upper (high velocity) and lower (low velocity) flow regimes for fluvial systems are shown in Figure 36.14. Grain sorting, cross-lamination and cross-bedding, asymmetric ripples and dunes, and a variety of erosional and depositional bed forms are formed (Chapter 36).

Current flow in lacustrine and marine water bodies is driven by differences in water temperature and by the process known as long-shore current movement. These environments are relatively low velocity in which planar beds and asymmetric ripples tend to form.

Wave action generates symmetrical and slightly asymmetric ripples. Strong surges of flow in opposing directions related to strong storm waves produce **hummocky cross-bedding**. **Barrier bars** are also known as **breaker bars**, indicating a genetic tie with breaking of waves. The bars are asymmetric, with the steeper side facing the shore, and consist of cross-stratifications and superposed ripples.

Tidal flow may be rapid or slow, depending on the configuration of the tidal channel. Current flow during the flood-flow stage tends to be higher velocity than during the ebb-flow stage, resulting in only partial reversal of the smaller bed forms during the ebb stage. Ripples form and, since flow at any given stage is unidirectional, tend to be asymmetrical. Sand waves and megaripples form during the flood stage and remain largely intact.

Settling of grains to the bottom of a lake or ocean below wave base and out of the way of currents generates stratification without superposed ripples. Such is the case at distal end of deltas and wherever airborne pyroclastic and pelagic debris of both biotic and nonbiotic derivation accumulates and settles.

Soft sediment deformation takes place during compaction and shearing of bedded sediments and reduction of porosity by fluid escape. A layer of water-saturated sand overlying a water-saturated bed of clay is an unstable situation. Freshly deposited sand has 45% porosity, whereas fresh clay has 70–90% porosity. This amounts to a density inversion, and the sand will locally "sink" into the mud, its space being replaced by an upward and inward flow of mud forming a complex load structure. If liquefaction were to occur in response to a seismic event, it would affect the mud layer to a greater degree than the sand because the permeability of muds is very low and pressure can be raised instantly. This would allow portions of the stronger sand layer to sink into the liquefied mud layer.

Crystal Movement and Plating in Magmatic Systems

Since most magmas crystallize in "fractions" (**fractional crystallization**—Philpotts, 1988, 1990; Shelley, 1992; McBirney, 1992), the coexistence of crystals and melt is common and sets the stage for a variety of kinetic processes involving the crystals and melt. Unlike surface water and atmospheric systems, crystal-bearing magmatic systems cannot be directly observed, except for surface lava flows. The results of differential motion between crystals and their coexisting melts are reflected in the rocks after complete consolidation, cooling, and erosional exposure. The modeling of magmatic systems based on these indirect criteria and on general physical principles applying to solids and fluids has been, and promises to continue to be, the focus of considerable investigation (Komar, 1972; Shaw, 1980; Hildreth, 1981; Marsh, 1981, 1984, 1987; Huppert et al., 1984; McBirney, 1984; Sparks and Huppert, 1984; Marsh and Maxey, 1985; Brandeis and Jaupart, 1986; Ross, 1986; McCallum, 1987; Turcotte et al., 1987; Ryan, 1990; Jaupart and Tait, 1990; Bergantz, 1991; Cashman and Bergantz, 1991).

The geometry of magma chambers has considerable influence on the specific attributes of magmatic rocks (e.g., cumulates are characteristic of dikes), but the fundamentals of rock-forming process are the same in all of them. For example, crystallizing magma in chambers shaped and oriented as dikes, sills, cylinders, cones, lenses, and spheres, and ranging in size from outcrop dimensions to batholiths, are all influenced by internal gravitational instability and commonly also by external factors such as melt replenishment and tectonic activity taking place during magmatic consolidation.

The Effect of Internal Gravitational Instability Crystal settling and rising is a function of crystal and melt composition. Crystals may be denser (typically olivine) or less dense (typically plagioclase) than coexisting melt, and as a result crystals may rise or sink in response to buoyant forces.

The viscosity of highly polymerized siliceous magma is considerably greater than H_2O-bearing siliceous magma and basaltic magma. Settling and rising velocities vary accordingly. Hot magma is less dense than cold magma, and a nonuniform heat distribution results in a buoyantly unstable condition leading to **convection**. Magma cools to environment at the walls and roof of magma chambers. Cooled magma may be denser and tend to sink into hotter magma. Well-organized convective cells may develop, ranging from two-cell convection with upwelling along a vertical medial zone, to multiple cells elongated in a vertical position (Marsh and Maxey, 1985; Marsh, 1988a), to flat-lying stacked convection cells (Turner, 1985). Viscosity differences may also lead to zoning of magmas of different composition within volcanic conduits (Carrigan and Eichelberger, 1990).

If the crystal content of a magma or a portion of a magma is more than about 50%, the crystal–magma system is "locked up," unable to convect and not likely to erupt (Marsh, 1981). The viscosity of the layer closest to the roof may become quickly crystalline and too viscous to convect, whereas in the interior of the magma chamber there may be active convection.

Flow differentiation may occur in response to convective motion and injection. **Viscous drag** is generated where a velocity gradient is established in which shear force, due to flow, is greatest where the drag is greatest. When crystals collide with each other in a shear zone, a **crystal-dispersive pressure** is generated as the crystals are forced past each other. Since the effect is directly proportional to the rate of shear, crystals tend to be repulsed from the high-shear-force border zones of magma chambers toward the inner or medial zone (Komar, 1972; Ross, 1986). Since there is dispersion of crystals in the shear zone, magma moves in to replace the volume occupied by the crystals, and the proportion of crystals to magma decreases in the zone of viscous drag.

The Effect of Hot Magma Replenishment Layer injection of fresh magma at the base of a floored magma chamber or by injection into the interior of a magma body (such as lateral injection into a sill, Shirley, 1987; Husch, 1990) creates additional instabilities. The replenishing magma is hotter but is likely to be denser than the partially crystallized magma (of the same composition) because it has not yet begun to crystallize.

Cooling by convective heat transfer to partially crystallized magma above, or above and below, results in crystallization and generation of a less dense magma. Such a magma might buoyantly rise, perhaps as upgoing plumbs of crystal-bearing magma (Huppert et al., 1984).

The Effect of Tectonics A crystal-bearing magma confined in a chamber may be able to expand like a balloon (Paterson, 1988) if the magmatic pressure exceeds the strength of the enclosing wall and roof rock. Magmatic pressure beyond that generated by negative buoyancy may be generated in a tectonically active region.

Extraction of melt from a crystal–magma system by filter pressing may occur if the crystals are not packed tightly, especially if their volume is less than 55%. If the crystals are deformed in the packing process, even more melt may be expelled. **Filter pressing** is a tectonic-related mechanism (Miller and Weiblen, 1990).

A body of magma containing greater than 55% crystals deformed in response to regional tectonic adjustments will experience shear. Fluid may be sucked ("pumped") into shear zones as dilation occurs. **Dilational pumping** is one way late magmatic dikes and veins form in granitic plutons (Chapter 33).

(a)

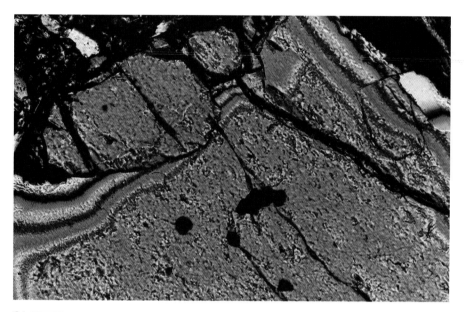

(b)

(c)

PLATE ONE

Birefringence as a function of thickness and orientation.

(a) Tapered sillimanite crystal within standard thickness section. Presenting birefringence is second-order blue for this orientation (near maximum), changing to first-order red, orange, yellow, and white as thickness decreases. X-polars. Bar scale = 0.12 mm.

(b) Tapered edges of olivine crystal. The major surface is fourth-order green, the deep green band is third-order, and the green next to blue is second-order. Since maximum birefringence for olivine is third-order green, thin section is too thick. X-polars. Bar scale = 0.12 mm.

(c) Tapered margin of thin section of quartzite. Crystals in extinction are either optic axis sections or other sections oriented with a principal vibration direction E–W. Interior of section (left) has a few quartz crystals presenting orange birefringence (especially lower left) indicating section is too thick (other orientations present yellow, white, and grays governed by crystal orientation). Central vertically oriented zone has pale yellow to cream-colored maximum presenting birefringence (correct thickness), whereas a maximum of white (right) indicates < 30 microns. X-polars. Bar scale = 0.5 mm.

PLATE TWO

Effect of additive or subtractive retardation obtained by superposing red-1 plate over anhydrite crystal.

(a) Anhydrite crystal with plucked area (center) reveals mounting epoxy and right-angle cleavage. Since presenting birefringence is second-order blue-green with patches of first-order red, anhydrite crystal is not oriented to provide maximum birefringence. Since maximum is about third-order green, this means (assuming 30 μm thickness) that the crystal is not oriented to provide maximum birefringence. Tapering indicated by change of colors to red, orange, yellow, and white (the epoxy is black).

(b) Same crystal position with red-1 plate inserted. All colors rise, indicating slow direction in plate (SW–NE) is coincident with slow direction in crystal. Second-order blue-green is now a mixture of third-order red and green, white edges are second-order blue, and the epoxy area is first-order red (the plate).

(c) Rotation of crystal 90° leaving red-1 plate in results in fall of colors (first order red patches become black, second order green becomes white, and white edges become first order yellow). All X-polars. All bar scales = 0.12 mm.

(a)

(b)

(c)

(a)

(b)

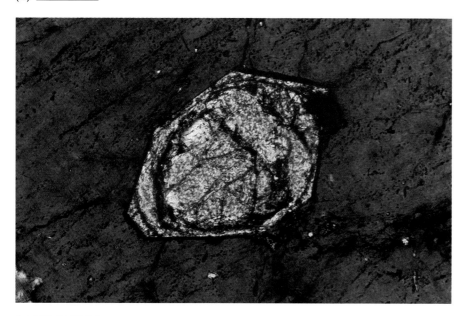
(c)

PLATE THREE

Determination of length fast andalusite using the red-1 plate, and an example of a relict sedimentary grain of zircon occurring in a magmatic rock.

(a) Elongate andalusite crystal presents cream-colored interference color, square andalusite crystal with symmetrical extinction is black, and another crystal presents a light gray interference color. X-polars. Bar scale = 0.5 mm.

(b) Same as (a) but with red-1 plate inserted. Elongate crystal becomes second-order yellow (colors rise), meaning that the fast direction in the red-1 plate (NW–SE) is coincident with the fast direction in the crystal, and the crystal is therefore length fast. The extinct prism cross section becomes magenta (first order "red") — that is, the birefringence of the red-1 plate alone. The light gray crystal goes to yellow, indicating a fall in color (red - grey = yellow); the slow direction in plate is coincident with fast in crystal. X-polars. Bar scale = 0.5 mm.

(c) Two-stage zircon from Predazzo monzonite pluton, northern Italy. Rounded interior zircon is an inherited detrital grain that has a magmatic overgrowth. X-polars. Bar scale = 0.12 mm.

PLATE FOUR

Birefringence as a clue to optimum orientation for conoscopic optical determinations, and examples of enhancement of textural relationships using the red-1 plate.

(a) Central muscovite presents a patchy second-order red and yellow birefringence — that is, about maximum for muscovite (find first-order red along thin margins)(note basal cleavage). Muscovite crystal (right) presents patchy second-order blue and yellow, meaning that its orientation cannot provide maximum birefringence (note lack of cleavage). Muscovite crystals (left and upper right) are basal sections (no cleavage showing), good for obtaining an optic axis figure. Note sphene crystals (center and left center). X-polars. Bar scale = 0.12 mm.

(b) Albite (near E–W) and pericline twin lamella of plagioclase (An_{52}) with red-1 plate inserted. Pericline twin lamellae are blue in this position. X-polars. Bar scale = 0.12 mm.

(c) Partial "swapping" relation between intergranular albite and two differently oriented crystals of alkali feldspar (see Figure 8.6 for explanation). Connections between albite and host have epitaxial associations. X-polars. Bar scale = 0.12 mm.

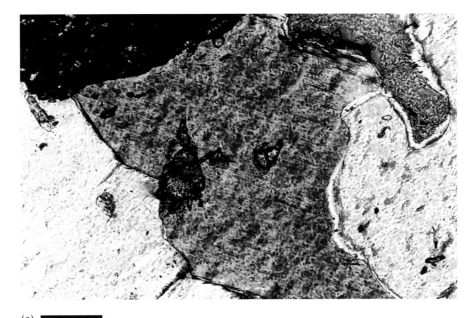

(a)

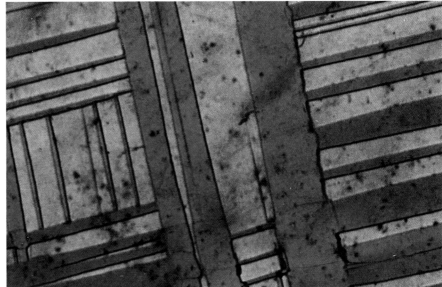

(b)

(c)

(a)

(b)

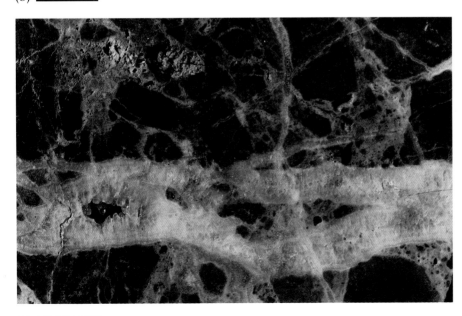

(c)

PLATE FIVE

Mantled feldspars, and an example of mineral staining as an aid in observing mineral distribution.

(a) Rapakivi texture in Wiborgite, Finland. Green oligoclase mantles pink alkali feldspar ovoids. Alkali feldspar ovoid has graphic-vermicular interlocking with matrix quartz (dark gray), biotite, and hornblende. Bar scale = 2 cm.

(b) Antirapakivi texture with pink alkali feldspar mantle on grey plagioclase. Whiter plagioclase crystal (left) has no mantle or a partial mantle. Minnesota pre-C, U.S.A. Bar scale = 1 cm.

(c) Vuggy epithermal quartz veins with associated adularia (stained yellow) in altered Alta andesite, Con Imperial Mine, Comstock Mining District, Nevada. Note the tendency for adularia to occur between andesite and quartz. (Courtesy D. Hudson.) Bar scale = 1 cm.

PLATE SIX

Examples of textural equilibrium and textural disequilibrium.

(a) Interleaved cluster of muscovite and biotite in quartz–mica schist as an example of textural equilibrium. X-polars. Bar scale = 0.5 mm.

(b) Penninite chlorite alteration of biotite. Location of chlorite at margins and along cleavage directions of biotite is indicative of textural disequilibrium, with biotite → chlorite. Sphene crystals (left). Osumi granodiorite, Japan. (Courtesy Ato Oba.) X-polars. Bar scale = 0.12 mm.

(c) Replacement of metamorphic olivine by serpentine as an example of textural disequilibrium. Calcite (lower left and upper left) has extreme birefringence. Interfaces between olivine crystals and along fracture planes in olivine were the locii of lateral replacement into olivine. Cream-colored birefringent mineral marking the medial replacement zone (lower right) is chrysotile. X-polars. Bar scale = 0.12 mm.

(a)

(b)

(c)

(a)

(b)

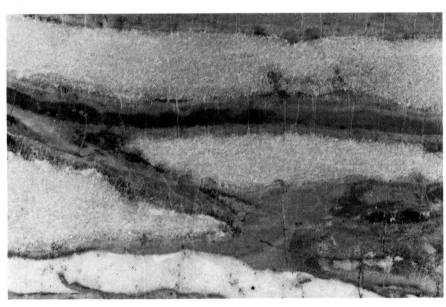

(c)

PLATE SEVEN

An example of dynamothermal metamorphism; use of the red-1 plate for approximating preferred crystallographic orientation of quartz; mineral zoning associated with thermal metamorphism.

(a) Synmetamorphic (dynamothermal) microboudinage of blue amphibole. Blue crossite was fractured and extended as darker crossite (uniaxial member of series) grew on fracture surfaces, followed by incomplete "soda-actinolite" (green-blue) as extension continued (Misch, 1969). (Shuksan blueschist, courtesy P. Misch.) PPL. Bar scale = 0.05 mm.

(b) Section of mylonitic quartzite perpendicular to kinematic X. Two populations of quartz orientation are revealed by insertion of red-1 plate. Since quartz is optically positive, the c axes of the blue crystals are generally NE–SW in the photomicrograph; c axes of yellow crystals are generally NW–SE (see Figure 6.3). Northern Snake Range, Nevada–Utah. X-polars. Bar scale = 0.5 mm.

(c) Calc–silicate zonation in metamorphosed muddy carbonate. Calcite marble (white crystalline) lenses zone to grossularite (pink), then to diopside dominant zones (green), and finally to darker brownish zones (not shown) that contain phologopite, biotite, cordierite, and quartz. White vein (bottom) consists of prehnite. House Range, Notch Peak area, Utah. Bar scale = 1 cm.

PLATE EIGHT

A complex sandstone-mudrock depositional contact, and examples of dolomitization.

(a) Contact between bentonitic mud rock of Chinle Formation and overlying Wingate Sandstone. Sinuosity of Wingate sand in mud cracks is a measure of compaction of Chinle mud as sediments were added above. Reduction zone (blue-green) lateral to sand fillings resulted from reduction of ferric iron in Chinle by percolating groundwater. Capitol Reef National Park, Utah. Lens cap for scale.

(b) Dolomitization (rhombs) of micritic limestone (calcite stained red). Wah Wah Range, Utah. (Courtesy J. Kepper.) X-polars. Bar scale = 0.12 mm.

(c) Partly dolomitized oosparite. Calcite stained pink. Oolites contain radial-fibrous calcium carbonate (center), sparry calcite with thin dolomite margin (upper left center), or saddle dolomite (upper right center and lower right). Matrix is sparry calcite. Cambrian limestone, House Range, Utah. (Courtesy J. Kepper.) PPL. Bar scale = 0.5 mm.

(a)

(b)

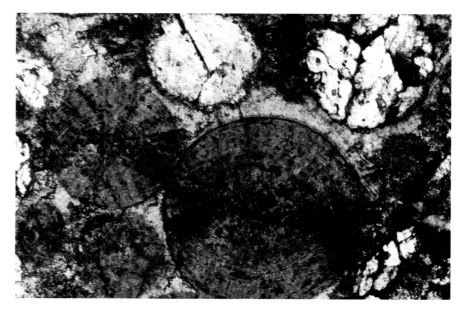

(c)

Deformation and Recrystallization of Solids

16

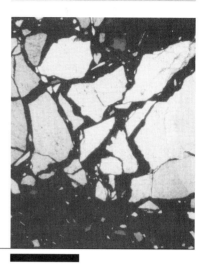

Micro mosaic breccia formed by brittle deformation of quartz crystal. Gossan zone of hydrothermal vein. Matrix (black) is limonite and jarosite. X-polars. Bar scale = 0.6 mm. Cucamunga Canyon, California/Nevada.

Rocks that evolve from other rocks do so by a variety of solid-state processes. The principal driving forces resulting in change are **heat** and **deviatoric stress**. Both mineralogical and textural changes record the new conditions and provide clues to the processes by which the new rocks are generated.

The Effect of Heat

As solids are heated, there is an increase in their molecular or ionic motion, and in this higher-energy state there is more chance for things to happen. An increase or decrease in the volume of a crystal lattice or of a rock in response to heating or cooling does not produce an observable texture. But much more can happen in crystals in response to heating. Recrystallization can occur, generating new crystals from old crystals of the same mineral type. Recrystallization also occurs when there is reaction between unlike minerals generating new crystals of a different type.

Heating promotes migration of crystal defects known as **point defects** and **dislocations**. Point defects in crystals include misplaced ions, lattice vacancies, and the presence of foreign ions. Dislocations are line defects. Defects form during initial crystal growth and in response to superposed deformation prior to or during heating. In either case, added heat is allowing the solid material to rearrange itself from a prior high-energy level to a lower-energy level attendant to the rearrangement or elimination of dislocations, a process known as one type of **recovery**. The various mechanisms of dislocation movement are collectively referred to as **dislocation creep**, and one of the most visible effects of this process, assuming some other conditions are appropriate, is formation of new crystals. This is **recrystallization**, or more specifically, **recrystallization-accommodated dislocation creep**. If it occurs in response to heating in a static environment, it is known as **annealing recrystallization**. Annealing is heating after solid-state deformation and the resulting recrystallization is another type of **recovery**. If deformation and heating are simultaneous, this is also a prime situation for recrystallization, and it is referred to as **dynamic recrystallization** or **dynamothermal recrystallization**. The new crystals will be obvious, but not the mechanism of recrystallization.

The Effect of Confining Pressure

Pressure applied to rock resulting in no change in shape is variously referred to as **confining pressure**, **non-directed pressure**, **lithostatic pressure**, **confining** stress, **geostatic pressure**, or **nondirected stress**. Change in volume of crystal lattices or rocks occur, but these would not be petrographically apparent. Without movement, crystals or noncrystalline materials of rocks cannot deform, and there is no record of a change in pressure. Confining pressure can be expressed numerically as **mean stress**. Stress or pressure is defined as a force divided by the area over which it is applied (dynes/cm^2).

The effects of confining pressure at the time of crystallization of minerals is more apparent. For example, with an appropriate starting chemical system, the mineral kyanite (specific gravity (s.g.) = 3.55) will form under relatively higher pressure conditions in preference to andalusite (s.g. = 3.16) even though the two minerals have the same composition. The higher-pressure condition must be favoring the formation of a denser crystal structure—thus one with a higher specific gravity (Figure 16.1a). Similarly, if andalusite is already in existence, subduction to higher pressure levels might result in the transformation of andalusite to kyanite. In reality, since volume for volume transformation of andalusite to kyanite is not energy efficient, growth of kyanite occurs somewhere in the rock as andalusite is destroyed somewhere else.

Denser minerals make denser rocks. If an oceanic crustal basalt (s.g. = 2.8) is subducted into the upper mantle, it can be expected to be converted into a granulite then into an eclogite (s.g. = 3.4). Eclogite is a chemically equivalent material containing denser minerals than those of the basalt (Figure 16.1b). The difference between the mineralogical content of basalt, garnet granulite, and eclogite is a parallel to the distinction between andalusite and kyanite.

The Effect of Deviatoric Stress

The most obvious physical record of pressure in rocks is tied to features such as folds, faults, and the shape of constituent grains. Since this is a record of **deformation** in rock materials, there must have been much more than just a volume change during their formation and the pressure involved could not have been equal in all directions. Change of shape of rocks and grains is **strain**. Strain is typically presented as a percent change of some linear dimension of a rock. Strain results from **deviatoric stress**, but it is loosely referred to as **differential pressure**, **load pressure**, **directed pressure**, **differential stress**, and **directed stress** as well. A deviatoric stress has principal stress directions that are (1) normal compressive ($\sigma 1$), (2) normal tensile ($\sigma 3$), and intermediate ($\sigma 2$).

From an energy point of view, with respect to a single crystal, a crystal may acquire or lose **strain energy** (density and configuration of dislocations) and **surface energy** (unsatisfied chemical bonds). **Mechanical energy** is transferred to strain and surface energy that is ultimately observed in the textures of rocks.

Deviatoric stress affects on rocks also depends on the value of confining pressure. At low confining pressure, rocks will "fracture" (**brittle** response), whereas at high confining pressures it is easier for them to "flow" (**ductile** response).

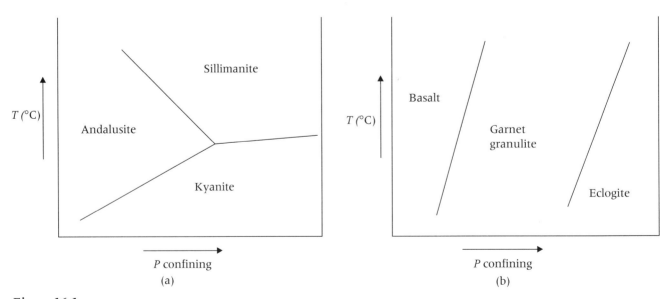

Figure 16.1
Qualitative representation of *P–T* conditions as they would influence crystallization of (a) Al-silicate polymorphs and (b) eclogite, garnet granulite, and basalt.

Deviatoric stress has a major effect on observable physical characteristics of rocks, and it also may have an effect on mineralogical reactions. Preference for kyanite to form in a relatively higher pressure environment does not seem to depend, per se, on the pressure being confining or differential. The level of pressure is important, not whether there is motion resulting in strain in the rock during growth of the kyanite.

If kyanite is considered as a product of reaction between other mineral phases, rather than as forming directly from andalusite, then a consideration of deviatoric stress becomes significant. For example, if the pressure level is appropriate for the formation of kyanite, the reactants that will produce kyanite may need to be mechanically brought together before the reaction can begin. Mechanical motion produced by deviatoric stress becomes a prerequisite for the production of kyanite in this situation.

Similarly, grain surfaces normal to high-stress directions are known to be more susceptible to dissolution, as long as there is at least some intergranular fluid. At pressure points between grains, there is more elastic energy since lattice bonds are compressed. The crystal is less stable at these points and therefore more soluble. **Pressure solution** leads to diffusional mass transfer to low-pressure sites where reprecipitation occurs. Pressure solution may be required to establish chemical gradients by which growth materials can be located in appropriate places and proportions for growth of new phases, such as kyanite. Once again, deviatoric stress has indirectly contributed to a favorable environment for the generation of kyanite.

The Effect of Time

Deformation of rock materials occurs in response to an appropriate level of deviatoric stress. On an atomic scale, such as within a crystal or in the polymeric complex of a glass, there must be movement of ions or molecules relative to each other. The observable effects of these motions in crystals are **dislocations**.

If the magnitude of deviatoric stress increases slowly, ions have time to adjust to the stress by moving. Diffusion of ions in solids is a very low-rate process. If there is a rapid increase in deviatoric stress, ionic motion is speeded up, but typically not enough to prevent breakage of chemical bonds. **Strain rate** is the rate at which rocks are forced to change in shape (and to a lesser extent, volume) as controlled not only by the magnitude of deviatoric stress but also by the level of temperature and confining pressure.

Strain rate typically is measured as percent shortening per second in laboratory experiments. Unfortunately, only clues and trends can be envisioned in natural rocks, since that is the realm of geological time. Except for seismic faulting, explosive volcanism, and meteor impact, most geological environments have very low strain rates.

Temperature and Pressure Within the Earth

The temperature of rocks increases with depth in the Earth. The rate of increase with respect to depth is the **geothermal gradient**. The gradient varies somewhat, depending on the heat flow existing along any selected vertical section of upper mantle and crust. The temperature of rocks along such a gradient ranges from a value corresponding to that at the Earth's surface to values at or near where rocks melt (600–700°C for water-bearing felsic rock materials, and 1100–1200°C for mafic rock materials). High temperatures may be generated locally as **frictional heat** along faults and more generally by local heating adjacent to magma intruded into shallower, cooler rock. Thus, the effects of elevated temperatures on rocks, in terms of their solid-state behavior, are realized at any pressure. Nevertheless, the converse of this is not true. Rocks under high confining pressure are not cold rocks except for local inversion of isotherms that have been modeled around "cool" oceanic crust plates subducting into the "hot" mantle.

Confining pressure also increases with depth. The geobaric gradient varies somewhat, depending on the densities of rock in the crust and upper mantle through which a gradient is calculated. Denser rocks generate steeper gradients. A range of pressure increase of 25 to 33 MPa/km is considered typical (MPa = 10 bars, 1 bar = 1 atm). If the mean density of rock is taken as 2.7 g/cm^3 the increase per kilometer depth is 270 bars (27 MPa).

As with the geothermal gradient, some momentary tectonic situations locally modify the geobaric gradient. A sudden tectonic event such as a fault is likely to generate local regions of reduced pressure (dilation) that may be the loci of melting if the rock is already near its solidus. Similarly, a tectonic overpressure associated with tectonism may lead to the crystallization of higher-pressure minerals than would be expected for an ambient confining pressure.

A rock brecciated a few kilometers from the surface may acquire and maintain open spaces. A clastic sedimentary rock with initial porosity will experience collapse of pore spaces as burial or subduction occurs. A rock under higher confining pressure has more strength and requires slightly greater stress for it to be deformed. Since temperature must rise as confining pressure rises, the weakening effect of heat on rocks overrides the slight gain in strength attendant to higher confining pressure.

Interrelation of Temperature, Confining Pressure, Deviatoric Stress, and Time

Strain in rocks is a function of temperature, confining pressure, and deviatoric stress, and it has two expressions: (1) the amount of strain, expressed as a percentage of some dimensional character, and (2) the rate at which the strain occurs. The amount of strain in relation to increasing deviatoric stress at constant temperature and confining pressure is shown in Figure 16.2a for two different rocks. One rock deforms (strains) along a straight-line curve to the **elastic limit**. At this point, if the deviatoric stress were returned to zero, the rock would return to its original shape on its own (recoverable strain). **Elastic strain**

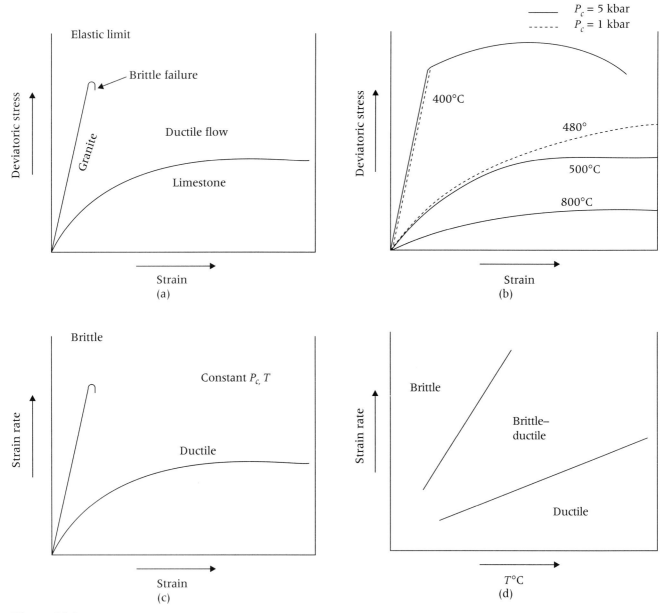

Figure 16.2
Affect of rock composition, temperature, confining pressure, deviatoric stress, and strain rate on rocks.

(a) Ductile behavior of limestone compared to brittle failure of granite. (b) Ductile behavior favored by higher temperature. (c) Higher strain rate favors brittle failure. (d) Brittle, brittle–ductile, and ductile regions as a function of temperature and strain rate (Heard, 1960; Griggs et al., 1960).

is proportional to deviatoric stress so far. Notice that the strain is different for a limestone than for a granite, indicating that deformation is in part dependent on rock type.

At the particular fixed condition of temperature and confining pressure shown in Figure 16.2a, the limestone will strain beyond its "elastic limit" and be permanently deformed. This condition is **ductile strain**, also known as "nonrecoverable strain" and "plastic strain." The granite experiences brittle failure beyond its elastic limit, but does this mean that granites cannot deform ductilely regardless of the magnitude of deviatoric stress?

The effect of temperature on strain is shown in Figure 16.2b. Heat weakens rock, and deformation can occur at lower values of deviatoric stress, which indicates that if the granite were hotter it would have experienced ductile deformation.

Strain rate is very significant in determining the style of rock deformation. If a granite is at its elastic limit as governed by temperature, confining pressure, and deviatoric stress, a sudden tectonic event may result in brittle failure rather than ductile deformation (Figure 16.2c). Even a nonsolid, such as viscous magma, has been known to "break" into blocks under certain explosive conditions (Walker, 1969), similar to the breaking of "silly putty." Conversely, a laborious tectonic event will allow ductile flow to occur (Figure 16.2c).

One of the useful concepts of strain rate is shown in relation to temperature in Figure 16.2d. Three fields can be generalized for rocks: (1) **brittle deformation** occurs at lower temperatures at any strain rate, and at somewhat higher temperatures if strain rate is rapid, (2) **ductile deformation** occurs at low strain rates if temperatures are elevated, and (3) an intermediate brittle–ductile region accommodates the behavior of different kinds of minerals and other rock-forming materials in a given rock. For example, a granite contains feldspars and quartz. The quartz behaves ductilely and the feldspar brittlely under the same environmental conditions, making this a transitional region.

A very high strain rate occurs along some faults, particularly if seismic waves are generated. Aseismic faults have a lower strain rate. A very low strain rate can be envisioned for the slow movement of rock involved in orogenic events.

The Effect of Fluid

Fluids are present in most rocks, at least in small amounts or tied up in hydrous minerals. This means that most "solid-state" processes are not strictly solid state.

It has been estimated that even mantle rocks contain on the order of 0.1 to 0.2 wt% H_2O. At the other extreme, porous sedimentary rocks have considerable pore fluid. Most metamorphic rocks form in the presence of at least an intergranular "film" of water. At very high temperatures rocks begin to melt, and the presence of even a little melt has marked effects on deformational behavior.

Deformation experiments have shown that as little as 0.01 wt% aqueous phase significantly reduces the deviatoric stress and temperature needed to produce strain. This reduction is called **hydrolytic weakening** (Chopra and Paterson, 1984; Tullis and Yund, 1989) and is thought to be due to $(OH)^-$ or H^+ ions somehow facilitating movement of dislocations, effectively rendering polycrystalline aggregates more susceptible to deviatoric stress.

Physical Expression of Rock Deformation

A summary of deformation mechanisms and their textural expression are presented in Table 16.1.

Brittle behavior on the thin-section and hand-specimen scale is characterized by fractured and faulted crystals (Figure 16.3) and crystalline aggregates. There may be only an incipient brecciation, or there may be comminution to aggregates of very small grains. Grains and grain fragments contribute strain by their rotation and translation. Movement of fragments past each other on a shear plane may produce open spaces in the rock as fragments slide up on each other (the Bagnold effect). Rocks formed by brittle mechanisms in fault zones are cataclasites (Chapter 23). Low-strain-rate deformation of "cold" rock occurs by brittle flow, typically resulting in deformed sedimentary rocks such as those in a mélange and in near-surface thrust sheets (Chapter 25).

Deformation of crystals and crystalline aggregates under high-temperature conditions usually takes place at slow strain rates, except locally at sites of deep-focus earthquakes. Some materials such as halite, gypsum, and ice deform ductilely at low temperature (Chapter 22).

The mechanical response to stress in crystals such as feldspar and quartz at elevated temperature is quite different than brittle behavior (Tullis and Yund, 1977; Debat et al.; 1978; Vidal et al., 1980; Schmid, 1982; Simpson, 1985; Mainprice et al., 1986). **Ductile flow** is characterized by dislocation creep in crystals. A number of intercrystalline and intracrystalline flow processes have been characterized based on electron microscopic observation of dislocations (Figure 16.4).

Table 16.1
Correlation of Pressure and Temperature Environment with Solid-State Deformation Processes and Rock Texture

Physical Environment: Low Pressure, Low to Medium Temperature

Geologic setting: brittle faults

Grain-scale processes: rigid-body rotation, cracking, fracturing, microfaulting, microcrushing, frictional grain boundary sliding, bending of crystals and twins, local melting

Textures produced: granulation, flattening or rotations by slip on cleavage, deformed crystals and twins, porphyroclasts, voids, patchy extinction, pseudosubgrains

Rock examples: gouge, breccia, slickensided surfaces, cataclasites, mud-matrix mélanges, pseudotachylite

Physical Environment: Medium Pressure, Medium Temperature

Geologic setting: brittle–ductile faults, "regional" brittle–ductile deformation

Grain-scale processes: brittle and ductile

Textures produced: depends on minerals present (e.g., quartz ductile, feldspar brittle)

Rock examples: mylonitic cataclasites

Physical Environment: High Pressure, High Temperature

Geologic setting: ductile faults, "regional" ductile deformation

Grain-scale processes: dislocation glide flow, dislocation creep, diffusion creep, recrystallization by nucleation and growth, recrystallization by subgrain rotation (recovery), pressure solution, hydrolytic weakening, cracking if high fluid pressure

Textures produced: progressive (undulatory) extinction, subgrains, deformation "bands," deformation lamellae, preferred orientations, reduction in grain size via recrystallization, increase in grain size via recrystallization, neoblasts, polygonalization, "ribbon" texture

Rock examples: mylonites, tectonites (schists and gneisses)

Dislocations may be linear and curvilinear defects, or point defects (lattice vacancies), but they may also be related to interstitial and substitutional impurities. The location of these imperfections can change as stress is applied, and new ones can form. This movement is intracrystalline and results in **crystal plasticity**, quite unlike the fracturing or rupturing of a crystal lattice under lower-temperature and/or high-strain-rate conditions.

The complex formation and movement of dislocations in crystals under conditions of stress at elevated temperatures results in zones of collection and zones of depletion of dislocations. For example, an increase in dislocation density in the form of a "pileup" of dislocations produces **progressive extinction** (undulatory extinction) (Figure 16.5a). This is one visible manifestation of intracrystal strain, as is the formation of **deformation lamellae** ("Fairbairn lamellae" of Groshong, 1988) in quartz (Figure 16.5b). A "mobilization" of dislocations may result in a reduction of dislocation density (recovery) in areas of crystals that become bordered by "tangles" of dislocations forming dislocation "walls" or boundaries. **Deformation bands** (Figure 2.15), **kinkbands** (Figure 16.5c), and **subgrains** (Figures 16.4, 16.5d) in crystals are formed in this manner (Wilson and Bell, 1979). **Quartz ribbons** (Figure 16.5e) are "ribbonlike" deformation bands (Wilson, 1975).

Other intracrystalline movements under relatively high temperature conditions includes **glide**, occurring along certain preferred crystallographic planes. The glide involves motion of dislocations along a **slip plane**. In some directions in certain minerals, such as plagioclase and calcite, glide produces **secondary twinning** (Figure 16.6) and is referred to as **twin gliding**. Twins produced in this manner, as opposed to growth twins, are known as **deformation twins** or **secondary twins**. In plagioclase, such twins typically are tapered at one or both ends (Figure 16.6).

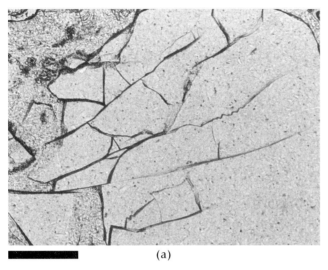 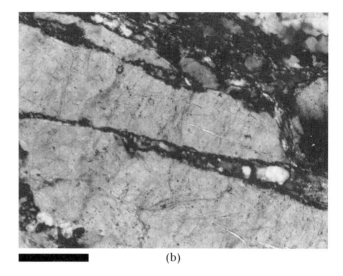

Figure 16.3
Brittle deformation.

(a) Fractures subradial to point of impact with another grain (presumed) on left side. Displaced quartz fragment and fragments of "dirty" glass matrix (left) in space-filling trydymite and late clear opal. Rhyolite tuff, Yellowstone National Park. PPL. Bar scale = 0.12 mm. (b) Displaced cleavage fragments of K-feldspar in finely recrystallized quartz. Brittle–ductile deformation. Mylonitic two-feldspar granite, Northern Snake Range, Nevada. X-polars. Bar scale = 0.06 mm.

Rotation of glide planes may occur in positions more normal to the maximum compressive direction, resulting in **preferred lattice orientation**. Gliding on the basal plane of quartz (Figure 16.7) is one explanation of preferred position of crystallographic c in some quartz-bearing tectonites (Figure 6.3 and Plate 7b). Since intracrystalline flow occurs at considerable confining pressure, voids do not form as they can in some brittle regimes. Consequently, movement of one grain relative to another must be accompanied by intercrystalline **grain boundary sliding**.

Another aspect of relatively high temperature deformation in rocks at low strain rates involves **recrystallization**. One way for an assemblage of crystals in

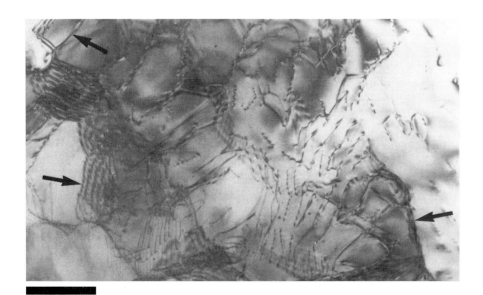

Figure 16.4
TEM bright-field micrograph of deformed and then annealed quartzite in the presence of H_2O showing well-formed subgrain boundaries (arrows) defined by high density of dislocations contrasting with clear areas with few dislocations. Bar scale = 0.7 μm. [Courtesy Richard A. Yund (Tullis and Yund, 1989).]

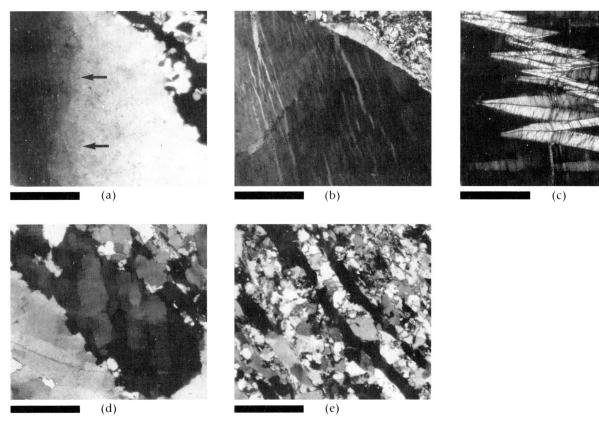

Figure 16.5
Textures resulting from ductile deformation.

(a) Progressive extinction in quartz. Transitional extinction zone has incipient subgrains (arrows). X-polars. Bar scale = 0.8 mm. (b) Probable deformation lamellae in quartz grain. X-polars. Bar scale = 0.2 mm. (c) Subgrains in the form of kinkbands in kyanite. X-polars. Bar scale = 0.8 mm. (d) Subgrains in quartz grains of quartzite. Okanogan Range, Washington State. X-polars. Bar scale = 0.8 mm. (e) Quartz "ribbons" in K-feldspar–quartz–plagioclase gneiss. Note subgrains within ribbons. North Cascade Range, Washington State. X-polars. Bar scale = 0.8 mm.

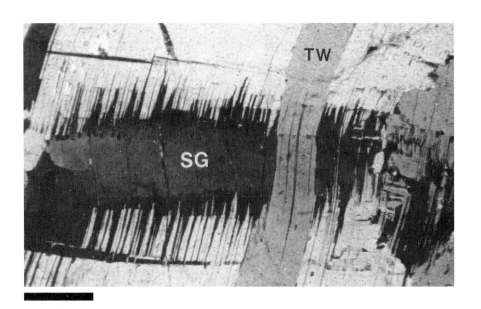

Figure 16.6
Secondary (mechanical) twinning in plagioclase localized along crystal flexure. Note weakly developed subgrains (SG) in plagioclase and secondary twin lamella in predeformation primary growth Carlsbad twin (TW). Charnokite. Adirondack Mountains, New York. X-polars. Bar scale = 0.5 mm.

a rock to respond to a high state of lattice energy and surface energy is to **recover** by forming relatively strain-free subgrains in crystals (Figure 16.5d) that have only slightly different orientation than the host crystal. Recovery involves a reduction of dislocation density, whereas an increase in dislocation density is **work hardening**. It is now thought that much recrystallization in rocks takes place by rotation of subgrain lattices through a process of accumulation of dislocations at subgrain boundaries. **Subgrain rotation recrystallization** may be very common in the recrystallization of minerals such as quartz (Figure 16.8a). A rise in temperature is the most common catalyst in the recrystallization processes, evidently allowing dislocations to clear out of certain regions in deformed crystals. It makes little or no difference if the high-energy state of crystals is acquired prior to or during heating—the result is effectively the same.

Growth of new silicate crystals at the expense of old crystals can also occur as a nucleation or growth process—that is, **nucleation recrystallization** producing new crystals known as **neoblasts**. If the new crystals occur along fractures and cleavages in minerals, it has been assumed that these are the high-energy sites at which nuclei are more likely to form (Figure 16.8b, c).

The formation of new crystals by either recrystallization mechanism generates more-or-less strain-free crystals that fit together into a **polygonal mosaic** (also known as **foam structure** and **comb structure**). Polygonalization is characterized by planar nonrational interfaces meeting at triple points with 120° angles (typically) between each of the interfaces (Figure 16.8c, d).

At a given temperature and deformational level, not all minerals behave either brittlely or ductilely. Quartz typically shows the effects of ductile flow, whereas at the same time feldspar deforms brittlely (Figure 16.3b). This is the basis of establishing the brittle–ductile transition region (Figure 16.2d).

High-energy sites are the driving force for any type of recrystallization. A deformed rock or an initially high energy rock such as a clastic sedimentary rock may recrystallize during a postdeformational heating related to intrusion of magma. Alternatively, deformation may accompany heating.

An example of recrystallization and deformation being more or less simultaneous is demonstrated by the movement of glacier ice. As a valley glacier makes a turn, there must be deformation of ice crystals because glacial ice is a solid. If there were no recrystallization, the ice would be a fine-grained comminuted mass of strained crystal fragments by the time the glacier reached its terminus. Regrowth of a crystal, initially strained in "conformity" with the flow direction change, will be a strain-free crystal in that new direction. Imagine this occurring on a collective scale with an unimaginable number of ice crystals. Such a process gets the glacier around the bend with a wealth of strain-free ice crystals.

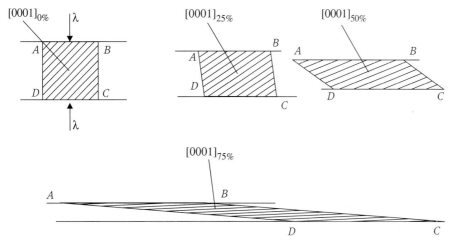

Figure 16.7
Demonstration of *c*-axis "rotation" in quartz by slip on basal plane in response to vertical compression (Hobbs et al., 1976).

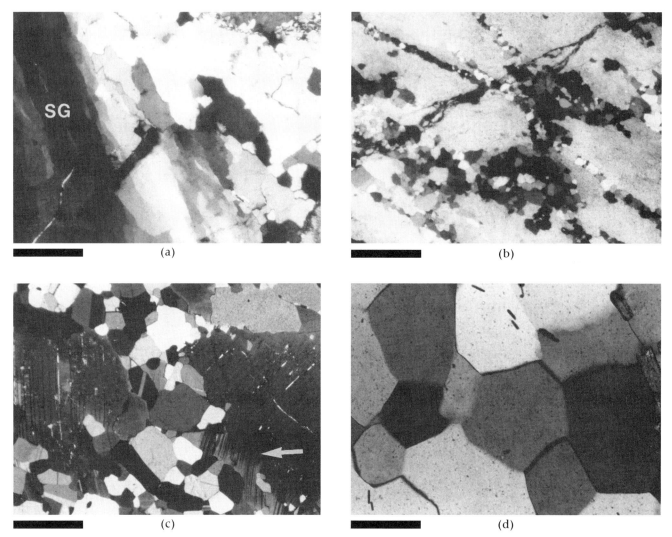

Figure 16.8
Ductile deformation leading to recrystallization.

(a) Partial recrystallization of large undeformed magmatic quartz grain of two-feldspar granite. Undeformed quartz (far left) grades to zone with subgrains (SG) and to mosaic of grains (middle and right) that may be rotated subgrains. Snowy Mountains, New South Wales, Australia. X-polars. Bar scale = 0.6 mm. (b) Rotated subgrains or neoblasts preferentially along cleavage-controlled fractures in quartz. Note marked deviation of new grains from host orientation. Metagranite, Thorvald Nelsen Mountains, Antarctica. X-polars. Bar scale = 0.6 mm. (Courtesy D. McLelland.) (c) Neoblasts of calcic plagioclase in magmatic plagioclase of same composition. Note relict magmatic zoning (arrow, upper left) and secondary twinning with marked curvature (arrow, lower right) in host plagioclase. Neoblasts locally form polygonal mosaic with triple points. Single albite twin is common in strain-free neoblasts. Metagabbro. X-polars. Bar scale = 0.6 mm. (d) Nearly perfect polygonal mosaic of quartz in quartzite. Note black rods of deerite. Laytonville, California. X-polars. Bar scale = 0.14 mm.

PART V

Rocks Formed by Magmatic Processes

Crystallization of magma generates crystalline magmatic rocks, and very rapid solidification of magma forms glassy magmatic rocks. A fixed volume of magma that consolidates as a closed chemical system results in magmatic rock that has a composition closely approximating the composition of that magma, minus excess aqueous phase and gases that are allowed to leave the chemical system. Magma that crystallizes rapidly or quenches to glass has little time for chemical exchange with its external environment, and even volatile phases may be trapped. Some internal textural heterogeneity develops in magmatic systems of unusual composition in which there is liquid immisibility (Philpotts, 1976).

Glassy and fine-grained rocks forming in (1) lava flows and domes, (2) small dikes and sills, and (3) the marginal chill zones of larger intrusive bodies are all reasonably representative of an initially homogeneous magma composition, regardless of magma source.

A large volume of magma convecting within a chamber as it cools and fractionally crystallizes leads to segregation of crystals from magma and unequal distribution of mineral phases. The composition of rock collected from the resulting rock body does not represent the initial composition of the magma. Depending on the site of sampling, a rock may be quite homogeneous and contain no overt textural evidence that it is a product of fractionation.

Zoned plutons contain magmatic rock members that are remarkably homogeneous. Although each zone consists of magmatic rock reasonably representing the magma from which it crystallized, no one rock type represents the composition of a presumed parent magma from which each rock zone was derived. Evidently, the fractionation process can lead to homogenization of magma batches that are emplaced in pluses at higher levels distant from the primary site of fractionation.

Magmatic rocks representing either (1) the composition of a "flash-injected" batch of homogeneous magma or (2) the composition of "apparently homogeneous" fractionates that do not represent initial magma composition are described as homogeneous magmatic rocks in Chapter 17.

Magmatic rocks that contain overt mineralogical and textural evidence for mechanical or chemical accumulation of mineral phases in fractionation systems are described in Chapter 18 as cumulate magmatic rocks. Accumulation of crystals of a given composition by differential flow of crystals and melt during a convective process, some other dynamic process, or some chemical mass transport process leads to the formation of rocks of unusual composition. Not only are these rocks not representative of the magma from which they evolved, some of them are monomineralic (dunites and anorthosites) and could not have been magmas anyway, in view of their extremely high liquidus temperatures.

Magmatic rocks containing overt evidence for mixing of magmas of different composition are described in Chapter 19 as mixed magmatic rocks. Magmas that have been compositionally modified by incorporation of wall rock are assimilation-hybrid magmatic rocks, discussed in Chapter 20.

17

Homogeneous Magmatic Rocks

Tertiary andesite dike in Cretaceous granodiorite. Carson Range, near Carson City, Nevada.

Textural Attributes of Homogeneous Magmatic Rocks

Grain size is a textural aspect of magmatic rocks that has proven to be very useful in classification (Chapter 9) and in characterizing environment of solidification. In general, the finer-grained magmatic rocks are **volcanic**, and the coarser-grained rocks are **plutonic** (MacKenzie et al., 1982). Although this is a convenient correlation for classification purposes and a general sense of petrogenesis, it is not specifically adaptable to those rocks known to be transitional between plutonic and volcanic on the basis of petrochemistry and structure, nor does it accommodate those occurrences of coarse-grained rock in the volcanic environment, such as spinifex ultramafic lavas (Pyke et al., 1973), and fine-grained rocks in the plutonic environment, such as aplite. Therefore, the discussion of magmatic rock types below focuses on grain size as a principal textural attribute rather than on assignment to their tectonomagmatic association on the basis of grain size.

Grain-size variation is another principal textural attribute that has not been a first-order textural classification parameter, except for some early classifications that made distinction between porphyritic and nonporphyritic rocks (Grout, 1940; Pirsson and Knopf, 1947). Recognition of grain-size variation as an important ingredient of the rock-forming process is on the increase (Marsh, 1988; Cashman and Ferry, 1988; Cashman, 1990), especially in view of the quantitative capabilities of automatic imaging analysis (Chapter 6).

The occurrence or absence of **phenocrysts** in magmatic rocks is particularly intriguing. Are phenocrysts good indicators of one-stage and two-stage crystallization or some other important aspect of crystallization or environment of crystallization (Chapter 14)? On the assumption that phenocrysts are petrogenetically important in some way, distinction between porphyritic and nonporphyritic magmatic rock is included in the outline of magmatic rock types given here.

Examples of Magmatic Solidification

Nonporphyritic (Aphyric) Glass

Glass without crystals (phenocrysts or microphenocrysts) includes some varieties of **pumice** (highly vesicular, Figure 17.1a), some **obsidian** (vesicular or nonvesicular, Figure 17.1b), and devitrified spherulitic

221

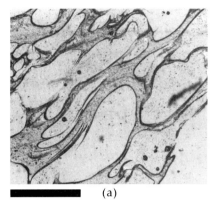 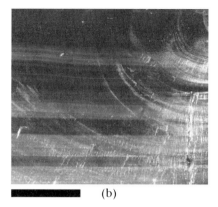 

Figure 17.1
Nonporphyritic glassy magmatic rocks.
(a) Pumice with somewhat flattened vesicles. PPL. Bar scale = 0.8 mm. (b) Flow-layered obsidian with conchoidal fracture. Bar scale = 1.5 cm. (c) Tachylyte chill margin (arrow) in vesicular pahoehoe Hawaiian basalt. Bar scale = 1.5 cm. (Courtesy Thea Robbins.)

obsidians. Obsidians have a wide range of composition and H_2O content, ranging from silicic rhyolitic to the basaltic glass **tachylyte**. A nearly crystal-free tachylyte rind is shown associated with a **pahoehoe** surface and a **vesicular** crystalline basalt interior (Figure 17.1c).

Porphyritic (Phyric) Glass

Porphyritic glass includes vitrophyry and other magmatic rocks that have a glass matrix. Some vitrophyrys form by dense compaction and welding of glass shards and pumice in ash flow tuffs; others; such as the glassy phonolite containing alkali feldspar phenocrysts (**kenyte**) shown in Figure 17.2a, represent two-stage crystallization–solidification. The phenocrysts have a spongy cellular texture that may be the result of partial dissolution when the crystal-bearing melt was erupted.

A vitrophyry fragment contained in an ash flow is shown in Figure 17.2b. The glass contains plagioclase phenocrysts and **trichites** (chainlike and hairlike crystallites).

The vesicular porphyritic basalt shown in Figure 17.2c contains plagioclase, olivine, and augite mi-

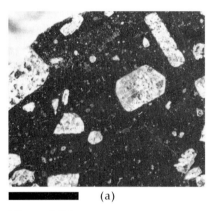 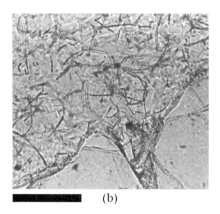 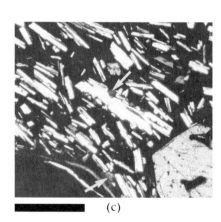

Figure 17.2
Porphyritic glassy magmatic rocks.
(a) Kenyte (phonolite) containing sanidine phenocrysts with glass-filled cells in glass matrix. Mount Kenya, Africa. Slab surface. Bar scale = 0.8 cm. (Courtesy J. Kurtak.) (b) Plagioclase phenocrysts (lower right) in glass containing chainlike trichite crystallites. Confusion Range, Utah. PPL. Bar scale = 0.07 mm.
(c) Skeletal olivine phenocryst (lower right) with smaller microphenocrysts of augite (arrows) and plagioclase in glass matrix. Note flow alignment of plagioclase. Vesicle (lower left). Mexico. X-polars. Bar scale = 0.8 mm.

crophenocrysts in a black glass matrix. The skeletal form of the olivine suggests growth in a strongly undercooled melt (perhaps modified by dissolution), and the preferred orientation of the plagioclase phenocrysts indicates flow of the crystal-bearing magma.

Nonporphyritic (Aphyric) Very Fined-Grained Glass-Bearing Magmatic Rocks

Very fine-grained (<0.1 mm) magmatic rocks with intercrystal glass ("intersertal" texture) are not considered porphyritic unless the volume of glass is large enough to isolate crystals. Late glass in small amounts (Figure 17.3a) represents a final stage of rapid heat loss to cold rock, the atmosphere, or to a water body in a volcanic environment. Rapid drop in pressure, attendant volcanic eruption, leads to vesiculation, raising of liquidii, and quenching remaining melt. The interior of a lava flow can be expected to be more crystalline and less glassy than basal or surface zones where quenching is extreme.

The exterior of the vesicular basalt shown in Figure 17.1c is fine-grained, glass-bearing, and nonporphyritic. If vesicles dominate, the rock is **scoreaceous** (Figure 17.3b) or **pumiceous**. A highly vesicular rock does not mean that there was a high percentage of dissolved gases with which the magma was overwhelmed during vesiculation. Since gas bubbles are mobile in magma, upward migration and accumulation are to be expected.

Porphyritic (Phyric) Very Fined-Grained Glass-Bearing Magmatic Rocks

Very fine-grained magmatic rocks with intercrystal glass can be porphyritic. Some glass-bearing basalts of the Columbia River Plateau have relatively large phenocrysts of plagioclase and olivine (Hooper et al., 1984). Glass-bearing rhyolite lava forming domes commonly have a few sanidine and quartz phenocrysts. The glass in these rocks is typically devitrified.

Nonporphyritic (Aphyric) Very Fine-Grained Rocks

Nonporphyritic very fine-grained basalts and basaltic andesites without glass are common. Very fine-grained silicic or alkaline rocks are most commonly porphyritic. The riebekite "granite" (microgranite) from Ailsa Craig, Scotland, used in the manufacture of curling stones is an exceptional occurrence.

Porphyritic (Phyric) Rocks with Very Fine-Grained Matrix

A two-feldspar granite porphyry with a very fine-grained (<0.1 mm) matrix is shown in Figure 17.4a. The matrix texture consists dominantly of K-rich

(a)

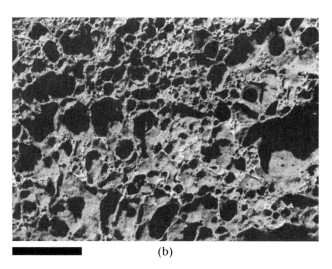
(b)

Figure 17.3
Nonporphyritic very fine-grained basaltic rocks with glass.
(a) Columbia River basalt containing plagioclase, clinopyroxene, opaque oxide, and acicular crystal-bearing glass (white arrow). Priest Rapids, Washington. PPL. Bar scale = 0.12 mm. (b) Scoreaceous basalt without vesicle flattening. Bar scale = 1 cm.

alkali feldspar and quartz in a quasi-graphic relationship, and locally this textural relation is clearly graphic (not shown). Graphic quartz–feldspar texture suggests that there was simultaneous crystallization of the two phases, with morphology influenced by the presence of aqueous phase (Fenn, 1986). Another two-feldspar granite porphyry with a quartz and alkali feldspar matrix is shown in Figure 17.4b. In this case, there is a side-by-side granular relation between these minerals.

The porphyry shown in Figure 17.4c has a more-or-less bimodal phenocryst grain-size distribution. A very fine-grained fraction serves as matrix to both phenocryst-size groups. It is tempting to interpret (see Chapter 14) this texture as resulting from three-stage crystallization. The contrast in grain size between all phenocrysts and the very fine-grained matrix strongly suggests that at least two stages of crystallization (and cooling) have affected this system. The larger phenocrysts are quartz and sanidine (not shown) as well as the microphenocryst fraction. The somewhat graphic intergrowth of sanidine crystals in the marginal zone of large quartz phenocrysts (Figure 17.4c) is very similar to the matrix of the porphyry shown in Figure 17.4a. It is tempting to speculate that if another eruption had occurred prior to

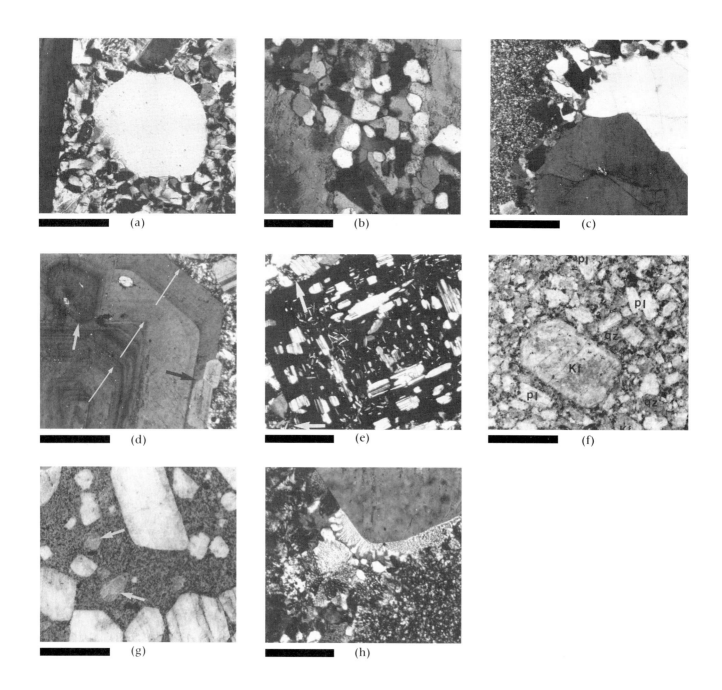

final crystallization of the system in Figure 17.4a, a very fine-grained matrix similar to that in Figure 17.4c would have formed.

The large plagioclase phenocryst shown in Figure 17.4d occurs with plagioclase microphenocrysts in a very fine-grained matrix, again suggesting, but not demonstrating, three-stage crystallization. **Synneusis** of two plagioclase crystals was succeeded by a coalescing overgrowth of plagioclase, completing crystallization of the large composite phenocryst (Vance, 1969). Some of the microphenocrysts are partially incorporated into this overgrowth plagioclase.

Another possibly three-stage rock is represented by the Judith Peak granite porphyry (Figure 17.4e). Alkali feldspar phenocrysts **poikilitically** include smaller oriented plagioclase crystals (Hibbard, 1965) equivalent in size and composition to the fine-grained microphenocrysts crowding the matrix. The third grain-size fraction consists of very-fined grained quartz and alkali feldspar.

The two-feldspar granite porphyry shown in Figure 17.4f contains only about 20% very fine-grained matrix. A fine- to medium-grained fraction of phenocrysts consists of plagioclase, alkali feldspar, quartz, hornblende, and biotite, suggesting that a final quenching event occurred very late in the crystallization history of this magmatic system. Some of the alkali feldspar crystals are very large, probably reflecting a relatively low nucleation rate for them rather than a separate stage of crystallization (see Chapter 14).

Figure 17.4
Porphyritic very fine-grained magmatic rocks.

(a) Two-stage "microgranite" porphyry with plagioclase (left), quartz (center), and K-feldspar (not shown) phenocrysts in very fine-grained matrix characterized by graphic and semigraphic quartz and K-feldspar. Toiyabe Range, Nevada. X-polars. Bar scale = 0.02 mm. (b) "Microgranite" porphyry with phenocrysts of plagioclase, quartz, K-feldspar, and biotite in fine-grained side-by-side granular matrix consisting of K-feldspar, quartz, and less plagioclase. X-polars. Bar scale = 0.02 mm. Lakeshore Mine, Arizona. (Courtesy W. P. Johnston, Hecla Mining Co.) (c) "Microgranite" porphyry with phenocryst of quartz (twinned), alkali feldspar (not shown), and plagioclase (not shown) in a very fine-grained granular matrix consisting mostly of quartz and alkali feldspar. Possible three-stage texture. Second stage represented by fine-grained crystal fraction (quartz, alkali feldspar, plagioclase), partly included with semigraphic relation in margin of large quartz (first stage) phenocryst. St. Flour, Massif Central, France. X-polars. Bar scale = 0.8 mm. (Courtesy Jean-Paul Couturié.) (d) Plagioclase phenocryst in two-feldspar "microgranite" porphyry. Synneusis contact of two plagioclase crystals (thick white arrow) followed by coalescing plagioclase overgrowth. Two small partially included plagioclase crystals (black arrow) probably indicate a later stage synneusis. Morphological changes can be traced from center of lower crystal (thin white arrow) through overgrowth portion having the sequence: edge (earliest crystal) → face → edge (end of second thin arrow) → face (final). Gillis Range, Nevada. X-polars. Bar scale = 0.8 mm. (e) K-feldspar phenocryst with oriented inclusions of poorly formed plagioclase and smaller nonoriented plagioclase. Probable three-stage texture. Larger phenocryst fraction consists of the K-feldspar and quartz (not shown). Second-stage fraction consists of microphenocrysts of plagioclase (same as inclusions) in very fine-grained matrix (third stage) (arrows). "Microgranite" porphyry. Judith Peak, Montana. X-polars. Bar scale = 0.8 mm. (f) Two-feldspar "microgranite" porphyry. Largest phenocryst is K-feldspar (left of center); smaller phenocrysts are plagioclase (pl), quartz (qz), hornblende (black), biotite (black), and more K-feldspar (Kf). Very fine-grained matrix (15%) is not readily visible in this slab curface. Probable two-stage crystallization. Stillwater Range, Nevada. Bar scale = 1.5 cm. (g) "Microgranodiorite" porphyry containing phenocrysts of plagioclase and quartz (arrows). Two-stage crystallization. Slab surface. Hell's Gate, British Columbia. Bar scale = 1.4 cm. (h) "Microgranite" porphyry with K-feldspar (upper right) and quartz phenocrysts (not shown) in very fine-grained K-feldspar–quartz matrix. Fine-grained vermicular–graphic K-feldspar–quartz intergrowths ("granophyric") tend to localize in plane of weak flow layering, oriented NNE–SSW. Slieve Gullion, Northern Ireland. X-polars. Bar scale = 0.2 mm.

The diorite porphyry in Figure 17.4g contains (1) large plagioclase and quartz phenocrysts, (2) a fine-grained fraction of hornblende, quartz, and plagioclase phenocrysts, (3) and a very fine-grained granular matrix consisting of quartz, K-feldspar, and oligoclase. The question of three-stage crystallization versus differential nucleation rates is raised once again.

The granite porphyry shown in Figure 17.4h has a matrix that is mostly very fine-grained granular, but locally coarsens into a vermicular–graphic ("granophyric") intergrowth of quartz and K-feldspar. Localization of the intergrowth is in the "pressure shadow" zone of the large alkali feldspar phenocryst (Figure 17.4h), as defined by a crude flow layering in the rock oriented NNE–SSW in the photograph. This relationship suggests localization of very late stage, probably H_2O-rich fluids (Chapter 33; Fenn, 1986).

Silicic lava flows (not ash flows) occur if the magma is relatively "dry" (Chapter 13) and capable of flow as sheets rather than freezing up into lava domes. Such rocks are typically porphyritic rhyolites with very fine-grained matrices (Henry, 1990).

Nonporphyritic Fine-Grained Rocks

The famous Westerly granite (Chayes, 1952; Tuttle and Bowen, 1958) is a myrmekite-bearing two-feldspar nonporphyritic fine-grained (0.1–1 mm) granite (Figure 17.5a). The plagioclase is oligoclase, not albite, meaning that the rock corresponds to the quaternary granite system rather than to the ternary system. Consequently, initial crystallization of the most calcic oligoclase (now cores of the larger plagioclase crystals, Figure 17.5a) was followed by direct crystallization of plagioclase alongside alkali feldspar and quartz as predicted from the phase relations in the quaternary system (Chapter 13).

Aplites are fine-grained leucocratic magmatic rocks (Figure 17.5b) most commonly containing oligoclase, K-rich alkali feldspar, quartz, and minor amounts of biotite and muscovite. Myrmekite is common in aplite (Figure 17.6c), and the alkali feldspar is typically microcline. Inversion of a monoclinic alkali feldspar to microcline in aplites and in "aplitic" granites such as the Westerly probably is facilitated by the generation of a late-stage aqueous phase (Eggleton and Buseck, 1980).

The texture of a nonporphyritic fine-grained diabase is shown in Figure 17.5d. Early crystallizing olivine and plagioclase are partly included in late-crystallizing clinopyroxene and hornblende. This oikocrystic ("ophitic") character of clinopyroxene and hornblende indicates their crystallization extended to a late stage with a relatively low nucleation rate.

Porphyritic Rocks with Fine-Grained Matrix

The matrix grain size of the porphyritic dolerite (diabase) shown in Figure 17.6a ranges from fine grained in the interior of the dike to very fine grained along the chill margin. Distribution of plagioclase phenocrysts may be controlled by flow differentiation (Ross, 1986).

The fine-grained matrix of a porphyritic diabase is shown in Figure 17.6b. The phenocrysts are orthopyroxene (not shown) set in a fine-grained matrix consisting mostly of plagioclase and clinopyroxene, with the latter tending to oikocrystically ("ophitic/subophitic") include the former. A small amount of very late, graphically intergrown quartz and K-feldspar occurs intergranular to plagioclase and clinopyroxene, a convincing demonstration of extreme fractionation of a basaltic system to a final small-volume granitic fluid (Bowen, 1928).

Nonporphyritic Fine- to Medium-Grained Rocks

Although most designations of "equigranular" are generalizations, not reflecting actual ranges in grain size, some rocks are clearly seriate inequigranular. The fine- to medium-grained granodiorite shown in Figure 17.7a, p. 229, a contains "blebs" of relatively small quartz crystals included in the rim zones of plagioclase and alkali feldspar. This textural relation suggests that there was shift in conditions, perhaps greater undercooling, promoting a higher rate of quartz nucleation.

Inequigranularity and an intergrowth relation of quartz with feldspar is even more apparent in the Isle of Skye (Scotland) granite shown in Figure 17.7b. Graphic quartz occurs in the outer portions of alkali feldspar crystals and as epitaxial extensions of quartz crystals. Evolution of this system from a side-by-side crystallization of the larger crystals of nonintergrown quartz and alkali feldspar to a late stage characterized by graphic intergrowth is a characteristic feature of this rock.

Variation in grain size from fine to medium may also occur as the result of very late intergranular crystallization. Intergranular hornblende and quartz in the cpx–opx–pl–ho diorite known as the "Academy Granite" of the Sierra Nevada (Figure 17.7c) is indicative of magmatic fractionation. The mantling of orthopyroxene and clinopyroxene by hornblende is a classic example of Bowen's "reaction" series (1928) in which mafic magmas generate less mafic melts as crystallization proceeds, including a progressive increase in the H_2O content of the evolving melt, some of which becomes partially incorporated into hydrous minerals such as hornblende.

Figure 17.5
Nonporphyritic fine-grained magmatic rocks.

(a) Two-feldspar granite with zoned plagioclase (arrow), microcline (Kf), quartz, biotite, and muscovite. Westerly, Rhode Island. X-polars. Bar scale = 0.5 mm. (b) Aplite dike in diorite. Pyramid Peak Range, Nevada. (Courtesy C. Sabine, University of Nevada, Desert Research Institute.) (c) Plagioclase, microcline, quartz aplite containing myrmekite (arrow). Wassuk Range, Nevada. X-polars. Bar scale = 0.5 mm. (d) Diabase with olivine (ol), hornblende (HO), clinopyroxene (CPX), and opaque oxide. CPX and HO have oikocrystic ("ophitic") relation with plagioclase laths. Sudbury, Ontario. X-polars. Bar scale = 0.5 mm.

Nonporphyritic Medium-Grained and Medium- to Coarse-Grained Rocks

Slab surface views of a diorite (Figure 17.8, p. 230; a), the two-feldspar Barre granite (Figure 17.8b), and the miarolitic one-feldspar Keyhole Canyon granite (Figure 17.8c) are shown as representatives of nonporphyritic medium-grained magmatic rocks.

Miarolitic cavities are prime textural evidence for late-stage separation of an aqueous phase (Jahns and Burnham, 1969) in felsic magmatic systems. Miarolitic cavities are most common in high-level plutons and do not occur in deeper plutons where they are precluded by high confining pressure.

Figure 17.6
Porphyritic fine-grained magmatic rocks.

(a) Dolerite (diabase) dike with larger plagioclase phenocrysts in medial zone. Chill margins (darker) locally have phenocrysts (black arrows). Cross-shear (between white arrows) resulted in displacement of phenocryst-bearing chill zone (bottom white arrow) with corresponding displacement and filling of extension fracture (upper white arrow). One-feldspar granite host (Figure 17.11c) was not completely crystallized at time of dike emplacement and late shear. Rockport, Massachusetts. (b) Matrix of diabase containing plagioclase and pyroxene phenocrysts (not shown) consists of oikocrystic ("ophitic") clinopyroxene–plagioclase and a local late-stage micrographic K-feldspar–quartz (white arrow). New Haven, Connecticut. X-polars. Bar scale = 0.12 mm.

The texture of one phase of the medium-grained Osumi granodiorite (Japan) has well-formed plagioclase against quartz and alkali feldspar (Figure 17.9, p. 231; a). In view of (1) high modal plagioclase and (2) plagioclase core compositions in the andesine range, the plagioclase probably was on its liquidus before quartz and alkali feldspar. When quartz and alkali feldspar began to crystallize, plagioclase continued to crystallize, zoning to more sodic compositions.

The pre-Cambrian nonporphyritic medium-grained two-feldspar granite from which the famous Vigeland sculptures (Oslo) were carved consists of microcline, oligoclase, quartz, and some biotite (Figure 17.9b). The possibility of subsolidus adjustments in mineralogy and texture is a consideration for granitic rocks that lack eustructural (well-shaped) crystals, well-developed plagioclase zoning, and have fully developed microcline, such as represented by this granite. A similar granite is shown in Figure 17.9c. This is the two-feldspar granite of Mount Rushmore, South Dakota, containing poorly formed microcline and quartz, poorly zoned plagioclase, but rather well-formed biotite and muscovite. The quartz vermicules in myrmekite have acquired a "digitated" relation with host oligoclase, suggesting that there has been some postmagmatic intracrystalline flow. Similar modified myrmekite occurs in some of the fine-grained New England granites (Chayes, 1952) and in the Stone Mountain granite of Georgia.

Well-formed alkali feldspar and nepheline crystals are larger than somewhat intergranular sodalite, aegirine–augite, barkevikite, and albite in the medium- to coarse-grained nepheline syenite shown in Figure 17.9, p. 231, d.

Porphyritic Rocks with Medium- or Coarse-Grained Matrix

Phenocrysts in coarser-grained rocks generally do not indicate two stages of crystallization. The most compelling evidence for this conclusion is that the

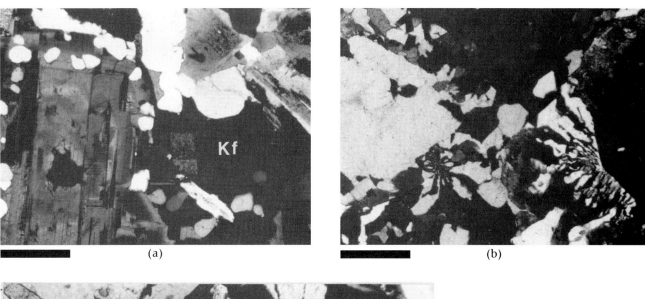

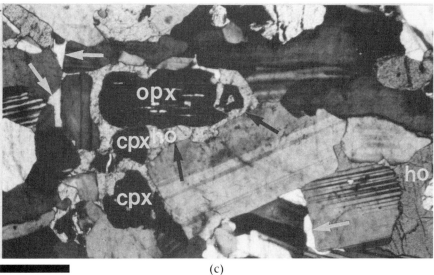

Figure 17.7
Nonporphyritic fine- to medium-grained magmatic rocks.

(a) Granodiorite with small grains of quartz (not in optical continuity) included in outer zone of plagioclase (left) and in K-feldspar (Kf), probably indicating simultaneous crystallization and possibly slightly accelerated rate of nucleation and crystallization. Sierra Nevada, California. X-polars. Bar scale = 0.7 mm. (b) Quartz in graphic–vermicular ("granophyric") relation with K-feldspar, probably representing late-stage simultaneous and possibly somewhat accelerated rate of crystallization. Early-formed quartz crystal (left) is optically continuous with quartz of intergrowth adjacent to it; early-formed K-feldspar (far right) is optically continuous with adjacent K-feldspar containing quartz. Isle of Skye, Scotland. X-polars. Bar scale = 0.7 mm. (Courtesy R. Watters.) (c) Early-formed orthopyroxene (opx), clinopyroxene (cpx), and plagioclase with late-forming intergranular hornblende (ho) and quartz (white arrows). Hornblende mantle on orthopyroxene and clinopyroxene (black arrows) indicates partial reaction relation. Gabbro. Sierra Nevada, California. X-polars. Bar scale = 0.5 mm.

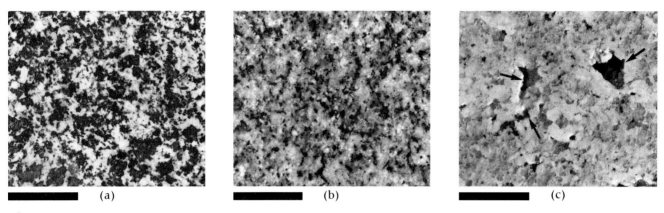

Figure 17.8
Nonporphyritic medium-grained magmatic rocks.

(a) Sierra Nevada diorite. Slab surface. Bar scale = 1.4 cm. (b) Two-mica, two-feldspar granite. Barre, Vermont. Slab surface. Bar scale = 1.4 cm. (c) One-feldspar miarolitic (arrows) granite. Note increase in crystal size of alkali feldspar and development of crystal faces facing miaroles. Keyhole Canyon, Eldorado Mountains, Nevada. Slab surface. Bar scale = 0.8 cm.

phenocrysts are typically alkali feldspar crystals that contain small plagioclase inclusions throughout the crystals (Figure 17.10, p. 232; a). This means that alkali feldspar did not nucleate first and enlarge before the many smaller crystals of plagioclase formed in a presumed second stage of crystallization. To be included in the alkali feldspar, the plagioclase must have nucleated early, before or at the same time as alkali feldspar. More likely (see Chapter 14), the alkali feldspar crystals are large due to low nucleation rate compared with plagioclase. Contrast of grain size between large alkali feldspar phenocrysts and medium- or coarse-grained matrices is especially marked in two-feldspar granites such as the Cathedral Peak Granite of Yosemite (Figure 17.10b).

Phenocrysts of alkali feldspar are accompanied by muscovite phenocrysts in some peraluminous two-feldspar granites, such as the granite of the Kern Mountains (Figure 17.10c, Best et al., 1974). Muscovite phenocrysts are also characteristic of one phase of the Leinster batholith in Ireland (Bruck, 1974). As with alkali feldspar phenocrysts, the muscovite must have nucleated early enough to allow relatively unimpeded growth and developed a morphology characterizing their crystal structure.

Nonporphyritic Coarse-Grained Rocks

The Half Dome granodiorite of Yosemite National Park (Bateman and Chappell, 1979) contains well-formed (eustructural) biotite and hornblende (Figure 17.11, p. 232; a). The Conway granite of New Hampshire contains alkali feldspar and radioactively darkened quartz (Figure 17.11b).

The Cape Ann granite (Figure 17.11c) contains perthitic alkali feldspar and quartz. This is Tuttle and Bowen's (1958) hypersolvus granite, crystallizing at the ternary granite minimum at low H_2O pressure. Some aqueous phase must have been present in this system to account for the formation of arfvedsonite.

Distinction Between Crystallization of Supercooled Magma and Devitrification of Glass

The **glass transformation temperature** marks the boundary between supercooled silicate melt and glass (Carmichael, 1979). The temperature, marking a sudden change in thermodynamic properties, varies according to composition and H_2O content. Above the glass transformation temperature, there can be **crystallization** of magma that has been **supercooled** to a state that is highly viscous and glasslike. Below the transformation temperature, glass is in the solid phase, but it too may become crystalline by a **devitrification process**. In the strict sense of the process, there must be true glass (below the glass transformation temperature) before there can be devitrification.

Textural distinction between crystalline aggregates formed by crystallization of supercooled magma, in contrast to devitrification, is not well defined.

17 Homogeneous Magmatic Rocks 231

Figure 17.9
Nonporphyritic medium-grained magmatic rocks.

(a) Osumi granodiorite (Hedaokawa type) with well-formed plagioclase and biotite and somewhat late-forming quartz (QZ) and K-feldspar (Kf). Osumi, Japan. X-polars. Bar scale = 0.12 mm. (Courtesy Noboru Ôba.) (b) Two-feldspar granite, in which alkali feldspar has inverted to microcline. Fredrikstad quarry (Vigeland sculptures). X-polars. Bar scale = 1 mm. (Courtesy H. Neumann, University of Oslo.) (c) Microcline (mic), plagioclase, quartz (qz), biotite (bt), muscovite (ms), two-mica (bt,ms) granite. Intracrystalline slip(?) has "digitated" vermicular quartz (arrow) in myrmekite unit. Mount Rushmore, Black Hills, South Dakota. X-polars. Bar scale = 0.5 mm. (d) Nepheline syenite containing early-starting alkali feldspar (medium gray laths), nepheline (white), sodalite (dark gray), aegirine–augite + barkevikite (black). Magnet Cove, Arkansas. Slab surface. Bar scale = 0.5 mm. (Courtesy V. Scheid.)

Figure 17.10
Porphyritic medium- and coarse-grained magmatic rocks.

(a) Alkali feldspar phenocryst containing preferentially oriented small plagioclase crystals. Two-feldspar granite. Sands Springs Range, Nevada. X-polars. Bar scale = 1.6 cm. (b) Cathedral Peak porphyritic (alkali feldspar) two-feldspar granite–granodiorite. Yosemite National Park, California. (Courtesy National Park Service.) (c) Two-feldspar peraluminous granite with phenocrysts of alkali feldspar (not shown) and muscovite (ms). Kern Mountains, Nevada–Utah. Hand specimen. Bar scale = 2.4 cm.

Spherulites are more or less spherical solids (Figure 17.12a) characterized by internal radial intergrowths of minerals, which in rhyolitic systems consist of acicular quartz (or tridymite) and alkali feldspar.

Lofgren (1971a, 1971b) generated spherulites from glass in devitrification experiments, but he also generated spherulites by crystallization in undercooled melts (Lofgren, 1974). This indicates that spherulites, per se, are not diagnostic of devitrification of glass, in contrast to crystallization of supercooled melt. Nevertheless, at least three textural clues are useful in making the distinction.

(1) Spherulites that have formed along perlitic cracks (Figure 17.12b) indicate that the spherulites result from devitrification of glass (below the transformation temperature) on the assumption that perlitic cracks can form only in glass, not in highly viscous supercooled melt (Marshall, 1961).

Figure 17.11
Nonporphyritic coarse-grained magmatic rocks.

(a) Biotite (short arrows), hornblende (long arrows), granodiorite. Half Dome, Yosemite National Park, California. Slab surface. Bar scale = 1.2 cm. (b) One-feldspar Conway Granite, New Hampshire, containing alkali feldspar, dark gray (radioactively bombarded) quartz, and arfvedsonite (black). Slab surface. Bar scale = 1.6 cm. (Courtesy J. B. Lyons.) (c) Cape Ann Granite. Coarsely perthitic alkali feldspar and quartz (center). An example of Tuttle and Bowens's (1958) "hypersolvus granite." Rockport, Massachusetts. X-polars. Bar scale = 0.8 mm.

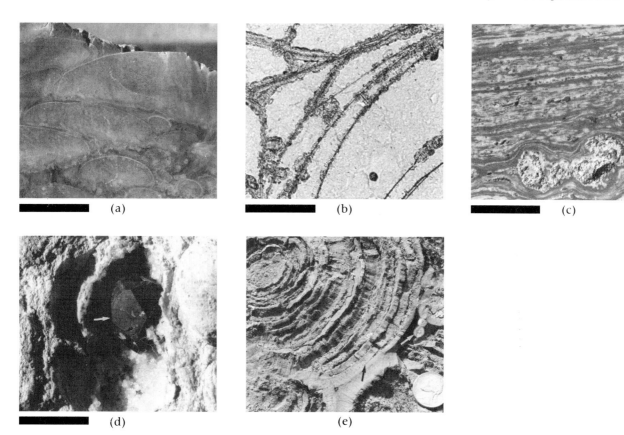

Figure 17.12
Comparison of devitrification textures formed in glassy magmatic rocks with textures resulting from crystallization of supercooled magma (see text).

(a) Nested spherulites in domal rhyolite dome resulting from rapid crystallization in supercooled lava. Garnet Hill, Eagan Range, Nevada. Slab surface. Bar scale = 1.6 cm. (b) Bilateral devitrification (axiolitic) along perlitic cracks in perlite. Rough appearance of glass due to refractions–reflections associated with inclined surfaces of frost pits (frosted glass slide) interfacing with mounting epoxy. PPL. Bar scale = 0.2 mm. (c) Flow structure in lava dome rhyolite. Dark layers are glass. Light layers consist of nonvesicular spherulites and vesicular spherulitic rhyolite. Larger pods of vesiculated spherulitic rhyolite (lower) acted as ridged obstacles to flow. Washington Hill, Virginia Range, Nevada. Slab surface. Bar scale = 0.8 cm. (d) Garnet (white arrow) and quartz in lithophysae of rhyolite. Rounded surface of unexposed lithophysal cavity (upper right). Garnet Hill, Eagan Range, Nevada. Bar scale = 1.5 cm. (e) Large lithophysal spherulite. Shells of nonvesicular spherulites (light gray) alternate with vesicular shells consisting of crystal-lined voids and spherulites. Note apparent expansion caused by vapor generation in vesicular zone between two black arrows. Note small spherulites (white arrow) in nonvesicular glass (black). "Obsidian flow," Yellowstone National Park. U. S. quarter for scale.

(2) Vesiculation hastens crystallization of magma (Chapter 13). Flow of rhyolitic magma during vesiculation and associated spherulitic crystallization is indicated by flow patterns of nonporous light-colored layers of rhyolite that are deflected by vesiculated (and crystallized) portions that become obstacles to flow (Figure 17.12c). This indicates that supercooled magma is present during spherulitic crystallization, therefore suggesting that the spherulitic crystallization took place in supercooled magma.

(3) Vapor-phase cavities larger than vesicles are commonly lined with precipitates or have concentric shells of precipitates. These are **lithophysal cavities** (Figure 17.12d) requiring "plastic" expansion of the

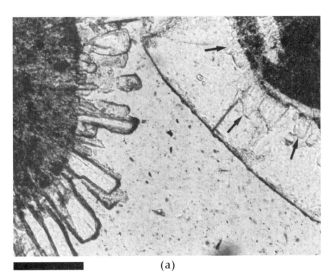 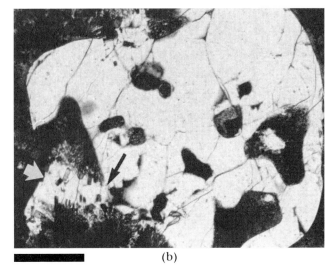

Figure 17.13
Extension of fine accicular spherulitic intergrowths of quartz (tridymite) and alkali feldspar (sanidine) to coarser laths of alkali feldspar and enclosing quartz.

(a) Vapor-phase cavity lined with alkali feldspar laths aligned with radial fibrous fabric of spherulite (left). Quartz has overgrown alkali feldspar (arrows) parallel to spherulite fabric (upper right). Indian Springs, Nevada. PPL. Bar scale = 0.12 mm. (Courtesy P. Purington.) (b) Crystallization of supercooled melt to quartz and laths of alkali feldspar (short white arrow). Note epitaxial attachment (black arrow) of this late-stage matrix quartz to large skeletal quartz phenocryst. Indian Springs, Nevada. Bar scale = 0.5 mm. (Courtesy P. Purington.)

host rhyolite. Extraordinarily large lithophysal cavities occurring at the "Obsidian Cliff" in Yellowstone National Park consist of porous shells alternating with spherulitic nonporous shells (Figure 17.12e). The apparent wedging of nonporous spherulitic shells (black arrow in photo) suggests non-porous spherulites formed first and were expanded along concentric planes where vapor phases concentrated (vesiculated) and crystallization continued. Since these **lithophysal spherulites** require expansion, it is likely that the process takes place in supercooled melt rather than rigid glass.

Lofgren (1971a,b) also generated micropoikilitic quartz in his devitrification experiments. Nevertheless, **micropoikilitic quartz** and similar intergrowths such as "granophyric" quartz and feldspar are thought to also be a product of crystallization in undercooled magmas of silicic composition (Smith, 1960; Swanson et al., 1989). Concentration of aqueous phase in the very latest stage of crystallization seems to aid in the production of quartz–alkali feldspar poikilitic and "granophyric" intergrowths. Relatively large well-formed crystals of alkali feldspar and quartz form beyond the spherulitic stage of crystallization, the latter being characterized by submicroscopic intergrowths of these minerals (Figure 17.13a). In this example, the larger crystals line gas cavities, suggesting that crystallization of the spherulitic portion took place in supercooled melt (not glass) that was evolving a vapor phase conductive to final growth of the larger well-formed crystals, much as crystal-lined miarolitic cavities form in granites. An epitaxial relation of late-forming (aqueous phase present) micropoikilitic quartz containing small laths of alkali feldspar to an earlier-formed (prior to saturation in aqueous phase) quartz phenocryst is shown in Figure 17.13b.

18

Cumulate Magmatic Rocks

Cyclically layered cumulate gabbro series consisting of modally graded plagioclase (lighter) and olivine + ilmenite + augite (darker). Newark Island, Nain, Labrador. (Courtesy Robert Wiebe, Franklin-Marshall College, 1988.)

Processes Leading to Mineral Concentration Not Representing Magma Composition

Solidification of magma involves either crystallization or rapid freezing to glass. Crystallization is a progressive process, strongly temperature dependent for most magma compositions, in which there are early-forming crystals followed by later-forming crystals. This process is **fractional crystallization**, and since silicate melts are liquid and mobile, mechanical concentration of crystals relative to magma may occur (Jackson, 1967; Irvine, 1979, 1982; McCallum et al., 1980).

There are other processes leading to concentration of minerals in rocks, operating in addition to, or without, mechanical accumulation of crystals. There can be infiltration of fresh magma into crystal mushes and chemical diffusion in an immobile magma leading to partitioning of crystalline phases. Segregation related to liquid immiscibility is another process favoring heterogeneity within magmatic rocks of special composition (Roedder, 1979). In all cases, the rocks formed do not represent the composition of starting magmas since they are either accumulations of crystals or concentrations of chemical components within initially "homogeneous" magmas. These are **cumulate magmatic rocks**, in the broad sense.

Mechanical Accumulation

Consider a few newly formed crystals suspended in their magma. If such a batch of crystal-bearing magma is not in motion, the principal force acting on the system is gravity. Crystals with density greater than the magma will tend to sink, their space being simultaneously filled with magma as they move through it. Crystals lighter than the magma will tend to rise as the heavier magma occupies original crystal space. Thus, **gravity settling** and **buoyant rising** of crystals are two mechanical processes leading to segregation of crystals in fractionally crystallizing magma.

Sinking or floating of crystals in magma is governed by the density and viscosity of the magma, the density of the crystalline phases, and the sizes and shapes of the crystalline phases. Minerals such as olivine, orthopyroxene, and clinopyroxene are typically denser than their host magma and tend to sink, whereas some plagioclase is less dense and tends to rise. As the ratio of crystals to magma increases, there is a progressive increase in the mechanical interaction between the crystalline phases, reducing the rates of sinking or rising.

A framework of crystals exists when a magma is more than about 60% crystallized, inhibiting movement of crystals. Where there is gravity-induced compaction, such as on the floor of a magma chamber, the ratio of crystals to magma can be even further increased, since geometric adjustment of crystals results in expulsion of intercrystal melt. The melt cannot be completely eliminated unless there is change in shape of the crystals themselves, such as by relocation of the melt (Chapter 27).

Convection within a crystal-bearing magma is another process that may result in mechanical segregation of crystals relative to melt. A batch of magma intruded into a chamber is subject to rapid development of thermal and compositional gradients. Heat is lost to the roof and walls, and crystallization changes the composition of the magma. The resulting thermal and compositional gradients produce density (buoyancy) gradients, and the stage is set for movement within the crystal-bearing magma. Convective motion produces differential flow of crystals and magma, resulting in local concentration of crystals.

The configuration of convection cells in magma chambers depends on factors such as the shape of the chamber, the rate of heat loss (warm versus hot wall rock), heat supply from nearby magma, and the possibility of resupply of hot magma into the chamber coupled with eruption of magma from the upper regions of the chamber. The simplest convective cell consists of an upward bore at the center, outward dispersion along the roof in opposing directions, downward motion along the walls, and inward convergence along the floor.

Elutriation can be expected to separate minerals of different densities, such as plagioclase in relation to olivine and pyroxenes. Flow in any direction is conducive to differential transport of crystals relative to magma, and accumulations of crystals may occur by capturing on surfaces of crystal mushes that can no longer convect.

The configuration of convecting cells in magma chambers may be much more complex than the two-cell model. Horizontally stacked cells may develop as a result of density difference attendant to both temperature and composition. Heat and solute diffusion (double diffusion) may occur and result in accumulation of crystals in layers. Vertically stacked subcells may lie in a larger two-cell configuration (Marsh, 1988a), providing for crystal capture on the chamber roof and accumulation on the floor (Brandeis and Jaupart, 1986).

The velocity of magmatic convection depends on the boundary conditions, but it has been estimated to be about 1 cm/yr (Morse, 1988). Magmas cannot behave like Newtonian fluids if the crystal content is much above 60%, meaning that they will be too viscous to convect.

Mechanical accumulation of crystals may be exaggerated by **filter pressing** in crystal-bearing magma systems. Consider a batch of magma containing a large percentage of crystals to which stress is applied. There is likely to be a deviatoric stress component, and the mass will change shape (Bingham flow). For example, a flattening of the mass would be gravity compaction, concentrating crystals (Figure 18.1) and expelling remaining magma in extension directions.

Concentration of feldspars by filter pressing may produce some monzonitic and anorthositic rocks. Stresses resulting in filter pressing can be related to intrusion (Miller and Weiblen, 1990) as well as to gravitational compaction in a stagnate system. Some **schlieren** in granitic bodies may result from differential movement of melt relative to early-forming mafic mineral phases (Figure 18.2). Not all schlieren in granitic rock results from concentration of mafic crystals: some undoubtedly represent dispersal of mafic minerals already concentrated in some other way.

Gravitational, convective, and nonconvective dynamic processes of crystal accumulation can be expected to occur in combinations (Marsh and Maxey, 1985). For example, the effects of settling and floating in response to gravity alone can be expected to be a factor in convecting systems that is itself gravity controlled. Upward flow of magma tends to retard or even eliminate settling. Downflow of magma cooled at a thermal boundary will tend to arrest rising of less

Figure 18.1
Tightly packed elongate plagioclase.

Well-formed crystals indicate growth occurred prior to compaction. Bent crystal (arrow) indicates some space accommodation by crystal deformation. Less than 5% intergranular opaque oxide + biotite + amphibole. X-polars. Bar scale = 0.5 mm.

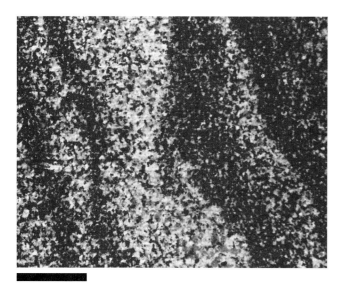

Figure 18.2
Possible mechanical segregation of hornblende relative to plagioclase–K-feldspar–quartz, forming schlieren. Granite de la Margeride, Massif Central, France. Bar scale = 3 cm. (Courtesy Jean-Paul Couturié.)

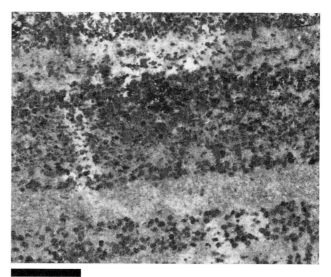

Figure 18.3
Rhythmic-layered gabbro of cumulate derivation. Layering defined by varying proportions of plagioclase (lighter) and pyroxenes (darker). Smartville ophiolite, Sierra Nevada, California. Bar scale = 1 cm.

dense crystals and accelerate sinking of denser minerals. Layers of heavier minerals over lighter ones may originate in this manner. Layering of cumulates more or less parallel to vertical chamber walls, with characteristics more or less identical to those on chamber floors, is good reason to suspect that convective processes can dominate over purely gravity-controlled sinking and floating (Parsons, 1987). Some layers that at first seem to be the result of gravity settling have been shown to be fundamentally related to multiple injection (McCallum et al., 1980; Husch, 1990; Wiebe and Snyder, 1993). Cumulate layering of mechanical origin may be typical of ophiolite complexes (Figure 18.3), but other segregating mechanisms may be involved as well.

Segregations within magmas generate structures that are similar to some clastic sedimentary rocks. Graded beds, cross-bedding, and scour-and-fill and slump structures are described (Parsons and Butterfield, 1981; Parsons and Becker, 1987; Wiebe, 1988; Conrad and Naslund, 1989).

Crystal Growth from Refreshed Magma

Crystallization of a batch of magma without segregation of crystals from melt in a closed system is in contrast to continued growth of mineral phases from melt whose original composition is maintained by passage of fresh melt along a crystal–melt boundary or by infiltration through a crystal mush (Figure 18.4).

Crystallization in an open system proceeds without fractionation, and monomineralic, or at least rocks of simple mineralogy, can be formed (Wager et al., 1960; McBirney, 1987). Such concentrations of certain mineral phases may have been preceded by mechanical accumulation of crystals, but crystals nucleated on a substrate can enlarge without any motion as long as fresh magma passes by.

Crystal Growth Controlled by Diffusion and Nucleation in Magma

Crystal growth can also be nurtured by chemical diffusion in magma rather than by convection of magma. There are indications that diffusion may be significant even in a relatively large magma bodies. For example, diffusion of alkali ions or alkali silicate complex ions coordinated with aqueous ions is thought to be the process resulting in K-feldspar enrichment in the roof zone of a granitic–syenitic magma system (Boone, 1962). Carten et al. (1988) point to a concentration of metals and volatiles in the apex of a stock prior to the onset of significant magmatic crystallization.

Calculation of mass transfer by way of diffusion in silicate melts is not so supportive of such a process

Closed System

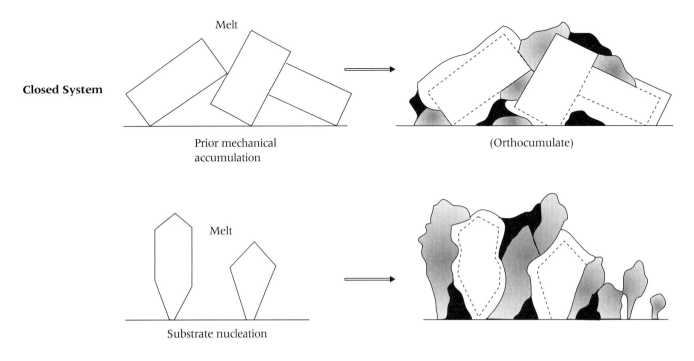

Open System

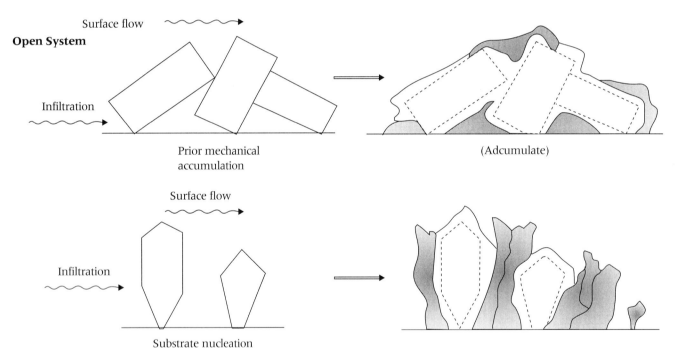

Figure 18.4
Representation of mechanical accumulation compared with chemical accumulation of mineral phases in magmatic rock. Fractionation occurs in closed systems, forming three mineral-phase orthocumulates in mechanical accumulations. Only two mineral phases form in open systems because of melt resupply.

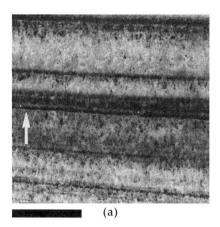  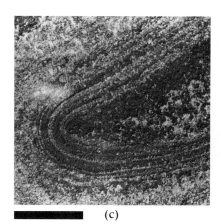

Figure 18.5
Mineral concentrations resulting from crystallization on substrate-producing cumulates of nonmechanical origin. Rhythmic layering probably governed by a diffusion–nucleation mechanism. Fischer Lake, Sierra Nevada, California.

(a) Comb-layered plagioclase–clinoproxene–hornblende diorite. Growth direction indicated by up arrow. Rock surface. Bar scale = 4 cm. (b) Plagioclase dominant layer in specimen shown in (a). Growth direction is right to left as indicated by enlargement of plagioclase in that direction (between arrows). X-polars. Bar scale = 0.8 mm. (c) Orbicular structure with core consisting of weakly layered diorite fragment. Bar scale = 3.3 cm.

---

forming magmatic rocks with unusual modal compositions. It has been calculated that within the time frame of the cooling and crystallization of a relatively large magma body, mass transfer by diffusion can be no more than a few meters (Shaw, 1974). However, if convective flow is in the same direction as diffusion, appropriation of certain elements to the roof zone of a magma chamber may be enhanced.

Diffusion may play a major role on a local scale. Layering is the trademark of cumulates in mafic bodies such as the Skaergaard body in Greenland and the Stillwater Complex in Montana. Layers measured in centimeters are laterally continuous and cyclically repeated in the direction of crystallization (Wager, 1959; McBirney and Noyes, 1979; Peterson, 1985; Boudreau, 1987; Sorensen and Larsen, 1987). Diffusion may play a major role in this in situ **oscillatory crystallization** model, a process somewhat analogous to the diffusion–supersaturation model of plagioclase zoning (Chapter 14). The effect of differing powers of nucleation of minerals may play hand-in-hand with diffusion. A delay in crystallization of a particular mineral phase has the potential of leaving a given layer depleted in that mineral.

The types of layering that may fit diffusion–nucleation models, at least in part, are those characterized by growth of crystals extending at high angle from a substrate. They include **harrisitic layering** (Donaldson, 1982), **comb layering** (Figure 18.5a,b), **orbicular layering** (Figure 18.5c; Brooks McKinney, personal communication), and layering in pegmatites (Duke et al., 1988; see Figure 33.7).

Crystallization of Cumulates

Crystallization in Mechanical Accumulations

Closed System Crystallization (Orthocumulates). Crystallization in a system containing mechanically accumulated crystals may be chemically and physically closed with respect to magma at large. After initial mechanical accumulation of a certain mineral phase or phases, fractional crystallization of intercumulus melt (not refreshed from without) leads to crystal zoning, nucleation of new mineral phases, and reaction with the early phases, partially converting, for example, olivine to orthopyroxene, orthopyroxene to clinopyroxene, or olivine to clinopyroxene. The final intercumulus mineral assemblage may be quite different from the earlier accumulated phases. These are the orthocumulates of Wager et al. (1960).

The textures of such rocks originating by both mechanical accumulation and crystallization of intercumulus melt are varied. Nucleation rates, growth rates, and attachment kinetics probably all dictate what the grain size and grain relations are going to be in these cumulates. For example, poikicrysts and oikocrysts result from low nucleation rates, compared with the mineral phases included in these large low-integrity crystals. Even if two or more phases are crystallizing simultaneously, there may be a tendency for a phase to include another if it has few seed crystals to grow on. The presence of a late-stage hydrous magma, leading to an escape of aqueous phase, may relate to finer grain sizes completely unrelated to the external thermal regime (Thy and Esbensen, 1982), much as it does in the production of aplite in granitic systems (Jahns and Burnham, 1969).

Open System Crystallization (Adcumulates). If mechanically accumulated crystals are nurtured by growth components that are continuously supplied by fresh magma or by diffusion of these components from nearby fresh magma, the character of the final rock may be quite different than the orthocumulates (Figure 18.4). Local crystal-scale fractionation is reduced or eliminated, and minerals may continue to grow more or less on their liquidi. As a result, growth zoning may be subordinate, and final intercumulus liquids need not produce different mineralogy than the original cumulates. These are the **adcumulates** of Wager et al. (1960).

Mesocumulates have intermediate characteristics between orthocumulates and adcumulates: that is, not all original intercumulus liquid is eliminated. **Heteradcumulates** are characterized by open system growth of large oikocrysts (Jackson, 1961) around cumulate phases of different composition. Lack of zoning in these oikocrysts indicates continual supply of fresh melt.

Crystal growth can be essentially an "add-on" and "fill-in" process in the formation of adcumulates. However, there are some additional possibilities. If the refreshing magma is at a temperature above the liquidus temperature of some of the already accumulated minerals, partial resorption may occur, producing irregular crystal morphologies. Plagioclase crystals included in oikocrysts may have these characteristics, and since they become enclosed in the other mineral, renewed crystallization cannot heal the morphological effects of resorption.

A mineral such as plagioclase can grow at a constant temperature, unzoned, in an open system, and since there is no undercooling to increase the rate of nucleation, the crystals can enlarge and the rock becomes coarse grained. This is one way unzoned plagioclase crystals can form without being at variance with the phase relations in the plagioclase series (Chapter 13).

In extreme cases, not only is zoning suppressed, but minerals of a specific composition may be able to enlarge until they interface with their neighbors. This is one way monomineralogic rocks such as dunite and pyroxenite can originate. The interfacing tends to produce polygonal mosaics (Figure 18.6), mimicking recrystallization in metamorphic systems. Interpretation of such textures as being adcumulates is by no means without question.

The discovery of komatiites, extrusive equivalents of peridotites, confirms the existence of ultramafic magmas. Crystallization of an ultramafic magma might result in local monomineralic polygonal mosaics of olivine or pyroxene, mimicking adcumulates. Therefore, an ultramafic layer in a layered mafic–ultramafic complex might represent either multiple injection (McCallum et al., 1980) of ultramafic magma or the formation of an adcumulate by the infiltration mechanism (Bédard, 1987).

Development of monomineralic rock by additive processes in an open system does not necessarily account for polygonal texture. Hunter (1987) suggests that solution reprecipitation may generate polygonal textures not only in monomineralic cumulates but in bimineralic and polymineralic rocks as well.

Figure 18.6
Pyroxenite adcumulate. Polygonal mosaic of unimodal clinopyroxene probably generated by postcumulus melt infiltration or diffusion. Note triple point (arrow) with nearly perfect 120° dihedral angles. Stillwater, Montana. X-polars. Bar scale = 0.5 mm. (Courtesy R. Neilsen.)

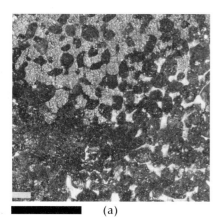 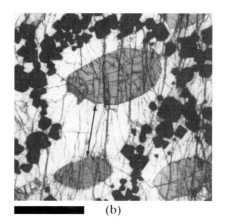 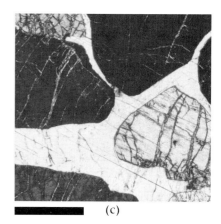

Figure 18.7

(a) Clinopyroxene grains with intercumulus bronzite oikocryst (upper) and plagioclase oikocrysts (lower right). Slab surface. Bar scale = 1.6 cm. (b) Cumulate olivine and chromite with partial replacement of olivine (chromite-free areas) by bronzite. Double arrow indicates relict grains of olivine in optical continuity. Stillwater, Montana. X-polars. Bar scale = 1.6 mm. (Courtesy M. Zientek.) (c) Part of a plagioclase oikocryst host to grains of bronzite. X-polars. Bar scale = 0.8 mm. Stillwater complex, Montana. (Courtesy R. Neilsen.)

If growth components supplied by the magma at large result in the growth of several phases, the development of oikocrystic textures may be extreme. Since near-liquidus temperature tends to be maintained, nucleation rates are minimal and certain minerals may be forced to enlarge on existing crystals, tending to surround other mineral grains. Oikocyrsts of orthopyroxene (Figure 18.7a), clinopyroxene (Figure 18.7b), and plagioclase (Figure 18.7c) are very impressive (Jackson, 1961). Reactive replacement generating clinopyroxene at the expense of olivine is indicated by a chromite-free zone (Figure 18.7b).

Crystallization with No Prior Mechanical Accumulation

In situ magmatic crystallization producing local (typically layers) rock compositions that do not directly reflect magma composition can form without any mechanical accumulation of crystals. For minerals to form in inordinate concentrations, in situ crystal growth must be fed by passage of fresh magma or by diffusion in magma, just as in adcumulates.

If a magmatic system is undercooled, accumulative growth may be rapid, yielding cellular crystals. Dendritic olivine crystals characterize the rock known as **harrisite** (Donaldson, 1982) and the generic **crescumulates** (Wager and Brown, 1968) that occur in association with other rocks known to have a component of mechanical accumulation.

Komatiite is similar to harrisite in that its **spinifex texture** consists of dendritic cellular crystals (Figure 14.6b) and it is of ultramafic composition. Undercooling of komatiite is realized by extrusion, but how are deep magmas undercooled to generate cellular crystals typical of the harrisites and crescumulates? Mingling of hotter magma with cooler magma can provide an undercooled condition in the hotter magma. Comb layering and orbicular structure may be linked to undercooling.

Crystal growth starting on a substrate, fed by refreshed solutions, yielding concentrations of elongate, commonly dendritic crystals, also occurs in the presence of high-temperature aqueous fluids in granitic pegmatite systems (Figure 33.4d) and in low-temperature brines of evaporites (Figure 38.5c).

Mixed Magma Rocks

Enclaves of more mafic microgranodiorite magma-mixed hybrid in granodiorite host. Host pluton contains hornblende-mantled quartz and pseudophenocrysts of boxy cellular plagioclase. Heusser Mountain, Nevada.

Magmas of different composition, different temperature, and different stages of crystallization mingle and mix with each other. If the identity of the participating magmas is still apparent, such as blobs of one magma in a host magma, there is **magma mingling**, whereas if the identity of the different magmas is not apparent, there is **magma mixing** on a crystal scale.

Several mineralogical and textural features of magmatic rocks at the hand-specimen–thin-section scale point to a process of mingling and mixing of magmas of different composition and temperature. At least three reasonable scenerios show juxtaposition of magma of different composition:

1. A basaltic magma is generated in one location, such as in the mantle, and is emplaced into siliceous rock somewhere else, such as in the continental crust. Siliceous magma forms in response to heating by the basaltic magma, and there is mingling and mixing (Huppert and Sparks, 1988).
2. Magma may inject an established magma chamber in which a magmatic system is already differentiating (Sparks et al., 1977; Eichelberger, 1980; Blake and Cambell, 1986; Emeleus, 1987; Wiebe, 1991). An ongoing state of magmatic differentiation and with episodic replenishment of magma with attendant mingling and mixing is well documented (Ballhaus and Glikson, 1989; Wiebe, 1991; Helz and Wright, 1992).
3. There may be a magmatic system that is closed to replenishment or invasion of any other magma. With fractionation and convection-dominating segregation, there may be internal mixing, especially if concurrent tectonic adjustments affect the magma chamber.

A mixing magmatic system consists of at least one magma, or partially crystallized magma, that is more felsic (or more mafic) than another (Eichelberger, 1975; Frost and Mahood, 1987). Either or both magmas may be (1) primitive, (2) evolved, or (3) hybrid by some prior magma mixing event (Hibbard and Watters, 1985).

Textural and Mineralogical Predictions

Flash injection of mafic magma at about 1200°C into a water-bearing felsic magma at about 700°C is as catastrophic as basalt lava flowing into the ocean. There is

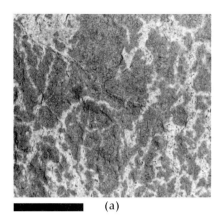  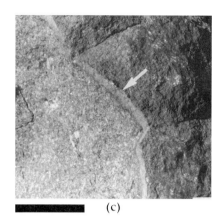

Figure 19.1
Magma mingling.

(a) Mingling between microgranodiorite (light) and more mafic micro diorite-granodiorite magma-mixed hybrid (dark). Relation ranges from "net-veining" of the more mafic member by the more felsic member, to enclaves of the more mafic in the more felsic. Fox Range, Nevada. (Courtesy of M. Leshendok.) Bar scale = 4 cm. (b) Investigation of magma mixing in porphyritic two-feldspar granodiorite Heusser Mountain stock, near Ely, Nevada. More mafic hydrid enclaves (two black arrows) and host pluton contain abundant textures (such as hornblende-mantled quartz) compatible with magma-mixing model. (c) More mafic microgranodioritic magma-mixed hybrid enclave with shrinkage cracks and convex surfaces exposed by weathering. Weathering rind (arrow) is not a chill facies against host granodiorite—little or no quenching would be expected against a magma of similar composition. Heusser Mountain, Nevada. Bar scale = 2.5 cm.

an immediate undercooling of the mafic magma (chill margins form) and a corresponding superheating of the felsic magma. The mafic magma freezes, forming lobate masses of basalt in the felsic fluid, analogous to basalt pillows forming in water. With heat loss to the environment from this mingling system, either rapidly as in a volcanic setting or slowly as in a plutonic setting, the felsic portion eventually solidifies. This is **mingling of magmas**, and rocks formed range from mafic dominant to felsic dominant.

If the mafic magma dominates, the felsic magma is relegated to an interlobe position, producing **net-veined complexes** (Figure 19.1a) (Marshall and Sparkes, 1984). Such complexes may originate by injection of more felsic magma into more mafic magma, as implied by the term "net-veined," but it can also be generated by the accumulative settling of globular and pillowlike masses of the more mafic magma in a more felsic magma, a spectacular example of which is shown by Wiebe (1991).

If the felsic magma dominates, the mafic lobes become mafic inclusions (enclaves) in the host rhyolite or granite (Figure 19.1b). The more mafic inclusions may not have chilled margins, especially if the thermal contrast between the mingling members is not large, but also if there has been "erosion" of inclusions in a mobile system. Crenate surfaces convex toward the host are typical (Figure 19.1c), evidently resulting from contraction of the rapidly solidifying material.

Inclusions may be strewn along the course of dikes injected into incompletely solidified, more felsic, host. Disaggregation of such dikes by "backdiking" leaves a trail of enclaves within a dominating host pluton of more felsic composition (Figure 19.2).

Mingling geometry can be expected to be the same whether or not the magmas contain crystals. However, on a grain scale, a more complex record of disequilibrium is recorded if crystals are present. For example, premixing crystals in the mafic magma might be relatively large compared with assemblage of small crystals appearing as a result of quenching of the mafic magma in the "cold" felsic magma. Prior crystals in the felsic magma should show the effects of dissolution resulting from reheating.

What are the conditions promoting mixing of magmas or crystal-bearing magmas on a crystal scale? Quenching of a significantly more mafic system locks

Figure 19.2
Back-diking (back-veining) in a 1-m-thick diabase dike (oriented east–west in lower half of photo) in Cape Ann Granite. The dike is discontinuous in host granite, indicating host pluton was mobile during dike emplacement. Felsic back-dikelets indicate granite system contained intercrystalline fluid when dike was fractured and extended. Rockport, Massachusetts.

Textures Compatible with Magma Mixing

Twenty-one textural relations and their interpreted mode of formation compatible with magma mixing are shown in Figure 19.3. No one of these textures is irrefutable evidence for magma mixing; they are presented here (see also Hibbard, 1991) in the spirit of possible, unconfirmed, indicators of magma mixing. Many of the textures are well known to occur in other rocks for which there is not a shred of evidence for magma mixing, but this does not preclude their development in mixing systems.

Magma-mixing occurs in volcanic and plutonic systems, and the processes leading to textural expression of mixing are fundamentally the same in both environments. Textural differences arise as the result of quenching in an eruptive volcanic setting, limiting "equilibration" of mixing textures, characteristic of the plutonic environment, in lieu of quenching of melt to very fine-grained or glassy matrices.

Alkali Feldspar Ovoids

Occurrence of ovoids of alkali feldspar–K-feldspar in magma-mixing systems is compatible with dissolution of pre-existing alkali feldspar–K-feldspar crystals in the more felsic system as a consequence of reheating by the more mafic (hotter) system. The ovoids may be mantled with plagioclase (Plate 5a, Figures 19.4, p. 249; 19.5, p. 249), mantled with biotite (Figure 19.5), or they may be nonmantled (Plate 5a, Figure 19.4) or partially mantled (Figure 19.5).

If they are nonmantled, renewed growth of alkali feldspar–K-feldspar after partial dissolution and after thermal stabilization in a plutonic mixing system can either (a) juxtapose the ovoids with matrix crystals (Plate 5a) or resume tabular growth.

Rapakivi Mantling

Rapakivi texture (Plate 5a, Figures 19.4, 19.5) has been proposed as the undercooling-induced nucleation of plagioclase on the growth surfaces of coexisting alkali feldspar–K-feldspar (Hibbard, 1981). These surfaces act as "emergency" substrates for plagioclase

continued on p. 249

up crystals into rock before they can be dispersed. With less thermal contrast between mixing end members, such as a dioritic (andesitic) magma mixing with a granodioritic (latite–andesite) magma, crystallization of the more mafic system is less rapid, allowing for mechanical stirring by such processes as convection and dynamic magmatic flow. With thorough mixing on a crystal scale, involving both disaggregation of crystal mushes into less crystallized magmas (Seamon and Ramsey, 1992) and diffusion between magmas (Blichert-Toft et al., 1992), crystals and melts of one system become juxtaposed with those of the other system. Since the systems are of different composition and temperature, the composite system is thrown into a state of intense disequilibrium, resulting in generation of an unusual assemblage of minerals and textures.

Figure 19.3
Pictoral representation of textures compatible with magma mixing. Textures shown evolving from initial stage through an intermediate stage to an "equilibration" advanced stage characterizing plutonic conditions. Volcanic equivalents would not have an "equilibration" stage.

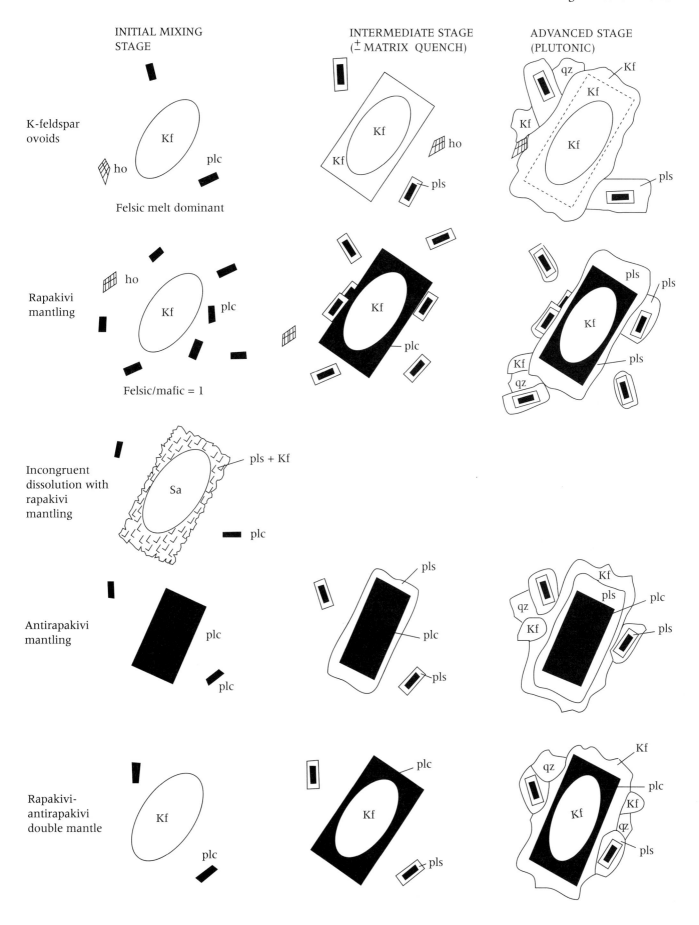

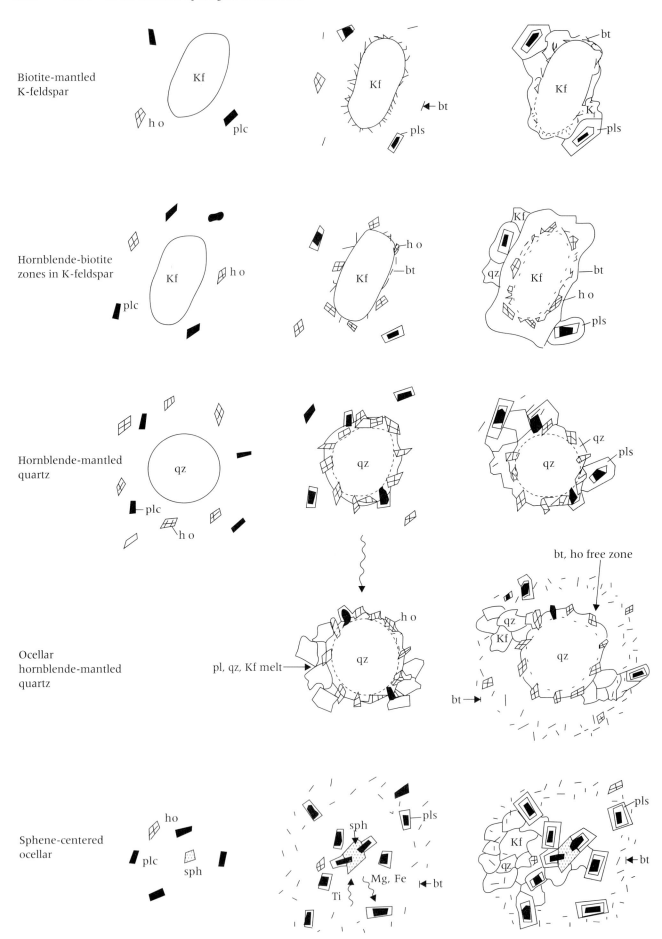

Figure 19.3 continued

19 Mixed Magma Rocks 247

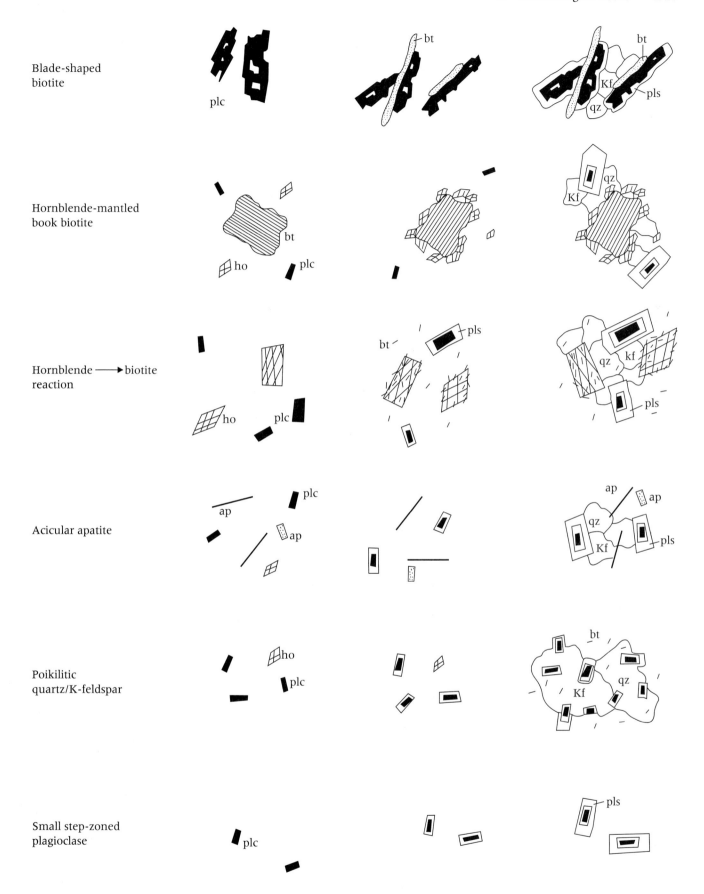

Figure 19.3 continued

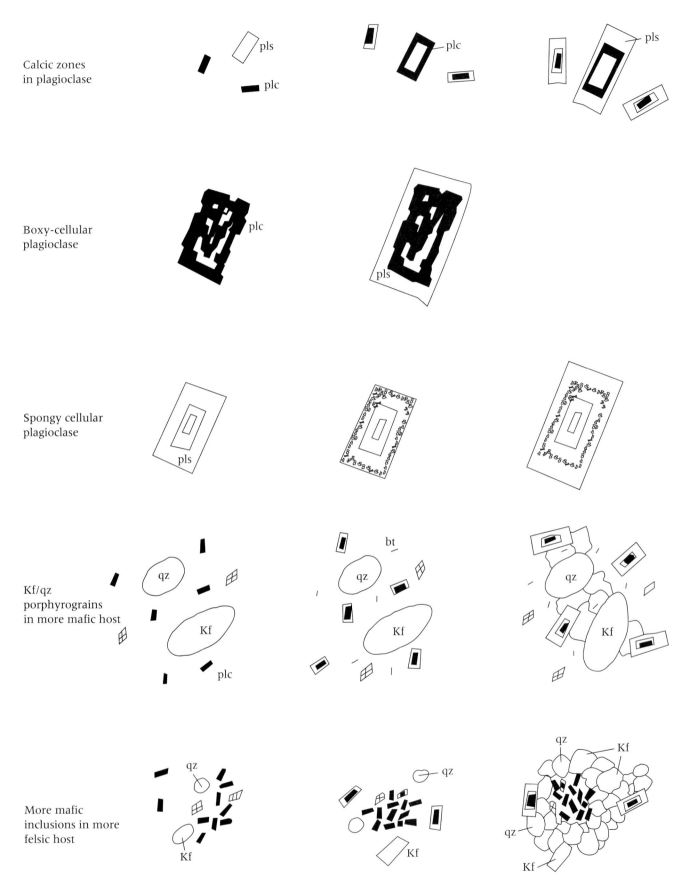

Figure 19.3 continued

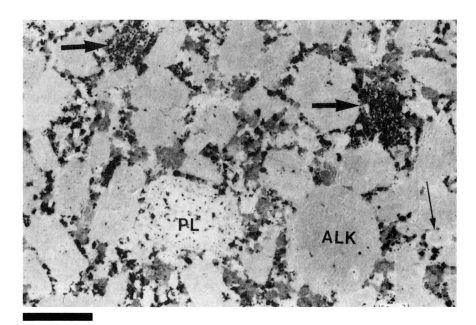

Figure 19.4; text p. 244
Two-feldspar granite with alkali feldspar ovoid (ALK), boxy cellular plagioclase (PL), oligoclase mantle on small alkali feldspar crystal (thin arrow), and two mafic-rich microdiorite enclaves (thick arrows). Slab surface. Northern Italy. Bar scale = 2.7 cm.

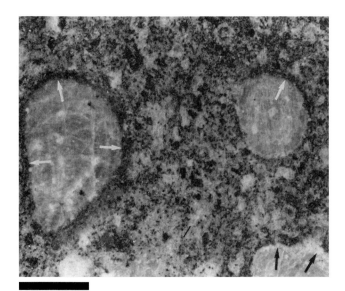

Figure 19.5; text p. 244
Ovoidal alkali feldspar pseudophenocrysts in magma-mixed microgranodiorite with partial biotite mantles (white arrows). Oligoclase mantle (black arrows) on alkali feldspar (lower right) (rapakivi texture). Note heterogeneous size and distribution of biotite and hornblende in host rock. Sylvania Mountains, Nevada. Slab surface. Bar scale = 0.5 cm.

growth, producing the mantles. Alkali feldspar–K-feldspar crystals must already exist in the more felsic member of the mixing system for mantling to occur, as they typically do in two-feldspar granite systems. Consequently, the texture is precluded in the granodiorite-hosted systems in which alkali feldspar–K-feldspar is a late intergranular phase, but it is also precluded in the one-feldspar-hosted systems in which alkali feldspar tends to crystallize simultaneously with quartz at a thermal minimum.

Incongruent Dissolution Rapakivi Mantling

Partial dissolution of an homogeneous alkali feldspar (such as a sanidine), derived from a more felsic system, may generate a mantle of sodic plagioclase (Stimac and Wark, 1992) (Figure 19.6). Dissolution yields a sodic plagioclase component and a K-feldspar component. The sodic plagioclase crystallizes as dissolution progresses, relegating the K-feldspar component to an intergranular melt. The morphology of the sodic plagioclase may be cellular, with a tendency toward lamellar habit. In this example, dissolution-generated sodic plagioclase was preceded by mantling of the sanidine with epitaxial coarse boxy cellular oligoclase (Figure 19.6a).

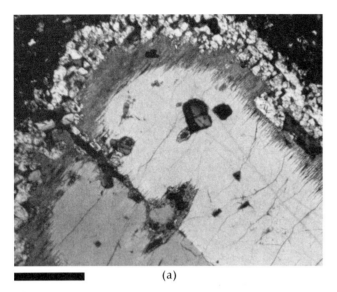 

Figure 19.6
Incongruent dissolution rapakivi mantling. Platoro cauldera, San Juan Mountains, Colorado. (Courtesy D. Rogers.)
(a) Sanidine phenocryst in rhyolite with outer, coarse, boxy cellular plagioclase mantle (growth) and inner, spongy, cellular plagioclase (incongruent dissolution derived). X-polars. Bar scale = 1.2 mm. (Courtesy D. Rogers.)
(b) Close-up of mantle textures of (a). K-feldspar (arrows) was generated in the melting process and resides in the spongy cellular oligoclase. X-polars. Bar scale = 0.1 mm.

Antirapakivi Mantling

Antirapakivi texture (Plate 5b) may form in mixing systems as a result of the juxtaposition of near-liquidus melt, rich in alkali feldspar–K-feldspar components, with plagioclase already crystallized in a more mafic system. Since the more felsic system becomes superheated if it was at its liquidus prior to heating by the hotter system, alkali feldspar–K-feldspar does not nucleate and begin to grow until there is postmixing loss of heat to the environment. In a plutonic setting, rate of heat loss to the environment is slow and nucleation of alkali feldspar–K-feldspar may be delayed. The nucleation rate of K-feldspar at or just below its liquidus is known to be slow (Fenn, 1977). Under these circumstances, plagioclase may serve as local sites of nucleation of alkali feldspar–K-feldspar, resulting in the mantling relation.

Rapakivi–Antirapakivi Double Mantles

It is not likely that the alkali feldspar component would be exhausted at the time plagioclase mantles are forming on alkali feldspar–K-feldspar crystals in the formation of rapakivi texture. Partial dissolution of pre-existing alkali feldspar–K-feldspar, forming ovoids, puts alkali feldspar–K-feldspar components back into the magma. When resumption of alkali feldspar–K-feldspar growth occurs, in response to loss of heat to environment, it may do so epitaxially on alkali feldspar–K-feldspar crystals already mantled with plagioclase, forming double mantles (Figure 19.7).

Biotite-Mantled Alkali Feldspar–K-feldspar

Ovoidal alkali feldspar–K-feldspar crystals may acquire a mantling of biotite in a mixing system (Figure 19.5). Crystallization of biotite in a mixing system is in part a consequence of juxtaposition of iron and magnesium of the more mafic system with potassium of the more felsic system. Such a biotite is "hybrogenic" (Hibbard, 1991). Concentration of hybrogenic biotite on and near ovoids of K-feldspar may result from a high potassium concentration adjacent to dissolving alkali feldspar–K-feldspar that reacts with iron–magnesium being stirred in from the more mafic magma.

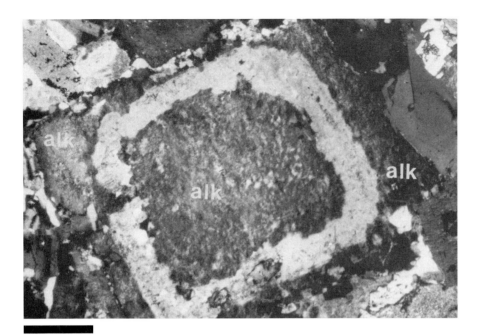

Figure 19.7
Double mantle relation with alkali feldspar core, oligoclase mantle zone, followed by outer alkali feldspar mantle. Ames monzodiorite, New Hampshire. X-polars. Bar scale = 0.12 mm.

Hornblende–Biotite Zones in Alkali Feldspar

Zones in alkali feldspar–K-feldspar phenocrysts containing biotite and/or hornblende (Figure 19.8) are also compatible with magma mixing. This texture is analogous to the feldspar double mantles shown in Figure 19.5. Let an alkali feldspar–K-feldspar become "mantled" with hornblende and biotite (especially hybrogenic biotite as shown in Figure 19.5) in a mixing environment. With heat loss to environment during the equilibration stage of mixing in a plutonic environment, the remaining alkali feldspar–K-feldspar grows epitaxially on the hornblende–biotite-mantled alkali feldspar–K-feldspar, forming a crude "double mantle." If a surge of more felsic magma dilutes the hybrid system, the alkali feldspar–K-feldspar with its zone of hornblende–biotite occurs as a relict mixing texture in the more felsic system (Figure 19.8).

Hornblende-Mantled Quartz

Another texture compatible with magma mixing is represented by an unusual relation of hornblende to ovoidal quartz (Figure 19.9a). The relationships parallel those of biotite and hornblende zoning ("double mantling") in alkali feldspar–K-feldspar.

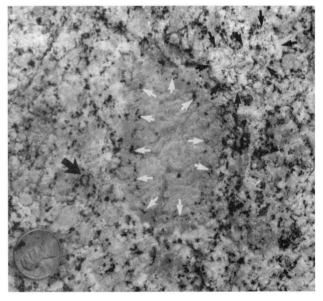

Figure 19.8
Alkali feldspar phenocryst with biotite–hornblende zone (semimantle) (white arrows). Host two-feldspar granite contains hornblende-mantled quartz (thick black arrow, center left) and boxy cellular plagioclase pseudophenocryst (small black arrows, upper right). Heusser Mountain, Nevada. Rock surface. Penny = 2 cm.

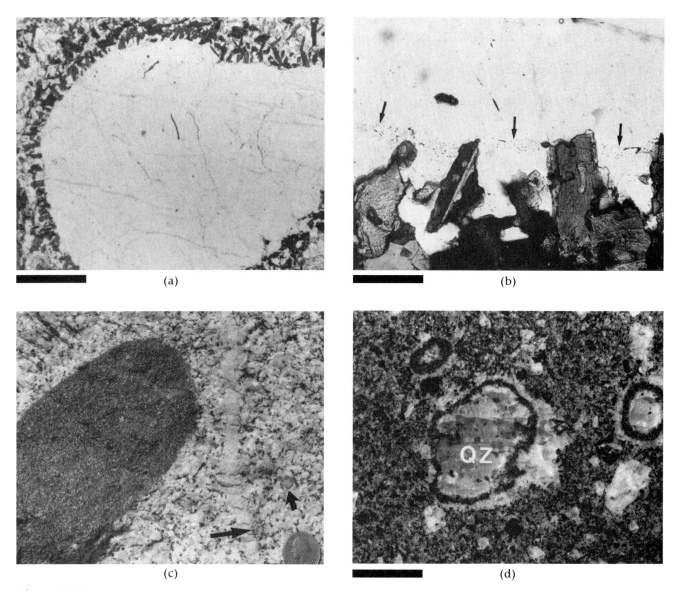

Figure 19.9
Hornblende-mantled quartz ovoids.

(a) Hornblende "locked" into rim portion of rounded quartz pseudophenocryst. Note tendency for hornblende crystals to orient perpendicular to a quartz surface existing prior to "locking" epitaxial quartz overgrowth. Heusser Mountain, Nevada. PPL. Bar scale = 1.2 mm. (b) Close-up of (a) showing hornblende locked in quartz up to a discontinuity marked by "impurities" in the quartz (arrows). X-polars. Bar scale = 0.14 mm. (c) Hornblende-mantled quartz (short arrow) in two-feldspar granite. Rounded microdiorite enclave (left) and obscure mini-enclave of microdiorite (long arrow). Postmixing, postmingling aplite dikelet (vertical). Heusser Mountain, Nevada. Quarter = 2.5 cm. (d) Biotite-mantled quartz ovoids with second mantle of quartz–plagioclase. Note exclusion of matrix biotite–hornblende from outer mantle zone, generating an ocellar texture. Slab surface. Granite Mountain, Shoshone Range, Nevada. Bar scale = 1 cm.

Ovoidal quartz in magma-mixing systems has the same origin as ovoidal alkali feldspar–K-feldspar. Pre-mixing quartz crystals in a more felsic system are subject to rounding by partial dissolution attendant to heating by the more mafic system. If the ovoidal quartz is mixed with a more mafic system crowded with small hornblende (or pyroxene) crystals, the quartz may "lock in" adjacent hornblende crystals as further growth of quartz occurs. The hornblende mantle is actually a cluster of hornblende crystals included in the rim portion of the quartz (Figure 19.9b).

The mechanically stable "locked-in" mantling relation would not be destroyed if there were to be a second or "rollover" stage of mixing engulfing the hornblende–quartz unit in other batches of hybrid magma or in a surge of the original more felsic magma (Figures 19.8, 19.9c).

Ocellar Hornblende-Mantled Quartz

Ocellar hornblende-mantled quartz is a relation analogous to the alkali feldspar–plagioclase double mantle (Figure 19.7) and the hornblende-biotite–K-feldspar "double mantle" (Figure 19.8). In this case, quartz ovoids are first mantled with "locked-in" hornblende. Subsequent to mantling, the quartz–hornblende unit is engulfed by a surge of more felsic magma followed by partial crystallization of a relatively hornblende-free assemblage of plagioclase, quartz, and K-feldspar. This is followed by a third stage or surge of mixing with more mafic "matrix" hybrid rock that contrasts sharply with the felsic "mantle," generating the ocellar texture (Figure 19.9d).

Sphene-Centered Ocellar Texture

An ocellar texture of possible mixing origin consists of biotite-free "spots" of sphene + plagioclase + quartz + K-feldspar + minor hornblende in a biotite-bearing granodioritic matrix (Figure 19.10). The texture has been tentatively explained as the result of a two-stage mixing involving magma capable of generating an abundance of sphene (Hibbard, 1991).

Blade Biotite

An unusual morphology of biotite is prone to form in the magma-mixing hybridization systems. The biotite is characterized by marked elongation in one direction, not the two directions characterizing platy biotite (Figure 19.11a,b). As previously discussed, mixing of an aluminosilicate system rich in Mg and Fe with another rich in K sets the stage for crystallization of

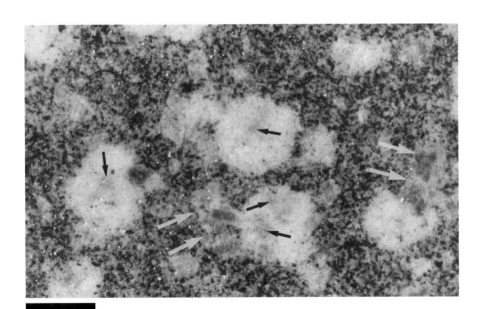

Figure 19.10
Sphene–centered ocellar texture. Ocelli consist of white plagioclase, light gray sphene (black arrows), and a few hornblende crystals (not shown). Irregular extensions of some ocelli contain larger plagioclase crystals (white arrows). Matrix consists of plagioclase, K-feldspar, quartz, and biotite. Austin Stock, Nevada. Slab surface. Bar scale = 1 cm.

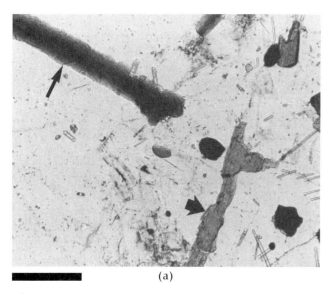

Figure 19.11
Blade biotite.

(a) Elongated basal section of biotite (long arrow) is similar to nonbasal section (short thick arrow). Note acicular apatite crystals (right half). Okanogan Range, Washington State. PPL. Bar scale = 0.14 mm. (b) Elongate biotite blades (short arrow) crisscrossing boxy cellular plagioclase (long arrow). Some biotite blades are basal sections. Quartz, alkali feldspar, and epitaxial, more-sodic, plagioclase fill cells. Willard Mining district, Nevada. X-polars. Bar scale = 0.5 mm. (Courtesy P. Muto.)

Figure 19.12
Hornblende crystals mantling premixing magmatic biotite. Andesite dike. Carson Range, Nevada. PPL. Bar scale = 0.5 mm.

biotite as a direct consequence of the hybridization. In some situations, growth of this "hybrogenic" biotite is physically restricted by a high density of pre-existing crystals, perhaps resulting in physical restriction and/or rapid growth in an undercooled environment. Unusual elongations have been documented for magnetite and halite in the rapid crystallization environment at fumaroles (Symonds et al., 1987).

Hornblende-Mantled Biotite

Well-formed prisms ("books") of biotite may form in the more felsic system prior to mingling and mixing. During the mingling stage, this biotite may be partially resorbed. Mixing with a hornblende-bearing more mafic system may result in attachment of small hornblende crystals to the larger biotite (Figure 19.12).

Conversion of Hornblende to Biotite

In a magmatic system capable of crystallizing "hybrogenic" biotite, reaction of potassium-bearing melt or hydrous melt with pre-existing hornblende is to be ex-

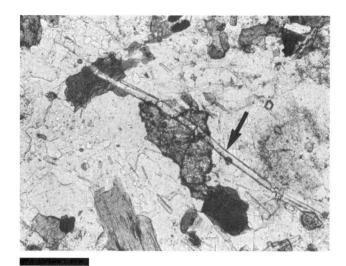

Figure 19.13
Needle apatite (arrow) in hybrid granodiorite. Note deformation of apatite and its apparent continuation through sphene (center) and biotite (upper left) indicating early crystallization. Heusser Mountain, Nevada. PPL. Bar scale = 0.14 mm.

ing nonmixing magmatic systems as well as forming during the deuteric-hydrothermal stage as "secondary biotite."

Acicular Apatite

Acicular apatite is common in rocks containing other textural features compatible with magma mixing (Figures 19.11a, 19.13). Evidently rapid growth in undercooled magma (Wyllie et al., 1962), in this case in a magma-mixing hybridizing system chemically capable of crystallizing apatite, effectively results in acicular, rather than stubby, prismatic growth of the apatite.

Poikilitic Quartz and Poikilitic K-feldspar

Poikilitic quartz and poikilitic K-feldspar texture (Figure 19.14) in the context of magma mixing result from late-stage crystallization of felsic melt or hydrous melt, after the crystallization of an abundance of quench-generated plagioclase, hornblende, biotite, and apatite. Since the more felsic system is reheated, or superheated, there may be only a few quartz and K-feldspar nuclei available for growth as the system slowly cools to the environment. Consequently, a few large crystals of quartz and K-feldspar are destined to include the earlier assemblage of relatively small crystals.

pected. Hornblende deriving from the more mafic system in a mixing system typically has small biotite crystals within and rimming its crystals. However, by no means is this a texture indicative of magma mixing even though it is characteristic of magma mixing. Such a textural relation is very common in fractionat-

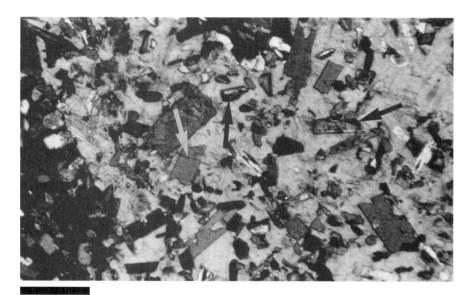

Figure 19.14
Poikilitic K-feldspar with biotite and plagioclase inclusions. Note needle apatite across biotite (white arrow), and steep zoning range of plagioclase (two black arrows). Granite Mountain, Shoshone Range, Nevada. X-polars. Bar scale = 0.5 mm.

Figure 19.15
Calcic growth zones (white) in plagioclase of andesite dike, probably representing two pulses of mixing. Carson Range, Nevada. X-polars. Bar scale = 0.12 mm.

"equilibration" stage of mixing in plutonic environments (Figure 19.14). Since the crystals are complete, not skeletal or otherwise cellular, the undercooling probably was no more extreme than that of interior portions of basalt lava flows in which small tabular noncellular crystals are characteristic.

Calcic Zones in Plagioclase

Abrupt (step) zoning to more calcic plagioclase (Figure 19.15) is also consistent with mixing of magmas deriving from the mixing of more mafic (calcic) magma with crystal-bearing more felsic magma. Pre-existing plagioclase growth surfaces are ideal substrate surfaces for nucleation and growth of the more calcic plagioclase. The calcic zone may be noncellular or cellular if undercooling is appropriately large. A step zone back to more sodic plagioclase, or progressive normal zoning to more sodic plagioclase from the calcic zone, occurs during the "equilibration" stage in slow-cooling systems.

Small Step-Zoned Plagioclase

Rapid nucleation of small plagioclase crystals from growth components deriving from a more mafic system is favored by rapid heat loss to a more felsic system. Such crystals may be quite calcic and typically step zone to more sodic plagioclase rims during the

Boxy Cellular Plagioclase

Cellular plagioclase crystals can result from a relatively high rate of growth and a low rate of nucleation (Lofgren, 1974, 1980). The undercooled environment, provided by the transfer of heat from the hot to the cool magma in a mixing system, may be ideal for boxlike cellular growth of more calcic plagioclase deriving

(a)

(b)

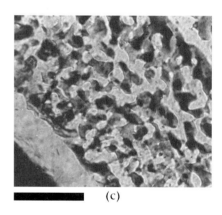
(c)

Figure 19.16
(a) Boxy cellular growth plagioclase (zoned) in granite of Heusser Mountain, Nevada (see Figure 19.8). Cells contain quartz and K-feldspar. X-polars. Bar scale = 1.6 mm. (b) Boxy cellular growth plagioclase occurring as epitaxial mantle (rapakivi texture) on alkali feldspar phenocryst (right). Newberry Mountains, Nevada. X-polars. Bar scale = 0.17 mm. (c) Spongy cellular plagioclase of probable dissolution melting origin. Noncellular rim against glass (lower left) formed by growth during "equilibration" stage postdating dissolution. Coso Mountains, California. X-polars. Bar scale = 0.07 mm. (Specimen courtesy C. Bacon, U.S. Geological Survey.)

from the more mafic system (Figures 19.11b, 19.16a, b). Zoning to more sodic compositions within the cells, and as an enclosing noncellular rim zone, is to be expected as "equilibration" is approached in slow-cooling systems.

Boxy cellular plagioclase may also occur as mantles on alkali feldspar–K-feldspar (Figure 19.16b). The beginning of plagioclase mantling in the rapakivi relation is typically boxy cellular, even if the bulk of the mantle is noncellular (Hibbard, 1981).

Occurrence of large boxy cellular plagioclase crystals in two-feldspar granites containing other textures compatible with magma mixing are shown in Figures 19.4 and 19.8. Such crystals are heterogeneously distributed through these plutons, unlike the more "homogeneous" distribution of co-existing alkali feldspar phenocrysts.

Spongy Cellular Plagioclase

Spongy cellular plagioclase (Figure 19.16c) also occurs in rocks with other textures compatible with magma mixing. In this case, the cellularity is the result of dissolution (and/or direct melting) attendant to reheating of plagioclase of the more felsic system as mixing occurs with the hotter more mafic magma. The spongy morphology of the cells contrasts sharply with boxy cellular growth plagioclase (Figure 19.16a,b). Spongy morphology has been produced experimentally by Tsuchiyama and Takahashi (1983) and Tsuchiyama (1985).

Spongy cellular texture is another example of a texture compatible with magma mixing but known to form in nonmixing systems. A rapid decrease in total pressure, as expected in rapid ascent of relative "dry" crystal-bearing magma to a volcanic environment, promotes crystal dissolution (Chapter 13).

Alkali Feldspar–K-feldspar Porphyrograins in Mafic Inclusions and Host Pluton

Mixing of more felsic magmas initially containing relatively large alkali feldspar crystals with more mafic systems results in a spectacular juxtaposition of mineralogy. Alkali feldspar phenocrysts in mafic inclusions (enclaves) (Figures 19.5, 19.17a,b) and in larger more mafic rock masses (Figure 19.17c) is typical. The phenocrysts may be rounded prior to mechanical mixing (Figure 19.4), or they may have little or no morphology modification attendant to reheating (Figure

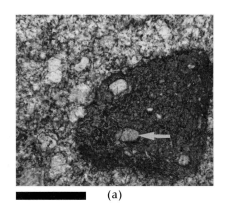 (a) 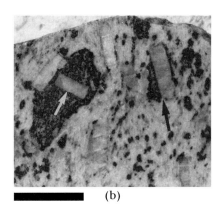 (b) 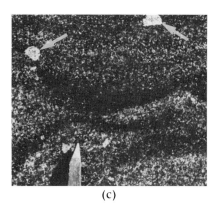 (c)

Figure 19.17
Alkali feldspar porphyrograins (phenocrysts) in mafic inclusions and schlieren.

(a) Alkali feldspar (arrow) in mafic inclusion is identical to alkali feldspar phenocrysts in host two-feldspar granite. Note partially included phenocrysts on upper left portion of inclusions. Rock surface. Sylvania Mountains, Nevada/California. Bar scale = 5 cm. (b) Inclusion and partial inclusion of alkali feldspar phenocrysts in mafic mini-inclusions. Flow orientation of inclusions and phenocrysts contrasts sharply with "misorientation" of phenocryst (white arrow) that was mixed with mafic system prior to observed flow. Phenocryst largely exhumed from inclusion during flow (black arrow). Wassua, Wisconsin. Slab surface. Bar scale = 1.3 cm. (c) Alkali feldspar phenocrysts (white arrows) in mafic schlieren in two-feldspar granite. Massif Central. Hammer point = 10 cm. (Courtesy J.-P. Couturié.)

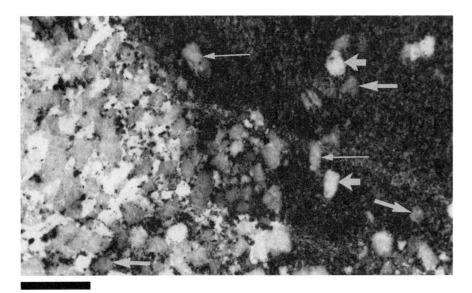

Figure 19.18
Mixing of two-feldspar granite with microdiorite. Crystals in dominating diorite (upper right) are alkali feldspar (thin long arrows), plagioclase (thick short arrows), and quartz (thick long arrows). Diorite becomes mini- and micro-inclusions (lower center–left) where minerals of the granite dominate. Slab surface. Okanogan Range, Washington State. Bar scale = 1.6 cm.

19.17b). This variation in morphology is probably a matter of timing and thermal contrast.

Fine-Grained More Mafic Mini- and Micro-Inclusions

Bits and pieces of microdioritic rock ("mini-" and micro-enclaves) in more felsic hosts is very common in magma-mixed hybrid systems. These inclusions are obvious if large (Figure 19.4) and obscure if small (Figure 19.9c). In well-mixed hybrid batches of crystal-bearing magma the inclusions texturally blend with the hybrid host (Figure 19.18). Many accounts of inclusions in magmatic rocks are interpreted as being a consequence of magma mixing (Smith, 1979; Bacon and Metz, 1984; Bacon, 1986; Eberz and Nicholls, 1988; Dorais et al., 1990; Vernon, 1991b; Barbarin, 1990; Barbarin and Didier, 1991).

Summary of Texture Compatible With Magma-Mixing

Progression from initial mafic and felsic magmas to a mafic-rich hybrid granitic rock to an alkali feldspar phenocryst-bearing microdiorite to microdiorite-"contaminated" two-feldspar granite is shown with a min-

Figure 19.19
Schematic representation of textural–mineralogical phenomena occurring when a crystal-bearing felsic system and a crystal-bearing mafic system mix on a grain scale. Textural–mineralogical units become part of hybrid system that is the source of the hybrid inclusion (within hand specimen) in resurgent host granitic magma into which most of the textural–mineralogical units have been dispersed.

19 Mixed Magma Rocks

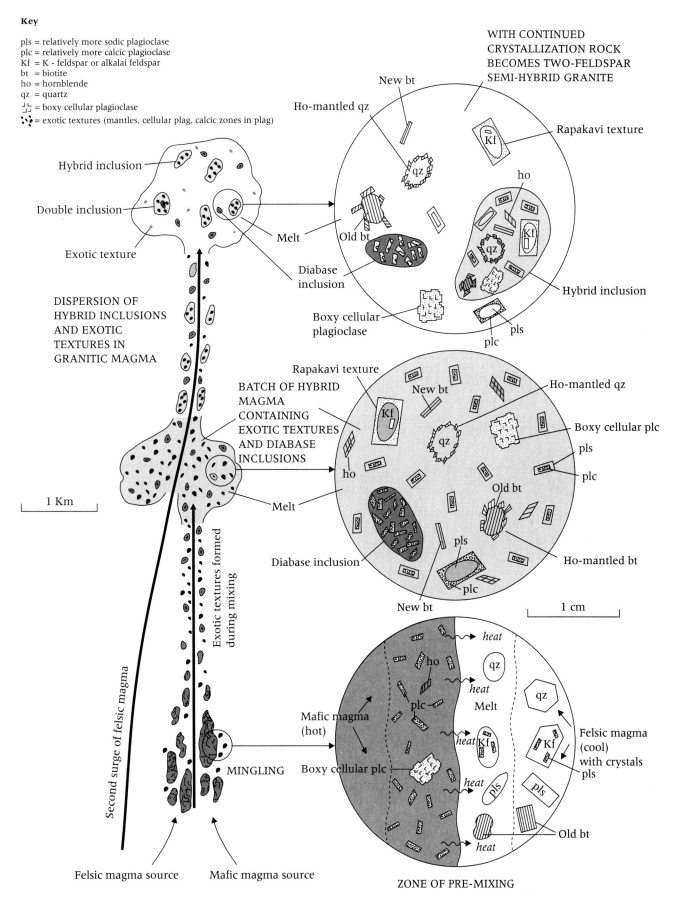

gling-mixing cartoon (Figure 19.19). An initial stage of mingling progresses into the first stage of mixing that generates a hybrid granitic rock rich in mafic minerals and exotic mixing textures and that contains some diabase inclusions. A second surge of felsic magma is depicted as leading to a second stage of mingling and mixing that generates a large volume granitic hybrid containing both diabase inclusions of the first stage and hybrid granitic inclusions of the second stage, in some cases as double inclusions.

20

Assimilation-Hybrid Magmatic Rocks

There seems no reason, therefore, to doubt that direct solution of foreign material in superheated magmas cannot be a factor of importance in petrogenesis.

(N. L. Bowen, 1922)

The degree to which such a concept [assimilation] now seems to be a fundamental violation of all that is known about the thermal relations of magma is a measure of the pervasive impact of Bowen's writing.

(A. R. McBirney, 1979)

Magma that comes in contact with pre-existing rock may be compositionally and texturally modified. If the magma simply freezes against the rock, there is no interaction and the magma is not modified. This is likely to be the case if the thermal disparity between magma and wall rock is large and the magma is quenched, solidifying to glass or a very fine-grained rock at the contact, with little chance for chemical interaction.

At the other extreme, there may be so much interaction that the distinction between an original mass of magma and the "foreign" materials becomes mineralogically and texturally indistinct, perhaps leaving only chemical signatures of assimilation. ". . . the effects of crustal assimilation by mantle-derived magmas are not easily demonstrated. The problem lies in the fact that the products of assimilation are not mineralogically conspicuous in igneous rocks" (Hess, 1989).

What Is Assimilation?

The term *assimilation* (also "contamination") indicates several possible processes that result in incorporation of rock into magma (Bowen, 1922; McBirney, 1979). Fragments of foreign rock incorporated into magma are **xenoliths**. Single crystals, disaggregated from rock and engulfed by magma, are **xenocrysts**. The contact between magma and the surfaces of xenocrysts, xenoliths, and the wall rock from which these are derived is where the processes resulting in assimilation may begin. The more contact between magma and rock there is, the more assimilation there can be, if assimilative processes are activated. Silicate melt in contact with minerals that have not crystallized from that melt presents a state of disequilibrium. **Dissolution melting** or **dissolution-reprecipitation** may occur across the melt–crystal interface if kinetic factors are favorable. Heat speeds up most reactions, and since there is melt present the temperature is inherently elevated.

Magmas may be initially saturated in aqueous phase, or they may become saturated as crystallization in the contact zone proceeds. Sooner or later, therefore, reaction of foreign minerals is likely to be with hydrous magma. There must be a transition from reactions of minerals with hydrous magma to reactions between aqueous phase and minerals when silicate melt is depleted; thus, there is a transition from reactions leading to assimilation to hydrothermal alteration within rock having nothing to do with assimilation.

In addition, assimilative processes include partial **direct melting** of rock if enough heat is supplied by the magmatic system. Melts generated in this way may mix with melts formed by dissolution melting, generating a system that has components of both assimilation and in situ magma mixing.

The Processes of Assimilation

The interaction between magma and rock may take any or all of five forms: (1) dissolution of crystals in contact with melt with which they are not in equilibrium (congruent dissolution), (2) dissolution at the crystal–melt interface accompanied by nucleation and growth of a new crystalline phase nearby (one type of incongruent dissolution), (3) dissolution at the crystal–melt interface accompanied by chemical diffusion back into the remaining crystalline phase (another variation of incongruent dissolution), (4) chemical exchange between rock and magma resulting in reactive replacement within the rock, concurrent with growth of the same mineral phase in the magma, and (5) direct melting in the rock at high-energy sites, especially where an aqueous phase is present or is generated. These various processes are shown schematically in Figure 20.1.

In the first case, there is a breakdown of crystal structure assisted by the attractive forces of adjacent molecular units in the melt. Crystals dissociate, dumping their chemical components into the adjacent melt without generating crystals of another composition. Congruent dissolution is most likely to occur if the melt is unsaturated in the components held in a mineral that is not part of a solid solution series, such as quartz.

This does not mean that any crystal will dissolve that is not in equilibrium with the melt. For example, an almandine garnet is not likely to dissolve in a granitic melt even if the melt is undersaturated with respect to the garnet. There are kinetic barriers to detachment of ionic units that cannot be overcome in this case, especially, for example, at the relatively low liquidus temperatures of a granitic melt containing an aqueous phase.

In the second type of interaction between magma and rock, dissolution of a mineral is accompanied by nearby nucleation and growth of a mineral phase in equilibrium with the melt (one type of incongruent dissolution). This type of dissolution is expected if a mineral is part of a solid solution series such as plagioclase and is unzoned. The new plagioclase crystal or an equivalent calcic growth zone on a pre-existing crystal in or facing magma will be more calcic than the beginning crystal. Melt produced will be more sodic, initially in equilibrium with the more calcic plagioclase, but gradually mixing with the ambient magma in which further compositional adjustment in the new plagioclase may be activated.

The third type of interaction is incongruent dissolution involving partial compositional modification of the mineral being dissolved. Unzoned plagioclase of the rock is chemically modified to a more calcic crystal in equilibrium with a more sodic melt. In this case, equilibrium is established by a two-way diffusion of Ca into the crystal and Na from the crystal into the melt (Tsuchiyama, 1985). This process ("coring" of metallurgists) would seem to be limited by the armoring effect of the calcic margin, arresting dissolution, at least in a relatively rapid cooling system.

In the fourth case of interaction, growth of new crystals is governed by chemical supply from the rock and the magma. A connecting fluid between rock and magma allows reactive replacement to occur on the rock side of the interface, generating a mineral that also grows into magma on the magma side of the interface (Figure 20.1).

In the fifth process, new melt is generated in the rock at high-energy sites within and between crystals. This is direct melting generating a melt independent of the ambient magma (Tsuchiyama and Takahashi, 1983). It was established by Bowen long ago (1922) that this process can only occur if the magma is hot enough with respect to the temperature required to initiate melting. For example, a basaltic magma may initiate melting in a rock that can generate a water-bearing granitic melt, such as a schist or a granitic gneiss, especially if aqueous phase is available from dehydration of micas and amphiboles contained in the rock or if aqueous phase is available from an external source. Conversely, direct melting in a mafic rock cannot be expected to be initiated by a relatively low temperature water-bearing granitic melt.

What actually occurs in a zone of assimilation is likely to involve several of the processes just described. Identification of process and dominance of one process

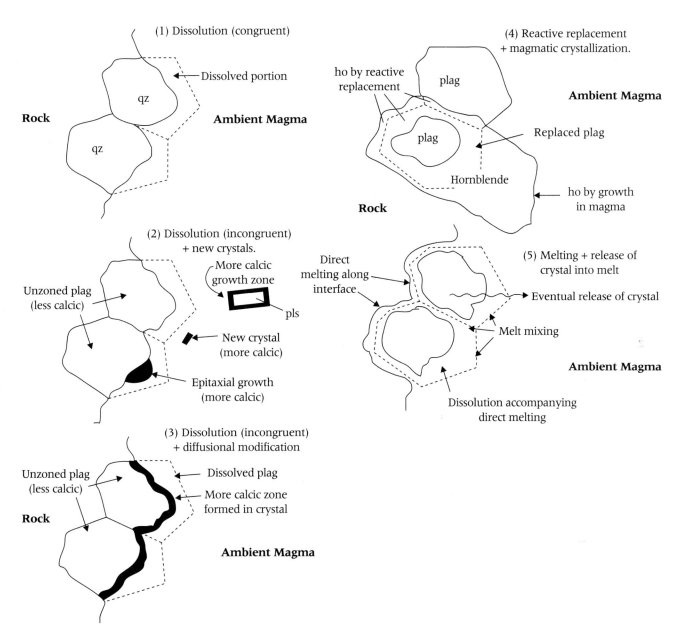

Figure 20.1
Five processes by which rock can become assimilated by magma. Processes 1–4 involve reaction between magma and interfacing minerals. Process 5 is higher temperature, direct melting shown occurring along grain boundaries, unavoidably accompanied by lower-temperature dissolution at magma–crystal interfaces.

over another cannot be expected to be a simple matter. Worse than that, there is likely to be controversy as to whether the interaction between rock and magma is an assimilation process at all. The following examples of assimilation are presented in the spirit that they seem to be compatible with a model of assimilation involving one or more of the five processes. Certainly, they are not clear-cut examples of assimilation, recognizable in routine petrographic analysis.

Textural Expression of Assimilation: Some Examples

Assimilation of Quartzite by Dioritic Magma

A cluster of three quartz crystals is shown in a thin section of quartz diorite (Figure 20.2). The crystals do not express hexagonal forms of quartz. They are locked together with continuous contact, unlike "loose" attachments typically resulting from synneusis of crystals in magmas, the principal origin of glomerophenocrysts. The quartz diorite is not porphyritic; there is no widespread distribution of quartz crystals the size of those occurring in the cluster.

This sample was collected very close to a contact with metasedimentary quartzites, and there are mesoscopic quartzite xenoliths of the quartzite in the quartz diorite. In view of these field data, it is likely that the quartz cluster shown in the photomicrograph is a microxenolith, disaggregated from a larger block of quartzite.

In addition, the contact between two of the quartz grains, shown by the arrow in Figure 20.2, is sutured if viewed with higher magnification (not shown). This suggests that the grain contacts are of metamorphic, not magmatic, origin. The external surfaces of the grains are irregular; this is not characteristic of free growth in a magma, nor resorption in a magma that typically generates distinctly rounded forms with embayments. Nor can this be a case of magmatic quartz metamorphosed in situ; the plagioclase has not recorded any such event.

If there had been further disaggregation, each quartz grain of the cluster would be isolated in the plagioclase-dominant quartz diorite. Although there is additional quartz in the rock (as labeled in the figure), this quartz has a tendency to be intergranular to plagioclase. Would a single crystal of assimilated quartz be recognized as such in a rock that already contains quartz? One clue would be that quartz diorites with about 10% quartz are not likely to have phenocrysts of quartz in a magmatic system that reaches quartz saturation only at a late stage. Large crystals of quartz would be anomalous, supporting either an assimilation or a magma-mixing interpretation. Furthermore, could the "indigenous" quartz itself be a product of assimilation, either as small grains with overgrowths or by dissolution of quartzite with the silica reappearing as intergranular quartz?

Figure 20.2
Microxenolith of quartzite consisting of three quartz grains in quartz diorite of French Lake pluton, northern Sierra Nevada. (Courtesy of Chris Osterberg). Contact between two grains (arrow) has sutured character (not shown). Additional quartz labeled qz. X-polars. Bar scale = 0.5 mm.

Assimilation of Tonalite by Granite

Assimilation of tonalite by granite has been interpreted on the basis of information beyond the scope of the thin section and hand specimen, just as it was in the previous example. The sample (Figure 20.3) has a medium-grained mafic-rich tonalite in association with a fine-grained two-feldspar granite.

"Bits and pieces" of the tonalite are "strewn" through the granite. A case for magma mixing cannot be made in the face of isotopic dates, indicating that the tonalite is Jurassic and the granite is Cretaceous (Hibbard, 1971). A case can be made for metasomatic "transformation" on the basis of the gross textural relations and some assumed chemical exchanges. If there had been deterioration of the tonalite by reaction of minerals with incoming solutions containing silica and alkalis, the bits and pieces of crystals and crystal clusters would be relicts of the tonalite, iso-

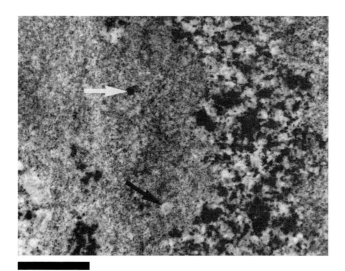

Figure 20.3
Medium-grained tonalite partly assimilated by fine-grained two-feldspar granite. Relicts of mafic minerals (hornblende + biotite) (white arrow) and plagioclase (black arrow) are isolated within granite portions. Okanogan Range, Washington. Slab surface. Bar scale = 2 cm.

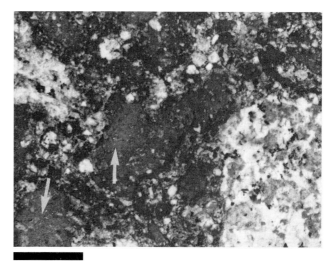

Figure 20.4
Two-feldspar granite partially melted and disaggregated by trachyandesite. Marginal portions of trachyandesite blobs (arrows) contain glass and crystals derived from the granite. Rattlesnake Gulch, northwest of Mono Lake, California (Kaczor et al., 1988). Slab surface. Bar scale = 1.4 cm.

lated (not mechanically strewn) by local conversion of tonalite to granite.

A case for assimilation of the tonalite by a granitic magma may at first seem to be unreasonable in the face of a more felsic system expected to assimilate a more mafic rock. Although direct melting of mineral phases in the tonalite is not likely, there is still the possibility of reaction leading to contamination of the granitic melt along the contact. Reaction leading to mechanical release of bits and pieces of mineral grains and grain clusters from the tonalite into the granitic magma would be an expected consequence of selective dissolution within the tonalite.

Assimilation of Two-Feldspar Granite by Trachyandesite

Detailed description of reactive melting (and direct melting?) and melt–crystal reactions involving quartz, plagioclase, and K-feldspar of a Cretaceous Sierra Nevada granite intruded by a Pliocene mafic plug (Kaczor et al., 1988), presents a fine example of partial assimilation. Bits and pieces of minerals and mineral clusters, derived from the granite, are strewn through glass derived from dissolution melting (and direct melting) of the granite (Figure 20.4). Direct melting of plagioclase and quartz is suggested by the presence of glass in spongy cells of the former and glass in fractures of the latter. Microcline of the granite is converted to sanidine, commonly rimmed with anorthoclase, representing reaction between melt and the alkali feldspar. Reaction of melt with plagioclase of the granite also produces rims of anorthoclase composition.

Globs of trachyandesite, containing calcic plagioclase, olivine, augite, and magnetite, are crudely stirred with the glassy and crystalline granitic materials (Figure 20.4). In thin section, relicts of the granitic minerals, along with their glasses, are strewn through very fine-grained, microporphyritic trachyandesite.

Assimilation involving dissolution melting and direct melting is compatible with a high-temperature system (the trachyandesite melt) emplaced into granitic rock where there was either an external source of H_2O and/or H_2O was derived from biotite and hornblende, reducing melting temperatures.

Assimilation of Carbonate-Bearing Rocks by Granodiorite

Chemical interaction between impure carbonate rocks and granitic magma is complicated by the presence of carbonate ion as well as an aqueous phase. What happens when calcium carbonate dissociates and the calcium and carbonate ions are available to become part of a granitic magma?

A rise in calcium in the magma can be expected to be reflected in eventual crystallization of a more calcic plagioclase or as more calcic zones on pre-existing plagioclase crystals (Bowen, 1922; McDowell, 1978). In addition, since hornblende is a common calcium-bearing mineral in granodiorites, an increased in calcium might be reflected in an increase in modal hornblende. However, if Mg and Fe are not available to form hornblende, a purer calcium mineral might be expected, such as wollastonite. If dolomite is assimilated, an additional source of Mg is available, and a mineral such as tremolite or diopside would be expected to form.

If new silicates form in the magma as the result of adding Ca and Mg, the proportion of quartz in the granodiorite should be reduced. In the extreme, the granodiorite magma could be desilicated to the point that it becomes syenitic or monzodioritic.

Growth of new clinopyroxene in siliceous magma interacting with carbonates is typical. Clinopyroxene may mantle hornblende (Figure 20.5a) or biotite (Figure 20.5b). The irregular interface between the clinopyroxene and the original hornblende and biotite suggests that there has been at least some volume for volume replacement. However, extensions of clinopyroxene off into quartz–plagioclase–K-feldspar space (Figure 20.5b) suggest that clinopyroxene growth components were available in the ambient magma. Conversion of biotite to clinopyroxene requires incoming Ca and outgoing K. Less of a chemical exchange is required to form clinopyroxene at the expense of hornblende. In both cases, the hydrous minerals have given way to the anhydrous clinopyroxene. If the granitic system was capable of generating hydrous phases, as indicated by initial presence of hornblende and biotite, why is the new phase anhydrous? The answer may relate to the presence of excess carbonate ion introduced along with the Ca and Mg. It has been suggested (Swanson, 1979) that the presence of carbon dioxide decreases the activity of aqueous phase in magmas. An overwhelming input of carbonate may then prevent crystallization of hydrous minerals, and clinopyroxene forms instead of clinoamphibole.

A hybrid zone forming in the granodiorite in contact with carbonate-bearing rocks is not only characterized by new clinopyroxene but by reduction or even elimination of quartz. How does this desilication occur without major generation of minerals such as wollastonite and perhaps nepheline (Watkinson and Wyllie, 1969), a silica-undersaturated equivalent of alkali feldspar?

One suggestion has been (Nabelek et al., 1988) that quartz in carbonate wall rock is consumed by reactions producing calc-silicate minerals. If magmatic fluids are interacting with this wall rock, there may be a chemical potential gradient established between the silica-rich granodiorite system and the quartz-poor calc-silicate metamorphic system. Transfer of silica to

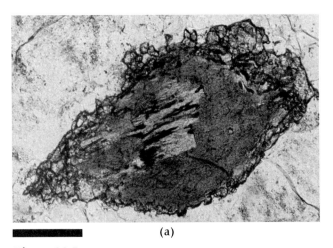
(a)

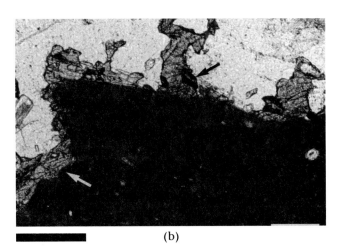
(b)

Figure 20.5
Clinopyroxene mantling mafic minerals of granitic rocks in contact with carbonate rock.

(a) Clinopyroxene on hornblende surrounded by alkali feldspar, plagioclase, and less quartz. Chlorite after biotite occurs within the hornblende. Osgood Mountains, Nevada. PPL. Bar scale = 0.17 mm. (b) Clinopyroxene on biotite. Irregular contact between clinopyroxene and biotite (white arrow) points to reactive replacement, whereas extensions of clinopyroxene into alkali feldspar–plagioclase–quartz assemblage from upper surface of biotite suggests growth into melt. Sphene crystals (black arrow) derive from Ti released from biotite and Ca introduction from carbonate wall rock, located a few centimeters away. Notch Peak contact, House Range, Utah. PPL. Bar scale = 0.17 mm.

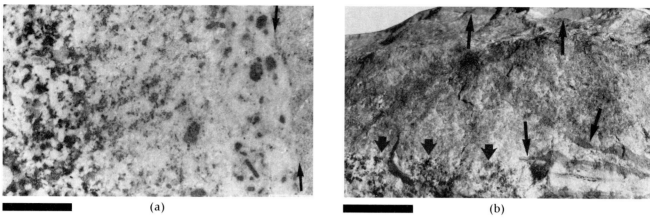

Figure 20.6
Mesoscopic characteristics of assimilation-modified granitic rocks in contact with carbonates.

(a) Hybrid zone between granodiorite (left) and calcite marble (far right) (contact shown with black arrows) is about 5 cm thick. Hybrid rock contains much less quartz than original granodiorite and has clinopyroxene, not occurring in unmodified granodiorite. Osgood Mountains, Nevada. Bar scale = 1.6 cm. (b) Fine-grained two-feldspar granite sill in layered calc-silicate hornfelses, Notch Peak, House Range, Utah. Upper contact with diopside-rich hornfels shown with two long arrows. Lower contact is with disrupted layers of diopside-rich hornfels (right) (two long arrows), and hybrid zone (left) (three short arrows). Hybrid zone characterized by clinopyroxene enrichment and reduced modal quartz, coincides with position of former lateral extension of layered hornfels. Lack of hybridization along contacts parallel to layering (upper contact and lower right contact) at diopsidic hornfels-granite contacts may be due to protective nature of diopside layers, breached along lower left contact zone where access to carbonate was enhanced. Bar scale = 2.5 cm.

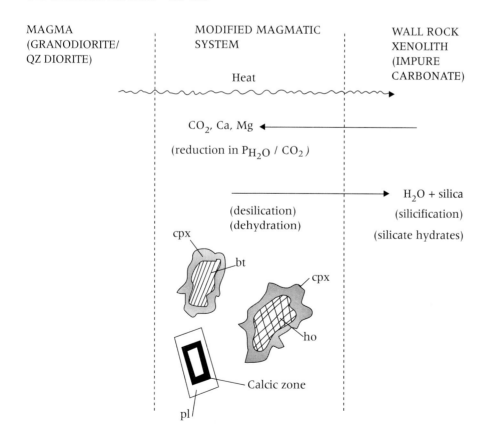

Figure 20.7
Pictoral representation of chemical exchange and mineral growth in assimilative contact zone between granitic magma and carbonate rock.

wall rock depletes the magmatic system in silica, resulting in final crystallization of a syenite or monzodiorite in the contact zone.

Clinopyroxene-enriched monzodioritic rock, interpreted as the result of assimilative modification is shown in Figure 20.6a, and incorporation of calc-silicate metamorphic rock into aplitic granite along a still contact is shown in Figure 20.6b. A schematic representation of the relationships shown in Figure 20.6a, which would generally also apply to the relations shown in Figure 20.6b, is given in Figure 20.7.

Assimilation of Biotite–Hornblende–Diopside–Plagioclase Hornfelsic Schists by Granodioritic and Tonalitic Magmas

The results of probable partial reaction between siliceous magmas capable of crystallizing hornblende and metamorphic rocks of sedimentary and volcanic parentage capable of generating biotite, hornblende, diopside, and plagioclase is shown in the photographs of slabbed hand specimens (Figure 20.8).

Partial conversion of the diopside to hornblende in a diopside granofels xenolith, is spatially related to an adjacent tonalite (Figure 20.8a). Complete conversion occurred at the interface with the magma, and a local reaction yielding hornblende occurred where fluids were able to penetrate the xenolith. Similarly, conversion of a biotite–hornblende–plagioclase (An_{50}) hornfelsic schist to coarse hornblende is shown in Figure 20.8b. In this case, penetration of fluids has been controlled in part by schistosity, although pods of quartz + feldspar lined with coarse hornblende appear isolated from the main body of granitic rock.

The metamorphic rocks were derived from sedimentary and volcanic rocks during regional metamorphism and were subsequently thermally metamorphosed at the time of emplacement of plutons ranging from two-feldspar granite to tonalite (Hibbard, 1971). The reactions indicated by the textural and mineralogical relations shown in the photographs must have taken place more or less simultaneously with the thermally induced metamorphic recrystallization since the magmas are responsible for both.

Although there is a clear genetic relation between magmatic and metamorphic systems in the sense that coarse hornblende is generated in the contact zones, the process of assimilation is not obvious. Has hornblende been incorporated into magma, or is this a case of infiltration of metamorphic rock by magmatic fluids, with coarse hornblende crystallizing only within the original confines of the metamorphic rock? The textural and mineralogical relations shown in Figure

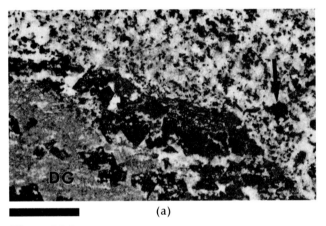
(a)

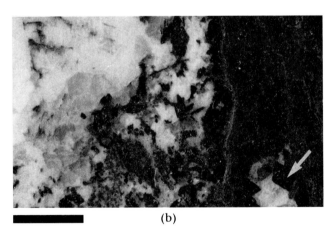
(b)

Figure 20.8
Mesoscopic views of assimilative reaction at magma–xenolith contacts.

(a) Hornblende growth at the expense of diopside along margin and within calc-silicate hornfels xenolith in contact with tonalite. Large hornblende (arrow) in tonalite may have been derived from the reaction zone. Chopaka Mountain, Okanogan Range, Washington. Bar scale = 1.4 cm. (b) Growth of relatively large hornblende in plagioclase–biotite–hornblende hornfelsic schist (HS) in contact with two-feldspar granite. Reaction occurs along planar structure in schist and within pods where there is fluid access. Hornblende tends to develop crystal faces toward felsic phases (arrow). Okanogan Range, Washington. Bar scale = 7 mm.

20 Assimilation–Hybrid Magmatic Rocks 269

Figure Figure 20.9
Growth, detachment, and incorporation of crystals in assimilation zones between hornfelsic rocks and Mesozoic granitic rocks, Okanogan Range, Washington.

(a) Large twinned hornblende crystal (center) has grown as poikiloblast in plagioclase–biotite hornfels (arrow) and with crystal faces in opposite direction facing granodioritic magma. X-polars. Bar scale = 0.5 mm. (b) Hornblende crystal with core zone containing inclusions of plagioclase derived from hornfels at early growth stage. Rim zone portions of crystal are complete, having grown in contact with magma, at least partly after detachment from hornfels side of the assimilation zone. PPL. Bar scale = 0.5 mm. (c) Detachment of plagioclase aggregates from plagioclase–biotite–hornblende hornfelsic schist (left). Grain contacts within plagioclase clusters (arrows) are relict polygonalized interfaces formed during metamorphic recrystallization. Magmatic host mineral is quartz, but it may also be K-feldspar. X-polars. Bar scale = 0.12 mm. (d) Clusters of relatively small hornblende crystals of assimilation origin (see text) between larger-zoned plagioclase. Hornblende is in quartz, but it may also be in K-feldspar of granodioritic host. PPL. Bar scale = 0.5 mm.

20.9 and Figure 20.10 suggest that both processes have occurred. Can these two processes be differentiated?

Growth of new hornblende at the expense of plagioclase, biotite, and metamorphic hornblende has resulted in the poikigranular portion of the large hornblende crystal in Figure 20.9a. Many of the plagioclase inclusions are smaller than those of the adjacent coarser-grained mosaic plagioclase, indicating that a reaction consuming plagioclase has occurred, such as

pl + bt + ho → hornblende + K (into fluid) + Na (into fluid)

Since this portion of the new hornblende crystal grew by reactive replacement, it is poikiloblastic.

The other half of a large hornblende crystal has grown in contact with ambient magmatic fluids and has developed crystal faces in that free-growth environment. Growth of both the poikiloblastic and magmatic hornblende is governed by equilibrium relations with fluids. Whether these fluids occur as thin films between metamorphic grains or as part of the magma (or hydrous magma) at the growth surface of hornblende is not a limiting factor in the growth of a homogeneous mineral phase. Presumably, there was either diffusional or infiltration connection of fluids between the metamorphic system and the magmatic system, establishing a uniform chemical environment for hornblende growth.

The hornblende crystal in Figure 20.9b has a poikiloblastic core and a solid rim in contact with large K-feldspar crystals of the magmatic system. The horn-

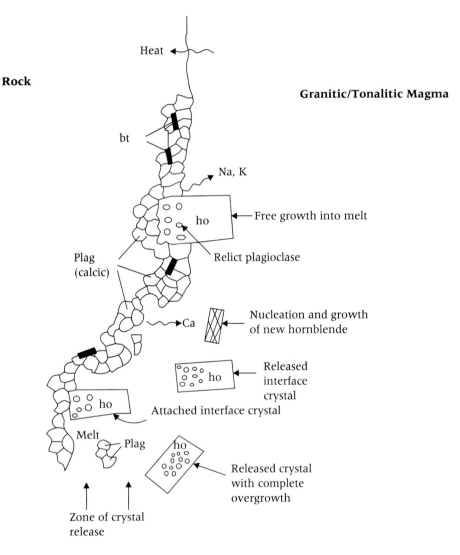

Figure 20.10
Pictorial representation of hornblende generated in assimilation zone between hornfels and magma. Portions of hornblende crystals forming by replacement of plagioclase–biotite hornfels is poikiloblastic. Detachment of hornblende into magma occurs as mosaic plagioclase of hornfels is melted or dissolved along grain boundaries and becomes detached as single or multiple polycrystal grains itself.

Calcic plag + Biotite ⟶ Hornblende + K (to melt) + Na (to melt)

blende texture indicates that it initially grew as a poikiloblast in metamorphic rock and was eventually released from that environment into the magma, continuing to grow in equilibrium with the melt, just as the solid half of the hornblende did (Figure 20.9a). How did this release take place?

A clue to detachment of crystals from an interlocking metamorphic texture with subsequent incorporation into adjacent magma is shown in Figure 20.9c. Here there are crystal clusters of plagioclase isolated in quartz of a granodiorite. Arrows on the photograph point to contacts within these pairs or clusters of plagioclase crystals that are interpreted as polygonal interfaces developed during metamorphic recrystallization.

One possible explanation of the relationships in Figure 20.9c is that there has been reactive replacement in which quartz takes the place of plagioclase, the plagioclase being in situ relicts. However, since there is no marked diminution of plagioclase crystals expected in a wholesale replacement process progressing inward along grain boundaries, the preferred interpretation is that plagioclase clusters have been mechanically released from the main plagioclase polygonal mosaic by minor replacement along plagioclase grain boundaries. This mechanical release is part of the process of assimilation, dumping plagioclase into the magma. If the release of plagioclase happens to be where there are poikiloblasts of hornblende, the release of hornblende is activated as well (Figures 20.9a,b).

An "overload" of hornblende in granitic magmas in contact with wall rock and associated xenoliths is shown in Figure 20.9d. The texture consists of an abundance of hornblende crystals in quartz and K-feldspar more or less intergranular to large plagioclase crystals. This is a textural anomaly, since if the magmatic system had been closed to modification attendant to assimilation a chemical system capable of crystallizing this much hornblende typically would take one of two textural forms: (1) relatively large, well-formed crystals or (2) relatively large crystals that become hosts to plagioclase laths as they grow, generating an ophitic relationship. The interpretation that the relatively small well-formed hornblende crystals in Figure 20.9d are of assimilation origin is supported by the restriction of this texture along the contacts with foreign rock, vis-à-vis Figure 20.8.

Notice that the hornblende crystals in Figure 20.9d are generally not poikigranular. This suggests that initial growth in a metamorphic environment with subsequent release, such as has been inferred for the hornblende in Figure 20.9b, has not occurred. Reactive dissolution of calcic plagioclase (as interpreted for Figure 20.8c) and biotite enriches the magma in Ca, Fe, Mg, and Al, just the ingredients needed for crystallization of an additional mafic mineral such as hornblende.

This example of assimilation has suggested that new crystals of hornblende form at the expense of foreign rock. The crystals are interpreted to have formed in the foreign rock, at the rock–magma interface, and out in the proximal magma. The relationships are sketched in Figure 20.10.

The Scale of Assimilation

The interaction between magma and foreign rock is energy consuming and therefore self-arresting unless there is excess heat. As Hess (1989) puts it: "Crystallization of 1 g of melt releases about 100 calories of energy. Because 300 cal/g are needed to melt the rock, in our example, about 3 g of melt must crystallize for every gram of rock melted." This means that if assimilation is occurring by direct melting, the magmatic system will soon freeze up and have no heat for further melting of wall rock.

If magmas were initially superheated (not a popular notion), there could be an extended period of assimilation by direct melting. Alternatively, if the supply of hot magma is large, as obtained by a relatively large volume of magma or by continual or repeated flow of magma through the contact zone, a lasting heat source can be achieved.

It takes less energy to activate reactions between magma and rock in comparison with direct melting. This means that the processes involved in assimilation can remain active somewhat longer, there being time for detachment of crystals from rock–magma interfaces and for stirring them into the magma before the system runs out of energy.

Batholithic magmatic bodies, thought to have acquired compositions in part through regional bulk assimilation of crustal rock on the basis of whole-rock and trace element chemistry (Myers et al., 1984; Ague and Brimhall, 1988), probably require processes other than, or in addition to, assimilation. Regional anatexis (direct melting) of crustal rock can occur if there is a sustained heat source, not an exhaustible one linked to a single batch of crystallizing magma. It is reasonable that mantle-derived basaltic magma can be the heat source for anatectic melting in the crust (Huppert and Sparks, 1988).

With portions of the continental crust "softened up" by large-scale partial anatexis, local assimilation of "migmatite" by an intruding magma becomes more feasible from an energy point of view. Alternatively, injection of magma into a migmatitic system in a dynamic environment might lead to the development of a "contaminated" magmatic system in which assimilation played no significant role.

Assimilation is only one way "conversion" of rock to magma can be made. Assimilation as used here means that a pre-existing batch of magma is contaminated by foreign rock by reactive and melting processes. The formation of anatectic magma and subsequent extraction of this magma from a rock mass is "conversion" of rock to magma, not modification of a pre-existing magma. If anatectic melt is extracted and mixes with another magma generated somewhere else, that is more appropriately considered as magma mixing.

Disappearance of wall rock from a structural point of view has been presented as a case for assimilation (Compton, 1955). The assimilation was envisioned to have changed the composition of trondhjemitic magma to that of a heterogeneous tonalite along the margins of the pluton where stoping was active. It was noted that xenoliths are uncommon in the contaminated zone, raising the question of rates and extent of assimilation. Would the presence of abundant xenoliths indicate that assimilation was arrested, whereas their absence indicates thoroughness (facility) of assimilation processes, leaving only inhomogeneous distribution of mafic mineral phases?

Assimilation as a means of providing "room" for emplacement of large magmatic bodies is not reasonable from an energy point of view nor necessary in the face of the new concept of regional dilation related to transpressional tectonics (Tikoff and Teyssier, 1992).

Part VI

Rocks Formed by Solid-State Processes

Solid-state processes lead to deformation and/or recrystallization of pre-existing rock. Rarely is a process strictly solid state because pre-existing rocks commonly contain hydrous minerals and at least a film of intergranular water inherited from water trapped in buried sediments or introduced along fractures and faults. The presence of even a minor amount of aqueous fluid (± carbonate) significantly increases rates of deformation, chemical reaction, and recrystallization, being an essential component of most "solid-state" processes.

Metamorphism in the broad sense includes all "solid-state" processes leading to modification or complete reorganization of pre-existing rock by processes including deformation at low temperature, recrystallization at low temperature, and deformation at extremely high pressure (meteor impact).

Metamorphism in the restricted sense is conveniently limited by mineralogical reactions occurring above about 200°C, identified by natural occurrence of indicator minerals, experimental reproduction, and thermodynamic calculation. This metamorphism generates phyllites, schists, granofelses, and hornfelses that are classically recognized as metamorphic rocks, leaving the other solid-state rock-forming processes distinct from metamorphism and classifiable on the basis of pressure and temperature conditions beyond the bounds of classical metamorphism.

Cataclasites and impactites are distinct solid-state rock types even though they result from deformation, as do many classic metamorphic rocks. Cataclasites result from brittle deformation along faults mostly at low temperature, and impactites form at pressures far above those of classical metamorphism.

Rock deformed and/or recrystallized at temperatures below the mineralogically established lower limit of classic metamorphism (200°C) can be characterized as follows. Low-temperature low-strain-rate deformation can be ductile or brittle–ductile, depending on the minerals involved. Ease of recrystallization characterizes some rock bodies such as salt intrusions and ice glaciers, forming the low-temperature recrystallized rocks. Deformation at low temperature involving both brittle and ductile processes generates low-temperature deformed rocks. These are the folded and sheared sedimentary rocks so common in the shallow region of compressive tectonics. Both of these process distinctions are well established, but the terminology is not, it being employed here for convenience of presentation.

21

Metamorphic Rocks

Glacier-polished surface on tightly folded Gile Mountain Formation schist. Hartford, Vermont. Bar scale = 11 cm. (Courtesy John B. Lyons, Dartmouth College.)

The Metamorphic Regime

Metamorphic rocks are the result of **static–thermal** and **dynamothermal** recrystallization of pre-existing rock at elevated temperatures (Tracy, 1987). Recrystallization occurs within a broad field defined by pressure and temperature (Figure 21.1). The upper limit of metamorphism is in that P–T region where minerals dissolve into magma. For most rock compositions there is a range of P–T conditions over which dissolution can occur. The position of the liquidus and solidus depends on the bulk composition of material being metamorphosed, including H_2O. Thus, there is an overlap of metamorphic recrystallization and magmatic phenomena. The position of this overlap range, with respect to temperature and pressure, depends on rock composition and the availability of H_2O.

The lower limit of metamorphism is partly fixed by the prevailing geothermal gradient at a given location for the period of metamorphism. There is a region of increasing pressure at very low temperatures that cannot naturally occur (Figure 21.1)—at depth, rock cannot be cold. At low pressure, the lower limit has been defined as a generalized P–T boundary below which recrystallizations, reactions, and deformations are considered to be nonmetamorphic. This is the domain of diagenesis in buried sediments, low-temperature brittle–ductile deformation, low-temperature recrystallization, and weathering. In highly reactive rocks of complex composition, such as tufaceous graywackes, the boundary between metamorphism and diagenesis has been placed at approximately 200°C at relatively low pressure on the basis of appearance of certain diagnostic minerals that form at the expense of certain diagenetic minerals (Coombs, 1953; Winkler, 1979; Turner, 1980).

Deformation of rock in fault zones generates cataclasites and mylonites. Cataclasites form mostly below 200°C except where frictional heat is concentrated, and they are not generally recrystallized (Chapter 23). Mylonites form at higher temperatures and are characteristically recrystallized, being a special variety of metamorphic rock (Bell and Etheridge, 1973; Maher, 1987). The dynamic aspect of cataclasites is metamorphic, but the lack of much reaction and recrystallization favors a separate classification (Chapter 9).

Chemical Limits of Metamorphism

Chemical migration is facilitated by intergranular fluids. If migration occurs on no more than approximately the scale of a hand specimen, metamorphism is essentially isochemical. Such small-scale

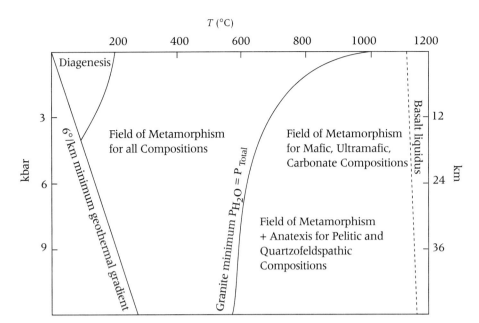

Figure 21.1
Field of metamorphism defined by temperature and confining pressure, between diagenesis and anatectic melting.

redistribution of chemical components is characteristic of metamorphism, in which segregation of some minerals with respect to others results in one type of **metamorphic differentiation**. If there is more involvement with fluids, say, beyond the scale of a hand specimen (Beach, 1980; Nishiyama, 1989, the process of chemical migration and exchange is appropriately characterized as **metasomatic metamorphism**. Overwhelming participation of aqueous fluids in which rocks are flushed with H_2O is a hydrothermal process that can also be characterized as hydrothermal metasomatism (Chapter 34).

Metamorphism may be overprinted by hydrothermal phenomena. Hydrothermal reactions occur in the same range of P–T as metamorphism. How can isochemical metamorphism and metasomatic metamorphism be distinguished from hydrothermal metasomatism? Mineral assemblages generated by hydrothermal reaction may not correspond to any igneous or sedimentary rock. For example, a rock containing mostly andradite garnet or sericitic mica has no igneous or sedimentary counterpart. Unusual mineralogical abundances are one indication that there has been major chemical exchange via large quantities of aqueous fluid.

Mineralogical distinction between isochemical metamorphism and hydrothermal metasomatism may not be possible since reactions may generate minerals unique to neither. Minerals such as quartz, K-feldspar, albite, epidote, clinozoisite, chlorite, muscovite, biotite, calcite, and dolomite can form by either process. Even an important isochemical metamorphic mineral such as andalusite has been shown to be a product of hydrothermal activity under special conditions (Figure 34.5c).

Protoliths of Metamorphism

If metamorphism is thorough, there may be no mineralogical or textural clues to the identity of the premetamorphic rock. However, if the bulk chemical composition of the metamorphic rock is comparable to the composition of some known rock, a protolith may be tracked.

A list of common protoliths and their mineral content is organized according to six chemical groups in Table 21.1. Many other less common rock types may be metamorphosed, including sulfide-bearing rocks and iron formations (Klein, 1973).

If metamorphism has not completely destroyed original texture, there is an improved chance of identifying the protolith. Features such as relict pebbles, ooids, graded bedding, phenocryst–matrix grain-size relation, magmatic crystal morphology, and fossils are useful. The first step in determining the protolith of a metamorphic rock is to identify the constituent minerals and their volume percentages. The mineralogical mode can then be converted to a quasi-chemical composition as shown in Chapter 10, followed by assignment to one of the six chemical groups (Table 21.1).

The second step is to identify relict minerals and textures. Of course, if the field relations are known, that may be very useful in identifying the protolith. Here are three examples.

(1) A rock contains 90% quartz, 8% chloritoid, and 2% graphite (Figure 21.2a). Relatively large ellipsoidal grains containing polygonalized quartz crystals have preferred dimensional orientation in the plane of schistosity. The estimated rock chemistry places this rock in the quartzofeldspathic system (Table 21.1).

Table 21.1
Chemical, Mineralogical, and Lithological Characterization of Metamorphic Protoliths

Composition Group	Chemical Content	Protolith Minerals	Protolith
Calcareous	Ca, CO_2 ($\pm$ Mg, SiO_2 H_2O)	Calcite, Dolomite, Quartz	Limestone, dolostone calc–arenite ($\pm$ siliceous)
Calcareous–argillaceous	Ca, Na, Mg, Fe, Al, SiO_2 ($\pm CO_2$, H_2O)	Calcite, Dolomite, Quartz, Kaolinite, Chlorite, Smectites, Glauconite, Biotite, Amphibole, Pyroxene, Plagioclase, Alkali feldspar, Muscovite, Illite, Albite	Calcareous mud rock, marl, argillaceous limestone/dolostone Calc–alkaline volcanic/plutonic
Argillaceous	K, Na, Al, Mg, Fe, SiO_2, H_2O	Quartz, Illite, Muscovite, Smectite, Kaolinite, Chlorite, Glauconite, Biotite	Non-calcareous mud rocks
Quartzofeldspathic	K, Na, Al, SiO_2 ($\pm$ Ca, Mg, Fe, CO_2, H_2O)	Quartz, Albite, Muscovite, Calcite, Alkali feldspar, Plagioclase, Leucite, Nepheline, Biotite, Na–pyroxene, Na–amphibole	Quartz arenite chert, arkose litharenite Alkaline volcanic/plutonic
Mafic	Ca, Mg, Fe, Al, Na, SiO_2 ($\pm H_2O$)	Plagioclase, Albite, Amphibole, Chlorite, Quartz, Na–Ca zeolites, Clinopyroxene, Orthopyroxene	Wackes, Litharenites Mafic volcanic/plutonic
Ultramafic	Mg, Fe, Ca, SiO_2 ($\pm H_2O$)	Olivine, Amphibole, Orthopyroxene, Clinopyroxene	Dunite Pyroxenite Harzburgite Lherzolite Hornblendite Peridotite

 (a) 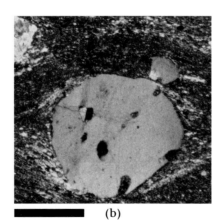 (b) 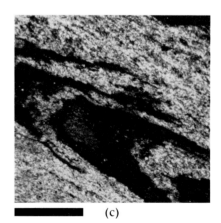 (c)

Figure 21.2
Metamorphic rocks containing relict protolith textures.

(a) Flattened and recrystallized quartz pebble in schistose matrix of quartz and chloritoid. Hardy Hill quartzite, New Hampshire. X-polars. Bar scale = 1.6 mm.
(b) Metarhyolite with schistose fabric diverging around quartz phenocryst and alkali feldspar phenocryst (upper left). Note dissolution embayments in quartz. Naciemento Range, New Mexico. X-polars. Bar scale 1 mm. (Courtesy D. McLelland.) (c) Cone-in-cone worm tube (Scalarituba) in hornfelsic greenschist. Reeve formation, Cisco Grove, California. PPL. Bar scale = 1.5 mm. (Courtesy D. Davis.)

Since no magmatic rock has such a high quartz content, a siliceous conglomerate is a likely protolith.

(2) A rock contains 30% quartz, 50% alkali feldspar, 10% muscovite, and 10% biotite (Figure 21.2b). The rock composition lies in both the argillaceous and the quartzofeldspathic groups (Table 21.1), and a number of sedimentary and felsic igneous rock protoliths are indicated. The rock has a bimodal grain-size distribution with relatively large quartz and alkali feldspar crystals occurring in a very fine-grained matrix. The quartz crystals have rounded embayments, similar to partially dissolved quartz phenocrysts in silicic volcanic rock, and the alkali feldspar phenocrysts tend to be tabular. The rock is weakly schistose, as defined by the preferred orientation of sericitic white mica. A rhyolite protolith is indicated, inferring that the larger crystals are relict phenocrysts.

(3) A fine-grained rock contains quartz, biotite, plagioclase, hornblende, and graphite. There is a weak schistosity and a curious arrangement of graphite (Figure 21.2c). Identification of the graphite forms as cone-in-cone backfillings caused by the worm Scalarituba confirms that this low-grade metamorphic rock had a sedimentary (litharenite) protolith.

Metamorphic Mineral Reactions

Objective of Determining Reactions

Chemical reactions involving minerals in metamorphic systems produce mineral assemblages that indicate the chemical and physical environmental conditions existing at the time of recrystallization. An estimation of these conditions is vital in a reconstruction of the metamorphic and tectonic history of an area.

Laboratory experimentation and thermodynamic calculation can provide such information. The experiments are simplified versions of what actually occurs, and the values of temperature, total pressure, partial pressure of aqueous and carbonate fluids, and other compositional aspects of fluids must be evaluated accordingly. Summary of such data are in Carmichael, Turner, and Verhoogan (1974), Turner (1980), and Winkler, (1979).

More recently, it has been suggested that the presence of the chloride ion in the fluid phase and extreme conditions of pH could allow metamorphic-type reactions to take place at drastically different temperatures than indicated by experimentation (Barnes, 1985). For example, although wollastonite might be expected to form at very low temperatures in the presence of solutions of specific composition, the field geologist does not find wollastonite in geological circumstance of low temperature. Either the occurrence of wollastonite has been overlooked in rocks recrystallized at low temperature, perhaps due to extremely small grain size, or these special chemical conditions do not originate in nature.

Equilibrium and Stability

The following reaction represents metamorphism of a siliceous dolostone:

$$5CaMg(CO_3)_2 + 8SiO_2 + H_2O$$
$$\text{(dolomite)} \quad \text{(quartz)}$$
$$= Ca_2Mg_5Si_8O_{22} + 3CaCO_3 + 7CO_2$$
$$\text{(tremolite)} \quad \text{(calcite)}$$

The reaction has three phases shown in equilibrium with three other phases. If the partial pressures of CO_2 and H_2O are kept equal in an experimental reproduction of the reaction, a curve representing the equilibrium can be shown (Figure 21.3) with respect to variable temperature and pressure.

A state of **equilibrium** is given at any point on the reaction curve. The curve is **univariant** in the sense that one variable, either P or T, can be changed provided the other changes in such a direction that the system maintains equilibrium (some point on the curve). To the left of any point on the curve, dolomite and quartz are the stable phases, and to the right tremolite and calcite are stable.

Tremolite and calcite are stable to the right of the equilibrium curve up to temperatures at which they participate in some other reaction to form a higher-temperature mineral assemblage. Does this mean that to the left of the reaction curve dolomite and quartz are stable until some other reaction takes place at very low temperature? The kinetics of such a reverse reaction may be so slow that a condition of **metastability** may exist "forever." This situation is common for many metamorphic mineral assemblages. In fact, most minerals in igneous and metamorphic rocks must be existing metastably at 1 atm and 25°C. However, for dolomite and quartz, stability is prevalent at the Earth's surface where quartz and dolomite readily crystallize.

Reactions Generating Metamorphic Mineral Assemblages

The following reactions have been chosen to illustrate some fundamental chemical and mineralogical relationships applying to the development of natural metamorphic mineral assemblages:

1. Kyanite → sillimanite

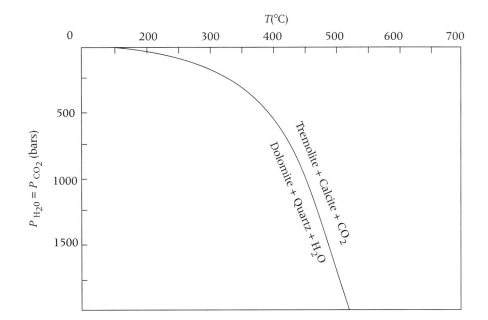

Figure 21.3
Univariant reaction curve for dolomite + quartz = tremolite + calcite. (Modified from Turner, 1968, p. 147.)

2. Muscovite + quartz → K-feldspar + andalusite + water
3. Calcite + quartz → wollastonite + carbon dioxide
4. K-feldspar + water → muscovite
5. Chlorite (Fe) + clinozoisite → almandine + chlorite (Mg) + quartz + water

Reaction 1: Kyanite → Sillimanite. The equilibrium curve for reaction 1 is shown in Figure 21.4 and extends from the aluminum silicate triple point to higher pressure and temperature. There are three instructive aspects of this reaction: (a) the uncertainty of experiment and thermodynamic calculation; (b) the slope of the curve is opposite to that of a sister reaction, andalusite → sillimanite; and (c) the textural expression of the reaction.

(a) Two triple points are plotted in Figure 21.4. They reflect the conflicting results of experimental and calculated determinations of an invariant equilibrium point such as this. It is a good reminder that application of such data to real rocks must be made with considerable caution.

(b) The kyanite–sillimanite univariant curve has a positive slope; that is, the equilibrium shifts in a similar direction with respect to changes in pressure and temperature. Conversely, the andalusite–sillimanite curve (Figure 21.4) has negative slope because kyanite has a small molar volume compared with sillimanite and sillimanite is denser than andalusite. This means that higher pressure favors kyanite in equilibrium with sillimanite, and lower pressure favors andalusite in equilibrium with sillimanite; this results in the negative slope.

(c) What is the textural expression of the kyanite → sillimanite reaction? If a rock already contained kyanite, a temperature increase, such as may be attendant to the emplacement of a magma nearby, might be expected to drive the reaction to the right. Will the kyanite recrystallize to sillimanite? Since the minerals are polymorphic forms with fixed composition, all the growth materials for sillimanite are in situ. Do neoblasts of sillimanite partly or completely occupy the sites of kyanite crystals? Microscopic observation typically reveals that sillimanite occurs with muscovite or quartz, not with the kyanite. Evidently, other reactions take place at lower temperatures before kinetic barriers to kyanite recrystallization to sillimanite can be overcome. Quartz and muscovite are typical minerals in kyanite schists. If the activation energies are right for the conversion of kyanite to sillimanite, it would be unlikely for quartz and muscovite to remain unaffected. The following reaction produces sillimanite very nicely:

$$[2K^+ + 3H_2O] + 4 \text{ kyanite} + 3 \text{ quartz} \rightarrow \text{sillimanite} + 2 \text{ muscovite} + [2H^+]$$

If $[2K^+ + 3H_2O]$ infiltrate (or potassium diffuses) along the interface between quartz and kyanite, there is likely to be growth of new crystals of sillimanite and muscovite and partial replacement of kyanite and quartz. Furthermore, if the liberated $[2H^+]$ comes in contact with original muscovite (or the new muscovite with which it is already associated), the following reaction may occur:

$$[2H^+] + 2ms \rightarrow 3qz + 3sil + [2K^+ + 3H_2O]$$

Now more sillimanite is associated with muscovite, and sillimanite also associates with quartz

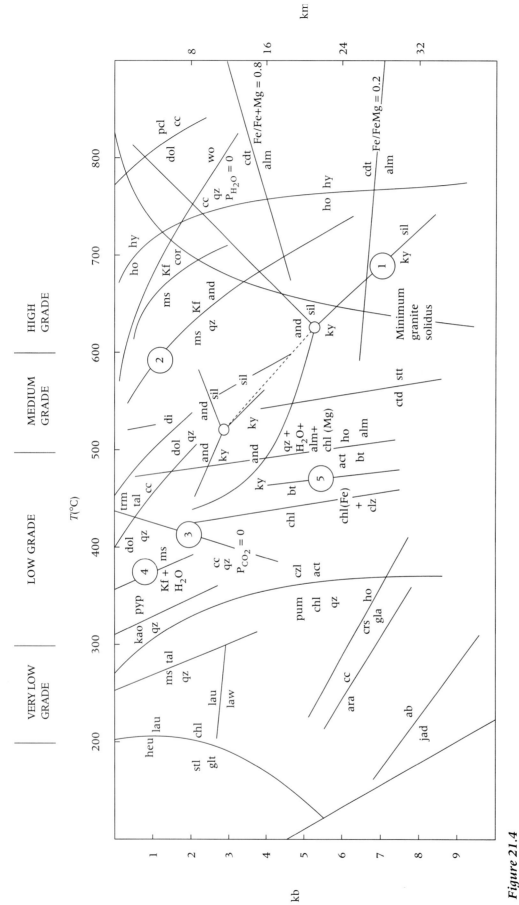

Figure 21.4
P–T field of metamorphism showing approximate location of some well-known reactions and grades of metamorphism. Reactions 1–5 referred to in text. Position of reactions 4 and 5 very approximate. (Data from Winkler, 1979; Turner, 1968, 1980.)

(Carmichael, 1969). Now there is a water and potassium source to participate in the first equation.

Similar simultaneous reactions may be involved in the following reactions:

calcite → aragonite
laumontite → lawsonite
andalusite → kyanite

Reaction 2: Muscovite + Quartz → K-feldspar + Andalusite + H_2O. Reaction 2 is a **dehydration reaction** typically occurring as rocks are recrystallized at higher temperatures, effectively driving out water tied up in hydrous minerals such as mica and chlorite. The reaction produces its own intergranular fluid even if there is no initial fluid present.

The equilibrium curve for this reaction is shown in Figure 21.4. The positive slope indicates that the forward reaction is favored by an increase in temperature *or* a decrease in pressure. This means that as reaction occurs, the partial pressure of water may increase, tending to inhibit or even arrest the reaction. If the water is free to escape the system, and that depends on other geological factors, the reaction will proceed as long as the temperature remains elevated.

The direct dissociation of muscovite occurs at higher temperature (Figure 21.4) according to the reaction

muscovite → K-feldspar + corundum + H_2O

The higher reaction temperature indicates that it takes more thermal energy to break bonds in muscovite if quartz is not there to assist. This is similar to the relative ease with which sillimanite can be formed at the expense of kyanite (reaction 1) as long as other phases, such as quartz and muscovite, are involved in the reactions.

The dehydration reaction producing pyroxene at the expense of amphibole is common in systems of appropriate composition that are being metamorphosed or remetamorphosed at high temperatures in situations allowing water to escape the system.

amphibole → pyroxene + H_2O

Reaction 3: Calcite + Quartz → Wollastonite CO_2. Calcite in association with quartz occurs in protoliths such as calcareous quartz arenites and siliceous limestones. In carbonate-bearing systems, the generation of CO_2 (or HCO_3^{-1}) as a process of decarbonatization is analogous to the production of water in dehydration reactions. In addition, since carbonate-bearing rock systems may also contain water, the ratio of CO_2/H_2O (mole fraction $CO_2/CO2 + H_2O$) becomes an additional chemical factor and can have a dramatic effect on mineral equilibria.

Since wollastonite can be identified by simple petrographic observation in a rock in which calcite and quartz have reacted, the presence of wollastonite may seem to be an excellent indicator of a certain minimum temperature reached during metamorphism. If the partial pressure of CO_2 in the intercrystalline fluid is low—that is, if the CO_2 is free to leave the system—or if the CO_2-bearing fluid is diluted with water, the reaction occurs at a certain minimum temperature, not much affected by the confining pressure (Figure 21.4). If the system is not open with respect to CO_2, and if there is no water dilution, the reaction will not take place until much higher temperatures are reached. As the partial pressure of CO_2 increases in this situation, the reaction becomes self-arresting even at the higher temperatures, but of course it will proceed if it is an open system. Thus, the presence of wollastonite per se is not a precise fix on temperature of reaction.

Suppose a reaction involves only calcite? Without participation of quartz in the reaction, the direct dissociation of calcite requires very high temperatures (Figure 21.4):

calcite → lime + CO_2

Similar decarbonatization and dissociation reactions involve dolomite (Figure 21.4):

dolomite + quartz → diopside + CO_2
dolomite → periclase + CO_2

Reaction 4: K-feldspar + H_2O → Muscovite. Reaction 4 is a **hydration reaction**. K-feldspar occurs in arkosic arenites and many igneous rocks. Most hydration reactions involve conversion of a mineral that formed at higher temperature to a lower-temperature equivalent. In general, if a protolith has a lower-temperature mineral assemblage, metamorphic recrystallization will generally produce a higher-temperature assemblage. If a higher-temperature assemblage already exists, as in igneous rocks or unaltered clasts derived from them, metamorphic recrystallization is likely to produce a lower-temperature assemblage.

If water is not available at the temperature at which this hydration reaction could take place, the K-feldspar survives as a metastable phase. As a result, the presence of K-feldspar in a "dry" metamorphic system would be useless as an temperature indicator of metamorphism. Even if the magmatic K-feldspar had recrystallized into a aggregate of new crystals, there would be no specific temperature indicated.

The same principles apply to other reactions involving hydration and destruction of originally high-temperature minerals:

forsterite olivine + H_2O → serpentine
clinopyroxene + H_2O → amphiboles
orthopyroxene + H_2O → talc
plagioclase + H_2O → clinozoisite + more sodic plagioclase

Hydration reactions commonly occur with decarbonization, as illustrated by the following:

dolomite + quartz + H_2O → talc + calcite + CO_2

Reaction 5: Chlorite (Fe) + Clinozoisite → Chlorite (Mg) + Almandine + Quartz + H_2O. Reaction 5 can be used to illustrate how some additional chemical adjustments may take place during reaction recrystallization. The production of water and retention of a chlorite indicates that there has been only partial dehydration. A similar adjustment is shown by the reaction

serpentine (hydrous) + quartz → talc (less hydrous) + H_2O

The chlorite in reaction 5 occurs on both sides of the equation. The stability of metamorphic minerals that are members of a solid solution series is controlled by the specific ratios of the substituting ions that become fixed in the crystal lattice at certain values of pressure and temperature. Chlorites, biotites, and plagioclases are good examples.

The Chl(Fe) is stable at lower temperatures than Chl(Mg) in the system shown. The smaller ionic radius of Mg^{2+} allows it to bond into higher-temperature structures. Thus, Fe/Mg ratios can be an indication of temperature of formation.

The anorthite content of plagioclase increases with metamorphic grade. This only works if there is anorthite or an "anorthite substitute" mineral such as calcite or clinozoisite available to supply the calcium for the higher An plagioclase. Albite may participate in mineralogical reactions over a very wide range of temperature as long as there is a deficiency in calcium. In that case, the sodium of albite remains fixed in albite at almost any temperature, since there is no other place for it to go unless there is OH^- available to form paragonite, or Fe and Mg are available to form sodic amphiboles or sodic pyroxenes. To sum up, if an anorthite substitute mineral is present, it may simply coexist with albite along with actinolite and chlorite in low-grade metamorphic rocks such as greenschists, but it will react with the albite with increasing temperature to form progressively more calcic plagioclase in rock such as amphibolite.

The temperature and pressure conditions favoring the coexistence of cordierite and almandine is another example of the effect of varying Fe/Mg ratio (Green, 1976). The position of the reaction curve is shown for two values of Fe/Fe + Mg in Figure 21.4. The presence of cordierite or almandine in a rock can now be said to be a function of temperature (elevated), pressure (elevated for almandine), and the Fe/Fe + Mg ratio, which extends the stability of cordierite to higher pressures.

Metamorphic Mineral Assemblages

Mineral grains comprising a fully recrystallized metamorphic rock are not likely to be the result of mineralogical reaction alone. For example, if a siliceous (cherty quartz or quartz) limestone were to be metamorphosed at relatively low P–T conditions, no reactions would take place and original calcite and quartz would simply recrystallize into new grains:

calcite + quartz → calcite (new crystals) + quartz (new crystals)

If the temperature during metamorphism were higher, wollastonite appears:

calcite + quartz → wollastonite + CO_2

What will the metamorphic rock contain if the wollastonite reaction occurs? It is very unlikely that quartz and calcite would be in exactly the proportions to produce a rock containing 100% wollastonite. In the case of a siliceous limestone, calcite will be in excess of that required to react with quartz to form wollastonite, and this excess will simply recrystallize to new crystals of calcite:

calcite + quartz → wollastonite
calcite (excess) → calcite (new crystals)

What would be indicated by the presence of all three minerals in the final metamorphic rock? One possibility (but not very likely) would be that the P–T reached corresponded exactly with the position of the equilibrium curve (for given conditions of fluid composition and partial pressure) where quartz, calcite, and wollastonite can coexist in thermodynamic equilibrium. Assuming escape of CO_2 generated, and assuming a kinetic inability for retrogressive reaction to take place as the system is cooled, the three minerals will co-exist. Of course, if such a rock is lying on a laboratory table, the reverse reaction has been precluded and the assemblage is metastable. It is more likely that co-existence of the three minerals has resulted from incomplete reaction, even though the temperature may have been well above the equilibrium curve.

Metamorphic reactions may also produce a mineral that is already in excess in the rock. For example, quartz is in excess in a mud rock containing quartz and lesser amounts of illite and chlorite. If the following reaction were to occur, new quartz will appear in the rock as a product of the reaction:

$$\text{illite} + \text{chlorite} \rightarrow \text{quartz} + \text{biotite} + \text{staurolite} + H_2O$$

Original quartz in the mud rock will simply recrystallize:

$$\text{quartz} \rightarrow \text{quartz (new crystals)}$$

The final metamorphic rock will consist of quartz from two sources, in addition to biotite, staurolite, and either illite or chlorite, depending on what was in excess. More likely, subsidiary reactions will take place in which potentially excess illite or chlorite would occur instead as muscovite (illite $\rightarrow$ muscovite) or new chlorite (chl$_1$ $\rightarrow$ chl$_2$). The distinction between quartz evolving by reaction and quartz inherited from the original sediment can be expected to be both mineralogically and texturally indistinguishable.

Metamorphism of a siliceous dolomite to a diopside-bearing metamorphic rock can be used to illustrate some additional principles relating to the production of metamorphic mineral assemblages. Consider the following reaction that might occur in a siliceous dolostone heated to an appropriate level (Figure 21.4):

$$\text{dolomite} + \text{quartz} \rightarrow \text{diopside} + 2CO_2$$

If dolomite is in excess, the final rock will consist of dolomite + diopside in a granular textural relation. Suppose the diopside crystals have relict tremolite in their cores. This mineralogical and textural variation has very important significance with respect to the metamorphic history of the rock. If the tremolite indicates incomplete reaction, an intermediate reaction may have occurred, such as

$$5 \text{ dolomite} + 8 \text{ quartz} + H_2O \rightarrow \text{tremolite} + 3 \text{ calcite} + 7CO_2$$
$$(+ \text{ dolomite excess})$$

By experiment or thermodynamic calculation, it can be shown that this reaction would have occurred at a lower temperature than that required to produce diopside. Physical introduction of an aqueous phase would be required to form tremolite since it is a hydrous mineral. Diopside could then be formed at higher temperatures via the reaction

$$\text{tremolite} + 3 \text{ calcite} \rightarrow 4 \text{ diopside} + \text{dolomite} + CO_2 + H_2O$$
$$(\text{dolomite excess maintained})$$

Another basic principle of mineral association can be illustrated here. Larnite (Ca_2SiO_4) is silica-deficient compared with wollastonite ($CaSiO_3$), as is monticellite ($CaMgSiO_4$) compared with diopside ($CaMgSi_2O_6$) (Burnham, 1959). Does an especially low content of quartz (or chert) in a siliceous limestone or siliceous dolostone favor the crystallization of the low-silica minerals? The answer is certainly no, because not one unit cell of any of these minerals will crystallize if the temperature is not high enough for their respective nuclei or "seed crystal" to form, and ratios of starting mineral phases are irrelevant.

Metamorphic Facies and Grade

Natural rock compositions may be too simple for their metamorphically recrystallized products to be readily useful as indicators of pressure and temperature. For example, recrystallization of a pure limestone, dolostone, or quartz arenite will yield new crystals of calcite, dolomite, and quartz respectively, each mineral being, by itself, stable over nearly the entire P–T range of metamorphism, and therefore not indicative of specific P–T conditions. Similarly, metamorphism of iron formations initially containing chert, jasper, quartz, hematite, and magnetite recrystallize without reaction, generating quartz–specularite–magnetite assemblages that are not indicative of what was high-grade metamorphism (Klein, 1973).

Complex chemical systems are capable of yielding diagnostic minerals or mineral assemblages that indicate a specific P–T subfield in which the metamorphism took place. The mafic, calcareous–argillaceous, and argillaceous chemical systems are particularly sensitive to changes in P–T, since many mineralogical reactions are possible.

The entire region of metamorphism as defined by P and T has been subdivided by the determination of many metamorphic reactions involving a wide variety of bulk compositions. A few of the more important reaction curves are shown in Figure 21.4, and many more reactions are listed in Table 21.2. Some of these reactions have been determined experimentally, others have been predicted on the basis of thermodynamic calculation, and still others have been proposed on the basis of observed natural mineral associations. On the basis of these reaction curves, the P–T region of

Table 21.2
Simplified Mineralogical Reactions Occurring at the Four Grades of Metamorphism for the Six Chemical Protolith Groups[1]

Calcareous System

Very low T
    cc = ara (high P)

Low T
    cc = ara (high P)
    3dol + 4qz + H_2O = tal + 3cc + $3CO_2$
    5dol + 8qz + H_2O = trm + 3cc + $7CO_2$

Medium T
    cc = ara (high P)
    cc + qz = wo + CO_2
    dol + qz + di + CO_2
    trm + 3cc + 2qz = 5di + $3CO_2$ + H_2O
    trm + cc = di + dol

High T
    cc = ara (high P)
    cc + qz = wo + CO_2
    dol = pcl + cc + CO_2
    trm + 11dol = 8fo +13cc + $9CO_2$ + H_2O
    3dol + di = 2fo + 4cc + $2CO_2$

Calcareous–Argillaceous System

Very low T
    cc = ara (high P)
    kao + cc = law (high P)
    lau + cc = prh + qz + $3H_2O$ + CO_2
    anl + qz = ab + H_2O
    kao + cc + qz + H_2O = lau + CO_2

Low T
    3law + cc = zo + CO_2 + $5H_2O$ (high P)
    pum + chl + qz = clz + act + H_2O
    dol + qz + H_2O = trm + cc + CO_2
    3dol + 4qz + H_2O = tal + 3cc + $3CO_2$

Medium
    2zo + 5cc + 3qz = 3gro + $5CO_2$ + H_2O (high P)
    prh + cc = gro + CO_2 + H_2O
    cc + qz = wo + CO_2
    dol + qz = di + CO_2
    trm + cc = di + dol
    4epi + qz = gro + 5an + $2H_2O$
    epi + cc + qz = gro + CO_2 + H_2O
    5ido + 4epi + 11qz = gro + 10di + $12H_2O$

High T
    ms + cc + 2qz = Kf + an (high P)
    cc + qz = wo + CO_2
    gro + qz = an + wo
    ms + qz = Kf + sil + H_2O

Quartzofeldspathic System

Very low T
    pl + H_2O = law + ab (high P)
    pl + H_2O = ab + lau
    pl + H_2O = ab + clz
    pl + H_2O = par + clz

Quartzofeldspathic System (continued)

Very low T
    alk + H_2O = ill + ab
    Kf + H_2O = ser

Low T
    pl + H_2O = clz + pl_s
    alk + H_2O = ms + ab
    Kf + H_2O = ms + qz

Medium T
    pl + H_2O = clz + pl_c
    alk + H_2O = ms + ab
    Kf + H_2O = ms + qz

High T
    ms + qz = Kf + ky/sil + H_2O (high P)
    ms + qz = Kf + and/sil + H_2O

Argillaceous System

Very low T
    ser + chl = bt + stl + qz + H_2O (high P)
    kao + qz = pyp + H_2O
    ill = ms
    Kf + H_2O = ms + qz
    3Kf + $2H^+$ = ms + 6qz + $2K^+$

Low T
    chl + ms + qz = alm + bt + H_2O (high P)
    chl + bt + qz = alm + bt_2 + H_2O (high P)
    pyp + chl(Fe) = cdt + qz + H_2O
    ms + chl(Fe) = bt + qz + cl(Al)
    Kf + H_2O = ms + qz

Medium T
    cht + ms = stt + bt + qz + H_2O (high P)
    ms + qz + chl(Fe) = alm + bt + ky + H_2O (high P)
    chl + qz = stt + chl(Mg) + H_2O (high P)
    chl + ms + qz = cdt + bt + and/sil + H_2O
    Kf + H_2O = ms + qz

High T
    ms + qz = Kf + ky/sil + H_2O (high P)
    stt + ms + qz = ky + alm + bt + H_2O (high P)
    bt + qz = hy + alm + Kf + H_2O
    ms + qz = Kf + and/sil + H_2O
    bt + and/sil = Kf + cdt + cor + H_2O
    bt + qz = Kf + cdt + hy + H_2O

Mafic System

Very low T
    clz + cl(Al) + qz + H_2O = law + chl (high P)
    act + mag + ab = gla/cro (high P)
    chl + ab + act = gla + ep + H_2O (high P)
    prh + chl = pum + act + ep (high P)
    lau = law + 2qz + $2H_2O$ (high P)
    ab = jad + 2 qz + $2H_2O$ (high P)
    heu = lau + ab + qz
    anl + qz = ab
    pyx + pl_c + H_2O = epi + chl + ab

Table 21.2   Continued

Mafic System (continued)

Low T
  law + chl = clz + qz + chl(Al) + H_2O (high P)
  pyx + pl_c + H_2O = alm + ho + chl + ep + pl_s (high P)
  pum + chl + qz = clz + act + H_2O (high P)
  prh + cl + qz = clz + act + H_2O
  pyx + pl + H_2O = ep + chl + pl_s + ho

Medium T
  pyx + pl_c + H_2O = pl_c + ho + alm (high P)
  pyx + pl_c + H_2O = pl_c + ho

High T
  ho + qz = cpx + hy + pl_c + H_2O

Ultramafic System

Very low T
  2fo + 3 H_2O = srp + bru
  2en + 2H_2O = tal + bru

Ultramafic System (continued)

Low T
  di + tal + H_2O = trm
  5srp + 2di = trm + 6fo + 9H_2O

Medium T
  5srp = 6fo + tal + 9H_2O

High T
  9tal + 4fo = 5anp + 4ho
  trm + fo = 5en + 2di + H_2O
  anp = 7en + qz + H_2O

[1]See Table 4.1 for mineral abbreviations.

metamorphism has been subdivided into subregions known as **grades** and **facies** (Figures 21.4, 21.5). The intent has been to relate the existence of naturally occurring mineral assemblages to the approximate P–T environment in which they form (Winkler, 1970, 1979; Mason, 1978; Turner, 1980).

In an attempt to relate facies and grade to a suite of metamorphic rocks generated along "paths" of increasing T and P, the **metamorphic facies series** have been invented. An argillaceous and mafic protolith composition are shown for three series in Figure 21.5. Indicator minerals are shown for both compositions along the (1) low-pressure "contact series," (2) the pressure and temperature-sensitive "Barrovian series," and (3) the low-temperature "alpine series." The idea is that a cycle of metamorphism affects protoliths that are part of a developing rock mass that is affected by varying conditions of P and T through time. Since orogenic tectonic events include juxtaposition of structural blocks of differing metamorphic evolution, and since synchronous emplacement of magmas can raise temperatures at high or low pressure, interpretation of rocks in the light of the facies series concept requires knowledge of these structural and magmatic events.

Texture of Metamorphic Rocks

The results of mineralogical reactions in metamorphic systems are recorded in texture (Spry, 1969; Vernon, 1975; Borradaile et al., 1982; Bard, 1986; Barker, 1990). Textural relations provide a **sense of reaction** or a **sense of nonreaction**. A sense of reaction is particularly evident if there are remnants of mineral reactants or morphological expression of their former presence.

Textures Providing Sense of Reaction

Microveins. Crystallization of a new mineral phase along a fracture or cleavage in an old phase, producing a microvein, is a clear indication of replacement and therefore of reaction. Microveins of serpentine in olivine (Plate 6c) and microveins of magnetite in serpentinized olivine (Figure 21.6a) are two such examples. Microveins containing neoblasts (Figure 21.6b) express reaction that did not involve mineralogical change. Replacement of calcic plagioclase by garnet along cleavage directions is shown in Figure 21.6c (Misch, 1964; Whitney and McLelland, 1983).

Pseudomorphs. Partial pseudomorphic replacement of garnet by chlorite (Figure 21.7) has preserved the dodecahedral form of the garnet. If reaction had gone to completion, the garnet may have been identified on the basis of the pseudomorph alone. Total pseudomorphic replacement of andalusite porphyroblasts by sericitic mica is very common. The rectangular staurolite section shown in Figure 21.37g has been completely replaced by chlorite and muscovite.

Remnant Mineral Grains. A few remnant grains of olivine occur in the nearly complete serpentinized rock shown in Plate 6c. The serpentine shown in Figure 21.6a has no relict olivine, but other patches of serpentine in the same rock do have remnant olivine grains.

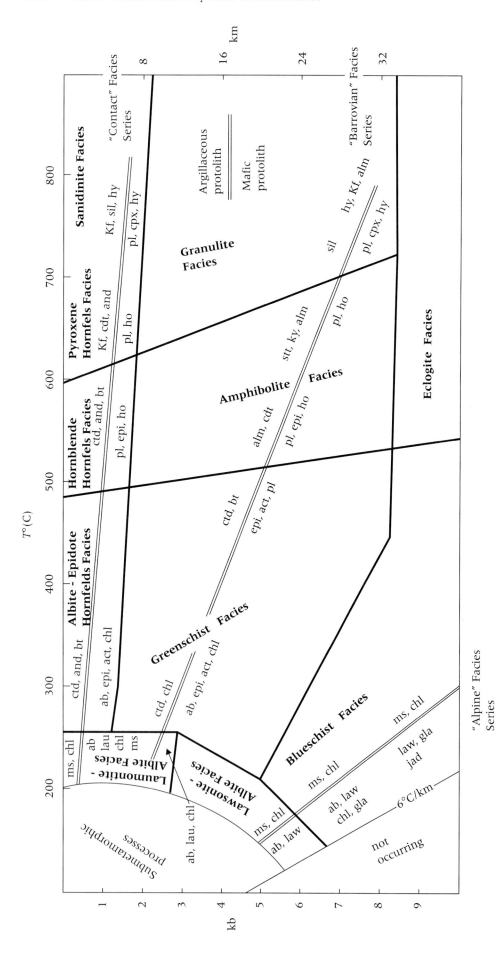

Figure 21.5
P–T field of metamorphism showing facies subfields and three facies series. (From Winkler, 1979; Turner, 1968.)

 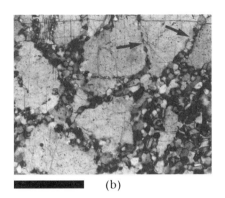 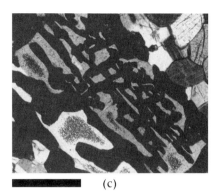

(a) (b) (c)

Figure 21.6
Microveins as indicators of sense of reaction.
(a) Magnetite veins formed in former olivine, the latter replaced by serpentine during alteration of granulite. Mantling of olivine by clinopyroxene (small arrow) followed by second mantle of garnet (large arrow) during granulite facies metamorphism (Whitney and McLelland, 1983). Adirondack Mountains, New York State. PPL. Bar scale = 1.8 mm. (b) "Microveins" (arrows) of recrystallized calcic plagioclase in relict magmatic calcic plagioclase. X-polars. Bar scale = 0.2 mm. (c) "Microveins" of garnet resulting from replacement of calcic plagioclase along cleavage directions. Same source as (a). X-polars. Bar scale = 0.2 mm.

The occurrence of one or two small grains of staurolite or andalusite in one thin section of a sericitized pelitic hornfels is not uncommon.

Mantles. Mantling texture is an including relation in which the included mineral is prominent, not remnant. Mantling in metamorphism results from reaction between an earlier mineral phase and adjacent minerals or fluids, producing a new phase. A cordierite mantle on almandine garnet is shown in Figure 21.8. The serpentinized olivine shown in Figure 21.6a has two mantles. The inner mantle is clinopyroxene. The outer mantle of garnet resulted from reaction along the clinopyroxene–plagioclase interface (Whitney and McLelland, 1983).

Reaction with a fluid phase generating a new mineral phase more or less independent of contiguous mineral grains is shown by the chlorite–garnet relation (Figure 21.7). Mantling can also originate by replacement of the interior of single crystals. Replacement of calcic cores of magmatic plagioclase related to deuteric or hydrothermal activity generates an epidote core "mantled" with remaining more sodic plagioclase.

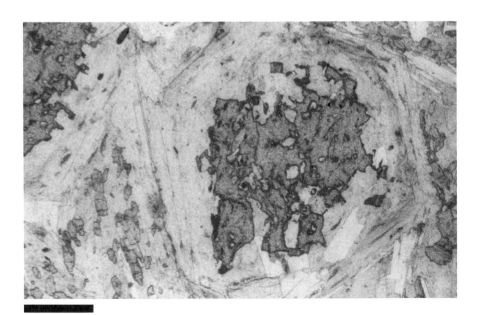

Figure 21.7
Sense of reaction indicated by partial pseudomorphic replacement of garnet by chlorite, outlining dodecahedral morphology. Blueschist of Franciscan Formation, California. PPL. Bar scale = 0.5 mm.

288   Part 6   Rocks Formed by Solid–State Processes

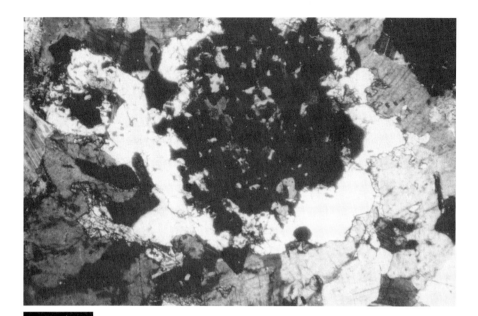

Figure 21.8
Sense of reaction indicated by mantling of almandine garnet by cordierite. Northern Cascades, Washington State. X-polars. Bar scale = 0.5 mm. (Courtesy J. Vance.)

Poikiloblastic and Porphyroblastic Texture. Poikiloblastic and porphyroblastic texture give a sense of reaction because reactive replacement must have taken place in order for the new crystal to grow. The hypersthene–cordierite hornfels (Figure 21.9a) has poikiloblastic hypersthene with quartz inclusions and poikiloblastic cordierite with quartz and biotite inclusions. Presuming that the protolith contained biotite or its equivalent, chlorite + muscovite, lack of biotite in hypersthene suggests that biotite was consumed in the reaction biotite → hypersthene + K^+, leaving excess quartz. The reaction generating poikiloblastic cordierite is less clear. In some cases, there may be textural evidence of mineral consumption beyond the

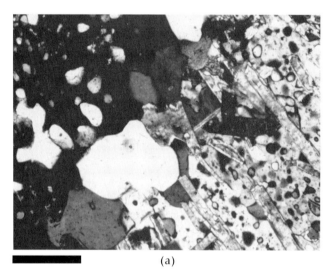
(a)

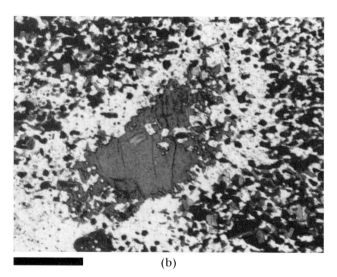
(b)

Figure 21.9
Including relations as indicator of sense of reaction.

(a) Poikiloblastic hypersthene (upper left) with residual inclusions of quartz, and poikiloblastic cordierite (lower right) with residual inclusions of quartz and boitite. Bushveld complex hornfels, South Africa. X-polars. Bar scale = 0.14 mm.
(b) Partly poikiloblastic horblende with included plagioclase. Absence of biotite in hornblende and in peripheral fine granular plagioclase zone indicates biotite was consumed in reaction generating hornblende. Hornfels. Okanogan Range, Washington State. PPL. Bar scale = 0.6 mm.

 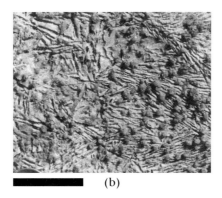 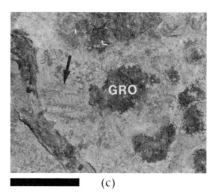

Figure 21.10
Textures indicative of static thermal metamorphic recrystallization.
(a) Randomly oriented tapered prisms of lawsonite in very low-grade metamorphosed litharenite. New Zealand. X-polars. Bar scale = 0.2 mm. (b) Random orientation of plumose groups of tremolite crystals associated with calcite (negative relief) and grossularite (positive relief). Notch Peak contact hornfels, House Range, Utah. Bar scale = 3.5 cm. (c) Trilobite cast (black arrow, left of center) in marble containing grossularite porphyroblasts (dark patches, GRO). Bar scale = 2.5 cm. Same location as (b).

physical limits of a new mineral. The porphyroblastic–poikiloblastic hornblende shown in Figure 21.9b indicates the reaction biotite + plagioclase → hornblende + Na^+. There is a biotite-free zone around the hornblende, indicating consumption of biotite (probably via a fluid phase) beyond the precise site of hornblende crystallization.

Porphyroblasts and poikiloblasts of garnet and staurolite are common in metamorphosed pelitic rocks (Figures 21.32a, 21.35a, 21.37a).

Textures Without Sense of Reaction

Polygonal Mosaic and Side-by-Side Granular Textures.
Polygonal mosaics (or "foam," "comb" texture) represent a low-energy, strain-free configuration (Chapter 16) in which there is no sense of mineralogical reaction (Figure 16.8d). Quartz (Figure 16.8d) and feldspar (Figure 21.6b), or quartz with feldspar, commonly form polygonal mosaics in higher-grade metamorphic rocks, although pyroxene and hornblende can participate in polygonal mosaics as well. Minerals such as muscovite and biotite rarely polygonalize, but tend to maintain their crystallographic expression with a side-by-side granular relation to minerals that are polygonalized (Figure 8.9c). The textural assemblage still represents both mineralogical and textural equilibrium.

An interleaving side-by-side textural relation between muscovite and biotite (Plate 6a) does not present a clear sense of reaction. There is no textural reason indicating that muscovite formed at the expense of biotite nor that biotite formed at the expense of muscovite. In cases where the interleaving grades to a mantled relation, such as the chlorite–biotite relation shown in Plate 6b, the reaction biotite → chlorite is indicated because the chlorite tends to mantle the biotite. Although mantling, per se, does not confirm reaction sense, fluid access to biotite, required for the chlorite reaction, is easiest around the rim of the crystal and along cleavage. From a strictly mineralogical point of view biotite → chlorite is a well-known retrogressive metamorphic reaction, and chlorite → biotite is a well-known progressive reaction.

Recrystallization and Physical Environment

Metamorphic recrystallization takes place in rock that is heated and in rock that is deformed as it is heated. Mineralogical assemblages are primarily determined by composition and level of heating, whereas the textural–structural character of metamorphic rock depends on whether the recrystallization was static thermal or dynamothermal.

Static (Annealing) Recrystallization

Metamorphic recrystallization in a static environment typically results in more or less random orientation of dimensional crystals. Such is the case with the lawsonite prisms that have grown in deeply buried gray-

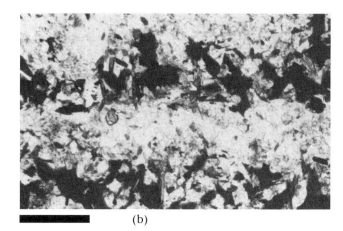

Figure 21.11
Thermal metamorphism in rock with inherited compositional layering. Dalradian of Scotland.

(a) Layered quartz–biotite schist. Bar scale = 1.5 cm. (b) Random orientation of biotite in biotite-rich layers. PPL. Bar scale = 0.7 mm.

wackes (Figure 21.10a) and the tremolite prism clusters shown in the calc–silicate hornfels of Figure 21.10b.

The sense of recrystallization under nearly perfectly static conditions is nicely exemplified by the trilobite fossil in a grossularite–diopside granofels (Figure 21.10c). Although this trilobite is a bit "under the weather," it is essentially undeformed and remained so as the calc–silicate mineral assemblage crystallized around it.

Static–thermal recrystallization does not mean that the rock will necessarily have a nondirectional fabric. Calc–silicate minerals have grown in an undeformed sedimentary rock consisting of interlayered carbonate and mud rock (Plate 7c). Preferred location of these calc–silicate minerals is controlled by compositional variation in the protolith, not as a result of metamorphic differentiation. Similarly, the layered quartz–biotite schist shown in Figure 21.11a has randomly orientated biotite crystals (Figure 21.11b) in pelitic layers that alternate with quartzitic layers. This is preferred location without preferred orientation.

Preferred orientation of crystals in a static or near-static metamorphic environment may originate by (1) mimetic recrystallization, in which crystal growth is controlled by prior grain-preferred orientation or (2) pressure solution redeposition (Figure 21.12).

Dynamothermal Recrystallization

Dynamothermal metamorphism is characterized by preferred orientation of crystals and dimensional grain shapes. Preferred orientation can originate by (1) mechanical rotation after crystal growth, (2) mechanical rotation during crystal growth, (3) pressure solution redeposition, and (4) intracrystalline glide (Figure 21.12). Preferred dimensional orientation is shown by calcite crystals that define a linear fabric (Figure 21.13a) and by micas that define a planar fabric (Figure 21.13b).

Preferred location forming in a dynamothermal environment is one expression of metamorphic differentiation. One major process of differentiation is pressure solution redeposition, shown in the formation of slaty cleavage (Figure 21.14a), crenulation cleavage (Figure 21.14b), and in schistosity (Figure 21.14c). Folding and crenulation of a prior cleavage or schistosity results in solution of quartz in the limbs of the folds or crenulations and reprecipitation of quartz in zones of reduced strain. Since pressure solution redeposition resulting in preferred location of minerals is related to strain in folding rocks, development of preferred orientation is likely to occur at the same time (Figure 21.14b).

Preferred location of minerals and textures defining planar fabric in metamorphic rocks provide marker beds in folded rocks (Figures 21.15a–d). All of these rocks have some preferred orientation of grains as well. A combination of inherited preferred location, metamorphically derived preferred location, and metamorphically derived preferred orientation is very common in metamorphic rocks.

Syndeformational Recrystallization

Recrystallization synchronous with deformation is fundamental to dynamothermal metamorphism. Although it is implicit that recrystallization takes place

continued on p. 295

21 Metamorphic Rocks 291

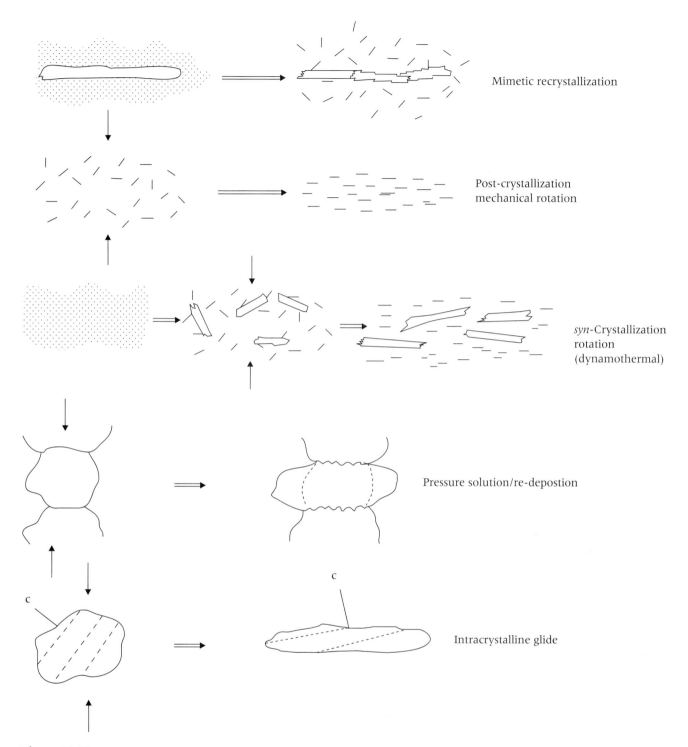

Figure 21.12
Mechanisms of preferred grain orientation in metamorphic rocks.

Figure 21.13
Preferred grain orientation in dynamothermally recrystallized metamorphic rocks.

(a) Preferred orientation of elongated calcite. Okanogan Range, Washington State. X-polars. Bar scale = 0.14 mm. (b) Preferred orientation of mica (not shown) defining continuous cleavage in phyllite. Lens cap provides scale. East of Conners Pass, Schell Creek Range, Nevada.

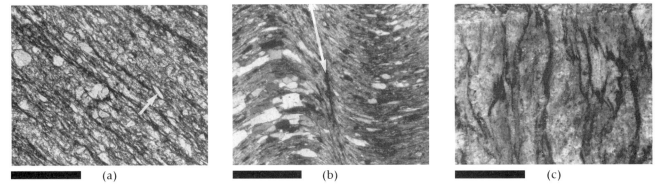

Figure 21.14
Preferred grain orientation and preferred grain location in dynamothermally recrystallized metamorphic rocks.

(a) Preferred alignment of mica grains (arrow), and concentration of micaceous minerals and graphite (dark anastomosing layers) due to pressure solution of quartz, and development of slaty cleavage in slate. PPL. Bar scale = 0.2 mm.
(b) Crenulation cleavage (arrow direction) resulting from deformation of prior schistosity. Concentration of muscovite-defining cleavage resulted from pressure solution of quartz that has reprecipitated along schistosity on either side of crenulation. X-polars. Bar scale = 0.8 mm. (c) Concentration of biotite in schistose quartzite at least partly related to pressure solution redeposition of quartz. Biotite has preferred orientation (not shown). Okanogan Range, Washington State. Slab surface. Bar scale = 1.6 cm.

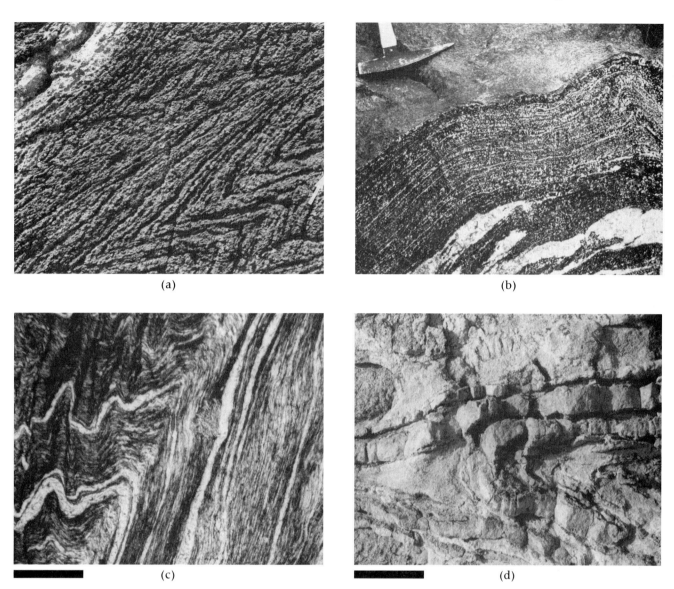

Figure 21.15
Dynamothermally recrystallized metamorphic rocks with preferred location (and preferred orientation, not shown) of minerals defining layers and folds.

(a) Quartz–tremolite interlayered with recrystallized calcite (negative weathering relief), defining chevronlike folds. Northern Snake Range décollement, Nevada/Utah. Scale is lens cap (arrow). (b) Amphibolite with plagioclase-dominating fine and coarse layers (white) alternating with hornblende (black), defining folds. West Townshend, Vermont. (c) Folding and compression of a portion of a graphite–biotite–quartz schist has resulted in pressure solution concentration of mica and graphite defining crenulation cleavage. Preferred location of quartz (white layers) in schist facilitates kinematic analysis of deformation. PPL. Bar scale = 1.2 mm. Dalradian, Sandend, Scotland. (d) Tight folds (Antler orogeny) of interlayered garnet–pyroxene calc-silicate and marble (weathered to low relief) from Sachse Monument pendant, Sierra Nevada, California. Bar scale = 4 cm. (Courtesy M. Lahren and R. Schweickert.)

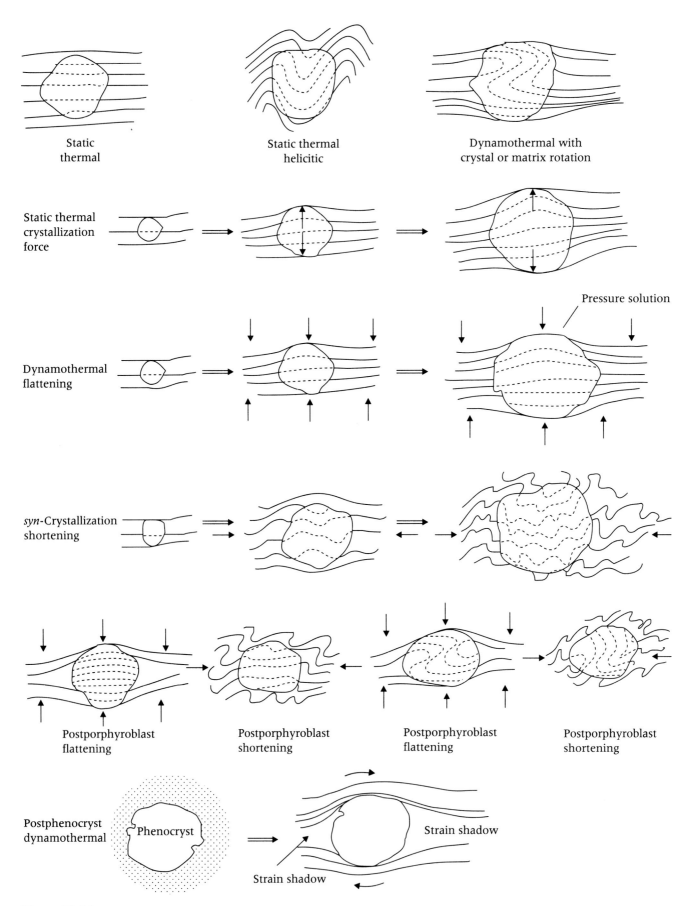

Figure 21.16
Timing of porphyroblast growth relative to deformation and inclusion of residual matrix minerals. Relict phenocryst shown for comparison.

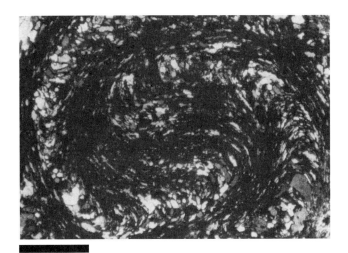

Figure 21.17
Textural evidence for dynamothermal recrystallization metamorphism. "Snowball" (spiral or sigmoidal pattern) garnet indicating either garnet or matrix rotation during growth of garnet. Proctorsville, Vermont. X-polars. Bar scale = 2 mm. (Courtesy J. Rosenfeld.)

Powell and Vernon, 1979), or rotation of the rock matrix via penetrative shearing (Lang and Dunn, 1990; Bell et al., 1986; Bell et al., 1992) had to have occurred during garnet growth (Figures 21.16, 21.17). Curvature of inclusion trains can also be generated by flattening during crystal growth (Zwart and Calon, 1977; Figure 21.16).

Another indicator of crystal growth during deformation is demonstrated by **syncrystalline microboudinage** (Figure 21.18). If a crystal is stretched to the point of rupture as it is growing, the fracture surfaces become sites of additional growth. If there is at least a slight change in fluid composition or temperature, this additional growth is optically distinct, appearing as external zoning as well as along the fracture surfaces. The crossite deformed in this way (Plate 7a) has partial overgrowth of an actinolitic blue amphibole (Misch, 1969). Similarly, zoning of omphacitic clinopyroxene in eclogite indicates stretching during growth (Figure 21.19a). Separation of biotite along cleavage may occur if the biotite crystal is oriented properly with respect to shear. Pre-extension biotite typically contains "dusty" graphite resulting from crystallization in carbonaceous pelitic rock. Separation generates potential space that may completely fill with graphite-free biotite, if biotite is still crystallizing (Figure 21.19b), or with biotite and quartz (Figure 21.37c). Stretching after crystal growth and thus generating quartz-filling extension fractures (Figure 21.19c) do not prove crystal growth exactly during deformation, but it does suggest that deformation is still active during the general period of recrystallization in the rock. Stretching during a later metamorphism is the alternative explanation of the texture.

during folding and penetrative shearing, there are limited textural indicators that any given crystal grew exactly as deformation was in progress.

Snowball texture is the most convincing evidence that crystal growth is syndeformational, at least in some circumstances. Inclusion patterns in garnets (and other minerals), typically defined by quartz and graphite, indicate that either rotation of growing porphyroblasts (Rosenfeld, 1970; Schoneveld, 1977;

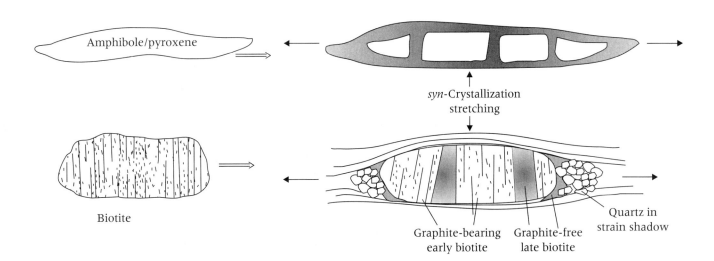

Figure 21.18
Representation of syncrystallization microboudinage in amphibole and biotite.

 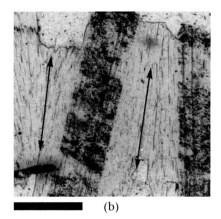 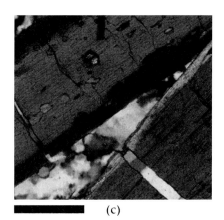

Figure 21.19
Syncrystallization microboudinage.

(a) Syncrystallization boudinage of omphacitic pyroxene, with overgrowths of sodic pyroxene of slightly different composition in fractures and exterior surfaces. Eclogite in Franciscan Formation, California. X-polars. Bar scale = 0.2 mm. (b) Graphite-bearing biotite extended perpendicular to cleavage with simultaneous epitaxial growth of graphite-free biotite. Slight rotation of cleaved plates results in mismatch of new biotite along irregular interface (arrows). PPL. Bar scale = 0.2 mm. (c) Zoned crossite with extension fractures filled with quartz derived from quartzite matrix. Ductile flow of quartz into fractures indicates mobility of quartz during deformation, but extension is postcrossite crystallization since there is no additional amphibole crystallization on fracture surfaces. Finney Creek, Northern Cascade Mountains, Washington State. X-polars. Bar scale = 0.2 mm. (Courtesy J. Vance.)

Other textures have been proposed as indicators of dynamothermal recrystallization, including "semipolygonal arcs" described by micas in schists containing microfolds (Figure 21.20), pressure solution–redeposition relationships in schists with crenulation cleavage (Figure 21.14b), and certain textures related to flattening during growth of porphyroblasts (Figure 21.16).

Semipolygonal arcs may be difficult to identify, but the concept of crystallization during folding is sound (Misch, 1969). Crystallization of mica following a pre-established fold is mimetic crystallization yielding undeformed crystals arranged to form a crude "arc" (Figure 21.20). Folding after crystallization necessarily results in bent mica crystals (Figures 21.20 and 21.21a). Syn-deformation crystallization generates slightly bent crystals that are arranged in semipolygonal fashion defining folds (Figures 21.20, 21.21b).

Sense of Shear

Deformation of rocks and minerals typically generates strain resulting from shear. The sense of shear may have textural expression (Simpson and Schmid, 1983; Simpson, 1983), and it becomes an important factor in reconstruction of tectonic events before, during, and after metamorphism.

Several grain-scale indicators of shear sense are sketched in Figure 21.22. The shear sense for sinuous inclusion trails in porphyroblasts may be difficult to determine, depending on whether the porphyroblast or the matrix rotated during growth (Bell et al., 1992). Morphology of strain shadow quartz in relation to porphyroblasts and matrix may be useful (Figure 21.22). Granular tails on porphyroblasts and metamorphic porphyroclasts are shown in Figure 21.22; p. 299. A variety of this texture are the mica "fish" (Figure 21.23a; p. 300) described by Lister and Snoke (1984). Fault-imbricated and -displaced grain fragments (Figure 21.22) are useful as well. In some cases fragments can be located some distance from their source (Figure 21.23b). Overturned marker "beds" or veins indicate sense of shear (Figure 21.23c).

Microscopic sense of shear must relate in some way to rock at the mesoscopic scale. In some cases this relation can be well documented. Crenulation cleavage has a clear sense of shear at the microscopic scale (21.14b). Crenulation cleavage in association with **shear band foliation** is shown in Figure 21.24;

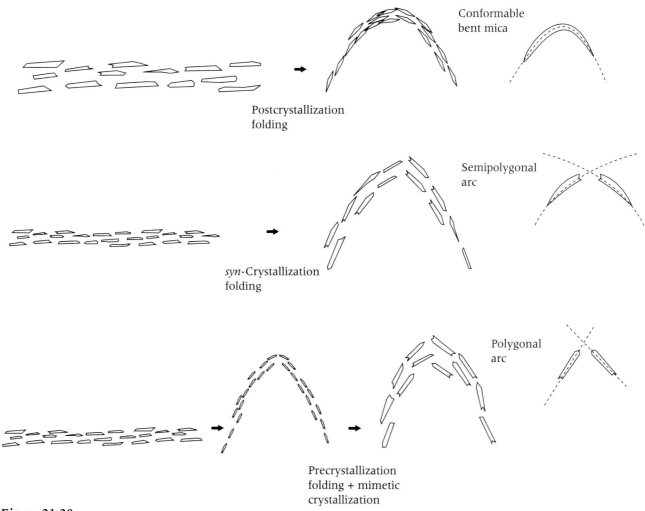

Figure 21.20
Representation of folds defined by mica in relation to timing of deformation.

p. 300. A shear "band" contains s planes marked by preferred orientation and location of minerals (Weijermars and Rondeel, 1984). The planes are tangential to c surfaces (Berthé et al., 1979) that define a spaced schistosity on the mesoscopic scale and the sense of shear. A microscopic example of the angular relation between s planes and c surfaces is shown in Figure 21.25; p. 301.

Crystallization Force or Compressive Shortening?

Porphyroblasts in dynamothermal metamorphic rocks commonly have divergence of matrix fabric in the plane of schistosity (Figure 21.26a; p. 301). Two fundamentally opposite processes could account for this textural relation: (1) Either the growing porphyroblast has exerted a **crystallization force** on its surroundings, "pushing aside" the matrix, or (2) there has been a **flattening** of the matrix with attendant pressure solution and **wrapping** of the micaceous fabric around the porphyroblast, either as it grew or after growth. Subtle patterns of inclusions may be useful in distinguishing "pushing" from "wrapping" (Misch, 1971; Fisher and Ehlers, 1982; Zwart and Calon, 1977), but the most important aspect of these relationships is whether deformation took place during or after porphyroblast growth. Proof that postcrystallization "wrapping" can occur is shown in Figure 21.26b in which relict alkali feldspar phenocrysts could not have grown during the development of planar fabric that diverges around them. Flattening associated with either pure shear or noncoaxial shear is consistent with porphyroblast growth during deformation.

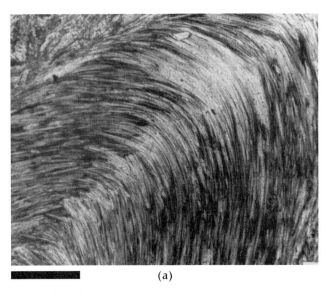
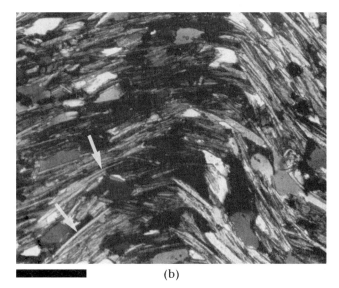

Figure 21.21
Timing of crystallization of mica defining folds relative to deformation.

(a) Postcrystallization folding (deformation) results in mica curvature conforming to fold. X-polars. Bar scale = 0.14 mm. (b) Semipolygonal arcs defined by slightly bent mica crystals (arrows) suggests crystallization during folding. X-polars. Bar scale = 0.14 mm.

Tectonites with Preferred Crystallographic Orientation

Metamorphic tectonites are characterized by preferred grain and crystallographic orientation (Turner and Weiss, 1963). In quartz-bearing rocks, such as granitic rocks and quartzites, intense deformation at elevated temperatures may induce several mechanisms of intracrystalline flow. Elongation and flattening of quartz grains in a mylonitic quartzite describe an ideal tectonite for which kinematic axes X, Y, and Z can readily be assigned (Figures 6.3, 21.27; p. 302. Gaudemer and Tapponnier, 1987). Basal glide on appropriately oriented quartz grains is a major contributor to shape change, being one mechanism by which grains acquire preferred orientation in metamorphic rocks (Figures 16.7, 21.12). In the process of dimensional preferred orientation, the c crystallographic is rotated to positions more normal to the foliation plane (see Figure 16.7).

A dominating linear fabric in a metamorphic granitic gneiss is shown in Figure 21.28; p. 303.

Metasomatic Metamorphism and Porphyroblasts

Metasomatism may lead to the formation of veins and segregations during metamorphisms. Some porphyroblasts are one expression of such metamomatism. Growth of K-feldspar in a metasomatic metamorphic setting may be represented by the porphyrograin shown in Figure 21.29; p. 303. The inclosed garnet, truncation of mica-defined schistosity, and lack of associated myrmekite point to porphyroblastic growth. Whether the K-feldspar represents dominantly in situ conversion of muscovite to K-feldspar or is the result of larger-scale potassium metasomatism is not likely to be determinable on the microscopic scale alone.

Reconstruction of Metamorphic History

Ingredients of a Metamorphic History

The character of a metamorphic rock is determined primarily by its protolith composition, thermal history, deformational history, and the degree to which fluid phases participate.

Progressive heating of rock in the metamorphic field results in reactions producing minerals that reflect that thermal rise, as long as the recrystallizing system is chemically reactive. Classic **progressive metamorphism** is detailed by Barrow (1893), Tilley (1924), Bowen (1940), and Kennedy (1949). Siliciclastic sedimentary rocks with high concentrations of

continued on p. 301

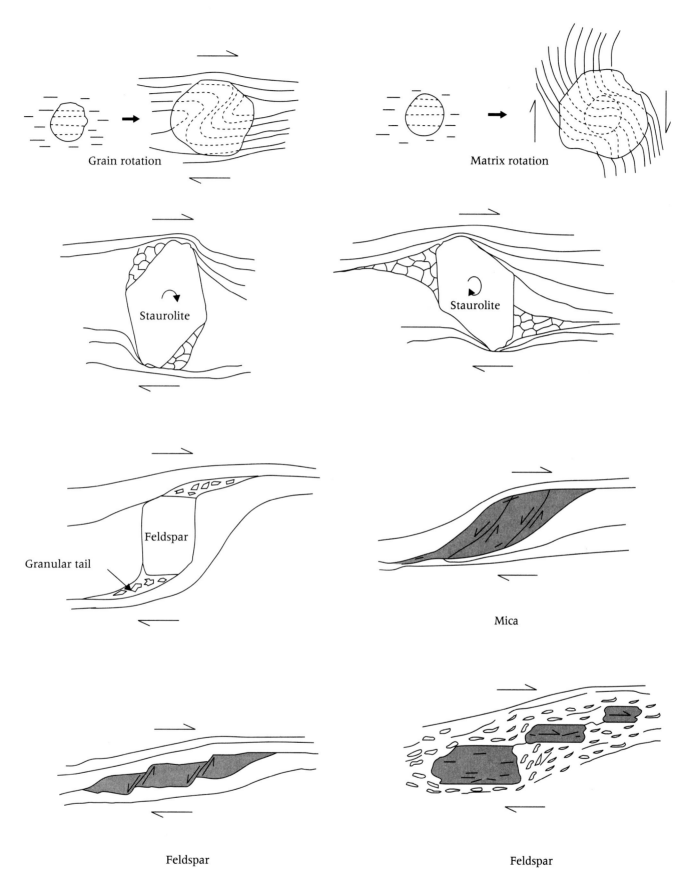

Figure 21.22; text p. 296
Representations of microscopic sense of shear indicators. (Primarily after Simpson and Schmid, 1983.)

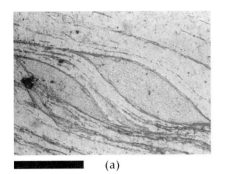 (a)
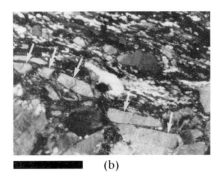 (b)
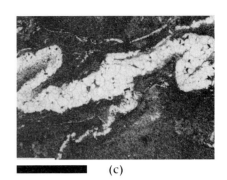 (c)

Figure 21.23; text p. 296
Sense of shear textures.

(a) Mica "fish" in mylonitic quartzite. Sinstral (left lateral) shear. Northern Snake Range décollement, Nevada/Utah. PPL. Bar scale = 0.2 mm. (b) Displaced cleavage fragments (arrows) of K-feldspar in sheared granite. In contrast, note ductile behavior of quartz (upper portion of photo). Sinstral shear. Northern Snake Range, Nevada. X-polars. Bar scale = 1 mm. (c) Folded quartzite layer in hornfelsic schist. Dextral (right lateral) shear indicated by overturned folds. Okanogan Range, Washington State. PPL. Bar scale = 2 mm.

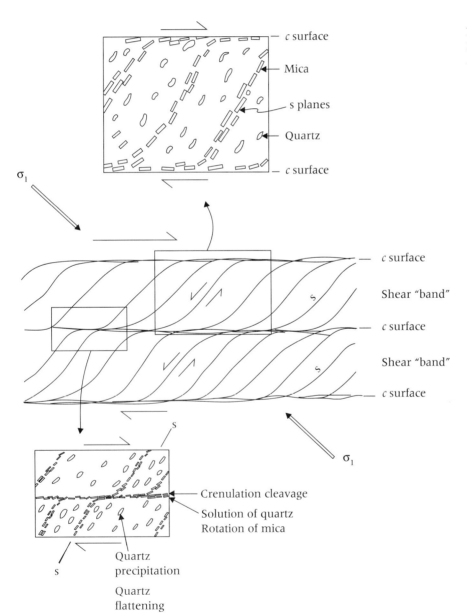

Figure 21.24; text p. 296
Representation of relationships between crenulation, shear bands, *c* surfaces, and s planes.

Figure 21.25; text p. 297
Trace of s-plane direction in quartzite layer at high angle to schistosity defined by horizontally oriented crossite crystals (see Figure 21.24). Finney Creek, Northern Cascade Mountains, Washington State. X-polars. Bar scale = 1 mm. (Courtesy J. Vance.)

hydrated minerals such as clays, micas, chlorites, and zeolites are chemically reactive. Igneous rocks are much less reactive, unless they are sheared and penetrated by aqueous fluids.

As recrystallization proceeds through the metamorphic P–T field, there is a general tendency for dehydration. If metamorphism reaches the pyroxene hornfels, granulite, or eclogite facies, anhydrous mineral assemblages are generated. Since a higher-temperature mineral assemblage is less hydrous at the same pressure level, and since recrystallization has reduced crystal strain, there is little reason for reverse

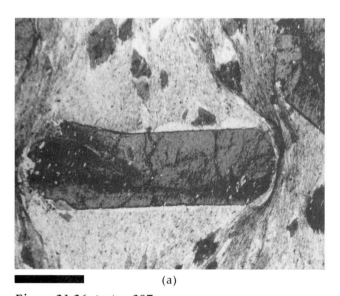
(a)

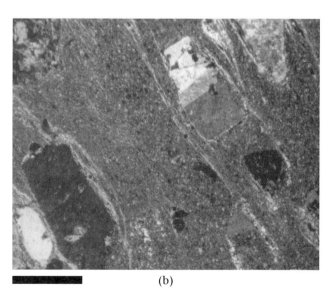
(b)

Figure 21.26; text p. 297
Crystallization force in comparison with dynamic compaction with pressure solution redeposition.

(a) Staurolite porphyroblast perpendicular to schistosity. Poststaurolite compression with accompanying pressure solution of quartz at ends of staurolite and of staurolite itself, is favored over crystallization force (see text). Burke, Vermont. PPL. Bar scale = 1.2 mm. (b) Divergence of micaceous matrix around relict alkali feldspar phenocrysts in metarhyolite cannot be due to crystallization force. Naciemento Range, New Mexico. X-polars. Bar scale = 1.2 mm. (Courtesy D. McLelland.)

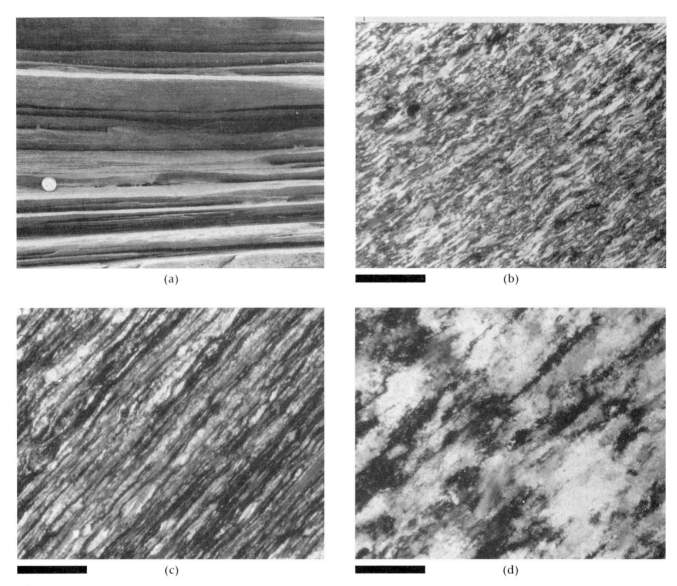

Figure 21.27; text p. 298
Dynamothermal metamorphism represented by preferred crystallographic orientation as well as grain dimension orientation.

(a) Streamworn mylonitic quartzite expressing marked continuity of planar–lenticular fabric domains. Northern Snake Range décollement, Nevada/Utah. Scale is U.S. quarter. See Figure 6.3 for crystallographic orientation of quartz. Bar scale for (b)–(d) = 0.6 mm. X-polars. (b) Section perpendicular to kinematic X, near parallel to two directions of quartz c axes (see Plate 7b). (c) Section perpendicular to kinematic Y and subparallel to quartz c axes. (d) Section perpendicular to kinematic Z and subperpendicular to two clusters of c-axis poles (see Figure 6.3).

 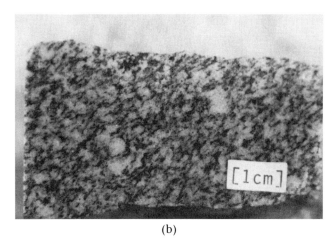

Figure 21.28; text p. 298
Dynamothermal metamorphism represented by granitic gneiss with dominating lineation. Pencil gneiss, Uxbridge, Massachusetts.

(a) Section parallel lineation marked by streaky mineral concentrations and preferred orientation of grains (not shown). (b) Section perpendicular lineation.

reactions to occur as temperature declines. The minerals remain metastable and only show the effects of **retrogressive metamorphism** if there is an episode of deformation and/or fluid introduction. This is precisely why an igneous rock or previously metamorphosed rock with an anhydrous mineral assemblage is not very reactive at initiation of metamorphism. Such rocks have formed in relatively "dry" magmatic or metamorphic systems.

The environment in which metamorphic recrystallization takes place may be static or dynamic, and it can vary as metamorphism proceeds. Dynamic environment is prominent in diapirism, ductile faulting, and in a wide variety of penetrative deformational processes associated with orogenesis. Thermal environment is prominent at magmatic contacts, in nonorogenic deep burial of sediments, and during late-orogenic regional heating related to emplacement of large magma bodies.

Synthesis of History Ingredients

The mineralogical and textural expression of thermal, dynamic, and fluid components of metamorphism can be approximated for any metamorphic rock (Zwart, 1962; Roper and Dunn, 1973; de Wit, 1976; Gillen, 1982).

A metamorphic rock may contain a record of more than one cycle of metamorphism. If the effects of the first metamorphism are completely eradicated by the second, reconstruction of metamorphic history may require observation of a suite of samples collected from an area, in which the mineralogical and structural aspects of both metamorphisms may be synthesized.

The Lewesian of Scotland is a fine example of **polycyclic metamorphism** (Figure 21.30a) on a regional scale. Although other rock-forming processes are involved, such as anatectic and injection

Figure 21.29; text p. 298
Augen-shaped K-feldspar porphyroblast in mica–feldspar–quartz schist. Note lack of myrmekite around perimeter of augen, included garnets, and termination of schistosity defined by mica within porphyroblast (arrows). X-polars. Bar scale = 1.2 mm.

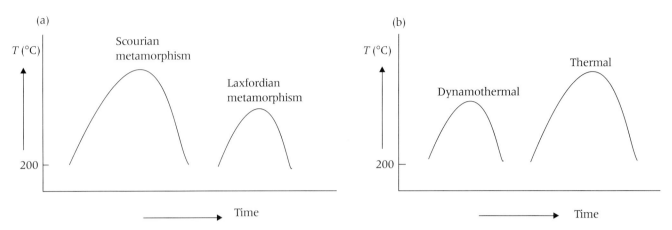

Figure 21.30
Time–temperature representation of polycyclic metamorphism.
(a) Two cycles dominated by dynamothermal metamorphism modeled after the Lewesian and Moine metamorphic terranes of Scotland. (b) Superposed thermal metamorphic overprint on earlier dynamothermally metamorphosed rocks, typical of magmatic pluton contacts in rocks of a previous metamorphism.

migmatization, the metamorphisms are fundamentally dynamothermal (Sutton and Watson, 1951, 1969; Ramsay, 1963). The earlier Scourian cycle is characterized by pyroxene-bearing facies and a characteristic structural direction. The later Laxfordian cycle is characterized by amphiboles and a different structural trend.

Static thermal metamorphism related to magmatic intrusion can involve wall rock that has had a prior metamorphic record. If the earlier metamorphism is completely unrelated to the magma emplacement, polycyclic metamorphism is indicated (Figure 21.30b).

Monocyclic metamorphism is characterized by a single cycle of thermal rise and decline. Although there may be sudden rises of temperature, as related to magma emplacement, and intensification or relaxation of dynamic components of the metamorphism, there is no major interruption of the metamorphic cycle. The following accounts of monocyclic metamorphism are primarily based on textural relations.

Static–Thermal Sequence. A single static–thermal cycle of metamorphism starting with unmetamorphosed sedimentary rock is indicated by the spotted hornfels in Figure 21.31a. The hornfels derived from a mud rock interlayered with carbonates in response to heating along the contact with a granitic stock. The matrix of the hornfels is mostly biotite and quartz without a trace of preferred orientation (Figure 21.31b). A temperature–time diagram shows the relations (Figure 21.31c).

Dynamothermal Sequence. A single cycle of dynamothermal metamorphism is represented by the garnet–mica schist in Figure 21.32a. The garnets are not porphyroblastic and are presumed to have crystallized along with the observed mica and quartz. Schistosity is presumed to have a dynamic origin, not mimetic after an earlier planar fabric. The temperature–time plot of this metamorphism is shown in Figure 21.32b.

Dynamothermal → Thermal Sequence. An early dynamothermal metamorphism followed by a late-stage thermal overprint related to magma emplacement is a common tale of regional orogenesis. The thermal stage of this metamorphism may result in a thermal "jump," but an overall single cycle of metamorphism is maintained since the magmatism is essentially a climactic phase of the orogeny.

Chiastolite porphyroblasts have grown across schistosity in the hornfelsic schist in Figure 21.33a, p. 307 as a result of late-stage contact metamorphism. Similarly, phyllitic siltstones contain randomly oriented chloritoid as a result of late-stage magmatic heating (Figure 21.33b). The relationships are summarized in the temperature–time diagram of Figure 21.33c.

Dynamothermal metamorphism may evolve into a thermal stage, without magma involvement, as the

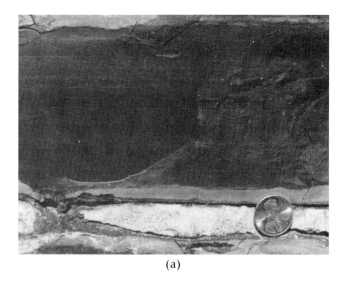
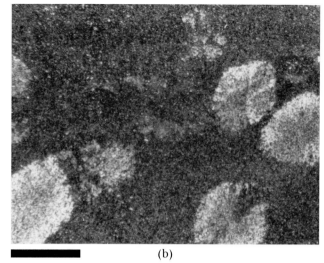
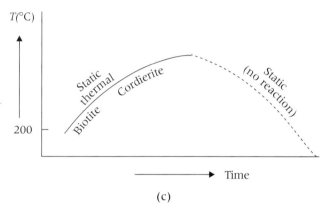

Figure 21.31
Metamorphic history: Single cycle of thermal metamorphism.

(a) Pelitic hornfels interlayered with calc–silicate hornfelses. Relict sedimentary bedding and soft sediment deformation (bottom). Notch Peak contact zone, House Range, Utah. Rock surface. (b) Hornfels shown in (a). Note cordierite poikiloblasts that have wedge-shaped twins. X-polars. Bar scale = 0.6 mm. (c) Metamorphic history plotted against temperature and time for the hornfels.

dynamic component of dynamothermal recrystallization subsides while the system is still hot and reactive. This may occur in medium- and high-temperature regional metamorphism or at low temperature during burial metamorphism (Figure 21.34a; Turner, 1968). Figure 21.34b, p. 308 applies to both low- and high-temperature environments.

Dynamothermal → Thermal → Dynamothermal Sequence. An early dynamothermal metamorphic stage can generate a planar fabric that is inherited by porphyroblasts crystallizing during a thermal stage at higher temperature. The thermal stage can either rep-

resent an episodic pause in the dynamic component or a high growth rate of the porphyroblast in relation to ongoing deformation. In either case, effectively renewed dynamothermal recrystallization can result in rotation of porphyroblast or matrix. Since the matrix consists of low-strain mica and quartz (Figure 21.35a; p. 308), recrystallization of matrix is indicated during rotation, effectively a second stage of dynamothermal metamorphism (Figure 21.35b).

Dynamothermal → Thermal → Dynamic Sequence. The schist shown in Figure 21.36a contains quartz, muscovite, chlorite, clinozoisite, graphite,

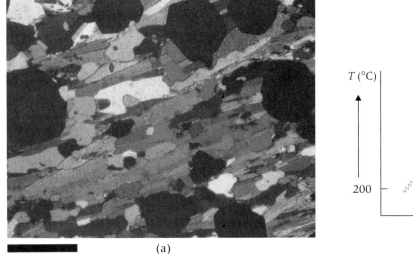
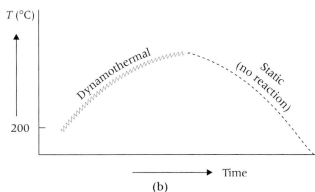

Figure 21.32
Metamorphic history: Single cycle of dynamothermal metamorphism.
(a) Almandine–muscovite–biotite–quartz schist. Littleton Schist, New Hampshire. X-polars. Bar scale = 0.6 mm. (b) Metamorphic history plotted against temperature and time for the schist.

sphene, and larger crystals of albite and lawsonite (not shown). An early stage of dynamothermal metamorphism generated a muscovite–chlorite–plagioclase schist. A second stage of thermal metamorphism generated albite and lawsonite porphyroblasts, in the same way that staurolite was generated in Figure 21.35a. Graphic inclusions in the porphyroblasts define the earlier planar fabric, as they so commonly do in porphyroblasts whose growth cannot utilize the carbon. A third stage was dominantly dynamic, occurring at reduced, perhaps even submetamorphic, temperatures. This stage is characterized by a strong penetrative shearing that rotated porphyroblasts and/or rotated (by penetrative shearing) matrix (Figure 21.36b; Johnson, 1990), as indicated by the nonalignment of graphite inclusions. Deformation without crystallization is indicated by strongly deformed (bent) muscovite and chlorite and by strained quartz (progressive extinction). Evidently, temperatures were too low for recrystallization to occur. The relationships between the various stages are shown in Figure 21.36c.

Dynamothermal → Thermal → Dynamothermal → Thermal Sequence. A staurolite–almandine schist (Gile Mountain Formation) from the East Burke area, Vermont (Woodland, 1963) has a mineral assemblage that has elegantly recorded a complex single-cycle metamorphic history. Textural relations are shown in Figure 21.37a–g, p. 310 based on the mineralogical and textural relations occurring in two specimens, and in Figure 21.37h, a summary metamorphic history.

An initial dynamothermal stage was probably characterized by mechanical rotation of crystals during recrystallization, converting shale to slaty rock according to the generalized reaction

illite + smectite + chlorite + quartz + carbon
→ illite + white mica + phengite
 + Mg–chlorite
 + quartz + graphite

The second stage was characterized by a series of progressive reactions generating graphite-bearing biotite, then almandine, followed by staurolite (which may include almandine, Figure 21.37a), according to the following generalized reactions:

quartz + muscovite + chlorite → biotite$_1$ + mobilized ions
quartz + muscovite + biotite$_1$ → almandine + chloritoid(?) + mobilized ions
muscovite + biotite$_1$ + chloritoid(?) → staurolite + mobilized ions

In each case of porphyroblast growth, there is textural evidence that either the "dynamic" component of dynamothermal metamorphism had subsided or the rate

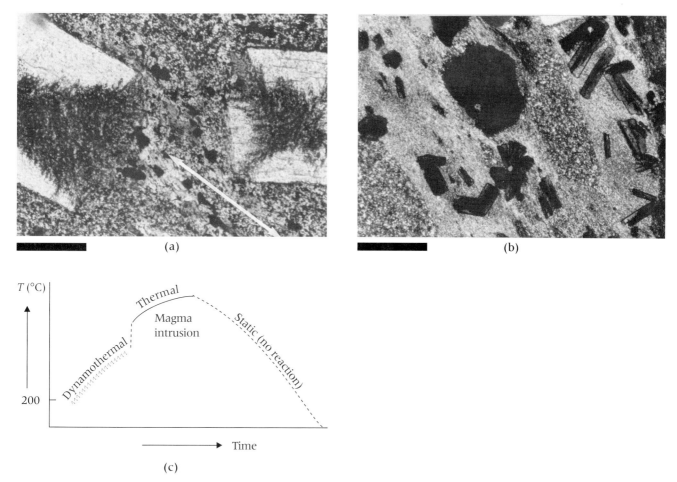

Figure 21.33; text p. 304
Metamorphic history: Single cycle of metamorphism with dynamothermal → thermal sequence.

(a) Muscovite–biotite–quartz hornflesic schist with chiastolite porphyroblasts that truncate weak schistosity (double arrow). Oslo, Norway. PPL. Bar scale = 0.2 mm. (b) Very low-grade meta quartz-rich litharenite with cleavage parallel to long axis of rock clasts. Randomly oriented chloritoid porphyroblasts have grown in the more argillaceous portions. X-polars. Bar scale = 0.6 mm. Candelaria Formation, Central Sierra Nevada. (Courtesy M. Lahren.) (c) Metamorphic history plotted against temperature and time for the hornfelsic schist and the hornfelsed metasediment.

---

of porphyroblast crystallization drastically exceeded strain rate. Biotite$_1$ porphyroblasts contain inclusion trails of graphite that define a very flat S pattern (Figure 21.37b–d) subparallel to schistosity, indicating minor biotite or matrix rotation during biotite crystallization. Graphite inclusions in almandine are radial zonal (Figure 21.37b), indicating static growth environment (Andersen, 1984). Graphite and quartz inclusion trails in staurolite define a crude direction subparallel to schistosity (Figure 21.37a), similar to biotite. Other staurolite crystals lie athwart crenulation cleavage (21.37g), clearly indicating postdeformation crystallization during this thermally dominated stage.

A second dynamothermal stage is indicated by (1) major pressure solution and reprecipitation of quartz in strain shadows (Figure 21.37a–c), (2) syn-crystallization microboudinage of biotite$_1$ forming graphite-free biotite$_2$ (Figure 21.37c), and (3) partial replacement of staurolite by biotite$_2$ (Figure 21.37d). At least partial recrystallization (grain enlargement) of quartz and muscovite occurred during this stage as

continued on p. 311

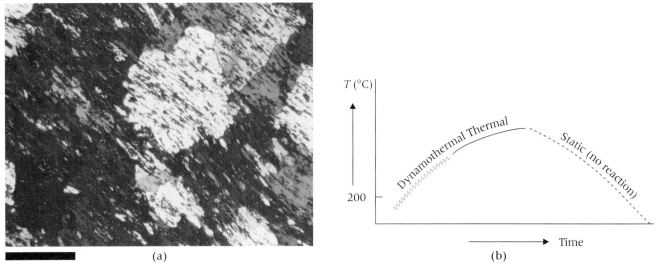

Figure 21.34; text p. 305
Metamorphic history: Single cycle of metamorphism with dynamothermal → thermal sequence.

(a) Very low-grade metatuff with planar fabric defined by elongate epidote and chlorite grains. Planar fabric parallel bedding may have been enhanced by pressure solution redeposition. Nonoriented albite poikiloblasts include the grains defining the earlier directional fabric. "Burial metamorphism," New Zealand. X-polars. Bar scale = 0.6 mm. (b) Metamorphic history plotted against temperature and time for the schist.

Figure 21.35; text p. 305
Metamorphic history: Single cycle of metamorphism with dynamothermal → thermal → dynamothermal sequence.

(a) Staurolite poikiloblast includes quartz and graphite (not shown) that define early directional fabric (double arrow). Staurolite crystallized across this early schistosity during a "thermal" stage attendant to either (1) reduced strain rate or (2) increased growth rate (staurolite). Subsequent rotation of staurolite (or rotation of matrix around staurolite) was accompanied by thorough recrystallization of matrix quartz and mica defining new schistosity (larger double arrow) in which the long axis of the staurolite lies. X-polars. Bar scale = 0.6 mm. (b) Metamorphic history plotted against temperature and time for the staurolite schist.

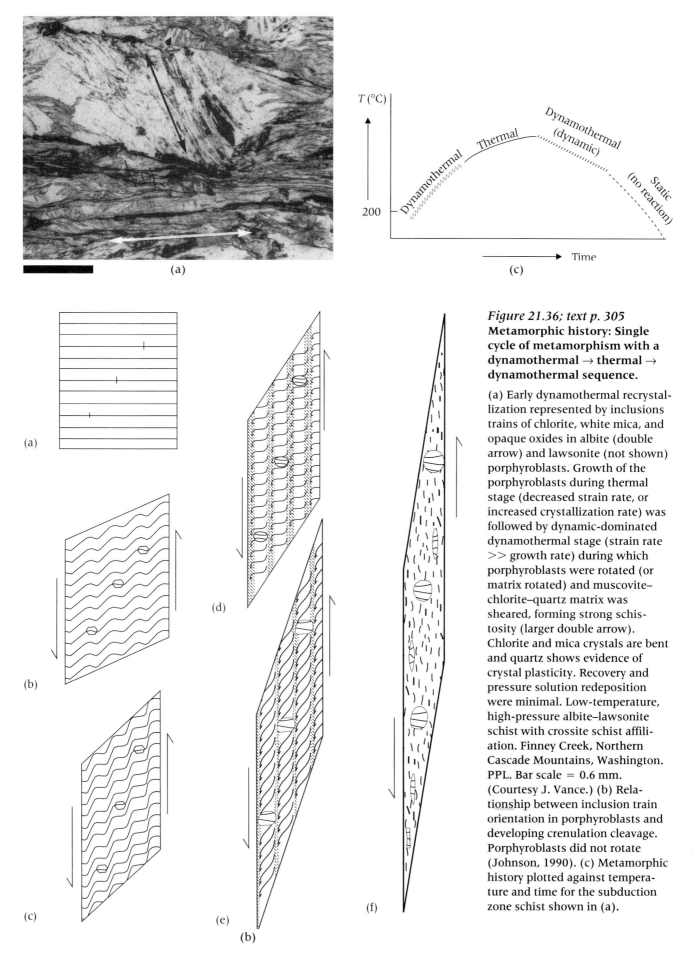

Figure 21.36; text p. 305
Metamorphic history: Single cycle of metamorphism with a dynamothermal → thermal → dynamothermal sequence.

(a) Early dynamothermal recrystallization represented by inclusions trains of chlorite, white mica, and opaque oxides in albite (double arrow) and lawsonite (not shown) porphyroblasts. Growth of the porphyroblasts during thermal stage (decreased strain rate, or increased crystallization rate) was followed by dynamic-dominated dynamothermal stage (strain rate >> growth rate) during which porphyroblasts were rotated (or matrix rotated) and muscovite–chlorite–quartz matrix was sheared, forming strong schistosity (larger double arrow). Chlorite and mica crystals are bent and quartz shows evidence of crystal plasticity. Recovery and pressure solution redeposition were minimal. Low-temperature, high-pressure albite–lawsonite schist with crossite schist affiliation. Finney Creek, Northern Cascade Mountains, Washington. PPL. Bar scale = 0.6 mm. (Courtesy J. Vance.) (b) Relationship between inclusion train orientation in porphyroblasts and developing crenulation cleavage. Porphyroblasts did not rotate (Johnson, 1990). (c) Metamorphic history plotted against temperature and time for the subduction zone schist shown in (a).

309

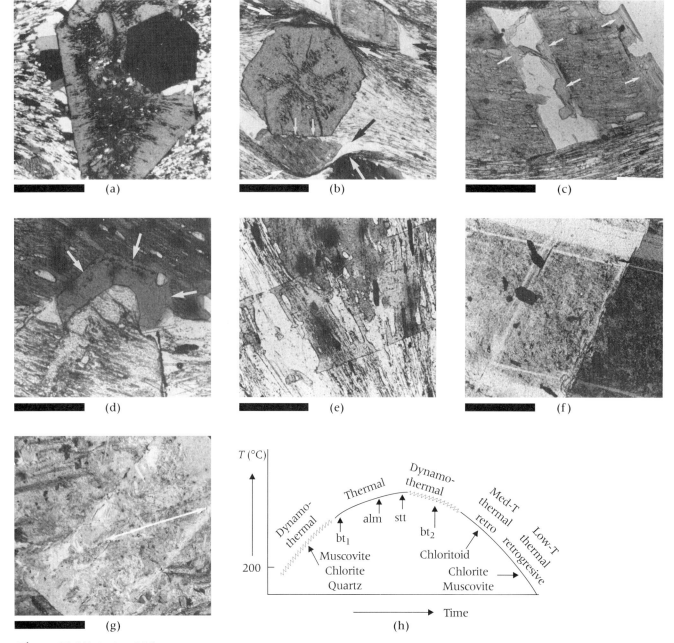

Figure 21.37; text p. 306
Metamorphic history: Single cycle of metamorphism with a dynamothermal–thermal–dynamothermal–high-temperature thermal retrogression–low-temperature thermal retrogression sequence. Gile Mountain Formation, Burke, Vermont (see text).

(a) Staurolite porphyroblast with partially included almandine porphyroblast in schistose matrix consisting mostly of muscovite and quartz. Note quartz in strain shadow beside rotated staurolite. X-polars. Bar scale = 0.8 mm. (b) Almandine and biotite porphyroblasts in schistose muscovite–quartz matrix. Note quartz in strain shadows associated with biotite (black arrows), quartz-free mica-rich zones resulting from pressure solution (white arrows), and probable dissolution surface in biotite compressed by almandine (small white arrows in garnet). PPL. Bar scale = 1.8 mm. (c) Biotite porphyroblast containing graphite-defining early-stage schistosity that has been extended perpendicular to cleavage. Simultaneous epitaxial growth of graphite-free biotite (white arrows) accompanied by quartz in the strain shadows. PPL. Bar scale = 0.2 mm. (d) Partial replacement of staurolite by graphite-free biotite (white arrow). Staurolite had partially replaced graphite-bearing biotite (top center) at an earlier thermal stage. PPL. Bar scale = 0.2 mm. (e) Medium-temperature retrogressive chloritoid (lower left to upper right) as skeletal replacement of biotite porphyroblast and matrix muscovite. PPL. Bar scale = 0.2 mm. (f) Chloritoid (light, left) replacement of graphite-bearing biotite with clear biotite zones (right) preserving the graphite and graphite-free distribution that formed during syncrystallization extension shown in (c). PPL. Bar scale = 0.2 mm. (g) Staurolite pseudomorphed by chlorite (light gray) and muscovite (white) had previously grown across crenulation cleavage. Inclusions trains in staurolite align (double arrow) with well-developed crenulations of matrix. PPL. Bar scale = 0.8 mm. (h) Metamorphic history shown in relation to time and temperature.

well. The replacement of staurolite by biotite$_2$ indicates thermal decline.

A second thermal stage is characterized by chloritoid crystallization in an environment that was completely static. Some chloritoid crystals lie across schistosity (Figure 21.37e); others that happen to lie across biotite have preserved both graphite-bearing biotite$_1$ and graphite-free biotite$_2$ (Figure 21.37f). This thermal stage is retrogressive in the sense that a lower-temperature mineral, chloritoid, has formed at the expense of biotite.

A third thermal stage consists of low-temperature retrogression, converting some almandine (not shown) and staurolite to chlorite and muscovite (Figure 21.37g). A summary of the metamorphic history is shown in Figure 21.37h.

Low-Temperature Recrystallized Rocks

Gornergletscher (foreground) with tributary glaciers issuing from Monte Rosa region. Zermatt, Switzerland (1971).

Ice, halite, and anhydrite are three important rock-forming minerals that recrystallize readily at low temperatures under conditions of low strain rate associated with co-axial compression or non-co-axial shear. Deformation mechanisms in glacier ice are ductile and dominated by recrystallization. Deformation mechanisms in salt diapirs and tectonic evaporites range from brittle to ductile, depending mainly on confining pressure. Recrystallization obscures or obliterates evidence of specific deformation mechanisms in both ice and evaporite-bearing rocks.

Recrystallization of carbonate rocks can also take place at low temperatures if strain energies are very high (Figure 23.1c). Recrystallization without mineralogical change also takes place during diagenesis, but in this case participation of a fluid phase is characteristic.

Glacial Ice

Transformation of Snow to Ice

Ice is a very common rock. Snow accumulates and soon transforms into ice, comparable to the precipitation of crystals in water bodies that accumulate and become certain sedimentary rocks. Ice that is converted into a glacier is subject to stress, deforming and recrystallizing very similar to the processes that change sedimentary rocks into metamorphic rocks.

Accumulation of snow produces bedded deposits to which the law of stratigraphic succession applies—deep ice is old ice. Pressure increases with depth, and since the ice is not perfectly confined, a deviatoric stress is generated. As a result, ice flows and we call it glacial ice. A section through an ice sheet is likely to be relatively simple compared with a section through ice confined in a valley, in which flow is complicated by the restricted nature of the valley and the joining of tributary ice with main valley ice. Nevertheless, rough bed topography beneath an ice sheet can be reflected in ice structure.

The conversion of snow to ice in the uppermost region of a glacier is a process with close parallel to diagenesis in sedimentary rocks. There are two categories of conversion: one is in polar regions where there is no freezing and thawing; the other involves diurnal generation of meltwater during the warm season on temperate glaciers.

Conversion of Dry Snow to Ice. Sedimentation of snow begins in the air. Snow crystals typically are delicate dendritic hexagonal plates. Mechanical breakage of delicate forms begins in their fall through air but becomes more active as crystals accumulate and are subject to gravity compaction.

Within an accumulation of snow, several additional processes now operate that result in the reduction of air spaces and change in the size and shape of snow crystals in dry snow. Snow is not far removed from its melting point, and as such it is in a relatively high energy state. Direct evaporation of a solid tends to reduce the size of crystals with large surface area, simultaneously creating **depth hoar** crystal forms in cells and on irregular surfaces of nearby crystals, thereby reducing surface area and free energy. Water vapor migrates upward into cold snow and then precipitates. This means that a temperature gradient has been established with warming downward from the surface. Below 10 m the temperature gradient is almost completely damped, and processes of snow conversion to ice operate isothermally.

There is also reduction of surface free energy by diffusion of water molecules over the surface of ice crystals, leading to conversion of cellular forms to noncellular grains. This reduction is particularly significant where the pressure at crystal contacts generates a surface film of water molecules that diffuse to sites of lower pressure, filling in high-energy configurations of crystals. Volume diffusion (intracrystalline diffusion) also begins as pressure is increased on the buried crystals and growth dislocations begin to move.

The gradual energy-reducing, stress-reducing change from flakes to grains reduces porosity. Rounded noncellular grains can now pack closer together and porosity is further reduced. Grain contacts now are particularly susceptible to bonding, a process to which the term **sintering** has been applied. Pore-filling precipitations, pressure solution reprecipitation, and bonding between grains are similar to diagenetic processes that make sedimentary rocks.

With further burial and increase in pressure, there is an increase of intragranular movements and pressure solution redeposition. There occurs some rupture of the intergranular bonds that have just formed. All of these processes facilitate even closer packing and reduction of porosity in the top 10 m or so of the snow pack. A **spongy cellular** ice is generated (Figure 22.1) that has a relative density of about 0.5.

With continuation of these processes and partial expulsion of air to the surface, a **sieve ice** is formed in which there is still significant porosity but marked reduction of their interconnections and of permeability. Transition from sieve ice to **bubbly ice**, in which permeability is lost, is shown in Figure 22.2. Below about 65 m in the ice sheet at Camp Century (Gow, 1969, 1971, 1975) and below about 115 m at the South Pole (Gow, personnel communication), there is a marked reduction in air spaces in bubbly ice (Figure 22.3). At this stage, air bubbles are completely trapped within the ice and now can only get smaller by being compressed under the ever-increasing overburden pressure. At much greater depths and pressures the bubbles disappear by diffusion of the gas molecules into the ice.

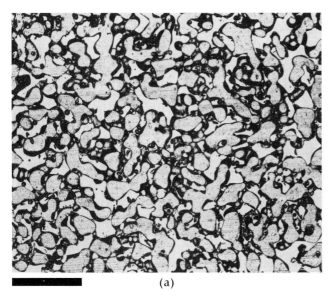
(a)

(b)

Figure 22.1
Spongy cellular grain structure in 12-year-old firn at 9 m depth having a density of 0.53 g cm^{-3}. Camp Century, Greenland. Bar scale = 4.3 mm. (Courtesy A. Gow (CRREL).)

(a) Uncrossed polars showing aniline impregnation (white), trapped air (black with white mottling), and ice (light gray speckled). (b) Same with X-polars. Ice grains typically are bicrystalline.

Conversion of Wet Snow to Ice. Snow that partly melts at the surface but survives one summer without being transformed into glacial ice is known as **firn**. With local melting there is an increased rate of rounding. With rounding there is tighter packing since grains can move together with fewer physical hangups. In addition, surface tension of water tends to pull grains together, further increasing packing and reduction of porosity. Porosity is also reduced by the filling of spaces with meltwater. Air escapes to the surface until trapped at the sieve ice stage at which pore interconnections are eliminated and permeability is nonexistent. Grain size increases by default as small grains are eliminated by melting. Actual increase in grain size occurs as grains are enlarged by growth on single crystals and by sintering of grains by refreezing meltwater in the daily cycle. Thus, grains consist of single crystals or several crystals in spongy cellular ice.

Below the zone of thawing and refreezing, recrystallization in firn is controlled by the same intercrystalline and intracrystalline processes that affect the transition zone between snow and ice in dry polar regions. There is a further increase in grain size and reduction in cellular crystal forms as energy-reducing mechanisms. Below about 10 m, polycrystalline grains have been eliminated because grain coalescence and porosity reduction has advanced to the stage at which definition of polycrystalline grains is lost. The ice is now an aggregate of single crystals sieved with isolated air spaces. At the stage the firn transforms into bubbly ice its porosity has been reduced to about 10%. Crystal sizes at the firn–ice transition zone are on the order of 1–4 mm in diameter, with a mean cross-sectional area of about 3 mm^2.

Deformation and Recrystallization of Deep Ice

Deformation of glacial ice occurs at depth as the weight of the overlying ice mass exceeds the strength of the ice. The driving force of ice flow is therefore gravity, and flow occurs because the ice is not confined within a ridged container. Glaciers are natural laboratories for observation of relatively simple ductile flow phenomena that has more complex analogs in dynamothermal metamorphic rock systems in which a variety of silicate minerals participate. However, in glaciers the main deformational driving force is gravity, and only the mineral ice is involved. In addition, strain rates are relatively high compared with those in silicate metamorphic systems.

The deep region below the firn-to-ice transition is characterized by (1) compression of air bubbles and their eventual elimination, (2) further reduction of energy in crystal aggregates by way of recrystal-

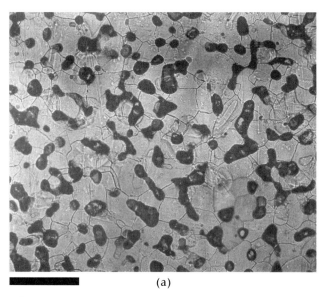
(a)

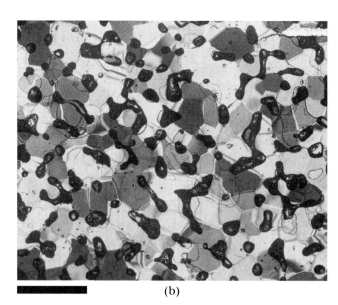
(b)

Figure 22.2
Transitional texture between sieved ice and bubbly ice in 118-year-old ice at 65 m having density of 0.82 g/cm^{-3}. Camp Century, Greenland. Bar scale = 4.3 mm. (Courtesy A. Gow [CRREL].)

(a) Uncrossed polars showing about 10% air bubbles. (b) Same with X-polars showing polygonal ice mosaic.

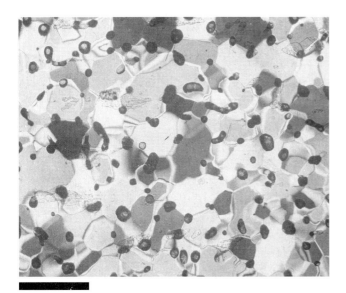

Figure 22.3
About 175-year-old bubbly ice at 90 m depth. Entrapped air bubbles are fewer, smaller, and rounder compared with 65 m ice in Figure 22.2. X-polars showing well-developed polygonal ice mosaic of ice crystals. Camp Century, Greenland. Bar scale = 4.3 mm. (Courtesy A. Gow [CRREL].)

Grain size generally increases with depth. The region near the base of the ice is warmer than within the ice above. This thermal gradient is an energy drive that results in further recrystallization and enlargement of crystals. Transitional textures are characterized by local crystal enlargement, producing larger crystals within aggregates of smaller crystals, analogous to porphyroblasts in silicate mineral metamorphic rocks. Polygonalization is not characteristic of these larger crystals. Evidently, deformation attendant to flow of the ice, including the effects of drag against the rock–ice bottom contact, tends to keep crystals in a relatively high energy state. Progressive extinction indicates lattice strain. Interpenetrating interfaces are more common than polygonal mosaics, even though crystal size has increased (Figure 22.4). At 1200–1300 m depth at Byrd Station, there is a reduction of grain size (Gow and Williamson, 1976). Evidently, localized deformation has resulted in recrystallization of larger grains into smaller grains in a low enough energy state to keep from enlarging by further recrystallization.

The movement picture in glacial ice is not as simple as it may seem in view of the monomineralic nature of the rock stressed in response to gravity alone. In particular, valley glaciers are partially confined by the valley walls, and there is convergence of tributary and main valley lobes. Since these glaciers are traveling at different velocities, there is a complex stress

lization, and (3) energy-increasing processes in crystals as a result of deformation related to "flow" of the glacier.

Air spaces in freshly formed ice become relatively smaller bubbles as they are compressed. At 100 m depth, ice contains about 2% air. This bubbly ice gradually changes to bubble-free ice as pressure continues to increase. At 300 m the bubble porosity is reduced from about 10% by volume at the firn-ice transition to 0.4%, equivalent to a bubble pressure of about 25 bars. The pressure at about 800 m depth exceeds the vapor pressure of ice, and ice can no longer dissociate to a vapor phase. As a result, air enters the ice lattice and locates at dislocation vacancies, forming what is known as a clathrate compound (Miller, 1969). Bubbles were found to be completely eliminated at 1100 m at Byrd Station, Antarctica (Gow, 1971, 1975).

If deformation is not dominant, recovery of ice crystals that have interpenetrating contacts and high dislocation densities allows grain boundary and dislocation migration to occur. The ultimate result is polygonalization, generating relatively strain-free crystals (Figure 22.3). Most commonly, the interfacial surfaces of polygonal crystals retain some curvature even though the low-energy shape of crystals yields intersections of the interfaces at triple points that make 120° angles with each other.

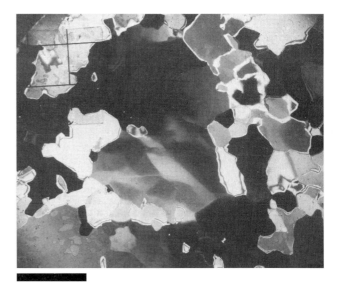

Figure 22.4
Coarse-grained, bubble-free ice with crude mosaic texture at 1833 m depth. Crystals have progressive extinction and what appear to be subgrains. Sample has several c-crystallographic maxima (not shown) encircling vertical axis. Byrd Station, West Antarctica. Bar scale = 1.7 cm. (Courtesy A. Gow [CRREL].)

picture at their intersection. In these dynamic zones, superposition of structures may occur. Folds may be generated that are subsequently sheared with transposition of planar fabric into new planar fabric with different orientation.

Shear is also important in ice sheets. With increasing depth, pure shear is followed by simple shear as ice moves downward and outward. At 95–130 m there is a linear fabric produced at Byrd Station (Gow, 1970). There is a crude flattening of grains and reshaping of some air bubbles into tubes. Since ice cores are not recovered with initial orientation known, such dimensional fabric cannot be directly related to the movement picture of the glacial ice.

The crystal structure of ice has a lot to do with its behavior in a deviatoric stress environment. Tetrahedral molecules are arranged hexagonally in ice. As a result, gliding on the basal plane is easiest, not unlike glide on the basal plane of quartz producing rocks such as mylonitic quartzites. If stress is applied to an ice crystal so that there is a component of shear in the basal plane, the crystal will deform, tending to bring the crystallographic c axis into a position normal to the shear plane (Figure 22.5) just as it does in quartz (Figure 16.7). In polycrystalline aggregates such as glacial ice, the rate of strain, as controlled by basal slip, is somewhat reduced by the crystals in the aggregate that are not oriented ideally for slip to occur.

Deformation of ice depends also on the presence of dislocations inherited from crystal growth and on dislocations produced during deformation. Dislocations allow for easier movement of planes of ions, resulting in increasing strain rates. As dislocations "pile up" there is a strain hardening and reduction of strain rate. Deformation of ice is at first elastic and then ductile as these intracrystalline creep mechanisms become dominant. Grain boundary sliding, grain boundary migration, and grain rotation all contribute to flow. As in silicate metamorphic systems, strain rate is time and temperature dependent. Relatively higher temperatures, such as at the base of an ice sheet, are conducive to higher flow rates: the longer the time frame, the greater the total strain.

Deformation of ice and recovery of high-energy intra- and intercrystal configurations should be expressed by lattice orientation. Petrofabric analysis of glacier ice reveals some not-so-simple relationships (Gow and Williamson, 1976). Uniaxial compression (pure shear) experiments produce small-circle girdles in which the c axes cluster in a ring around the axis of compression. Simple shear experiments produce two maxima, one perpendicular to the plane of shear and a weaker one at about 20° to that plane. Direct observations of thick glacial ice show that initial fabrics are random, becoming nonrandom at depth. Large crys-

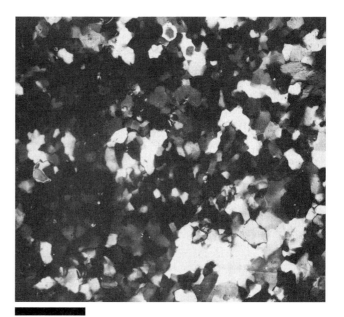

Figure 22.5
Bubble-free polygonalized ice mosaic at 1384 m depth. Abundance of grains in extinction reflects single-maximum fabric with c axes perpendicular to plane of shear, which is nearly the plane of view. Byrd Station, West Antarctica. Bar scale = 1.7 cm. (Courtesy A. Gow [CRREL].)

tals tend to orient with multiple maximas, whereas smaller crystals may yield single-maximum fabrics.

Salt Diapirs, and Evaporites in Thrust Belts

Deformation and Recrystallization of Halite at Low Temperature

Simple shear experiments have produced cataclastic flow in halite aggregates at 25°C and low confining pressure (Chester, 1988). Distributed fractures, sliding along planar surfaces, extension fractures, flattened grains, and alignment of grain boundaries generating a planar fabric were produced. The natural fabric shown in Figure 25.8 is similar to some of those generated in the experiments.

Partial recrystallization was observed by Hiraga and Shimamoto (1987) and Shimamoto (1989) in presheared compacted samples at 200 MPa (2 kbar), 100 MPa deviatoric stress, and 25°C. Shearing of these specimens at various confining pressures produced foliation defined by straight and parallel grain boundaries and more recrystallization. The deforma-

22 Low-Temperature Recrystallized Rocks 317

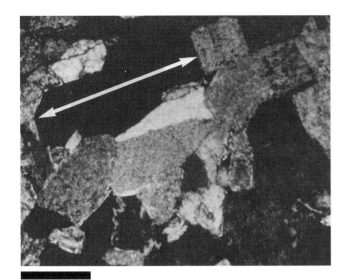

Figure 22.6
Recrystallized anhydrite with slight preferred orientation (double arrow) of blocky anhydrite crystals, in fold limb of Oakwood salt dome, East Texas. Bar scale = 0.7 mm. (Courtesy M. Jackson.)

tion mechanism was presumed to be intercrystalline glide, especially by slip on the dodecahedron, but confirmation was precluded by pervasive recrystallization. Triple points and S-C mylonitic fabrics were also produced.

Dynamic recrystallization accompanying gliding, tensional separations, and grain boundary diffusion have been documented in natural rock salt of salt glaciers (Talbot, 1981). Porphyroclastic texture formed, similar to the texture in sheared rock salt from a Texas salt diapir (Figure 25.8; Dix and Jackson, 1982). In the diapir, recrystallized anhydrite occurs in fold limbs (Figure 22.6), very similar to recrystallized anhydrite in sheared chickenwire anhydrite observed by Spötl (1989) in an alpine thrust complex.

Coupling of brittle and ductile (crystal plastic) processes—that is, their simultaneous operation—is emphasized by Chester (1988). Evidently, brittle and ductile behavior and accompanying recrystallization are typical of rock salt deformed at low temperature and variable confining pressure.

From Bedded Halite to Salt Domes to Salt Glaciers

Halite melts at about 800°C, ensuring that flow observed in salt diapirs and salt glaciers must be in a solid-state process. Salt moves because it is plastic under conditions of differential stress and low temperature.

Diapirs are initiated because of gravitational effects produced by a heavy overburden overlying a more buoyant salt bed, and in some cases in response to tectonism. If a 1-km-thick source bed is activated, the distance traveled to the base of a salt dome is between 3 and 15 km. The salt diapir rises between 2 and 15 km and may flow into an overhanging mushroom-shaped bulb another kilometer or so. Thus, the total flow of salt is between 5 and 30 km. This means that an extended period of deformation and recrystallization occurs in the formation of a salt diapir.

Grain size of halite in diapirs is 0.5–5 cm and tends to be finer-grained in shear zones. Coarsening in

Figure 22.7
Recumbent folds in active Kuh-e-Namak namakier (salt glacier), SW Iran. The folds are sheathlike and are bounded by subhorizontal shear zones. Flow is characterized by dynamic rotation, recrystallization, and solution mass transfer. (Courtesy C. Talbot.)

pressure shadows is the result of diffusional mass transfer. At depth, fluid inclusions weaken salt, thereby lowering its elastic limit.

Emergent salt diapirs forming salt glaciers in Iran suggest that halite can flow ductilely at surface temperatures. The salt column cannot support its own weight in an extrusive dome, spreading laterally in the form of **salt glaciers**. In Iran (Talbot, 1979, 1981; Talbot and Jackson, 1987; Jackson et al., 1990) glaciers are 2 km wide, 50–100 m thick, and 1–3 km long. Recumbent folds of rock salt are characteristic of high-strain zones (Figure 22.7)

Evaporites in Tectonites

Spötl (1989) describes chaotic mélanges containing fragments of shale, siltstone, sandstone, and anhydrite in a clayey halite matrix. There has been severe tectonization, including diapirism, gravitational sliding, and alpine thrust tectonics. Isoclinal folding of rock salt and recrystallization of anhydrite are documented.

Malavielle and Ritz (1989) describe ductile deformation of evaporites in shallow décollements. Mylonitic foliation, stretching lineations, and shear bands have formed, as well as a variety of folded structures in anhydrite-bearing rock salt beds. Anhydrite nodules, siliciclastic rock fragments and boudins provide pressure shadow locations for gypsum crystallization. Borchet and Muir (1964) show many examples of ductile behavior involving evaporites, including

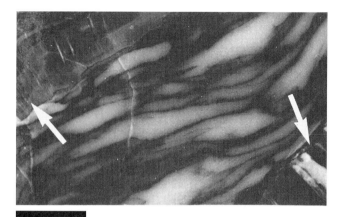

Figure 22.8
Ductile deformation of anhydrite between brittlely deformed carbonate beds containing calcite in brittle fractures (arrows). Fabric is characterized by stretched anhydrite nodules containing recrystallized anhydrite. Saint Ghislain Formation, Belgium. Bar scale = 2.6 cm. (Courtesy J. M. Rouchy.)

boudinaged halite beds in less competent carnallite and complex folding in various evaporites.

Ductilely deformed and recrystallized anhydrite (Rouchy et al., 1987) generates a tectonic lamination (Figure 22.8). The anhydrite is ductile, whereas enclosing carbonate beds deform brittlely. Somewhat blocky recrystallized anhydrite similar to that shown in Figure 22.6 is formed.

23

Cataclasites

Slickensided surface (dip slip) in Silver City Fault zone. Footwall consists of brecciated meta-andesite cemented with hydrothermal calcite and quartz. Lucern Cut, Silver City, Nevada.

The Regime of Brittle Faulting

Thorough brecciation of rock occurs primarily by (1) meteor impact, (2) explosive volcanism, and (3) faulting. Rocks resulting from brittle deformation of preexisting rock in fault zones are **cataclasites** (Wise et al., 1984).

Fracturing, cracking, granulation, rigid-body rotation, frictional sliding are all involved in the formation of cataclasites (Engelder, 1974; Hadizadeh and Rutter, 1983; Blenkinsop and Rutter, 1986; Rutter, 1986; Knapp et al., 1987). Since these brittle phenomena require an increase in volume, cataclasites form in shallow faults where a volume increase is mechanically possible.

Cataclasites are generated in faults of any structural type, including the shear at the base of gravity slide blocks. The faulting may be seismic or aseismic. Relatively incohesive **fault breccia** (visible fragments >30%) and **gouge** (visible fragments <30%) form within a few kilometers of the Earth's surface, are either included as cataclasites in the broad sense (Wise et al., 1984), or distinguished from cohesive cataclasite rock as being incohesive fault debris (Sibson, 1977).

Faulting at depth, where the confining pressure and temperature are elevated, generates mylonites. Mylonites form in the ductile regime and are described in Chapter 21 as variants of dynamothermal metamorphic rocks, occurring on a local scale at high strain rates compared to "regional" dynamothermal metamorphism.

Penetrative deformation, such as bulk rock deformation forming slaty cleavage, characterizes low-temperature deformed rocks that form where there is relatively low-temperature upper crustal tectonics, such as portions of foreland thrust belts, and subduction accreted mélange wedges (Chapter 25).

Some brittle faults may contain pseudotachylite. The origin of the pseudotachylite in brecciated rock is not always clear. For example, the type location of fault-generated pseudotachylite in the Vredefort Dome, South Africa, is now thought to have resulted from shock metamorphism (Reimold, 1990). Furthermore, magma injected into fault zones, either during the faulting or after, can be misidentified as friction glass. Nevertheless, friction melting associated with high strain-rate faulting has been identified (Maddock, 1983). Short of melting, there may be formation of extremely fine-grained **crush microbreccia**, typically accompanied by **slickensided surfaces**.

Frictional heating must temporarily bring local portions of brittle faults into the regime of metamorphism, but since the strain rate of faults that generate frictional heat must be high, there is brittle rather than ductile deformation, and since the thermal inertia is low (rapid dissipation of heat) there is little or no time for crystal growth and formation of mylonites (Bell and Etheridge, 1973). Nevertheless, minerals such as quartz in the high-strain zone of cataclasites may have some optical indications of crystal plasticity. Carbonate and evaporite minerals can deform plastically at lower temperatures than quartz and they may even recrystallize (Chapter 22).

The behavior of rock in a fault zone depends on strain rate as well as temperature. During a seismic fault event (earthquakes generated), opposite sides of a fault move past each other at rates on the order of 10–100 cm/sec, lasting for tens of seconds, whereas movement on aseismic faults is by creep, typically on the order of millimeters to several centimeters per year. Generation of pseudotachylite requires a high strain-rate; thus, it is expected to occur on faults acquiring offset by seismic faulting but not by aseismic creep.

Renewed movement on brittle faults, such as stick-slip faults or reactivated faults, may be ductile if there is an intervening period of veining along the shear planes (Stel, 1981). For example, precipitation of quartz may result in crystal plastic behavior in a shear zone that was dominated previously by brittle behavior of feldspars.

Textural Characteristics of Cataclasites

A brittle fault zone may consist of no more than a layer a few millimeters thick of extremely dense comminuted rock crush microbreccia with only moderately fractured rock on either side. These microbreccias may have grooved and "polished" slickensided surfaces (Figure 23.1a,b). Rocks with multiple slickensided surfaces are known as **slickenites** and can be considered as an additional variety of cataclasite.

Recrystallization of carbonate in a slickenite formed in a calc–arenite is shown in Figure 23.1b,c). Temperatures could not have been elevated for very long in view of the completely unchanged character of micritic fossils and glauconite pelloids adjacent to the calcite that formed by very localized recrystallization of the micrite. There is no indication that calcite veining preceded or accompanied the shearing.

Brecciation is characteristic of cataclasites, occurring adjacent to slickensided microbreccias (Figure 23.2a) or distributed throughout a shear zone in which there is no single plane of slickensides (Figure 23.2b). Although typically without directional fabric

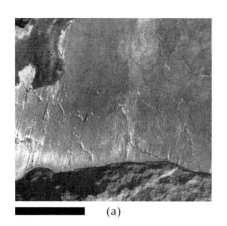

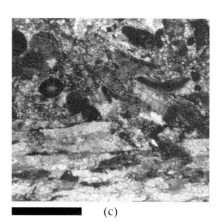

(a)   (b)   (c)

Figure 23.1
Slickensided crush breccia.

(a) Extremely smooth, well-striated surface on 3–4-mm-thick microbreccia formed from metavolcanic rock. Bar scale = 2 cm. (Courtesy R. Schweickert.) (b) Stepped slickenside surfaces in bioclastic limestone. Curtis Formation, Vernal, Utah. Bar scale = 4 cm. (c) Dynamically recrystallized calcium carbonate in rock shown in (b), in which new calcite blades are elongate in direction of shear (lower portion of photo). Sharp contact of calcite blades with glauconite pellets, calcareous ooids, calcareous biclasts, and preshearing matrix sparrite, indicates very little frictional heat was generated. X-polars. Bar scale = 1 mm.

Figure 23.2
Mesoscopic fault breccias.

(a) Nondirectional breccia (hammer head) associated with slickensided surface (hammer handle). Osgood Mountains, Getchell Mine area, Nevada. (b) Calcite-bearing shear zone in Cretaceous sandstone. Lithons of sandstone are separated by cataclasite. Mendocino, California. Bar scale = 6.8 cm.

(Sibson, 1977; Wise et al., 1984), cataclasites in fault zones may be arranged between lensoid breccia fragments known as **lithons**. A directional rock fabric is generated (Figure 23.2b) and is one variety of foliated cataclasites (House and Gray, 1982; Chester et al., 1985; Wojtal and Mitra, 1986).

Breccia fragments may dominate over matrix in cataclasites and related rocks (Figure 23.3a,b). Typically, cataclasites consist of both visible rock fragments and a submicroscopic very finely comminuted matrix (Figure 23.4a–d). **Microcataclasites** occur along microfaults (Figure 23.5) and have the same

Figure 23.3
Brecciated rock with a large fragment/matrix ratio.

(a) Breccia zone in quartzite cobble. Note lack of milling of breccia fragments. Titus Canyon Formation, Sierra Nevada. Bar scale = 1.2 cm. (Courtesy R. Schweickert.) (b) Quartzite monolithic breccia formed at base of gravity slide block. Sacramento Pass, Snake Range, Nevada. Bar scale = 1.2 cm.

322   Part 6   Rocks Formed by Solid–State Processes

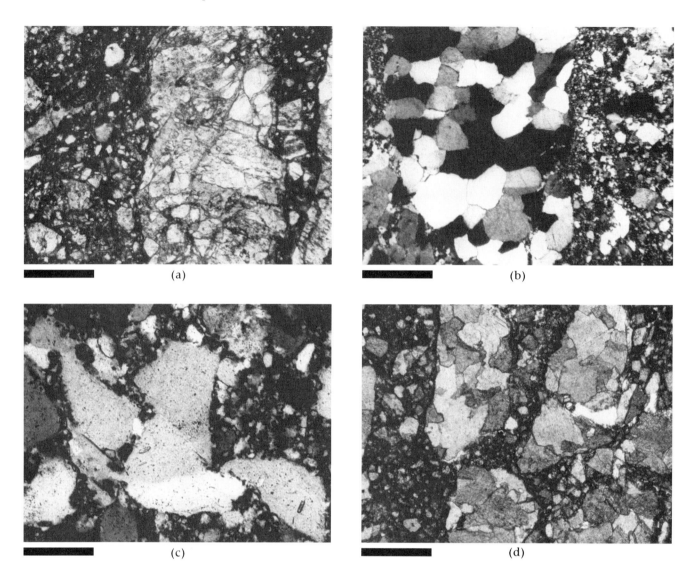

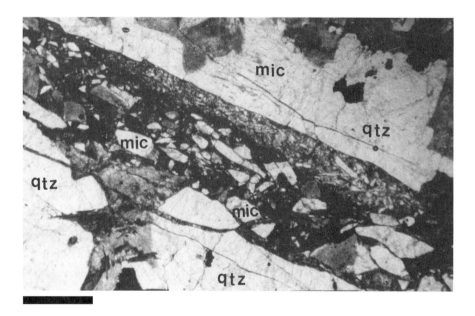

Figure 23.5
Microcataclasite zone in two-feldspar granite containing fragments of quartz (QTZ), microcline (MIC), altered plagioclase (dusty grains), and comminuted crush microbreccia matrix (black). Oracle Granite, San Manuel Mine, Arizona. PPL. Bar scale = 1 mm.

Figure 23.4
Typical cataclasites.

(a) Cataclasite fromed in Cretaceous two-feldspar granite along Holocene fault. Genoa, Nevada. PPL. Bar scale = 0.6 mm. (Courtesy C. Osterberg.) (b) Cataclastic Eureka quartzite. Note relatively undisturbed quartzite texture in large breccia fragment (left of center). Ely, Nevada. X-polars. Bar scale = 0.6 mm. (c) Same as (b) showing disaggregated grains of quartzite and shear displacement of two grains. X-polars. Bar scale = 0.14 mm. (d) Cataclasite from fault zone in Bonnaza King Formation (limestones). Nevada Test Site. PPL. Bar scale = 0.6 mm. (Courtesy J. Caskey.)

breccia fragment–matrix relation that characterize mesoscopic cataclasites.

Pseudotachylites occur as dikelets and as complex networks enclosing brecciated rock (Philpotts, 1964). Pseudotachylites of fault origin may contain skeletal and spherulitic crystals indicating some primary crystallization of melt (Maddock, 1983). Glass may be totally devitrified, and as such is difficult to distinguished from recrystallized dense gouge (Wenk, 1978).

Veining is typical in cataclasites (Figure 23.2b). Increase in volume attendant to fragmentation generates open spaces in which precipitates may form at any later time.

24

Impactites

Rainy day at Meteor (Barrenger) Crater, Arizona. Crater covers 1 km² and is 1200 m in diameter, 170 m deep to the topographic floor and twice that amount to the base of the breccia lens. Tilted rim rock (right) is Kaibab Limestone, stratigraphically underlain by Coconino Sandstone within crater. An iron meteorite calculated to be 60 m in diameter, weighing 1 million metric tons and moving at a velocity of 15 km/sec, impacted 30,000 years ago.

Rocks impacted by fast-moving extraterrestrial bodies are drastically changed to new rock types by a nearly instantaneous process known as **shock metamorphism**. High-pressure shock waves generated by such impacts may last only a fraction of a second, but in that brief time they produce extraordinary changes in the target rocks. **Impactites** are the rocks that result from shock metamorphism. Some impactites contain glass resulting from melted target rock (**impact melts**). Large impacts can produce large volumes of magma, and such events represent yet another source of magma in the terrestrial setting that eventually crystallizes into magmatic rock.

Impacts and Natural Shock Waves

Since the 1960s, geologists have become increasingly aware that impacts of large extraterrestrial objects have played a significant role in Earth history (French, 1968, 1990; Roddy et al., 1977; Grieve, 1987; Melosh, 1989). More than 130 terrestrial structures have been identified as the sites of such impacts. The structures range from <100 m to >200 km in diameter and from a few thousand years to nearly 2 billion years in age (Mark, 1987).

Recognition of large impacts as an important geological process has come about chiefly because of one factor: large impact events generate intense, transient, high-pressure stress waves (**shock waves**) that produce distinct and permanent deformation features in the target rocks. Impact as a rock-forming process is important in the petrologic context of this textbook, but the possible effects of impacts on the biological evolution of the Earth are far more significant. Large impacts may temporary modify the surface environment, causing some of the major extinctions observed in the geological record (Silver and Schultz, 1982; Sharpton and Ward, 1990). In particular, there is now a large body of evidence that at least one major impact occurred 65 million years ago, at the same time as the major extinction of dinosaurs and other species that

Figure 24.1
Pressure–temperature diagram showing widely separated fields of normal crustal metamorphism (lower left) and shock metamorphism (lower right) (modified from Grieve, 1990) (pressure plotted on logarithmic scale). Pressure–temperature curves for the formation of several high-pressure minerals (coesite, stishovite, diamond) are shown for reference, as are some high-temperature melting and decomposition reactions observed in shock-melted rocks.

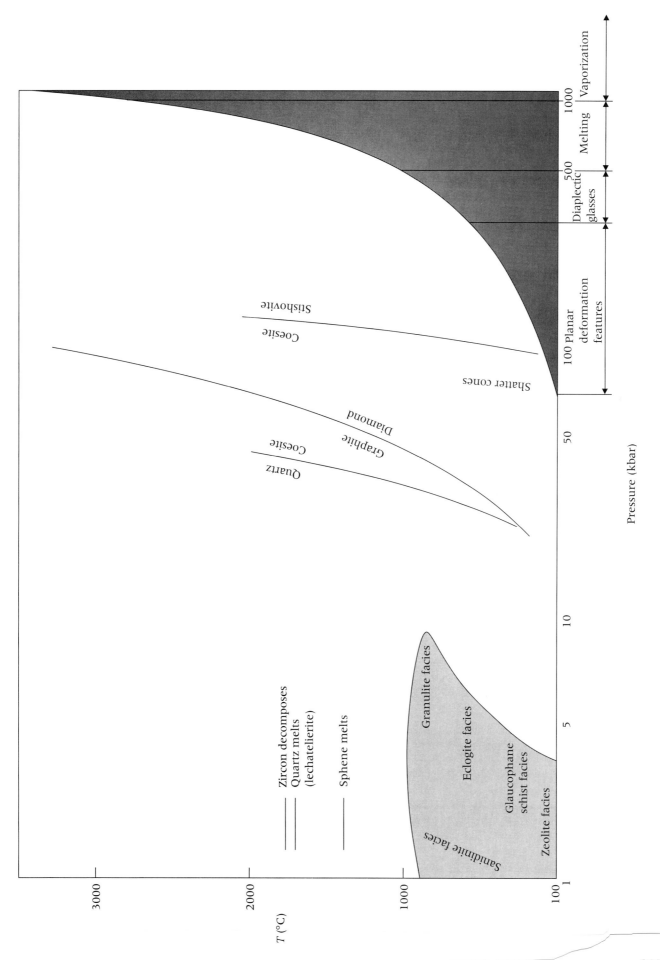

mark the Cretaceous–Tertiary boundary. There is still active debate about the extent to which meteorite impacts have affected the Earth (Officer and Drake, 1983; Alvarez, 1986; Hallam, 1987), but the importance of large impacts in the history of the Earth has been generally accepted.

Extraterrestrial objects strike the Earth at high velocities, typically about 10–30 km/sec for asteroidal bodies and up to perhaps 65 km/sec for comets. Impacts at such velocities release immense amounts of kinetic energy. The impact of a meteorite only a few meters across releases as much energy as an atomic bomb. Collisions with larger objects (5–10 km in size) are rarer but more lethal. Such an event releases more energy in an instant than the whole Earth releases (through heat flow, volcanism, and earthquakes combined) over tens to hundreds of years.

Upon impact, virtually all the kinetic energy of the impacting object is transmitted into the target rocks by intense shock waves. These shock waves typically have peak pressures ranging from a few tens of kilobars to a few megabars, much greater than pressures produced by normal geological processes (Figure 24.1). As the shock waves radiate from the impact point and pass through the target rocks, they interact with the Earth's surface to excavate an impact crater that may be 10–20 times the diameter of the original projectile (Roddy et al., 1977; Melosh, 1989).

Shock Waves and Shock Metamorphism

In the process of forming an impact crater, shock waves also subject large volumes of the target rocks to transient pressure–temperature conditions that are totally beyond those of conventional geological processes (Figure 24.1; French, 1968; French and Short, 1968; Stöffler, 1972, 1974; Dence and Robertson, 1989; Grieve, 1990).

Energy deposited by the most intense shock waves can raise the target rock temperatures to thousands of degrees, high enough to melt or even vaporize large volumes of rock after the shock wave has passed. At lower shock-wave pressures, distinctive deformation effects are produced instead. Reaction times are short, from microseconds to minutes, so that chemical equilibrium is rarely attained in shocked rocks. Such high-pressure minerals as coesite, diamond, and stishovite (Chao et al., 1962) occur in some impacted rocks. Their production is due to the high pressures (Figure 24.1) and rapid reactions times produced during the impact.

Diamonds occur in both impacted terrestrial rocks and in some silicate-rich (stony) meteorites. These are not engagment-ring-size stones; instead they occur as small (less than 1 mm) grains typically detected by x-ray diffraction. Theories for formation of such diamonds range from nonshock models, in which diamonds are formed by gravitational pressures within bodies larger than the moon (Carter and Kennedy, 1964; Lewis et al., 1987), to shock models in which the diamonds are formed by shock waves generated in terrestrial impacts (Lipschutz and Anders, 1961; Gilmour et al., 1992) and in nonterrestrial impacts on asteroid parent bodies (Lipschutz, 1964).

High-pressure shock waves typically generate enough heat in the target rocks to produce such reactions as the melting of sphene, melting of quartz to **lechatelierite**, and decomposition of zircon (Figure 24.1). All of these transformations have been detected in impactites.

Impact-produced shock waves acting on the target rocks produce distinctive rock deformation and melting effects that characterize shock metamorphism. These effects provide unique evidence for extraterrestrial impact events in the geological record. Shock-metamorphic features can be conveniently divided into those formed at three different shock pressures (B. French, personal communication): low (about 10–100 kbar) (shatter cones and breccias), intermediate (100–500 kbar) (microdeformation effects in mineral grains), and high (>500 kbar) (high-temperature melting and vaporization). However, the development of shock features is a complex and gradational process. These pressure limits are only approximate, and the presence of a given shock-metamorphic feature does not necessarily specify an exact pressure of formation.

Various descriptive terms have been applied to shock-metamorphosed rocks. Impactites have distinctive shock effects, particularly shatter cone structures and planar deformation features in quartz crystals. Such shock features have been reproduced in numerous dynamic laboratory experiments and nuclear explosions. With some exceptions (such as the high-pressure minerals coesite and diamond) shock-metamorphic features have not been identified in any nonimpact environment. In particular, definite shock effects have not been identified in explosive volcanic rocks nor in rocks subjected to intense tectonic deformation. Witness the recent debate over deformation features in explosive volcanic rocks (see Carter et al., 1986; Izett and Bohor, 1987; Alexopoulos et al., 1988; Sharpton and Schuraytz, 1989.) It is now widely accepted that the occurrence of such features as shatter cones and planar features in quartz is definite

Figure 24.2
Typical shatter cones from two impact structures.

(a) Small cones in limestone from the Haughton Structure, Northwest Territories, Canada. (Courtesy R.A.F. Grieve, Geological Survey of Canada.) Bar scale = 3 cm. (b) Large shatter cone in Mississagi Quartzite from the Sudbury structure, Ontario. Hammer (lower left) is 40 cm long. (Courtesy B. M. French, NASA Headquarters.)

---

evidence for high shock pressures and, therefore, extraterrestrial impact (French, 1990).

Low-Pressure Shock Effects (about 10–100 kbar)

The first distinctive shock effects form at peak shock pressures of roughly 10–100 kbar. The lower part of this interval overlaps the range of pressures existing during tectonic deformation. At these shock pressures, the target rocks develop megascopic **shatter cones** (Figure 24.2 a,b) and pseudotachylite breccias (Figure 24.3). Both of these features tend to occur beneath and around the floor of the original crater, where the shock-wave interactions were not intense enough to excavate and disperse the target rock after the shock effects had formed (Melosh, 1989).

Pseudotachylite breccias are compatible with impact but are not conclusive evidence of impact. Similar breccias of fault origin are well known (Chapter 23). In both cases, very fine-grained black rock serves as a matrix for breccia fragments. In some cataclasites, the pseudotachylite matrix contains glass formed

Figure 24.3
Netlike pseudotachylite breccia in crystalline basement rocks of the Vredefort Structure, South Africa. The pseudotachylite vein cuts coarse granite and contains numerous rounded granite inclusions in the very fine-grained black pseudotachylite matrix. Hammer scale. (Courtesy W. U. Reimold, University of the Witwatersrand (Reimold, 1990).)

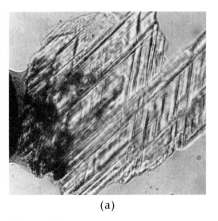  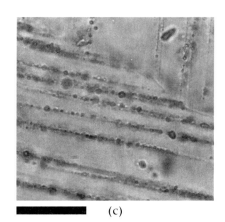

(a) (b) (c)

Figure 24.4
Shock-produced planar deformation features in quartz.

(a) Quartz grain (0.27 mm in diameter) from the Cretaceous–Tertiary boundary west of Trinidad, Colorado. The grain contains two well-developed sets of fresh, continuous planar deformation features. PPL. (Courtesy G. A. Izett, U.S. Geological Survey (Izett, 1990).) (b) Two sets of planar deformation features in quartz grain in granite from Sudbury, Ontario, Canada. The original planar features have been decorated with small fluid inclusions that preserve the orientation of the original planes. The quartz grain is about 1 mm across. (Courtesy B. M. French, NASA Headquarters.) (c) Close-up view of inclusion-decorated planar features in quartz from the Charlevoix structure, Quebec, Canada. PPL. Bar scale = 0.18 mm. (Courtesy R. A. F. Grieve, Geological Survey of Canada.)

from friction-induced melting. In the impact setting, pseudotachylite breccias form several kilometers below the original crater floor, by shock waves whose original pressures had dropped to relatively low levels (French and Nielsen, 1990).

Shatter cones (Dietz, 1968) resemble some non-shock features, such as cone-in-cone structures that form in sediments, but cones of shock origin occur in all kinds of sedimentary and crystalline rocks. The axes of shatter cones in sedimentary rocks may be oriented at any angle to the original bedding, indicating their origin is unrelated to sedimentary processes. In addition, the cones themselves point inward and upward when rocks are restored to their pre-impact positions, indicating a shock-wave source from above.

Intermediate-Pressure Shock Effects (100–500 kbar)

Higher shock pressures produce microscopic deformation features in individual mineral grains. These features include (1) multiple sets of distinctive **planar deformation features** in quartz and feldspar (Figure 24.4a–c); (2) amorphous glassy phases (**diaplectic** or **thetomorphic glasses**), which are formed without actual melting and which retain the shape of the original quartz or feldspar grains (Figure 24.5); (3) high-pressure mineral phases, including coesite and stishovite produced from quartz.

Planar deformation features in quartz (Alexopoulos et al., 1988) have become the most widely recognized microscopic shock effect. The features have been observed in numerous young and old impact structures as well as in quartz grains from Cretaceous–Tertiary boundary clays (Figure 24.4a). Planar deformation features occur in multiple sets and are relatively stable. They are preserved even in geologically old structures, although the original planes may be replaced by arrays of fluid inclusions that preserve the original orientations of the planes (Figures 24.4b,c).

Diaplectic glasses, where they are preserved in their original isotropic state (Figure 24.5), are equally impressive evidence for impact, chiefly because of the higher shock pressures (200–400 kbar) required for their formation.

Some of the high-pressure minerals formed by shock (e.g., coesite) also occur normally in deep-seated rocks subjected to high pressures, but their presence in near-surface crustal rocks, such as in the Coconino Sandstone at Barringer Crater, Arizona

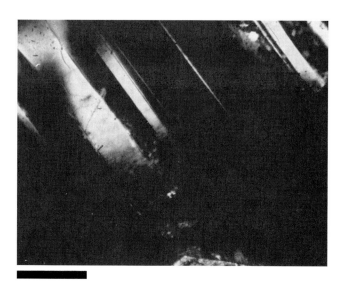

Figure 24.5
Glassy feldspar (maskelynite), produced by shock waves acting on plagioclase crystals in anorthosite from the Manicouagan structure, Quebec, Canada. Parts of the crystal remain birefringent with multiple twin lamellae (top), which disappear in the dark isotropic parts of the crystal (bottom). X-polars. Bar scale = 0.08 mm. (Courtesy R. A. F. Grieve, Geological Survey of Canada.)

(Chao et al., 1962), is strong evidence for impact. Furthermore, since shock-metamorphic changes occur in milliseconds to minutes, chemical disequilibrium is characteristic. Accordingly, a shocked sandstone may contain original quartz as well as coesite, stishovite, and silica glass (lechatelierite).

High-Pressure Shock Effects (500–1000 kbar)

Shock waves deposit some of their energy in the target rocks in the form of heat, raising the temperature of the rocks. High-pressure shock waves (>500 kbar) deposit so much energy that the rock's temperature is suddenly raised to thousands of degrees, far above the normal melting points of most rocks and minerals (about 700–1400°C). As a result, the target rock melts, and some may even be vaporized.

This process of impact melting produces two types of unusual rocks. Small bodies of melt, ejected from the crater, cool rapidly to form glasses characterized by quench textures, compositional diversity, and turbulent flow structures (Figure 24.6a). The impact origin of such glasses is indicated by unusual mineralogical reactions which occur far above the thermal range of normal geological processes (Figure 24.1), including melting of quartz to silica–glass (lechatelierite) (T = 1713°C), melting of sphene (T = 1395°C), and the decomposition of zircon to the oxide baddeleyite (T = 1775°C).

Glass also occurs in glassy breccia (Figure 24.6b). Glasses of impact origin may contain lechatelierite (Figure 24.6c), but they may also contain small crystals of minerals such as pyroxene (Figure 24.6c) similar to many glassy mafic igneous rocks of nonimpact origin.

Globules of glass occur in **suevite** of the Nördlinger Ries Crater in Germany (Figure 24.6d). Suevite is a tufflike breccia in which the centimeter- to decimeter-sized glass "bombs" are thought (Stöffler, 1966; Hörz, 1982) to be the result of meteorite impact. The glass was produced by the melting of impacted sedimentary and igneous rocks and was aerodynamically sculptured during ballistic flight. Lack of sorting between the glass "bombs" and fine-grained ashy matrix is attributed (Hörz, 1982) to turbulent motion in a gaseous environment of positive pressure (>1 atm) produced in the ejected debris as part of the excavation of the impact crater. In some respects, suevite is similar to igneous ash flow tuffs that contain pumice fragments. However, the glass fragments in the suevite are not pumiceous. Instead, they are dense and slightly vesicular, and they have a composition reflecting a mixture of impacted sedimentary and granitic rocks. Furthermore, the presence of planar features in quartz occurring in the suevite (Stöffler, 1966) is strong evidence for impact origin.

Larger, more extensive bodies of impact melt are thought to have collected inside craters where they cool more slowly and crystallize to bodies of igneous rock (Dence, 1971; French and Nielsen, 1990). Such rocks differ from those crystallizing from nonimpact magmas in their unusual chemical characteristics (even allowing for extensive assimilation of wallrock by "normal" magma), Ni and Fe enrichment, lack of phenocrysts, and lithic inclusions that have shock features. In addition, rocks resulting from crystallization of impact melts cannot have shock features themselves since they crystallize after impact.

Extraterrestrial Impactites

Rock impacted by meteorites on the surface of extraterrestrial bodies such as the moon and asteroids generates impactites in more or less the same manner as on Earth (Heiken et al., 1991).

For example, the meteorite known as "Petersburg" consists of a rock type known as **howardite**

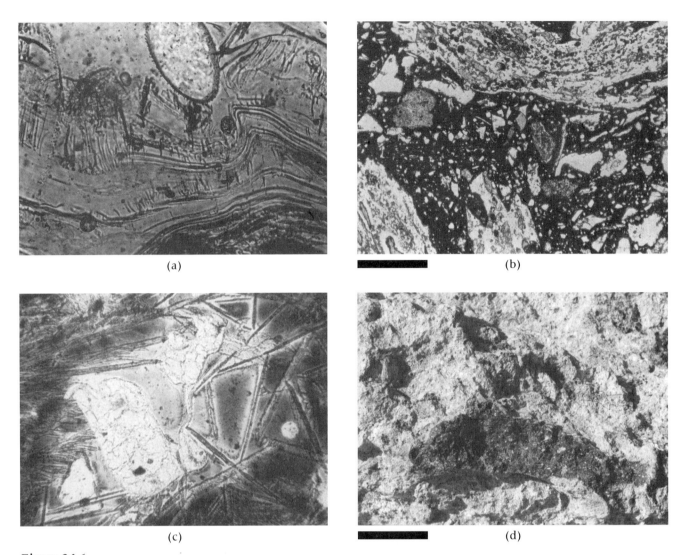

Figure 24.6
Impact melts.

(a) Fragment of shock-produced glass from Clearwater Lakes, Quebec, Canada, with compositional variability and clear, deformed streaks of silica glass (lechatelierite) formed by melting of quartz grains. PPL. Field of view is about 2 mm long. (Courtesy M.R. Dence, Royal Society Canada.) (b) Glass-rich, fragmental impact breccia (Onaping Formation) from Sudbury, Ontario, Canada. The large, flow-layered glass fragments, now recrystallized to chlorite, quartz, and other minerals, resemble pumice fragments of pyroclastic origin. (Courtesy B.M. French, NASA Headquarters.) Bar scale = 0.9 mm. (c) Crystal-bearing glass of impact origin from the Tenoumer, Mauritania crater. A patch of clear lechatelierite (center), probably derived by shock-melting of quartz, co-exists with a dark brown glass containing acicular quench crystals of pyroxene. PPL. Field of view is about 2 mm long. (Courtesy B.M. French, NASA Headquarters.) (d) Suevite (tufflike breccia) from the Ries Crater, Germany. Flattened "bomb" (black) is a bubbly glass object that has come to rest in a fine-grained gray to yellowish gray breccia matrix. This matrix contains quartz crystals with planar features (Stöffler, 1966). Target rock is Mesozoic limestones, arkose, and shale. Impact occurred 15 m.y. BP. Bar scale = 2.5 cm. (Courtesy Richard Schultz.)

 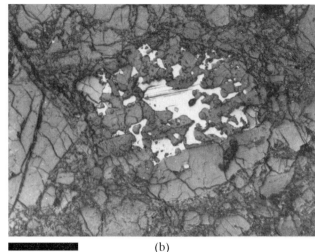

Figure 24.7

(a) Petersburg polymict (fragments of different composition) brecciated howardite. Fragment of recrystallized eucritic basalt (short arrow) contains pigeonite (high relief), plagioclase (white), and a few oxide grains (black). Oikicrystic (ophitic) textured eucritic basalt fragment (long arrow) is not recrystallized. Matrix contains fragments of pyroxene, plagioclase, and minor amounts of ilmenite, chromite, and troilite. PPL. Bar scale = 1.5 mm. (Courtesy R. Schultz.) (b) Kamacite (Co- and Ni-bearing iron) intergranular between pyroxene crystals and as poikilitic inclusions in pyroxene. Reflected light. Bar scale = 0.1 mm. (Courtesy R. Schultz.)

(Figure 24.7). This is a **stony meteorite** (containing silicate minerals) of the **achondritic** type (lacking the spherical granules characterizing **chondrites**). Howardite is an impact breccia containing mainly orthopyroxene and plagioclase (Figure 24.7a) but also some ilmenite, chromite, and Ni–Co iron known as **kamacite** (Figure 24.7b). The petrogenesis of the howardite is interpreted to have involved magmatic differentiation, impact on the surface of an asteroid, recrystallization (metamorphism), refragmentation by further impact, ejection from parent body (asteroid collision?), and finally entrance into the terrestrial environment as a meteorite (Richard A. Schultz, personal communication; Duke, 1965; Duke and Silver, 1967; Bunch, 1975).

Low-Temperature Deformed Rocks

Franciscan Formation bedded cherts in a highly disrupted zone of angular parallel folds and small shear zones. Folding occurred by flexure and layer-parallel slip at low to moderate temperature. Marine Penninsula, north of San Francisco, California. Bar scale = 0.5 m. (Courtesy G. LaFreniere, Willamette University, Oregon.)

Low-Temperature Low-Strain-Rate Deformation

Rocks deformed within a few kilometers of the Earth's surface are both tectonic and nontectonic. Folding, cleavage formation, penetrative shearing, and faulting of sedimentary rocks are characteristic of orogens, and some of these structural features are formed in sediments and sedimentary rocks as the result of gravity-induced sliding. Deformation restricted to brittle fault zones is discussed in Chapter 23.

Siliciclastic rocks deform both brittlely and ductilely at low temperature (Blenkinsop and Rutter, 1986; Wu and Groshong, 1991). Many brittle textures are preserved in these rock types. On the other hand, carbonate rocks can deform ductilely at low temperature and low strain rate. Calcite is very susceptible to solution and reprecipitation (Green, 1984) in dynamic environments, and it may also deform by intracrystalline glide and creep at very low strain rates (Groshong, 1988), tending to eradicate evidence of earlier or contemporary brittle deformation. Major recrystallization is characteristic of glacial ice and evaporites (Chester, 1988) involved in deformation, discussed as a separate rock-forming process in Chapter 22.

Deformation is not limited to sediments and sedimentary rock. Volcanic lavas, pyroclastics, and epiclastic pyroclastic equivalents are formed in the shallow low-temperature region and are subject to deformation, especially magmatic arc structural settings. Deep level crystalline plutonic and metamorphic rocks must first be elevated and exhumed before they can be caught up in uppermost crust tectonic activity.

Deformation Mechanisms

There are many indicators of strain in deformed rocks. They can be grouped conveniently as **brittle** or **ductile** on a rock scale (hand specimen, outcrop), reflecting, accordingly, brittle and **plastic** behavior at the grain scale (Chapter 16). Record of deformation in rocks is not limited to textural attributes. For example, **reflectance fabrics** defined by vitrinite in coals and coal-bearing rock can be an indicator of tectonic history (Levin and Davis, 1989).

Deformation is either localized or distributed. In low-temperature-deformed rocks such as in fold–thrust belts (Caskey and Schweickert, 1992; Evans and Neves, 1992) and in accretionary wedges (Lucas and Moore, 1986; Lash, 1989; Boone et al., 1989; Talbot, 1989; Brandon et al., 1988; Brandon,

Figure 25.1
Contrast in deformation style within interbedded cherty limestone and shales. More competent chert beds bend and buckle brittlely, generating a fold (left center) due to gliding in incompetent shale bed above and another below (white arrows). Accumulation of shale in hinge zone of fold (long arrow), although appearing to be the result of ductile deformation, is largely the result of cataclastic flow. Transition zone between Glenerie Formation and Esopus Formation, Appalachian Mountains, New York. Long arrow represents four decimeters.

1989) deformation ranges from dominantly cataclastic to dominantly ductile, depending on rock type and strain rate, and varies from localized to distributed (Figure 25.1).

Comparison of the processes involved in rocks deformed locally along faults and by distributive deformation are listed in Table 25.1.

Brittle Fracture

Rock grains may slide past each other (frictional sliding), and they may rotate into new positions. This happens readily in soft sediment deformation and in loosely cemented rock with pores filled with air or water. If the rock consists of interlocked grains, deformation at low temperature is likely to begin with **microcracking** (Kranz, 1983). A mineral grain may crack along cleavage directions (Figure 25.2) or nonspecified directions. Cracking results in a volume increase, and the new grain fragments may slide and rotate, the accumulative effect being **granulation** (**comminution**). Chemical dissolution may also "loosen up" rock on a grain scale, permitting grain boundary sliding and rotation in appropriate dynamic settings.

Cataclastic flow involves change of shape of grain aggregates involving microcracking of individual grains, sliding of grains past each other, and grain rotation. Grains involved in cataclastic flow are not deformed internally. Lattice strain and dislocation creep are not characteristic of cataclastic flow, although glide may occur in minerals such as calcite.

Table 25.1
Mechanisms of Deformation Occurring in Local and Distributed Deformational Settings

Dominant Deformation	Local	Distributed
Cataclastic (± some recrystallization)	Brittle faulting ± frictional heat [cataclasites]* [± pseudotachylite]	Cataclastic flow distributed shear [sheared rock] [folded rock]
Recrystallization (± crystal plasticity)	Fault zones [folded evaporites in nappes/folds]	Distributed dynamic recrystallization [glacier ice] [glacier salt] [diapir salt]
Ductile (crystal plasticity)	Ductile faulting [mylonite]	Distributed dynamic recrystallization [tectonites]

*In brackets, examples of rock produced by the deformation.

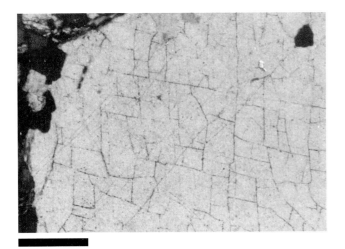

Figure 25.2
Cleavage-controlled (rhombohedral) brittle fracture in quartz. Camelback Mountain, near Phoenix, Arizona. X-polars. Bar scale = 0.5 mm.

minerals when deposited. Compaction results in preferred orientation and expulsion of vast amounts (70–80%) of water.

Transgranular microfracturing and **microfaulting** are also brittle processes contributing to deformation of rock. The fractures and faults may be localized, but they may also be **spaced** on a small scale, appearing "penetrative" on a mesoscopic scale. Pervasively sheared rock with intershear oblate-shaped rock fragments is characteristic of large volumes of deformed sedimentary rock, related to both gravity sliding and tectonic activity.

Grain cracking and rock fracturing–faulting result in volume increase. Dilatant openings are commonly filled with minerals such as quartz and calcite, precipitating from solutions that migrate into these zones of reduced pressure. Microveins and mesoscopic veins (Figure 25.4a,b) are formed.

Cohesive strength of rock typically is reduced by the presence of intergranular or intrafracture water. Both the chemical aspect (**hydraulic weakening**) and the mechanical aspect (**pore water pressure** and **hydraulic fracturing**) contribute to rock weakness.

Rotation of clays, chlorites, and micas takes place in response to either gravity compaction of soft sediment (Curtis et al., 1980; Agar et al., 1989), mechanical shortening related to gravity-induced deformation such as slumping, or tectonic movements (Weaver, 1984). Grains may be bent (Figure 25.3a) or kinked (Figure 25.3b) as a result of compaction. Mud rocks have a more or less random orientation of platy

Ductile Deformation

Ductile behavior is related to intracrystalline **dislocation creep**, **glide**, and **dynamic recrystallization**, processes reflecting **crystal plasticity** of constituent minerals (Chapter 16). Crystal plasticity is favored by

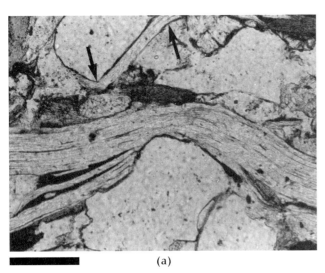
(a)

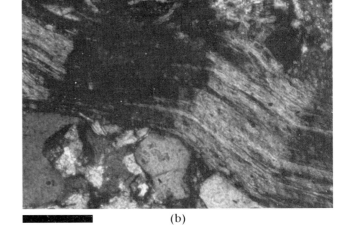
(b)

Figure 25.3
Deformed mica in compacted siliciclastics.

(a) Bent muscovite in sublitharenite due to compression against ridged quartz grain. Bent ends of smaller muscovite crystal (arrows) also indicate shortening in the vertical. Portland, Connecticut. PPL. Bar scale = 0.15 mm. (b) Chlorite with kinkbands. X-polars. Bar scale = 0.15 mm.

Figure 25.4

(a) Fragmented limestone (light gray, actually light brown, fragments) and mud rock (dark gray). White patches and veinlets are recrystallized limestone, probably aragonite of a high pressure subduction stage later recrystallized to calcite (J. A. Vance, personal communication). Mélange complex northwest of Deadman Bay, San Juan Island, Washington. Bar scale = 1.6 cm. (b) Calcite-filled fractures in limestone involved in brittle extensional tectonics. Upper plate, Northern Snake Range décollement, Nevada. Scale = U.S. quarter.

elevated temperature and confining pressure, but in the low-temperature regime there are generally early signs of crystal plasticity, especially twin gliding in calcite and progressive and patchy extinction in quartz, occurring concurrently with cataclastic behavior (Wu and Groshong, 1991). Identification of processes in very fine-grained rocks is difficult and may be precluded in totally recrystallized rocks.

Diffusional Mass Transfer

A high-pressure point at a grain contact, such as in grain-supported sediments and sedimentary rocks, results in crystal lattice strain. This portion of the crystal is in a higher-energy state and is more susceptible to dissolution if there is an intergranular liquid present. Diffusion of solute to lower-energy sites at the ends of the crystal in the direction of extension (**strain shadows–pressure shadows**) leads to reprecipitation (Figure 25.5a,b). This is **pressure solution redeposition** and is one form of diffusional mass transfer. **Sutured contacts** between grains characterizes strain accommodated by solution redeposition (Figure 25.11a). Dissolution of vein quartz related to the formation of crenulation cleavage in a phyllitic shale is shown in Figure 25.6

Deformation of Soft Sediments

Sediments deposited in forearc, backarc, and foredeep basins are eventually deformed, as are sediments scraped off subducting plates. Deformation of sediment accumulations begins in response to differential compaction, dewatering, and slope instability. Grain cementation converts these soft sediments to sedimentary rocks, preserving some of these soft sediment deformational structures. Folds originating from gravity slumping, load casts resulting from differential compaction and expulsion of water, and even incipient cleavage may be preserved. Scaly fabric containing oriented platy grains may form in accreted trench-fill muds as the result of both distributed fabric collapse resulting from compaction and shear-related compaction (Cowan, 1982; Brandon, 1989). Lack of cleavage and pervasive cataclasis in these **olistostromal** (chaotic mixture of blocks and muds) mélanges is an indication that the extreme grain-size variation in such rocks was acquired by submarine falls and slides, not by tectonic deformation. Some mélanges are characterized as flow of stiff hydraulic slurries. Grain rotation and grain boundary sliding dominate in **hydroplastic faults** and **sedimentary volcanoes** (Talbot and Brunn, 1989).

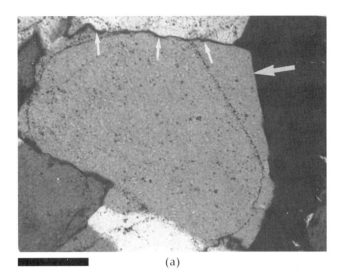

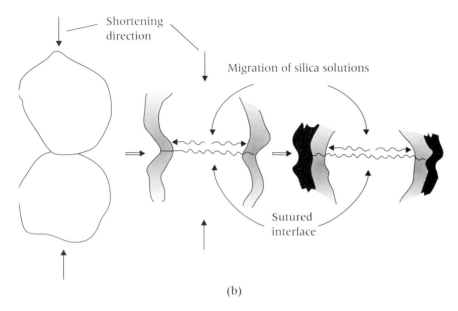

Figure 25.5
Pressure solution at quartz grain contacts.

(a) Pressure solution generating serrated solution surface on quartz clasts (small arrows) with redeposition of quartz in "pressure lows" as overgrowth (large arrow), converting quartz arenite to orthoquartzite. X-polars. Bar scale = 0.14 mm. (b) Sketch showing sequence of solution and redeposition shown in (a). Note how redeposited quartz is itself dissolved as shortening progresses.

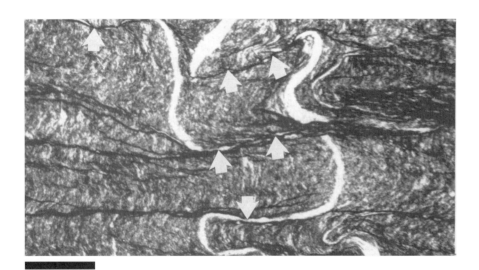

Figure 25.6
Quartz microvein parallel to continuous cleavage, perpendicular to later crenulation dissolution cleavage in phyllitic shale. Note thinning or disappearance of quartz due to dissolution (arrows). Preble formation, Golconda, Nevada. Bar scale = 1 mm. (Courtesy Nevada Bureau of Mines and Geology.)

Deep burial of sediments without much deformation can lead directly to metamorphism, known as **burial metamorphism**, which is marked by very low-grade mineral reactions, not by development of schistosity. Alternatively, in a dynamic environment such as an accretionary wedge, sediments being modified by diagenesis and brittle deformation grade into zeolite facies metamorphism (Ernst, 1975).

Deformation of Sedimentary Rock

Nontectonic and Tectonic Deformation

Most observable low-temperature deformation is in sedimentary rock because this is the most abundant rock type at and near the Earth's surface. Deformational features in nonmetamorphic sedimentary rocks can be easily traced into metamorphosed sedimentary rocks (Borradail et al., 1982; Ramsay and Huber, 1987).

Nontectonic deformation includes blocks of rock falling, sliding, and slumping on oversteepened basin margins, ending up, for example, in olistostromes and diamictites. Subaerial gravity sliding typically occurs where there is oversteepening related to normal faulting and at the front of advancing thrust sheets. Brittle deformation at the base of slide blocks produce breccias with little or no directional fabric (Figure 23.3b; Grier, 1983).

Tectonic deformation of sedimentary rocks at low temperatures occurs in subduction melanges (Figures 25.4a, 25.7a), miogeocline foreland thrust belts, the interior of orogens, and in diapiric salt (Figure 25.7b). Deformation is characterized by well-defined fault zones as well as distributive faulting. Distributive faults may be spaced (Figure 25.8) or pervasive shears. Scaly fabrics, characterized by curviplanar surfaces, typically have slip surfaces (Lash, 1989; Brandon, 1989). Kinematic reconstruction of sheared rocks at low temperatures is aided by the presence of movement indicators such as fossils, reduction spots, ooids, and veins (Vroliūk and Sheppard, 1991).

Structures of Tectonic Deformation

Folds. At low temperature and pressure, folding of bedded siliciclastic sedimentary rocks involves mainly cataclastic flow and diffusional mass transfer (pressure solution). Flexural slip folds maintain constant bed thickness by slip along bedding planes. Flexural

(a)

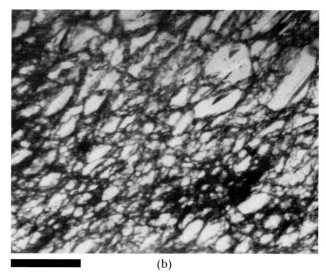
(b)

Figure 25.7

(a) Deformed (accreted) ocean floor ribbon chert with shaley partings (darker). Light zones in chert define recrystallization along fracture planes (note conjugate pattern, conjugate spaced cleavage). Orcus Chert, Deadman Bay, San Juan Island, Washington. Bar scale = 1.2 cm. (b) "Foliated" rock salt with porphyroclasts of halite in comminuted and probably recrystallized halite matrix. Oakwood Dome, East Texas. Slab surface. Bar scale = 4 cm. (Courtesy M.P.A. Jackson, Bureau of Economic Geology, University of Texas at Austin.)

Figure 25.8
Closely spaced high-angle step faults in strongly indurated Triassic siltstone. Sierra Nevada, California. Rock surface. Bar scale = 2.3 cm. (Courtesy R. Schweickert.)

flow folds have thinning of fold limbs, in part the result of pressure solution but including other grain-scale flow processes of a collectively ductile character.

Competent beds inclosed by incompetent beds, such as a bed of limestone or sandstone in shale, beds of ribbon chert in shale (Figure 25.7a), or sandstone in limestone (Figure 25.9), contribute to disharmonic folding. Brittle fragmentation of the competent bed leads to dismemberment and block rotation in the mobile host. Deformation also generates crush breccia in fold hinges and **kink folding**. Kink folding consisting of angular parallel folds in a highly disrupted zone related to subduction is shown in the frontispiece photo.

Cleavage. Development of cleavage in folded rock is very common. In the low-temperature, low-pressure regime cleavage forms by rotation of ridged grains and by concentration of residues by way of pressure solution. Cleavage may also form in sediments by compaction with or without contributing pressure solution.

Cleavage is either continuous (distributed) or zonal (spaced). **Slaty cleavage** is continuous on the mesoscopic scale and begins to form in shales (Figure 25.10), mudstones, siltstones, and tuffs. It is best developed in very low-grade metamorphic rocks such as slates and phyllites (Figure 21.13b) in which the cleavage is accentuated by preferred orientation of new crystals resulting from recrystallization. Refolded slaty cleavage results in the development of **crenulation cleavage**, characterized by internally undisturbed microlithons between cleavage domains in which there has been concentration of certain minerals, such as mica, and by dissolution of others, such as quartz (Figures 21.14b, 21.15c).

Spaced cleavage includes **dissolution cleavage** (including crenulation cleavage), **crack-like disjunctive cleavage**, and **differentiated zonal cleavage** (Borradaile et al., 1982). Dissolution cleavage results from the concentration of residues by the same process that forms **stylolites** in carbonate rocks (Figure 25.11a), namely by solution perpendicular to the compression (shortening) direction (Figure 25.5b, 25.11b). There may be thinning or even disappearance of vein material caught up in the formation of dissolution cleavage (Figure 25.6). An example of disjunctive spaced cleavage with a tendency for conjugate geometry in deformed ribbon chert is shown in Figure 25.7a.

Dissolution cleavage also forms, in part, by mass transfer of material to the pressure shadow zone

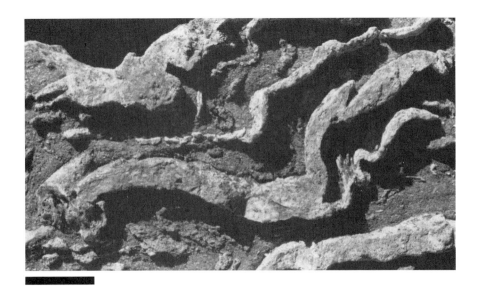

Figure 25.9
Disharmonic folding with brittle disaggregation of quartzitic beds in recrystallized limestone matrix (in low relief). Northern Snake Range décollement, Nevada. Bar scale = 2 cm.

Figure 25.10
Bedding parallel continuous cleavage in slaty shale. Coal Canyon, West Humboldt Range, Nevada. Bar scale = 4 cm.

(**strain shadow**) at the ends of grains. Reprecipitation in these places generates **mica beards** at the ends of quartz grains and quartz overgrowths.

Fractures, Faults, and Joints. Fractures that are more widely spaced than those of crack-line cleavage are joints. Jointing represents brittle fracture, with propagation direction shown by plumose markings. Except for a single joint surface, the scale of mesoscopic jointing is beyond that of hand specimens. However, if the joint contains mineral precipitates, the resulting vein may be collected in hand specimens. Vein fillings of quartz and carbonates are common in the low-temperature regime, generating from pressure solution or from infiltrating solutions.

Extension fractures (or dilated joints) form perpendicular to the least principal stress direction. A tensile fracture forms if the least principal stress is tensile, as it is in a stretching situation such as in very viscous lava (Figure 25.12). Pore-fluid pressure can

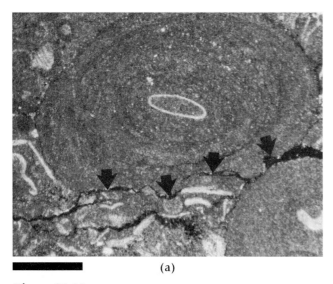
(a)

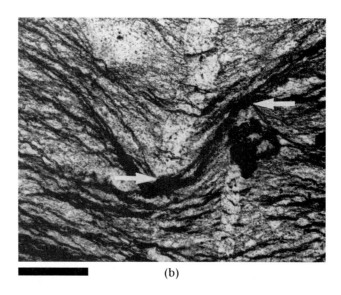
(b)

Figure 25.11
Stylolitic solution cleavage.

(a) Micritic peloid with dissolution surface (arrows) and concentration of insoluble residues (black). Deep Creek Range, Nevada. PPL. Bar scale = 1.2 mm. (Courtesy J. Kepper.) (b) Pressure solution, "stylolitic cleavage" marked by insoluble residues. Note shortening (between arrows) indicated by displaced vertical quartz microvein. Valmey Formation, Golconda, Nevada. PPL. Bar scale = 0.6 mm. (Courtesy D. Davis, Nevada Bureau of Mines and Geology.)

Figure 25.12
Extension cracks in obsidian flow rock. Mono Craters, California. Bar scale = 1 cm. (Courtesy R. Schweickert.)

generate tensile fractures. Extension fractures are commonly associated with boudinage and dismemberment in disharmonic folding (Figure 25.9).

A fault is a surface along which there has been displacement as a result of sliding, typically generated during noncoaxial strain in the brittle regime. Displaced and rotated rock fragments characterize tectonic breccias such as occur in subduction mélanges (Figure 25.4a) and in other cases of brittle faulting (Chapter 23).

Brittle meso- and microfaults may be filled with mineral precipitates. Reactivation of faulting may crack through these vein fillings to be filled again. **Crack-seal veins** are useful in establishing sense of shear and other reconstructions of brittle deformation (Cox and Etheridge, 1983).

Cooling cracks and **desiccation cracks** are extensional features forming in the brittle regime. Forceful injection or passive intrusion of sediments along extension fractures of any origin may form **clastic dikes**.

Brittle Deformation or Ductile Deformation?

How a rock behaves mechanically and chemically in the low-temperature regime depends a great deal on grain composition. A folded mass of rock may appear to have flowed ductilely, but the folding process may have involved mainly cataclastic flow and pressure solution redeposition if most of the constituent minerals have not been able to deform plastically.

The presence or absence of crystal plasticity would seem to be a sure way to isolate the brittle regime from the ductile regime with due consideration of the mineral species involved. However, there are at least four limitations. (1) Certain types of plastic deformation occur in some minerals even at low temperature. Brittle and ductile processes may complement each other, occurring simultaneously

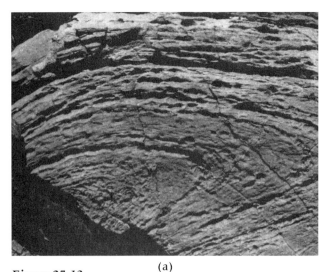
(a)

(b)

Figure 25.13
Apparent ductile deformation of bedded carbonates.
(a) Folded limestone. Silurian Hills, California. Hammer scale (lower left corner).
(b) Folded mylonitic limestones and dolostones in the high-strain zone of the Northern Snake Range décollement, Nevada.

(Rutter, 1972). (2) The nature of deformation may be obscured by recrystallization. (3) Many rocks in the low-temperature region are extremely fine grained, and since the products of deformation may be gouges and other extremely fine-grained materials, observation of crystal deformation may be impossible (Rutter, 1986). (4) Interlayered rock types deform differently than monolithic rocks. For example, gypsum may act as a lubricant in shale deformation, reducing strength and generating ductility (Jordan and Nuesch, 1989).

In view of these limitations, deformation mesoscopically characterized as **homogeneous flow** may not be specifically identified as cataclastic flow or ductile flow. Even folded carbonates that appear to have behaved ductilely (Figure 25.13a,b) may have had a component of cataclastic flow (Rutter, 1972) that has been overridden by plastic deformation and recrystallization. Furthermore, the role of solutional mass transfer in the change of shape of grains and rock masses, particularly carbonates, may not be apparent.

In the face of these limitations, Groshong (1988) proposed a classification of **deformation mechanism associations**, starting with the weakest "rock" and lowest temperatures.

1. Hydroplastic deformation and cohesionless, nondestructive particulate flow. Strain may take place in grains such as carbonate pelloids, ooids, and fossils.
2. Brittle behavior (cataclasis) characterized by fracturing or faulting of framework silicates occurs in association with pressure solution but with development of only 0–2% intracrystal "plastic" deformation, such as glide features and weak progressive extinction.
3. Significant crystal plastic deformation that cannot, however, produce a grain-orientation fabric. Fracturing, faulting, and pressure solution may still occur. Phyllosilicates may change species, but there is no general recrystallization for most rock materials.

These three mechanism associations characterize the **low-temperature deformational regime**. In this regime sedimentary rocks maintain their sedimentary character. The threshold to metamorphism and the formation of tectonites takes place at elevated temperature that allows preferred crystallographic orientation of minerals such as quartz and calcite (Schmid et al., 1987).

PART **VII**

Rocks Formed by Magmatic and Solid-State Processes

Solidification of magma and deformation of solids occur simultaneously in several geological circumstances. Although these rocks have mineralogical and physical attributes indicating that both magmatic and high-T solid state processes occurred, there is no comprehensive classification by which such rocks can be treated as a generic rock group separate from those of magmatic and metamorphic rocks.

Incorporation of xenoliths and autoliths into magmatic bodies is traditionally characterized as a phenomenon rather than a rock-forming process. Viewed from the perspective of a hand specimen of some magmatic rock containing rock inclusions measured in centimeters, magmatic breccias have a distinct physical appearance for which distinct formation process can be defined. The inclusions in magmatic breccias are typically subangular, but they may be rounded as well. The included fragments may be foreign to the magma (xenoliths), distantly related to the magmatic system (cognates), or derived from solidified portions of the magma itself (autoliths).

Magmatic systems are characteristically mobile as a consequence of their liquid state. Flow alignment of early-forming crystals is recognized as a phenomenon within the fundamental context of magmatism. Since flow of solids is also a well-documented process, leading to the formation of dynamothermal metamorphic rocks, for example, there is likely to be juxtaposition of magmatic flow and solid flow in magmatic systems that are (1) nearly but not completely solidified and (2) involved in localized forceful emplacement or participating in large-scale compressive tectonics.

It has been increasingly realized that vast volumes of granitic gneisses are the result of both magmatic and metamorphic processes, not simply the result of metamorphism of previously crystallized granites or thorough recrystallization of silicic volcanic rock or siliclastic sedimentary rock. Formation of granitic gneisses along the borders of plutons was recognized long ago as a gneiss-forming process occurring during pluton emplacement. The process is now recognized as occurring deep into the interior of some large plutonic bodies. *Dynamomagmatism* is proposed as a term characterizing this particular juxtaposition of rock-forming processes.

Migmatites are already recognized as rock types and they have their own classification. Co-existence of both magmatic and metamorphic (including metasomatic metamorphic) rock-forming process is the trademark of migmitization, but unlike dynamomagmatism, the metamorphic (solid-state) components are fundamentally inherited from pre-existing rock, not developed in situ by crystallization of magma.

Processes leading to migmatization can be characterized on the hand-specimen scale if the juxtaposition of magmatic and metamorphic components is on a scale measured in centimeters, which it is locally in most migmatite complexes. Migmatizing processes are varied but are chiefly the result of either magmatic injection or partial anatectic (in situ) melting. Injection migmatites are characterized by intrusion of exogenous magma and local assimilation of the pre-existing host rock. Anatectic migmatites are characterized by in situ dissolution melting that typically leads to self-injection. If melt is extracted from an anatectic system the solid residue becomes a rock type itself, occurring either as restite mineral accumulations from which melt has been entirely extracted or as local concentrations of minerals remaining in the migmatite as melanosomes.

Lacking chemical and isotopic knowledge of magma source in relation to the pre-existing rock in migmatites, identification of specific migmatizing processes is tentative at best. Similarly, knowledge of geochemical fertility (no melt extraction) or infertility (melt extracted) of restite mineral assemblages, typically occurring as inclusions rafted along in magmatic systems, is necessary for at least tentative identification of inclusions as fragments of restite (or not) carried away from a site of partial anatexis.

26

Magmatic Breccia

Diorite inclusion swarm in granodiorite host. Pyramid Peak Range, California. Lens cap scale (arrow). (Courtesy C. Sabine.)

Angular and rounded rock fragments in magmatic rock bodies range in size from microscopic grains to enormous slabs of lava crust in surface lava flows and slabs of roof and wall rock in chambered magma. Brittle rock fragmentation followed by incorporation in magma generates inclusions. In a broad sense, inclusion-bearing magmatic rock is **magmatic breccia** regardless of the ratio of inclusions to host rock.

Brittle Fragmentation

Fractured rock is formed from rock that behaves brittlely in an appropriate stress regime. If the strain rate is high, even magma itself can be fractured (Walker, 1969), but fracturing leading to formation of magmatic breccias occurs either before magma is emplaced or by low-strain processes during emplacement. An exception is the interface region between magma in volcanic conduits and pyroclastic eruption where there can be explosive generation of blocks of rock. Blocks derived from conduit walls and temporary plugs in the conduit, and by caldera collapse may be engulfed by effusive lava (magmatic breccias) or caught up in pyroclastic flows and surges as **lag breccias** (Lipman, 1976; Druitt, 1985). Fragmentation of submarine silicic domes can occur by autobrecciation of magma quenched against water (Pichler, 1965).

Crystallization of magma or quenching of magma to glass leads to contraction and generation of thermal stresses. Such stresses are fundamentally tensile and are relieved by fracturing. If only the margin of a magma body has solidified, such as the glassy–crystalline crust on lava or the chilled facies along an intrusive contact, stress builds up in that rock, eventually being relieved by fracturing, perhaps preceded by an initial ductile stretching.

Fracturing within a body of partly crystallized magma may also occur if there is a high rate of shear. Fracturing of a magma containing more than about 65% crystals can occur if the strain rate is appropriately rapid. Thus, fracturing and fragmentation may occur within a stressed magma body, not just in the quenched surface of a lava flow, lava lake, or the wall zone of a magma chamber.

Fracturing of wall or roof rock completely foreign to the magma may occur prior to or during magma emplacement. Pre-existing fractures, originating by many possible means, including tectonic extension, gravitational collapse, and release of stored strain, may be extended by **magma fracturing**. Pressurized magma may extend the fracture and create new fractures in the process.

Fracturing leading to fragmentation may not be entirely brittle. Stressed hot rock may deform in the brittle–ductile region if some of the contained minerals,

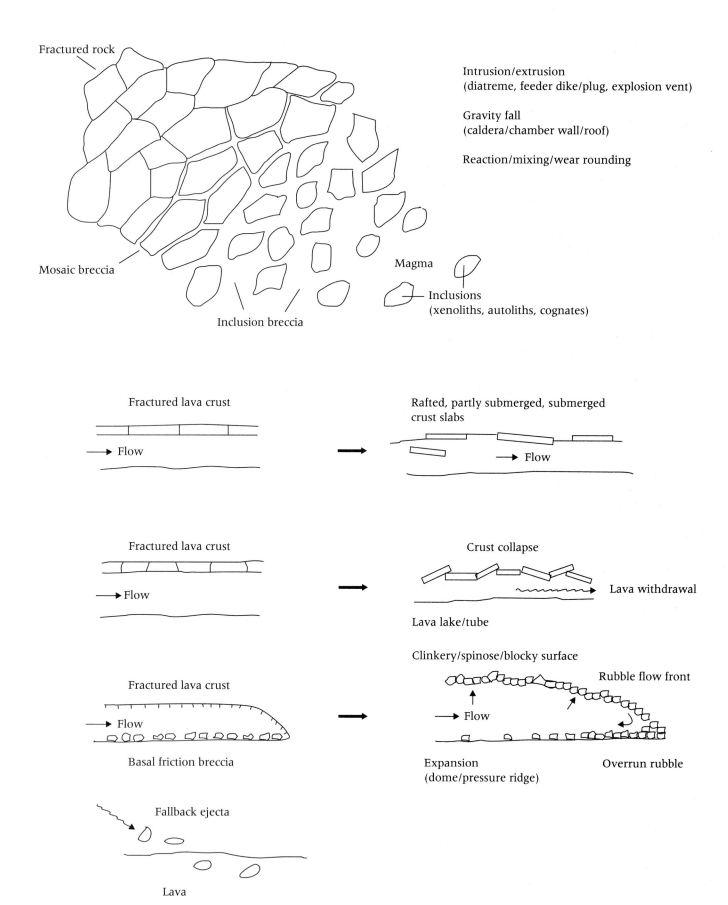

Figure 26.1
Representations of circumstances in which magmatic breccias form.

such as quartz, deform ductilely while others, such as feldspars, deform brittlely.

Incorporation into Magma

Fractured rock may be injected by magma-forming **agmatitic injection migmatites**. The rock fragments form a mosaic breccia in which the individual blocks have been displaced only slightly from their neighbors without significant rotation. At the other extreme, rock fragments may be incorporated and then dispersed through magma, forming isolated inclusions (enclaves). Inclusions formed within a magma system itself are **autoliths** (**cognate inclusions**), and they are coeval in the sense that they are geologically of the same age as the host magma, (Long and Wood, 1986; Irving, 1987). Inclusions derived from rock that has no genetic link (or no overt link) to the magma system are **xenoliths** (Nixon, 1987; Schneiderman, 1989).

Incorporation of autoliths is best illustrated by looking at the solidified crust of a lava flow or lava lake. Contraction fractures generate blocks of crust. Lateral flow of lava will separate blocks and lava fills in between the slabs. Total incorporation may take place by (1) gravity settling, since rock is denser than its melted equivalent, and/or (2) by turbulent flow, an extreme example being the overriding of lava crust at a flow front. A similar incorporation may occur by roof collapse attendant to lava withdrawal. Several possibilities of fragmentation and incorporation are shown in Figure 26.1. Similar mechanisms of fragment incorporation may take place in chambered magma (Figure 26.2), especially in a convecting and crystallizing system, but documentation of these processes is difficult in comparison with observable surface flow of lava.

Mingling of magmas of different composition generates net-veined complexes and concentration of more mafic inclusions in more felsic magma. Hotter magma quenching against cooler magma tends to yield spherical or oblate spheroidal inclusions of magma or crystal-bearing magma.

Examples of Magmatic Breccia

Autoliths

Initiation of fragmentation of lava crust caused by shrinkage and an expanding interior is shown in Figure 26.3a. Further development of these partly desegregated rock fragments generates a blocky surface known as **aa lava** flow. The tops and bottoms of plateau basalt lava flows are characterized by such blocky fragments that are loosely attached to each other (Figure 26.3b).

Fragmented rhyolite engulfed by lava of approximately the same composition is shown in Figure 26.4a,b. Presumably, the relationship evolved by an explosive episode that fragmented consolidated rhyolite, allowing fresh magma to enter the breccia zone. The fragment shown in Figure 26.4b is a glassy equivalent of the microporphyritic host rock. Devitrification spherulites occurred in this rock after brecciation and fragment incorporation, to account for the nearly perfect coincidence of spherulite rims with fracture surfaces.

Fragmentation within supraplutonic magmatic systems is shown in Figure 26.5. The graphic alkali feldspar–quartz fragment in Figure 26.5a occurs in a granite porphyry containing phenocrysts of quartz and alkali feldspar. Assuming that the simultaneous growth mechanism proposed by Fenn (1986) is correct, namely that such intergrowths are late stage and form in hydrous melts, this intergrowth unit is most likely a fragment (not a phenocryst cluster) of a previously crystallized portion of the granitic system, forming in a batch of magma that evolved to the late-magmatic stage, later engulfed by a fresh batch of parent magma.

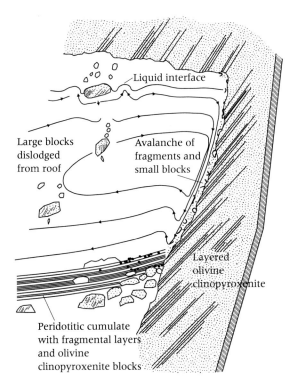

Figure 26.2
Gravity-assisted fragmentation of earlier crystallized roof and wallrock that forms a chamber in which a new convecting magma resides. (From Irving, 1987.)

Figure 26.3
Autobrecciation associated with lava flows.

(a) Fractured and extended basalt forming partial blocks at surface of lava flow. White portions ar lichen covered. Near Cedar Breaks, Utah. Bar scale = 10 cm.
(b) Thick basal columns resting on blocky basalt flow surface of underlying flow. A thin basal breccia at base of columns is not readily visible. Columbia River basalt. Washington State (Long and Wood, 1986). Hammer scale located with arrow.

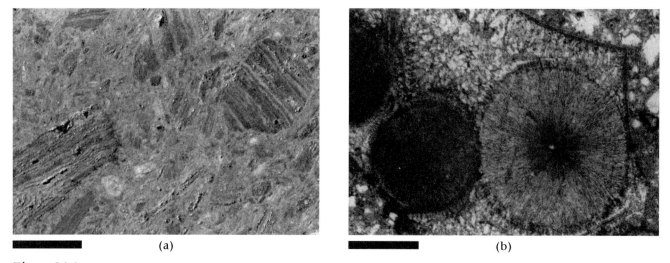

Figure 26.4
Autobrecciation associated with near-vent flows.

(a) Laminated flow rhyolite fragments in nonlaminated engulfing rhyolite of same composition. Garnet Hill. Ruth, Nevada. Slab surface. Bar scale = 1.2 cm.
(b) Devitrified rhyolite fragment engulfed microporphyritic rhyolite. Spherulite surfaces tangential to fracture surfaces suggests postfracturing devitrification. Indian Springs, northwestern Nevada. PPL. Bar scale = 0.6 mm. (Courtesy P. Purington.)

Figure 26.5
Mineral fragmentation in supraplutonic porphyry.

(a) Probable fragment of earlier pegmatitic graphic alkali feldspar–quartz included in and partly disaggregated by later pulse of rhyolite magma that forms a very fine-grained matrix. Dragoon Mountains, Arizona. X-polars. Bar scale = 1.2 mm.
(b) Microbrecciated plagioclase crystal in phenocryst-dominated two-feldspar granite porphyry. Intercrystalline melt flowed into dilated fractures forming microveins. Gillis Range, Nevada (Hibbard, 1986). X-polars. Bar scale = 0.6 mm.

Very late-stage fragmentation is indicated by the microbrecciated plagioclase crystal shown in Figure 26.5b. A presumed explosive event, probably related to sudden release of volatiles, resulted in fracturing of plagioclase, K-feldspar, and quartz phenocrysts. The co-existing interphenocryst melt, amounting to only about 20% by volume, redistributed into microveins in the fractured crystals. In this case the breccia consists of crystal fragments, not rock fragments.

A few examples of fragmentation and formation of inclusions and breccias in plutonic systems are shown in Figure 26.6 and the frontispiece photo. The

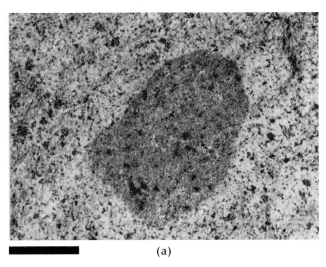

Figure 26.6
Fragmentation in plutonic systems.

(a) Cognate inclusion of microdiorite in fine-grained granodiorite. Note similarity of mafic mineral textures. Mount Davidson. Virginia City, Nevada. Rock surface. Bar scale = 3 cm. (b) Mosaic breccia consisting of hybrid diorite–granodiorite (Endicott Diorite) inclusions separated by one-feldspar granite (Conway Granite). Belknap Mountains, New Hampshire.

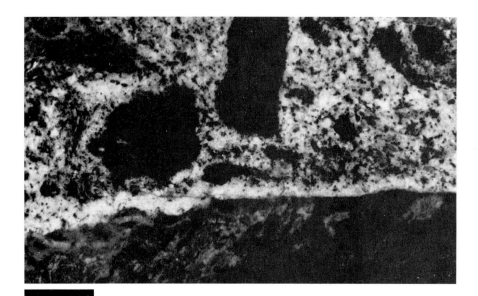

Figure 26.7
Hornfelsic schist xenoliths adjacent to larger diopside–hornblende layered granofels xenolith in granodiorite. Assimilation has increased the hornblende and biotite content of the granodiorite. Chopaka Mountain. North-central Washington State. Slab surface. Bar scale = 1 cm.

cognate nature of the rounded inclusion shown in Figure 26.6a is suggested by textural and mineralogical similarity between the inclusion and host. The inclusion is slightly finer-grained and somewhat more mafic, but the similarities are apparent. In addition, the rounded form of the inclusion suggests that fragmentation and inclusion may have taken place when the inclusion was not completely crystallized, facilitating rounding by disaggregation.

In contrast, the host-dominated breccia shown in the frontispiece photo is characterized by cuspate inclusion morphology. A cognate and/or mingling relation is indicated by mineralogical and textural similarities. The inclusion-dominated breccia shown in Figure 26.6b is characterized by angular blocks of diorite in a more-or-less **mosaic breccia**. The marked mineralogical (color) contrast between the diorite and the invading granite and the apparent fracturing of a completely crystallized diorite at first suggest that the two rocks are genetically unrelated. A coeval relation is indicated, nevertheless, by the presence of evidence for magma mixing and mingling. The diorite is actually heterogeneous, ranging from a granodioritic phase dominated by poikilitic K-feldspar and quartz (Figure 19.14) to true diorite. Fracturing may have occurred before complete crystallization of the granodioritic–dioritic hybrid, as indicated by infiltration of quartz and K-feldspar equivalent fluids through the fracture surfaces. Therefore, in this case, fracturing occurred in a not completely crystallized magma. The angularity of breccia fragments suggest a high rate of strain, perhaps related to a sudden tectonic or eruptive event.

Xenoliths

Intrusion of magma into rock that is in no way related to the generation or crystallization of the magma is likely to generate xenoliths or xenolithic breccias (Figure 26.7). Any rock type may be incorporated into the magma. The shape of the xenoliths reflect both the mode of fragmentation and possible assimilative modifications (Chapter 20) and the inherent fabric characteristics. Breakage is typically controlled by bedding, schistosity, or jointing.

27

Dynamomagmatic Gneisses

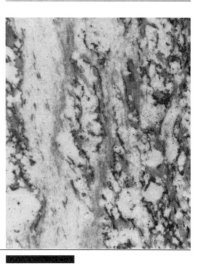

Border phase of Papoose Flat two-feldspar granite (Sylvester et al., 1978). Anastomosing veinlets of aplite (left) result from fluid relocation in deforming late-magmatic system. Inyo Mountains, California. Bar scale = 1.2 cm.

Granitic Gneisses

Quartzofeldspathic gneisses are relatively common in crystalline continental crust. They occur as large-body homogeneous "orthogneiss" bodies (frontispiece photo; Figure 27.1), (Sylvester et al., 1978; Clarke and Lyons, 1986) and as migmatites in which commonly there are portions consisting of relatively homogeneous gneiss (Figure 27.2).

The most common gneiss compositions are two-feldspar granitic, granodioritic, tonalitic, and quartz dioritic. One-feldspar granitic gneisses, gabbroic (amphibolitic) gneisses, and charnockitic granitic gneisses are less abundant.

Relatively homogeneous granitic gneisses may form either with or without melt present. Melt can be present in at least four scenarios: (1) partial anatexis of supracrustal siliciclastic rocks in a homogenizing dynamic environment (Clarke and Lyons, 1986), (2) partial anatexis of nongneissose magmatic granitic rock or previously formed homogeneous granitic gneiss, in a dynamic environment, (3) homogenization of injection migmatite or injection–anatectic migmatite in a dynamic environment, and (4) late-stage deformation of crystallizing granitic magmatic systems (Waters and Krauskopf, 1941; Hibbard, 1987; Paterson et al., 1989; Bouchez et al., 1992).

Melt-absent formation of relatively homogeneous granitic gneisses are represented by (1) dynamothermal metamorphism of siliciclastic supracrustal rock (classic origin of "paragneiss"), (2) dynamothermal metamorphism of siliciclastic supracrustal rocks accompanied by metasomatic exchange on a large scale (classic "granitization"), (3) superposed metamorphism of a prior nongneissose granitic rock (Borradaile, 1976) or previously formed homogeneous granitic gneiss.

Relatively homogeneous gneisses formed by both magmatic and metamorphic processes in a dynamic environment during the late stages of magmatic consolidation are here designated as **dynamomagmatic gneisses**.

The Late-Magmatic Dynamomagmatic Environment

Deformation of a crystal–magma system takes place at near-solidus conditions, brittle in an explosive volcanic environment and ductile (or brittle–ductile) in plutonic systems at advanced stages of crystallization where (1) confining pressure is relatively high,

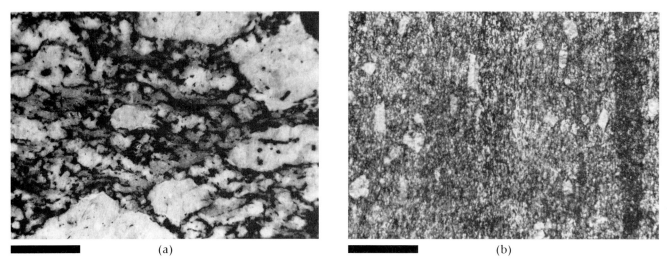

Figure 27.1
Dynamomagmatic gneisses interpreted to have experienced fluid-accommodated strain in part.

(a) Two-feldspar granite gneiss with shape-modified alkali feldspar phenocrysts. South Bergell. San Martino, Italy. Bar scale = 1.6 cm. (b) Same as (a) with diorite inclusion elongate parallel to gneissosity. Diorite may also have been partly liquid during deformation. Bar scale = 6.5 cm.

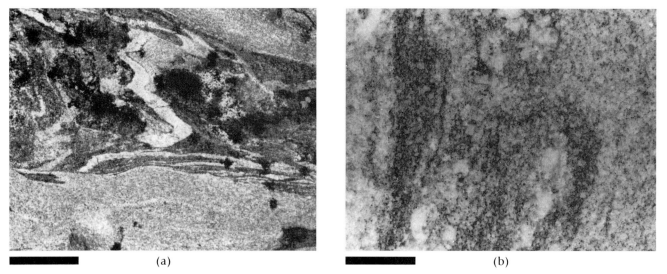

Figure 27.2
Migmatitic gneisses with possible dynamomagmatic components.

(a) Heterogeneous granitic gneiss with homogeneous (across the bottom) injection components containing dynamomagmatic textures (not shown). Okanogan Range, Washington State. Bar scale = 10 cm). (b) Homogeneous fine- to medium-grained two feldspar gneissose granite (right) contains fluid relocation textures (not shown). Felsic–mafic complex (left) may be the segregated equivalent of the homogeneous portion, occurring during late-stage deformation. Okanogan Range, Washington State. Bar scale = 2.3 cm.

(2) temperatures are above the solidus, and (3) there is slow-strain-rate penetrative deformation. These conditions are favorable to the development of dynamomagmatic gneisses.

The character of flow in rock systems containing a melt phase must vary in part according to the amount of intercrystalline melt present relative to the crystalline phases. Experiments suggest (Molen and Paterson, 1979) that if a system contains greater than about 30% melt there is no stress-transmitting crystalline superstructure, and strain is completely accommodated by movement of melt and rafting of crystalline phases in suspension. In systems with less than about 30% melt (critical melt fraction), and in which a crystal framework exists, deformation-related flow can be expected to have textural expression in the crystalline phases and to redistribute intercrystalline fluids in relation to deviatoric stresses (George, 1978). Even small amounts of melt can have major effects on the tectonic behavior of a terrain (Hollister and Crawford, 1986; Dell'Angelo and Tullis, 1988).

During deformation, late-stage magmatic fluids can be expected to locate preferentially in zones normal or subnormal to the least principal compressive stress (direction of maximum extension) having migrated from zones normal or subnormal to the greatest principal stress direction. The low thermal inertia of a small fraction of intercrystalline fluid relative to that of the crystals may in part limit the redistribution (Marsh, 1981). Additionally, limited redistribution can be expected if local (micro) quenching of a melt saturated in aqueous phase occurs as the fluid phase enters lower-pressure regions in a system affected by deviatoric stress. Nevertheless, fluid relocation forming mesoscopic aplites (frontispiece photo) occurs, grading to microscopic fluid relocation features.

Since deformation occurs in the presence of a crystalline superstructure, solid-state processes must take place simultaneously with fluid relocation. Temperatures are high (near solidus) and, if the local ratio of fluid to crystals is very low, many of the crystal plastic deformation mechanisms and the recrystallization characteristic of dynamothermal metamorphism occur. The presence of quartz and feldspars, for which thresholds for crystal plasticity are different, result in brittle behavior for the feldspars and ductile behavior for the quartz.

Dynamomagmatic Textures

Two-feldspar (calc–alkaline) granitic systems are capable of producing a larger variety of diagnostic relocation textures than one-feldspar granitic or quartz dioritic, trondhjemitic, or tonalitic systems for the following reasons: (1) Both plagioclase and K-feldspar are present, and distribution of later-crystallizing K-feldspar relative to earlier-crystallizing plagioclase (more calcic) is texturally distinct because it is mineralogically distinct. (2) Myrmekite and microaplite are likely to occur in aqueous-phase-bearing calc–alkaline systems. Both are easily observed where they locate in the deforming crystal–melt system. (3) In calc–alkaline systems there is a relatively large separation of liquidus and solidus, providing an extended period of crystallization in the plutonic environment during which strain is active.

The significance of myrmekite in its association with late-crystallizing magma or hydrous magma has been discussed in the context of late-magmatic crystallization from hydrous melt (Hibbard, 1979). Although still a controversial subject (Vernon, 1991) the presence of myrmekite along with other crystallized products of relocated late-stage fluids is important in the present context because it is optically distinct and aids in locating where these late-stage fluids accumulate.

Fluid Relocation Textures

Microveins. Extension fractures in feldspars can result from bulk flow of a plutonic magmatic system nearing completion of crystallization. Since quartz is behaving plastically at this time, the overall deformational style is brittle–ductile, describing the different behavior of feldspars compared to quartz under the same conditions.

In two-feldspar granitic systems microveins (Hibbard, 1980, 1987; Bouchez et al., 1992) containing quartz, K-feldspar, and myrmekite commonly occur in either plagioclase or alkali feldspar as fracture fillings oriented perpendicular to the extension direction (Figure 27.3a) as marked by elongated crystals, quartz ribbons, and recrystallization trains of quartz. Directional rock fabric may be well developed (Figure 27.3a) or subtle (Figure 27.3b).

If a crystallizing granitic system experiences a thermal quench (due to very shallow emplacement and/or a pressure quench), intercrystalline melt is undercooled and nucleation rate is elevated. In this case, relocation of late-stage fluids occurs in microfractures crystallizing to fine granular assemblage of quartz, K-feldspar, oligoclase, and myrmekite (commonly) that connects to the fine-grained matrix of the porphyry containing the same minerals (Figure 27.3c).

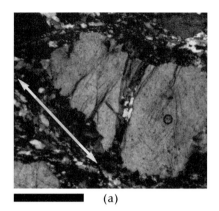  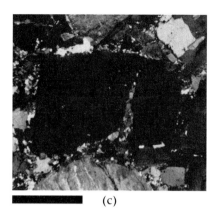

Figure 27.3
Microveins resulting from fluid relocation during dynamomagmatic flow.
(a) K-feldspar–quartz microveins in extension fractures within relict magmatic alkali feldspar. Fracture planes are at high angle to fine-grained recrystallized flow fabric of adjacent quartz (double arrow). Deformation during late-magmatic stage continued into postmagmatic stage generating a strong mylonitic fabric. Silver Creek two-feldspar granite. Northern Snake Range, Nevada. X-polars. Bar scale = 0.8 mm. (b) Cracked alkali feldspar crystals (arrows) filled with K-feldspar and quartz. Note weak gneissose fabric (double arrow). "Colonial Rose". Lac Du Bonnet, Manitoba, Canada. Slab surface. Bar scale = 1 cm. (c) Microveins containing quartz, K-feldspar, oligoclase, and myrmekite in fractured plagioclase (center) and K-feldspar phenocrysts of granite porphyry. "Homogeneous cracking" occurred when there was about 20% remaining melt. No gneissose fabric was generated. Gillis Range, Nevada. X-polars. Bar scale = 1.8 mm.

Local Concentrations of K-feldspar and Microaplite. Volumes of K-feldspar faced with lobes of myrmekite locate in the pressure shadow zones of plagioclase (Figure 27.4a) and early-crystallizing alkali feldspar crystals (Hibbard, 1987) and as lensoid forms not obviously related to zones of reduced fluid pressure (Figure 27.4b,c). The apparent lack of direct relation of late-stage fluid location to the pressure shadow zones associated with rigid feldspar crystals can be explained by the limitations inherent in viewing two-dimensional sections (Figure 27.4d).

Fluids capable of crystallizing a quartz, K-feldspar, oligoclase, myrmekite (commonly) assemblage characterizing microveins in calc–alkaline systems may also locate in zones of reduced pressure. These fine-grained aggregates are conveniently categorized as **microaplite** (Figure 27.5).

In the plutonic environment, the microaplite is formed in response to something akin to a pressure quench as proposed by Jahns and Burnham (1969). In their model, separation of aqueous phase leads to the formation of pegmatites. In the present context, separation of aqueous phase leads to the final crystallization of K-feldspar that occurs in microveins (Figure 27.3a) and localized K-feldspar (Figure 27.4d).

Healed Microbreccia

Multiple fractures and extension results in microbreccia healed with additional growth of the mineral involved. Healing with more sodic plagioclase presents good contrast with more calcic plagioclase fragments (Figure 27.6).

Intracrystalline Deformation

The crystalline components of a deforming crystal–magma "mush" also record the deformation responsible for relocation of intercrystalline fluid. Deformation textures become more prominent as the abundance of crystals increases and mechanical interaction between them increases. In addition, as melt and hydrous melt are consumed, there is less accommodation of strain by displacement of fluid, promoting intracrystalline strain. Crystal deformation in the form of fracturing and microbrecciation of feldspar has been described. In addition, feldspars are reshaped by shear and rotation. Shearing generates crystal fragments that conform to the gneissose fabric (Figure 27.1a). Earlier magmatic flow alignment of phenocrysts may also contribute to gneissose fabric that is accentuated by

 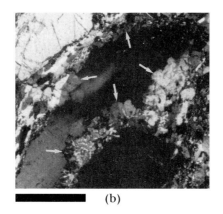 

1. Section level results in apparent deficiency of relocated fluid
2. Section level reveals fluid relocation enclosing crystal
3. Section level reveals fluid relocation in pressure-shadow
4. Section level does not reveal presence of crystal

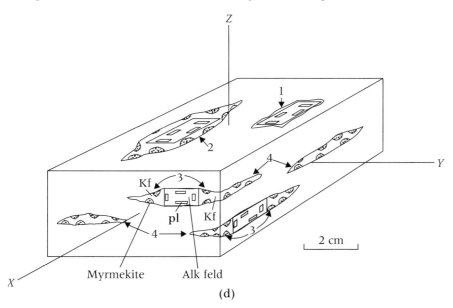

Figure 27.4
Relocation "pools" of K-feldspar and myrmekite.

(a) Pool of K-feldspar (Kf) faced with myrmekite lobes (arrows) in pressure shadow position adjacent to plagioclase crystal (at extinction). Papoose Flat diapiric two-feldspar granite pluton. Inyo Range, California. X-polars. Bar scale = 0.8 mm. (b) Lensoid "pool" of K-feldspar with riming myrmekite (arrows). Papoose Flat pluton. X-polars. Bar scale = 0.8 mm. (c) K-feldspar (center) with facing myrmekite lobes. Two-feldspar granite zone in migmatite (see Harme, 1965). Southern Finland. X-polars. Bar scale = 0.2 mm. (d) Three-dimensional view of the location of myrmekite-lined K-feldspar in relation to early-forming alkali feldspar or plagioclase crystals in deformed two-feldspar granitic rock (kinematic axes shown). Relations are compatible with late-stage fluid relocation to pressure shadow zones, even though view #4 would not reveal the actual relationship.

Figure 27.5
Late-magmatic fluid relocation represented by microaplite containing quartz, K-feldspar, oligoclase, and myrmekitic oligoclase, located between quartz crystal (lower left) and K-feldspar crystal (upper right). Papoose Flat granite gneiss. Inyo Range, California. X-polars. Bar scale = 0.12 mm.

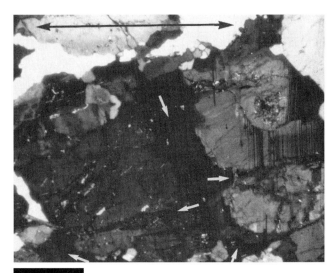

Figure 27.6
Microbrecciated plagioclase healed with late-stage oligoclase (arrows) and quartz in microvein (far left). Gneissose fabric direction shown with double arrow. Papoose Flat gneissose granite. X-polars. Bar scale = 0.5 mm.

continuation of flow into the late stage when fluid relocation and crystal reshaping are active.

Crystal Plasticity and Recrystallization

A stressed two-feldspar granitic system near its solidus is a prime candidate for intracrystalline ductility and recrystallization as it is in high-temperature metamorphic systems containing no melt. Quartz is particularly susceptible to the development of progressive extinction, subgrains, ribbon texture, deformation lamellae, crystallographic preferred orientation, and recrystallization (Figure 27.7). Any or all of these may associate with the brittle textures characterizing feldspars. Thus, dynamomagmatic rocks have components of both magmatism and dynamothermal metamorphism in which the deformation is both ductile and brittle.

If deformation continues beyond the melt-present stage, dynamothermal metamorphic features dominate and the critical textural evidence for dynamomagmatism may be scarce or even eliminated.

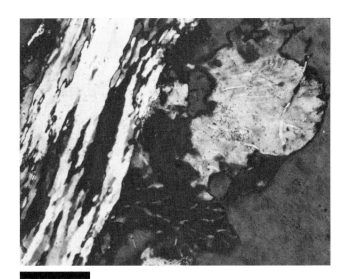

Figure 27.7
High-strain zone containing ribbons and dynamically recrystallized quartz (left) adjacent to relatively undeformed K-feldspar "pool" (right) with facing myrmekite lobes. Deformation in brittle–ductile transition continued beyond dynamomagmatic stage. Silver Creek gneissose two-feldspar granite. Northern Snake Range, Nevada. X-polars. Bar scale = 0.12 mm.

28

Anatexites

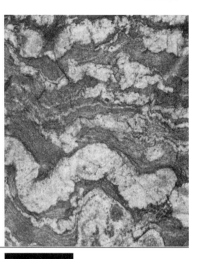

Ptygmatic quartz-rich pegmatitic layering in granitic gneiss as possible representative of partial anatectic melting. Note mafic-rich borders on the leucosome. Sawatch Range, Colorado. Scale = 3 cm.

Melting of Rock

Causes of Melting

Rocks begin to melt if the solidus temperature for a given rock composition is exceeded. The solidus temperature varies with temperature, total pressure, rock composition, and the amount of dissolved H_2O. This means that melting in a rock just below its solidus can be initiated by (1) further heating, (2) reduction in total pressure, or (3) infiltration of H_2O.

Heating of rock above its solidus can occur if there is heat transfer from magma or hot diapiric intrusions. The temperature of the hot material must exceed the solidus temperature of the rock; a basaltic magma can initiate melting in a silicic crustal rock containing hydrous minerals, but a relatively lower-temperature aqueous-phase-bearing silicic magma cannot initiate melting in a rock of basaltic (or graywacke) composition. Internal heating of silicic crustal rock occurs as a consequence of radioactive decay. Heating can also occur by depression of rock to deeper crustal levels, increased heat flow related to crustal thinning, or by development of convective mantle plumes. Total pressure is higher at deeper levels, and effective heating is somewhat countered by the raise of solidi (Chapter 13).

Subducted lithosphere is heated, although a reversal of isotherms may occur initially, leading to partial melting and rise of magma. The zone of partial melting is likely to retain some melt. This melt eventually crystallizes, and a rock of mixed metamorphic and magmatic character is born that is known as **anatectic migmatite**.

Heating of rock leading to melting also occurs in fault zones (Chapter 23) and in meteor-impacted rock (Chapter 24), where glass-bearing breccias are formed. **Friction melting** and **impact melting** are transient processes of restricted occurrence.

Melting activated by reduction of confining pressure is **adiabatic melting**. This occurs beneath spreading centers where there is upward mantle convection, reduction in pressure, and melting of mantle rock to form basaltic magma. Dilatant zones associated with any lower crustal and upper mantle faulting may be sites of melt accumulation (Figure 28.1).

Infiltration of aqueous fluids potentially lowers the solidus, and melting may be initiated with no change in temperature or total pressure. Metamorphic water, liberated during regional high-temperature recrystallization, can participate in local melting. This **ultrametamorphism** is a major processes of migmatization and source of granitic magma.

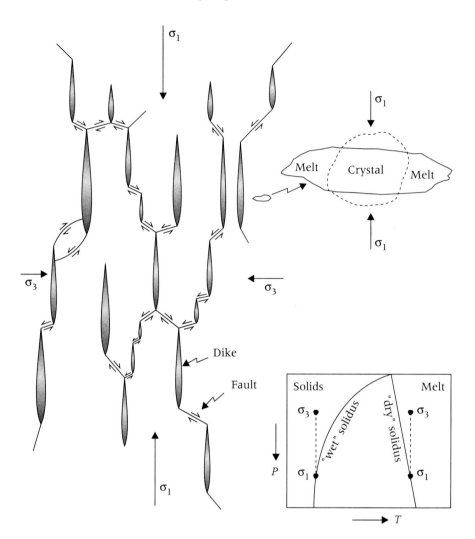

Figure 28.1
Location of anatectic melt generated in rock affected by deviatoric stress. Melt generated at high-pressure points migrates to low-pressure "shadows" around grains and in lenses normal to extension. Melting is favored in "dry" rocks and precluded in "wet" rocks if affected only by a drop in pressure, whereas "wet" systems melt at lower temperature than "dry" systems if there is heating. (Based on Shaw, 1980.)

Location of First Melt

Melting of rock begins at high-energy sites in crystalline aggregates. Pools of melt form at grain intersections in experiments (Mehnert et al., 1973; Jurewicz and Watson, 1984; Hollister and Crawford, 1986), with partially melted crystals becoming convex toward the melt. Interconnection of these melt pockets generates a permeability when there is a considerable fraction of melt and/or if there is deformation within the rock that may concentrate melt in certain zones. Melt may then be extracted from the anatectic system.

When melts nucleate, the energy barrier to further melting is reduced, since the kinetics of phase change along a melt–crystal interface is energy conserving in comparison to melt nucleation. Preferential location of aqueous phase, such as along microfractures and where hydrous minerals are located, can also influence the location of first melt, since this is where the solidus is depressed the most.

Anatectic melting is likely to occur in a tectonically active region because this is where there is magmatic and diapiric intrusion, extensional strain, and structures formed allowing infiltration of aqueous phase. In a strained rock, melting is favored at grain contacts normal to sigma 1 (the direction of greatest principle compressive strength) because at these high energy sites (with aqueous phase present) there is depression of the solidus. Melts so generated will tend to migrate to zones normal to sigma 3, in this case to **pressure shadow** zones at the "ends" of grains. On a larger scale, extensional fracturing parallel to planes normal to sigma 3 is the preferred location of melt (Figure 28.1; Shaw, 1980). This is adiabatic melting.

Composition of Melt Produced

The composition of melt produced reflects the composition of the original rock if the chemical system is closed, with allowances made for possible introduction of aqueous phase. The metamorphic equivalents of mud rocks and wackes are particularly susceptible to melting because they contain hydrous minerals and yield relatively low-temperature melts. Sedimentary rocks are progressively dehydrated during metamorphism, but retention of micas and amphiboles during high-pressure metamorphism sets the stage for ultrametamorphism in which aqueous phase liberated from these hydrous mineral phases lowers the solidus.

The composition of anatectic melt also reflects the degree of melting. Partial melting generates a melt composition that is not equivalent to the rock composition, just as late-crystallizing melts in magmatic systems are not the same composition as the initial magma. For example, first melt in mica–feldspar–quartz schist derived from a mud rock has a composition near the granitic cotectic, yet the remaining rock has a mineral assemblage unlike any granite.

Mineralogical Composition of Anatexites

Rocks formed by partial melting of rock contain minerals generated by (1) metamorphic reactions and (2) reactions that simultaneously yield new crystals and silicate melt.

Metamorphic mineral assemblages derived from three bulk compositional systems likely to be involved in widespread anatexis of geoclinal accumulations involved in orogenesis are

A. Muscovite + biotite + quartz (from illite-bearing mudrock)
B. Muscovite + biotite + quartz + oligoclase (from feldspathic or calcareous montmorillonite-bearing mudrock)
C. Biotite + hornblende + plagioclase (from tufaceous graywacke)

With continued heating, or reheating, the following reactions may take place:

(A) muscovite + biotite + quartz →
K-feldspar + cordierite ± almandine
± sillimanite + H_2O

The specific mineral assemblage on the right-hand side depends on the composition of the biotite, but it will also include whatever is in excess on the left-hand side. If quartz and muscovite are in excess, further heating can generate melt as follows:

ms + qz → sillimanite + H_2O + melt
(Kf + qz potential)

Starting with assemblage (B), the following reaction may apply (Dougan, 1979; Clark and Lyons, 1986):

(B) bt + sil + qz + oligoclase ± Kf →
alm + cort + Kf + pl (more calcic)
+ H_2O + melt (equivalent to Kf
+ qz + albitic pl)

In this case a "minimum" melt is generated, being at a lower temperature than the eutectic melt generated from assemblage (A). In both (A) and (B), the high-pressure polymorph kyanite might be generated instead of sillimanite, and if it is, the melting temperature must be higher since at higher pressure it takes more energy to dissociate the water-bearing micas.

Assemblage (C) is somewhat variable, depending on the initial composition of the tufaceous graywacke and the pressure level of metamorphism. Accordingly, biotite or pyroxene may be present. If the composition of the rock just prior to the appearance of melt consists of biotite, hornblende, and plagioclase, the following reaction applies:

(C) biotite + hornblende + plagioclase →
pyroxene + pl (more calcic) + almandine/
cordierite + H_2O
+ melt (equivalent to sodic pl + qz + Kf)

The final system consists of hornblende, pyroxene, plagioclase, garnet, cordierite, and localized silicic melt.

In all three examples, there is generation of aqueous phase from hydrous minerals that immediately dissolves into the melt as it forms. Unless there is infiltration or diffusion of additional aqueous into the anatectic system, any substantial volume of melt will be undersaturated with respect to the aqueous phase. This is because premelting metamorphic systems may be relatively "dry," especially if there has been high-temperature metamorphism, as, for example, in the pyroxene hornfels and granulite facies.

Further melting, attendant to additional heating, incorporates more and more of the crystalline phases into the melt phase, thus changing its composition. If melting is not complete, refractory mineral phases remain, such as garnet, cordierite, sillimanite, hypersthene, and calcic plagioclase. Reactions (A), (B), and (C) show these refractory minerals being generated along with melts. However, metamorphism can generate the same minerals well out of the field of beginning of melting, especially if there is a deficiency of aqueous phase. Garnet–mica schists, cordierite hornfelses, and pyroxene hornfelses without evidence of melting are common. How can refractory phases formed by way of high-grade metamorphic reactions be distinguished from the same phases generated in (1) anatectic melting reactions and (2) by direct crystallization from a peraluminous magma? The genetic association is typically equivocal, being based solely on mineralogical and textural relations. Field relations may be very useful, as well as trace element geochemistry.

Can Anatectic Migmatites Be Identified?

On the basis of melting experiments, there is no doubt that granitic magma can be generated on a grand scale by melting of continental crust, but petrographic characterization of anatectic migmatites is in a state of development. Crystalline cores of continents are characterized by migmatites, but just what portion of these migmatites represent in situ anatectic melting is not obvious because several other rock-forming processes are likely to occur in an environment conducive to anatectic melting.

Migmatizing processes include (1) partial melting in a static environment, (2) partial melting in a dynamic environment in which melt is relocated and forms fluid relocation textures and self-injection features mimicking those of dynamomagmatism (Chapter 27), (3) chemical and mechanical partitioning on a local scale without melting ("metamorphic differentiation"), (4) open-system metasomatism accompanying metamorphic process without melting ("metasomatic migmatization"), and (5) injection of magma from an external source forming "injection migmatites."

Knowing that water-bearing silicic crustal rock melts at geologically observable levels is the inspiration for trying to identify small-scale features resulting specifically from anatectic melting.

An Example of Partial Anatexis

An example of partial melting in a rock chemically capable of generating garnet, cordierite, and orthopyroxene is shown in Figure 28.2. Interpretation of the mineralogical and textural relations in this rock attempts to distinguish those processes resulting from reaction with and without involvement of anatectic silicate melt in a single rock-forming event.

A paragneiss containing an assemblage of oligoclase, biotite, almandine garnet, and ilmenite was thermally heated by a 25-m.y. tonalite in the Mt. Pough region, Northern Cascades Washington (J. Vance, personal communication). Reactions generated cordierite, hypersthene, quartz, oligoclase, and magnetite–ilmenite, based on interpretation of texture. Cordierite mantles almandine (Figure 28.2a). Hypersthene and quartz tend to mantle cordierite, occurring against cordierite crystal faces (Figure 28.2b).

  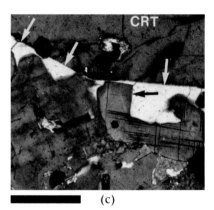

(a)   (b)   (c)

Figure 28.2
Reaction and melting in paragneiss of Mt. Pough area, Washington.

(a) Cordierite mantle on almandine garnet. Note deteriorated aspect of almandine and development of cordierite crystal faces. Note also how cordierite has extended to the left beyond the original confines of almandine. X-polars. Bar scale = 0.5 mm. (b) Hypersthene (upper left, black arrows) and quartz (lower right, white arrows) partial mantles on cordierite. Almandine probably located above or below plane of thin sectioning. X-polars. Bar scale = 0.5 mm. (c) Quartz against cordierite (CRT) crystal faces (white arrows) and oligoclase-rimmed plagioclase crystal faces (black arrow). Small opaque grains are ilmenite, high relief grains are hypersthene, and biotite is located lower center and right. X-polars. Bar scale = 0.12 mm.

Plagioclase zones from An_{50} to An_{40} to oligoclase rims facing quartz (Figure 28.2c).

The following reaction showing the generation of cordierite (see also Gribble, 1970; Ashworth and Chinner, 1978) and hypersthene is based on SEM microprobe analysis:

$$4 \text{ almandine} + \text{biotite} \rightarrow$$
$$3 \text{ cordierite} + 0.8 \text{ hypersthene} + 1.5 \text{ ilmenite/magnetite} + 0.2 \text{ quartz}$$

The development of cordierite crystal faces against quartz, and sodic plagioclase crystal faces against quartz (Figure 28.2c), are interpreted to mean that there was melt generated during the reaction that simultaneously produced cordierite and hypersthene. The melt allowed for freedom of growth and development of crystal faces. The melt is now represented by quartz and oligoclase rims on the plagioclase, but it initially also contained potassium. The rock now contains a mineral assemblage and a texture that is both metamorphic and magmatic by way of anatexis. These relationships are schematically summarized in Figure 28.3.

A similar textural relation involves K-feldspar and cordierite having crystal faces toward quartz (Vernon and Collins, 1988). Other parts of the same crystals texturally participate in polygonal aggregates. If the polygonalization has resulted from metamorphic recrystallization, as it most probably has, this is another example of the coexistence of metamorphic and magmatic textures in an anatectic system (see also McLellan, 1983).

Migmatization

Partial anatexis leads to the formation of anatectic migmatite. If there is extraction of this melt, the residual compacted mineral assemblage is **restitite** (Figure 28.4) (Chapter 29).

Migmatite is "mixed rock" in the sense that both metamorphic and magmatic components are, or seem to be, present (Sederholm, 1923; Mehnert, 1971; Ashworth, 1976, 1985; Yardley, 1978; Gupta and Johannes, 1982; Atherton and Gribble, 1983; McLellan, 1988). Local migmatization can occur in contact zones. Here the environment may be more or less static, such as at Mt. Pough, or dynamic if the intrusive is diapiric. Regional migmatization occurs at convergent plate boundaries, and the environment is dynamic.

Dominantly static isochemical anatexis is characterized by local development of felsic mineral assemblages, in a host that maintains its metamorphic fabric. Partial anatexis in rock that is undergoing dynamothermal metamorphism acquires new fabrics ranging from well-organized layered segregation of felsic and more mafic residual components (Figure 28.4) to chaotic mixing of the melt and residual solid phases.

In a dynamic environment—that is, one with a deviatoric stress, anatectic melt is redistributed by shear. Melt finds itself in contact with crystalline phases with which it may be out of chemical equilibrium, especially if the original rock was compositionally heterogeneous. Melt may locally inject its co-existing restites, forming microveins and dikelets. There may be transport of both the melted and unmelted portions of the migmatite in the form of **migmas** and **migmatitic diapirs**.

The portion of a migmatitic system that was melt is more felsic because first melts contain "minimum" components. Partitioning of felsic and more mafic components probably is aided by diffusion and small-scale convection in the melts, by chemical exchange via intergranular aqueous phase, and by mechanical mass transport processes.

The felsic portion of migmatites is the **leucosome**, and the more mafic portion is the **melanosome**. These terms are descriptive and do not apply only to anatectic migmatites. If the leucosome is shown to have been derived from the melanosome, it is a **neosome** and the melanosome is the **paleosome**. If anatexis is demonstrated, the leucosome–neosome is anatexite, and the melanosome–paleosome is the restitite (Figure 28.4).

Crystallization of In-Situ Anatectic Melt

Is has been shown for the Mt. Pough rock (Figure 28.3) that the mineral assemblage crystallizing from anatectic melt can be different than the mineral assemblage of the premelting metamorphic rock.

Suppose the original rock was quartz–muscovite schist. The melting reaction is

$$\text{muscovite} + \text{quartz} \rightarrow \text{sillimanite} + H_2O + \text{melt (KF + QZ potential)}$$

The original rock did not contain K-feldspar, but the reaction suggests that K-feldspar could appear in the neosome. Will the melt crystallize to K-feldspar + quartz, or will there be a reaction between the melt and sillimanite, as the system cools, producing muscovite once again? This may well depend on a number of factors: (1) rate of cooling, (2) possible mechanical separation of melt from sillimanite-bearing restite,

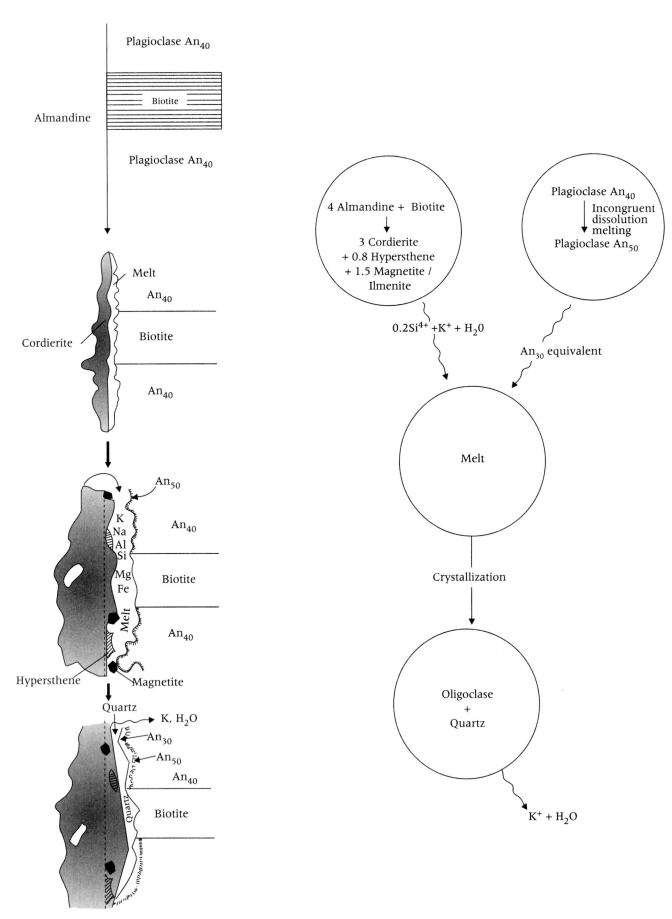

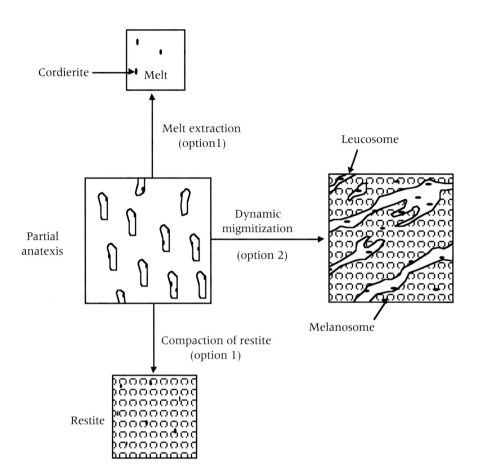

Figure 28.4
Pictorial representation of two different fates of partially melted rock: (1) extraction of partial melt forming batch of magma and restitite and (2) anatectic migmatization.

and (3) undefined kinetic factors involved in the reverse reaction. The availability of aqueous phase for the reverse reaction is not a problem. Since an aqueous phase generated from muscovite in the partial melting process is a solidus-lowering factor for melting in the first place, it is in the melt and automatically available to participate in the reverse reaction.

Melt generated by anatexis must eventually crystallize, but what actually occurs? Can magmatic textures develop in small fractions of anatectic melts just as they do in large batches of magma? Why should crystallization in a melt pocket in a developing anatectic migmatite be significantly different than crystallization in a large magma chamber? It would seem that in-situ anatectic melts (1) are not able to fractionate in view of their near-cotectic composition, and (2) lack the capability for significant gravity and convective crystal–melt segregation. Nevertheless, within the confines of restricted composition and volume, compositional zoning of crystals, magmatic growth morphology, and intergrowths such as graphic and vermicular textures are likely to occur.

If the environment is dynamic during anatexis, melt relocation textures may develop, mimicking those forming at the latest stages of magmatic crystallization (Chapter 27). Furthermore, crystallization of anatectic melt containing dissolved water can be expected to generate a late-stage hydrous magma. A dynamic environment would favor segregation of aplites, pegmatites, and quartz veins evolving from such late- and postmagmatic fluids (Figure 28.5).

Magma Bodies of Suspected Anatectic Origin

There are many good reasons to claim that many granitic magmas have an anatectic source. High strontium isotope ratios and the occurrence of relict restite

Figure 28.3
Schematic representation of reaction and melting in Mt. Pough contact rock. Aqueous phase and potassium are shown migrating from the zone of anatexis.

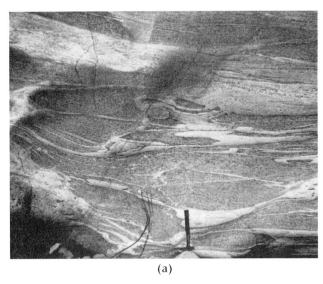
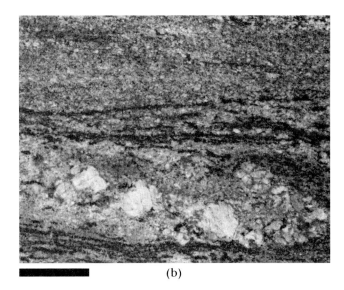

Figure 28.5
Possible results of at least some anatectic melting.
(a) Quartz dioritic (darker) melanosome with trondhjemitic (white) leucosome. Skagit gneiss. North Cascades, Washington. (b) Granitic gneiss with localized pegmatitic K-feldspar and quartz. Pre-Cambrian. Minnesota. Bar scale = 4.5 cm.

(Figure 29.1a,b) are especially indicative, but most compelling is the realization that the volume of granitic magma is much too large to have been generated by other processes. Fractional crystallization of mafic magma cannot yield large volumes of silicic magma. Assimilation of crustal rock, modifying mafic magma composition, is limited by heat supply and the mechanics of "stirring in" the silicic contaminant. Anatexis of the sialic crust requires heating, but there are no mechanical restrictions, and heating on a regional scale at convergent plate junctions is to be expected (Hutton and Reavy, 1992).

Extraction of anatectic melt from restite is another matter (McKenzie, 1985). Buoyant rise of melt accompanied by gravity compaction of restite or dynamic "filtering pressing" may both contribute to the extraction and accumulation of anatectic magma. Granitic plutons with "bits and pieces" of restite seem to have a logical link to anatectic migmatites (Larsen and Poldervaart, 1961; Couturié, 1969; Flood and Shaw, 1975; Clemens and Wall, 1984; Chappell et al., 1987).

Restitites

Garnet amphibolite of possible restite derivation, included in diorite porphyry of laccolithic intrusive. Mount Hillers, Henry Mountains, Utah (Hunt et al., 1953). Bar scale = 1.6 cm.

Partial melting of rock (anatexis) leaves an unmelted **restite** mineral assemblage. Separation of melt and compaction of the remaining mineral assemblage generates rock known as **restitite** (Figure 28.4).

Identification of Restitite

Bulk chemical and isotopic composition of some granitic rocks indicate derivation from continental crust, presumably by anatectic melting. This means that there should be considerable amounts of restite in the zone of anatexis (McKenzie, 1984; Scambos et al., 1986). Petrographic criteria indicating restite origin are not well established but is in a state of development (Barbry, 1991; Montel et al., 1991). The main reason for this is that there are several processes by which mineral assemblages of restite character can form, including metamorphic differentiation, metasomatic metamorphism, and the many processes associated with injection migmatization. Nevertheless, the premise of mechanically accumulated leftover mineral phases in anatectic zones is sound, and similarity to other processes does not preclude future petrographic characterization.

Melanosomes in migmatites are dominated by mafic minerals. In anatectic migmatites melanosomes are the restite of the melting process. Since extraction of melt from anatectic migmatites is to be expected in appropriate dynamic environments, there should be a complete series of rocks containing restites, ranging from the anatectic migmatites to anatectically derived bodies of granitic (and silicic volcanic) rock containing relict restite mineral and lithic inclusions. In this broad-spectrum context, restitites occur as melanosomes in anatectic migmatites and as inclusions in magmatic bodies.

Genesis of Restitite Minerals

Crystalline phases remaining after an episode of anatectic melting reflect (1) the bulk composition of the premelting rock, (2) how much of the rock was melted, and (3) the nature of reactions taking place between minerals and between minerals and melt.

(1) Partial melting of mantle rock generates melt of mafic composition, whereas melting of pelites generates melt of felsic composition. (2) Beginning of melting in any source rock yields compositions near a cotectic. Typically the composition of this melt is the equivalent of quartz, sodic plagioclase, and K-feldspar. Even a mafic volcaniclastic rock can generate a small amount of felsic melt. As the ratio of melt to

crystalline phases increase, the composition of the melt becomes more and more like the bulk composition of the original rock.

(3) Restitite consists of (a) unmelted and unreacted minerals that occurred in the premelting rock (resistors), (b) new minerals that formed during subsolidus (metamorphic) reactions as conditions of anatectic melting are approached, and (c) new minerals generated by reactions that simultaneously produce anatectic melt (incongruent melting). Since it is not feasible as yet to identify the specific source of any given mineral in the context of these three options, restitite, in the broad sense, is any mineral assemblage that remains after anatectic melting.

Identification of restitites, even in the broad sense, is complicated by an additional factor. If anatectic melt is completely extracted, the remaining minerals assemblage is the restite. However, partial melt extraction leaves melt in the system that eventually crystallizes. If the melt is segregated in situ, lenses of leucosome are distinct from restitic melanosome. But if the melt is mechanically remixed with restite, the sense of restitite as a rock type is less meaningful, and anatexite is more appropriate.

Further complications obscure the definition and identity of restitites. Two stages of anatectic melting were proposed by Zeck (1970). The first is a regional anatexis generating anatectic migmatites. After in situ crystallization of the melt, a new melting event took place in migmatite intruded by mafic magma. The challenge, in this example, is to identify what was restite at each stage of melting.

Another problem is in distinguishing xenocrystic restites and direct magmatic crystallization in peraluminous magmatic rocks. The importance of minerals such as garnet, cordierite, spinel, corundum, sillimanite, and calcic plagioclase as single-crystal xenocrysts of restite derivation is stressed by Chappell et al. (1987). Crystallization of garnet and cordierite directly from peraluminous magma, derived from a pelite anatectic source, is emphasized by Flood and Shaw (1975), Miller and Stoddard (1981), Phillips et al. (1981), Clemens and Wall (1984), and Clemens (1989). Anatectic melt may remix with its own restite (Clark and Lyons, 1986). Mobility may obscure the textural and mineralogical aspects of the restite–melt equilibrium.

Suspect Restitites

Pelitic Source Rocks

Tracy and Robinson (1983) describe anatectic melting in which the restite is not distinctly separated into restitite. Pelitic gneisses were formed metamorphically, containing biotite, sillimanite, quartz, plagioclase, minor K-feldspar, cordierite, and garnet. Melting generated new well-formed garnets occurring in veins with K-feldspar. This observation is an example of the importance of texture, besides mineralogy, in recognition of anatexites and restitites. It also is a reminder that typical restite minerals such as garnet and cordierite are also typical metamorphic minerals.

Calcareous and noncalcareous pelites melt to form garnet and cordierite-bearing restitites and melt (Clark and Lyons, 1986). The following equation is suggested:

$$\text{biotite} + Al_2SiO_5 + \text{quartz} + \text{feldspar} \rightarrow \text{garnet} \pm \text{cordierite} + Kf + \text{melt}$$

In this case, mixing of the anatectic melt with its own restite precludes formation of restitite as a discrete rock, forming instead a "migma" of pluton size and shape containing restitic mineralogy. This is one way an anatectic migmatite can evolve into a granitic pluton without regional extraction and collection of melt (see also Flood and Vernon, 1978).

Partial melting of pelites on a subducting plate generates restite and melt, the latter rising from the subduction zone (Mazzone and Haggerty, 1989). The restite is metamorphosed to corundum–garnet–spinel rock with continued subduction. Intrusion of kimberlite from a deeper source penetrated the subduction complex, included metamorphosed restitite, and brought the inclusions to high crustal levels.

Metamorphism of pelites to quartz, K-feldspar, sillimanite, cordierite, biotite, and garnet gneisses preceded probable partial melting in response to heating by an intrusion of monzogabbro (Grant and Frost, 1990). Granulitic restitite containing K-feldspar, cordierite, orthopyroxene, corundum, spinel, and garnet are generated along with a melt that yields quartz, K-feldspar oikocrysts, biotite, and cordierite. The granulites occur as wallrock and as inclusions in the monzogabbro. The leucosome occurs in streaks and patches, tending to be localized in ductile shear zones.

Melting of biotite pelites generating restite granulites and melt is described by Barbey et al. (1991). The restitite is a garnet–kyanite gneiss with some residual quartz, K-feldspar, and sodic plagioclase. In this case, there is thought to have been extensive pre-anatexis metamorphic differentiation forming quartzofeldspathic lenses (leucosomes). This is a good reminder of the possibility of more than one process being involved in the formation of anatexites and restitites.

Surmicaceous enclaves in the Velay Granite of the Massif Central are interpreted to be restitite (Figure 29.1; Couturié, 1969). Folded pelitic gneisses containing biotite, sillimanite, quartz, and sodic plagioclase

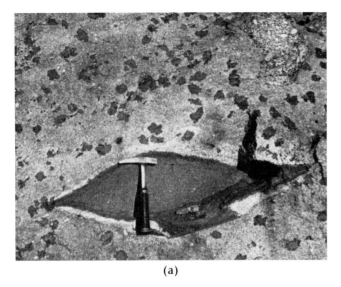 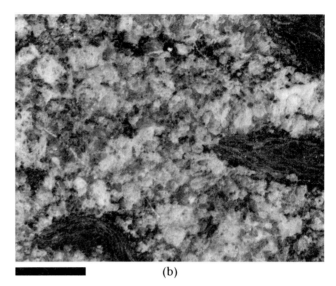

Figure 29.1
Restite enclaves in cordierite-bearing (not shown) Granite du Velay, Massif Central, France. (Courtesy J.-P. Courturié.)

(a) Cordierite-bearing pelitic hornfelses with biotite reaction rims (large lensoid fragment and many smaller enclaves). Enclave of granite porphyroïd de Margeride (upper right). (b) Cordierite-bearing biotite "schist" inclusions (surmicaceous enclaves) in Granite du Velay. Bar scale = 1 cm.

were partially melted, forming restitite containing biotite, cordierite, spinel, sillimanite, garnet, and melt. The host granite also contains cordierite providing a genetic link to the melt fraction. Reaction between the restitite inclusions and host magma generated biotite envelopes on many of the inclusions (Figure 29.1a)

Mafic-Intermediate Source Rock

Metamorphism of graywacke to amphibolite and then partially melted generates granulite restitite and melt (Nesbitt, 1980). It is thought that contemporaneous metamorphism to granulite facies without melting generates an aqueous fluid that participates in nearby fluid-present melting, forming the granulites of restite origin. These conclusions emphasize that granulites can be both metamorphic and restitic.

Melting of tonalitic gneisses during granulite facies metamorphism generated trondhjemitic and granitic melts with restitic mafic salvages containing clinopyroxene, orthopyroxene, amphibole, biotite, and garnet (Cartwright and Bernicont, 1986). The melts went on to form leucogneisses and partly fractionated to pegmatites.

Partial melting of greenstone in the contact aureole of gabbroic plutons is described by Beard (1990). Melting at approximately 900°C and 3 kbar generated pyroxene hornfels restitite containing clinopyroxene, orthopyroxene, and calcic plagioclase. The melt fraction was partly extracted from the anatectic system, leaving magmatic-looking diorite mixed with hornfels, expelling a supposed tonalitic or trondhjemitic melt.

Mantle Source Rock

Deformed and recrystallized spinel–herzolite xenoliths in basanite are interpreted to be restitites by Fabriès et al. (1987). Diapiric rise of mantle rock leads to partial melting driven by adiabatic decompression.

Injection Migmatites

Lit-par-lit injection migmatite. Two-feldspar granite in biotite–hornblende hornfelsic quartzite. Note biotite reaction rims (two white arrows). Heusser Mountain, Nevada. Bar scale = 1 cm.

Petrographic criteria delineating what is injected magmatic rock from host rock is clear in those cases where there is minimal mingling between the two. In cases where there is accompanying assimilation, anatectic melting, and/or metamorphic differentiation the sense of injection migmatization as a distinct rock-forming process is lost. This means that characterization of injection migmatization is inherently structural, not petrographic. Dikelets of granite in a schist host indicate the process, not the textural and mineralogical attributes of the participants.

Source of Magma

Exogenic Source

The classic view of magmatic injection forming "arteritic" migmatite (Sederholm, 1967) is intrusion of magma from a source external to the injected host rock. The injection may parallel planar discontinuities, such as bedding, foliation, cleavage, and joint sets forming **lit-par-lit gneisses** (frontispiece; Figure 30.1), or follow anastomosing fracture patterns forming **mosaic breccia** (Figure 30.2) and **agmatitic injection migmatite**.

The ultimate source of magma is not relevant; it can be from an established magma chamber or defiltering from a zone of partial anatexis. Composition of the magma typically is two-feldspar granitic and granodioritic, but trondhjemitic, quartz dioritic, and leucocratic equivalents of all of these, including aplites and pegmatites (Duke et al., 1988) are common.

Variation of this classic exogenic model includes (a) injection of earlier crystallized portions of a magmatic system, such as a chill phase, by magma arriving from elsewhere within the system, and (b) injection of magma into a fractured, but not totally crystallized, magma with which it is mingling (Chapter 19).

In all cases there is likely to be transition of injection migmatite to inclusion-bearing magmatic bodies. In the classic exogenic case the inclusions are xenoliths. In case (a) they are cognate inclusions, and in case (b) they are coeval inclusions in mingling magmatic systems.

Indigenous Source

Indigenous magma sources include (1) dilational pumping of late-stage intercrystalline melt into late-stage fractures in plutonic rock, generating dikes and dikelets of aplite, pegmatite, and other leuco-

 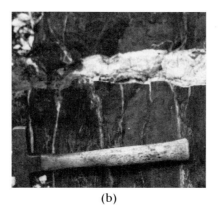 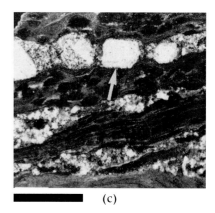

Figure 30.1
Lit-par-lit injection migmatites.

(a) Two-feldspar granite interlayered with siliceous–argillaceous hornfels. Note local discordance and hornfels termination (right). Heusser Mountain, Nevada. Bar scale = 4 cm. (b) Aplite injections controlled by layered fabric in metagabbro, extending from a larger discordant aplite–pegmatite dike. Okanogan Range, Washington State. (c) Injection of alkali feldspar phenocryst-bearing granitic magma parallel and discordant to spotted hornfelsic schist and quartzite (bottom). Dynamic environment has resulted in compression of host rock around phenocryst (arrow) with relocation of adjacent melt, isolating the phenocryst in hornfelsic schist. Schist xenoliths partly rotated in injected–relocated granitic fraction. High-strain wallrock zone of Papoose Flat pluton. Inyo Mountains, California. Bar scale = 1.7 cm.

granitic rock variants, and (2) mobilization of anatectic magma in a dynamic environment. In both cases, the source of magma is local, and the injection is immediate.

In case (1) a few indigenous dikelets in an otherwise homogeneous magmatic rock body does not constitute migmatite. Fluid relocation in a dynamomagmatic system does not have much injection migmatitic character either, although mesoscopic partitioning of aplite in the form of discontinuous dikelets in granite host does occur (Chapter 27 frontispiece photo).

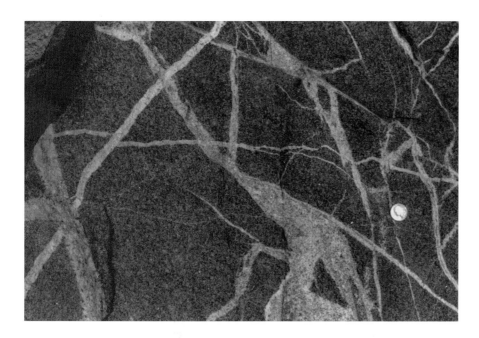

Figure 30.2
Agmatitic injection migmatite consisting of leucogranite granitic dikelets in diorite host. Pyramid Peak Range, Sierra Nevada. Scale = U.S. quarter.

In case (2) a few seams or dikelets of mobilized anatectic melt do not constitute injection migmatite. At the other extreme, a large proportion of anatectic melting may lead to bulk mobilization (such as in a diapir) and loss of migmatitic character. Cases (1) and (2) and their variations are considered phenomena transitional between crystallization of homogeneous magma (Chapter 17) and injection migmatization, and between anatectic migmatization (Chapter 28) and injection migmatization.

Classic Injection Migmatites

Scale of Observation

Intrusion of magma or magma containing a separated aqueous phase (hydrous magma) into wall or roof rock on a mesoscopic and macroscopic scale has been described from many continental crystalline complexes; (Osborne, 1936; Runner, 1943; Lovering and Goddard, 1950; Buddington, 1948, 1959; Pitcher and Berger 1972; Olsen, 1982). Injection of silicate melt into foreign rock ranges from local contact injection associated with small-body magma (mesoscopic), to laccolithic silling (macroscopic), to sheets of magmatic rock parallel to nonfolded or domed metasedimentary sections (macroscopic), to syntectonic partly concordant lenses of magma in folded metasedimentary terrain (regional macroscopic).

Structural Relations

Concordance and Discordance. Injection along parallel sets of planar discontinuities such as bedding and schistosity generates the lit-par-lit relation. Slabs of quartzite interlayered with granitic rock occur in the marginal part of the Main Donegal pluton (Pitcher and Berger, 1972). If the parallel planar structures are folded, the injections may in part follow the fold geometry. If injection is concurrent with folding, concentration of magma in fold hinges generates **phacoliths** (Buddington, 1959). Domal concordant structures are characteristic of the Black Hills injection complex (Runner, 1943; Duke et al., 1988). Injection may be both concordant and discordant on the scale of a hand specimen (Figure 30.1) or larger.

Injection into fractured rock generates mosaic breccia and anastomosing dikes (Figure 30.2), or concordant dikes connecting to discordant feeder dikes (Figure 30.1b).

Proportions of Magma and Rock. A minimum of injection leaves the host rock intact, being best described as rock containing a few dikes, sills, and veins. With greater volume of injected magma, migmatitic character is acquired. At one extreme, injection migmatites consist of rock isolated by a subordinate volume of magma (Figure 30.2); at the other, the injected material dominates, leaving only thin screens of host rock (Osborne, 1936; Runner, 1943).

Forceful and Passive Injection. Pressurized magma propagates fractures and dilates layered structures. Forceful injection of crystal-bearing magma into wallrock along the Papoose Flat diapiric pluton (Sylvester et al., 1978) has resulted in local deviation of foliated rock around alkali feldspar phenocrysts (Figure 30.1c). Injection of aplite–pegmatite systems into schistose rock along the border of the Calamity Peak pluton is thought to have resulted in forceful distension and plastic deformation (Duke et al., 1988).

More-or-less passive injection can take place during folding of layered rock, perhaps explaining the development of phacoliths in the hinge zone of folds (Buddington, 1959).

Associated Processes

Metamorphism, metasomatic metamorphism, assimilation, and hydrothermal activity are processes likely to accompanying magmatic injection (Read, 1957). Metamorphic recrystallization and local metasomatic adjustments can hardly be avoided as a result of the thermal rise attendant to magma injection. The spotted hornfelsic schist shown in Figure 30.1c has resulted from this contact thermal metamorphism.

Local assimilation can be expected as well. Development of a mafic-rich border phase is shown in the frontispiece photo. Such contamination is difficult to distinguish from melanosome selvages lining leucosomes in anatectic migmatites (Olsen, 1982). If the invading magma is significantly hotter than the lowest melting temperature of the host rock, partial anatexis can occur. A blending with anatectic migmatization consists of local mixing of anatectic melts with exogenous injected magma.

Some of the biotite–quartz–feldspar migmatitic gneisses of the Pre-Cambrian Idaho Springs Formation in the Front Range of Colorado (Olsen, 1982) are injection migmatites in which the injected magma has reacted with host.

PART VIII

Rocks Formed by the Mechanical Interaction of High-Temperature Aqueous Fluids or Gases with Rocks or Magma

Aqueous fluids, including steam and supercritical aqueous phase, and gases such as carbon dioxide and sulfur dioxide, can evolve directly from H_2O-bearing magma and indirectly by the magmatic heating of meteoric and connate water. If the lithologic and structural environment is favorable, these fluids may be confined as they evolve and become pressurized. An explosive potential is created that may lead to violent disaggregation of magma and fragmentation of rock. Such explosive activity is the main process in pyroclastic eruption and hydrothermal brecciation.

Pyroclasis is directly related to volcanism, although an initial association with magma is quickly terminated at the eruptive stage when magma becomes particles of glass in the eruptive column. Rocks are formed from pyroclastic debris by processes that are more akin to those of clastic sedimentation than magmatism. Since the initial association is magmatic and since the erupted materials are very hot, pyroclastic eruption is an igneous process.

Separation of aqueous phase from shallow magmatic systems sets the stage for eruption into the atmosphere. The eruptive stage is followed by depositional stages, including air fall sedimentation, sedimentation of air fall tephra in water bodies, ash flows, and debris avalanching. Post-eruption epiclastic reworking of these primary pyroclastic deposits is in the realm of clastic sedimentation, completely separated from all igneous activity even though the clasts may be entirely of igneous derivation.

Explosive activity beneath the Earth's surface is characterized as hydrothermal brecciation. Typically, there is no direct association with magma, and fragmentation affects previously solidified magmatic rock as well as nonmagmatic rock. This type of eruption is smaller scale than pyroclastic eruption and does not lead to sedimentation processes. Hydrothermal brecciation includes the formation of diatremes, breccia pipes, and geyser-related fragmentation. Fluids may evolve from magmas directly beneath the site of fragmentation, or they may be meteoric water heated by nearby magma or a cooling magmatic rock body.

Pyroclastic Rocks

Rounded dacitic pumice fragments (artificially stacked) of Plinian type eruptions at Mount St. Helens in 1980. Diameters of grains range from 4–6 cm.

Formation of pyroclastic rocks involves processes that for the most part can be directly observed in present-day active volcanism. There are not many rock-forming processes to which the principle of uniformitarianism ("the present is the key to the past") can be so successfully applied.

Formation of pyroclastic rocks involves an initial stage of explosive eruption followed by accumulation of mineral, crystalline rock, and glassy rock fragments. Eruption occurs subaerially (Figure 31.1), subaqueously or partly subaqueously, and through rock saturated with groundwater (Figure 31.2). Specifically, pyroclastic deposits form in five volcanic settings: (1) subaerial eruptive and depositional, (2) subaerial eruptive–subaqueous depositional, (3) phreatosubaerial eruptive–subaerial depositional, (4) subaqueous eruptive and depositional, and (5) subglacial eruptive and depositional (Williams and McBirney, 1979; Fisher and Schmincke, 1984; Cas and Wright, 1987).

The rock equivalent of pyroclastic deposits is volcanic **tuff**. Distinction between unconsolidated pyroclastic debris and coherent rock (tuff) for modern pyroclastic activity is of little practical consequence. Completely incoherent pyroclastic ashes are likely to be removed by wind and water action, leaving the more coherent materials as rock or semicoherent rock. Ancient tuffs are more likely to be coherent since postdepositional diagenetic-like processes or metamorphism contribute to lithification.

The Eruptive Stage

Vesiculation

Magma arriving at the Earth's surface extrudes either as lava or pyroclastic debris. Silicic magma typically contains dissolved aqueous phase. With rapid emplacement of such a magma to or just below the Earth's surface, aqueous phase must separate from silicate melt since its solubility in the magma is greatly reduced at reduced pressure (Chapter 13). Separation of volatile phases in magma is **vesiculation**. The phase separation involves a large increase in volume since aqueous phase forms relatively large molecular units. Retardation of the separation process resulting from confinement in the near-surface conduit leads to generation of an overpressure. In effect, a bomb is created that ultimately may result in explosive volcanism and the generation of pyroclastic debris.

Low-silica mafic magmas contain less H_2O and are less susceptible to explosive disintegration. In addition, because of relatively low silica content and reduced silicate polymerization, the viscosity of mafic

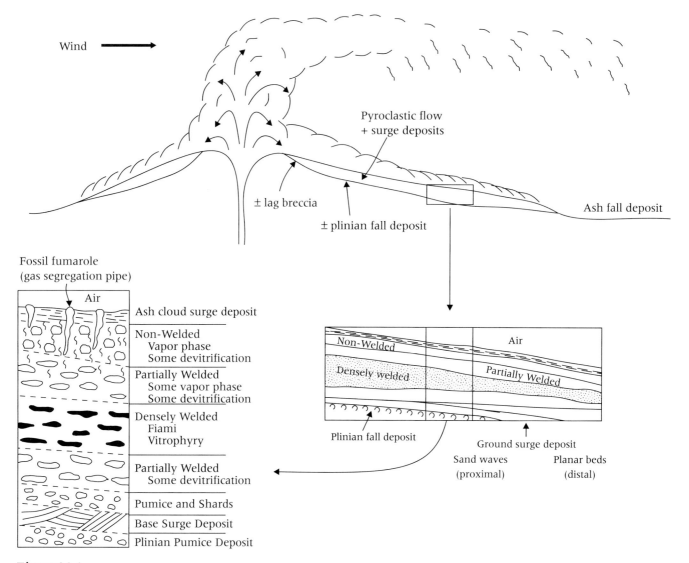

Figure 31.1
Pictorial representation of subaerial pyroclastic eruption showing zonation in welded pyroclastic flow.

magma is lower and the magma is more likely to find paths to the surface before there is much heat loss, crystallization, and overpressurization.

Eruption into the Atmosphere

Eruption into the atmosphere produces an **eruption column** consisting of frothy vesiculated magma, previously crystallized would-be phenocrysts, and fragments of crystalline rock derived from the volcanic conduit (Druitt, 1985; Calvache and Williams, 1992). The magma quickly quenches to glass as it is desegregated in the eruptive jet. It is common practice to distinguish **lithic fragments** that are crystalline and **vitric fragments** that are glass; however, both are rock fragments in the broad sense. **Juvenile** fragments form directly from magma and are typically glassy, whereas lithic fragments may be **cognate** (formed from crystalline rock of same volcanic eruption) or **xenolithic** ("accidental," formed from rock of prior volcanic eruption). All rock fragments formed by pyroclastic eruption are **pyroclasts** or **tephra**. The size of tephra ranges from **ash** (< 2 mm) (Figure 31.3a) to **lapilli** (2–64 mm) (Figure 31.3b) to **blocks** and **bombs** (> 64 mm).

Vitroclastic particles include **pumice** fragments (Figure 31.4) and ash-sized fragments (Heiken, 1972) consisting of **glass shards** (Figure 31.5), resulting from fragmentation of pumice and irregularly shaped glass particles, with or without attached crystals

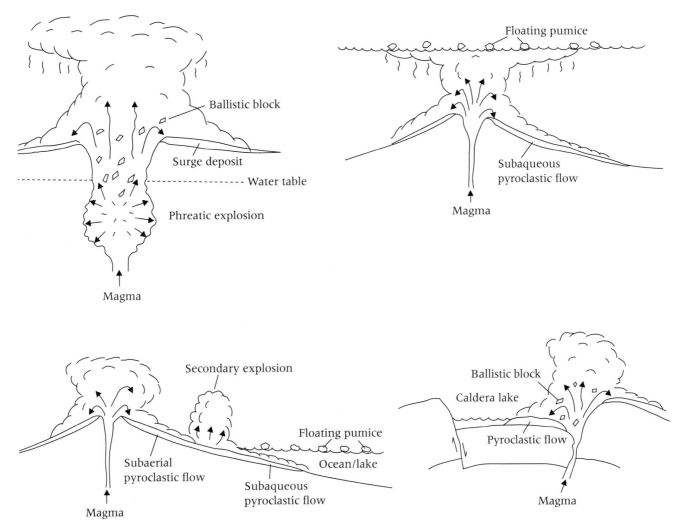

Figure 31.2
Pictorial representations of phreatomagmatic, submarine phreatomagmatic, and subaerial-submarine pyroclastic eruptions.

(Figure 31.3a). Although silicic magma is noted for its H$_2$O content and pyroclastic disintegration, basaltic magma can also form basaltic pumice, **achneliths** including droplets of basaltic glass (Pele's tears), and threads of glass (Pele's hair) (Decker and Christiansen, 1984). More commonly, lumps of mafic magma are thrown into the air as lava fountains. Magma bombs, magma lumps, and magma droplets acquire a frozen exterior rind, but the larger ones are still molten inside, flattening and agglutinating on impact with the depositional surface.

Accretionary lapilli consist of lapilli-sized particles that develop by cementation of ash on a water nucleus, resulting from rain passing through an ash cloud or eruption through water or water-bearing rock. An erupting magma commonly carries **intratelluric crystals** that form in the magma during a pre-eruption stage of magmatic cooling and partial crystallization. The force of eruption may separate crystals from magma. Partial separation of plagioclase and hornblende from melt is shown by the glass attachments in ash from Mount St. Helens (Figure 31.3a). The force of explosion may generate deformation features in quartz, feldspar, biotite, and other minerals that are similar to shock deformation textures in impactites.

An eruption column consists of a lower part (a few kilometers) that has a density greater than air. Heating of incorporated air and fallout of larger pyroclasts tends to reduce the density, taking the form of an expanding convective plume. The density of the plume is lower if the gas content is high. A lower gas content, or a gas with a higher proportion of carbon dioxide relative to water vapor and other gases, results

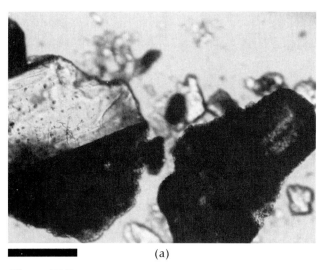 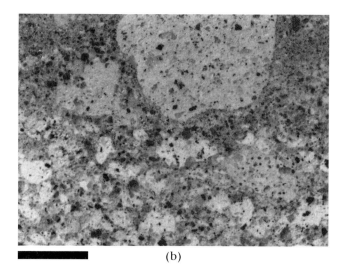

Figure 31.3

(a) Volcanic ash containing plagioclase partially enclosed by glass (left) and hornblende partially attached to glass (right). Mount St. Helens eruption May 18, 1983. PPL. Grain in oil. Bar scale = 0.12 mm. (b) Lapilli tuff dominated by fragments of porphyritic rhyolite. Possible surge deposit. Big Indian Creek, Nevada. Bar scale = 1 cm.

in a denser column, promoting gravitational column collapse. Such a collapse results in a flow of hot pyroclastic debris known as an **ignimbrite**.

The upper part of an eruptive column has a lower density than the atmosphere and can transport fine particles to great heights. Sorting of fines from coarse pyroclasts occurs in this way. Sorting of crystals from glass fragments takes place during the fall in the lower part of the eruption column. This glass winnowing can lead to higher concentrations of crystals in the pyroclastic deposit as compared with the original crystal-bearing magma prior to eruption. **Crystal tuffs** are formed in this way.

Eruption may also take the form of collapse of a lava dome rather than an eruptive column. Collapse results from oversteepening, leading to both avalanch-

Figure 31.4
Welded tuffs with pumice fragments.

(a) Pumice fragment compressed against sanidine crystal (arrow). Note compaction of glass shards above and below pumice fragment. Bishop Tuff, California. PPL. Bar scale = 0.12 mm. (b) Devitrified pumice fragment in densely welded tuff has resulted in light color of fragment. White Pine Range, Nevada. Rock surface. Bar scale = 1 cm. (c) Black vitrophyric fiamme formed by collapse of pumice fragments, dense welding, and possible rheomorphic flow. Bishop Tuff, California. Bar scale = 1 cm.

 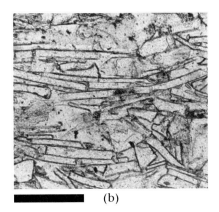 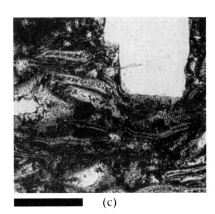

Figure 31.5
Glass shards in tuffaceous rocks.
(a) Nondevitrified glass shards (in extinction) in association with twinned quartz phenocryst (left). Pervasive calcite has filled voids in what was probably a porous ash fall permeated by groundwater originating in limestone. Few shard contacts suggests partial replacement of glass by calcite. White Pine Range, Nevada. X-polars. Bar scale = 0.12 mm. (b) Possible aeolian accumulation of glass shards. Note lack of delicate shard architecture. Horse Springs Formation, Muddy Mountains, Nevada. PPL. Bar scale = 0.12 mm. (Courtesy S. Castor, Nevada Bureau of Mines and Geology.) (c) Devitrified shards in welded tuff. Sanidine phenocryst (top). PPL. Bar scale = 0.12 mm. White Pine Range, Nevada.

ing of loose debris of all sizes and collapse of the walls of volcanoes and calderas (Lipman, 1976).

Eruption Through Water-Bearing Rock, Beneath Glaciers, and into Water Bodies

If magma is erupted into porous rock containing groundwater (phreatic zone) or directly into standing water bodies (Wohletz and McQueen, 1984; Branney, 1991), the explosivity of the eruption is increased because of the participation of the nonmagmatic water that is turned to steam. These **phreatomagmatic** eruptions result in a higher degree of fragmentation because there is fracturing related to rapid quenching of magma contacting water as well as explosive fragmentation. Eruption beneath glaciers can also produce pyroclasts. The ice is melted, leading to interaction between the magma and the water (Furnes et al., 1980).

The Transportive and Depositional Stages

Erupted pyroclasts travel in air, air and water, or water only. The movement of pyroclasts occurs as (1) pyroclastic falls, (2) pyroclastic surges, and (3) pyroclastic flows. All of these modes of transport may occur during a single pyroclastic event, and there may also be accompanying lava flows, debris avalanches, and lahars.

Pyroclastic Falls

Pyroclastic air falls are characterized by tephra that cover highs and lows of topography with layers of relatively constant thickness. Falls originate from the lower and upper portions of eruption columns and from turbulent clouds following pyroclastic flows.

Hawaiian and **strombolian** falls are basaltic and only mildly explosive. Scoria fragments are most common, but there are also ballistic bombs and a variety of achneliths (Decker and Christiansen, 1984). Association with lava spatter and lava flow is characteristic. **Vulcanian** falls are stratified ashes containing large bombs and blocks with breadcrust surfaces. Eruptions are basalt or andesite–basalt, occurring as cannonlike explosions. **Plinian** air falls consist of silicic pumice blocks (tens of centimeters), smaller pumice fragments (frontispiece photo), glass shards, and crystals. They originate from highly explosive events related to highly viscous siliceous magma. Nonbedded texturally homogeneous deposits are common, but there can be stratification and reverse-graded bedding with relatively large pumice fragments toward the top of beds.

Figure 31.6
Pumice flows with multiple lobular fronts; Mount St. Helens eruption 1980. (Lipman and Mullineau, 1981; courtesy U.S. Geological Survey MSH–PP1250 105.)

Fall of subaerially erupted tephra into water leads to good sorting and well-bedded or well-laminated deposits. Commonly there is sorting of crystals relative to glass fragments and reverse-graded bedding as a consequence of the tendency of pumice fragments to float.

Phreatomagmatic eruptions through water-saturated sediments or rock and through water bodies (such as crater lakes) leads to **Surtseyan** or **phreatoplinian** air falls (Branney, 1991). **Tuff rings** and **tuff cones** are common, and the tephra is characteristically fine grained and commonly accretionary. **Hyaloclasitite** deposits result from such eruptions. Fallout from submarine volcanic eruptions tends to generate bimodal size distributions in which pumice diameters are two to three times as large as lithic fragments deposited at the same stratigraphic interval (Cashman and Fiske, 1991).

Palagonitic alteration of basalts is an indicator of phreatomagmatic eruption. Palagonite is an orangish, isotropic-to-weakly-birefringent material produced by alteration of sideromelane (basaltic glass).

Pyroclastic Surges

A **pyroclastic surge** is a ring-shaped inflating cloud consisting of particles dispersed in a matrix of hot volatiles that moves outward at hurricane velocities as a turbulent density current from the base of a vertical explosion column. Surge is caused by release of large volumes of steam or carbon dioxide capable of fluidizing particles (Wohletz and Sheridan, 1979). Surges may be associated with subaerial pyroclastic falls, subaerial pyroclastic flows, phreatomagmatic eruptions, and even subaqueous eruptions. Surges differ from flows in that surge occurs as pulses in which kinetic energy rapidly decays, whereas flows maintain kinetic energy over longer periods of time. Surges tend to mantle topography like air falls rather than flow around topography.

Fludization may lead to enrichment of crystals. Sand waves, massive beds, and planar beds are formed as a turbulent surge moves along. Reverse-graded bedding is common as laminar grain flow occurs in the surge mass containing multisized grains. Larger particles will seek lower-shear-stress regimes, tending to moving to the surface. Cross-stratification, dunes, antidunes, pinch and swell structures, and chute and pool structures are all possible in this phase of surge deposits. The massive-bed stage represents deposition that is not turbulent or quiescent, generating rocks with little or no sense of bedding. The planar-bed stage represents lower velocity and waning stages of the surge.

Early-stage ground surge deposits consist of vesiculated and nonvesiculated rock fragments, crystals, and unidirectional bedforms, enriched in denser components. Later-stage ash cloud deposits consist of elutriated particles, accretionary lapilli, and many other characteristics similar to air fall deposits. Pyroclastic surge deposits may precede pyroclastic flows, in which case they are capped by the pyroclastic flow. Alternatively, surge deposits may have no associated pyroclastic flow.

Subaqueous eruptions similar to subaerial surges are common in the marine environment and in crater lakes. Tephra restricted to the water body is temporarily suspended, sorted, and followed by sedimentation on the flanks of submarine volcanoes, on the ocean floor, or on the crater floor. Coarser and denser materials settle out earlier, except that pumice fragments tend to float. Turbidity currents commonly redistribute the eruptive debris after sedimentation.

Pyroclastic Flows

Pyroclastic flows are generated from the lower portions of eruption columns in either a subaerial or subaqueous environment (Smith, 1960; Ross and Smith, 1961; Fiske and Matsuda, 1964; Fisher, 1966, 1979; Sheridan, 1979; Smith, 1979; Wilson and Walker, 1982; Cole and DeCelles, 1991). Subaerial flows may flow into the ocean and caldera lakes.

Pyroclastic flows have a lower gas–solid ratio (poorly inflated) than do surge blasts and air falls. Gravity collapse of this relatively dense tephra–gas–air mixture leads to flows that are topographically controlled, tending to thicken in valleys and depressions. Sorting in pyroclastic flows is poor due to high particle concentration, not to turbulence (unlike pyroclastic surges). Flow is dominantly laminar, with drag resistance between the flow and the ground creating a low-velocity zone.

Flows range from basaltic–andesitic, dominated by block and ash or scoria. Silicic calc–alkaline magmas generate pumice flow deposits (Figure 31.6) that are also known as ignimbrites.

Ignimbrites typically consist of several flow units. Zonation in single flow is shown in Figure 31.1. A **cooling unit** consists of several flows whose eruptions occur within a short time frame, forming a layer of pyroclastic flow debris that cools as a unit. A single flow may be only few meters thick, whereas a cooling unit may be as much as 600 m thick. Accumulation of flow material results in the flattening of pumice tephra (Figure 31.4). Partings within cooling units may result from momentary air falls, being marked by concentrations of lump pumice, crystals, ash, and even reworking of surface ash by water or wind.

Lag breccia may form at the proximal end of a pyroclastic flow. Segregation of larger or denser fragments generated by collapse of blocks in the eruptive vent cannot be transported in the flow. Breccia consists of dense lithic fragments and poorly vesiculated clasts. Lithic-dominated pyroclastic breccia flows related to vulcanian-type eruption may precede more typical pumiceous pyroclastic flows (Calvache and Williams, 1992).

Welding Within Pyroclastic Flows. The high rate accumulation of hot eruptive debris in ash flows and in some air falls is conducive to heat retention and consequent sintering of compacted tephra. **Sintering** is the nonmelt process of bonding between solid particles. **Welded tuffs** occur in the central zones of ash flows (Figure 31.1) and ash flow cooling units. Glass shards and pumice fragments are compacted and sintered together into a rock that has **eutaxitic structure** (streaked appearance) (Figure 31.4c), mimicking that derived by flow of lava.

The porosity of pumice and nonwelded tuff is reduced and finally eliminated as fragments are flattened and internally and externally welded. Draping of shards and pumice fragments (Figure 31.4a) over pre-existing crystals is a common consequence of compaction. Partial welding generates rock that is somewhat darkened (salmon pink is typical), somewhat vitreous, and containing relict vesicles. Fracturing through compacted pumice fragments rather than around them is a fair test of at least partial welding.

Elimination of porosity of pumice fragments generates a dark (purplish brown is typical) welded tuff. With extreme deformation and welding, obsidian-like (black) **fiamme** are produced (Figure 31.4c). Dense welding converts all tephra to glass known as **vitrophyre**. Since tephra commonly contain early-magmatic crystals, porphyritic glass characterizes vitrophyre.

Flattened pumice fragments assume disk shapes, but if there is unidirectional flow as a result of secondary flow known as **rheomorphism** (Wolff and Wright, 1981; Branney et al., 1992), fragments become triaxial ellipsoids and rodlike forms, commonly with tension cracks that break fiamme into boudinlike pieces. Rheomorphic tuffs are similar to silicic lava flows (Henry and Wolff, 1992), but whether the lava-like flow occurs as a continuation of pyroclastic flow (primary flow) as well as after halt of the pyroclastic flow (secondary flow) is of some debate (Chapin and Lowell, 1979; Wolff and Wright, 1981).

A high rate of accumulation of air fall tephra minimizes radiative cooling. Hot fragments are buried on impact, and high porosity precludes rapid heat loss by conduction. Welding within these air falls may occur as a result, demonstrating that welding does not prove pyroclastic flow (Sparks and Wright, 1979).

Vapor-Phase Crystallization in Pyroclastic Flows. Precipitation of minerals such as cristobalite, tridymite (Figure 31.7), and alkali feldspar as druses in pore spaces within the partially welded and nonwelded upper layers of pyroclastic flows (Figure 31.1) results from upward migration of high-temperature hydrothermal fluids. These fluids derive from cooled gases trapped in the flows and by devitrification of water-bearing glass. A continuous transition of these higher temperature fluids to lower-temperature fluids characterizes fumarolic activity (Chapter 35). **Gas segregation pipes** have concentration of crystal and lithic fragments and depletion of ash resulting from gas streaming. These **fossil fumaroles** are characteristic of the upper portion of some pyroclastic flows (Sheridan, 1970; Cas and Wright, 1987).

Devitrification of Welded Tuffs. Devitrification of glass generates silica minerals, feldspars, and water (Chapter 17). The glassy portions of welded tuffs that may devitrify are compacted shards (Figure 31.5c) and pumice fragments (Figure 31.4b). Devitrification results in lighter color of the materials, which may have been originally dark or even black (compare Figures 31.4b, c). Nonwelded glass shards and pumice fragments in the nonwelded and partially welded zones of pyroclastic flows may devitrify as well.

Devitrification textures are characterized by radial–fibrous intergrowths of silica minerals and feldspars, extending inward from shard boundaries, or as spherulites with no obvious textural relation to shard or pumice form, especially when welding is so thorough that the form of shards and pumice fragments is obliterated.

Figure 31.7
Vapor-phase tridymite replacement of devitrified glassy matrix of rhyolite tuff. X-polars. Bar scale = 0.05 mm. Yellowstone National Park, Wyoming.

Epigenetic Deposition and the Transition to Clastic Sedimentary Rocks

Airborne tephra falling into water bodies make an immediate transition from pyroclasis to clastic sedimentation. Blankets of tephra on the ground are particularly susceptible to erosion by wind and water, providing clasts for epigenetic sedimentation. Fluvial–lacustrine redeposition is characterized by well-developed bedding (Figures 31.8a,b), although in the

(a)

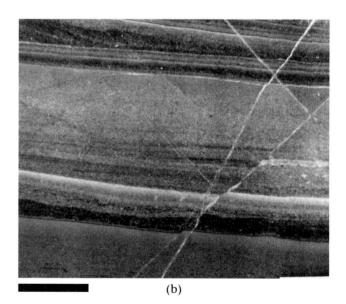

(b)

Figure 31.8
Epiclastic rocks containing pyroclastic debris.

(a) Well-bedded fluvial beds containing nonvesiculated rhyolite clasts, pumice fragments, and perlite fragments in an ash matrix. Keystone Junction, Ruth, Nevada. Scale = lens cap (center). (b) Lacustrine air fall or fluvial–lacustrine laminated tuffaceous rock with graded bedding (up is toward top of photo). Slab surface. Sutro Tuff, Virginia Range, Nevada. Bar scale = 2 cm.

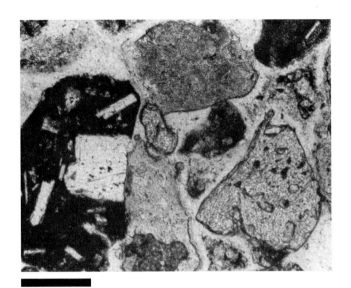

Figure 31.9
Fluvial volcaniclastic arenite containing a variety of lithic fragments, including porphyritic basalt (left) and nondevitrified pumice, along with glass-free or partially glass-free crystals of plagioclase, augite, and biotite. Rock also contains a few diatoms (see Figure 37.7a,b,c) and fossil leaves (see Figure 37.1a). PPL. Bar scale = 0.14 mm. Hunter Creek Formation, Verdi, Nevada.

absence of cross-beds the distinction from gravitational bedding in water bodies receiving air fall tephra is not always clear.

The glass shards shown in Figure 31.5b may have been redeposited by wind or water since they have few of the delicate forms that characterize untransported shards (Figure 31.5a). Large-scale cross-bedded volcanic arenites of the Bear Valley formation (Anderson, 1971) are wind deposited, containing glass shards and fragments of quartz and feldspar crystals.

Redeposited pyroclastic debris may superficially resemble pyroclastic flows. The occurrence of fossil fumarole pipes and color changes attendant to welding are sure signs of pyroclastic flow, not redeposition.

Epigenetic sedimentation may include nonpyroclastic volcanic rock fragments mixed with tephra (Figure 31.9). Rock fragments may be related to an eruption producing tephra from the walls of the eruptive conduit, or they may be "accidently" incorporated into pyroclastic materials during epigenetic sedimentation that occurs in regions of prior volcanism. Ash falls onto accumulating clastic sediments in fluvial–lacustrine environments may occur as very thin beds (Figure 31.10) up to beds measured in many

(a)

(b)

Figure 31.10
Air fall (probably water reworked) ash tephra interlayered with clastic sediments.

(a) Volcaniclastic litharenite with very thin vitric ash beds (horizontal across center of photo and lower left). Clastics are interbedded with thick-bedded diatomite. Hunter Creek formation, Chalk Bluffs, Reno, Nevada. Rock surface. Bar scale = 0.5 cm. (b) Ash bed (6–8 mm, between arrows) in Pleistocene Lake Lahontan deltaic sediments. Mullin Creek, Pyramid Lake, Nevada. Lens cap scale (lower left).

Figure 31.11
Andesitic agglomerate of probable lahar origin. Floriston, California. Scale = U.S. quarter.

centimeters. **Tephrochronology** is very useful in identifying times of pyroclastic eruption, which may have occurred far from the site of ash fall, and useful in establishing stratigraphic relations of nonvolcanic clastic sediments with which the ashes are interbedded.

Lahars consist of volcanic debris in a mud slurry. Deposits are non-sorted and largely non-stratified (Figure 31.11). Lahars may occur during eruption, especially if there is melting of ice and snow on the flanks of a volcano, or later as volcanic debris is saturated with rainwater or meltwater. Pumice lahars may resemble pyroclastic pumice flows, but the characteristic zones in flows resulting from welding, vapor-phase crystallization, and late-stage fumarolic activity are lacking in the lahars.

Hydrothermal Breccias

Hydrothermal explosion breccia containing autoliths of perlitic glass, nonperlitic devitrified glass, and xenolithic fragments of Ely Limestone (weathered out), porphyritic hornblende monzonite (Lane Monzonite), and Rib Hill Sandstone. Matrix consists of tridymite-bearing devitrified pumiceous glass, indicating that vesiculation of magma was more-or-less contemporaneous with explosive eruption. Breccia occurs along vertical cylindrical contact between Tertiary rhyolite intrusive into Upper Paleozoic Ely limestone. U.S. quarter scale (white arrow lower right). Garnet Hill region, Egan Range, Nevada.

Explosive Rock Brecciation

Separation of aqueous phase in magmatic systems is preceded by crystallization of anhydrous magmatic minerals such as quartz and feldspar, effectively increasing the H_2O content of remaining intercrystalline melt. Intrusion of H_2O-bearing magma or partially crystallized magma typically results in lowering of confining pressure. Both factors lead to saturation and separation of an aqueous phase. Since there is a substantial increase in volume accompanying separation of aqueous phase from magma, there must be a corresponding physical response. A body of granodioritic magma containing 2.7 wt % H_2O will potentially expand nearly 50% upon completion of crystallization at a depth of 2 km (550 bars) (Burnham and Ohmoto, 1980). The separating fluid is typically aqueous rich, and it may be in part of meteoric origin in shallow magmatic systems, but in some alkaline magmatic systems such as carbonatites and kimberlites, CO_2 is likely to be the dominant volatile phase.

If aqueous phase dominates, it is a hydrothermal fluid, occurring as bubbles at first, coalescing to form larger volumes of fluid that may be "plumbed" into fractures forming in the largely crystallized magmatic body and in wall rock where the fluid is free to escape the magmatic system.

If the increase in volume is accommodated by progressive diffusion of bubbles through magma to leakage sites provided by fractured rock, there is no buildup of pressure and no sudden release of mechanical energy. Alternatively, if the increase in volume is accommodated by an increase of internal pressure resulting from lack of fluid leakage into nonfractured or otherwise impermeable rock, the magmatic–hydrothermal system may literally explode.

If failure occurs in the throat of a volcano in the presence of magma with violent expulsion of hydrothermal fluid, small globules of magma, would-be phenocrysts, vesiculated glass, and finely crystalline rock, the event is pyroclastic. If explosion occurs at the postmagmatic stage, rock fragmentation occurs, generating **hydrothermal breccia**, also known as hydrothermal eruption breccia. Transition between pyroclasis and hydrothermal brecciation is to be expected. Blocks of bedded pyroclastics in diatremes are derived from essentially contemporary pyroclasis at the surface (Hearn, 1968; McCallum et al., 1976),

Figure 32.1
Meter-size blocks of siliceous sinter ejected by the hydrothermal explosion at Porkchop Geyser, Yellowstone National Park, on September 5, 1989, witnessed by eight Park visitors. (Fournier et al., 1991). (Courtesy of Robert Fournier, U.S. Geological Survey.)

whereas the hydrothermal brecciation in the diatreme is linked to magma at depth.

Hydrothermal breccia may also originate by extreme heating of meteoric water. An explosive geyser-like eruption, generating sinter block ejecta, was witnessed by eight visitors to Yellowstone National Park on September 5, 1989 (Fournier et al., 1991; Figure 32.1). Thus, the hydrothermal fluid involved in explosive brecciation ranges from purely magmatic H_2O to magmatic H_2O mixed with adjacent meteoric water to purely meteoric water heated by a nearby magmatic body.

Hydrothermal Breccia Bodies

The character of rock in hydrothermal breccias is a clue to the depth at which explosive release of aqueous (or CO_2) phase occurs. **Diatremes** are pipelike breccia zones assumed to be associated with magmatism. Some diatremes consist of diamond-bearing kimberlite indicative of a very deep magmatic source. Other diatremes contain rock fragments characteristic of the walls of the pipes and/or fragments of rock (including pyroclastics) that characterized the uppermost levels in the pipes and the surface. Such rock fragments subside in the pipe immediately following an explosive event.

Breccia pipes are hydrothermal breccias typically occurring in roughly cylindrical vents, but their form can also be dike-like or irregular (Bryant, 1968). They are common in porphyry systems (Chapter 34), where they occur in structurally favorable places in the host pluton and in roof rock. The size of breccia pipes and similar bodies ranges from a few centimeters to many meters in diameter or thickness.

Breccia fragments occur in eruptive conduits, but they may also be ejected from the vent onto the surface, as they did at Yellowstone (Fournier et al., 1991). A temporary seal of a hydrothermal vent, such as a precipitate of siliceous sinter, may become part of the breccia assemblage in subsequent explosions. Fragments range from angular to subrounded, but some may be milled in the eruptive process and become well rounded, as they did in the igneous complex of the Isle of Rhum (Hughes, 1960).

Hydrothermal breccias are especially characteristic of epithermal precious metal ore deposits (Hedenquist and Henley, 1985; Vikre, 1985; Nelson and Giles, 1985; Bonham, 1988; Nicholson, 1989). The breccias associated with porphyry systems form at greater depths and are more likely formed by hydraulic fracturing rather than by explosion. Breccias of these higher-pressure environments are less likely to have rotated and displaced fragments. Fragment separation without rotation generates one type of mosaic breccia, and fragmentation without separation forms crackle breccia. Reactive replacement along fractures in crackle breccias generates stockwork breccias (Laznicka, 1988).

Examples of Hydrothermal Breccia

Brecciation of welded tuff is shown in Figure 32.2a. It is likely that entrapped meteoric water resulted in this explosive eruption because explosive breccias resulting from sudden release of volcanic gases postdating welding are not common. Fragments of quartz crystals and devitrified glass indicate explosive activity within a Yellowstone rhyolite (Figure 32.2b). There has been postbrecciation open-space precipitation of opal in both of these breccias.

Angular breccia fragments of rhyolite supported by a finely comminuted rock flour containing hematite, quartz, and cinnabar occurs at the Buckskin mine (Figure 32.3a), site of a volcanic-hosted epithermal precious metal deposit. Brecciation of a hydrothermally altered calcareous shale into a mosaic of fragments is shown in Figure 32.3b. This mosaic

(a)

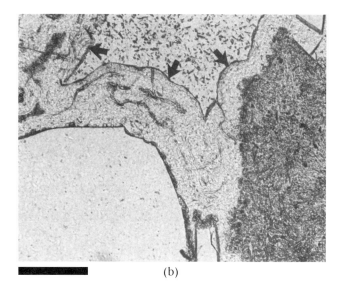
(b)

Figure 32.2
Hydrothermal breccias with extensive open space.

(a) Open-space precipitation of opal on quartz and welded tuff fragments. White Pine Range, Nevada. Bar scale = 1.2 cm. (b) Opal (arrows) precipitate on fractured quartz crystals (lower left and upper right), tridymite-bearing (not shown) devitrified glass (right). Open space (upper center) contains carborundum grinding compound embedded in epoxy. Rhyolite breccia. Yellowstone National Park, Wyoming. PPL. Bar scale = 0.14 mm.

(a)

(b)

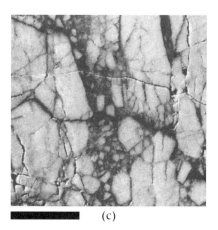
(c)

Figure 32.3
Gold-bearing hydrothermal breccias.

(a) Rhyolite fragments in finely comminuted rock flour matrix containing hematite, quartz, cinnabar, and gold. Note silicified earlier hydrothermal breccia (top right) that has become a fragment itself in the younger breccia. Buckskin Mine, Esmeralda Country, Nevada. (Courtesy J. Tingley, Nevada Bureau of Mines and Geology.) Bar scale = 0.6 cm. (b) Mosaic hydrothermal breccia consisting of weakly silicified and mineralized (Au) pilot shale that was decarbonized prior to mineralization. Dark material between fragments is supergene hematite after primary sulfides. Bar scale = 2 cm. Alligator Ridge, Nevada. (Courtesy H. Bonham, Nevada Bureau of Mines and Geology.) (c) Variation from breccia with rotated fragments to mosaic and crackle breccia. Matrix contains hematite, limonite, jarosite, and gold. Hydrothermal brecciation of Rib Hill sandstone, Star Pointer Mine, Robinson Mining District, Nevada (Smith et al., 1988). (Courtesy M. Houhoulis, Nevada Magma Copper Company.) Bar scale = 2 cm.

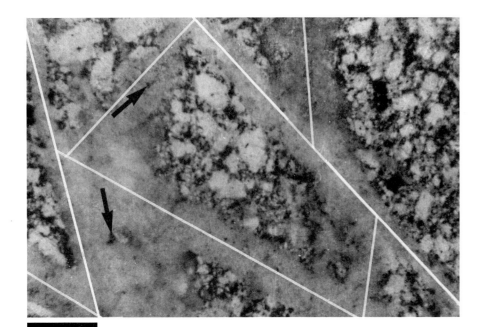

Figure 32.4
Quartz stockwork in granite resulting from reactive replacement in crackle breccia (crack orientation shown with white lines). Note residual fragments of granite in the veins (black arrows). Christmas Mine, Arizona. (Courtesy D. Hudson.) Bar scale = 1 cm.

hydrothermal breccia is associated with a Carlin-type gold deposit (Chapter 34). A gold-bearing hydrothermal explosion breccia, locally having characteristics of mosaic and crackle breccia, is shown in Figure 32.3c. The host rock in this case is a quartz arenite. Reactive replacement of granitic rock by quartz along fractures of a crackle hydrothermal breccia can generate a quartz stockwork (Figure 32.4).

If hydrothermal brecciation takes place in the presence of magma, the breccia fragments are encased in a pumiceous glass matrix (frontispiece). In this case, explosion occurred simultaneously with magma vesiculation, providing a direct link between separation of aqueous phase from magma and explosive eruption, in the portion of a system that is subsurface, not pyroclastic.

PART IX

Rocks Formed by Precipitation and Reactive Crystallization Involving High-Temperature Aqueous Fluids or Gases

A much greater volume of rock is formed by low-rate interaction of aqueous fluids with pre-existing rock than by the high-rate explosive mechanisms discussed in Part VIII.

Hydrothermal alteration is a rock-forming process of major importance because of the direct and potential link to the formation of ore deposits. Hydrothermal interaction with rock consists of direct precipitation from aqueous solutions and reaction of aqueous fluids with pre-existing rock, forming new minerals in the place of old minerals. The aggregate of these "alteration" processes is not commonly thought of as a rock-forming process, no more than the products of weathering are thought of as distinct rock types. Instead, the rocks are whatever they were before alteration plus the imprint of the alterations. Such rocks are named according to the protolith and the type of alteration, such as "propyllitically altered andesite."

An ore sample consisting of siltstone only slightly modified by permeating hydrothermal solutions that deposit "invisible" gold is a product of hydrothermal alteration just as much as an ore sample containing mostly coarse crystalline sphalerite and galena. The former is likely to be referred to as an altered siltstone, whereas the latter is not likely to be considered a rock type. Similarly, veins containing, for example, fluorite or quartz are traditionally thought of as "minerals occurring in veins."

Classification of hydrothermally altered rocks on the basis of rock-forming processes must include both protolith rock-forming processes and "alteration" processes. In this way, altered rocks stand as rocks themselves, consisting of their total assemblage of primary and secondary mineralogical and physical attributes, not on what they were prior to hydrothermal modification nor exclusively to the effects of alteration.

Similarly, veins can be classified as rocks in the sense that they are aggregates of crystals. Quartz veins are also quartz rocks in this sense. The classification of hydrothermally altered (and weathered) rocks and vein rocks is considered in Chapter 10.

Of much less abundance, but nonetheless fascinating occurrence of rock formed by aqueous fluid precipitation and reaction, are pegmatites and fumarolites. Pegmatite enjoys full status as a rock type, whereas fumarolites are typically referred to as "fumarolic sublimations."

Pegmatites formed in association with magmatism are magmagenic pegmatites. The association with magmatism is reflected in their traditional classification as a variety of magmatic rock. In detail, these pegmatites are not magmatic any more than ash flow tuffs are magmatic. Both are derived from magmatic systems; they are igneous, but the actual rock-forming processes involved in the formation of pegmatites is precipitation from high-temperature aqueous phase, not crystallization of magma.

A link back to the processes described in Part VIII, involving explosive release of volatiles, is provided by the observation that the pocket zone of large pegmatites contain ruptured crystals that seem to indicate a final explosive event in the evolution of some pegmatite systems.

Fumarolites consist of minerals resulting from reaction with rock as well as direct sublimation on rock surfaces. They are the gaseous analogue of hydrothermal activity in the sense that direct precipitation and reactive replacement takes place for both.

33

Magmagenic Pegmatites

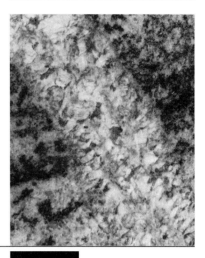

Pegmatitic vein of granitic composition without associated aplite crosscutting granitic gneiss. Bar scale = 4 cm.

Pegmatitic Texture

Very coarse-grained quartz–feldspar rocks are texturally pegmatitic. Large crystal size indicates low nucleation rate in an environment conducive to growth of large crystals. Such an environment arises in H_2O-bearing granitic magmas during the very latest stages of crystallization if the evolving hydrous magma is allowed to accumulate in dilatant fractures or some other structurally favorable chamber.

Pegmatites of granitic composition occurring in granitic plutons and in adjacent wall rock are relatively common. They have been the focus of considerable interest and research partly because of their gem mineral content and former industrial use of giant muscovite crystals (Cameron et al., 1949; Jahns, 1953, 1955).

Granitic pegmatites also occur in anatectic, metasomatic metamorphic, and hydrothermal systems. Distinction among granitic pegmatite generated by these various processes is not always possible even if the field relations are known. This chapter focuses on granitic pegmatites associated with aplite because these pegmatites are thought to have evolved specifically from hydrous silicic magma. These are **magmagenic pegmatites**, and most are of granitic composition.

Crystallization of H_2O-Bearing Granitic Magma

Stages of Crystallization

One way to describe the consolidation of H_2O-bearing granitic magmas is as follows (Hibbard, 1980). Early magmatic crystallization occurs up to the time of saturation in aqueous phase. For most granitic magmas this represents 85% or more crystallization of a given volume of magma. Late magmatic signifies the appearance aqueous phase, pictured as silicate melt containing bubbles of aqueous phase (hydrous magma). The aqueous phase is prone to concentration in reduced pressure zones, in some cases due to dilatant pumping, occurring as dikes, lenses, and pods adjacent to silicate melt. Complete separation of aqueous phase from silicate melt marks the postmagmatic stage, in which the fluid can move on to form hydrothermal systems in fractured portions of the host pluton and into fractured, or otherwise permeable, wallrock.

Late-Magmatic Fluids

The final 15% or so of fluid in a consolidating H_2O-bearing two-feldspar granitic magma contains mostly the equivalent of sodic plagioclase, quartz, K-feldspar, and aqueous phase. Several magmatic environments have an approach to such a composition, among them being (1) diffusion of alkalis and aqueous phase to the roof zone of chambered magmas (Boone, 1962; Carten et al., 1988; Candela, 1991), (2) fractional crystallization with accumulation of silicic magma in the roof zone of diabase sills (Ernst, 1960), (3) accumulation of centimeter- and decimeter-scale pod-shaped masses, including those with miarolitic cavities, within otherwise homogeneous granitic or microgranitic bodies (Jahns and Burnham, 1969; McMillan, 1986), (4) in dikes and veins within the parent pluton as a consequence of dilational pumping (Hibbard, 1980), (5) as passive or forceful injections into wallrock (Duke et al., 1988), and (6) as highly evolved rhyolites (Congdon and Nash, 1991).

Indigenous Microveins

Microveins in granitic rocks containing late-stage minerals can be considered to be the beginning of larger aplite–pegmatite systems rather than their distal terminations (Hibbard, 1980). Crystallization of a granitic magma in a static environment eventually generates an intergranular hydrous magma containing all the ingredients of an aplite–pegmatite system.

If crystallization proceeds in this static intergranular environment, final crystallization is (1) as rim zone additions on existing crystals of quartz, plagioclase, and alkali feldspar, and (2) as replacements of some phases by others. Local replacement of sodic plagioclase by K-feldspar is typical, the partial replacement of oligoclase in myrmekitic relation with vermicular quartz being one such example (Figure 33.1; Hibbard, 1979). The final postmagmatic aqueous-rich fluid either remains in miarolitic cavities (Figure 17.8c) or microcavities (Sprunt and Brace, 1974), if the pluton is shallow and under relatively low confining pressure or dissipates along intergranular or fracture pathways. Deuteric alteration of hornblende, biotite, and feldspars is possible at this stage.

A dynamic environment existing during the late-magmatic stage can generate structures ranging from microcracks (Figures 27.3a,c; 27.6) to penetrative shear zones. Relegation of late-magmatic fluids into microcracks forms microveins (Figure 33.2). Dikelets, pods, lenses, and pressure shadow pools of relocated fluids characterize the dynamomagmatic environment (Chapter 27). In all cases, timing of mechanical events is critical in determining the mineralogical character of these small-scale late-magmatic features. If fracturing does not occur until the postmagmatic stage, only deuteric–hydrothermal fluids will be available to move through and concentrate in these structures.

Late-magmatic microveins commonly contain quartz and K-feldspar (Figure 33.2). In two-feldspar granitic systems, late plagioclase typically crystallizes on earlier-formed plagioclase crystals that may or may not contain vermicules of quartz (myrmekite) (Hibbard, 1979, see also Phillips, 1974; Vernon, 1991). In one-feldspar granitic systems, late-stage intergranular "microveins" of albite are characteristic (Plate 4c; Figure 8.7a). The process or processes by which sodic plagioclase, perthitic and nonperthitic K-feldspar, and quartz become separated in microveins and mesoscopic aplite–pegmatite systems is not as yet well understood (Burnham and Nekvasil, 1986). However, it is certain that partitioning takes place in hydrous magmas that are evolving to hydrothermal fluids (Jahns and Burnham, 1969; Cerney, 1971; Burnham,

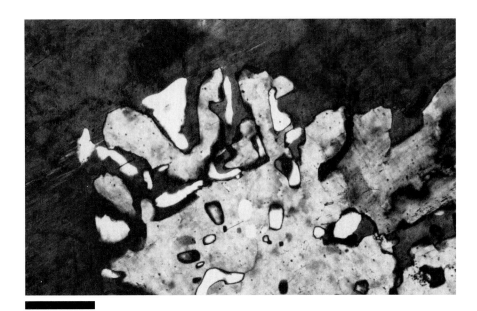

Figure 33.1
Partial replacement of myrmekite by K-feldspar (dark grey) in two-feldspar granite. Note tendency for isolation of quartz vermicules (white) from oligoclase host (mottled gray-white), indicating reaction along grain boundaries. Colville batholith, Washington State. X-polars. Bar scale = 0.12 mm.

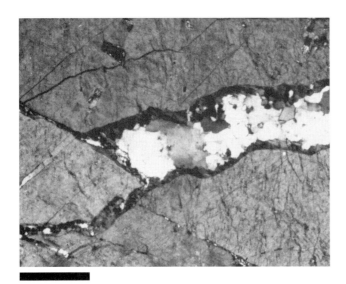

Figure 33.2
Microveins of postmagmatic quartz and K-feldspar in microfractured and microfaulted alkali feldspar crystal, analogous to mesoscopic pegmatite. New K-feldspar (in extinction) epitaxial on host crystal (light gray) with quartz in medial zone. Granite of Lac Du Bonnet, Manitoba, Canada (see Figure 27.3b). X-polars. Bar scale = 0.12 mm.

1979, 1986; Burnham and Ohmoto, 1980). There is some indication that Cl content of these late-stage fluids may be a factor in the transport capacity of aqueous-rich fluids (Burnham, 1986; Czamanske et al., 1991).

Mesoscopic Aplite–Pegmatite Systems

Magmagenic granitic pegmatites typically occur in association with aplite. Aplite–pegmatite dikes, lenses, and pods are common in granitic plutons (Figures 33.3a,b) and even large-zoned pegmatites occurring in metamorphic host rock without apparent association with magmatic rock have a marginal aplite zone (Cameron et al., 1949). Small pegmatite pods and pegmatite dikes in granites may appear to occur without aplite association (frontispiece). Similarly, aplite dikes may seem to be without pegmatite (Figure 17.5b). A general field rule is that where you find aplite you will eventually find pegmatite, and vice versa.

The experiments of Jahns and Burnham (1969) provided an explanation for the association of aplite and pegmatite. Sudden relocation of separated aqueous phase is envisioned as a pressure quench, accelerating nucleation rates of the hydrous magma, forming aplite as a product of melt crystallization. It is thought (Burnham, 1986) that very small aqueous

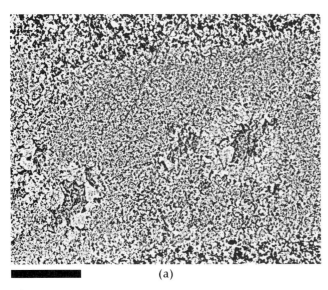
(a)

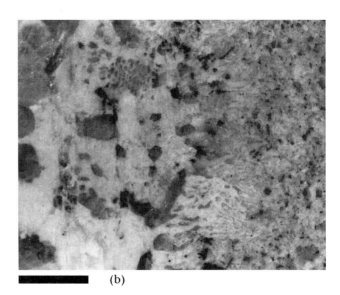
(b)

Figure 33.3
Aplite-pegmatite associations.

(a) Aplite–pegmatite dike in granodiorite. Pegmatitic quartz and K-feldspar locates in pods along medial zone of dike where latest aqueous-rich fluids were relegated as aplite crystallized. Slab surface. Osgood Mountains, Nevada. Bar scale = 2 cm. (b) Transition from aplite (right) to pegmatite (left) marked by graphic quartz-feldspar zone (center vertical). Pine Creek, Bishop, California. Slab surface. Bar scale = 1.5 cm. (Courtesy Union Carbide Corporation.)

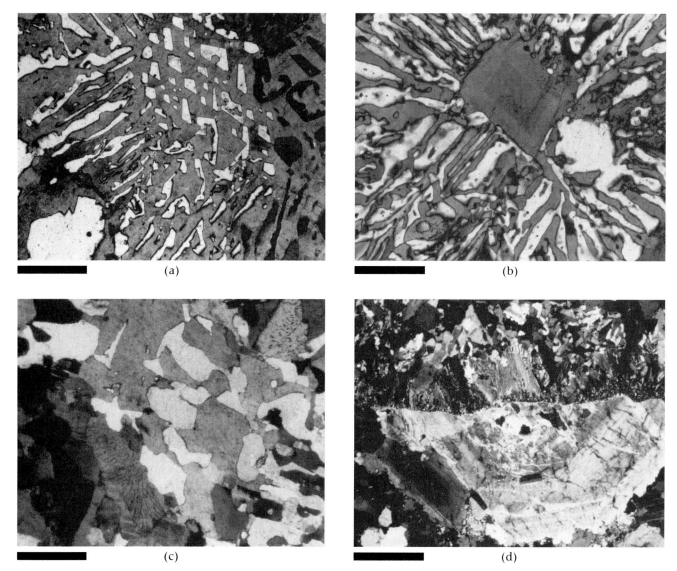

Figure 33.4
Graphic intergrowths.

(a) Micrographic quartz in K-feldspar. Note somewhat vermicular morphology of longitudinal sections of quartz (left) in contrast to diamond-shaped cross sections (upper right). Aplitic granite near Contact, Nevada. X-polars. Bar scale = 0.5 mm. (Courtesy Leo Sheehan.) (b) Longitudinal sections of graphic quartz (see (a)) in K-feldspar radiating from quartz-free zoned K-feldspar core crystal. Presumably, simultaneous crystallization followed initial crystallization of K-feldspar. Aplitic granite. Heusser Mountain, Nevada. X-polars. Bar scale = 0.5 mm. (c) Pegmatite consisting of graphic–vermicular intergrowth of quartz with K-feldspar and oligoclase associated with myrmekite. X-polars. Bar scale = 0.5 mm. (d) Portion of 3-cm aplite dikelet emplaced in fractured granodiorite. Initial crystallization consists of very fine-grained aplite on fracture surface of host-rock plagioclase, followed by somewhat plumose micrographic K-feldspar–quartz oriented perpendicular to this substrate, and finally fine-grained granular aplite in medial zone of dikelet. Note similar sequence in Figure 33.3b. Diamond Mountains, Nevada. X-polars. Bar scale = 0.5 mm.

Figure 33.5
Aplitic granite with local centers of coarse feldspar, quartz and tourmaline-lined miarolitic cavities. Sierra Nevada, California. Glacially polished rock surface. Bar scale = 1 cm.

phase bubbles capable of exchanging Na for K cause rapid nucleation of melt and growth of sodic plagioclase in association with K-feldspar and quartz in the form of the aplite. As bubbles of K-bearing aqueous phase are expelled from the region of aplite crystallization, they concentrate in favorable structural sites, where large crystals of feldspar and quartz are formed. The aqueous phase in which these crystals form is enriched in the silicate structure-breaking components Li, B, P, F, and H, reducing nucleation rates (Candela, 1991; see also Fenn, 1977; Swanson, 1977) and promoting growth of large crystals.

Quartz coprecipitates graphically (Figure 33.4) with K-rich feldspar (Figures 33.4a,b) or with plagioclase (Figure 33.4c). Fenn (1986) has developed a convincing genetic model for graphic texture. He suggested that simultaneous crystallization of quartz and K-feldspar results in the intergrowth, controlled by growth kinetics rather than by cotectic equilibrium relations in the bulk system.

The characteristic K-feldspar of pegmatite is coarsely twinned microcline perthite, with or without graphic quartz. It is generally agreed that twinned microcline must have a monoclinic predecessor. Since the presence of aqueous phase seems to promote inversion of monoclinic K-feldspar to maximum microcline (Eggleton and Buseck, 1980), it appears that the aqueous-phase–melt environment characterizing the late-magmatic stage means that a monoclinic phase formed first, inverting to microcline, probably as exsolution took place.

The grain-size contrast between aplite (fine-grained granular) and pegmatite may be marked by a transition zone characterized by graphic and vermicular intergrowths of quartz and K-feldspar (Figure 33.3b). A core of aplite-size plagioclase may evolve to a graphic K-feldspar–quartz "mantle" (Figure 33.4b). A similar relation is shown in Figure 33.4d. In this case, an exceedingly fine-grained aplite crystallized against wallrock, in this instance a sheared plagioclase crystal. From this aplitic substrate, micrographic quartz–K-feldspar crystallized toward the medial zone of a 3-cm dike, from both sides. The medial zone of the dikelet is "coarse" aplite, in a sense equivalent to pegmatite.

Medium-grained masses of leucogranitic rock (Figure 33.5), bearing tourmaline and miarolitic cavities as evidence of the presence of an aqueous phase, may represent the "failed" development of an aplite–pegmatite association. Many grain sizes of leucogranitic rocks (alaskite, aplitic granite) fit between fine-grained aplites and very coarse-grained pegmatites, even though the occurrence of aplite and pegmatite without intermediate grain-sized rock is very common.

The grain-size contrast between aplite and pegmatite does not necessarily correspond to the distinction between crystallization in melt (the aplite) and crystallization in aqueous phase (the pegmatite). Micrographic and microvermicular quartz-feldspar intergrowths occur in variety of magmatic rocks (Figures 33.4d; 17.4a,c; 17.6b; 17.7b). Granophyre, as a rock type, is characterized by these micrographic–microvermicular intergrowths, including granophyres in the upper region of diabase sills (Ernst, 1960) and as crosscutting veins in mafic cumulates (Czamanske et al., 1991).

If pegmatites are the result of crystallization in aqueous-phase solutions, pegmatite bodies should appropriately carry hydrothermal rather than magmatic terminology. Pegmatite "dikes" should be pegmatite veins. In reality, the common association of aplite with pegmatite in dikes leaves the choice of nomenclature optional since the aplite portion of the system represents crystallization of silicate melt, and "dike" is appropriate.

Pegmatite Zoning

Pegmatites characteristically have a simple zonation (Jahns, 1955; Jahns and Burnham, 1969; Figure 33.6), whereas some are rhythmically zoned (Rockhold et al., 1987; Duke et al., 1988; Figure 33.7). The wall zone of large pod-shaped pegmatites is characterized by muscovite crystals measured in decimeters (Cameron et al., 1949). Intermediate zones are characterized by giant crystals of feldspar, including microcline perthite,

394   Part 9   Rocks Formed by Precipitation and Reactive Crystallization

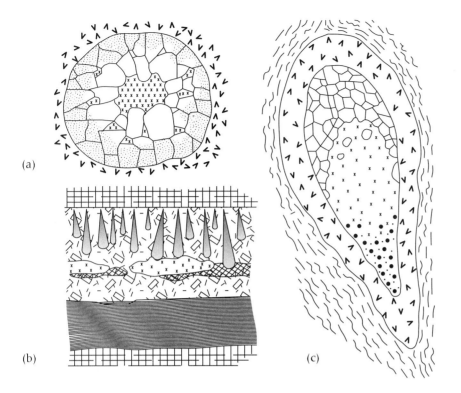

Figure 33.6
Mesoscopic sketches of pegmatite.

(a) Zoned miarolitic pod in granite. (b) Zoned pegmatite dike with aplite footwall (thickness about 1 meter). (c) Zoned pod-like pegmatite with granitoid (may be aplitic) margin (scale = meters). From Jahns and Burnham (1969).

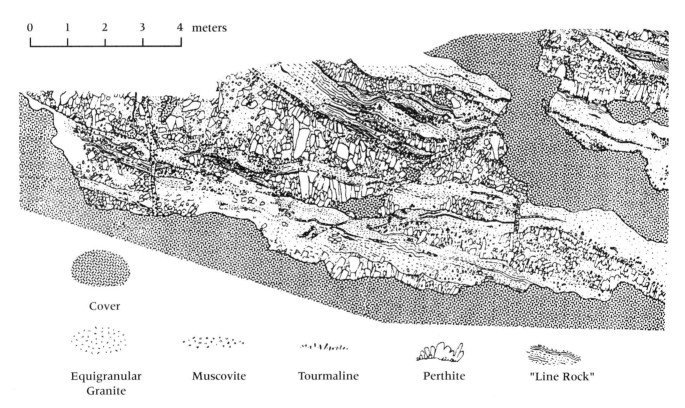

Figure 33.7
Complex layered and lensed association of pegmatite and aplitic rock. Calamity Peak pluton, Black Hills, South Dakota. (Courtesy James Papike [Duke et al., 1988]).

graphic quartz–microcline perthite, and cleavelandite (coarse thin-tabular albite).

Quartz dominates in the core of large-zoned, pod-shaped pegmatites (Figure 33.6), nicely mimicked by zoned miarolitic pods (Figure 33.3a) and zoned microveins (Figure 33.2) in which quartz is located in the central region.

A pocket zone generally is located within the quartz core of large pod-shaped pegmatites. This zone has its parallel in miarolitic cavities (Figure 17.8c). The pocket zone is characterized by a fascinating suite of minerals, including beryl, tourmaline, topaz, lepidolite, spessartine garnet, fluorapatite, spodumene, autunite, and columbite (Cameron et al., 1949; Foord et al., 1989).

Ruptured crystals in the pocket zone may be the result of explosive release of volatiles at the final stage of evolution of a pegmatite that is sealed by relatively impermeable wallrock (Jahns and Burnham, 1969; Foord et al., 1989).

Hydrothermal Rocks

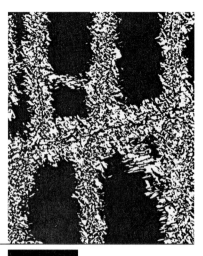

Cleavage-controlled sericite replacement of orthoclase. X-polars. Bar scale = 0.12 mm.

No demonstration of the action of hot water in the geological environment is more impressive than the **geysers** of Yellowstone National Park. The rock-forming capability of these eruptions and associated nonerupting **hot springs** is demonstrated by incrusting precipitations at the vents in the form of **siliceous sinter**. Some hot springs at Yellowstone are fed by **geothermal waters** that have passed through limestone, dissolving calcium carbonate and reprecipitating it in the form of magnificently terraced **travertine deposits**.

Holocene surface hydrothermal activity is linked to hydrothermal rock-forming processes that operate at depth within the Earth's crust. The results of hydrothermal activity at depth are recorded as mineralogical replacements, or partial replacements, of pre-existing rock, generating "hydrothermally altered rocks." Precipitation of minerals in fractures and other open spaces, or precipitation in "potential space," as, for example, the filling of a fracture as it dilates, is also an important process of hydrothermal activity and is traditionally considered an "alteration" process.

Estimation of composition and source of hydrothermal fluids is important in reconstruction of hydrothermal rock-forming process as well as to the genesis of associated ore deposits.

Hydrothermal Solutions

Source of Solutions

There are five fundamental sources of H_2O: meteoric, ocean, connate, metamorphic, and magmatic (Figure 12.8). The temperature of connate, metamorphic, and, of course, magmatic water is already elevated. Meteoric and ocean water can become heated (hydrothermal) by mixing with hot water from another source or by contact with hot rock. Which source or combination of sources is involved in hydrothermal rock-forming process in any given circumstance can be evaluated in the light of **oxygen isotopic ratios** (Taylor, 1979).

Oxygen consists of the stable isotopes ^{18}O and ^{16}O. This slight mass difference results in slightly different chemical and physical behavior in a variety of geological environments. Chemical bonds formed by lighter ions are very slightly weaker than those formed by heavier ions, meaning that ^{16}O is slightly more mobile and reactive than the heavier isotope. Evaporated water has a lower $^{18}O/^{16}O$ ratio than the water from which it evaporated. In reverse, condensation of water vapor in the form of rainfall generates a slightly higher $^{18}O/^{16}O$ ratio than its parent vapor because the lighter isotope tends to remain in the more mobile gaseous state. The ratio in meteoric water decreases with increasing altitude, higher latitudes, and in meteoric water with increasing distance from the ocean.

Meteoric and connate water in contact with most rocks is depleted in ^{18}O because ^{16}O is chemically more active and less likely to be tied up in crystal lattices. It follows that the $^{18}O/^{16}O$ ratio is relatively high in sedimentary rocks. Magmatic aqueous phase has a higher $^{18}O/^{16}O$ ratio than seawater, and the ratio is also higher in meteoric water and meteoric water that has interacted with rock. This means that the oxygen isotopic ratio of water at Yellowstone and of older rocks known to have formed in a hydrothermal environment should indicate an H_2O source, although discrimination between meteoric and magmatic fluids at low latitudes is less successful compared with higher altitudes and latitudes. Magmatic water signatures are found in deep environments associated with magmatism. It is now realized on the basis of oxygen isotope analysis that dilution of magmatic and metamorphic water occurs in hydrothermal systems as deep as several kilometers, and that the waters we see at hot springs and geysers are dominantly, if not entirely, meteoric. This situation is deceptive because the magmatic source of water vapor exhalations in active volcanic vents seems obvious. The physical relation is there, and the association of the water vapor with other fumarolic gases is highly suggestive of a direct connection to magma (White and Hedenquist, 1990).

Since meteoric water can be heated in the vicinity of a magma body or a magmatic rock body that is still cooling, the fact that the water is hot does not indicate source. The nature of solutes in hydrothermal waters may not be diagnostic either. Chloride, sulfur, and metal ions do not necessarily indicate magmatic source, in spite of what is apparent at active volcanic vents. Such ions can be dissolved out of wallrocks by heated meteoric water.

Composition of Hydrothermal Solutions

Hydrothermal solutions contain solutes that reflect source and interaction with rock. Rainwater is essentially free of dissolved substances, whereas connate water trapped in the pores of marine sediments is saline, and fluids emanating from a magmatic system typically contain H_2O, CO_2, SO_2, and lesser amounts of other gases. Groundwater, heated by a nearby magma, may infiltrate glassy tufaceous volcanic rock, dissolving silica out of the unstable glass. If geothermal water travels through carbonate rock, dissolution of carbonate results.

Determination of the composition of **fluid inclusions** in ore minerals and gangue minerals demonstrates (Bodnar et al., 1985) that hydrothermal solutions involved in most ore-forming processes are typically saline. Sodium and potassium chlorides are derived from ocean water, marine connate water, exhalations from magmatic systems, and from the dissolution of evaporites contained in some sedimentary sections. Rainwater inland from coastal regions may contain saline aerosols that impart an early salinity to the meteoric water. Nevertheless, some low-temperature hydrothermal systems have low salinities.

It was recognized long ago that metal sulfides such as sphalerite and galena are essentially insoluble in water, even if the water is very hot. An apparent paradox was realized in that geological relationships dictated that deposition of metal sulfides must be from H_2O-dominated fluids. Since fluid inclusion data 0commonly indicated the saline nature of these fluids, the connection between the presence of the chloride ion and increased solubility became obvious.

Formation of **complex ions** increases the solubility of metal cations in saline aqueous solutions (Helgeson, 1964; Romberger, 1988). Sulfur occurs in solution in forms such as SO_4^-, HS^-, HSO_4^-, H_2S, and possibly $NaSO_4^-$ in saline solutions. Various soluble complexes of these ions and chloride ions with metal cations are possible, among them are $AgCl_2^-$, $AuCl_2^-$, $Au(HS)_2^-$, $PbHS^+$, $PbHS_2^+$, $PbOH^+$, and HgS_2^{-2}.

Silver and the base metals (Cu, Pb, Zn) are most efficiently transported by Cl^- complexes under acid-oxidizing conditions. For example,

$$PbS(s) + 2H^+(aq) + Cl^{-1}(aq) = PbCl_2(aq) + H_2S(g)$$

Although gold forms a soluble chloride complex ion as well, it is also mobil under reducing conditions where hydrogen sulfide is present, forming the soluble gold bisulfide ion:

$$Au(s) + 2H_2S(aq) = Au(HS)_2^{-1}(aq) + H^+(aq) + 0.5H_2(g)$$

Gold may also form soluble complexes with arsenic and tellurium (Romberger, 1988).

It is possible that some hydrothermal fluids consist of colloidal dispersions rather than ionic chemical solution. Occurrence of extremely small particles of gold and cinnabar in opaline silica rock (opalite) suggests that colloidal silica may have been important in the transport and fixation of metals. The chalcedony so common in agate may evolve from colloid-bearing fluids as well.

Precipitation of Minerals from Hydrothermal Solutions

Precipitates form in aqueous solutions (1) if they are cooled by heat loss to environment, reducing solubility and stability of complex ions, (2) if there is a pressure reduction leading to adiabatic boiling and loss of volatile phases, (3) if fluid composition changes by diffusion of chemicals into the fluid from wallrock, and (4) if there are changes in pH, salinity, and temperature resulting from mixing of waters of different composition and temperature.

(1) Cooling a solution saturated in NaCl results in halite precipitation. Silica dissolved by geothermal waters precipitates as opal upon cooling, perhaps requiring even more cooling if there is a colloidal suspension of silica. Similarly, cooling of a solution containing the complex ion $Al(OH)_4^-$ results in precipitation according to the following equation because the stability of the complex decreases with decreasing temperature:

$$K^+ + Al(OH)_4^- (aq) + 3SiO_2(aq) \rightarrow KAlSi_3O_8 + 2H_2O \quad \text{K-feldspar}$$

In copper systems the stability of Cu–Cl complexes decreases with falling temperature as well.

(2) Rise of a hydrothermal solution along a conduit results in pressure decrease, leading to adiabatic boiling (Reed and Spycher, 1985). A good example of this is the heating of semiconfined meteoric water by hot rock or magma beneath the Yellowstone vents. As the water is heated in solution chambers, convective overturn is inhibited by the narrow conduits leading to the surface. As a result, the pressure of the semiconfined system increases until vaporization pressure exceeds confining pressure. With vaporization (boiling), there is an enormous increase in volume and geyser eruption issues, and the process starts all over again.

Boiling leads to escape of volatile compounds dissolved in the water at lower temperature. For example, loss of CO_2 can result in the precipitation of calcite if there is calcium dissolved in the water. Even very slight changes of pressure can result in the precipitation of calcium carbonate as travertine at hot springs and speleothems in caves:

$$Ca^{2+}(aq) + 2HCO_3^-(aq) + H^+(aq) \rightarrow CaCO_3(s) + H_2O(aq) + CO_2(g)$$

Escape of H_2S attendant to boiling can result in precipitation of mercury, gold, and realgar according to the equations:

$$HgS(H_2S)_2(aq) \rightarrow HgS + 2H_2S(g) \quad \text{cinnabar}$$

$$8Au(HS)_2^-(aq) + 6H^+ + 4H_2O \rightarrow 8Au + SO_4^{2-} \quad \text{bisulfide} \quad \text{gold}$$
$$+ 15H_2S(g)$$

Gold produced in this oxidation reaction ($S^{2-} \rightarrow S^{6+}$) is typically associated with pyrite and arsenic sulfides:

$$2AsS_3^{3-}(aq) + 6H^+ \rightarrow As_2S_3 + 3H_2S \quad \text{realgar}$$

(3) Diffusion of Fe from wall rock into copper-bearing hydrothermal solutions may lead to the precipitation of chalcopyrite:

$$Cu^+ + Fe^{2+} + 2H_2S + \tfrac{1}{4}O_2 \rightarrow CuFeS_2 + 3H^+ + \tfrac{1}{2}H_2O$$

(4) Mixing waters of different composition can cause dilution and precipitation. For example, the soluble gold chloride complex ion is unstable if there is dilution with freshwater low in chloride ion and reduction to native gold occurs:

$$2AuCl_2^-(aq) + H_2O \rightarrow 2Au(s) + 0.5O_2 + 4Cl^- + 2H^+$$

Hydrothermal Interaction with Rock

Direct Precipitation or Reactive Replacement?

Open spaces in rocks can be incrusted or filled with precipitates if hydrothermal solutions have access to these spaces. Potential space can be the site of crystallization as well if it is occurring as the "space" is generated. This "space" can be generated by (1) mechanical dilation in rock, such as along a fault, as precipitation occurs, or (2) if there is separation and localization of an aqueous phase in magmatic systems that becomes the site of further, now hydrothermal, crystallization.

If there is reaction of the hydrothermal solution with host rock rather than direct precipitation, new minerals (or new crystals of the same minerals) are produced at the expense of old ones by reactive replacement. These relations are sketched in Figure 34.1.

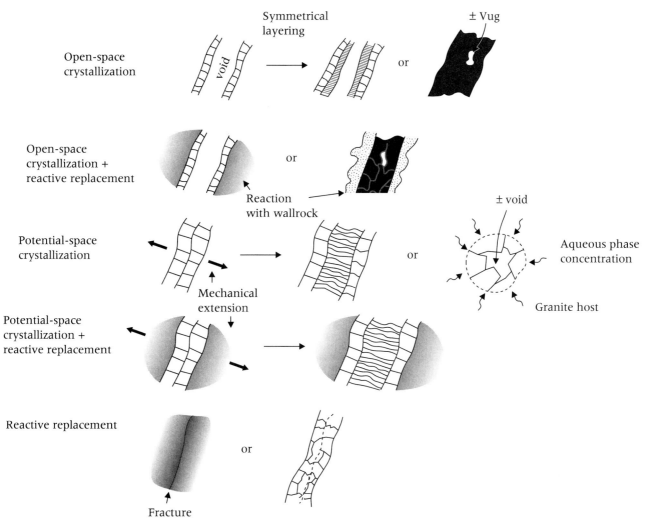

Figure 34.1
Representation of open space, potential space, and reactive replacement hydrothermal precipitation shown relative to a fracture or fault plane. Rock surfaces serve as substrate for crystallization and/or access for permeation of fluids.

Open-Space and Potential-Space Precipitation. Pre-existing open spaces provide rock surfaces that serve as substrates for precipitation of minerals from hydrothermal solutions. Most cavities occur near or at the Earth's surface where confining pressure is low. The volume of precipitates is limited by the size of the cavity, except at hot springs where precipitation is in direct contact with the atmosphere and the volume of precipitates is limited only by the availability of the hydrothermal solution and its solute.

The size and shape of open spaces are many. Vesicles are isolated, whereas spaces between breccia fragments are interconnected. Precipitation in a planar structure such as a fault or fracture generates veins. If the surface configurations of opposing rock surfaces along the course of a vein "fit together," extension-generating open space or potential-space precipitation is indicated. Precipitates in veins, such as quartz, may have little or no textural or chemical relation to substrate surfaces. More commonly, **symmetrical layering** (Figure 34.2a) develops, characterized by matching layer pairs of mineralogically and texturally identical material that has deposited inward toward a medial zone. **Comb structure** is a variety of symmetrical layering in which crystals, such as quartz, terminate with crystal faces toward the medial zone. Incomplete filling of an open space may leave faceted crystals lining vugs (Plate 5c). **Reniform** (radiating crystals terminating in kidney-shaped mass), **botryodial** (radiating crystals forming "bunch of grapes"), and **colloform**

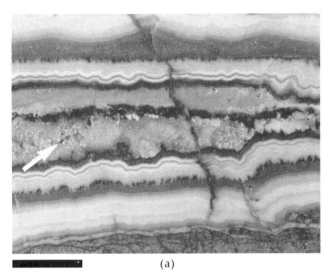 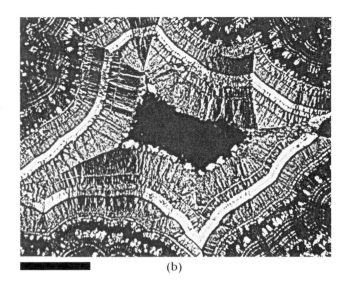

Figure 34.2
Open-space precipitation.

(a) Symmetrical layering consisting of chalcedony and a medial zone of vuggy (arrow) quartz. Slab surface. Bar scale = 2.3 cm. (b) Peripheral opal at extinction contains small quartz crystals. Central region consists of concentric layers of chalcedony and an inner quartz-lined void (black, center). X-polars. Bar scale = 1 mm.

(rhythmic layers with rounded surfaces) morphologies are also indicative of free precipitation into a fluid. Symmetrical layering with surfaces convex toward the central cavity is characteristic (Figure 34.2b).

Brecciation of any origin generates open space because there is an inherent volume increase. Brittle faulting and hydrothermal explosion breccias are prime candidates for infiltration of hydrothermal solutions. **Cockade structure** consists of breccia fragments that have been partly or completely "cemented" by precipitates, which may or may not have symmetrical layering. Vugs may be present if the confining

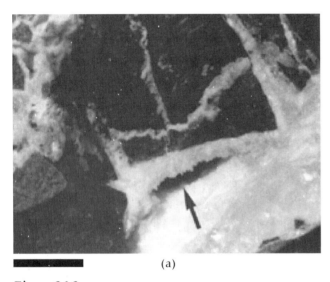 

Figure 34.3
Breccias.

(a) Cockade structure with limestone fault breccia fragments and calcite "cement." Note remaining open space lined with calcite crystals (arrow). Treasure Hill, Hamilton, Nevada. Slab surface. Bar scale = 1 cm. (b) Rubble breccia consisting of silicified limestone fragments coated by sphalerite. Bar scale = 1 cm.

pressure is not high (Figure 34.3a). **Mosaic breccia** is characterized by fragments that "fit" together, as a result of (1) minor expansion without significant rotation, (2) reactive replacement along a network of fractures (**crackle breccia**), or (3) mechanical repacking of angular fragments. **Stockworks** are a variety of mosaic breccia formed chiefly by reactive replacement in crackle breccia.

Open spaces occur in limestones that have had a history of dissolution. **Sanding** is the solutional removal of intergranular carbonate in siltstones and sandstones that in some cases leads to silicification and deposition of ore minerals such as the Carlin-type gold deposits (Radtke et al., 1980). **Karsting** involves the formation of caves and **collapse breccia**, which are also potential sites of hydrothermal precipitation (Figure 34.3b).

Deformation of earlier deposited material followed by renewed crystallization on fracture, fault, or breccia surfaces characterizes potential open-space precipitation. Such a process is analogous to the **crack-seal** textures occurring in low-grade metamorphic rocks (Cox and Etheridge, 1983) (Chapter 25). Deformation of earlier formed quartz and pyrite, followed by precipitation of argentite and a second generation of quartz is shown in Figure 34.4. This particular textural relation does not prove that the early quartz and pyrite grew in open space, but it is likely in view of the occurrence of vuggy quartz in nearby veins (Plate 5c).

Reactive Replacement. Minerals and amorphous rock-forming materials can be converted to new minerals by a more-or-less volume-for-volume recrystallization involving chemical reaction. For example, replacement of limestone by sphalerite can take place according to the following reaction:

$$\underset{\text{calcite}}{CaCO_3(s)} + H_2S(aq) + ZnCl(aq) \rightarrow \underset{\text{sphalerite}}{ZnS(s)}$$
$$+ Ca^{+2} + H_2CO_3(aq) + Cl^-$$

Reactive replacement can be intragranular or transgranular, and it may be partial or complete. Intragranular pseudomorphic replacement preserves the morphology of the original grain. It may be complete, as shown by the total conversion of plagioclase to montmorillonite (Figure 34.5a), sanidine to alunite (Figure 34.5b), plagioclase to andalusite plus quartz (Figure 34.5c), calcite to quartz (Figure 34.5d), or there may partial replacement such as by cleavage-controlled sericitization of orthoclase (frontispiece).

Since hydrothermal alteration is characteristically an open chemical system, a replacement mineral assemblage represents both indigenous chemical composition as well as chemical introductions. For example, the reaction plagioclase → andalusite + quartz, based on the observed minerals within the boundaries of the relict plagioclase crystal (Figure 34.5c), cannot be a closed chemical system, nor a system open only to pure aqueous phase, because there is too much silica and no in situ accounting for sodium and calcium of the original plagioclase.

Transgranular reactive replacement occurs along fractures, faults, or some other pre-existing structure. Pseudomorphic replacements may occur within host rock on either side of the transgranular replacement zone. Mismatch of vein walls is indicative of at least some reactive replacement (Figure 34.6a). Conversion of a magmatic mineral assemblage to a hydrothermal mineral assemblage along a transgranular fracture is shown in Figure 34.6b. In this case, reactive replacement has generated albite, epidote, and chlorite at the expense of plagioclase and hornblende. Tourmaline in the reaction zone underscores the chemically open hydrothermal system even though most of the alteration mineralogy can be locally accounted for by the following reaction:

$$\text{plagioclase} + \text{hornblende} + H_2O \rightarrow \text{albite} + \text{epidote} + \text{chlorite}$$

Reactive replacement along joints in a blue-gray limestone has generated a medial zone of quartz and

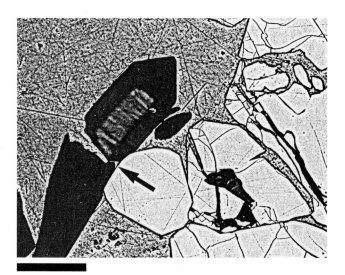

Figure 34.4
Faceted quartz crystal (black) formed in vug with pyrite (white), with subsequent precipitation of argentite (light gray) during or after a brittle fracture event. Note argentite in pyrite fractures (right) and ruptured quartz crystal (center, left) resulting from compression against a pyrite crystal acting as an anvil (arrow). Reflected light. Bar scale = 0.12 mm. From fissure vein, Comstock Lode, Nevada. (Courtesy D. Hudson.)

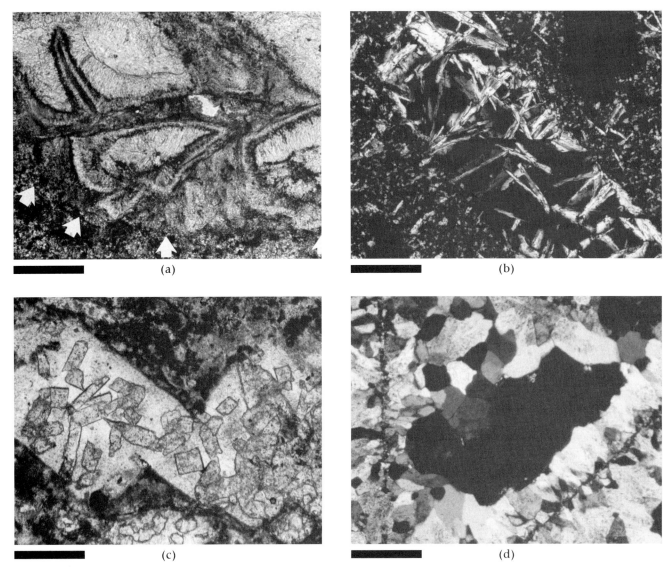

Figure 34.5
Pseudomorphic replacements.

(a) Calcic montmorillonite after plagioclase phenocryst in Alta Andesite, Comstock Lode, Nevada. Former position of plagioclase crystal faces indicated with arrows. PPL. Bar scale = 0.12 mm. (Courtesy D. Hudson.) (b) Alunite crystals (mostly plucked during thin sectioning) after plagioclase or sanidine crystal in Mickey Pass ash flow tuff. Gabbs Valley Range, Nevada. X-polars. Bar scale = 0.5 mm. (Courtesy D. Hudson.) (c) Andalusite crystals (high relief) with quartz as replacement of plagioclase. Green Talc Mine, Hawthorne, Nevada. PPL. Bar scale = 0.5 mm. (Courtesy D. Hudson.) (d) Intercrystalline quartz precipitation initially controlled by calcite crystal faces (vertical trace, left; inclined trace, lower right). Dissolved calcite only partly replaced by quartz, leaving open-space interior (black, center). X-polars. Bar scale = 0.5 mm.

bilateral zones of carbon-free (bleached) rock containing newly formed tremolite and recrystallized calcite (Figure 34.6c). Reactive replacement may be controlled by fossil or fossil-cast morphology, such as the local quartz replacement of colonial coral (Figure 34.7a), and by crystal grains, such as quartz clasts (Figure 34.7b) and quartz phenocrysts (Figure 34.7c) on which there is epitaxic crystallization at the expense of matrix minerals.

The mode of access of hydrothermal fluids into rock may not be as obvious as a fault- or fracture-controlled vein. Infiltration along grain boundaries may

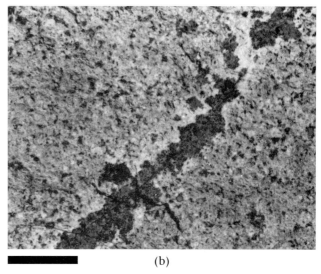
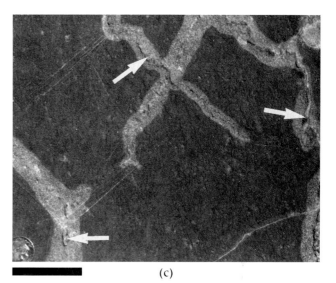

Figure 34.6
Reactive replacement in vein structures.

(a) Transgranular vein in quartz monzonite porphyry. Mismatch of vein walls in quartz phenocryst indicates at least some replacement. Therefore, some of the quartz in the microvein may be derived from the phenocryst. Vein beyond confines of quartz phenocryst contains sulfides (not shown). Mission, Arizona. X-polars. Bar scale = 0.5 mm. (Courtesy ASARCO.) (b) Replacement of plagioclase and hornblende in Mt. Davidson granodiorite by albite (white), chlorite, and tourmaline (black). Slab surface. Virginia City, Nevada. Bar scale = 1 cm. (c) Replacement of blue-gray limestone along fractures perpendicular to bedding results in tremolite and recrystallized calcite in bilateral bleached zones with a medial zone of quartz (arrows). Notch Peak contact zone, House Range, Utah. Dime scale = 1.8 cm.

leave little or no textural expression, other than the replacement phenomena at a site within the rock. Local partial replacement of polygonalized metamorphic plagioclase by quartz (Figure 34.8) is indicated by the isolation of plagioclase grains from the polygonal mosaic and reduction in size of these partially replaced grains.

Open- (and Potential-) Space Precipitation with Reactive Replacement. Several combinations of precipitation in open or potential space with reactive replacement are shown in Figure 34.1. Generation of open (or potential) space by brecciation of a micritic limestone can be the locus of quartz precipitation (Figure 34.9). In this case, reaction of the quartz-precipitating fluids with the micritic carbonate generated relatively large calcite crystals along the fluid–micrite interface, indicating that the fluid (and heat) were the catalysts for the carbonate recrystallization, with no significant change in mineral composition. Thus, reactive replacement ranges from simple recrystallization without mineralogical change to complex mineralogical reactions producing new mineral phases.

The quartz veining shown in Figure 32.4 represents a similar multiprocess episode involving hydrothermal fluids. Fracturing generated crackle breccia, in which there was little or no fragment rotation. However, some dilation of fractured rock is inevitable, allowing infiltration of hydrothermal fluids and at least some open- (or potential-) space precipitation of quartz. Reactive replacement of the host granite by quartz along the fractures is suggested by the "diffusive" contacts between quartz and granite.

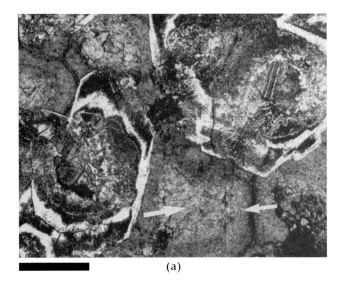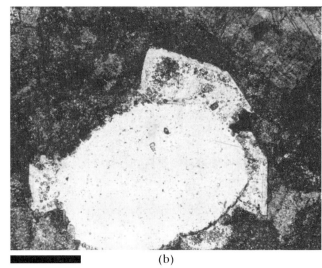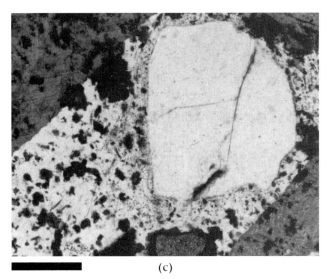

Figure 34.7
Hydrothermal quartz replacements.

(a) Colonial coral partly silicified with quartz (white) along cell walls. Cells contain sparry calcite (larger white arrow) and micrite (smaller white arrow). X-polars. Bar scale = 0.5 mm. Bisbee, Arizona. (Courtesy P. Murphree.) (b) Faceted quartz crystal epitaxial on detrital quartz in limestone. Silicified contact zone, Bisbee, Arizona. Field association with hydrothermal system points to hydrothermal origin of quartz overgrowth in preference to diagenetic quartz overgrowth. X-polars. Bar scale = 0.12 mm. (Courtesy P. Murphree.) (c) Quartz (and calcite) replacement in porphyry. Quartz is epitaxial on relict quartz phenocryst. X-polars. Bar scale = 0.12 mm. Buckskin Range, Nevada. (Courtesy D. Hudson.)

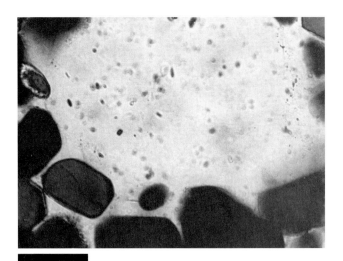

Figure 34.8
Grain boundary reactive replacement by quartz (white) in polygonalized plagioclase mosaic of hornfels. Note reduction in size of partially replaced plagioclase grains. X-polars. Bar scales = 0.05 mm. Okanogan Range, Washington.

Figure 34.9
Open-space (or potential-space) precipitation plus reactive replacement. Quartz precipitation in micritic limestone fault breccia with reactive replacement at solution–limestone interface generating new calcite (arrows). Salina, Utah. X-polars. Bar scale = 0.12 mm.

Zoning and Paragenetic Sequence

Precipitation of minerals from hydrothermal solutions, and replacement reactions resulting from chemical interaction between rocks and hydrothermal fluids, are controlled by temperature, pressure, rock composition, and fluid composition. At a given level in the crust, or for only minor changes in depth, pressure is essentially constant because pressure gradients are generally not steep, particularly when systems are hydrostatic.

Consider first temperature as a controlling factor in the distribution of minerals generated by hydrothermal activity. When a pluton is intruded into the shallow crust, it is necessarily hotter than enclosing wallrocks. Heat from the intrusion is conducted into the adjacent wallrocks and a thermal front progresses outward. When the pluton has crystallized and cooled to the temperature of immediately enclosing wallrocks, both the pluton and the wallrocks begin cooling as heat is conducted away. This process produces a thermal front that collapses on the intrusion during later stages of cooling.

If mineral stabilities are dictated by temperature alone, then it may be seen from line A–B (Figure 34.10, Beane, 1982) that K-feldspar will be stable at highest, and kaolinite at lowest, temperature. Thus at any given time, K-feldspar would be expected to occur closest to the intrusion and kaolinite most distal with muscovite in between. This spacial arrangement of minerals is called **zoning**. Zoning may also be recorded by the distribution of ore minerals. Minerals such as arsenopyrite, wolframite, chalcopyrite, and molybdenite crystallize in places where temperatures are relatively high, whereas lower-temperature minerals such as argentite, cinnabar, stibnite, and marcasite preferentially crystallize farther from the heat source.

During later stages of cooling, when isotherms are moving toward the intrusion, there are likely to be progressively lower temperatures with time at a given position in the system. As temperatures decrease, K-feldspar becomes unstable, first with respect to muscovite and then with respect to kaolinite. Thus K-feldspar might be expected to be replaced with muscovite and muscovite replaced by kaolinite. This change in mineralogy with time is called a **paragenetic sequence**, which establishes a new (overlapped or telescoped) zoning arrangement. Such an overlapping relation can apply to ore minerals as well as alteration minerals.

Another scenario in which to view thermal effects is with respect to a vein through which fluids flow. The fluid may be either hotter or colder than the rock through which the vein passes. If the fluid is hotter, such as one derived from a magma or one heated by a magma, the heat is conducted from the fluid into the

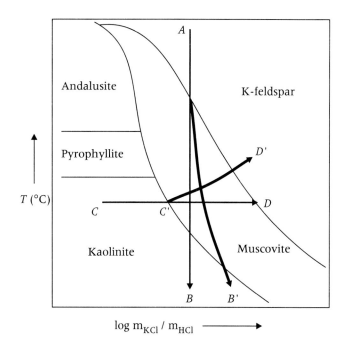

Figure 34.10
Fields of stability of andalusite, pyrophyllite, kaolinite, muscovite, and K-feldspar with respect to temperature and concentration of KCl relative to HCl. Vertical reaction path A–B indicates progressive passage of a fluid with constant composition into stability fields of K-feldspar, muscovite, and kaolinite with drop in temperature. Horizontal path C–D indicates the possibility of isothermal transition of kaolinite to muscovite and then to K-feldspar stability as the ratio m_{KCl}/m_{HCl} increases (equivalent to decrease in H⁺ activity or increase in pH). Path A–B′ indicates both temperature- and reaction-controlled compositional changes. Path C–D′ indicates possible increase in temperature attendant to meteoric fluids entering cool margin of cooling pluton, then moving toward hotter interior. Modified from Beane (1982).

adjacent rock, producing a thermal gradient. According to Figure 34.10, K-feldspar should appear closer to the vein while kaolinite would be stable farther out in the enclosing rock (Figure 34.11a). Similarly, with acid conditions, andalusite (Figure 34.5c) or pyrophyllite may be the stable phases at high temperature instead of muscovite or K-feldspar (Figure 34.10). Andalusite may form at the expense of K-feldspar by the following reaction:

$$2H^+ + 2KAlSi_3O_8 = 2K^+ + Al_2SiO_5 + 5SiO_2 + H_2O$$
$$\text{K-feldspar} \qquad \qquad \text{andalusite}$$

Alternatively, if the hydrothermal fluid is colder than wallrock, such as groundwater entering into a still-hot pluton or heated wallrock, then heat will be conducted into the fluid from the hot rock, and temperature will increase going outward from the vein, thus stabilizing kaolinite closer to the vein and K-feldspar farther out.

Compositional changes may also result in zoning and paragenetic sequence. If K-feldspar or muscovite is changed to kaolinite, then potassium must be transferred to some other place because there is no potassium in kaolinite. Because the compositions of the minerals are different, materials related to changes in mineralogy must be either removed from, or contribute to, some other phase or chemical medium.

Quartz is a common source for SiO_2 related to mineralogic changes because it is present in many geologic media. If K-feldspar is converted to muscovite by potassium loss:

$$3KAlSi_3O_8 + H_2O \rightarrow KAl_3Si_3O_{10}(OH)_2 + K_2O + 6SiO_2$$

Here, not only is SiO_2 released to the surroundings, but so too is K_2O. Potassium oxide does not occur as a rock-forming mineral, so it must be incorporated into some other phase. In shallow crustal environments, soluble cations such as potassium and sodium are commonly added to, or derived from, an aqueous phase which appears in the above reaction as H_2O. In such a case, the reaction could be rewritten as:

$$3KAlSi_3O_8 + 2H^+ \rightarrow KAl_3Si_3O_{10}(OH)_2 + 2K^+ + 6SiO_2$$

The conversion from K-feldspar to muscovite accompanied by formation of quartz is accomplished by reacting the feldspar with an acidic solution into which potassium is dissolved.

The role of K^+ and H^+ in the above reaction is reflected in the ratio KCl/HCl on the horizontal axis of Figure 34.10. Changes in mineralogy can occur at constant temperature, producing zoning (space related) and paragenetic sequence (time related), if chemical components are added to, or subtracted from, the system. Many mineral changes occur in the shallow crust in response to variations in solution composition because water, with its dissolved constituents, is essentially ubiquitous in this regime.

The mineral sequence K-feldspar → muscovite → kaolinite can reflect an increase in K^+/H^+ of an aqueous solution at constant temperature. One of the most common causes of an increase in this cation ration is through progressive decrease in dissolved H^+, corresponding to an increase in solution pH.

As an example, if K-feldspar is infiltrated by an acidic solution with low dissolved potassium, the ratio K^+/H^+ is low and the solution is in equilibrium with kaolinite (point C, Figure 34.10). The acidic solution would react with the K-feldspar to form kaolinite according to:

$$2KAlSi_3O_8 + 2H^+ + H_2O \rightarrow Al_2Si_2O_5(OH)_4 + 4SiO_2 + 2K^+$$

Kaolinization of K-feldspar uses up H^+ and contributes K^+ to the solutions. Thus the solution moves from point C along the line C–D (Figure 34.10). When the solution composition has changed to point C′, muscovite rather than kaolinite becomes stable according to the equation:

$$3KAlSi_3O_8 + 2H^+ \rightarrow KAl_3Si_3O_{10}(OH)_2 + 2K^+ + 6SiO_2$$

The net result of these reactions is a zone containing kaolinite, succeeded by one containing muscovite, and terminating in a zone in which K-feldspar is unaltered. Such zoning can occur along a vein or moving away from a vein into an adjacent wallrock (Figure 34.11).

Now consider the case in which a solution has moved along a vein producing the spatial mineral sequence kaolinite → muscovite → K-feldspar. Another volume of solution starting out with composition C begins moving along the vein. So long as the solution passes through the zone in which K-feldspar was converted to kaolinite by the earlier solution, then this solution will not react with the rock because the solution composition (point C) is compatible with kaolinite. But when the solution reaches the zone in which K-feldspar was altered previously to muscovite, the solution begins to react with the muscovite, converting it to kaolinite. As more and more solution moves through the rock, the kaolinite and muscovite alteration zones advance simultaneously in the direction of

Figure 34.11
Schematic illustration of mineralogical changes attendant to: (a) temperature-controlled hydrothermal reactions in two-feldspar granitic system generating an endogenic hydrothermal fluid, and (b) reaction-controlled changes in fluid composition as acidic meteoric water penetrates cooling pluton.

(a) Mineral Assemblage Dominated by Thermal Gradient

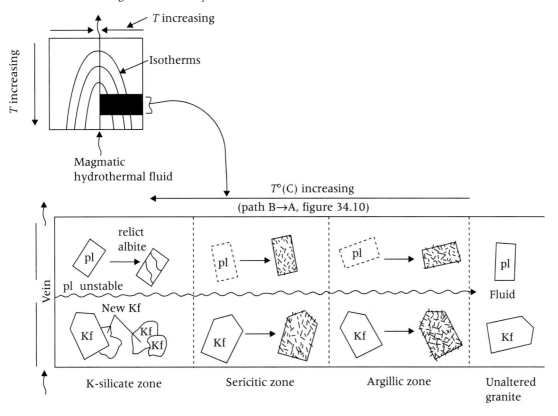

(b) Mineral Assemblage Dominated by Reaction-controlled Change in Fluid Composition

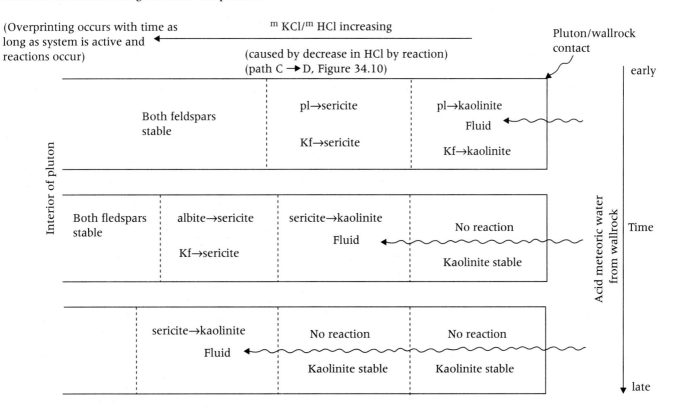

solution flow. These effects are called **overprinting** and they are diagrammed in Figure 34.11b.

In reality, if meteoric H_2O is entering a cooling crystallized pluton, the fluids will be heated as they penetrate toward the hotter interior of the pluton, that is, along path C–D' in Figure 34.10. Similarly, path A–B is more realistically drawn as path A–B' (Figure 34.10) because the effect of reaction-controlled change in fluid composition as well as temperature is likely to apply.

Sales and Meyer (1948) emphasized the continuous nature of zoning and paragenetic sequence during a hydrothermal event (see also Wallace, 1968; Nielsen, 1968). Lovering (1949) emphasized the overlapping of hydrothermal events.

Distinction Between Hydrothermal Replacement and Weathering Reactions

Mineral assemblages generated during primary hydrothermal precipitation and reaction at low temperature may be very similar to some of the mineral assemblages resulting from precipitations and reactions in the weathering zone. For example, montmorillonite and alunite can form in either environment.

Open-space precipitation and reactive replacement characterize the weathering zone as well as hydrothermal "alteration," meaning that textures associated with these processes are approximately equivocal. On the other hand, oxidation reactions occurring during weathering are typically related to the position of a former water table—a geological configuration not characterizing primary hydrothermal activity. Additionally, occurrence of minerals in fractures extending down through primary mineral assemblages may also be related to weathering processes and aid in the distinction between primary and secondary processes. However, these spatial relationships do not preclude the presence of relict primary minerals and textures in the weathering zone above the water table that may be mineralogically compatible with weathering.

Some attempts at radiometrically dating minerals of questionable ancestry seem to be successful (Ashley and Silberman, 1976; Arehart et al., 1992). If the age of crystallization of a mineral, such as alunite $[KAl_3(SO_4)_2(OH)_6]$, is demonstrably younger than associated minerals that are identified as being primary, a secondary origin is indicated. Textural distinction between primary and secondary alunite may also be possible. Secondary alunite (isotopically identified) preferentially occurs as fine-grained aggregates in veins, whereas primary alunite tends to be coarser grained (Ashley and Silberman, 1976).

Oxidation per se is not indicative of weathering as distinct from primary hydrothermal alteration. For example, the latter part of the main hydrothermal stage at the Carlin gold deposit in Nevada is characterized by oxidation of sulfides and organic compounds, brought on by boiling, loss of H_2O and CO_2, and production of H_2SO_4 and attendant acid leaching and oxidation (Radtke et al., 1980). Stable sulfur and oxygen isotope analysis of alunite in such deposits can discriminate between these primary acidic conditions and those obtaining during secondary, lower-temperature processes.

Although oxides and carbonates such as cuprite and malachite are likely to be secondary after a mineral such as chalcopyrite, occurring as secondary minerals in the weathering zone, it is well known that sulfides (unoxidized) may be secondary as well. Chalcocite is a very important secondary sulfide in porphyry copper deposits, and argentite may be partly secondary in some epithermal veins such as the Comstock lode of Nevada (Figure 34.12; Ramdohr, 1981).

Metamorphosed Hydrothermal Deposits

Hydrothermally altered rocks may be subsequently metamorphosed. The lead–zinc–silver deposit at Broken Hill, New South Wales, Australia is an interesting example. The ore minerals occur in regionally

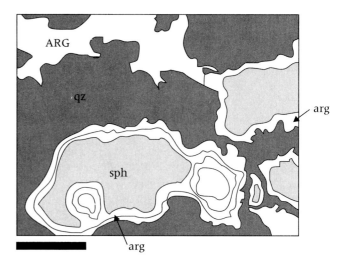

Figure 34.12
Textural distinction between primary hydrothermal ("hypogene") argentite (ARG) associated with quartz (qz), calcite, and sphalerite (sph), and secondary ("supergene") argentite (arg) redeposited as a layered incrustation on sphalerite. Gould and Curry Mine, Comstock Lode, Nevada. Drawn from reflected light micophotograph (Ramdohr, 1981). Bar scale = 0.1 mm.

metamorphosed amphibolite–garnet, locally granulite grade rocks. Metamorphism has increased the difficulty in establishing whether the deposit is of hydrothermal or sedimentary origin (Guilbert and Park, 1986).

Hydrothermal Systems

Historically, hydrothermal activity has been studied as a process that forms ore deposits. This case is clearly one of economics driving scientific investigation. From a noneconomic point of view, there is far more hydrothermal activity as a rock-forming process than that strictly associated with deposition of ore minerals.

The pressure and temperature regimes in which hydrothermal activity occurs have proven to be quite useful for classification purposes (Lindgren, 1933). As with most rock-forming processes, there is now a trend to characterize hydrothermal activity in the context of more inclusive physical and chemical parameters, including fluid composition, fluid source, and the dynamics of fluid flow as contributors to **hydrothermal systems** (White and Hedenquist, 1990).

Lindgren classified hydrothermal ore deposits according to conditions of temperature and depth (pressure). His "epithermal," "mesothermal," and "hypothermal" designations have proven to be very useful in some respects but somewhat misleading in others. The chief problem is that thermal rise associated with emplacement of magma can occur at any depth, making strict correlation of the thermal environment with depth unrealistic.

The elegant presentation of Sillitoe (1973) (Figure 34.13) demonstrates the importance of the system approach by showing the relation of surface volcanism, subvolcanic magmatism, and plutonic magmatism to a single ore-generating hydrothermal system ranging from what Lindgren would have viewed as separate conditions of "hypothermal," "mesothermal," and "epithermal" hydrothermal activity.

For our purposes hydrothermal rock-forming processes are classified according to many of the traditional classification parameters, recast in the context of hydrothermal systems with emphasis on fluid and heat sources. Accordingly, seven "systems" are established (Figure 34.14), representing in a general way most environments in which hydrothermal activity is a rock-forming process: (1) plutonic intramagmatic, (2) volcanic intramagmatic, (3) skarn–pluton, (4) porphyry–pluton, (5) nonpluton vein–replacement, (6) epithermal, and (7) exhalative–geothermal. Some typical features of each type of system are sketched in Figure 34.15.

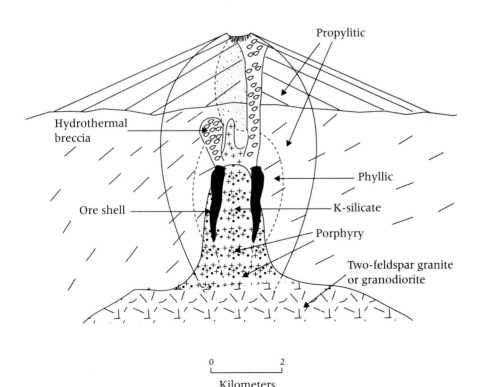

Figure 34.13
Schematic relation of surface, shallow, and deep zones in porphyry copper hydrothermal system. Modified from Sillitoe (1973).

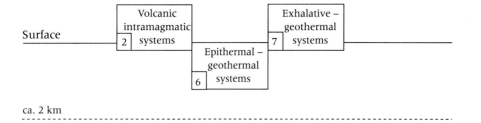

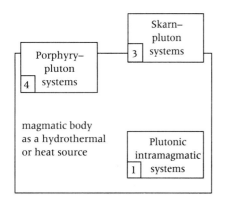

Figure 34.14
Schematic placement of seven generalized hydrothermal systems in which hydrothermal fluids are important as rock-forming processes. Systems shown in relation to depth, volcanic, and plutonic magmatism.

Plutonic Intramagmatic Systems

Water-bearing magmas must evolve a separate aqueous phase at the end stages of crystallization if there is more H_2O than can be incorporated into magmatic hydrous minerals.

Late-stage fracturing of granitic systems leads to migration of aqueous-rich fluids into veinlets and microveins where minerals such as quartz and K-feldspar are precipitated (Hibbard, 1980). Granitic pegmatite represents more extensive localization of a final H_2O-rich fluid phase capable of precipitating quartz and feldspars (Jahns and Burnham, 1969). Both the veins and pegmatite in the form of pods and dikes represent a transition between magmatic crystallization and precipitation in hydrothermal solutions.

A final aqueous phase separated from silicate melt can leave cavities in pegmatites and magmatic rocks (Chapter 33). There is typically a **pocket zone** in the inner quartz zone of granitic pegmatites that is lined with a variety of minerals that precipitate from a hydrothermal fluid. **Miarolitic cavities** occur in granites (Figure 17.8c), albite porphyries (McMillan, 1986), and other types of magmatic rocks that evolve a final hydrothermal fluid. If this fluid is shunted into fractures, veins are formed instead of the cavities.

Residual H_2O-rich fluids that remain in plutons after crystallization of quartz and feldspars are hydrothermal solutions capable of reacting with the magmatic minerals. This intrapluton reaction is **deuteric alteration**, typically converting plagioclase to sericite (Figure 34.16a; p. 413), hornblende to secondary biotite (Figure 34.16b), and primary magmatic biotite to chlorite (Plate 6b).

Volcanic Intramagmatic Systems

Vesicles are vapor gas bubbles that form in magma and are preserved in the magmatic rock. Vesicles are commonly well-developed spheres or oblate spheroids in basalts (Figures 17.2c, 17.3b) and in **pumice** (Figure 17.2a). Silicic lava flows and near-surface intrusions typically have **vapor-phase cavities**, representing concentration of H_2O-rich fluids. The cavities commonly cluster in lenses or layers, that are lined with crystals (Figures 17.12d, 17.13a). Concentric shells of crystalline material (including devitrified glass) may loosely fill the cavity, becoming **lithophysae** (Figure 17.12e). Lenticular and lithophysal cavities also occur in ash flow tuffs (Chapter 31). All of these features are open or potential spaces and therefore sites of **vapor-phase crystallization**.

continued on p. 413

Figure 34.15
Pictoral representation of typical geologic relationships occurring in the generalized hydrothermal systems shown in Figure 34.14.

(1) Intraplutonic Hydrothermal Systems (magmatic H$_2$O)

Quartz/feldspar veins

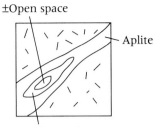
Pegmatite

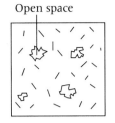
Miarolitic cavities

Deuteric reaction

(2) Intravolcanic Hydrothermal Systems (magmatic ± meteoric H$_2$O)

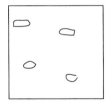
Vesicles in lava

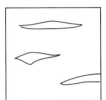
Lensoid cavities in lava

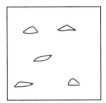
Cavities in tuffs

(3) Skarn–Pluton Systems (magmatic ± meteoric H$_2$O)

(4) Porphyry–Pluton Systems (magmatic + meteoric H$_2$O)

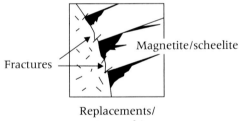
Replacements/ metamorphism

Stockworks/ breccia pipes

Veinlets/ disseminations

(5) Non-Magmatic Vein/Replacement Systems (meteoric/metamorphic/magmatic H$_2$O possible)

Bed replacement/ remobilization

Fissure veins

continued

412 Part 9 Rocks Formed by Precipatation and Reactive Crystallization

(6) Epithermal-Geothermal Systems (meteroic/ocean/connate ± magmatic H$_2$O)

(7) Exhalative-Geothermal Systems (meteoric/ocean ± magmatic H$_2$O)

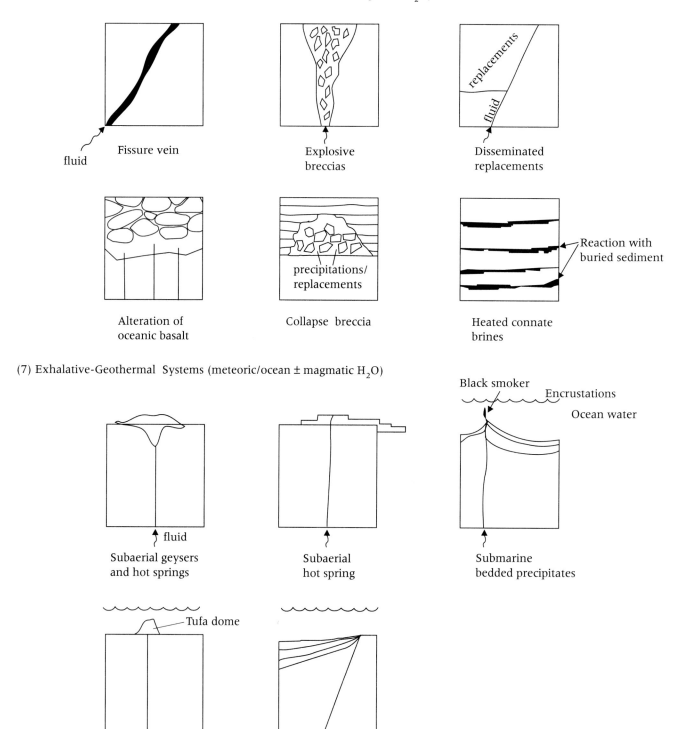

Figure 34.15 continued

34 Hydrothermal Rocks 413

Figure 34.16, text p. 410
Deuteric alteration.

(a) Plagioclase with coarse sericitic alteration (center) and biotite with chloritic (white) alteration (lower left). Springer stock, Imlay, Nevada. X-polars. Bar scale = 0.5 mm. (b) Biotite replacement of hornblende crystal in two-feldspar granite. Heusser Mountain, Nevada. X-polars. Bar scale = 0.12 mm.

Vesicles containing precipitates are **amygdules** (Figure 34.17). Amygdules may form immediately after magma consolidation if they can be accessed by the intramagmatic (endogenic) hydrothermal fluids. Amygdaloidal precipitates can also form from meteoric waters heated by cooling lava, permeating through the vesicular rock. Amygdaloidal precipitates of native copper along with chlorite, epidote, quartz, prehnite, pumpellyite, and laumontite occur in the Keweenawan volcanics of northern Michigan. The copper is derived from hydrothermal reaction with basalt at depth, probably moving up along fractures to the highly permeable vesicular tops of buried basalt and in interbedded conglomerates (Stoiber and Davidson, 1959).

Precipitation of hydrothermal phases may occur in hydrothermal breccias formed by explosive release of water vapor (Chapter 32). Such breccias are common in volcanic rocks and can be the site of subsequent precipitations of opal (Figure 32.1a,b).

Skarn–Pluton Systems

Faults and fractures occurring during the latest stages of crystallization of magmatic bodies are sites of dilatant pumping of indigenous hydrothermal fluids. At Iron Springs, Utah (Mackin, 1968) residual aqueous phase, through a process of deuteric solution, dissolved iron from magmatic biotite and hornblende as it passed along crystal interfaces and through

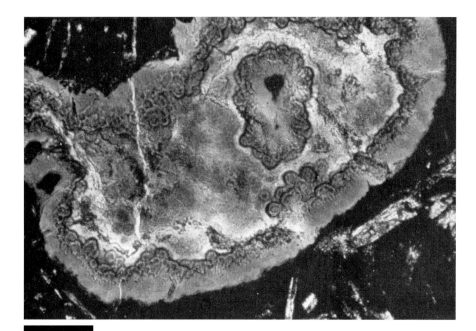

Figure 34.17
Amygdule containing smectite, heulandite, and thompsonite in glassy basalt. X-polars. Bar scale = 0.12 mm.

microcracks on its way to the fracture plumbing system. Much of the iron relocated as magnetite replacements of limestone wallrock forming one type of **skarn** (Figure 34.15). Magnetite on joint surfaces bordered by low-iron "selvages" in host quartz monzonite porphyry (Figure 34.18) suggests a dominantly magmatic source for the H_2O.

Mineralization in carbonate rocks adjacent to crystallizing magmatic bodies may also consist of scheelite, powellite, base metal sulfides, gold, as well as jasperoid and secondary dolomite (Einaudi, 1982). Heating of impure carbonate rock results in metamorphic recrystallization characterized by calc–silicate mineral assemblages, variously intermixed with hydrothermal minerals. As a result, these deposits are also known as "igneous-metamorphic." Temperatures range from about 300 to 800°C for these skarn–pluton systems, reflecting possible meteoric input on the low-temperature end to high temperatures in the magmatic range. Pressures range from moderate to high, depending on the depth of pluton emplacement.

The relationships in Figure 34.19 are particularly illustrative of the timing of thermal contact metamorphism and hydrothermal activity. The carbonate wallrock in this case consists of lensoid carbonate in argillaceous matrix, inherited from sedimentation. These impure carbonates are particularly susceptible to development of mineral assemblages indicating grade of metamorphism. Grossularite–idocrase–diopside and cordierite–biotite hornfels characterize the region close to the granite contact, followed by a tremolite–clinozoisite and biotite hornfels zone, and finally by very weakly metamorphosed limestone–argillite tens of meters from the contact (Figure 34.19). The grossularite occurs as "strata-bound" crystals a centimeter or less in thickness in the stratigraphic section (Plate 7c). With the exception of the grossularite, all mineral assemblages, both calcareous and pelitic, have been replaced by andraditic garnet (Figure 34.20a) in the zone of skarnification (Figure

Figure 34.18
Magnetite (black arrow, left) encrusted on joint surface with adjacent iron-depleted "selvage" zone extending to unaltered quartz monzonite porphyry (contact marked by white arrows). Features result from deuteric release of iron-bearing ore fluid (Mackin, 1968). Iron Springs, Utah. Slab surface. Bar scale = 1.2 cm.

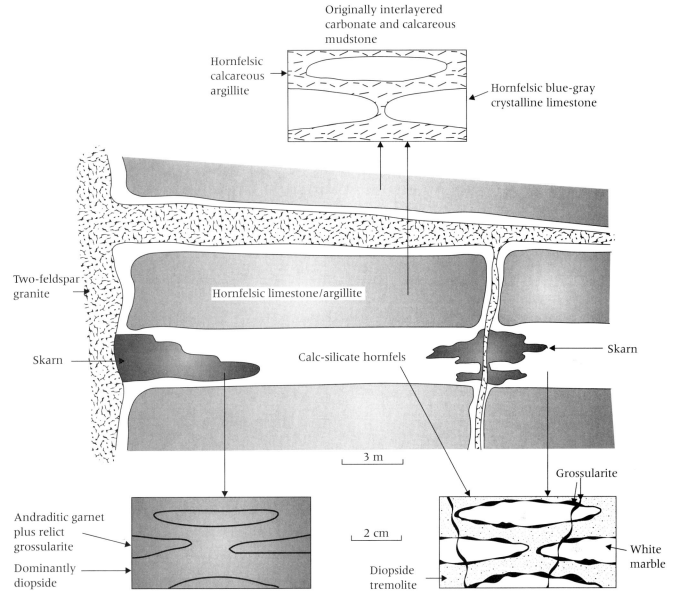

Figure 34.19
Localized skarn in contact zone between Notch Peak two-feldspar granite and interbedded calcareous–argillaceous. Marjum Formation, House Range, Utah.

34.20b). This particular skarn contains some scheelite and pyrrhotite (Gehman, 1958), suggesting that the hydrothermal fluid responsible for generating andraditic garnet was derived at least partly from the adjacent magmatic system.

Porphyry–Pluton Systems

Another close genetic tie between hydrothermal activity and magmatism is represented by **porphyry hydrothermal systems** (Titley and Hicks, 1966; Titley, 1982). In contrast to the skarn systems, concentration of hydrothermal fluids and associated mineral precipitates and alteration assemblages is centered within the magmatic system, although typically there is hydrothermal mineralization and alteration in wallrock as well (Lowell and Guilbert, 1970; Guilbert and Park, 1985) (Figure 34.15). The reference to "porphyry" in these hydrothermal systems derives from the close spacial and genetic relation to high-level porphyritic magmatic plutons. The ore fluids have both magmatic and meteoric components (White, 1974). Vein mineral

 (a)
 (b)

Figure 34.20
Granitiferous skarn shown in Figure 34.19.

(a) "Stratabound" grossularite garnets (gro) of thermal metamorphic stage enveloped in andraditic garnet (adr) of skarn stage. View on relict bedding surface. PPL. Bar scale =1 mm. (b) Skarn zone (dark) in thermal metamorphic contact rock (lighter). Adit (far left) is on pluton–wallrock contact. U.S. dime scale (arrow).

---

assemblages in porphyry systems also occur as **disseminations**, consisting of individual grains and microveins in hydrothermally altered rock adjacent to veins and veins in **stockworks**.

Mineralization in porphyry systems is dominantly either copper, molybdenum, or tin. At Climax, Colorado, veins containing quartz and molybdenite (Figure 34.21a) are considered to be more or less contemporaneous with aplite porphyry and rhyolite porphyry magmatism (Wallace et al., 1968), and at Henderson, another porphyry **molybdenum deposit**, similar relationships apply (Carten et al., 1988).

Veins containing quartz, orthoclase, pyrite, and chalcopyrite characterized the primary zone in porphyry copper deposits (Lowell and Guilbert, 1970).

Hydrothermal ("secondary") orthoclase associated with veins is shown blending with the host quartz monzonite porphyry in Figure 34.21b. Such an occurrence of hydrothermal K-feldspar typically involves partial replacement of plagioclase (Figure 34.21c). Complete replacement of original magmatic mineral assemblages generates irregular mosaics of interlocking K-feldspar crystals (Figure 34.21d) quite unlike the well-formed morphologies of alkali feldspar phenocrysts crystallized during the magmatic stage.

The morphological characteristics of K-feldspar in the **K-silicate** or **potassic alteration** zones of porphyry copper systems are more like those occurring in granitic pegmatites. In fact, gradation to pegmatites in a porphyry copper deposit was described by Gilluly

Figure 34.21
Veins and replacements in porphyry systems.

(a) Molybdenite-rich vein (lower left to upper right) cut by molybdenite–quartz vein (vertical, right) in quartz porphyry. Note gradational silicification into host along vertical vein (white arrows), and quartz phenocrysts (black arrows). Slab surface. Climax, Colorado. Bar scale = 2 cm. (Courtesy AMEX.) (b) Replacement veins of quartz (white arrow, left), and chalcopyrite (black arrow) plus quartz (right) with "secondary" K-feldspar, pervasive into adjacent quartz monzonite porphyry host. Slab surface. Bar scale = 1 cm. From primary ore shell, Ajo, Arizona. (c) Hydrothermal replacement ("potassic" alteration) of plagioclase (white) by K-feldspar (dark gray) in mineralized two-feldspar granite. X-polars. Bar scale = 0.12 mm. Hub Mine, Snake Range, Nevada. (d) Secondary K-feldspar alteration in primary zone of mineralized Snoqualamie granodiorite. Note interlocking interfaces of the K-feldspar (orthoclase) crystals, typical of hydrothermal, not magmatic, crystallization. X-polars. Bar scale = 0.5 mm. (e) Quartz pyrite vein in host porphyry at Sierrita-Esperanza, Arizona. X-polars. Bar scale = 1 mm.

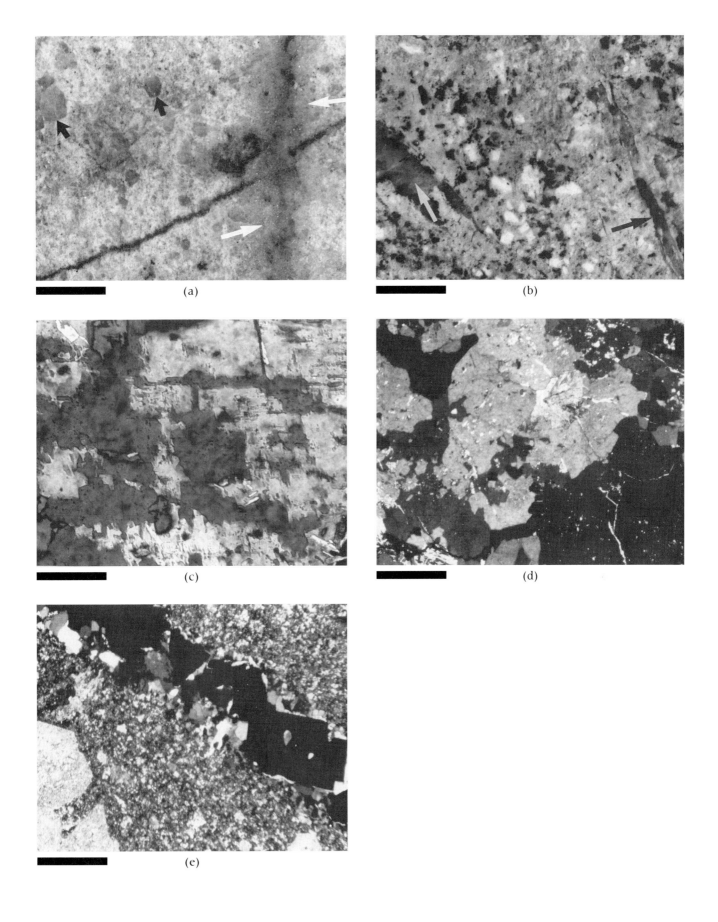

(1946). A vein containing quartz and sulfides without K-feldspar in a porphyry copper deposit is shown in Figure 34.21e.

Since there is participation of meteoric water in porphyry systems, and since there is direct association with magmatism, temperatures of formation are 200–800°C. Since porphyry systems are hosted by plutonic magmatic systems, pressures are moderate to high. In terms of Lindgren's classification these are "mesothermal" hydrothermal deposits.

Non-Pluton Vein–Replacement Systems

Fault breccias, fissures, and replacements containing hydrothermal precipitates and reaction products may occur without, or without noticeable, relation to igneous activity (Figure 34.15). In some cases, suggestion of relation of hydrothermal activity to magmatism is inferred based on oxygen isotope ratios (Beaty et al., 1989). Hydrothermal deposits in this generally nonmagmatic category include **vein deposits**, such as the "Cordilleran veins" of Guilbert and Park (1985), and **lateral–vertical secretion deposits** thought to be due to "remobilization" of metals in sedimentary and metamorphic rocks.

Vein mineralization without direct genetic tie to magmatic rocks is very common. Veins may occur in nonmagmatic rock or in magmatic rock with which there is no genetic tie. An example of the latter are the quartz veins and bordering alteration in plutonic granodiorite (Figure 34.22). A close spatial relation does not necessarily mean that the host pluton is a supplier of the hydrothermal fluid nor even of heat that drove the system (Lockwood and Moore, 1979).

The Coeur d'Alene silver and base metal veins of Idaho are examples of hydrothermal activity that is not obviously related to magmatic activity. Most of the mineralized veins occur in brecciated quartzites and argillites.

Veins occurring in metamorphic rock far from plutonic rocks may contain mineral assemblages reflecting the chemistry of the host metamorphic rock, suggesting an origin by lateral–vertical secretion. This may apply to the Yellowknife and Porcupine-Timmins gold districts in Canada and the Mother Lode in the Sierra Nevada (Guilbert and Park, 1985).

Epithermal–Geothermal Systems

Hydrothermal rock-forming processes at shallow depths (<ca. 2 km) are characterized by low pressure and low to moderate temperatures (ca. 50–350°C). These are the general constraints of Lindgren's "epi-

Figure 34.22
Quartz veins with lateral sericitic alteration of granodiorite localized by parallel fracture set. Pyramid Peak Range, Sierra Nevada, California. (Courtesy C. Sabine.)

thermal" hydrothermal vein systems, except for a somewhat wider range in temperature, and the term **epithermal** has been retained (Guilbert and Park, 1985).

Precious metal epithermal systems have been studied in great detail, but there are many other regions in which hydrothermal activity takes place within the pressure and temperature constraints of the epithermal that are less spectacular and in most cases less thoroughly studied (Figure 34.15).

In the comprehensive view of epithermal systems there are **vein** and **replacement** deposits. Stratabound replacement deposits may involve deeply buried connate water that is heated because of the burial or because of nearby magmatic activity. Water is typically meteoric (or seawater) in epithermal systems, but mixtures with magmatic water components are common if there is proximal volcanism. Many of the epithermal vein complexes are more or less contemporaneous with volcanic and supraplutonic magmatic systems on the basis of radiometric age dates of unaltered magmatic rock and the vein minerals. Even a difference in age between magmatism and hydrothermal activity as much as 1.5 m.y. is ascribed to repetitive hydrothermal events (Silberman, 1983). Thus, the epithermal gold–argentite–quartz bonanza veins of the Comstock Lode, Nevada, are indirectly tied to shallow igneous activity both on the basis of isotope studies and spatial association with andesitic volcanics and a high-level granodioritic pluton (Thomp-

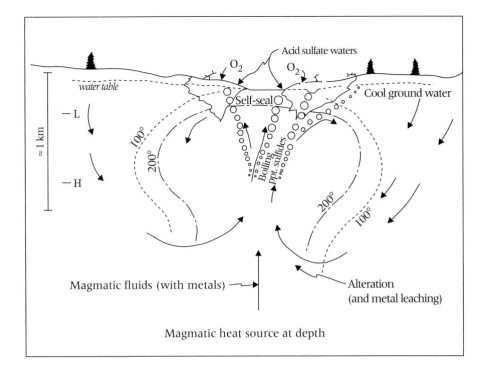

Figure 34.23
Schematic diagram of boiling hydrothermal system. Ascending hot waters begin to boil at depth in accord with pressure reduction. Mixing of meteoric water with possible magmatic fluids indicated. Magmatic heat source can be independent of amount of magmatic fluid input. Modified from Reed and Spycher (1985).

son, 1956). On the other hand, epithermal ceragyrite–carbonate veins at Treasure Hill, Nevada, are hosted by Paleozoic limestones with no nearby indications of magmatic activity (Humphrey, 1960; Figure 34.3).

The prime control in precipitation of ore minerals in precious metal epithermal veins is boiling induced by pressure drop occasioned by ascension of fluids along the vein structures (Figure 34.23). Low-sulfur systems (Figure 34.24) are characterized by jasperoidal silicification and quartz–adularia veins (adularia–sericite type) (Figure 34.25; Plate 8a), whereas high-sulfur systems (Figure 34.26) are characterized by argillic alteration and the presence of primary (hypogene) alunite (acid–sulfate type) (Hayba et al., 1985). Low-sulfur systems are presumed to involve hydrothermal fluids that are neutral or weakly alkaline, with low salinities in most deposits. High-sulfur systems are characterized by high acidity and dominance of the sulfate ion over the chloride ion, accounting for the precipitation of primary alunite. In both cases, hydrothermal activity extends to the surface hot spring environment, and vein mineralization may be accompanied by mineralization in stockworks and breccias (Silberman and Berfer, 1985; Henley, 1985; Bonham, 1988).

Perhaps the most economically important replacement-type epithermal hydrothermal deposit occurs as "easily overlooked" gold disseminations in carbonaceous siltstones. The "Carlin-type" ore deposits (Radtke et al., 1980; Ilchik, 1990) and similar disseminated precious metal deposits (Lovering and Heyl, 1974) are characterized by micron-size gold (Arehart et al., 1993) in rocks that have few visible minerals and very little veining. Carbon-bearing sediment host rocks are typical, and jasperoid (Lovering, 1972), such as that shown in Figure 34.27, is common. Alunite-veined supergene oxide ore resulting from weathering (Chapter 40) is shown in Figure 40.24a.

Epithermal hydrothermal mineralization in collapse breccias in carbonates may or may not have magmatic affiliations (McCormick et al., 1971). Epithermal replacement deposits in bedded sediments containing base metals may result from circulating heated connate brines. Such **lateral secretion** or **sedimentogenic** processes form the **strata-bound** "Mississippi Valley type" lead–zinc ore deposits. They are epigenetic deposits in the sense that the metals were not deposited at their present site during sedimentation. The Mississippi Valley–type deposits are a fine example of hydrothermal activity with no known link to magmatism. On the other hand, selective replacement in bedded carbonates produces fluorite "coon-tail" ore in the Illinois–Kentucky District (Grogan and Bradbury, 1968; Figure 34.28), but in this case magmatic association is inferred because of the presence of lamprophyry dikes and kimberlite pipes.

The epithermal region includes the alteration of pillow basalts and the upper part of the underlying sheeted dike zone along oceanic ridges. Circulation of ocean water through these basalts is driven by heat contained in recent eruptions and magma temporarily stored at depth.

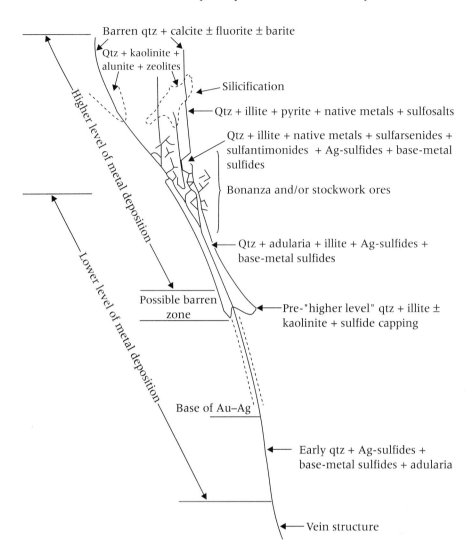

Figure 34.24
Schematic cross section showing fault-controlled vein structures in quartz-adularia, low-sulfur "epithermal" metal-bearing system. From Silberman and Berger (1985).

Exhalative–Geothermal Systems

Hot springs and geysers provide a direct look at hydrothermal activity. Exhalation of hot aqueous solutions and associated gases may be subaerial, sublacustrine, and submarine (Figure 34.15).

Subaerial hydrothermal precipitations are either siliceous or calcareous, depending on the rocks through which the fluids travel in reaching the surface. **Siliceous sinter** consists chiefly of porous opaline rock that has partly or completely inverted to cristobalite and tridymite (opal-CT) and chalcedony (Fournier, 1985). A variety of sinter occurring at Steamboat Hot Springs in Nevada (Figure 34.29a) is shown in Figure 34.29b, and the temporal relation of sinter to cinnabar and a final coating of opal facing open space is shown in Figure 34.29c. **Geyserite** is a typically botryoidal siliceous sinter that accumulates at geyser orifices. It is generally recognized that certain active Holocene siliceous hot springs precipitating siliceous sinter containing gold, cinnabar, and stibnite are equivalent to the surface expression of many

Figure 34.25
Adularia crystallized in open space on rock surface in epithermal vein system. Comstock Mining District, Virginia City, Nevada. Bar scale = 0.8 cm. (Courtesy H. Bonham, Nevada Bureau of Mines and Geology.)

34 Hydrothermal Rocks   421

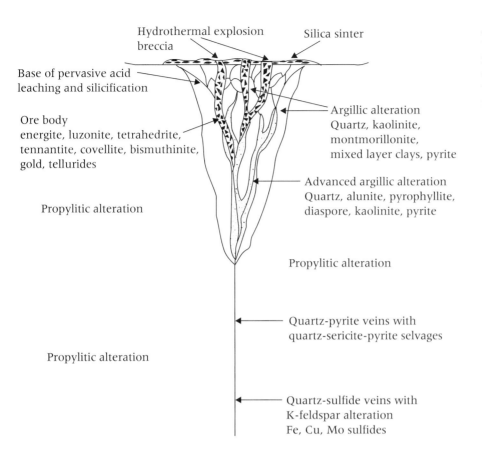

Figure 34.26
Schematic cross section showing fault–fracture control of quartz alunite, high-sulfur epithermal-bearing system. From Silberman and Berger (1985).

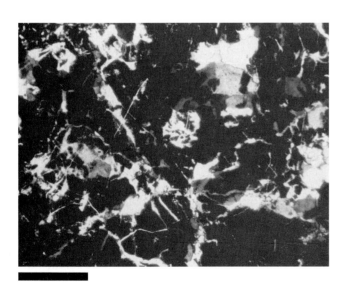

Figure 34.27
Jasperoid containing quartz (white) and hematite-bearing quartz (black). X-polars. Bar scale = 0.5 mm.

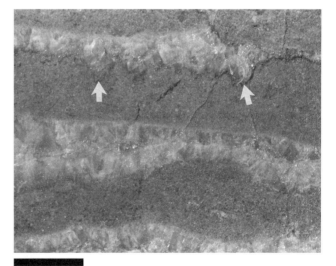

Figure 34.28
Coarse crystalline fluorite replacements in bedded carbonates forming "coon-tail ore." Dark layers contain rhombs of carbonate and fluorite. Note stylolite control of fluorite replacement (arrows). Illinois–Kentucky Mining District. Slab surface. Bar scale = 2 cm.

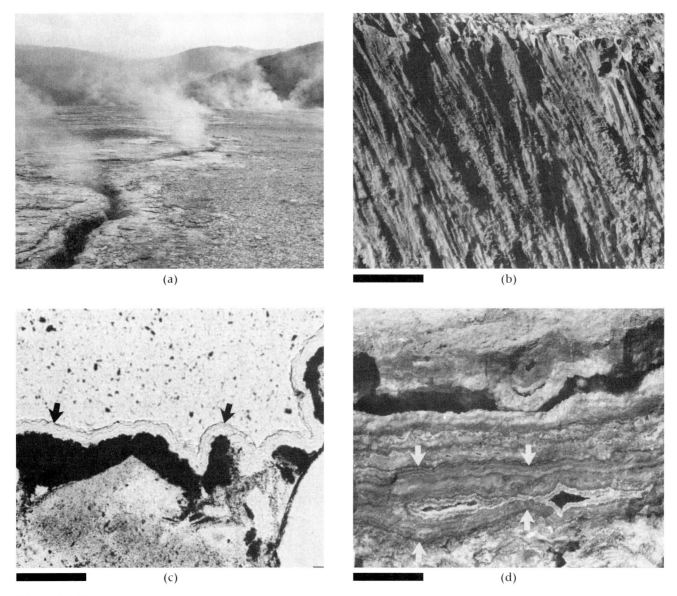

Figure 34.29
Siliceous precipitates in surface or very near surface hydrothermal environment.

(a) Steamboat Springs, Reno, Nevada. (b) Variety of siliceous sinter. Broken surface. Steamboat Springs, Reno, Nevada. Bar scale = 2 cm. (c) Silicified rhyolite (lower left, and upper right) with relict quartz phenocryst (lower right), coated toward open space (top) with cinnabar (black) followed by opal (black arrows). PPL. Bar scale = 0.12 mm. Steamboat Springs, Reno, Nevada. (d) Chalcedonic hot spring sinter with multiple zones of symmetrical layering, each typically with open spaces in medial zone. Cinnabar-bearing chalcedony (arrows). Slab surface. Bar scale = 0.1 cm. McLaughline gold mine, California. Lehrman (1986). (Courtesy J. Tingley, Nevada Bureau of Mines and Geology.)

epithermal precious metal deposits, such as the one at the McLaughlin Mine in California (Figure 34.29d).

Calcareous hot spring precipitates are **travertine**, forming the spectacularly terraced and cascaded deposits at Mammoth Hot Springs, Yellowstone National Park (Figure 34.30). Pools on terraces are confined by a rim of calcium carbonate precipitate around the outer terrace margin. The rim results from a somewhat greater rate of CO_2 loss at that exposed location and consequent higher rate of precipitation.

Sublacustrine hydrothermal precipitations necessarily involve mixing of hydrothermal solutions with

Figure 34.30
Terraced and cascading travertine. Mammoth Hot Springs, Yellowstone National Park.

lake water (Chapter 39). Precipitation of aragonite at sublacustrine hot spring sites forms **tufa mounds** (Figure 39.4a,b). Some of the precipitates are strictly chemical, whereas others have biotic associations. Hydrothermal fluids, issuing up along a fault zone, may react with surface and near-surface sediments, resulting in the generation of strata-bound mineral deposits (Robinson and Smoot, 1989).

Submarine hydrothermal precipitations result in the formation of economically important sulfide deposits as well. Activity of this type occurs at midocean ridges and accretionary plate boundaries, varying from **black smokers**, to back-arc basin **metalliferous brines** emanating along faults (Figure 34.31), to alteration of seafloor pillow basalt (Rona et al., 1983)

and palagonization of basalt tephra (Hay and Iijima, 1968; Jakobsson and Moore, 1986).

Finely bedded and laminated base metal sulfide and iron oxide rocks characterize chemical precipitations in this marine environment, such as those presently being deposited on the floor of the Red Sea and the Salton Sea. This type of submarine hydrothermal activity is **penecontemporaneous** with sedimentation.

"Banded" siliceous iron formations (Figure 36.8a) may represent accumulation of muds whose metal content may be supplied by exhalative-type discharges or by scavenging saline solutions discharging at the surface (Gross, 1980). The role of biotic activity in the precipitation of bedded iron oxides is probably significant in some cases.

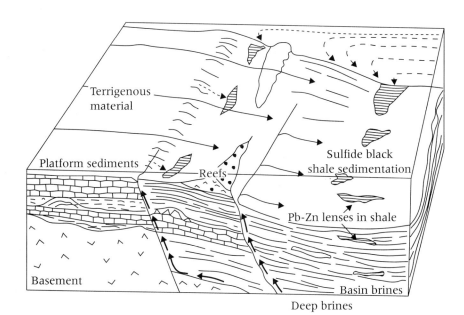

Figure 34.31
Schematic representation of base metal deposition in black shales along continental side of back-arc basin. Hydrothermal brines rise along fault structures. From Leblanc and Billard (1978).

Fumarolites

Fumarole exhalations, Santiaguito dome, Guatemala. (Courtesy Richard Stoiber, Dartmouth College.)

Gases and droplets of condensed gases (vapor) associated with volcanism are exhaled from fissures and other forms of vent structures known as **fumaroles** (Figure 35.1). The gases and vapors typically travel along fractures through tephra accumulations, ash flows, and debris avalanches. They vent into the atmosphere, into groundwater-bearing rock, or into standing bodies of water. Subaerial incrustations occur in the throat of fumaroles and on rock surfaces at orifices. Incrustation in nonwelded volcanic ash may result in **fumarole mounds** if the surrounding loose ash is removed by erosion (Sheridan, 1970). "The Pinnacles" in Crater Lake National Park, Oregon result from the same process.

Association of hot magmatic gases with hydrothermal liquids is common since both fluids evolve from consolidating magmatic systems. There is a clear connection between fumaroles and acid sulfate–type epithermal precious metal deposits. Near the surface, at reduced pressure, gases may coexist with liquids and erupt with them. Where gases and vapors dominate, **sublimation** can occur, precipitating amorphous substances and minerals on rock surfaces near vent orifices (Zies, 1929; Stoiber and Rose, 1974; Naughton et al., 1976; Keith et al., 1981; Symonds et al., 1987; Kodosky and Keskinen, 1990). If gas exhalations are directed into water, direct deposition of solids from the gaseous state does not occur and there are no sublimates. Rocks containing sublimates and related acid–rock reaction minerals are **fumarolites**.

Fumarole Exhalations

Source of Gases and Vapors

Heated fluids may come directly from magma, forming fumaroles, or they may be derived locally within pyroclastic deposits forming **rootless fumaroles**. Exhalation of gases and vapors through rock material containing pore water results in compositional shifts and cooling. Rootless fumaroles may contain heated meteoric water vapor component as well as gases trapped in the pyroclastic deposits. Devitrification of water-bearing glass can also contribute hydrothermal vapor to the fumarolic gas assemblage, especially in ash flow tuffs.

Composition and Temperature of Gases and Vapors

Gaseous exhalites range from a few hundred degrees to as much as 900°C. Since heat loss to the atmosphere is immediate, gases and vapors cool as they move away from the vent (Figure 35.2). The temperature of

Figure 35.1
Fumarolic vent with high values of Hg and Au. Vent is lined with quartz crystals, host limestone is silicified. Fern Mine, Yellow Pine district, Idaho (Cookro et al., 1988). (Courtesy M.L. Silberman, U.S. Geological Survey.)

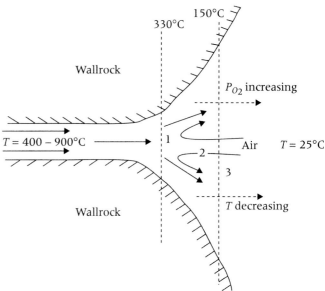

Figure 35.2
Schematic cross section of a fumarolic opening. As the gas leaves the vent, at position (1), the sulfur gas begins to quench. At (2), liquid H_2SO_4 aerosols form in the stream, and at (3) hydrated compounds become stable. Essentially from Stoiber and Rose (1974).

gases and vapors also change with time as volcanic systems cool.

Low-temperature (<150°C) **solfataras** are dominated by meteoric water and may occur far from volcanic centers. Fumarolic incrustation at temperatures less than 250°C were studied at Mount St. Helens by Keith et al. (1981) a few months after the May 18 eruption. Fumarole temperatures measured in June were from 75°C to 394°C soon after the March 1986 eruptive cycle of Mount St. Augustine, Alaska (Kodosky and Keskinen, 1990). Sublimates and other incrustation were sampled from 500 to 800°C fumaroles at Merapi Volcano, Indonesia (Symonds et al., 1987; Symonds, 1993).

The composition of fumarolic exhalations is governed by (1) the source magma composition, (2) mixing with hydrothermal vapors derived from meteoric water, (3) the stage of magma degassing as a temperature and pressure dependent factor, (4) the amount of degassing as a time-dependent factor, and (5) the degree of oxidation as determined by gas–atmospheric reactions.

Volcanic gases consist mostly of H_2O and CO_2 but include SO_2, SO_3, Cl_2, HCl, H_2, CO, HF, and a variety of very fine particulate substances (aerosols) such as sulfur and metal- and semimetal-bearing compounds. Rock-forming elements can be added to gas as it ascends through fractures. Prolonged degassing leads to loss of CO_2 and accumulation of more "soluble" components such as S and Cl in the magma.

Fumarole Incrustations

Sublimates and Reaction Precipitates

Fumarolites consist of incrustations resulting from sublimation and acid–rock reactions (Figure 35.3). Direct crystallization from the gas phase is **sublimation**. Sulfuric acid liquid aerosols commonly form in the gas stream landing on rock surfaces where they react with minerals and glass (Dethier et al., 1981).

Incrustations are not all crystalline. Amorphous colloform and botryoidal materials are common, some of which may partially crystallize in situ after initial deposition. Amorphous silica and fine crystalline quartz are characteristic of some fumaroles such as the one in Figure 35.1. Distinction between precipitates forming directly as sublimates and those resulting from reaction of acid hydrothermal liquids with rock

Figure 35.3
Fumarolites.

(a) Sublimates (yellows and white) in Crater Caliente, Santinguito dome, Guatemala. (Photo by W. Rose Jr., courtesy R. Stoiber.) (b) Salammoniac (NH_4Cl) and other sublimates on lava flow from Arenal Volcano, Costa Rica. (Photo by J. Carr, courtesy R. Stoiber.)

and previously deposited incrustations is not an easy task.

Yellow incrustations are not necessarily pure native sulfur. Mixtures of sulfur with other minerals can generate yellow-white or orange-yellow incrustation. Hydrated chlorides and sulfates of Fe and Al are yellow. Yellows are also characteristic of some amorphous materials containing Al, Cl, Fe, S, and H_2O (Keith et al., 1981). Reddish incrustations are typically tinted with iron oxides or hydroxides. Solfataric fumaroles yield sulfur coatings, but these are low-temperature fumaroles that release mostly air and steam.

Reactions of gases with each other can generate particulate solids in the gas stream and sublimates on rock surfaces. Sulfur may be generated by the following reactions:

$$SO_2(g) + 2H_2S(g) \rightarrow 3S(s) + 2H_2O(l)$$
$$H_2S(g) + 0.5O_2(g) \rightarrow S(s) + H_2O(l)$$

Thinardite may form by reaction of sodium chloride vapor with sulfur trioxide:

$$2NaCl(g) + SO_3(g) + H_2O(g) = Na_2SO_4(s) + 2HCl(g)$$

Na and K compounds are sufficiently volatile as halide or oxyhalite gas species to precipitate as sublimates. Ca and Mg compounds are not typically volatile enough to form sublimates, which is one reason for the general lack of carbonates in fumarolites. Another reason is the high solubility of carbonate in acid hydrothermal liquids that characterize the fumarolic environment.

Reaction incrustations involve hydrothermal processes. Reaction of primary minerals and glass in the presence of hydronium, hydroxyl, and other ions forms secondary phases. Reactions involving a liquid phase may be dominant in most fumaroles, contributing solid phases that dominate the fumarolites. Condensation of acid vapors on rock results in leaching of elements and precipitation of incrustations. For example, a volcanic rock containing magnetite may react with sulfuric acid to form hematite and sulfur:

$$\underset{\text{magnetite}}{6Fe_3O_4} + H_2SO_4(l) \rightarrow \underset{\text{hematite}}{9Fe_2O_3} + H_2O(l) + \underset{\text{sulfur}}{S(s)}$$

Plagioclase feldspars commonly yield gypsum and kaolinite as shown by the reaction of pure anorthite with sulfuric acid:

$$\underset{\text{anorthite}}{CaAl_2Si_2O_8} + H_2SO_4(l) + 3H_2O(l) \rightarrow \underset{\text{gypsum}}{CaSO_4 \cdot 2H_2O}$$
$$+ \underset{\text{kaolinite}}{Al_2Si_2O_5(OH)_4}$$

Plagioclase may also yield smectite or allophane. Reaction of hydrogen fluoride with the anorthite component of plagioclase in the presence of H_2O may generate fluorite and kaolinite:

$$2HF(g) + CAl_2Si_2O_8 + H_2O(g) = CaF_2(s) + Al_2Si_2O_5(OH)_4$$

Reaction of albite or the albite component of plagioclase or alkali feldspar with fluorite and hydrochloric acid can generate halite, anorthite, and quartz:

$$\underset{\text{fluorite}}{CaF_2} + \underset{\text{albite}}{2NaAlSi_3O_8} + 2HCl = \underset{\text{halite}}{2NaCl} + \underset{\text{anorthite}}{CaAlSi_2O_8} + \underset{\text{quartz}}{4SiO_2} + 2HF$$

Zoned Assemblages

As gases cool they become saturated with sublimate phases that fractionate from the gas in the order of their equilibrium saturation temperatures. A sublimate zoning is common. Here are some examples.

At Izalco volcano, El Salvador (Stoiber and Rose, 1974; Hughes and Stoiber, 1985) the highest-temperature incrustations (400–900°C) are white and consist mostly of the mineral thenardite (Na_2SO_4). At about 300°C golden brown incrustations consists of specular hematite (Fe_2O_3), tenorite (CuO), and scherbinite (V_2O_5). Another white zone occurs at 150–250°C, containing halite and fluorite, with minor sylvite and salammoniac (NH_4Cl). A dull red-pink zone at 100–200°C is dominated by anhydrite. At less than 100°C, yellow-orange incrustations consists of ralstonite (NaMgAl (F,OH) · H_2O), chloraluminate ($AlCl_3$ · $6H_2O$), earthy hematite, and cristobalite. Yellow-to-white zones at the lowest temperatures consist of sulfur, gypsum, and earthy hematite.

At Merapi Volcano, Indonesia, the following sequence of sublimates is described by Symonds et al. (1987). With decreasing temperature from 800 to 500°C a zone of cristobalite and magnetite is followed by a zone consisting of acmite, halite, sylvite, and pyrite. The next zone at lower temperature consists of aphthitalite (K–Na sulfate), sphalerite, galena, and a Cs–K sulfate, followed by Pb–K chloride plus Na–K–Fe sulfate, and finally Zn, Cu, and K–Pb sulfates.

The morphology of precipitates collected in silica tubes placed into high-temperature vents at Merapi can be studied in some detail (Symonds, 1993). Crystal habit varied with the amount of supersaturation, ranging from complete crystals at low undercoolings to skeletal and, finally, dendritic crystals at large undercoolings. Morphological anisotropy is indicated by elongated halite and magnetite crystals (Figure 35.4). This anisotropy is interpreted (Symonds, 1993) as resulting from dissimilar growth rates on different crystal faces at larger undercoolings than those favoring normal morphologies of these isometric crystals. There must be incomplete equilibrium between gases and the silica-tube sublimates because of rapid cooling, kinetically retarding sublimation reactions.

(a)

(b)

Figure 35.4
Sublimates collected in silica tubes placed into active fumarolite, Merapi Volcano, Indonesia (Symonds, 1993). (Courtesy R. Symonds, U.S. Geological Survey Cascades Volcano Observatory, Vancouver, Washington.)

(a) Elongated halite resulting from undercooled sublimation crystallization. SEM. Bar scale = 24 microns. (b) Normal halite morphology (cubes) with coatings of bladed K–Ca sulfate crystals. SEM. Bar scale = 120 microns.

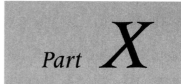

Rocks Formed by Mechanical and Chemical Processes Involving Low-Temperature Aqueous Fluids

Part X

The common denominator for rocks formed by mechanical and chemical processes involving low-temperature aqueous fluids is the participation of surface and near-surface water in the rock-forming processes. Groundwater and surface water are inherently low temperature, but there is transition to higher-temperature water where there is mixing with geothermal water. The rock-forming processes range from dominantly chemical to mechanical plus chemical.

Mechanical accumulation of clastic particles in the surface environment leads to burial and compaction accompanied by permeation of aqueous solution through connected pores. Direct precipitation of minerals and reaction with the clastic particles is at first a process of cementation, followed by continued or renewed chemical activity known as diagenesis. With deep burial and higher temperature, diagenesis grades to low-grade metamorphism.

The clasts of sedimentary rocks resulting from extensive sediment transportation are chiefly silicic because grains composed of chert and quartz are particularly stable, both mechanically and chemically in dynamic surface-water environments. Therefore, clastic sedimentary rocks are dominantly siliciclastic.

Bioclastic sediments originate in those water environments where organisms thrive. Diagenetic processes obscure the distinction between initial biotic precipitations and mechanically reworked biotic and diagenetic precipitates characteristic of bioclastic rocks. For purposes of discussion and topic organization, all rocks with any biotic associations, whether they be organic precipitates, inorganic precipitates, minerals generated by reactive replacement, or bioclastic reworkings, are considered biosedimentary rocks.

Evaporites and other rocks forming by inorganic precipitation represent the accumulation of crystals in water bodies (salinas and playas) much as crystals accumulated in magma bodies. Although the viscosities are very different, the precipitation of crystals and settling to the floor of evaporite basins is a process not unlike that of gravity settling in magmatic systems. Reworking of evaporites along marine and lacustrine shorelines represents a transition to clastic sedimentation on a very local scale. The minerals of evaporites are not likely to survive extensive transport and redeposition due to their solubility and mechanical weaknesses.

Chemical precipitation in water bodies, where there is little or no evaporation-driven supersaturation, is a rock-forming process generating low-temperature aqueous precipitative rocks. Recognition of evaporation as a rock-forming process and classification of rocks in evaporite systems is well established, whereas rocks formed as a result of nonevaporative chemical precipitation are poorly defined. Precipitations caused by mixing of aqueous solutions of different composition, by mixing of aqueous solutions of somewhat different temperature, or by gas loss related to agitation are sound chemical phenomena but not readily distinguished as distinct rock-forming processes in the geological record. The reasons for this include (1) the inevitable participation of organisms in systems involving carbonate that are more conveniently classified as biochemical sedimentary rocks, and (2) precipitation in pores of pre-existing sediments and rocks where the process is more conveniently considered diagenesis of clastic sedimentary rock, as a continuation of the processes forming the clastic sedimentary rocks rather than a new rock-forming process. Nevertheless, a few rocks can be defined as resulting from low-temperature precipitation that are neither evaporites nor associated with biotic processes.

Formation of weathered rocks is essentially the low-temperature analogue of hydrothermal alteration, although there are some mechanical processes, such as the shrinking and swelling of clays, that do not have an equivalent in the hydrothermal regime. In the broad sense, weathering includes all those processes occurring in the vadose (above the water table) and phreatic (water-saturated) zones on continents and in the phreatic zone beneath oceans and lakes.

36

Clastic Sedimentary Rocks

The new and the old. Cretaceous sandstones (dipping left) truncated by wave action. Subsequent emersion of wave-cut surface was followed by fluvial deposition of clastic sediments (including granite boulders, 0.5 m) derived from nearby Santa Lucia Range. Recent exhumation by storm waves has left boulders as lag on irregularly sculptured terrace surface. Garapada Beach, Big Sur, California.

Source, Transportation, and Accumulation of Sediments

Clastic sedimentary rocks are formed from accumulations of mineral and rock particles. Several dynamic processes are capable of removing and transporting clastic particles from sites of weathering to places of accumulation. These sedimentology processes are directly linked to the petrographic attributes of clastic sedimentary rocks.

Dynamic subaerial and shallow subaqueous environments include (1) gravity-driven downslope motion (mass wasting), (2) winds resulting from atmospheric thermal and pressure differentials (eolian), (3) gravity-driven running water (fluvial), and (4) tide and wind-driven wave action (nearshore). Subaqueous dynamic environments include (1) gravity-driven sedimentation of temporarily suspended sediment onto the floor of water bodies (pelagic sedimentation), (2) thermal-driven current flow, and (3) gravity-driven turbid flow of water-saturated sediments (turbidity flow).

Burial and lithification of sediment accumulations form rock that reflect (1) composition and (2) the mode of deposition. Composition is a reflection of **provenance**, the composition of the source rock terrain, and all the factors that affect physical and chemical development of mineral and nonmineral grains in this terrain. For example, a volcanic provenance (Anderson, 1971) is indicated by a lithic grain of basalt (Figure 36.1a), a granitic source is indicated by the grain of K-feldspar containing myrmekite (Figure 36.1b) and twinned microcline (Figure 36.1c), a metamorphic provenance is indicated by the lithic grains of quartzite (Figure 36.1d) and crenulated schist (Figure 36.1e), and a previously formed sedimentary rock is indicated by the recrystallized limestone fragment (Figure 36.1f).

Siliciclastic sediments are more common than carbonate fragments, and certainly more common than evaporite fragments (Hardie and Eugster, 1971; Weiler et al., 1974), because they are more stable in the surface environment. Calc–arenites containing carbonate ooids, pisoids, peliods, intraclasts, or fossil hash are discussed in Chapter 37 because of the close association with biogenic carbonate rocks.

Some siliciclastic sediments are more stable than others (Goldich, 1938), but the unstable ones are converted to other silicate minerals that are stable during initial weathering and during sedimentation. For example, chert (microcrystalline chalcedonic quartz), quartz, muscovite, iron oxyhydrates, chlorites, clay minerals, and even K-feldspar are common sediments

431

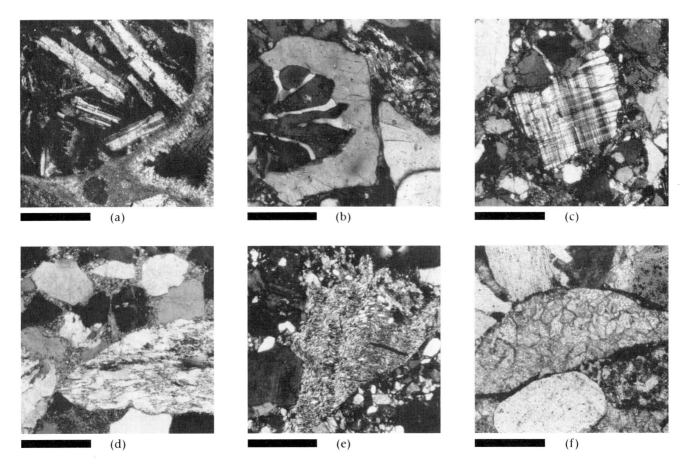

Figure 36.1
Grain composition and texture as indicators of provenance.

(a) Litharenite with porphyritic basalt lithic grain (upper left). Isopachous calcium carbonate cement does not completely fill pores. X-polars. Bar scale = 0.2 mm. Pleistocene sandstone, Lovelock, Nevada. (b) Arkose containing myrmekite-bearing K-feldspar grain derived from granitic rock. Muscovite, plagioclase, quartz, and hydrobiotite (?) also present. X-polars. Bar scale = 0.2 mm. (c) Twinned microcline in arkose indicating granitic source. Note angularity of microcline and quartz grains. Chlorite–calcite matrix. X-polars. Bar scale = 0.8 mm. Rico Formation, San Juan Mountains, Colorado. (d) Quartzite grain containing very fine-grained neoblasts indicates metamorphic source. Quartz wacke also contains muscovite grain (upper left) and monocrystalline quartz grains with sericitic mica matrix. X-polars. Bar scale = 0.8 mm. (e) Crenulated mica schist lithic grain in feldspathic wacke indicates metamorphic source. Also contains plagioclase, alkali feldspar, muscovite, and quartz set in brownish claylike matrix. Connecticut River Valley. X-polars. Bar scale = 0.8 mm. (f) Limestone clast containing sparry calcite indicates carbonate source. Litharenite also contains chert lithic grains and recycled quartz (bottom center) from eolian quartz arenite. Iron Springs Formation, Utah. PPL. Bar scale = 0.2 mm. (Courtesy P. Goldstrand.)

stable in surface environments. Plagioclase, amphibolespyroxenes, biotite, and sulfides are chemically unstable and readily convert to other minerals. The most abundant sediment is clay because most igneous and metamorphic rocks are dominated by feldspars, yielding clay minerals by reaction with aqueous solutions in the weathering environment.

The dynamics and duration of transport determine grain size, grain-size sorting, and grain morphology, as well as small-scale structures. Grain size is a good indicator of the energy of deposition. Clay-size particles reflect low-energy dynamics because otherwise the clays would not be deposited until such a low-energy environment was reached. Conglomerates indicate a high-energy environment because the finer particles have been removed from the gravel, such as in the surf zone of lakes and oceans and in channels of high-gradient streams. The importance of energy as it relates to grain size is underscored by the fact that grain size is a major parameter in the classification of clastic sedimentary rocks.

Textural maturity and **mineralogical maturity** are measures of how long an initial assortment of clastic grains has been in the sedimentation regime. The degree of sorting, rounding, and decomposition of chemically unstable clasts depends on environment and how long that environment has acted on the sediments. Weathering after deposition also contributes to mineralogical maturity. If grains are derived, at least in part, from a previously formed sedimentary rock, there is **grain recycling** (Goldstrand, 1991) and **acquired maturity** (Kidder and Swett, 1989). Abraded secondary overgrowths on spherical eolian quartz grains is an indicator of recycling. High content of resistor minerals, such as zircon, tourmaline, and rutile, is an indication of mineralogical maturity.

One objective in studying clastic sedimentary rocks is to generate a picture of sedimentation as it relates to both orogenic and nonorogenic regions. It is only natural that these rocks be considered as part of depositional environments because in that context a more comprehensive picture of sedimentation process is achievable. For example, in a fluvial system gravelly channels are bordered by sand bars, and both are adjacent to the silty and muddy flood plain. It makes more sense to study conglomerates, sandstones, and mudstones in the context of a fluvial system than it does to isolate the lithologies as self-contained subsystems. Each subsystem generates a rock type: conglomerate, sandstone, siltstones, and mudstones, but it is the **facies change** between these rock-forming materials that generates the complete picture of sedimentation and geological history (Reading, 1978; Blatt et al., 1980; Scholle and Spearing, 1982; Boggs, 1987; Greensmith, 1989; Einsele, 1992).

Compaction and Diagenesis

Accumulation of clastic sediments is the first stage in the evolution of clastic sedimentary rock. Lithification necessitates **compaction** of these sediments, expulsion of pore water (or air) (Figure 36.2), and **cementation** of the grains. **Diagenesis** converts stacks of sediments into rock. The initial composition of clastic grains may be modified by reactive replacement, and pore spaces are filled with precipitates. Absence of cement at grain contacts indicates cementation took place after sedimentation and compaction. Secondary pore spaces generated by postdiagenesis solution may not be lined with precipitates.

Early diagenesis begins at the time of sediment accumulation. Diagenesis is a weathering process as well, such as the cementation of grains with sparry calcite (Figure 36.1a) and the chemical reaction of seawater with biotite generating glauconite (Hughes and Whitehead, 1987) on the seafloor. Since early diagenesis affects grains as they are being accumulated, chemical processes are active as compaction and dewatering are in progress.

Late diagenesis is dominated by mineral reactions and solution reprecipitation after elimination of primary porosity. Reactive replacement of earlier cements (Figure 36.3a) and of grains (Figures 36.3b) occurs in response to infiltrating groundwater in both continental and submarine settings (Morad, 1988). Transition to hydrothermal alteration occurs if these waters are heated by nearby magmas or mix with magmatic waters.

Compaction alone can produce some lithification as a result of rigid grains such as quartz being imbedded into ductile particles such as clays, chlorites, and micas. There is more surface bonding with this grain interface geometry than the point contacts between spherical grains. For example, friable sandstones are characterized by lack of cementing material and lack of much ductile intergranular material.

Diagenetic cements include opal, chalcedony, quartz, hematite–limonite, and carbonate minerals.

Figure 36.2
Sedimentary structures in lacustrine delta sediments of Pleistocene Lake Lahontan. Ball-and-pillow structure associated with dewatering plume (center), overlain by later, truncating ripple-laminated silts. Scale = U.S. quarter (lower right). Clark Station, east of Reno, Nevada.

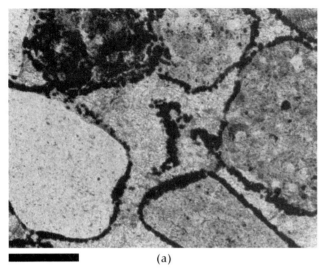 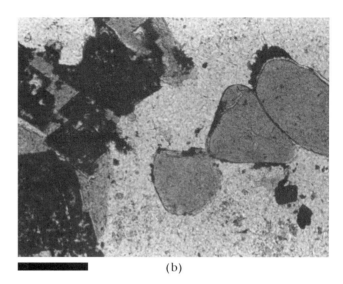

Figure 36.3
Replacement cements in clastic sediments.

(a) Rounded chert clasts (right) with recycled eolian quartz clast (left) in hematite-bearing matrix (black) largely replaced by diagenetic sparry calcite (note patch of matrix in calcite, center). Litharenite of fluvial origin. Canaan Peak Formation, Utah. PPL. Bar scale = 0.12 mm. (Courtesy P. Goldstrand.)
(b) Glauconite as replacement of pelloidal carbonate (?) in quartz arenite. Local diagenetic replacement of glauconite by siderite, which is now oxidized to hematite (left). PPL. Bar scale = 0.12 mm.

Cementing of clastic grains by chemical precipitation in situ (**authigenesis**) and by introduced mineral phases (**allogenesis**) is the normal course of lithification. Primary depositional (detrital) intergranular muds can recrystallize to form micaceous minerals (Figure 36.4). Introduction of carbonate into the pore spaces of sandstones may be without any sign of reaction with primary grains, the surfaces of quartz grains being clearly attributed to either spherical rounding in an eolian environment or to fracture (Figure 36.5a).

Siliceous cements are common in siliciclastic sediments. Although quartz is the most abundant siliceous cement, opal may precede the quartz and evolve to quartz during diagenesis. Nontaxic crystals of quartz on eolian-rounded quartz grains can serve as cement, as shown in Figure 36.5b. Quartz epitaxic on rounded quartz grains is defined by (1) a preserved surface containing minute particles of other materials (such as hematite) and (2) the faceted character of the overgrowth quartz (Figure 36.6). Crystal faces are incongruous with the abrading process characterizing clastic sedimentation. Quartz overgrowths do not luminesce when bombarded with electrons generated in a cathode tube. Activators such as rare earth ions and dislocations in original quartz grains generate luminescence.

The source of silica serving as overgrowths may not be obvious (Sanderson, 1984; Mazzullo and Magenheimer, 1986; Grutzeck, 1986). An original opal cement might be the source, but there may also be grain-scale in situ redistribution via pressure solution reprecipitation. Luminescence may not differentiate between introduced and in situ silica. Whatever the source of silica, the epitaxic quartz overgrowths may

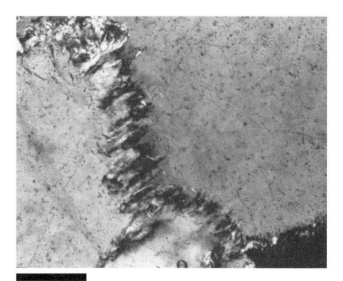

Figure 36.4
Sublitharenite with micaceous "cement" probably recrystallized from clay matrix during diagenesis. X-polars. Bar scale = 0.12 mm.

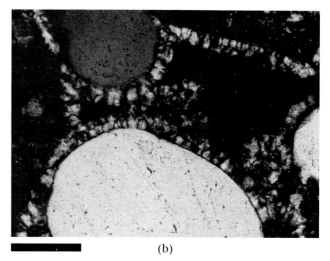

Figure 36.5
Recycled eolian quartz grains in sublitharenite of the Pine Hollow Formation, Utah. (Courtesy P. Goldstrand.)

(a) Fractured spherical quartz grain associated with subrounded chert grains (left) with diagenetic sparry calcite cement. X-polars. Bar scale = 0.12 mm.
(b) Spherical quartz with chert (upper right and left) grains and isopachous polycrystalline quartz cement. X-polars. Bar scale = 0.12 mm.

(1) completely eliminate porosity, (2) be followed by introduction of other pore-filling cements, or (3) be partly replaced by other cements. Quartz overgrowths followed by introduction of iron oxyhydrates into pore spaces is shown in Figure 36.7. Faceted quartz with adhered kaolinite crystals is shown in Figure 36.6b.

The relations in Figure 36.7 are more complex, pointing to the following genetic model. The sphericity of the original quartz grain indicates eolian derivation. The first cycle of diagenesis resulted in faceted overgrowths of new quartz. Subsequent release from a first-cycle quartz arenite results in another

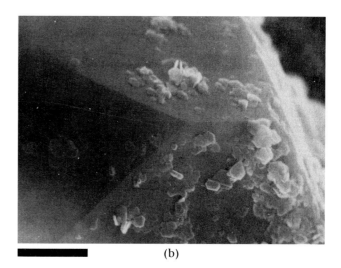

Figure 36.6
Epitaxic quartz overgrowths on quartz clasts.

(a) Rounded quartz grains defined by dark line of impurities with faceted quartz overgrowths of probable pressure solution redeposition origin. Clayey iron oxyhydrate "cement" is secondary and partly removed. PPL. Bar scale = 0.12 mm. Arkosic quartz arenite, Canyon de Chelly, Arizona. (b) SEM view of faceted quartz overgrowth on grain of St. Peter Sandstone. Kaolinite crystals on faceted quartz are thought to be postdiagenetic. Bar scale = 5 μm. (Courtesy M. Grutzeck.)

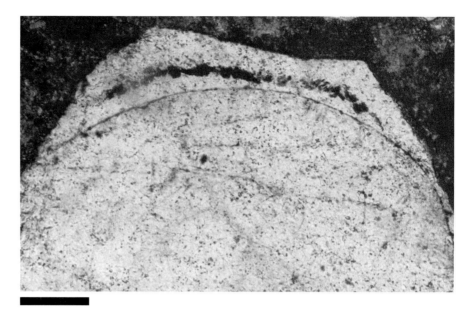

Figure 36.7
Probable doubly recycled quartz clast. Spherical wind-abraded clast, outlined by thin impurity zone, received epitaxic overgrowth of quartz during diagenesis of eolian sandstone (Entrada?), followed by formation of new grain that was rounded during second stage of transportation. Deposition in clayey iron oxyhydrate matrix (cf. Figure 36.3a) was followed by a second diagenetic epitaxic quartz overgrowth in sublitharenite. A third stage of diagenesis in represented by carbonate cement. Pine Hollow Formation, Utah. PPL. Bar scale = 0.12 mm. (Courtesy P. Goldstrand.)

rounding event that eliminates facets. Redeposition in a hematite–limonite bearing clay occurred during a second sedimentation cycle. A second cycle of diagenesis resulted in overgrowth of more quartz, epitaxic on first-cycle overgrowth quartz, but it included some of the iron oxide–bearing clay. Quartz growth beyond the limits of this limonitic clay resulted in the present facets. A third stage of diagenesis is represented by the carbonate cement, in this case resulting from infiltration of carbonate solutions deriving from an overlying marl formation.

Ferruginous cherty arenites are associate with **banded iron formations** (Figure 36.8d) (or **banded ironstone formations**; Young, 1989). These ironstone-formations are **granule ironstone formations** that probably represent mechanical reworking as well as diagenetic modification of the banded ironstone formations in which chert, hematite, and magnetite are the principal constituents (James and Trendall, 1982, Simonson, 1987). Hematite- or magnetite-bearing chert grains are typically ellipsoidal with a chert cement (Figure 36.8b), and there are typically many

Figure 36.8
Iron formations and ironstones.

(a) Granule containing cherty quartz and well-crystallized hematite. Matrix is cherty quartz. Taconite. Biwabek, Michigan. PPL. Bar scale = 0.12 mm. (b) Featureless greenalite granules (black) with quartz matrix. Note siderite rhomb in one granule (lower left). X-polars. Bar scale = 0.3 mm. Biwabik Formation. (Courtesy B. French.) (c) Greenalite (black) largely replaced by minnesotaite (gray) with quartz matrix (white). PPL. Bar scale = 1.5 mm. (Courtesy B. French.) (d) "Banded" taconitic (siliceous facies) iron. Light layers are red jasper (hematite + quartz); dark layers are rich in magnetite. Sedimentary origin indicated by smoothly laminated structure. Michigan Banded Iron Formation. Slab surface. Bar scale =1 cm. (e) Ferrugination of oolite. Quartz nuclei in calcareous ooids replaced by hematite. Matrix contains quartz (clear) and carbonate (mottled) grains. PPL. Bar scale = 0.5 mm. Clinton Formation (Silurian). New York State.

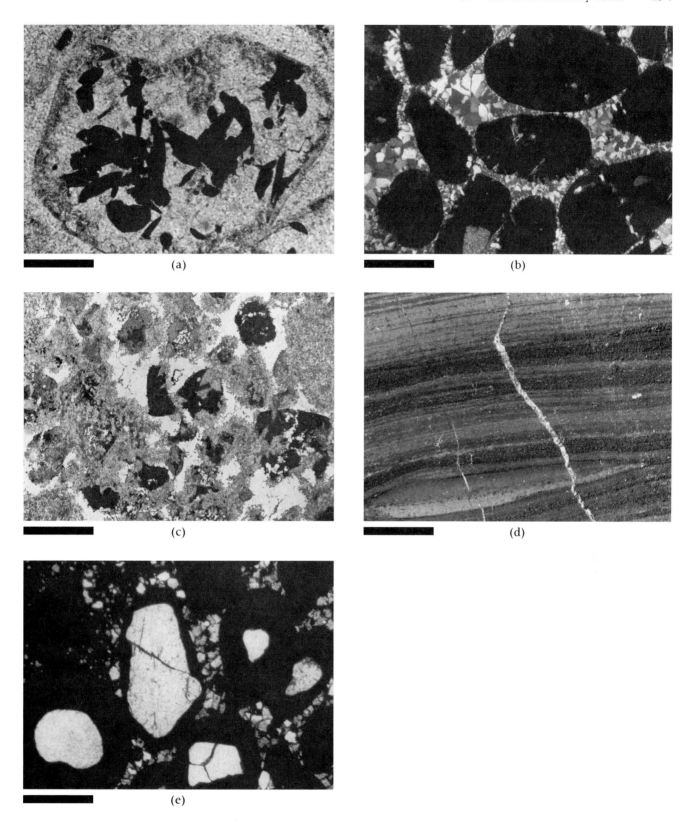

other textural and structural features characteristic of clastic sedimentary rocks. Very low-grade metamorphism of these granule formations generates greenalite [$Fe_9Fe_2Si_8O_{22}(OH)_{12} \cdot 2H_2O$](Figure 36.8b) and then the iron talc minnesotaite [$(Fe,Mg)_{11}(Si,Al,Fe)_{16}(OH)_{11}O_{37}$] (Figure 36.8c) and stilpnomelane (French, 1973). Still higher metamorphism converts this mineral assemblage to other iron-bearing silicates, including grunerite and clinopyroxene (Klein, 1973).

Ooidal ironstones are characterized by ferric oxide ooids (Figure 36.8e), of which the Clinton Formation (Silurian) of North America is a prime example. There are occurrences of ooids in the Clinton Formation that are clearly the result of ferruginization of calcareous oolites (Kimberley, 1979; Dimroth, 1976), but lack of relict calcareous ooids in the bulk of these ironstones indicates other processes are dominant. There are indications that reworking of lateric soils, which themselves may contain ferruginous ooids and pisoids, generates some of the ooidal ironstones (Madon, 1992), but such a genesis does not seem to be widespread. A model of intrasedimentary growth of berthierine microconcretions, followed by transformation to chamosite, in turn followed by diagenetic replacement of the chamosite by hematite and goethite, may have general applicability to ooidal ironstones (Young, 1989; Cotter, 1992). **Berthierine** is an iron-rich layered silicate of the serpentine group that may have precipitated directly from seawater or pore water under anoxic conditions. **Chamosite** is an iron-rich chlorite with an end-member composition ($Fe_5^{2+}Al$)(Si_3Al)$O_{10}(OH)_8$. It forms from berthierine during burial diagenesis. Partial replacement of ooidal chamosite by ferric oxides is shown in thin section by Cotter (1992).

Physical Characteristics of Siliciclastic Sedimentary Rocks

Physical characteristics of siliciclastic rocks at the mesoscopic and microscopic scale consist of (1) small-scale structures formed by soft sediment deformation, deposition, and erosion, and (2) textural attributes such as grain size, grain shape, grain location, and grain orientation (Collinson and Thompson, 1989).

Sedimentary Structures

Soft Sediment Deformation. Sediments saturated with water are prone to differential compaction as the sedimentary pile thickens. Sands deposited on muds are particularly unstable, and "balls" and "pillows" of sand form as local subsidence occurs. In extreme cases differential subsidence forming a **ball-and-pillow** structure leads to separation of the sediment masses from their original beds, forming **pseudonodules**. **Load casts** are the curved surface depressions in the underlying bed that may be the only preserved evidence of a ball-and-pillow structure. Upward flow of the water-saturated underlying sediments may form a **flame structure** typically in association with a ball-and-pillow structure (Figure 36.2). In extreme cases of instability, **sand volcanoes** form by extrusion of water-saturated sediment. Convoluted and chaotic structures may be due to liquefaction caused by the passage of a seismic wave.

Mud cracks (Figure 36.9) are a form of soft sediment deformation that occur as the result of dessication and shrinkage. Sediments of the following depositional cycle may fill these cracks (Plate 8a). **Syneresis cracks** are subaqueous shrinkage cracks formed as the result of water being expelled from gelatinous sediments. **Rainpits** may even be preserved on mud rock surfaces. Rainpits indicate that the sediment was extremely soft and saturated in water. Evidently, a drying stage after rainfall solidified the impressions that in some cases were preserved by burial beneath the next influx of sediment.

Deposition of sediments on a sloping surface, such as on continental shelves and on the margins of forearc and backarc basins, may lead to slumping. Downslope flow of sediments or partly consolidated sediment generates a variety of contorted fold structures.

Figure 36.9
Obstacle scour (lower right) in sand upstream (left) and lateral to cobble-sized rock in Holocene wash. Slack-water stage mud in adjacent pool has dissicated, forming mudcracks. Mullin Creek Wash, Pyramid Lake, Nevada. Scale = U.S. quarter (far left).

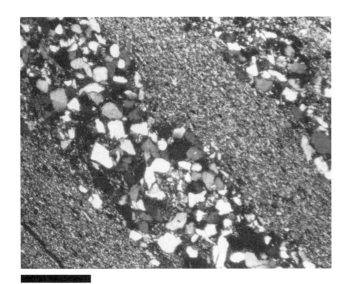

Figure 36.10
Nongraded, laminated silt and clay-sized clasts in slaty rock. Note subangular to subrounded morphology of quartz silt. X-polars. Bar scale = 0.5 mm.

Bioturbation has a major effect in destroying soft sediment structures such as laminations. Burrowing organisms may occur in siliciclastic sediments, but they are more common in carbonates.

Depositional Structures. **Planar bedding** (>1 cm) and **planar lamination** (<1 cm) are the most common depositional structures of clastic sedimentary rocks. A laminated mudstone–siltstone is shown in Figure 36.10. Mud rocks may be shaly. **Fissility** of shales depends on clay content and orientation. Flocculation into clumps results from interaction of clay minerals with ions in solution, generating nonfissile mudrocks.

Ripples are more or less regularly spaced undulations formed in silts and sands. If the distance between the crest and trough of ripples is greater than about 30 cm, the structures are **dunes** and **sand waves**. The crest of ripples may be **peaked** (Figure 36.11a) or **rounded**, and they may be **symmetrical** (Figure 36.11a) or **asymmetrical** (Figures 36.11b,c). Ripples may be **straight, bifurcated, sinuous, linguoid**, or have a combination of attributes (Figures 36.11b,c). Ripples form subaqueously and subaerially. Oscillation motion of waves in water bodies tends to form symmetrical ripples. Directional flow of water and wind form **current ripples** that tend to be asymmetrical. Ripples that form in water are a good indication of shallow conditions of sedimentation. They may be preserved on the upper side of strata, or as casts on the underside of successive strata. **Cross-bedding** and **cross-lamination** are depositional structures indicative of a dynamic environment. Stratification with low slope typically truncates strata of higher slope, with the latter tending to be asymptotic with underlying strata (Figure 36.12). Sedimentation characterized by ripples generates **ripple cross-laminations** (Figure 36.2). Cross-stratification in subaqueous and subaerial environments indicates paleocurrent or wind

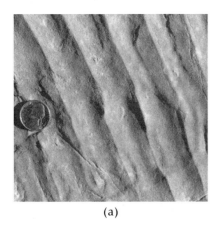
(a)

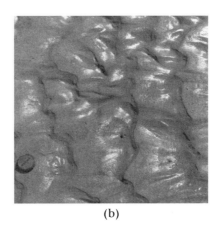
(b)

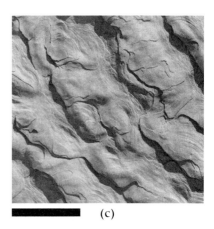
(c)

Figure 36.11
Ripple marks.

(a) Straight symmetrical ripples with peaked crests (rounded crests may falsely be observed) probably resulting from shallow-water oscillation on tidal flat silts. Scale = U.S. nickel (left). Moenkopi Formation, Canyonlands National Park, Utah. (b) Asymmetric, rounded crest, current ripples with sinuous to linguoid morphology. Flow direction is right to left (note steep lee side). Muddy bed of De Chelly River, Canyon de Chelly National Monument, Arizona. Scale = lens cap (lower left). (c) Sinuous to linguoid flow ripples similar to those shown in (b), preserved on upper surface of sandstone bed in Curtis Member of Stump Formation (Jurassic). Flow direction right to left. Vernal, Utah. Bar scale = 2 cm.

Figure 36.12
Cross-bedded Dakota sandstone. Note truncated foreset beds (arrows) that are asymptotic to truncation surface on underlying beds. Cortez, Colorado. Scale = U.S. quarter (lower left).

velocity regimes, such as on the inside of meander loops, where they are known as **point bars**. **Barrier-island bars** are typical along marine coasts where there is redistribution of deltaic sediments.

Graded bedding can develop where there is sedimentation of clasts of variable size. **Normal grading** starts with coarser clasts at the bottom and grades to the finest clasts at the top (Figure 36.13). In most cases this gradation reflects differing settling velocities in a batch of rapidly suspended clasts of mixed sizes, such as in turbidites. **Reverse grading** can develop on beaches, during migration of wind ripples, and in turbidite flows where there is rapid deceleration. **Kinetic sieving** is responsible for some reverse-graded bedding. This process occurs during grain flow, especially on the lee side of dunes, as smaller particles filter down through the larger grains.

Imbrication bedding is defined by elongate clasts that lie at an angle to bedding, inclined in an upstream direction. This position is acquired for certain river-deposited gravels. Alignment of elongate clastic particles can also parallel flow direction under appropriate conditions.

The development of planar beds, ripples, dunes, and antidunes in shallow water is related to flow velocity and grain size (Figure 36.14). In general, at lower velocities (**lower flow regime**) ripples form if the clastic grains are fine sands and silts, and dunes form if they are medium and coarse sand size. At very low flow velocities, planar beds may form with coarse sand. The **upper flow regime** is charac-

direction. Depositional surfaces of cross-stratifications viewed perpendicular to direction of flow may be trough shaped, characterizing the festooned morphology of eolian sands.

Sand bars are depositional features typically formed in fluvial and marine marginal environments. The position of bars in fluvial systems reflects lower-

 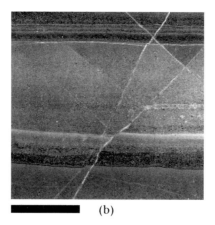 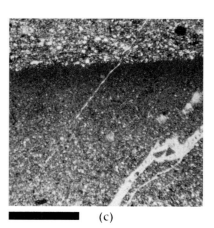

(a)   (b)   (c)

Figure 36.13
Graded bedding and laminations.

(a) Slaty laminated siltstone–shale with stratigraphic younging to the right indicated by coarser fraction on finest fraction. PPL. Bar scale = 0.8 mm. (Courtesy R. Schweickert.) (b) Normal graded beds in waterlain, air-fall (?) tuff, younging upward. Finer fractions at top of beds are lighter colored. Sutro Tuff, Virginia Range, Nevada. Slab surface. Bar scale = 1.8 mm. (c) Microscopic view of a portion of (b) showing contact surface between finer-grain fraction (chlorite-rich) (dark) and overlying coarser fraction (mostly plagioclase and calcite). Younging upward. PPL. Bar scale = 1.8 mm.

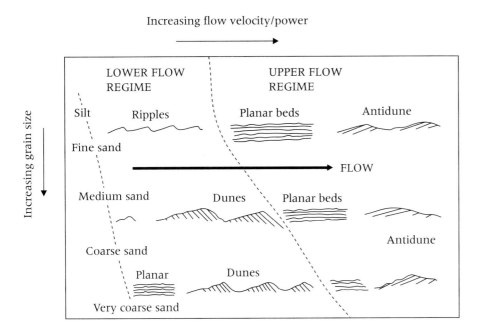

Figure 36.14
Bedforms in relation to grain size and flow velocity in shallow water. Modified from Blatt et al. (1980).

terized by planar beds and by **antidunes** if velocities are extreme.

Erosional Structures. Physical features resulting from local excavation can be preserved in the stratigraphic record. **Sole marks** are casts occurring on the bases of coarser-grained rocks resting on mudstones. Specifically, **scour marks** include **obstacle scours** (Figure 36.9) that consist of excavated sediments on the upstream side of cobbles and **flute casts** that are scooplike forms related to turbidity flow. **Toolmarks** are scour marks that relate to grooving by larger grains in softer sediment and to bouncing and skipping of grains. **Furrows** and **rills** are small-scale erosional channels. **Channel scars** and **slump scars** are larger-scale erosional features associated with lateral fluvial erosion, estuarine channels, and submarine channels. These features are preserved by channel filling and burial.

Siliciclastic Textures

Grain size and **grain shape** are the two most genetically significant primary depositional textural attributes of clastic sedimentary rocks. The **Udden–Wentworth grain-size scale** (Figure 36.15) places the cutoff between clay and fine silt at 0.0039 mm, between coarse silt and fine sand at 0.0625 mm, between coarse sand and granules at 2 mm, and between granules and pebbles at 4 mm. In general, the size of clastic grains reflects (1) sediment source (granite generates coarser grains than shales), (2) reactive replacement modification related to weathering (coarse feldspars become fine clays), (3) strength and hardness of the grains (quartz crystal grains maintain size easier than cleavable or softer minerals), (4) energy level of the erosive environment (channel deposits are gravels, whereas overbank deposits are silts and muds), and (5) duration of dynamic environment in which the grains are transported (long travel distances lead to reduced grain size).

The morphology of grains reflects many of the same parameters that apply to grain size. Mechanically weak grains and chemically reactive grains do not survive long enough to be rounded during transportation. Unreactive, mechanically strong clastics such as chert, polycrystalline quartz, and monocrystalline quartz can become extremely rounded and even spherical (Figure 36.16) if they participate in a long-term abrading environment. If initial angular grains are elongate or flattened, their worn equivalents are likely to have dimensional shape even though corners and edges are rounded. Spherical grains of feldspars are rare (Figure 36.16c) in view of their propensity to cleave.

Angular and subangular clastic grains typically reflect rapid, short transport distance sedimentation. In the extreme, intraclasts are generated more or less in situ, having had little chance for rounding (Figure 36.17a). Angularity of silt-sized clasts may in part be related to suspension-like transportation in muds, with marked reduction in grain to grain collisions, just the opposite of fluvial, beach, and eolian

	Name	Millimeters	Micrometers	φ
GRAVEL	Boulder	4,096		−12
	Cobble	256		−8
	Pebble	64		−6
	Granule	4		−2
SAND	Very coarse sand	2 ———	———	−1
	Coarse sand	1		0
	Medium sand	0.5	500	1
	Fine sand	0.25	250	2
	Very fine sand	0.125	125	3
MUD	Coarse silt	0.062 ———	62 ———	4
	Medium silt	0.031	31	5
	Fine silt	0.016	16	6
	Very fine silt	0.008	8	7
	Clay	0.004 ———	4 ———	8
		↓	↓	

Figure 36.15
Udden–Wentworth grain-size scale for clastic sedimentary rocks.

(a) (b) (c) (d)

Figure 36.17
Subangular clasts.

(a) Shaly siltstone intraclasts of probable rip-up origin in fluvial sandstone. Kayenta Formation. Hickman Bridge Trail, Capitol Reef National Park (Circle Cliffs), Utah. Bar scale = 2 cm. (b) Quartz and feldspar silt-sized clasts with muscovite in chloritic–carbonate matrix. Angularity of quartz due to short transportation distance and/or suspension in muds during transport. Note compaction deformation of large muscovite crystal. Fountain Formation arkose. Colorado. X-polars. Bar scale = 0.12 mm.

environments. The subangular character of quartz and feldspar grains in the arkosic siltstone shown in Figure 36.17b probably relates to rapid sedimentation and suspension like flow in mud.

Cement Textures

The infilling and replacement nature of cements dictates that their textural attributes are determined in part by the morphological characters of the primary clasts. Cements may simply mold against and between the clasts (Figure 36.1a). They may be relatively coarsely crystalline, such as sparry calcite (Figures 36.3a, 36.5a), or the materials may be submicroscopic, such as micrite (Figure 36.16b), chert, or hematite. Growth of crystals on pre-existing crystal substrates can be epitaxic or nontaxic, and faceted or nonfaceted crystals may be produced.

Compositional Characteristics of Siliciclastic Sedimentary Rocks

Grain Composition

Composition of primary grains can help determine the source, terrain, and history of the weathering and sedimentation processes of a siliciclastic sedimentary rock. Classification of siliciclastic sedimentary rocks into **conglomerates**, **sandstones**, and **mud rocks** is

Figure 36.16
Spherical and rounded clasts.

(a) Well-rounded lithic pebbles in coarse sand matrix. Tendency for bimodal grain-size distribution. Fluvial origin. (b) High-sphericity quartz grains of eolian derivation in carbonate matrix. X-polars. Bar scale = 1 cm. (c) Unusually spherical plagioclase grain in porous volcanic litharenite with isopachous carbonate cement. Pleistocene fluvial origin. Lovelock, Nevada. X-polars. Bar scale = 0.12 mm. (d) Rounded chert grains in finer quartz grain matrix. Chert (possibly derived from Vinini Formation) litharenite from Diamond Peak Formation, White Pine Range, Nevada. X-polars. Bar scale = 1 mm.

based on grain size. Subtypes of these major categories are determined chiefly by clast composition even if diagenetic cement is abundant.

Mud rocks are dominated by clays and quartz (Potter et al., 1980). Muds containing carbonate form **marls**, marking transition to carbonate-dominating rocks. Mixtures of muds and silts or sands generate **wackes**, considered to be a variety of sandstone rather than of mud rock since the mud fraction typically is matrix in relation to the larger clast fraction. "Graywacke" is a field name applied to dark, poorly sorted wackes that typically contain volcanic rock fragments and that may contain such minerals as chlorite and zeolites.

Sandstones are also known as **arenites**. Nomenclature is governed by clast composition. **Quartz arenites** contain more than 90% quartz grains. **Litharenites** contain a substantial fraction of rock fragments (Goldstrand, 1991). Most common rock fragments are chert, quartzite, siltstone, schist, andesite, basalt, and in some circumstances limestone and dolostone. **Sublitharenites** are quartz-rich arenites with less than 20% lithic fragments. **Arkoses** contain a substantial fraction of feldspar grains. **Subarkose** contains less than 20% feldspar and at least 75% quartz grains. **Lithic arkoses** and **feldspathic litharenites** express intermediate compositions (Folk, 1974).

Conglomerates are classified on the basis of the ratio of clasts (granules, pebbles, cobbles, boulders) to matrix (sand, silt, clay). **Orthoconglomerates**, also known as **oligomict conglomerates**, contain clasts of a single composition (typically quartz) set in less than 15% matrix. **Petromict conglomerates** contain mixed clast lithologies. **Paraconglomerates**, also known as **diamictites**, contain more than 15% matrix and are derived from glacial deposits (Eyles and Eyles, 1989). The composition of the clasts reflects the terrane that was glaciated.

Rock Color

The color of siliciclastic sedimentary rocks is fundamentally a reflection of the composition of grains and cements. However, composition is determined by environment of deposition as well as provenance. Furthermore, composition is determined by diagenetic and epigenetic processes.

For example, white quartz arenites, such as the Coconino Sandstone, reflect the very high content of quartz grains and secondary silica cement. The redbrown Wingate and Entrada sandstones reflect introduction of a relatively minor amount of iron oxyhydroxides. Pure white portions of these eolian quartz arenites occur locally where they have not been infiltrated by postdepositional iron-bearing solutions.

The "red rock" country of the Colorado Plateau (Baars, 1983) is characterized by Permian and Triassic oxidizing environment, converting the ferrous iron in primary minerals into ferrous iron that forms secondary oxyhydroxides or adsorbs on clay minerals. The oxidizing environment either produced the secondary iron minerals, perhaps in marine tidal flat deposits and in flood plain deposits, or maintained them if they were already present at the source region or along the course of transportation.

In contrast, Cretaceous marine marginal and epicontinental marine environments of the Colorado Plateau (Mancos, Mesa Verde, and Tropic) were characterized by reducing conditions. The sandstones in these stratigraphic sections are light gray, not white nor stained by iron oxyhydroxides, and the mud rocks are medium to very dark gray. Free carbon is abundant, occurring in black shales, greenish gray mudstones, and coals formed in stagnant lagoonal waters.

Varicolored rocks typically represent a mixture of reddish or purplish oxidized and greenish gray reduced material. Although primary environment may vary between oxidizing and reducing, as it probably did during the formation of the Morrison Formation of the Colorado Plateau, apparent juxtaposition can be generated by partial reduction of oxidized materials (Plate 8a) or by incomplete oxidation of reduced materials. Red-brown zones in fluvial and eolian sandstones that cross primary stratification boundaries may be the result of selective introduction of oxidized iron compounds in rock that had no materials indicative of oxidizing or reducing primary environment.

Siliciclastic Depositional Environments

Although the complexities of sediment transport and deposition are enormous in view of interactions between gravity, wind, standing water, moving water, and airborne pyroclastics, generalization of depositional environment is useful for understanding how siliciclastic rocks form. The following is a general summary of depositional environments for siliciclastic rocks. For more detail the reader is referred to Reading (1978), Reineck and Singh (1980), Allen and Allen (1990), Davis (1992), and Einsele (1992).

Dominantly Subaerial Environment

The effect of wind action on an accumulation of sediments of variable size is to leave the larger grains behind and winnow out the clay- and silt-size particles.

Fine sand accumulates because it cannot be suspended in the winds and dispersed, and the larger clasts have no means to accumulate other than as a passive residuum. Minerals other than quartz (and the gypsum sands of White Sands, New Mexico) may have initially been converted to clays; if not, they eventually are reduced by both chemical and mechanical processes to clay- and silt-sized particles that can be removed.

The final product is an accumulation of fine quartz sand. In the process of sorting and dune migration, the sand particles become spheres as a consequence of grain-to-grain collisions. The grains become microscopically pitted, which gives them a frosted appearance. This **eolian** environment generates sandy **deserts (ergs)** characterized by magnificent cross-stratification (Figure 36.18).

The finer particles removed from their association with the sand grains must eventually drop from suspension. Dispersal of the finer sediments is inevitable unless there is an unusually abundant source of very fine sediments. Glacial till contains an abundance of rock flour as a result of grinding between ice-bound rock and bedrock. Since wind erosion is not prevented by vegetation in freshly glaciated terrain, these fines may be separated from the coarse till fraction, forming **loess deposits** downwind from the region of glaciation.

The subaerial environment also includes accumulations of rock debris in response to gravity alone. **Talus** and **gruss** accumulations are dry mass-wasting deposits. Since these deposits occur on slopes, they are not likely to be preserved in the stratigraphic record without being reworked by other agents of erosion.

Subglacial Environment

Rock debris entrained in glacial ice accumulates as till, forming various types of moraines. **Tillite** or **diamictite** is lithified till. Till deposits are very poorly sorted, although transition to water-worked deposits is a consequence of entrance of glaciers into the ocean or lakes and glaciofluvial systems. Kames, eskers, and outwash deposits are a consequence of fluvial reworking.

Distinction between diamictites, subaerial debris flows, and subaqueous debris flows may require the presence of striated rock fragments or lateral facies changes and associated subenvironments.

Fluvial Subaqueous–Subaerial Environment

This environment involves the transportation of sediments by running water on the continents. The environment is partly subaerial since sediments deposited by fluvial processes on flood plains and alluvial fans are exposed to the atmosphere prior to final burial and lithification.

In three geomorphic systems running water is the dominant means of sediment transport: (1) alluvial fan, (2) pediment–bahada, and (3) river.

Alluvial fans typically form in semi-arid-to-arid climates where streams, or intermittent streams, issue from an elevated land mass into a basin. Fans also form in humid climates as outwash deposits in proglacial settings. The classic case of fan development is along a range front generated by normal faulting, such

Figure 36.18
Large-scale, festooned, eolian cross-beds with two well-developed truncation surfaces (upper left). De Chelly Sandstone (Permian). Canyon de Chelly National Monument, Arizona.

 (a)
 (b)

Figure 36.19
Fanglomerate.

(a) Tectonic block consisting of poorly sorted Tertiary conglomerate (rudite) of alluvial fan derivation. Boulders up to several meters in diameter. Sacramento Pass, Snake Range. Nevada. (b) Same formation as (a), showing possible clast imbrication oriented from lower left to upper right (some tectonic rotation likely). Note angular to subrounded clasts and lack of sorting at scale shown. Scale = lens cap (lower left).

---

as in the Basin and Range Province of the western continental United States. Episodic flash flow of water down canyons transports rock debris of all sizes to the base of the range, where abrupt change in slope and partitioning of discharge water into distributaries result in deposition. The clastics range from boulders to clays and are poorly sorted (Figure 36.19a,b), and since the transportation mechanism is very short-lived there is little opportunity for rounding. Channel deposits on the fan become clast-supported conglomerates containing a pebbly sand matrix. Channeled flow may change to sheet flow as the channel becomes choked with debris, eventually leading to establishment of a new channel elsewhere on the fan. Distal parts of fans are characterized by finer-grained rocks deriving from finer sediments deposited in braided distributaries and on sheetflood surfaces. Both fining and coarsening upward occur in this spontaneous high-energy environment. Trough-filled cross-stratification can form as deposition succeeds temporary channel erosion and the carrying power of the fluvial system fluctuates.

Deposition in basins adjacent to ranges can be extensive. Fans coalesce and generate a continuous surface of deposition reaching down to the lowest region of the basin, where there may be a playa lake. Erosion into the range may generate a rock-cut surface of low slope known as a **pediment**. Clasts originating in the range are transported across pediments to the **bahada** zone of thick sedimentation in the basin. If there is a frontal fault, its location at the lower end of a pediment, marking the beginning of the bahada, may be obscured. The veneer of sand and gravel on pediments is not likely to be preserved in the stratigraphic section, but the fan-type sediments accumulated in the basins from extensive conglomerates and mixed sediment clastic rocks.

Rivers are responsible for major deposits of clastic sediments in all climates. Gravels accumulate in the river channel (Figure 36.20a), bordered by sand bars and flood plain silts and muds. Layers of sand over gravel are formed by lateral shifting of the river channel that commonly contains gravel, generating a layer of gravel that is continuously covered by sand accumulating on the inside of meander loops. Sand deposition in a gravel-lined channel is shown in Figure 36.20b. Gravels over sands (Figure 36.20c) and sand lenses in gravels (Figure 36.20d) result from reverse of lateral shifting of a fluvial channel and by seasonal fluctuations of discharge in which local deposits of sand during low water are modified or removed during high water. Channel sands and gravels overlying

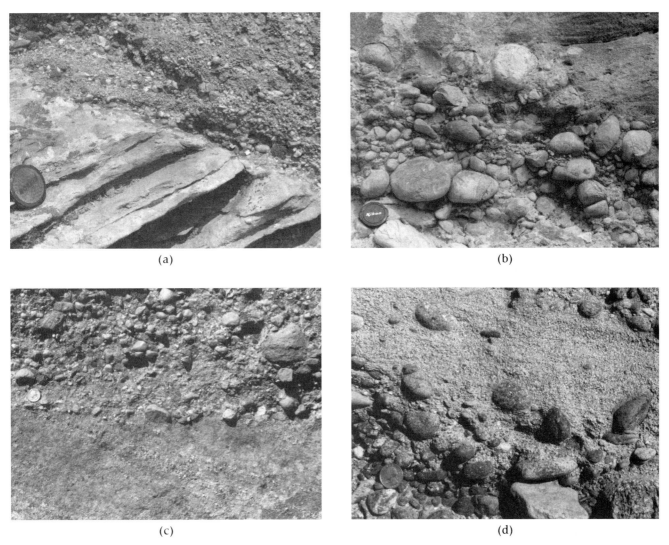

Figure 36.20
Fluvial sandstones and conglomerates.

(a) Channel gravel facies of Shinarump Conglomerate deposited on erosion surface truncating cross-stratification of eolian de Chelly Sandstone. Rim of Canyon de Chelly, Arizona. Scale = lens cap (left). (b) Sandstone and conglomerate derived from sand deposited in channel of underlying, well-rounded gravels. Duchesne River Formation, Vernal, Utah. Scale = lens cap (lower left). (c) Conglomerate truncates weakly stratified (inclined to the right) sandstone. North Horn Formation. Manti, Utah. Scale = U.S. quarter (left). (d) Early Pleistocene pebbles and cobbles with bedding-parallel bar sands. West Reno, Nevada. Scale = U.S. quarter (lower left).

silts and muds of contiguous flood plains (Figure 36.21a) represent channel shift in either a short-term (alluvial fan) or long-term aggradation mode. Deposition preserves the flood plain deposits that otherwise would be removed by strictly lateral erosion. Occurrence of fluvial sandstones and conglomerates on tidal flat mud rocks (Figure 36.21b) probably resulted from deposition on an erosion surface cut into mud rock, not by shifting of fluvial channels across active tidal flats.

Sorting of sediments generally is not as good as that occurring in eolian and beach environments, where the sorting processes are more constant. Abrupt change in clast size is common, and cross-stratification commonly consists of gravels mixed with sands (Figure 36.22).

Flooding results in reduced flow velocity since there is much more friction with the ground surface in proportion to the volume of water passing over this

Figure 36.21
Fluvial sandstones on siltstones.

(a) Flood plain horizontally laminated siltstone overlain by a lens of weakly cross-stratified channel (?) sandstone (white), in turn overlain by cross-stratified bar (?) sandstone. Kayenta Formation. Colorado National Monument. Bar sand bed is about 0.5 m thick on left side. (b) Shinarump Conglomerate Member of Chinle Formation here consisting of bar sands about 3 m thick, overlying Moenkopi tidal flat shales and overlain by bentonitic member of Chinle Formation. Capital Reef National Park, Utah.

surface. Gravels in fluvial channels, do not get out of these channels. However, sand may be raised onto the edge of the flood plain where it is quickly deposited as natural levees as flow velocity plummets. Silts and clays are carried in suspension onto the flood plain until they too eventually drop out where the carrying power of the water is further reduced. Overbank deposits are characterized by well-developed beds and laminations (Figures 36.21a, 36.23). Mud cracks form as the floodwaters recede (Figure 36.9) and the water-saturated sediments dry out, to be preserved in the stratigraphic section by the next blanket of wet overbank sediments.

Sublacustrine–Subaerial Environment (Marginal Lacustrine)

Marginal lacustrine deposits include (1) deltas with their associated beaches, swamps, channels, depositional plains above and below water level, (2) beaches not associated with deltas, and (3) swamps with or without associated beaches. Lacustrine swamps are sites of accumulation of organics with variable amounts of muds and silts. Other environments that

Figure 36.22
Cross-bedded, pebbly sands of fluvial Salt Wash Member of Morrison Formation (Jurassic–Cretaceous). Near Notom, Utah. Scale = lens cap (left).

Figure 36.23
Well-bedded and laminated siltstone of flood plain origin. Kayenta Formation, Capitol Reef National Park. Scale = lens cap (lower left).

are both sublacustrine and subaerial during the course of their existence include **playa** and supraglacial deposits. Playas are more appropriately discussed as an evaporite system, and supraglacial deposits such as kames are not likely to be preserved in the stratigraphic record. Sediments in the marginal lacustrine environment are siliciclastic if input is by fluvial systems. These sediments commonly intertongue with carbonates as they did in Lake Unita.

Foreset lacustrine deltaic stratification with truncating topset strata is shown in Figure 36.24a. The Green River delta facies in Utah consists of distributary channel sands (Figure 36.24b) in greenish gray locally maroon, probably interdeltaic mudstones (Figure 36.24c). Evolution to an open lacustrine environment is nicely illustrated by the carbonate and carbonaceous sediments covering the deltaic sediments (Figure 36.24b).

Beaches may form along the margins of deltas, but they form along any lacustrine shoreline where wave action works on bedrock or previously deposited sediments. The gastropod grainstone shown in Figure 37.14a was formed as a lacustrine beachrock without an associated delta. The composition of beachrock also reflects rock adjacent to the lake and to terrane reached by fluvial systems that empty into the lake. The beachrock shown in Figure 36.25 contains granules and gravel of volcanic rock fragments derived

(a)

(b)

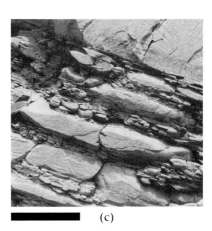

(c)

Figure 36.24
Lacustrine deltaic and interdeltaic facies.

(a) Sands and pebbly sands of delta foreset beds truncated by topset bed of similar characteristics. Mullin Creek delta built into Pleistocene Lake Lahontan. Scale = lens cap (lower left). Pyramid Lake, Nevada. (b) Bar sandstone with low-angle cross-stratification, interbedded with thin clayey siltstones (delta facies), overlain (middle line of photo) by lacustrine claystone, siltstones, and a bed of limestone (white, ca. 1 m thick), representing interdeltaic facies. Green River Formation, Willow Creek Canyon, Utah (Ryder et al., 1976). (c) Green-gray, finely laminated, mudstone with curved partings probably controlled by slight grain-size differences and orientations of platy minerals. Interdeltaic facies of Green River Formation. Bar scale = 4 cm.

Figure 36.25
Lacustrine beachrock. Pleistocene Lake Lahontan, Nevada.
(a) Bedded coarse sandstone, locally conglomeratic, exhumed along bedding weaknesses. (b) Close-up of rock shown in (a). Grain-supported, well-sorted granules of silicic ash flow quartz mixed with andesitic and basaltic grains derived from flows and intrusives. Calcite cement. Bar scale = 1 cm.

from bordering ranges. The excellent sorting capability of wave action is well illustrated by the granules shown in Figure 36.25b.

Sublacustrine (Open Lacustrine)

Rocks of the sublacustrine environment typically are rocks of biogenic affinity, such as limestones, marls, and diatomites (Chapter 37). Air fall pyroclastics may also accumulate in open water (Figure 36.13b,c). Interbedding and intertonguing with rocks of marginal lacustrine derivation are to be expected.

Marine Marginal Subaqueous–Subaerial Environments

Marginal marine environments are even more complex than the marginal lacustrine setting because of the effects of tides and subduction tectonics along some continental margins. The subenvironments of the marine marginal are deltaic or nondeltaic. The environments tidal flat-tidal channels, beach, lagoon, and barrier island may occur in association with deltas or develop along coastal regions where there is relatively minor fluvial input. Estuarine, subaerial delta, subaqueous delta are characteristically deltaic. The **sabkha** environment is dominantly evaporitic and glaciomarine deposits are relatively subordinate. **Deltas** develop on tectonically passive coasts where rivers empty into the sea. Sedimentation occurs as a result of sudden reduction in flow velocity. As sedimentation progrades seaward, the upper delta becomes part of the fluvial system, characterized by channel flow and susceptible to flooding and deposition of overbank sediments. Distributary channel flow is characteristic of the intertidal zone. If the incoming river water is less dense than the colder seawater, suspended sediments are carried great distances before they flocculate and finally drop onto the distal region of the delta system.

Distal delta front sand, distributary mouth sand bars, and lagoonal accumulation of organics are characteristic of the Late Cretaceous deltas along the Mancos seaway of what is now the Colorado Plateau (Hill, 1983). Sandstones with coal beds overlying deep-water shales are characteristic (Figure 36.26). The sandstones represent building of the delta, the coals represent temporary lagoons behind barrier islands, and the shales represent the open marine sediments on which the delta progrades.

Marine transgressions can generate beach sands from rock by wave action without fluvial input and delta formation. These mainland **beach** deposits lack barrier bars and lagoons. Vast beds of sandstone form in this way, such as the Dakota Sandstone (Figure 36.12), the Gallop Sandstone, and the Tapeats Sand-

Figure 36.26
Open marine shales overlain by lagoonal coal beds (not showing), subsequently buried by barrier island sands (two sequences). Foreground is surface on Salt Wash Member of Morrison Formation. Beyond is Tununk Shale overlain by cliff-forming Ferron Sandstone (accentuated by shadows), followed by deep-water Blue Gate Shale overlain by Emery Sandstone (ridge), all members of Mancos Shale Formation. Cainville, Utah.

stone. The Tapeats Sandstone rests with marked nonconformity on Precambrian metamorphic rocks in the inner gorge of the Grand Canyon (Figure 36.27). Presumably there were relatively high-energy, wave-dominated coasts.

Tidal flats occur along low-relief coasts where strong wave action is lacking or behind barriers on high-energy coasts. The flats may be associated with estuaries, bays, and deltas, consisting of muds, silts, and local marshes. Movement of the sediments occurs toward the mainland during flood tides and toward the ocean with the ebb tide. Mud cracks and ripples are characteristic. Bedding is well developed (Figure 36.28a). Evaporitic conditions may develop, resulting in the occurrence of halite crystals in the sediments (Figure 36.28b) and gypsum beds and secondary veinlets (Figure 36.28c) (Peterson and Pack, 1983). If seawater flow is restricted, especially in arid-to-semi-arid regions, tidal flats evolve into **sabkhas**. Shaly beds in evaporite sections represent episodic tidal flooding (Figure 36.29).

Open Marine Environment

Sedimentation on the ocean floor without exposure to subaerial processes occurs in epicontinental seaways, foreland basins, along passive continental shelves and slopes, in the trench of subduction zones, in forearc and backarc basins, and on abyssal plains.

Epicontinental seaways are characterized by accumulation of fine sediments in relatively shallow oceans. For example, the Late Cretaceous seaway extending from what is now the Gulf of Mexico to the Arctic Ocean across the Colorado Plateau region was the site of deposition of the Mancos Shale (Figure 36.30). This shale derives from fine sediments carried by suspension to deeper water adjacent to marine marginal deltas forming as the result of clastic input into a **foreland basin** associated with the Sevier orogenic highlands to the west in western Utah and Nevada.

The continental shelf along passive continental margins receives siliciclastic sediments where rivers issue from the continent. Sedimentation in open water beyond the limits of deltas results in

Figure 36.27
Marine transgressive Cambrian Tapeats Sandstone (foreground) and shown nonconformable with underlying Precambrian metamorphic rocks (upper center). Grand Canyon, Arizona.

 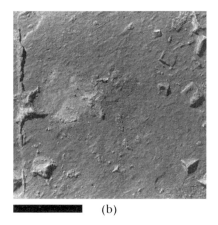 

Figure 36.28
Siltstones of tidal flat environment.
(a) Rhythmically well-bedded siltstones of Summerville Formation with angular unconformable relation with underlying Entrada Sandstone, overlain by fluvial pebbly sandstones of Salt Wash member of Morrison Formation. West of Bullfrog Marina, Utah. (b) Halite casts in Moenkopi Formation indicate ephemeral evaporitic conditions in tidal flat environment. Capitol Reef National Park, Utah. Bar scale = 2 cm. (c) Secondary "coconut" gypsum veins (subhorizontal) and "sugary" gypsum in vertical fracture (lower right) in Summerville Formation (Petersen and Pack, 1983). Possibly derived from gypsum bed in upper part of section. West of Hanksville, Utah. Scale = U.S. nickel (lower left).

accumulations of pelagic clay-sized particles on **continental shelves**, **continental slopes**, and deep **abyssal plains**. Episodic rapid transport of sediments in the form of density currents down submarine canyons generates **turbidites**. Where terrigenous siliciclastic sediment input is minimal or lacking, carbonates dominate.

Tectonically active marine environment sedimentation is especially well represented by the generation of **mélanges** formed in the **accretion wedge** of subduction zones (Ernst, 1971; Brothers and Grapes, 1989). The Late Jurassic and Cretaceous Franciscan Formation formed in this manner along the Pacific margin of North America (Figure 36.31). The Francis-

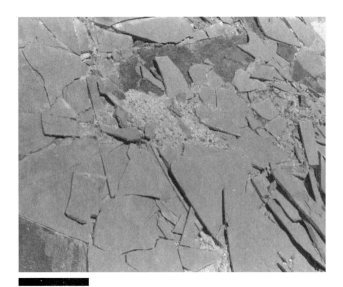

Figure 36.29
Shaly interbeds in anhydrite and gypsum of Arapien Formation, as a result of fluvial flooding into evaporite system. Mayfield, Utah. Bar scale = 1 dm.

Figure 36.30
Finely laminated mudstone with curved-surface partings. Open epicontinental seaway environment. Mancos Shale. Mesa Verde National Park. Scale = U.S. nickel (lower left).

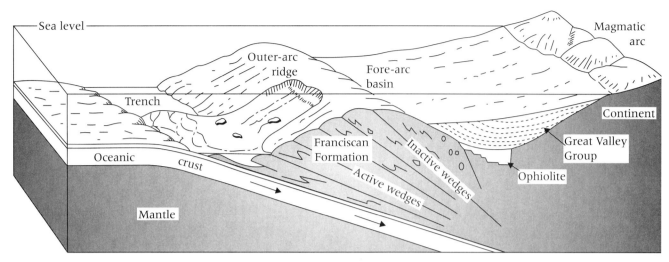

Figure 36.31
Accretional prism setting of Franciscan Formation shown diagramatically in relation to plate tectonic setting. (Courtesy B. Page: Page, 1977.)

can mélange consists of blocks of rock in a pervasively sheared matrix accumulated in a submarine trench environment. Wackstone ("graywacke") is most common (Figure 36.32) in the **subduction complex**, consisting of mineral assemblages derived from exotic volcanic terrains on the Pacific Ocean side of the subduction zone and/or from the arc on the continental side. There are also well-bedded portions (Figure 36.33) consisting of mudstones, conglomerates, thin- and thick-bedded sandstones of partly turbidite origin (Bachman, 1978).

Open marine sedimentation also occurs in forearc and backarc basins. Turbidites are common. Sedimentation on abyssal plains consists of pelagic siliciclastics (clays), diatomes and radiolarians (siliceous oozes) (Chapter 37), and planktonic calcareous foraminifera (carbonate ooze) (Chapter 37), as well as input of volcanic ash.

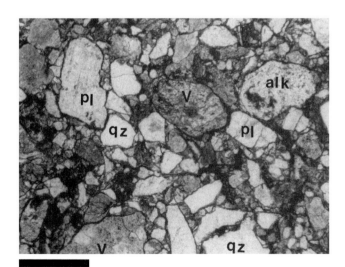

Figure 36.32
Franciscan volcanic litharenite ("graywacke") of subduction complex. Contains altered plagioclase, quartz, white mica, chlorite, and lawsonite. Pacheco Pass, California. PPL. Bar scale = 0.5 mm.

Figure 36.33
Bedded turbidites of Cretaceous Franciscan Formation. Mendocino Coast, California.

37

Biosedimentary Rocks

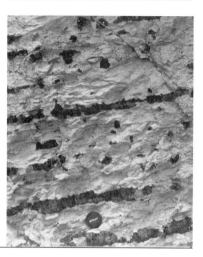

Bedded and nodular chert (black) in Ely Limestone, Eagan Range, Nevada. Scale = lens cap (bottom center).

Biota and Rock-Forming Process

How do biogenic materials contribute to the rock-forming process? Coral reefs represent a rock-forming process dominated by biotic activity that generates calcareous framestone, whereas the occurrence of a few brachiopod fragments in an otherwise nonfossiliferous limestone *appear* to indicate no more than an incidental biotic participation in the rock-forming process.

Biota-dominating rock forming processes include environments conducive to accumulation of vegetation (swamps, lagoons), growth of organic-generated precipitates (coral reefs), bacterial reduction of iron in the formation of iron formations (La Bergs et al., 1987; Nealson and Myers, 1990; Duhig et al., 1992), and secretions of sticky substances by algae on stromatolite surfaces that trap and bind passing clasts.

Gravity settling and sedimentation of postmortem pelagic biota and mechanical accumulation of benthonic fossil fragments are clastic sedimentary processes that are not directly dependent on the biogenic nature of the clasts. Nevertheless, the size and shape of the bioclasts, as determined by the initial biogenic structure and by different chemical response to environment during transport, must influence in part the attributes of the sediment deposit and the resulting rock.

Mechanical accumulation of carbonate bioclasts in shallow water generally occurs in the same environments that are favorable for growth of carbonate-secreting organisms. Consequently, distinction between in situ biota and transported postmortem bioclastic material is difficult. A sparsely fossiliferous limestone does not necessarily mean that there was only incidental input of biotically generated carbonate in comparison to chemical precipitation of calcium carbonate. It is well known that calcareous fossil exoskeletons are readily destroyed by algal borings, by diagenetic micritization, and by diagenetic or epigenetic recrystallization to sparite. This means that a limestone containing few or no fossils does not preclude an origin from an accumulation of dominantly biogenic material. Nevertheless, direct precipitation of calcium carbonate from aqueous solutions in the absence of biota or indifference to the presence of biota is well known (Chapters 38 and 39).

It would seem that a classification dividing "biochemical" sedimentary rocks from "chemical" sedimentary rocks would be the basis of a useful rock classification focusing on process. Carbonate rock associated with bedded halite, gypsum, and anhydrite in evaporite settings lends itself quite well to such a division. However, most limestones do not originate in evaporite settings and are generated where biota thrive. Even though direct participation by organisms may be minimal or lacking, the physical association of lime-secreting biota and purely mechanical or inorganic chemical accumulation is omnipresent and process separation is not practical. Therefore, rocks included in this chapter are those forming in environments where biotic activity occurs, regardless of the degree of participation of biota in the formation of the rocks.

Carbonate sedimentary rocks are by far the most abundant of the biochemical rocks (Bathurst, 1975; Scholle, 1978, 1983; Adams et al., 1984; Scoffin, 1987; Greensmith, 1989; Tucker, 1991). There are transitions to other biochemical rocks, such as cherts and phosphorites, and to nonbiochemical siliciclastic sedimentary rocks. Gradations form as the result of (1) juxtaposition of sediments from more than one source, (2) diagenetic infiltration of calcium carbonate-bearing solutions into the pores of siliciclastics where precipitation and reactive replacement occur, and (3) in situ development of biota in siliciclastic sediments, such as pelycepods in tidal-flat silts. Thus, even the distinction between carbonate and siliciclastic sedimentary rocks must be generalized where both types of materials accumulate.

The Remains of Organisms in Biosedimentary Rocks

What Is Preserved?

Any or all of the following may be preserved (Horowitz and Potter, 1971) and are indicators of the former presence of biological activity: (1) organic remains of flora, fauna, (2) calcareous (calcite–aragonite), siliceous (opal-A) and phosphatic (carbonate–fluorapatite, **francolite**) precipitates associated with algae, invertebrate exoskeletons, and vertebrate endoskeletons, (3) skeletal parts (overall form and/or internal structure) replaced by new minerals or new crystals, such as calcite after aragonite, chert after micrite, and dolomite after calcite, (4) chamber fillings that have no direct relation to skeletal parts, such as silicic silts and clays in calcareous shells, (5) impressions of organic or skeletal form in adjacent sediments, such as bivalve molds in silts, and (6) tracks, trails, and burrows (**trace fossils**), including fecal pellets or their micritized equivalents.

Organic remains consists of materials such as carbon, kerogens, resins, and hydrocarbon combustibles. The form of the organism may be preserved, such as stems, leaves, pollen, spores, logs, and roots, or the decayed products are more or less dispersed as a consequence of sedimentation, deformation, and diagenetic processes. **Humus** is accumulations of the larger plant forms. **Sapropel** is ooze or sludge composed of plant remains (mostly algae), macerating and putrefying in the anaerobic environment at the bottoms of lakes and oceans. Finely divided organics deriving from planktonic algae and foraminifera accumulate subaqueously in low-oxygen environments, thus contributing to the carbon content of black shales and dark limestones. **Kerogens** are the organic remains in oil shales that can be converted into petroleum.

Examples of Plant Remains (Kingdom Planta)

Leaf imprints and residual carbon in a volcanic litharenite are shown in Figure 37.1a, and silicified wood cells are shown in Figure 37.1b. Carbonaceous material in a silty bed of the Green River Formation (Figure 37.2) is closely associated with the famous oil shales of that formation.

A lignitic mud rock containing coal lamellae is shown in Figure 37.3a. **Bituminous coal** beds in the Blackhawk Formation of Utah are shown in Figure 37.3b. This coal is associated with distributary sand bars of a marine deltaic system. The woody character of a coal from the Franciscan Formation of California (Figure 37.3c) is similar to the coalified twigs of a Permian Bangladesh coal (Figure 37.3d) except for the presence of **vitrinite** in the latter. Vitrinite is a homogeneous-looking coal maceral that has a middle level of reflectance as a measure of burial temperature (Bostick, 1979). A **maceral** is any organic constituent of coal that has distinct physical and compositional properties (a nonmineral unit, Chapter 7). **Phyterals** are specific plant forms recognized by characteristic morphologies. **Fusinite** is a high-reflectance maceral, in part of fire origin (Figure 37.3e).

Coalification requires decomposition of organics by the enzymes generated by fungi and bacteria (van Krevelen, 1981; Ward, 1984; Ross and Ross, 1984). A low-oxygen environment such as a bog is required so that not all the organics are converted to carbon dioxide, and the rate of accumulation must exceed the rate of decomposition. **Peat** accumulation is followed by transformation to **lignite** and coal as burial occurs. The process is mainly dehydration at the early stages, followed by devolatilization of oxygen and carbon–oxygen compounds as the **rank** of the coal increases.

 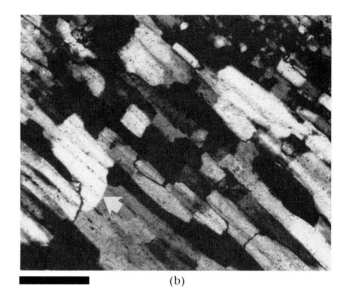

Figure 37.1
Fossil flora.

(a) Casts and residual carbon remains of Pliocene popular leaves in volcanic litharenite. Sandstone of Hunter Creek (informal name; Coal Valley Formation of Everden and James, 1964). Verdi, Nevada. Bar scale = 1.7 cm. (b) Silicified wood. Quartz probably after original replacing opal-A. Although quartz is generally confined to individual cells, some crystals bridge cell boundaries (arrow). X-polars. Bar scale = 0.12 mm.

Palynomorphs are insoluble microfossils, such as pollen and spores, that are useful in dating certain terrigeneous sedimentary rocks. Palynomorphs are best collected from fine-grained material (mudstone or coal) and are highly resistant to destruction, except under oxidizing conditions. Because of their resistance to destruction, polynomorphs can easily be recycled during erosion of older strata and redeposition into younger strata. **Phytoclasts** are fragments of organic matter occurring in shales and sandstones that have retained some sense of plant structure.

Plants or Animals?

Single-celled microorganisms containing chloroplasts and capable of photosynthesis originally were classified as plants. One-celled animals were defined as having flagella used for locomotion. Problems with classification of single-celled organisms such as bacteria, blue-green algae, and fungi arise mainly because some varieties of these organisms have both plant and animal attributes. At present, five kingdoms are recognized: (1) Planta, (2) Animalia, (3) Monera, (4) Protista, and (5) Fungi.

Kingdom Monera. **Red algae** are benthic algae that precipitate high-Mg calcite within and between cell walls, generating a skeletal structure. Some red algae are coralline (Figure 37.4a,b) and some of these have articulated branches (Figure 37.4b). More algal skeletal carbonate is produced in the photic zone than in deep water. This indicates that photosynthesis is involved in the process of precipitation and that algae have plant attributes. In an aqueous solution saturated with cal-

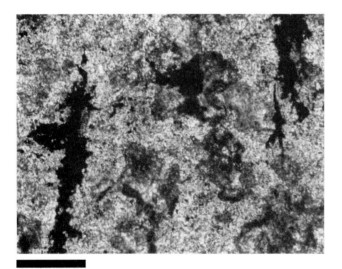

Figure 37.2
Organic matter (black) in cherty limestone bed of deltaic facies of Green River Formation. Uinta Mountains, Utah. PPL. Bar scale = 0.12 mm.

Figure 37.3
Lignite and coal.

(a) Very thin coal seams (arrows) in lignitic Menefee Formation (Cretaceous). Mesa Verde National Park. Rock surface. Bar scale = 1.8 cm. (b) Bituminous coal beds (dark) derived from flora accumulated in swamps adjacent to large rivers on marine marginal delta plain. Episodic burial by fluvial bar sands and thin mud interbeds indicate unstable basin conditions and shifting of river course. Upper portions of Cretaceous Blackhawk Formation. Castle Gate, Utah. Lower coal bed is 1.5 m thick. (c) Log wood with preserved cellular structure containing fluorescing resinous material little changed from plant resins. Sub-bituminous coal from Franciscan Formation (Cretaceous). Baker's Beach, San Francisco. Reflected light. Bar scale = 0.13 mm. (Courtesy N. Bostick.) (d) Coalified woody mush (light) containing massive, medium reflectant vitrinite layers (dark) formed by coalification of twigs. Reflected light. Bar scale = 0.13 mm. Permian coal. Bangladesh. (Courtesy N. Bostick.) (e) Vitrinite (lower gray portion) with high-reflectant fusinite (white) of fire origin. Reflected light. Bar scale = 0.18 mm. Permian coal. Bangladesh. (Courtesy N. Bostick.)

cium carbonate, utilization of carbon dioxide in the photosynthetic process reduces the concentration of bicarbonate leading to the precipitation of the calcite. Nevertheless, that some algae can grow in dark waters as well indicates that there must be an additional energy source. Red algae are encrusters and cementers especially in bioherms and biostromes, but they may be intertidal as well.

Cyanobacteria (blue-green algae) contain chlorophyll pigments that serve as a catalyst in photosynthesis. The well-known variety *Halimeda* (Figure 37.4c) generates calcareous plates that occur in branching structures held together by organic material. Decomposition of the organic portions leads to detachment of the plates. Plates consist of **utricles** that contain organic filaments shrouded in clusters of

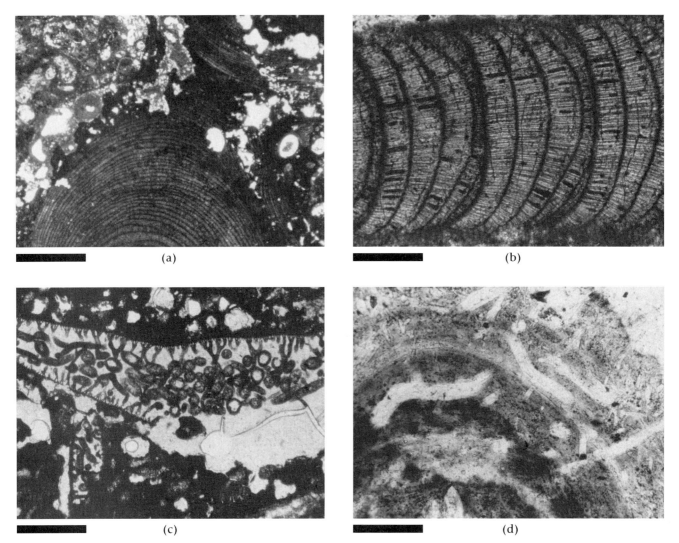

Figure 37.4
Calcium carbonate structures formed by algae.

(a) Transverse section of structure formed by branching coralline red algae *Neogoniolithon*. Cockburn Town Reef (Pleistocene). San Salvador, Bahamas. (Courtesy A. Berger and D. Zenger.) X-polars. Bar scale = 1 mm. (b) Longitudinal section of cellular structure formed by coralline red algae. Curved segmented branches were articulated. PPL. Bar scale = 0.12 mm. Recent sand from shoal of coral rubble at Molasses Island, Outer Reef Tract, Florida Keys. (Courtesy N. Silberling.) (c) Plate that consisted of micron-sized aragonite generated by green algae *Halimeda* has been diagenetically replaced by sparry calcite (white). Tubelike uticles, some of which cross within the thickness of the thin section, were originally filled with plant tissue (filaments) and are now mostly micritic carbonate. Branching uticles characterize rim of plate. PPL. Bar scale = 0.5 mm. Key Largo Limestone (Pleistocene) from Windley Key quarry, Florida Keys. (Courtesy N. Silberling.) (d) Algal borings generating filament tubes in phosphatic pisolite. PPL. Bar scale = 0.4 mm. (Courtesy K. Nichols and N. Silberling.) (e) Characteristic bulbous laminated structure of stromatolitic limestone. Bar scale =1.7 cm. (f) Oncoid with concentric calcium carbonate precipitated on limestone fragment (white). Lacustrine oolitic Middle Pliocene limestone. Hot Creek Range, Nevada. Bar scale = 1 cm. (g) Pisoids formed by algal-accreted calcium carbonate on brachiopod (also gastropod and echinoid plate) fragments. PPL. Bar scale = 1 mm.

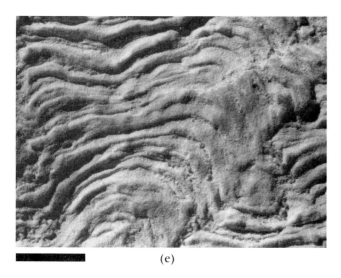

(e)

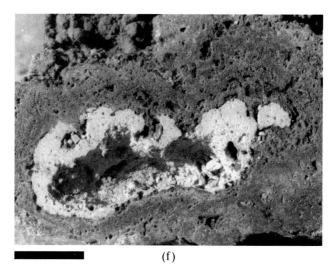

(f)

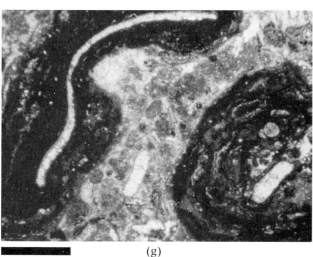

(g)

aragonite needles. The tubelike utricles commonly fill with cryptocrystalline aragonite, and the carbonate superstructure is typically micritized. Generation of aragonite within a blue-green algal cell is by the following reaction (Thompson and Ferris, 1990):

$$CH_2O \text{ (formaldehyde)} + O_2 + OH^- \rightarrow HCO_3^- + H_2O$$

then:

$$Ca^{2+} + HCO_3^{-1} + OH \rightarrow \underset{\text{(aragonite)}}{CaCO_3} + H_2O$$

Some algae are capable of boring into bioclastic debris and other forms of carbonate and phosphatic material (Figure 37.4d). These are **endolithic borings** that are instrumental in conversion of the internal structure of skeletal parts into structureless micrite.

Deposits of calcium carbonate in the form of stromatolites and some varieties of tufa and travertine, also known as algal reefs, algal bioherms, and algal mats, have a genetic link to algae (Walter, 1976). There are at least two types of genetic association: (1) uptake of carbon dioxide by algae causes precipitation of calcium carbonate (typically aragonite needles), and (2) mucilaginous algal filaments provide sticky surfaces to which calcium carbonate grains and crystals become attached and entrapped. **Stromatolites** are characterized by laminated structure and bulbous form (Figure 37.4e). They consist of both micrite and sparry calcite, the latter growing on nuclei trapped on the sticky material. The bulbous forms are an expression of enlargement of the stromatolite structure by accretion of particles where they are mechanically most stable, such as on low-slope surfaces and in cavities, and by erosion where attachments are less effective. **Oncoids** (Figure 37.4f) and some **pisoids** (Figure 37.4g) are the individual grain equivalents of stromatolites. Like stromatolites, accretionary layers of calcium carbonate material in the presence of algae are characteristic. A similar association is represented by phosphatic particles associated with filamentous microphytes (Soudry, 1987).

Sublacustrine reefal, domal, pinnacle, encrusting, and bedded deposits of calcium carbonate are thought to have a genetic link to both inorganic and organic processes (Scholl and Taft, 1964; Emeis et al., 1987; Straccia et al., 1990). Precipitation preferentially occurs at spring orifices where water nearly saturated in calcium carbonate issues into a lacustrine environment that may not be saturated. The presence of blue-green algae in such environments is not unexpected, especially if the water is warm. The presence of algae increases the chances of precipitation that otherwise may not have occurred.

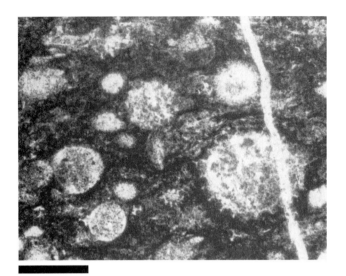

Figure 37.5
Section of radiolarian, diagenetically altered from opal-A to cherty quartz. Ferrugenous chert (red-brown) of Franciscan Formation. Near Muir Beach, California. PPL. Bar scale = 0.12 mm. (Courtesy G. LaFreniere.)

hot spring association (Figure 34.30) typically have brightly colored algal surfaces, although the extent of participation in precipitation of calcium carbonate is difficult to evaluate.

Kingdom Protista. **Radiolarians** were originally classified as animals, but they are now in the kingdom Protista along with foraminifera, diatoms, and coccolithophores. Radiolaria are spherical and bell-shaped tests that consist of a latticelike network of siliceous (opal-A) bars and spines (Figure 37.5). The organism is planktonic, ranging from Cambrian to Holocene, and surviving tests are a major contributor to siliceous oozes and the formation of chert, especially in the Paleozoic when calcareous pelagic organisms had not yet evolved.

Algae- and cyanobacteria-assisted precipitation may also occur at subaerial spring orifices: such is the case in the vicinity of sublacustrine tufa mounds and tufa pinnacles at Pyramid Lake, Winnemucca Lake, and Mono Lake in the Great Basin of the western United States (Chapter 39). Travertine deposits in the

Foraminifera are the main contributors to calcareous oozes in the Carboniferous and Permian. They were originally in the kingdom Animalia but now are in the kingdom Protista to allow for both plant and animal attributes. Forams were initially marine benthic organisms during the Paleozoic, and planktonic forms did not appear until the Mesozoic. **Fusulinids** are large forams with well-developed chambered skeletons and shapes similar to rice grains (Figure 37.6a,b).

Diatoms are both marine and lacustrine. Their siliceous (opal-A) frustules are bivalves that fit together like a pillbox. The **centric** variety (Figure 37.7a,b) is planktonic (pelagic), whereas the **pinnate**

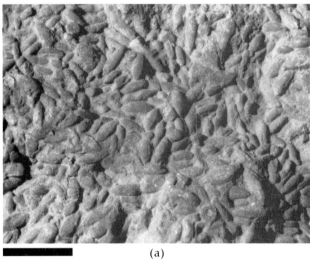
(a)

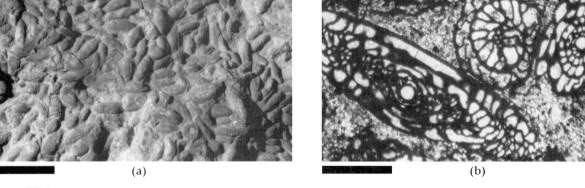
(b)

Figure 37.6
Fusulinid foraminifera.

(a) Weathered rock surface with fusulinids standing in relief. Bar scale = 1.5 cm.
(b) Longitudinal (lower left) and transverse (center, right center) sections of *Schwagerinid*-type fusulinid with skeletal architecture characterized by numerous chambers and plicated septa. Dark (porcellaneous) chamber walls contrast with white sparry calcite chamber fillings. Matrix contains quartz silt in micritic carbonate. PPL. Bar scale = 1 mm. Confusion Range, Utah. (Courtesy J. Kepper.)

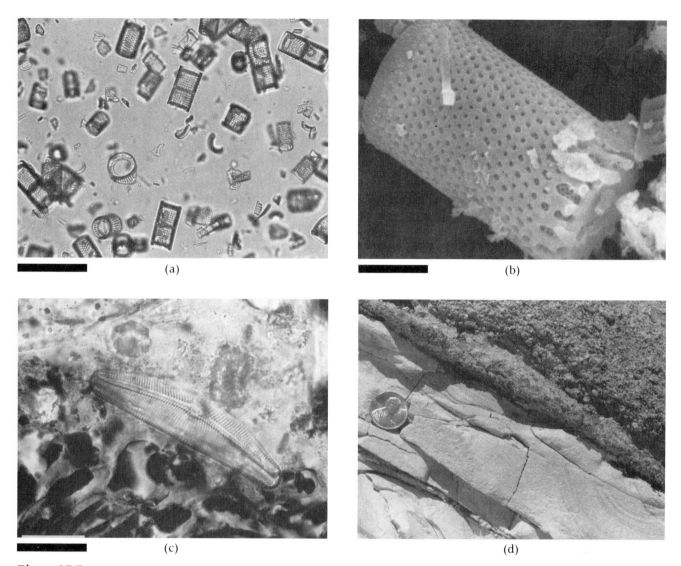

Figure 37.7
Diatoms of the Coal Valley ("Hunter Creek") Formation. Reno, Nevada.

(a) Centric diatom frustules dispersed in an oil emersion. PPL. Bar scale = 0.05 mm.
(b) SEM view of centric diatom frustule. Bar scale = 0.002 mm. (Courtesy J. Firby.) (c) Pinnate diatom frustule adjacent to pumice fragment (lower left) in volcanic litharenite. PPL. Bar scale = 0.02 mm. (d) Diatomite (note conchoidal fracture) overlain by volcaniclastic bed resulting from rapid deposition into Tertiary lake. Scale = U.S. penny (left).

(Figure 37.7c) type is mostly benthic. Diatoms have been a significant contributor to the formation of siliceous oozes since their first appearance in the Late Mesozoic. Dissolution of the opal-A frustules leads to the formation of chert. Lacustrine diatoms are favored by influx of volcanogenic glass (Figure 37.7c), which is much more soluble than silicate minerals, releasing silica for use in building diatom frustules. High concentrations of diatoms form the rock known as **diatomite** (Figure 37.7d), typically occurring in lacustrine environments where there is air fall and fluvial input of volcanic vitric clasts. Mixing with siliciclastic sediments generates diatomaceous arenites and diatomaceous wackestones.

Like algae, diatoms secrete sticky substances that trap sediments and provide nucleation sites for growth of seed crystals in environments that have the potential to precipitate calcium carbonate, such as travertine (Emeis et al., 1987).

Coccolithophores are the principal constituents of **chalk**, but they are also contributors to the formation of calcareous shale, calcareous mudstones, and marls.

Figure 37.8
SEM views of calcareous nannoplankton in chalk. (Courtesy A. Wetzel, 1989.)

(a) Disk-shaped coccoliths in nannoplankton hash. Bar scale = 0.004 mm.
(b) Asterolith (center), coccolith (upper left), and nannoplankton hash. Bar scale = 0.004 mm.

The calcareous tests are submicroscopic and shaped like shields (Figure 37.8a) and stars (Figure 37.8b). **Coccoliths** are mostly marine nannoplankton that have the ability to photosynthesize as well as to propel themselves with flagella. Coccoliths have also been contributors to the accumulations of calcareous oozes since the Early Jurassic when they first appeared.

Kingdom Animalia

Sponges (phylum Porifera) are a major contributor to siliceous oozes and therefore to the formation of chert. Along with radiolarians, the siliceous remains of sponges were the principal source of silica in the Paleozoic when their concentration was not diluted with planktonic foraminifera that had not yet evolved. The soft organic part of sponges has ostia (pores) (Figures 37.9a,b) that superficially resemble the zooecia of bryozoans. Sponge architecture includes the presence of either calcium carbonate or opaline silica **spiculites** that are either overlapping or fused into a crude endoskeleton. Sponge spicules are shaped like rods, anchors (Figure 37.9b), and tuning forks representing mechanical breakup of somewhat more elaborate forms. Siliceous spicules are more likely to be preserved than calcareous spicules, but the siliceous variety may be replaced by calcium carbonate, microquartz, or pyrite (Figure 37.9c). Availability of dissolved sulfide for the generation of pyrite may be the result of bacterial sulfate reduction (Coniglio, 1987).

Corals (phylum Cnidaria) have massive calcareous exoskeletons initially composed of fibrous calcite or aragonite oriented perpendicular to secreting surface of tissues and therefore at high angle to skeletal walls. They range in age from Ordovician to Holocene and occur mostly in marine environments as solitary organisms (Figure 37.10a) or as participants in colonies, where they constitute **boundstone** and **framestone**. Corals thrive in warm water, suggesting that calcium carbonate is more easily extracted from the aqueous environment by the coral organism when the solubility of calcium carbonate is at a minimum. Localized coral reef is a **bioherm**, and extension of bioherms forming a layer of coralline limestone is a **biostrome**. Sparite (Figure 37.10b) and micrite (Figure 37.10c) are common fillings of coral chambers.

Earthworms (phylum Annelida) are very important as a factor determining the textural attributes of carbonate rocks, since they are a major participant in **bioturbation**. Bioturbation is effective in destroying many primary sedimentary characteristics—for example, rendering laminated sediments to weakly laminated or massive morphology. Worms are soft-bodied organisms and are rarely preserved, but worm burrows and tracks may be well preserved. These features are trace fossils (also **ichnofossil**), since only the organism's activity, not the organism itself, is preserved. Borings of worms such as **scolithos** may be so closely spaced that the fecal sediment filling the tubes dominates the rock texture and structure (Figure 37.11).

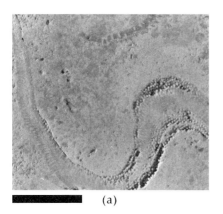 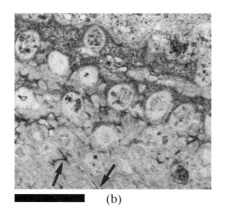 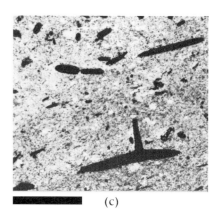

Figure 37.9
Sponges.

(a) Wall section of silicified siliceous sponge (*Actinocoelia*, Finks) (Finks, 1960) in nodular upper Kaibab limestone. Preserved pores (ostia) resemble bryozoa zooecia. Near Jacob Lake Junction, Arizona (Rigby, 1977). Slab surface. Bar scale = 2 cm. (b) Microscopic view of (a) showing relict ostia and more or less in situ anchor-shaped spiculites (arrows). Completely silicified. PPL. Bar scale = 0.05 mm. (c) Pyritized sponge spicules in silicified shale of Cow Head Group (Ordovician), Newfoundland. Bar scale = 0.4 mm. (Courtesy M. Coniglio, 1987.)

Trilobites (phylum Arthropoda) are very important fossils in determining the specific age of Lower Paleozoic marine strata. Hard parts typically have weakly laminated submicroscopic calcium carbonate structure. This material, or its diagenetically replaced equivalent, may be present in rocks, but the cast of these hard parts are just as diagnostic (Figure 37.12a). Silicification or dolomitization may result in differential solution, leaving fossilized trilobites in relief relative to matrix limestone (Figure 37.12b). Sections of skeletons commonly generate a diagnostic "shepherds hook" geometry (Figure 37.12c). Intraclasts composed of trilobite fragments encased in micritic calcium carbonate are shown in Figure 37.12c.

  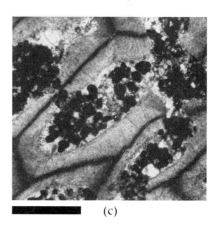

Figure 37.10
Corals.

(a) Silicified rugose corals exposed by artificial etching of limestone. Devonian limestone, Nevada. Bar scale = 2.5 cm. (b) Skeletal architecture of coral viewed on oblique section. Septa oriented NE–SW, trace of tabulae plates are NW–SE. Chambers were diagenetically filled with microspar. PPL. Bar scale = 1.8 mm. (c) Nearly transverse section across coral septa. Septa consist of amorphous material (black) faced with fibrous carbonate (aragonite?) oriented perpendicular to septa walls. Chambers contain pelletal calcium carbonate (black spheres) and microspar (white). Bisbee, Arizona. PPL. Bar scale = 0.8 mm.

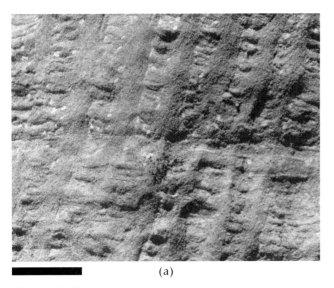

Figure 37.11
Earthworms.

(a) Longitudinal section of *Scolithos* burrows filled with fecal silty calcium carbonate. Burrows are perpendicular to lamellae in silty limestone, standing in relief on weathered surface as do silty laminae in host rock. Ely Limestone, White Pine Range, Nevada. Bar scale = 2 cm. (b) Transverse section of burrows shown in (a). Bar scale = 1.5 cm.

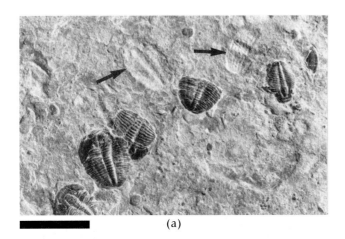
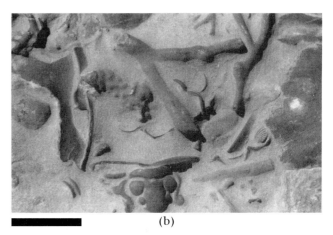
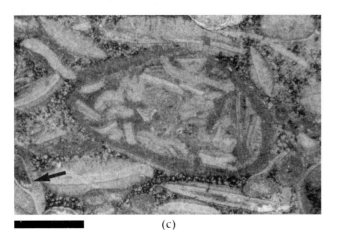

Figure 37.12
Trilobites.

(a) Trilobite skeletons and trilobite casts (arrows). *Elrathia kingii*. Wheeler shale (Middle Cambrian). House Range. Utah. Bar scale = 1.8 cm. (Courtesy J. Firby.)
(b) Trilobite *Glabella* (bottom center) with branching bryozoas (upper center) and brachiopods (right center). Devonian. Nevada. Bar scale = 1 cm. (Courtesy J. Firby.)
(c) Trilobite fragments occurring mostly in intraclasts that are in a muddy carbonate matrix. Note "shepherd's crook" (arrow). Large intraclast (center) contains many trilobite fragments. Deep Creek Range, Utah. Bar scale = 0.5 mm. (Courtesy J. Kepper.)

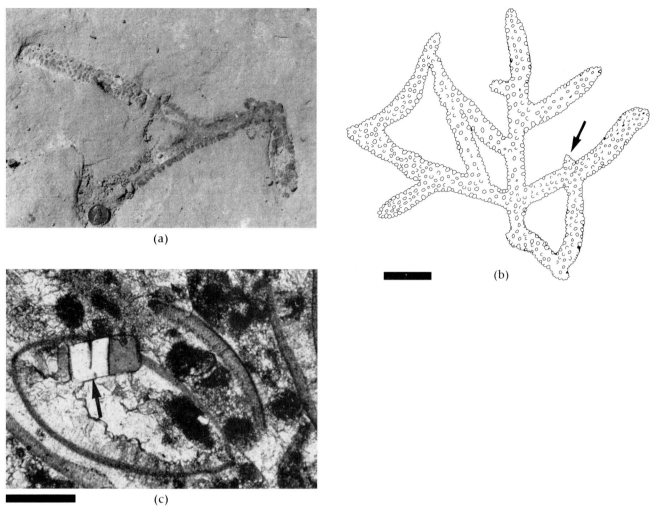

Figure 37.13
Crustaceans.

(a) Trace fossil burrows of shrimp *Orthomorpha*. Bifurcating burrows with knobby external surface occur in barrier island Upper Cretaceous Cliff House Sandstone (Donselaar, 1989). Chaco Canyon National Monument, New Mexico. Scale = U.S. penny (bottom). (b) Burrow complex drawn from a photo. Note uninterrupted intersection of two burrows (arrow) and constant burrow diameter, both features dissimilar to plant branches. Same location as (a). Bar scale = 4 cm. (c) Ostracod *Eukloedenella* in Lockport Limestone (Middle Silurian). Sherrell, New York. Note diagenetic twinned albite crystal (arrow) grown across thin-walled (micritized) bivalve. Microspar (white), micrite (speckled), and peloidal micrite (black) occur within bivalves and as matrix. X-polars. Bar scale = 0.12 mm. (Courtesy D. Zenger.)

Crustaceans (phylum Arthropoda) include crayfish, crabs, lobsters, shrimp, barnacles, and garden sowbugs. Trace fossil burrows of the shrimp **Ophiomorpha** (order Decapoda) are shown in Figure 37.13a,b (Kues, 1982). These burrows were originally misinterpreted as plant stem remains (see Chapter 8) because of their branchlike morphology. The nodular surfaces of the preserved burrows are probably the surfaces of well-packed fecal pellets, of which the modern analogue of this shrimp crustacean (*Callianassa major* Say) produces about 450 per day (Weimer and Hoyt, 1964).

Ostracods (order) are bivalved crustaceans, similar to those of pelecypods except for the diagnostically thin shell wall (Figure 37.13c). One valve tends to overlap the other, and the two valves commonly occur in their articulated position, unlike the valves of pelecypods and brachiopods. Ostracods are crustaceans (superclass) taxonomically grouped with trilobites, crabs, and lobsters. Shell microstructure consists of fine prismatic crystals perpendicular to shell surfaces.

Gastropods (phylum Mollusca) are both marine and freshwater. A Pleistocene lacustrine grainstone

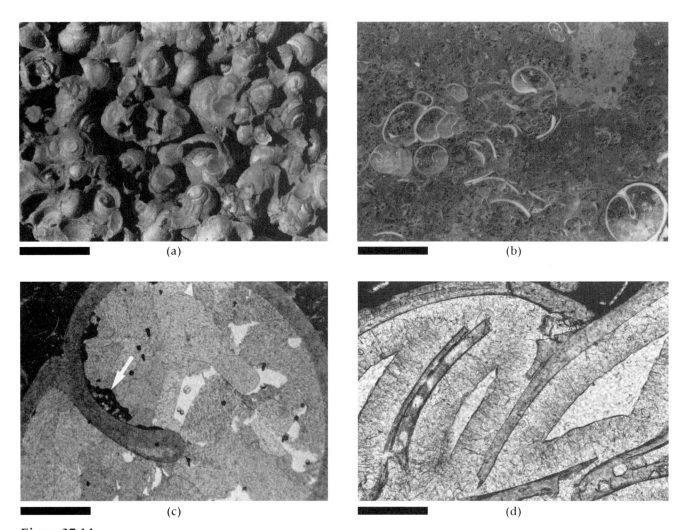

Figure 37.14
Gastropods.

(a) Planorbid freshwater Pleistocene to sub-Recent gastropod *Vorticifex* (*Carinifex*) *newberryi*. Grainstone beachrock from Winnemucca Lake, Nevada. Bar scale = 1 cm. (b) Transverse (lower right) and longitudinal (left) sections of freshwater conispiral gastropod *Viviparus*. Slab surface. Flagstaff Limestone (Paleocene), Wasatch Plateau, Utah. Bar scale = 1 cm. (c) Transverse section of gastropod from location (b) showing lack of microstructure in wall (dark), microsparite infilling (light gray), and peletal micritic mud (black) in chamber (arrow) and as matrix. White is plucked sparite. PPL. Bar scale = 1 mm. (d) Silicified freshwater gastropod consists of weakly laminated calcium carbonate shell walls that are fractured and displaced. Radial fibrous opaline silica is perpendicular to wall and wall fragments, terminating inward to clear cherty silica centers. Biomicrite. White Pine Range, Nevada. PPL. Bar scale = 0.05 mm.

composed of a planorbid (coiled in a plane) gastropod is shown in Figure 37.14a. Transverse and longitudinal sections through lacustrine coiled gastropods of the Flagstaff Limestone are shown in Figure 37.14b. Gastropods lack the septa (chamber partitions) so characteristic of cephalopods, and as a result filling of the shell tubes with sediment and diagenetic precipitation of microspar (Figure 37.14c) or silica minerals (Figure 37.14d) is common.

Cephalopods (phylum Mollusca) have a straight or planorbid shell divided into chambers by septa that appear as sutures on the shell surface. **Belemnoids** and some varieties of **Nautiloids** (Figure 37.15a) and **Ammonites** are straight shelled. A coiled ammonite with sutures characterized by a few lobes and saddles (agoniatitic) is shown in Figure 37.15b. Chamber walls are initially aragonite but easily recrystallize to microspar. Unlike coiled gastropods, some coiled ceph-

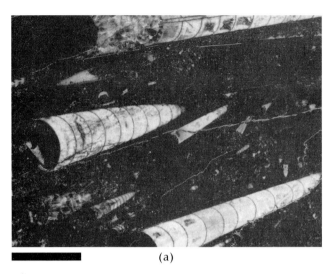
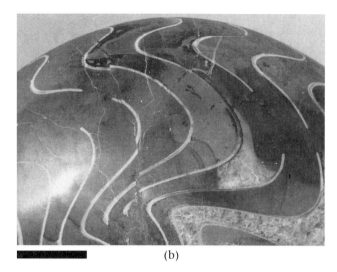

Figure 37.15
Cephalopods, Erfoud, Morocco. (Courtesy M. Owens, Comstock Rock Shop.)
(a) Straight-shelled (orthoconic) Silurian nautiloid with orthoceratitic (straight) septa. Bar scale = 1.7 cm. (b) Coiled Devonian ammonite (*Goniatites*) with simple fluted goniatitic sutures (septa) characterized by pointed lobes and rounded saddles. Discontinuous sutures probably due to local fragmentation of septa prior to or during chamber filling. Bar scale = 1.7 cm.

alopods are involute in that the last spiral covers all previous ones. Cephalopods are exclusively marine.

Pelecypods (phylum Mollusca) are bivalved organisms (Figure 37.16a) whose valves are mirror images of each other, unlike brachiopods. They are characterized by cross-laminar shell structure (Figure 37.16b) that is initially fibrous aragonitic but commonly is diagenetically recrystallized to microspar (Figure 37.16c). Pelecypods are both marine and freshwater, being important contributors to post-Paleozoic limestones.

Brachiopods (phylum Brachiopoda) have valves of unequal shape. The valves of inarticulate types are held together without hingement, whereas articulate brachiopods have teeth and sockets for articulation. Cross-lamellar shell microstructure is common, and since this characteristic is very similar to skeletons of

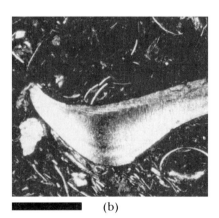
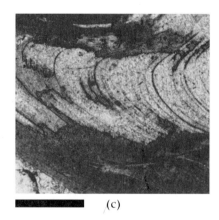

Figure 37.16
Pelecypods.

(a) Marine pelecypods in sandy siliclastic matrix. Merced Formation (Pliocene), San Francisco Bay, California. Bar scale = 4 cm. (b) Portion of pelecypod valve with primary cross-lamellar fibrous microstructure. Note extinction region. X-polars. Bar scale = 0.8 mm. Freshwater biomicrite. White Pine Range, Nevada. (c) Primary laminated carbonate (aragonite) skeletal microstructure (dark) selectively replaced by microspar calcite (white). Coquina. St. Augustine, Florida. Bar scale = 0.2 mm.

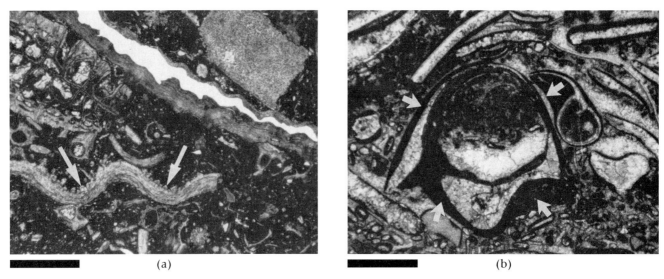

Figure 37.17
Brachiopods.

(a) Laminar calcium carbonate microstructure and ornamented surface (arrows) in "corrugated" brachiopod valve (lower left). Separation parallel to laminar structure has left void (white) in larger-value section. Bryozoa (upper left) and echinoderm plate (upper right). Biomicrite. PPL. Bar scale = 0.5 mm. (b) Section through skeletal architecture of productid brachiopod (within arrows). Wall structure appears amorphous (black). Interior of bivalve contains micritic carbonate mud (mottled dark), sparry calcite (light). PPL. Bar scale = 1 mm.

Figure 37.18
Bryozoa.

(a) Silicified branching bryozoa standing in relief on weathered limestone surface. Zooecia may have been filled with sparry calcite prior to weathering. Denay Formation (Devonian). Roberts Mountain, Nevada. Bar scale = 2 cm. (b) Section through bryozoa that has fibrous calcium carbonate walls and zooecia that have been diagenetically filled with micritic mud (black) followed by microsparite (white). Biomicrite. PPL. Bar scale = 0.12 mm.

pelecypods and bryozoa, fragmented debris is difficult to identify. Inarticulate brachiopods may be phosphatic, but others are calcareous, as are all the articulate types. Shell ornamentations may be useful in identification (Figure 37.17a). If the bivalves are still joined, interior fillings of sediment are common, as well as diagenetic microspar (Figure 37.17b). Brachiopods are marine only and were most abundant in the Paleozoic.

Bryozoans (phylum Bryozoa) are marine, calcareous colonial animals that form encrustations, branches (Figures 37.12b, 37.18a), or fenestrates (screens) (Figures 37.17a, 37.18b). The organic part of the animal resided in the **zooecia** that are boxlike tube pore cells. Sediment and sparite fillings of zooecia are typical (Figures 37.17a, 18b). Wall microstructure is fibrous and laminated (Figure 37.18b), very similar to brachiopods.

Echinoderms (phylum Echinodermata) occur in the fossil record as body plates and stem plates. Body plates are characteristically perforated unless replaced

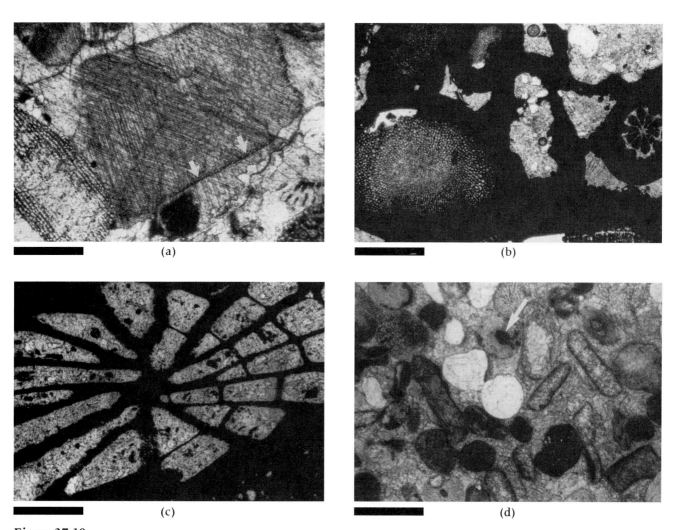

Figure 37.19
Echinoids.

(a) Porous echinodermal plate (left) adjacent to plate largely replaced with sparry calcite crystal (two twin sets) that extends beyond limits of plate. Vestige of plate porosity remains along margin of plate (arrows). PPL. Bar scale = 0.5 mm.
(b) Porous echinodermal plate (left) and echinoid spine (right) partly filled and replaced by hematite. Clinton Formation. New York State. PPL. Bar scale = 0.5 mm. (c) Echinoid spine structure replaced by hematite. Cells contain opal. PPL. Bar scale = 0.12 mm. Clinton Formation. New York State. (d) Echinoid spine (white arrow) and plates (upper left), well-rounded quartz grains (center), microspar intraclasts with micritic margins (center right), and pelletal micrite (very dark spheres) in sparry calcite matrix. Calcareous litharenite at interbedded zone between Diamond Peak and Ely Limestone Formations. White Pine Range, Nevada. PPL. Bar scale = 0.5 mm.

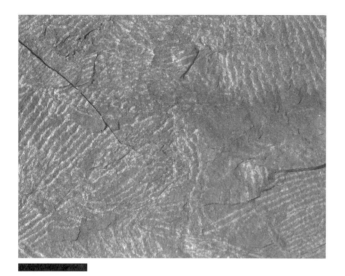

Figure 37.20
Carbonaceous films of bushlike dendroid *Dictyonema flabelliforme* graptolite. Tremadocian (Ordovician), Oslo, Norway. Bar scale = 1 cm.

by microsparite (Figure 37.19a). Stem sections have a radial skeletal architecture (Figures 37.19b,c) that may have a central hole (Figure 37.19d). Echinoderms are exclusively marine.

Graptolites (class Graptolithina, included in phylum Hemichordata) are marine invertebrate animals that range from Cambrian to Pennsylvanian. They typically occur in black shales, where they appear as relatively reflectant carbonaceous imprints on bedding surfaces. Most graptolites are planktic, resembling small hacksaw blades, several of which may be attached at a common point. Some graptolites are benthic, where they existed as complexly branched colonies that have bushlike aspects (dendroids) (Figure 37.20).

Conodonts (phylum Conodonta) are toothlike phosphatic microfossils (Figure 37.21) that represent parts of conodont "animals" whose composite characteristics are a matter of some conjecture. The hard-part microfossils are composed of laminated (Figure 37.21a) carbonate–fluorapatite [$Ca_{10}CO_3(PO_4)_6$] (francolite).

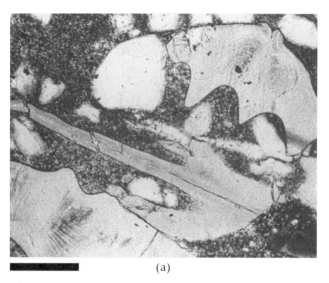
(a)

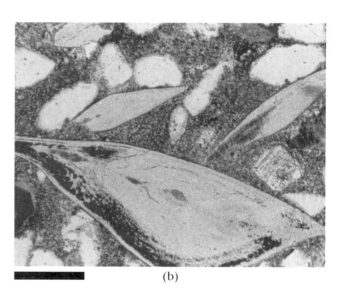
(b)

Figure 37.21
Early Mississippian (Kinderhookian, *Siphonodella sandbergi* Zone) conodont *Siphonodella* sp. Basal phosphatic lag sandstone of upper tongue of Cottenwood Canyon Member of Madison Limestone at Baker Mountain, Beartooth Mountains, Montana (Sandberg and Klapper, 1967). (Courtesy C. Sandberg.)

(a) Three specimens: (1) longitudinal section of one-quarter (anterior part) of large specimen (upper right), showing well-developed growth lines (carbonate fluorapatite); upper (oral) surface at bottom; (2) diagonal lower (aboral) view (lower left) of anterior end of specimen showing basal groove along part of platform at left and entire free blade at right; (3) latitudinal cross section (center) showing complete growth lines, carina flanked by two rostral ridges on upper (top) surface, and suture line extending downward into basal groove on lower surface. PPL. Bar scale = 0.12 mm. (b) One large and two small specimens in lower (aboral) view. Dark, mineralized (organic-carbon infiltrated) areas on large-specimen highlight growth lines, anterior end at left. Narrow line on upper-right specimen is basal groove of pseudokeel. PPL. Bar scale = 0.12 mm.

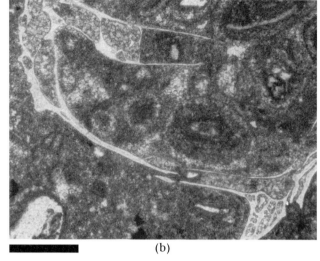

Figure 37.22
Fish (Vertebrate).

(a) *Goshvicthys (Parvis) knightia.* Eocene Green River Formation. Kremmer, Wyoming. Bar scale = 2.5 cm.
(b) Fish bone, possibly part of gillarch pharyngeal (T. Lugaski, personal communication) in Middle Pliocene lacustrine limestone. Hot Creek Range, Nevada. PPL. Bar scale = 0.5 mm. (c) Cycloid fish scale. Mowrey Shale (Cretaceous). Vernal, Utah. Bar scale = 5 mm.

Fish (subphylum Vertebrata) occur as skeletal imprints on bedding planes (Figure 37.22a), as bone fragments (Figure 37.22b), and as teeth and scales (Figure 37.22c). Hard parts are phosphatic and as such may be difficult to distinguish from fragments of inarticulate brachiopods and conodonts. Extraction of phosphate from upwelling marine waters by fish, inarticulate brachiopods, and conodonts leads to the postmortem accumulation of phosphatic debris forming **phosphorite**.

Submicroscopic Biogenic Material

Very fine-grained and amorphous-appearing materials important in formation of rocks closely associated with biotic activity are chiefly calcareous, siliceous, and phosphatic. Accumulation of calcareous planktonic organisms such as foraminifera and coccoliths are major constituents of **calcareous ooze**. Siliceous ooze consists of sponge spicules and radiolaria plankton, along with diatoms in younger accumulations. **Phosphatic oozes** contain fish bone, brachiopod, and conodont fragments.

Micrite (Figure 37.23a,b) is the very fine-grained form of calcium carbonate occurring as matrix (Figures 37.14c, 37.15a, 37.17a) and in a variety of clasts (Figure 37.23b). Specifically, micrite can occur as (1) calcareous replacements of submicroscopic, microscopic, and mesoscopic skeletons (Figures 37.16b, 37.17b), (2) inorganic precipitates, particularly as cements (difficult to verify), (3) **fecal pellets** and tube fillings, (4) **pelloids** of other derivation (Figures 37.10c, 37.13, 37.14c, 37.19d), (5) **intraclasts** (Figure 37.23b), (6) crystals produced within tissues of calcareous algae (Figure 37.4c,d), and (7) very fine clastic

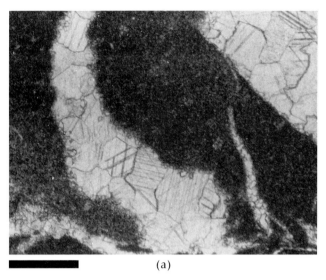
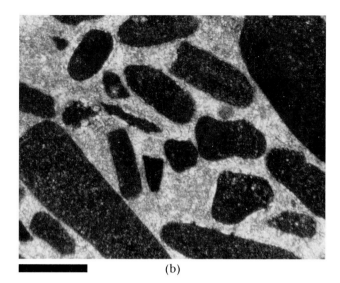

Figure 37.23
Micritic limestones.

(a) Micrite locally recrystallized to sparite. PPL Bar scale = 0.5 mm.
(b) Intramicrite with rounded micrite intraclasts (dark) in micrite matrix (light). PPL. Bar scale = 0.5 mm.

---

fragments deposited below wave base. Micrite appears dark in thin section due to its semi-opaque behavior in transmitted light. Since micrite is very fine grained, it is in a high-energy state and is particularly susceptible to recrystallization to sparry calcite (Figures 37.16c, 37.23a), with or without dolomitization. There is some indication that modern aragonite muds can convert directly to microspar without going through a micrite stage (Lasemi and Sandberg, 1984).

Phosphatic ooids, pisoids (Figure 37.24), and peloids (**phosphopeloids**) are less common than calcareous ooids and peloids but evolve through the same sort of processes (Baturin, 1982; Slansky, 1986). Phosphatic nodules are larger and form primarily during weathering processes and, thus, during the formation of phoscrete (Chapter 40).

Recycled Biogenic Materials

Biogenic material of any size, ranging from micrite to large shell fragments, may be released from rock and redeposited to become a constituent of new rock. Recycling of quartz sand grains is particularly common since quartz is chemically stable. Recycling of calcareous biogenic materials, although chemically less resistant, occurs if transport and sedimentation are not prolonged.

Calcareous **allochems** include intraclasts, ooids, pelloids, and skeletal fragments. Equivalent phosphorite allochems (Figure 37.24) also occur. Any of these particles may be of primary origin or recycled from pre-existing rock. Recycled phosphorites are well known (Soudry, 1987; Kidder and Swett, 1989; Abed and Al-Agha, 1989).

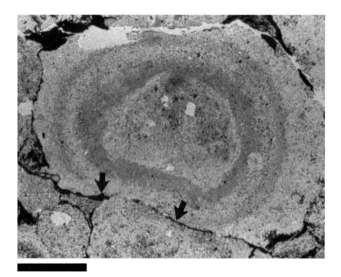

Figure 37.24
Pisolitic phosphorite. Note pressure solution surface (arrows) truncating growth zones in large pisoid. Bar scale = 0.12 mm.

Nonbiogenic Material in Biosedimentary Rocks

Primary Precipitates

Environments favoring organically induced precipitation of calcium carbonate, opal-A, apatite, or hematite can also be favorable for inorganic precipitation of these same materials. Nucleation and growth of minerals or formation of gels can be totally dependent on chemical factors, particularly supersaturation. Nevertheless, there can be a passive role played by organic and skeletal remains that promotes inorganic precipitation. For example, shell fragments provide a substrate for nucleation and growth of crystals that has nothing to do with the earlier biological process (Maliva and Siever, 1988). Additionally, in the case of stromatolites, bacteria and algae excrete sticky mucus that entraps mineral and skeletal particles that act as nuclei for crystallization (Emeis et al., 1987). Biotically induced precipitation on diatoms is described by Chafetz et al. (1991), and bacterially induced precipitation of carbonate minerals is discussed by Buczynski and Chafetz (1991). Nucleation sites for silica precipitation can be provided by woody organic matter, leading to the formation of silicified wood (Maliva and Siever, 1988).

Percolation of fluids through porous sediment accumulations after the period of initial deposition and compaction can lead to precipitations that are early diagenetic. The composition of these percolating solutions may be similar or quite different from the aqueous bodies in which sedimentation took place.

One of the best examples of inorganic precipitation in an environment also conducive to organic precipitation is the formation of **calcareous ooids**. Ooids are spheres less than 2 mm in diameter, consisting of concentric laminae of carbonate (Figure 37.25a). Modern ooids have radial fibrous aragonite in the layers, whereas in older ones this structure is diagenetically replaced by sparite or micrite, destroying the radial fibrous habit. There may be a quartz grain or shell fragment, or other such nucleus, serving as a substrate for nucleation and growth (Figure 37.25b). Silicification of calcareous ooids may preserve some of the concentric structures but not the radial fibrous texture (Figures 37.25c,d). Several reports describe primary precipitation of **phosphatic ooids** (Swett and Crowder, 1982) and pelletal–grain coating francolite (Abed and Al-Agha, 1989). Primary precipitation of dolomite spherulites (Gunatilaka et al., 1987) and dolomite cement (Lasemi et al., 1988) and conversion of limestones to dolostone by dolomitization (diagenetic) are also inorganic processes.

Silica minerals may form directly from precipitation in aqueous solutions. Silica in these solutions may have ultimately derived from dissolution of siliceous organisms, but it may also derive from the illitization of smectites for which there is no biotic link. An **authigenic** quartz crystal grown by replacement of micrite (Molenaar and Jong, 1987) is shown in Figure 37.26a, and an authigenic quartz overgrowth on a clastic quartz grain is shown in Figure 37.26b. Crystallization postdating the primary depositional stage is clearly shown by the authigenic albite crystal that has replaced part of an ostracod valve as well as a matrix (Figure 37.13).

An indirect clue that cherts can be the result of direct precipitation of silica, probably siliceous gels, is the observation that the silica of Precambrian cherts cannot have been supplied by sponges, first appearing in the Cambrian, nor radiolarians, first appearing in the Ordovician, nor diatoms that did not show up before the Jurassic. Of course, it is possible that there were siliceous organisms in the Precambrian for which diagenetic processes have left no trace. In fact, many younger cherts that have had a substantial history of diagenesis do not contain sponge spicules, radiolarians, or diatoms.

Siliciclastic Particles

The most obvious nonorganic materials in biochemical rocks are siliciclastic sediments. Pelagic clays dispersed at the distal end of deltas come to rest with calcareous and biosiliceous oozes. In anoxic aqueous environments, organically derived carbon occurs in shales of siliciclastic derivation. Some benthic organisms such as pelecypods thrive in siliciclastic silts and sands (Figure 37.16), although there can be clastic reworking of fossils into sediments not representing the primary depositional environment of the organisms. Transported skeletal debris, or reworked rock containing fossils, can mix with siliciclastic sediments in any ratio since the organic phase of development has terminated and the life-choking effect of clay-sized particles is not a limiting factor.

Diagenesis

Any process that modifies the physical and chemical attributes of sedimentary materials after initial deposition or crystallization is **diagenetic**. Early diagenetic processes may not be distinct from syngenetic processes (bioturbation, immediate cementation), but with burial the superposed nature of diagenesis is more apparent. Biomineralizing organisms (Lowenstam, 1989; Carter, 1991), such as conodonts that

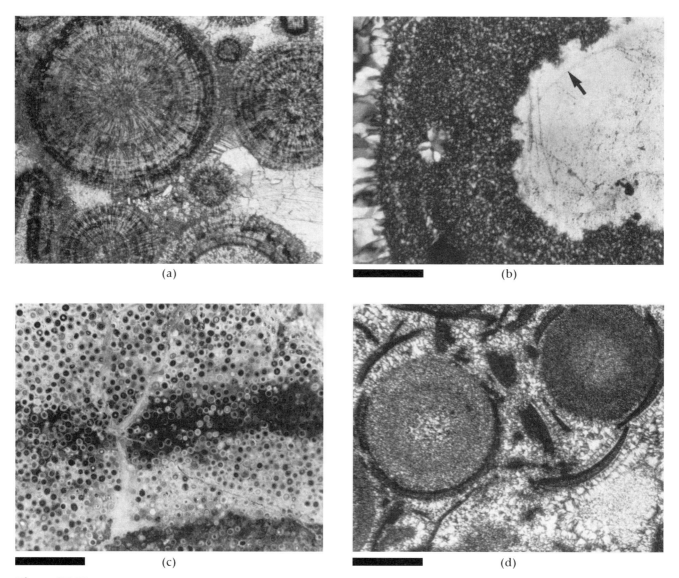

Figure 37.25
Silicified calcareous oolites.

(a) Calcareous ooids with concentric layers and radial fibrous microstructure. Sparry calcite matrix. Diameter of larger ooid (upper left) is 1.5 mm. Bear River, Indiana. X-polars (b) Well-rounded probably eolian quartz clast marked by impurity surface (arrow) was probably overgrown by additional quartz (epitaxial) in redeposited environment. Second reworking in high-energy oolite-forming environment resulted in partial dissolution of quartz and precipitation of ooidal calcium carbonate. Diagenetic cherty silicification of carbonate portion of ooid culminated with a coarser-grained chalcedonic quartz matrix (left). X-polars. Bar scale = 0.12 mm. (c) Completely silicified oolitic limestone. Slab surface. Bar scale = 1 cm. (d) Local detachment of marginal corticals of calcareous ooids occurred prior to silicification. Microcrystalline quartz (chert) interiors of ooids without preservation of primary microstructure. Chertified rim corticals (dark) with matrix of coarser chalcedonic quartz (light portions). X-polars. Bar scale = 0.5 mm.

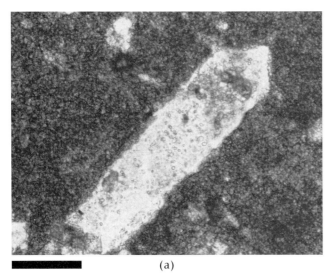
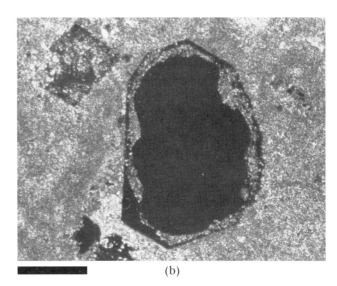

Figure 37.26
Silica diagenesis in micrite.

(a) Micrite containing a well-formed authigenic quartz crystal. Note remnants of micrite in crystal. PPL. Bar scale = 0.12 mm. Wah Wah Range, Utah. (Courtesy J. Kepper.) (b) Micrite with quartz clast modified by carbonate replacement and overgrown with faceted authigenic quartz. Authigenic dolomite (?) rhomb (upper left). X-polars. Bar scale = 0.12 mm. Biomicrite from Diamond Peak–Ely Limestone contact zone. White Pine Range, Nevada.

---

generate carbonate fluorapatite, are a source of chemicals that participate in diagenesis of the sediments in which the remains of the organism locate. In some cases, chemical migration leads to diagenesis of interbedded sediments that do not contain the organism.

Diagenetic changes include (1) reduction in porosity by infilling, (2) formation of secondary porosity by mixing in aqueous fluids undersaturated in a component of the rock, (3) reactive replacements, generating authigenic quartz, feldspar, and dolomite, and (4) obliteration of biota remains.

Diagenesis of Calcareous Rocks

Silicification. Silicification of limestones occurs as submicrocrystalline and microcrystalline: (1) primary pore-filling precipitates, (2) filling of pores formed by dissolution, (3) replacements of micritic and sparry carbonate matrix, and (4) replacements of calcareous skeletons (Maliva and Siever, 1988), calcareous intraclasts, and oolites (Figure 37.25). From a purely chemical point of view, silica can be expected to replace carbonate if the pH is less than 7–8. The most obvious sources of silica are radiolarians, diatoms, and siliceous sponges. Bedded cherts (Coniglio, 1987) and nodular cherts (Maliva and Siever, 1989) occurring in limestones and calcareous mud rock have formed by a combination of these processes.

Silicification of calcareous rocks may involve no more than a few crystals of authigenic quartz (Figure 37.26a) in micrite, and as much as complete replacement of the carbonate rock with chalcedony or quartz. Complete diagenetic conversion of calcareous ooids to microcrystalline quartz with retention of primary texture are shown in Figure 37.25c,d. Calcareous ooid cortices may be broken prior to silicification as a result of compaction (Figure 37.25d).

Calcification. Bioclastic grainstones are commonly cemented with calcium carbonate (Figures 37.14a, 40.12). Early diagenetic replacement of aragonitic material and micrite by microspar is common (Figure 37.16c) and there can be solution of calcium carbonate followed by infilling of another form of carbonate. Reactive replacement may preserve much of the primary architecture of organisms (Evans and Ginsburg, 1987).

Dolomitization. Stratigraphic sections of carbonates commonly include beds of dolomite (dolostone) as well as limestones. Two very significant observations have set the origin of dolomites apart from limestones. (1) Although some skeletal fauna consist of dolomite, examples of fauna that generate dolomite skeletons directly (dolomite biomineralization), are nearly nonexistent and a direct link of organic activity to the formation of dolomites is essentially absent. (2) Some bedded dolomites have portions that are limestone. Contact surfaces between dolomite and this limestone at some angle to the bedding structures indicate post-depositional conversion of limestone to dolostone. Both of these observations strongly point to the secondary nature of dolomite in dolostones and faunal remains (Zenger, 1972; Hardie, 1987). The reaction can be shown as follows:

$$2CaCO_3 + Mg^{2+} \rightarrow CaMg(CO_3)_2 + Ca^{2+}$$
$$\text{(calcite)} \qquad\qquad \text{(dolomite)}$$

Dolomite produced in this reaction has the stoichiometric ratio of Mg/Ca = 1, which forms a highly ordered crystal structure. Such an ordered dolomite has not been produced in laboratory experiments at low temperatures equivalent to depositional environments. Stoichiometric dolomite is easily synthesized above 100°C. Evidently the higher-energy environment allows ordered ionic positioning. Conversion of calcite crystals in calcareous muds and limestones to dolomite must be at low temperatures if it occurs at or near the surface during early diagenesis. Disordered calcian dolomite is to be expected in this case. With deep burial (Gawthorpe, 1987) and/or introduction of hydrothermal fluids, dolomitization is at a higher temperature, there is more time for possible reorganization of originally disordered structures, and stoichiometric dolomite is formed.

In the low-temperature depositional environment, where the kinetic barrier to ordering applies, a disordered dolomite ("protodolomite") may form containing up to 56 mole % calcium and 44 mole % calcium. Primary Holocene calcian dolomite (weakly ordered) spherulites have molded around quartz sand grain tidal bar deposits (Gunatilaka et al., 1987), and primary calcian dolomite cement, as indicated by void and endolithic boring fillings, has formed in supratidal carbonates in the Bahamas (Lasemi et al., 1988).

The process of secondary **dolomitization** is controlled by physical access (Wilson et al., 1990) of magnesium-bearing solutions into limestone and by the reactivity of the various forms of calcium carbonate in the limestone. Brines generated in salinas and sabkhas may become enriched in magnesium, mix with normal meteoric phreatic water, and infiltrate limestones in supratidal environments. Dolomites formed in this way characterize back-reef and near-shore locations rather than open marine settings. It is now thought that normal seawater, slightly modified by sulfate reduction, organic matter oxidation or evaporation, and perhaps heated to as much as 200°C (Wilson et al., 1990), can dolomitize limestones. Additionally, the source of magnesium in diagenetic generation of dolomite may be from interbedded mud rocks after deep burial (Gawthorpe, 1987).

Textural expression of dolomitization is impressive, especially if samples are stained with reagents that distinguish calcium carbonate and dolomite. Clear dolomite rhombs have formed at the expense of micritic calcium carbonate (stained red) (Plate 8b). Selective dolomitization in an oosparite is shown in Plate 8c. Contrast between a thin rim of dolomite around the sparry calcite of an ooid (perhaps resulting from replacement of a micritic rim) and an adjacent ooid that is completely replaced by dolomite and adjacent ooids that have no dolomite is an indication of the delicate relation between original aragonite, secondary sparite, and dolomitization. These are cases of dolomite forming by reactive replacement, not by filling in primary or secondary voids, nor by diffusional cation exchange through crystal lattices. In general, reactive replacement is indicated by the juxtaposed position of dolomite crystals relative to original aragonite or calcite crystals; that is, new crystals form simultaneously with dissolution of old crystals. Lattice exchanges are known to be very slow at low temperature and would generate pseudomorphs, not crosscutting crystals. Reactive replacement is also indicated by incorporation of inclusions that had characteristic distribution in, for example, the cortices (layers) of calcareous ooids and in stylolites. Reactivity may be controlled by (1) mineral type (aragonite is more reactive than calcite), (2) grain size (micrite is more reactive than sparite), and (3) grain shape and orientation (concentric layers may react differently than radial fibrous, radial columnar, or random crystal orientations). Faunal skeletons may be selectively dolomitized. The calcite wall structures of brachiopods, trilobites, and oysters are less susceptible to diagenetic dolomitization than the aragonitic wall structures of pelecypods and gastropods.

Dolomitization can be accompanied by an episode of **dedolomitization**. Limestones may be generated from dolomites with sucrosic texture (Theriault and Hutcheon, 1987). In an oxidizing meteoric water setting, iron-bearing dolomite can be converted to iron oxides and rhombs of calcite. Magnesium leaving the system may participate in dolomitization nearby where chemical conditions are different.

Ferruginization. Some ironstones are generated by diagenetic reactions leading to the replacement of calcareous rocks with iron silicates and iron oxides. A convincing model of oolitic limestone diagenetic replacement has been presented by many investigators, including Kimberley (1979) and Dimroth (1976), applying to the Clinton-type iron formation of the Appalachian region. Groundwater containing organic-decay materials is strongly reducing. Leaching of iron by such a fluid from organic-bearing muds interbedded with oolitic carbonate is thought to lead to replacement of ooids by a variety of minerals, including siderite, goethite, pyrite, chamosite [$3(Fe,Mg)O \cdot (Al,Fe)_2O_3 \cdot SiO_2 \cdot nH_2O$], and hematite, forming the oolitic ironstones. Organic limestones contain sulfur that can combine with iron leached from interbedded mud rocks to form pyrite, occurring as pyritized sponge spicules (Coniglio, 1987) and pyritized algal mats (Schieber, 1986).

Phosphatization. Dissolution of fish bones and conodont fragments associated with carbonate sedimentation are a source of phosphate (francolite) that may participate in diagenesis of the carbonates (Bentor, 1980; Baturin, 1982).

Diagenesis of Siliceous Rocks

Bedded and nodular cherts typically evolve by a maturation-like diagenesis process starting with biogenic opal-A and changing to an opal-CT porcellanite and finally to cherty quartz (Williams and Crerar, 1985; Maliva and Siever, 1988). Radiolarians, diatoms, and sponge spicules generate opal-A that collects as postmortem siliceous ooze. Opal-CT lepispheres in the porcellanites are spherulitic aggregates of blade-shaped opal-CT crystals (Figure 40.11).

Biogenic siliceous debris also occurs in siliciclastic muds, particularly as basin accumulations of pelagic sediments. Diagenetic migration of silica from protoshale to protochert is one way that bedded cherts may form. Such a mechanism avoids the need for a depositional mechanism by which mud rocks of normal siliceous content repeatedly alternate with beds of chert (Murray et al., 1992).

Diagenesis of Carbonaceous Rocks (Coalification and Kerogenization)

Coalification is at first a process of dehydration and later a process of devolatilization releasing carbon–oxygen compounds. Classification of coals is by **rank**, based on a scalar property observed as optical **reflectance** (Bostik, 1979). Low-rank coals, such as soft brown, hard brown, and subbituminous coals, have low reflectance. The high-rank **bituminous** and **anthracite** coals have moderate to high reflectance. **Vitrinite** (Figure 37.3d) is moderately high reflecting, whereas **intertinite** has higher reflectance. The variety of intertinite, **fusinite**, is shown in Figure 37.3e.

Oil shale contains **kerogens**, which consist of fossilized insoluble organic matter formed by maceration of organic debris generating a sludge known as **sapropel**. Algae, pollen and spores, and a wide variety of plant remains are the primary constituents. Kerogens can be converted to petroleum by distillation, making these shales a potential source of petroleum. The shales may be carbonate or silica rich.

Rock Type and Depositional Environment

Thin sections and hand specimens of biochemical rocks cannot be used exclusively to reconstruct the depositional environment. However, they are a clue to the nature of deposition and, in the context of stratigraphic relations, are essential to a complete analysis of a depositional terrain.

Faunal organisms depend on phytoplankton for food, and since phytoplanktons require light, most fauna live within 10 m or so of the surface of water bodies. The subaerial environment does not supply much biotic material to rock formation because of highly oxidizing conditions that drastically reduce the amount of material that can be retained. A few bones here and there and organics generated in the A horizon of soils is about all that can be contributed. Except for lacustrine and marine salinas and similar evaporite environments, shallow freshwater or ocean water is the place where biota flourish and is the location of sedimentation leading to the formation of biochemical sedimentary rocks.

Lacustrine Environment

The lacustrine environment includes the "open" region of the depositional basin and the "marginal" subenvironments including beaches and associated eolian deposits, swamps, bogs, and deltas (Figure 37.27). Marl lakes are characterized by the presence of calcium carbonates and the association of biota and hydrocarbons. Lake Flagstaff and Lake Unita (Paleocene) in Utah are good examples (Figures 37.14b, 37.28).

Some lakes are supportive of life that generates biosiliceous rocks. Lakes receiving particles of volcanic glass can maintain a thriving community of freshwater diatoms because the glass is a form of silica that is unstable and yields free silica for diatom growth.

Marine Environment

The marine environment consists of the **marginal marine** and **open marine** zones (Figure 37.27). Marginal marine includes the beach and barrier bar, supratidal eolian, tidal flat, lagoon and estuarine, and deltaic subenvironments.

The beach, barrier bar, and supratidal eolian subenvironments are high energy. In a relatively siliciclastic-free, calcium-carbonate-dominating environment, carbonate sands are generated where surf and wind interact with carbonate clasts and carbonate rock. This is where **grainstones** are generated, including beachrock, **bioclastites**, **oolitic limestones**, **coquinas**, **calcareous eolianites**, and **intraclastites** with

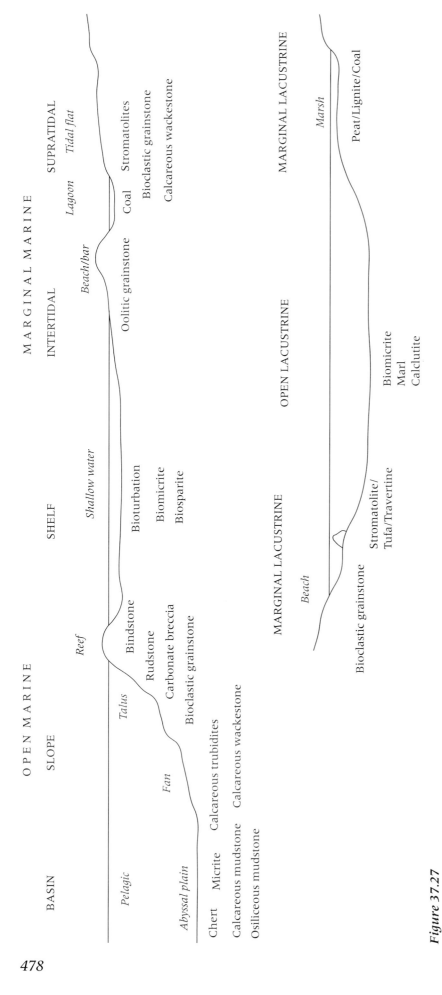

Figure 37.27
Diagram showing generalized location of biochemical rock types with respect to lacustrine and marine environments.

Figure 37.28
Lacustrine marly limestone with thin siliciclastic interbeds. Claron Formation. Bryce Canyon, Utah.

variable amounts of admixed siliciclastics. Fine sediments are winnowed out in this high-energy setting. Cements may only partially fill pores and are typically **meniscus**, meaning that cementation occurs locally by capillarity along grain contacts (Figure 40.12), or **pendulous** cement that occurs on the under surface of grains where solutions attach in response to gravity.

The tidal-flat subenvironment, as the name indicates, is coastal marine where relief is very low. There are both siliciclastic and carbonate tidal flats, as well as transitions from the sabkha environment of evaporite systems. Tidal flats are integrated systems including the subtidal, intertidal, and supratidal zones as well as tidal channels. The energy of this environment, except for times of storms, is much less than that of the beach environment. Even during times of storms, the energy of waves may be dissipated across the relatively extensive surface of the flats, in contrast to the crashing of waves against a sea cliff along a coastal mountain range. The sediments produced in the carbonate tidal-flat environment reflect this lower-energy environment, including laminated carbonate muds and a large variety of **calcareous packstones** and **calcareous wackestones**. Shrinkage cracks form in the intertidal and supratidal zones. Ripples and cross-laminations, and even graded beds, occur in the subtidal and intertidal zones. Many of these features may not be preserved in the rocks that eventually form from tidal-flat sediments. Storm activity, bioturbation, and fluvial-like tidal channel activity are likely to be equally as important in reshaping sediment deposits. Meandering channels erode and redeposit sediments, generating laminations, cross-beds, and graded beds. The intertidal zone can be favorable for growth of blue-green algae, forming algal mats and columnar stromatolites.

The lagoon and estuarine subenvironments are particularly favorable for the accumulation of floral debris and eventual generation of coals. Swamps associated with rivers and lagoons on and adjacent to deltas also are sites of coal formation.

The open marine environment is divided into shelf, slope, and basin subenvironments. The shelf environment is where biotic activity is maximum. The shelf has relatively low energy, shallow water, and in tropical-to-subtropical latitudes the water is warm. The inner shelf is transitional to bays, lagoons, tidal flats, and evaporite basins. Above wave base, there is an accumulation of coarser clastic sediments forming grainstones, such as oolitic limestones, that may mix with finer particles eventually to form packstones and wackestones. Deposition of ice-rafted glacial debris may occur and mix with the fauna already established (Eyles and Lagoe, 1989).

The middle shelf is below wave base and is dominated by muds and the eventual formation of micritic limestones, which can contain variable quantities of chert (frontispiece photo) and siliciclastics sediments (Figure 37.29). Bioturbation tends to destroy laminar structure. The outer shelf is characterized by development of coral bioherms and biostromes. Reefal carbonates include both allochthonous and autochthonous components but are dominated by corals and the debris derived from them. Rocks derived from these materials are **boundstones, framestones**, and packstones.

The marine slope environment consists of the fore reef and basin margin portions. It is characterized by

Figure 37.29
Marine shelf Ely Limestone (Pennsylvanian) with sandy (quartz) laminations. Bar scale = 2 cm. (Pennsylvanian). Eagan Range, Nevada.

gravity-transported debris, including reefal material and **shelf limestones**. The debris takes the form of talus breccia, slump blocks, and turbidites. Admixed siliceous and calcareous pelagic particles add to the complexity of rocks formed in this environment.

The ocean basin environment is characterized by **turbidites** along the basin margins and accumulations of nannoplanktons and microplanktons elsewhere. The abundance of pelagic sediments depends on the rate of supply relative to the rate of dissolution. Deep ocean water is undersaturated with respect to calcium carbonate: consequently there is a depth, the **carbonate compensation depth** (CCD), below which carbonates dissolve. The CCD varies from place to place and from ocean to ocean in accord with temperature and carbon dioxide content. The CCD is at about 4000 m in equatorial water and gets shallower at higher latitudes.

Sedimentation on the ocean floor is characterized by well-developed beds and laminations since it is a low-energy environment. Homogenization of these structures by bioturbation does not occur because there are few benthic fauna at these depths. Settling of pelagic fauna in the turbidite zone generates a wide variety of rock types. Cherts are common in some turbidites, particularly those involving mélanges at convergent plate boundaries (Figures 25.7a, 37.30). Carbonate material from the continental shelves and slopes can be involved in turbidity currents (Bustillo and Ruis-Ortiz, 1987). Carbonate turbidites and carbonate-bearing siliciclastic turbidites (Hesse, 1987) may contain nodular chert, since rapid burial of radiolarians and sponge spicules favors their entrapment and availability to participate in diagenetic processes. Calcareous turbidite packstones, turbidite lime mudstones, and radiolarian pelagic lime mudstones are characteristic of this region.

Figure 37.30
Vertical beds of ocean-floor ribbon chert of an accreted terrane. Orcas Chert, Roche Harbor, San Juan Island, Washington. Scale = U.S. quarter (center).

38

Evaporites

Megapolygons in Holocene salt pan at Four Mile Flat, Fallon, Nevada. Efflorescent hollow antiformal ridges (tepee structure) are due to repeated cycles of expansion–contraction related to thermal fluctuations or by expansion alone, forming pressure ridges. Bar scale represents about 0.5 m.

Precipitation of Evaporite Minerals

Chemical Conditions

Ocean water and alkaline lake water containing dissolved salts will precipitate minerals if the concentration of the solutes exceeds chemical saturation. There are many different kinds of dissolved salts in natural waters, and precipitation of **evaporite minerals** is governed by phase relations in these solutions as crystallization proceeds and concentration of various solutes increases. Concentrations are increased by long-term evaporation and by short-term changes in diurnal and seasonal temperatures.

Since the formation of **evaporites** occurs at the Earth's surface, pressure is close to 1 bar. Slight changes in atmospheric pressure have little or no effect on evaporite mineral equilibria. However, variation in the partial pressure of carbon dioxide is a major factor controlling the presence or absence of carbonate minerals, and these commonly constitute part of the evaporite stratigraphic record.

A **brine** is a low-temperature aqueous solution in which the concentration of dissolved salts is extreme. Since silicates, oxides, and sulfides are not very soluble in relatively cool water, dissolved compounds in the ocean, lakes, and rivers are mostly carbonates, sulfates, and chlorides. The ionic content of some of these waters is shown in Table 38.1, and the varieties of minerals that can precipitate from brines derived from them are given in Table 38.2.

Inspection of the chemical composition of ocean water, in comparison with continental lake and river water, reveals some major differences (Table 38.1). Ocean water has a very uniform composition, dominated by sodium and chloride ions but also containing Mg, K, Ca, and SO_4. Composition of continental waters varies according to source. The principal sources are (1) meteoric runoff water, groundwater, and some spring water, (2) geothermal brines, (3) diagenetic brines, and (4) volcanogenic brines.

Some continental waters are characterized by the presence of bicarbonate, whereas others are more like seawater. As a result, the distinction between marine and nonmarine evaporites on the basis of mineral assemblages is not always clear (Hardie, 1984; Lowenstein, 1989). Even marine evaporites are not strictly the result of evaporation of seawater, since evaporite basins occur along the continental margins where there must also be input of continental water.

Table 38.1
Analysis of Sea, Lake, and River Waters (in parts per million)

Ion	Seawater	Lakes[1]			Rivers[1]	
		A	B	C	D	E
Cl^-	18,800	112,900	1,960	208,020	15	113
Na^+	10,770	67,500	1,630	34,940	11	124
SO_4^{2-}	2,715	13,590	264	540	41	289
Mg^{2+}	1,290	5,620	113	41,960	7.6	30
Ca^{2+}	412	330	10	15,800	34	94
K^+	380	3,380	134	7,560	3.1	4.4
HCO_3^-	140	180	1,390	240	101	183
SiO_2			1.4		5.9	14

1(A) Great Salt Lake, UT, (B) Pyramid Lake, NV, (C) Dead Sea, Israel, (D) Mississippi River, Baton Rouge, LA, (E) Colorado River, Yuma, AZ.

Source: Krauskopf (1979).

Order of Precipitation

The sequence of minerals that precipitate directly in oceanic and continental waters largely depends on the solubilities of the various compounds in relation to the complex solutions from which they evolve (Sonnenfeld, 1984; Logan 1987). A very soluble salt will crystallize late relative to another that is less soluble if the ionic abundances are equivalent. But if the more soluble salt is in greater abundance, it may crystallize earlier despite the solubility difference.

The concentrations of calcium and bicarbonate ions in seawater are very low compared with sodium and chloride ions (Table 38.1), yet the first mineral to precipitate in evaporating seawater is calcium carbonate (Figure 38.1) because its solubility is much lower than that of halite. Similarly, the concentration of the sulfate ion in seawater is much less than the chloride ion but greater than bicarbonate. The solubility of gypsum (and anhydrite) in pure water is much lower than that of halite, but it is greater than calcium carbonate. Thus, gypsum (or anhydrite) precipitates after calcium carbonate but before halite in seawater (Figure 38.1).

The sequence of crystallization of evaporite minerals in brines can be predicted on the basis of experiment, thermodynamic calculation, and observation of the order of crystallization in stratigraphic sequences. For seawater, the sequence has been found to start with low-Mg calcite or aragonite, followed by gypsum (or anhydrite), then halite, and finally by a wide variety of chlorides known as **bitterns** (Figure 38.1). This is the general sequence, but in reality there is overlap with simultaneous crystallization of phases.

The sequence in continental waters varies according to the particular composition of the water, but very commonly there is either a deficiency in the chloride or calcium ion and several sodium carbonate minerals appear, of which **trona** ($Na_2CO_3 \cdot NaHCO_3 \cdot 2H_2O$) is the most common. Evaporites of meteoric water parentage are well represented in the Eocene Green River Formation (Eugster and Hardie, 1975; Ryder et al., 1976).

Hypersaline conditions are not conductive to the development of a high-density fauna and flora; this is no heartwarming environment for clams and snails. Nevertheless, the activity of certain bacteria, algae, and salt marsh vegetation may have an effect on the sequence of crystallization of evaporite minerals. For example, if a water body becomes anoxic, anaerobic bacteria may flourish, degrading carbonate and sulfate ions (Heydari and Moore, 1989). If these ions are not available, certain carbonate and sulfate minerals cannot precipitate. A deficiency of calcium carbonate minerals can be related to such degradation. The absence of gypsum–anhydrite is a well-known result of **sulfate reduction**.

Dissolution of Evaporite Minerals

Progressive precipitation of evaporite minerals in natural systems typically is interrupted by temporary conditions favoring partial dissolution of crystals during the primary formation of an evaporite. The stability fields of evaporite minerals are in the general region characterizing surface temperatures and one atmosphere pressure. Some evaporite systems are active at the brine–atmosphere interface, where solutions and crystals are affected by nighttime and daytime temperatures and by the longer-term annual temperature conditions characterizing summer and winter at temperate latitudes. Since the solubility of a salt such as

Table 38.2
Minerals That Occur in Marine and Continental Evaporites

Mineral	Formula	d	Solubility $T\ (°C)/g/L$
Anhydrite	$CaSO_4$	2.963–2.98	20/2.98 30/2.09 100/1.619
Aragonite	$CaCO_3$	2.93	25/0.0153 75/0.0190
Bischofite	$MgCl_2 \cdot 6H_2O$	1.56–1.604	Cold/1670 Hot/3670
Bloedite (astrakhanite)	$Na_2Mg(SO_4)_2 \cdot 4H_2O$		
Borax	$Na_2B_4O_7 \cdot 10H_2O$		
Calcite	$CaCO_3$	2.71	25/0.014 75/0.018
Carnallite	$KMgCl_3 \cdot 6H_2O$	1.602	19/645.0 Disintegrates in fresh water at 25°C
Celestite	$SrSO_4$	3.971	0/0.113 30/0.114
Colemanite	$Ca_2B_6O_{11} \cdot 5H_2O$	2.42	
Dolomite	$CaMg(CO_3)_2$	2.87	18/0.32
Epsomite (reichardtite)	$MgSO_4 \cdot 7H_2O$		
Glauberite	$Na_2Ca(SO_4)_2$		
Gypsum (karstenite)	$CaSO_4 \cdot 2H_2O$	2.317–2.33	0/2.41 100/2.22
Halite	$NaCl$	2.135–2.164	0/357.0 100/391.2
Hexahydrite	$MgSO_4 \cdot 6H_2O$		
Ikaite	$CaCO_3 \cdot 6H_2O$		
Kainite	$K_4Mg_4Cl_4(SO_4)_4 \cdot 11H_2O$		
Kernite	$Na_2B_4O_7 \cdot 4H_2O$	1.91–1.93	
Kieserite	$MgSO_4 \cdot H_2O$		
Langbeinite	$K_2Mg_2(SO_4)_3$		
Magnesite	$MgCO_3$		
Mirabilite	$Na_2SO_4 \cdot 10H_2O$		
Nahcolite	$NaHCO_3$		
Phillipsite	$KCa(Al_3Si_5O_{16}) \cdot 6H_2O$		
Polyhalite	$K_2MgCa_2(SO_4)_4 \cdot 2H_2O$	2.775–3.0	Incongruent
Sal ammoniac	NH_4Cl	1.527	0/107.4 85/425.4
Siderite	$FeCO_3$		
Strontianite	$SrCO_3$		
Sylvite	KCl	1.984–1.99	20/347.0 100/567.0
Thenardite	Na_2SO_4		
Trona	$Na_2CO_3 \cdot NaHCO_3 2H_2O$		

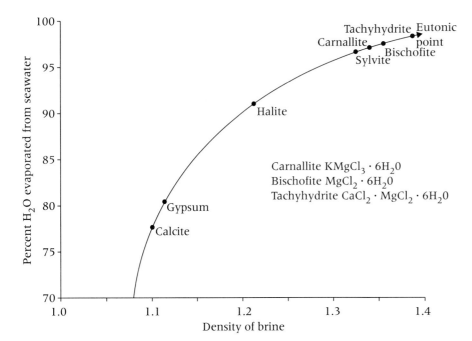

Figure 38.1
Precipitation path of an anoxic brine to the eutonic point. From Sonnenfeld (1984).

halite increases with temperature (Chapter 12), halite precipitated in cool water would be subject to dissolution as the water is warmed during the daytime and during the summer.

Another way that dissolution can occur is in response to changes in concentration. Flooding with dilute solutions can occur in evaporite systems, especially by storm waves and spring tides along the coast and by cloud bursts in the continental environment. Longer-term dilution of solutions typically relates to wetter seasonal (annual) conditions during which rainfall and meteoric inflow increase. Minerals that were close to equilibrium with brine now find themselves far out of equilibrium with the dilute solutions, and since the solutions are undersaturated with respect to the minerals, dissolution occurs. As a result, a sequence may consist of calcite (aragonite), gypsum (anhydrite), halite, and ending with more gypsum (anhydrite). The final gypsum (anhydrite) indicates a dilution of the waters at a late stage.

Dissolution of a complex mineral such as carnallite ($KMgCl_3 \cdot 6H_2O$) simultaneously generates sylvite if the solution is saturated in KCl (Wardlaw, 1968). This dissolution is incongruent.

Flooding may also be the reason for the scarcity of certain minerals in evaporite sequences, which on the basis of experimental and calculated prediction should be present. Potassium-free salts such as bischofite ($MgCl_2 \cdot 6H_2O$) represent extreme fractionation. Since this mineral is rare, repeated replenishment by flooding, keeping the brine somewhat less concentrated, is indicated.

Modification of Evaporite Mineral Assemblages by Reaction

The observed mineral assemblages in evaporites may not directly reflect the composition of the brine even if flooding and incongruent dissolution are considered. After burial, reactions between minerals and waters derived from additional sources may result in preferential abstraction or addition of ions and generation of new minerals, falsely indicating that the original brine was of different composition. Introduction of groundwater or hydrothermal water into a marine evaporite sequence can be responsible for such a paradox.

Reaction of evaporite minerals with its own late-stage brine, without introduction of new solutions, can also alter the mineral assemblage. The sequence of minerals produced from ocean water depends not only on the composition of this solution but on physical activity as crystallization proceeds. If crystals are mechanically removed or in some way physically restricted from the solution from which they just precipitated, these crystals cannot react with the fluid. Conversely, if they remain in contact with the solution, reactions may occur.

Dolomitization of limestone is an example of reaction of crystals with their own solutions. Abstraction of magnesium from solutions that have previously precipitated calcium carbonate converts calcite and aragonite to dolomite (Chapters 14, 37).

Carbonates as Evaporites

Only about 0.03% of the total dissolved solids in seawater is carbonates. The concentration of calcium ions is about 0.01 M, and that of the carbonate ion is 0.0003 M. The ion product indicates that seawater is drastically supersaturated in calcium carbonate, but since calcium ions also associate with sulfate ions, and carbonate ions partly associate with magnesium ions, the activity product is much less, although seawater is still nearly saturated in calcium carbonate.

Stratigraphic sections containing bedded evaporites commonly also contain carbonates. Since carbonate rocks are generated in environments in which biota flourish, how can carbonates be formed in the evaporite environment that is so chemically hostile to biological activity? A major portion of carbonate rocks associated with evaporites are at the base or top of the stratigraphic section, and they also occur as lateral facies extending away from what was the central brine pool. In any of these cases, the environment is likely to be much less saline, and there is a "normal" relation between biota and the formation of carbonates. Thus, the basal carbonate means that gypsum and halite saturation has not been reached. The lateral carbonates indicate that a gypsum- or halite-saturated brine pool is localized, and the capping carbonates signal termination of gypsum–halite brine phase of the evaporite system.

Some carbonates are interbedded with gypsum (anhydrite) and halite. In this case, there must be a change in conditions that are either (1) more conducive to biotic activity and/or (2) favor precipitation of calcium carbonate in place of gypsum or anhydrite. Although nonskeletal algae, bacteria, and some ostracods (*Cyprideis* sp.) (Decima et al., 1988) are relatively well adapted to high-saline environments, major involvement of biota in precipitation of calcium carbonate requires an environment that is temporarily (cyclically) much less saline.

The second possibility is that there is occasional or cyclic inflow of carbonate-bearing groundwater, spring water, or river water, favoring precipitation of calcium carbonate over gypsum and halite.

Physical Configuration of Evaporite Systems

Accumulation of precipitates requires that (1) water contain a high concentration of dissolved substances and (2) there be repeated replenishment of this water to an evaporative system.

Evaporation is promoted by heat, low humidity, and wind. These are the conditions of arid and semiarid climates characterizing certain inland basins and subtropical continental margins. The supply of dissolved substances is by inflow of marine or continental waters, and it is these same waters that resupply the system.

Continental evaporative systems receive their water in the form of surface runoff, groundwater discharge, and by subaqueous or nearby subaerial hot springs. Marine evaporative systems may also receive some meteoric and hydrothermal water along the continental margins. Direct rainfall is not a supplier of salts and has only a minor effect as a dilutant to brines or as a solvent for temporarily exposed surface crusts.

Inland basins contain one or more of the following: (1) a perennial salt lake, (2) an ephemeral **playa** or **salina lake**, and (c) a **sabkha** (mostly subaerial). A composite sketch of these settings is shown in Figure 38.2. Accordingly, there may be crystallization of evaporite minerals in brine, at a desiccation surface, or in brine-soaked sediments. Clastic particles are likely to be abundant if the basin is bordered by mountain ranges providing high-energy fluvial and sheetwash flooding.

Evaporative systems supplied by ocean water range from groundwater-driven sabkhas to shallow **salinas**, to **lagoons**, and finally to **deep-water basins** (Hanford, 1981; Hovorka, 1987; Lowenstein, 1988). In each case, there is proximity to the sea and a connection to hydrodynamic resupply of saltwater. The connection may be over a subaqueous barrier that may be no more than a ridge over which there is free flow of seawater (Figure 38.2).

Islands and shoals may be enough of a barrier for broad inland seas to become evaporative systems. Such continental margin lagoons and estuaries are an ideal setting for development of laterally extensive evaporites.

If the barrier is subaerial, and therefore totally restrictive, drawdown by evaporation lowers lake level with respect to sea level and there is seepage of seawater through the barrier to the basin in response to the head difference and in spite of the density difference between brine and seawater. Salinas form in this way and are equivalent to inland playas (Figure 38.3).

Continental waters may contribute to resupply on the landward side of the evaporative basin. As brines develop, there is lateral encroachment into the less dense groundwaters, which are then unable to inflow to the system. Evaporites transitional between the marine and continental types occur where the inflow of seawater is strongly restricted and the inflow of continental surface water is dominant.

Coastal sabkhas are fed by salty groundwater seeping into porous supratidal sediments. Capillary

486   Part 10   Rocks Formed by Mechanical and Chemical Processes Involving Low-Temperature Aqueous Fluids

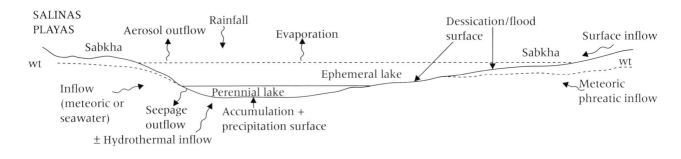

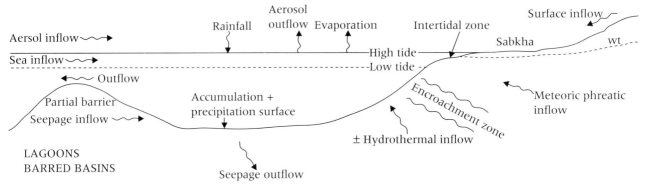

Figure 38.2
Diagrammatic composite of evaporite depositional environments.

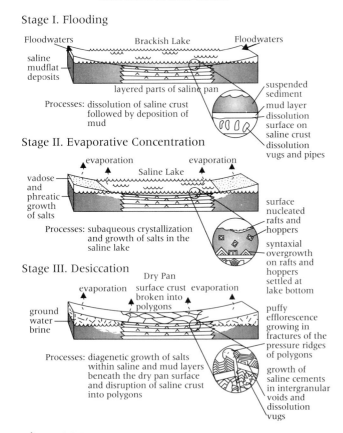

Figure 38.3
Saline pan cycle showing dissolution and growth of halite. From Lowenstein and Hardie (1985).

action and evaporation at the surface yield evaporite minerals. There need be no barrier in this case, but sand supply in this mostly subaerial environment must be low or the salt-encrusted surface is covered, suffocating the sabkha system.

Textural Interpretation of Evaporite Genesis

Characterization of depositional environment of "ancient" evaporites can be done by comparison with the mineralogical and textural characteristics of Holocene evaporites (Lowenstein and Spencer, 1990).

Evaporite textures can be classified as being either primary (syngenetic, syndepositional) or secondary (early diagenetic, late diagenetic, epigenetic, postburial) (Hardie et al., 1985; Lowenstein, 1987). Direct precipitation of minerals in a brine is a primary process. Growth of crystals on the brine floor or in underlying muds may be temporarily interrupted by dissolution or replacement processes. Both dissolution and replacement processes are considered a normal part of the primary depositional process (Figure 38.3). Postburial secondary processes include recrystallization and deformation.

There are also chemical means of interpretation of primary and secondary processes. For example, small amounts of bromine replace chlorine in all chlorides.

Figure 38.4
Halite rock containing cumulate halite cubes that have fluid inclusion zones in cores. Reverse-graded bedding defined by increasing size of crystals upward. Dark layer at base is microcrystalline polyhalite. PPL. Bar scale = 5 mm. Salado evaporite, New Mexico–Texas. (Courtesy T. Lowenstein.)

The bromine is absorbed by chlorides in a fixed relation corresponding to the bromine content of the brine. As the content of bromine in the brine increases, so does the content in halite that is forming from the brine. Departure from systematic bromine content of halite, particularly low-bromine contents, suggests that dissolution and regrowth have occurred, likely related to inflow dilution and a flushing out of the bromine.

Crystal Growth in Brines

In brines there is free growth of crystals either (1) at the brine–air interface, (2) totally within the brine, or (3) on the brine floor.

Crystallization of halite at the brine–air interface generates hopper cubes and "rafts" of connected hoppers (Figure 38.3) (Arthurton, 1973; Lowenstein and Hardie, 1985). Hopper crystals reflect rapid crystal growth in a "free" physical environment and therefore are strong evidence of primary growth in brine. An inverted halite hopper or hopper raft (s.g. = 2.1) can float as a "boat" in the brine (s.g. = 1.2) from which it has crystallized. Turbulence results in filling of hoppers with brine and breaking of surface tension between the brine and the crystals. Foundering of crystals and rafts occurs sooner or later, and cumulate halite is generated on the brine floor surface (Figure 38.4) much as gravity-driven cumulates are accumulated to form some layered magmatic rocks.

Once a halite crystal or raft has foundered, it becomes an ideal substrate for further crystal growth projecting upward from the brine floor. Nucleation on other materials covering the floor, such as muds, evaporite, or siliceous clastics, may also occur. In any case, continued growth produces either **chevron halite** (Figure 38.5a,b) or **cornet halite** (Figure 38.5b). Gypsum typically grows upward with an **inverted chevron** (herringbone) pattern (Figure 38.5c) (Rouchy, 1980). The chevron pattern of halite results from cube corners or edges being oriented in an initial upward position on the brine floor. If the cube faces are oriented upward, cornets are produced. The inverted chevron of gypsum reflects an initial vertical orientation of the twin plane. Evidently, the growth rate on a twin dovetail is greater than on domelike surfaces of the opposite end of a twinned crystal; thus, upward growth of "inverted chevrons" are favored for gypsum. If the brine floor consists of water-soaked mud, the gypsum crystals may sink and tilt as they get larger and heavier.

Vertical elongation of both halite and gypsum reflects competition for space. The chevron pattern within halite and gypsum crystals is defined by (1) fluid inclusions, producing a cloudy aspect, (2) small inclusions of other evaporite minerals crystallizing simultaneously with the host crystal (e.g., gypsum in halite), and/or (3) siliciclastic sediments that might happen to be present. Slower growth rates result in fewer inclusions. Cooler nighttime conditions slow growth rates, and fewer inclusions develop. The alternation of inclusion-rich zones with inclusion-poor zones as controlled on a daily time scale is an impressive reminder of the extreme difference in the rate at which some textures in evaporite minerals form in comparison with those in plutonic magmatic and metamorphic systems.

There is some question about the primary growth of anhydrite crystals. Anhydrite is thermodynamically stable at somewhat higher temperatures than gypsum, but still within the range of natural conditions. Experiments always produce gypsum, but it is possible that the kinetics of anhydrite crystallization are not compatible with short-lived experiments. Crystallization of anhydrite in brine and into brine from the brine pool floor has not been texturally documented.

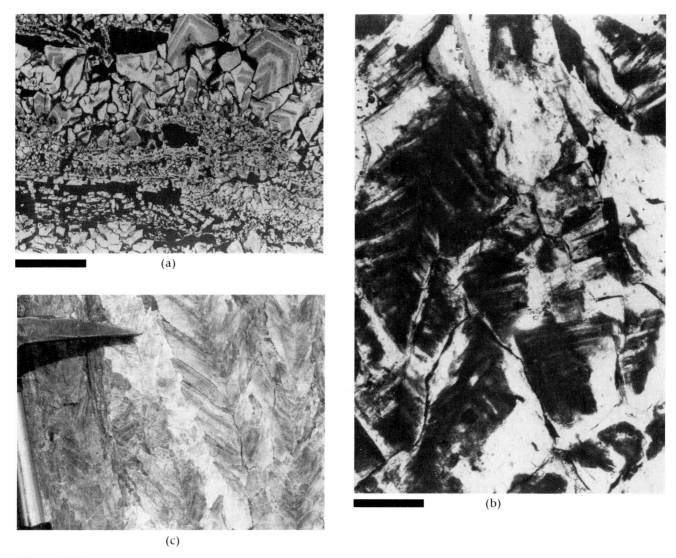

Figure 38.5
Primary growth on brine floor.

(a) Larger halite crystals are vertically oriented chevrons and cornets that grew upward from saline lake bottom. Fluid inclusion zones (light gray) can be traced stratigraphically across crystal boundaries from one crystal to the next. Interlayered with fine-grained cumulates consisting of halite rafts, plates, and hopper cubes. Dark areas are voids. Salina Omotepec, Baja California, Mexico. PPL. Bar scale = 10 mm. (Courtesy T. Lowenstein.) (b) Vertically oriented halite chevrons (upward-directed edges and corners) and cornets (upward-directed cube faces) (lower right). Dark zones in crystals have abundant fluid inclusions. Salado evaporite, New Mexico/Texas. PPL. Bar scale = 3 mm. (Courtesy T. Lowenstein.) (c) Primary twinned selenite gypsum. "Inverted chevrons" point upward along growth direction. Messinian of Murcia Basin, southern Spain. (Courtesy J. M. Rouchy.)

Crystal Growth in Brine-Soaked Sediments

Brines occur in the pore spaces of sediments in evaporite systems. Although the permeability varies with the nature of the sediments, these brines are connected with less concentrated fluids through the intergrain pores providing an open chemical system. Just as in surface-water bodies, there must be resupply of brine to sites of crystallization, either by flow of the solvent or by diffusion in the solvent if there is to be significant enlargement of crystals. Connections may be lateral, to depth, or to the surface. Although these routes of transfer are needed for crystal growth, they are the same avenues along which there can be inflow of diluting fluids and dissolution of crystals.

If the interpore brines are connected to the surface where there is no brine pool, capillary action toward the surface results in evaporation, supersaturation, and precipitation. In this case there must be a flow of brine inward from a source located laterally or from below. Connection to the surface where there is a brine pool or a saline lake provides a means of resupply by downward flow or diffusion.

Growth of evaporite minerals in muds is no longer in free space, although there may be primary voids close to the brine floor or a desiccation surface. There must be an interaction between the new crystals forming and the grains of the sediments. This interaction can be of three types: (1) the sediments are displaced as the crystals grow; (2) the sediments are incorporated and/or chemically replaced as the crystals grow; (3) the crystals grow in localized open spaces.

Displacive growth of evaporite minerals (Gornitz and Schreiber, 1981; Rouchy et al., 1987) is detected by the divergence of sediment lamellae around crystals (Figure 38.6a). Growth of evaporite minerals in muds just below a desiccation surface has been shown to exert lateral pressure, resulting in buckling of the drying sediments along upturned antiform (**tepees**) boundaries between **megapolygons** (Kendall and Warren, 1987) (frontispiece photo). This is evidence that the growth of these minerals generates a "force of crystallization."

Some inclusion of mud grains can be expected, just as there is in many metamorphic porphyroblasts. If inclusion is parallel to all the growth faces, growth in a completely enclosing mud is indicated. The shape of displacive crystals may be irregular, as indicated by muds that follow crystal irregularities, as opposed to precipitates in irregularities formed by later dissolution.

There is the possibility that host materials are "compressed" or "draped" around a pre-existing crystal rather than being displaced by crystal growth. Draping of muds over crystals of gypsum on the floor of a brine pool or on a desiccation surface could occur during a flood. However, divergence of lamellae around crystals would not be both above and below the crystal as in the case of displacive growth, unless compaction also deformed layering beneath the crystal. Of course, there may be no sense of displacement if the muds are too soupy to establish structures. In that case a relatively inclusion-free displacive crystal in a structureless rock could be interpreted as one grown replacively. Additionally, there is the possibility that the draping geometry results form soft sediment deformation during or just after crystallization.

Replacive growth and **incorporating growth** do not have the displaced lamellae beds that characterize displacive growth. Clay minerals, quartz grains, and other grains constituting the host sediment, including carbonate muds, are included and/or replaced by reaction. If reactive replacement is extreme, the crystals may be free of inclusions of the original materials. There may be minor (Figure 38.6a) or major (Figure 38.6b) inclusion of the matrix in crystals that transect layering.

Open-space growth occurs wherever there is a supersaturated brine in open fractures, prior dissolution cavities, and other such open spaces. Desiccation cracks in muds are prime sites of precipitation of evaporite minerals. Crystal growth in muds and dried muds can be partly incorporative and space filling (Figure 38.6c). Space filling by sylvite in a halite framework is shown in Figure 38.6d.

Dissolution of Crystals

Dissolution of crystals may occur at the surface or in the subsurface, with three possible effects: (1) voids are formed in the crystal, (2) there is crystallization of the mineral of the same composition in the voids closely following the formation of voids, or (3) there is crystallization of a new mineral without voids formation.

If voids are formed (1), they will contain fluid as the dissolution process occurs, but they will be left vacant if there is a drying out on a desiccation surface (Figure 38.7a). If new crystals of the same type fill the "voids" (2), such as the inclusion-free halite in "dusty" halite, there may have been dissolution followed closely by reprecipitation as solutions become more concentrated again after initial dilution. It might even result from diurnal temperature changes.

Replacement of one mineral by another does not generate voids because there is simultaneous crystallization of new mineral (3) as the ionic units of the old crystal are taken into the reactive solution.

The timing of these dissolution, dissolution–reprecipitation, and reactive replacement processes may

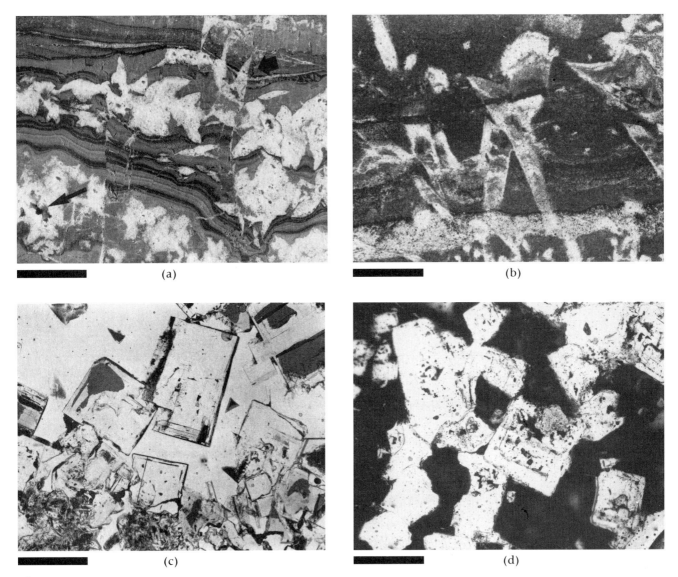

Figure 38.6
Displacive and replacive crystal–mud relations.

(a) Organic-rich, fine-grained, laminated limestone containing calcite pseudomorphs after gypsum. Displacive growth of gypsum in soft sediment indicated by divergence of lamellae around crystals. Some replacive growth indicated by crystals lying across lamellae (thick black arrow). Voids in calcite (thin black arrow) may be the result of volume reduction in the reaction gypsum fi calcite. Slab surface. Bar scale = 1.5 cm. Visean of Belgium. Belgium Geological Survey core. (Courtesy J. M. Rouchy and E. Groessens.)
(b) Anhydrite rock with vertically oriented, mud-incorporative gypsum pseudomorphs (now anhydrite). Mud layers extend laterally through several pseudomorph crystals. Mud (dark) consists of anhydrite, magnesite, and minor siliciclastic sediment. PPL. Bar scale = 5 mm. Salado evaporite, New Mexico/Texas. (Courtesy T. Lowenstein.) (c) Halite in muddy halite layer just beneath the surface of the modern salt pan of Saline Valley, California. Randomly oriented halite cubes have incorporated mud (black, and messy material lower left). Primary voids (adjacent to cubes). PPL. Bar scale = 5 mm. (Courtesy T. Lowenstein.) (d) Muddy halite rock with cube halite framework filled (cement) with zoned sylvite crystals and minor mud (black). McNutt Potash Zone, Salado Formation. Compare with modern halite framework without infilled sylvite (c). PPL. Bar scale = 7 mm. (Courtesy T. Lowenstein.)

be during the primary development of the evaporite bed (early diagenetic) or any time later, even as an epigenetic event. The only requirement is that dilute or reactive solutions be in contact with pre-existing crystals. Accordingly, early diagenetic dissolution may be the result of surface flooding, such as provided by storm waves or high-volume fluvial discharge, or simply by rainwater. Filling of the voids with mud at a waning stage of fluvial flooding is common. An exposed dissolution surface is likely to have an array of morphologies, including serrate ridges and jagged pinnacles. Late diagenetic and epigenetic timing of dissolutions is more likely to be subsurface, resulting from vadose or phreatic inflow or, less commonly, hydrothermal inflow. The evaporites at this later time are more indurated, whereas the activity during early diagenesis is in soft sediment.

Corrosion surfaces truncating exposed crystals of halite (Figure 38.7b) or gypsum represent dissolution of crystals down to a more-or-less planar, but in detail an irregular, surface—a **dissolution erosion surface**. Dissolution is confirmed because growth zones, as defined by fluid or solid inclusion, are truncated. It is common to have deposition of clastics or precipitation of new evaporite minerals on these "erosion" surfaces.

Dissolution may also occur along fractures or any other pre-existing structure that provides fluid access, either creating or enlarging voids of many shapes, including pipes and microkarst pits. If the evaporites are consolidated and contain mudstone interbeds, intraclasts may be formed as a result of dissolution of evaporite minerals below such a bed, resulting in collapse and fragmentation.

(a)

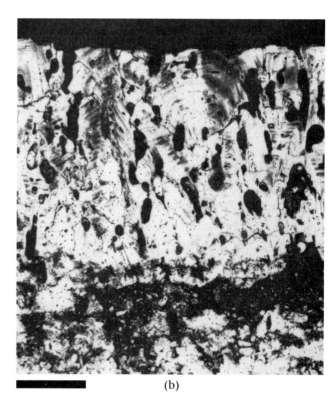
(b)

Figure 38.7
Dissolution features in evaporites.

(a) Dissolution voids (V), partly parallel to chevron growth zones in halite. PPL. Bar scale = 1 mm. Surface crust, near Tell Dafana, shore of Lake Manzala, eastern Nile delta, Egypt. (Courtesy T. Lowenstein.) (b) Halite crust (depth < 0.5 m) with vertically oriented solution cavities (dark tubular areas) between vertically oriented chevrons and cornets of halite (clear areas with faint inclusion zones). Uppermost surface is a smooth horizontal dissolution surface that truncates some crystals. Dark granular layer at bottom consists of gypsum crystal sand, siliciclastic mud, and insect larval cases, all cemented with clear poikilotopic halite (halite poikigrains, Chapter 6). PPL. Bar scale = 1.5 cm. (Courtesy T. Lowenstein.)

Reactive Replacement of Evaporite and Associated Minerals

Any evaporite stratigraphic section that becomes buried also is susceptible to late diagenetic and epigenetic replacement reactions.

Conversion of Gypsum to Anhydrite. The conversion of gypsum to anhydrite requires an environment that allows the following dehydration reaction to take place:

$$\underset{\text{(gypsum)}}{CaSO_4 \cdot 2H_2O} \rightarrow \underset{\text{(anhydrite)}}{CaSO_4} + 2H_2O$$

The textural relation shown in Figures 38.6b and 38.8a indicate that anhydrite has formed at the expense of gypsum. Thermodynamic calculations indicate that gypsum is more stable at low temperatures than anhydrite, and since laboratory experiments always produce gypsum the chances of there being a gypsum precursor to anhydrite are high, both in the diagenetic and epigenetic development of anhydrite-bearing evaporite rocks.

The formation of **anhydrite nodules** is very typical. Nodular form seems to be the result of several conditions existing at the time of reaction. The general nodular shape is typically controlled by the size and shape of the original gypsum crystals, as well as by the distribution of impurities within these crystals. The nodules consist of aggregates of anhydrite crystals, reflecting conversion from relatively pure portions of larger gypsum crystals. Zones containing impurities in the gypsum and siliciclastic sediments between gypsum crystals remain in place or become concentrated, giving visual definition to the nodular form. Since the process can occur in soft sediments soon after growth of gypsum crystals, soft sediment deformation is likely to modify the form of the nodules. Displacive growth of the anhydrite can further concentrate impurities in the internodule position. In some cases a ptygmatic (intestine-like) **enterolithic** chain of nodules form, giving the false impression of epigenetic folding. An even distribution of nodules with respect to their internodule impurities typically forms a **"chicken-wire"** structure (Figure 38.8b,c).

Since growth of gypsum in shallow muds of the sabkha environment is common, the formation of anhydrite nodules is commonly taken as evidence that there was a sabkha in existence. However, it is likely that circulation of brines capable of taking on water, **hydroscopic brines**, can occur in the subaqueous environment where gypsum lies in bottom sediments,

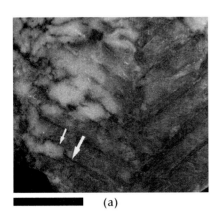 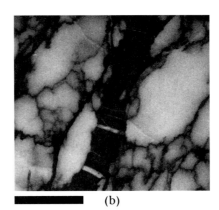 

(a) (b) (c)

Figure 38.8
Nodular anhydrite.

(a) Partial replacement of twinned selenite gypsum crystal by anhydrite that was itself replaced by gypsum (white). "Impurity" zones parallel to selenite growth surfaces (large arrow) become displacively distorted (small arrow) during anhydritization and gypsification, producing "chicken-wire structure." Slab surface. Bar scale = 3 cm. Messinian of the Murcia Basin, southern Spain. (Courtesy J. M. Rouchy.) (b) Anhydrite with "chicken-wire" structure. Slab surface. Bar scale = 2 cm. Visean of Belgium. Belgium Geological Survey core. (Courtesy J. M. Rouchy and E. Groessens.) (c) "Chicken-wire" structure consisting of fine granular gypsum nodules, presumably a replacement of nodular anhydrite that replaced primary coarse crystalline gypsum. Slab surface. Arapien Formation, Mayfield, Utah. Bar scale = 1.5 cm.

forming anhydrite nodules (subaqueous anhydritization). Furthermore, dehydration of gypsum crystals, forming porous plastic masses of very fine granular anhydrite, may be deformed into "nodular slugs" during deep burial diagenesis (Hardie et al., 1985).

Conversion of Anhydrite to Gypsum. Reaction of water with anhydrite forms gypsum. Such a reaction is to be expected in any situation that brings aqueous solutions in contact with anhydrite, regardless of whether the anhydrite is primary or has formed from a gypsum precursor. Influx of groundwater into a sabkha environment would make the conversion as an early diagenetic event. With deep burial of an evaporite sequence, either already containing anhydrite or experiencing conversion of gypsum to anhydrite at depth, subsequent re-emergence, attendant to tectonic uplift and erosion, brings this anhydrite into contact with groundwater where the secondary gypsum forms epigenetically.

Replacement of anhydrite crystals by gypsum is evidenced by pseudomorphic preservation of rectangular anhydrite sections and by relics of anhydrite. Larger relics of anhydrite in alabastrine (fine granular) gypsum also occur. In many examples of conversion of anhydrite to gypsum, the anhydrite is presumed to have originally formed from gypsum. For example, the nodular forms in the large selenite gypsum crystal shown in Figure 38.8a consist of an aggregate of small gypsum crystals that have very likely replaced an aggregate of anhydrite crystals. It is likely that an early diagenetic nodular anhydrite has become a nodular gypsum during late or postdiagenesis. The "chicken-wire" occurrence of nodules in Figures 38.8c was probably anhydritic, but the nodules are now fine-grained gypsum.

Replacement of nodular anhydrite is not limited to gypsum. Silica from a presumed nearby sponge spicule source has replaced anhydrite nodules and formed quartz geodes (Chowns and Elkins, 1974). Laths of anhydrite that were in the nodules are silicified, and small inclusions of anhydrite in the quartz are typical.

Conversion of Carbonates to Gypsum. Although bedded carbonates, especially dolomite, are common in evaporite sequences, the age relations may not be obvious. In some cases textural relations clearly indicate replacement of carbonate by evaporite minerals. For example, gypsum has replaced coarser-grained recrystallized limestone as shown in Figure 38.9a. Perfect crystals of gypsum have grown in micritic carbonate (Figure 38.9b). The reaction would be

$$CaCO_3 + SO_4^{2+} + 2H_2O = CaSO_4 \cdot 2H_2O + CO_3^{2-}$$
$$\text{(calcite)} \qquad\qquad\qquad \text{(gypsum)}$$

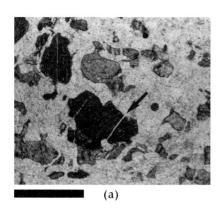 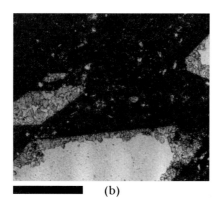 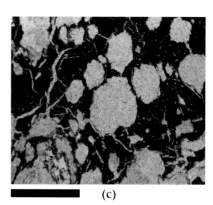

(a) (b) (c)

Figure 38.9
Replacements of carbonates by gypsum.

(a) Fine granular gypsum replacement of coarse crystalline calcite. Note microvein of gypsum in calcite grain (arrow). PPL. Bar scale = 0.8 mm. Arapien Formation, Mayfield, Utah. (b) Well-formed gypsum crystals in micritic limestone, pseudomorphed by calcite. Voids resulted from plucking during preparation of section. PPL. Bar scale = 1.8 mm. Messinian of Carboneras Basin, southern Spain. (Courtesy J. M. Rouchy.) (c) Nodular and microvein, fine-granular gypsum (white) in micritic limestone (black). Crude "chicken-wire" structure in lower left portion of photo. Gypsum may have replaced anhydrite that replaced primary selenite gypsum, or gypsum has replaced anhydrite replacements of carbonate (see Figure 38.10d). PPL. Bar scale = 1.8 mm. Arapien Formation, Mayfield, Utah.

Conversion of the gypsum back to carbonate (Figure 38.9b) could be by the reverse reaction or by dissolution of the gypsum and precipitation of calcite in these voids. In either case, these are diagenetic processes that occur in response to influx of solutions containing either the sulfate or carbonate ion. An epigenetic origin is possible if the appropriate rocks and solutions are available.

The textural relation of gypsum to carbonate shown in Figure 38.9c indicates that the gypsum postdates the carbonate. This, however, does not necessarily mean that the reaction carbonate → gypsum was direct. The gypsum may have replaced nodular anhydrite that had previously replaced the carbonate or that had replaced an earlier gypsum that replaced carbonate. The sequence of replacement may have been carbonate → gypsum → anhydrite → gypsum.

Conversion of Carbonate to Anhydrite. Carbonates can be replaced directly by anhydrite. Rectangular sections of anhydrite in micritic carbonate are shown in Figure 38.10a. Partial replacement of calcitic ooids by anhydrite (Figure 38.10b), partial replacement of micritic calcite pelloids by anhydrite (Figure 38.10c), and local replacement of micrite by anhydrite (Figure 38.10d) point to the reaction

$$CaCO_3 \text{ (calcite)} + SO_4^{2+} \rightarrow CaSO_4 \text{ (anhydrite)} + CO_3^{2+}$$

The reaction suggests that anhydrite is the stable phase instead of gypsum, which is at odds with experimentation that always produces gypsum first.

Conversion of Anhydrite to Calcite. Replacement of anhydrite laths by sparry calcite is shown in Figure 38.11. The anhydrite previously replaced micritic carbonate (Figure 38.10a). The double replacement texture suggests that reaction of carbonate with sulfate-bearing solutions was followed by reaction with carbonate-bearing solutions. Another possibility is that the anhydrite may have been dissolved and followed by crystallization of sparry calcite in anhydrite space.

Conversion of Gypsum to Calcite. Reaction producing sparry calcite pseudomorphs of perfect gypsum crystals is shown in Figure 38.9b. As with the calcite replacement of anhydrite (Figure 38.11), gypsum may have been dissolved and followed by precipitation of calcite in gypsum space.

Conversion of Gypsum to Anhydrite and Halite. Gypsum crystals may be pseudomorphically replaced by halite and anhydrite (Borchert and Muir, 1964; Hovorka, 1992; Schreiber and Walker, 1992). The reaction is

$$\underset{\text{gypsum}}{CaSO_4 \cdot 2H_2O} + Na^+ + Cl^- \rightarrow$$
$$\underset{\text{halite}}{NaCl} + \underset{\text{anhydrite}}{CaSO_4} + 2H_2O$$

One possibility is that this reaction is a dehydration type in which the brine is hydroscopic. Since gypsum can be expected to crystallize prior to halite from seawater, primary gypsum is subject to reaction with a brine that is becoming saturated in halite. Alternatively, the anhydrite may have initially replaced gypsum. Dissolution of anhydrite provided pseudomorphous space for later halite precipitation. Since gypsum and halite have no common ion, the dissolution–reprecipitation process is attractive. The earlier development of anhydrite by reactive replacement accounts for its presence in the pseudomorph along with halite.

Replacements Involving Polyhalite. Well-formed gypsum crystals can be pseudomorphically replaced by polyhalite and anhydrite (Stewart, 1949):

$$\underset{\text{gypsum}}{5CaSO_4 \cdot 2H_2O} + 2K^+ + Mg^{2+} + 2Ca^{2+} \rightarrow$$
$$\underset{\text{polyhalite}}{K_2Ca_2Mg(SO_4)_4 \cdot 2H_2O} + \underset{\text{anhydrite}}{CaSO_4}$$

A reaction producing polyhalite at the expense of well-formed gypsum crystals without in situ formation of anhydrite is shown in Holser (1966), suggesting the reaction

$$\underset{\text{gypsum}}{2CaSO_4 \cdot 2H_2O} + 2K^+ + Mg^{2+} \rightarrow$$
$$\underset{\text{polyhalite}}{K_2Ca_2Mg(SO_4)_4 \cdot 2H_2O} + 2SO_4^{2-}$$

Polyhalite may form at the expense of halite as indicated by finely crystalline polyhalite occurring along cleavages and at cleavage intersections in halite (Schaller and Henderson, 1932). Since the two minerals have nothing chemically in common, the replacement must involve total removal and total addition:

$$\underset{\text{halite}}{NaCl} + 2K^+ + Mg^+ + 2Ca^{2+} + 4SO_4^{2+} \rightarrow$$
$$\underset{\text{polyhalite}}{K_2Ca_2Mg(SO_4)_4 \cdot 2H_2O} + Na^+ + Cl^-$$

Perhaps influx of dilute solutions results in dissolution of halite and more-or-less simultaneous precipitation of the sulfate, which became polyhalite instead of

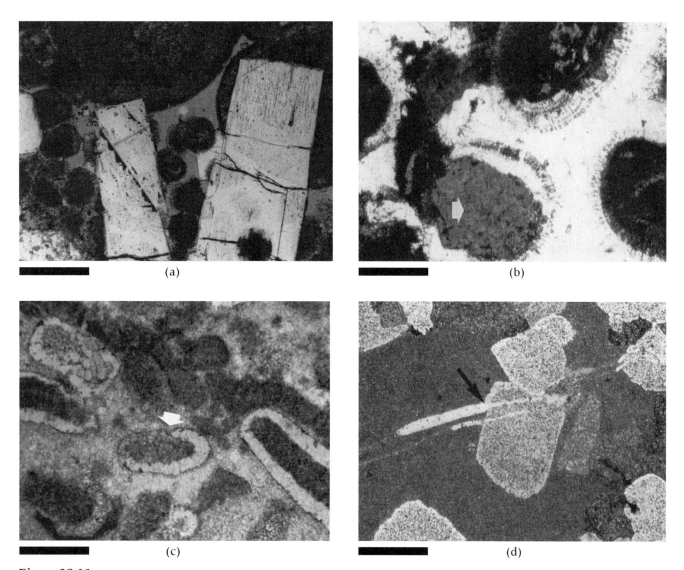

Figure 38.10
Anhydrite replacements of carbonates.

(a) Well-formed anhydrite crystals as replacements of micritized oolitic grainstone and calcite cement. X-polars. Bar scale = 0.25 mm. Smackover Formation, southeast Mississippi. (Courtesy E. Heydari and C. Moore.)
(b) Coarsely crystalline anhydrite with partial replacement of calcite ooids. Minute calcite inclusions define radial and concentric structure of ooids. Quartz nucleus (arrow) is resistant to replacement. Bar scale = 0.12 mm. Devonian of Belgium. Belgium Geological Survey core. (Courtesy J. M. Rouchy.) (c) Partial replacement of micrite pellets by fine granular anhydrite. Replacement, not overgrowth, is illustrated by lateral termination of anhydrite against micrite (arrow). Matrix is fine-granular anhydrite. Bar scale = 0.2 mm. Visean of northern France. (Courtesy J. M. Rouchy.) (d) Poikigrains (poeciloblasts) of anhydrite in micritic limestone. Replacement process has been complete only in thin clear rim zone and in optically continuous anhydrite microvein (arrow). Bar scale = 0.5 mm. Visean of Belgium. Geological Survey Belgium borehole. (Courtesy J. M. Rouchy.)

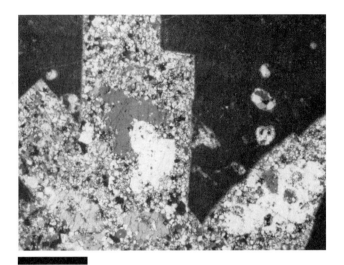

Figure 38.11
Well-formed anhydrite crystals in micritic limestone pseudomorphed by coarse and fine granular calcite. Compare Figure 38.10a. X-polars. Bar scale = 0.5 mm. Smackover Formation, southeast Mississippi. (Courtesy E. Heydari and C. Moore.)

gypsum or anhydrite because the original brine was enriched in potassium and magnesium.

Relatively large acicular polyhalite crystals in fine-granular anhydrite (Schaller and Henderson, 1932) illustrates the difficulty of interpretation of certain textures. Have the slender prisms of polyhalite replaced the finely crystalline anhydrite, or have they replaced gypsum that later recrystallized to anhydrite?

Reactive Replacement Involving Continental Evaporites. Reactions indicated by textural relations also apply to continental evaporite minerals. Smith and Haines (1964) show a bed of borax overlain and underlain by trona. Local replacement of borax by trona is indicated along these contacts, suggesting the reaction

$$Na_2B_4O_7 \cdot 10H_2O \rightarrow Na_2CO_3 \cdot NaHCO_3 \cdot 2H_2O$$
$$\text{borax} \qquad \qquad \text{trona}$$
$$+ BO_3^{2-} + 6H_2O + 3H^+$$

The textural relation shown in Figure 38.12 indicates replacement of colemanite by gypsum. The following reaction is indicated:

$$Ca_2B_6O_{11} \cdot 5H_2O + 2SO_4^{2+} + 2O_2 \rightarrow$$
$$\text{colemanite}$$
$$2CaSO_4 \cdot 2H_2O + 6BO_3^{2-} + H_2O$$
$$\text{gypsum}$$

Evaporite Clastics and Associated Stromatolites

Since there is a beach environment located between subaqueous evaporite environments and subaerial environments (Figure 37.2), the formation of clasts from evaporite minerals and rocks is to be expected. Evaporite sequences that are shoaling upward—that is, evaporite beds and lamellae that progress into the surf zone—are susceptible to breakup in this high-energy environment. Additionally, subaerial evaporites, as in the sabkha environment or on desiccation surfaces, are subject to mechanical reworking during storms.

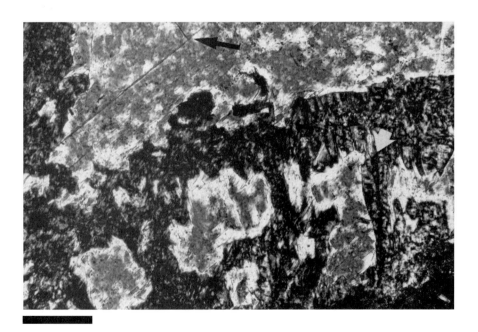

Figure 38.12
Partial replacement of colemanite (gray) with lower birefringent thin rims (white) by alabastrine gypsum (black). Note right-angle cleavage intersection (near top, black arrow), "islands" of colemanite in gypsum, and microvein of gypsum in colemanite (lower right, stubby white arrow). X-polars. Bar scale = 0.12 mm. Horse Spring Formation. Southern Nevada. (Courtesy S. Castor, Nevada Bureau of Mines and Geology.)

Desiccation cracks aid in the formation of rip-up clasts. Sands and larger grains are generated from individual minerals and laminated rock. Ripples and cross-laminations are typical hydrodynamic structures. Planktonic fauna such as forams may mix in with these clasts, becoming a significant part of the arenites and rudites that eventually form (Hardie and Eugster, 1971). Such lithoskeletal grainstones and packstones are rather common. A **gypsrudite** is shown in Figure 38.13, in which derivation of clasts is from laminated gypsum. Windblown silts and sands are derived from the clasts initially formed by wave action. Foreign particles may also be blown in, contributing to the clastic sediment accumulation.

There may also be additional chemical precipitation on clastic grains. Pisoids, peloids, and ooids of halite (Weiler et al., 1974), gypsum, and anhydrite occur. The intertidal zone of marine evaporites (Figure 37.2) is commonly inhabited by algae. Sticky mucilaginous mats, up to 1 foot thick, secreted by cyanobacteria (blue-green algae) become traps for clastic sediments. Stromatolites are formed in this manner, but there may also be nonbiogenic precipitation of carbonate contributing to the formation of these structures. Delicate laminites consisting of carbonates and gypsum occur. Halite also occurs in algal mats. Evidently, crystals physically trapped in the mats are sites of continued crystal growth.

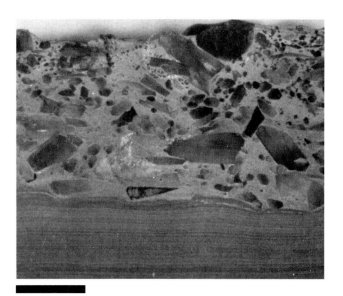

Figure 38.13
Gypsrudite (upper) with clasts of gypsum in gypsum sand. Deformed soft sediment contact with finely laminated gypsum (lower). Slab surface. Bar scale = 2 cm. Messinien of Psematismenos Basin, Cyprus. (Courtesy J. M. Rouchy.)

Low-Temperature Aqueous Precipitative Rocks

The "Needles" at Pyramid Lake, Nevada. Alignment of tufa mounds extend from north shore southward into the lake in two directions along fault lines. Holocene spring activity occurs locally at the base of exposed tufa mounds. (Courtesy Cordell T. Gray.)

Precipitation of minerals in aqueous solutions independent of pre-existing rock and biotic activity occurs in response to a shift in chemical equilibrium (Chapter 12). A shift to saturation and supersaturation can occur if there is (1) evaporation of water, (2) a change in temperature, (3) a change in total pressure, (4) a change in the activity of carbon dioxide, or (5) a change in solute concentration. Precipitation resulting from evaporation forming evaporites is described in Chapter 38. Precipitation caused by the other four factors in the low-temperature environment can be caused by mixing of solutions from different sources. Change in temperature also can be diurnal and seasonal. Loss of CO_2 can be caused by solution agitation and relocation of solutions from zones of higher pressure to lower pressure in the groundwater system as well as by photosynthesis.

Although the chemical processes leading to precipitation are well known, those causing the accumulation of mineral precipitates in rock-forming processes are not because precipitations occur in conjunction with other rock-forming process, such as (1) hydrothermal activity (Chapter 34), (2) weathering (Chapter 40), and (3) diagenesis of clastic and biochemical sediments (Given and Wilkinson, 1985, Chapters 36 and 37).

Other than evaporite systems, there are very few geological circumstances where abiotic low-tempera-

ture chemical precipitation leads directly to the formation of rock without the involvement of other rock-forming processes. However, if there were none, there would be no need for this chapter.

The Thinolite Tufa Story

One of the best examples of nonbiotic precipitation in aqueous solutions at low temperature, leading to the formation of rock, is the formation of thinolite tufa.

Thinolite tufa consists of a directional network of what appear to be crystals (Figure 39.1). The crystals are morphologically similar to those of aragonite (Figure 39.2) and gaylussite ($Na_2CO_3CaCO_3 \cdot 5HO_{20}$) (King, 1878). The thinolites (Russell, 1885) actually consist of rounded groups of calcite grains, some with concentric structure, clustered in the skeletal "crystals" (Figure 39.3). Because calcite has no morphology matching the thinolites (they are not scalenohedrons) and it is granular in the skeletal thinolite units, a pseudomorphic relationship is indicated.

A recent suggestion that the mineral **ikaite** ($CaCO_3 \cdot 6H_2O$) (monoclinic) may be the primary precipitate solves both the morphological and compositional problem (Shearman et al., 1989).

Ikaite is known to form only in cold water such as on the floor of the Ika Fjord of Greenland (Pauly,

Figure 39.1
Thinolite tufa. Pyramid Lake, Nevada, U.S. quarter scale.

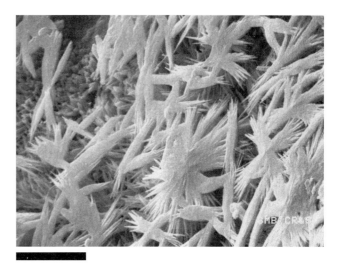

Figure 39.2
SEM view of aragonite from warm spring travertine system, southwestern Colorado. Morphology of aragonite clusters vaguely similar to thinolite form. Bar scale = 4 μm. (Courtesy Henry S. Chafetz.)

1963). The tie-in with cold water fits the occurrence of thinolite tufa in Pleistocene lake basins of the Great Basin of Nevada and California. Tufa mounds, pinnacles, and domes are now exposed along the shores of Pyramid Lake (Figure 39.4) and Mono Lake, which are the remnants of deep-water Pleistocene Lake Lahontan and Lake Russell respectively. Alignment of tufa mounds at the north end of Pyramid Lake (frontispiece photo) indicates formation of thinolite tufa where there was and still is sublacustrine discharge of water along fault structures.

Assuming that the sublacustrine springs contained an appropriate content of calcium and bicarbonate, supersaturation occurs in water near 0°C and ikaite precipitates as long as the crystallization of calcite is depressed. Bischoff et al. (1992) suggest that the presence of orthophosphate is a likely inhibitor of calcite crystallization that does not affect precipitation of ikaite. The concentration of orthophosphate may have been relatively high in cold waters of the deeper Pleistocene lakes where anoxic conditions controlled organic decay.

Since ikaite is unstable in warm water, pseudomorphic replacement by calcite is not unexpected in response to climate changes through the Pleistocene and into the Holocene.

Precipitation Resulting from Mixing of Aqueous Solutions

Mixing of Two Low-Temperature Waters

Mixing of two relatively cool waters occurs (1) along the interface of two groundwater flow systems, (2) along the encroachment zone between continental groundwater and the phreatic zone beneath lakes and oceans (Hardie, 1987), (3) in lake and ocean water receiving fluvial inflow (Sholkovitz, 1976), and (4) by

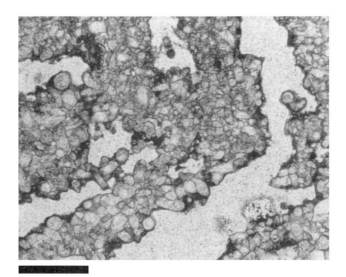

Figure 39.3
Granular, zoned calcite, locally with micritic mud (dark matrix) pseudomorphic after skeletal structure of presumed ilkaite. PPL. Bar scale = 0.5 mm. Thinolite tufa, Pyramid Lake, Nevada.

Figure 39.4
(a) Tufa dome. Exposed interior of tufa dome characterized by radial thinolite and concentric layering defined by differing sizes of thinolites and some nonthinolitic dendritic tufa. Pyramid Lake, Nevada. Hammer scale (bottom center). (b) Row of tufa pinnacles (looking east), north shore Pyramid Lake, Nevada (see frontispiece photo).

mixing of deeper marine water with shallower marine water attendant to upwelling.

Runnells (1969) and others have pointed out that solubility of carbonates is a nonlinear function of the partial pressure of CO_2. This means that mixing of two waters that are in equilibrium with respect to CO_2, but in which the partial pressure of CO_2 differs, generates an undersaturated solution that has the potential to dissolve carbonate rock because the equilibrium curve is concave downward (Figure 39.5a). Differing contents of NaCl in carbonate solutions can also cause dissolution because the equilibrium curve for such compositions is also concave downward. Conversely, the solubility curve for the $CaSO_4$–$CaCl_2$–H_2O system is concave upward (Figure 39.5b). This means that mixing of waters with different contents of $CaCl_2$, both saturated in gypsum, depresses the solubility of gypsum and leads to precipitation. This phenomenon is called the **common ion effect** (Chapter 12).

Precipitations may also occur if low-temperature solutions of differing pH are mixed. Substances like $Al(OH)_3$ and $Fe(OH)_3$ possess both acidic and basic properties, passing through a minimum in solubility as a function of pH. Mixing of continental acid waters poor in dissolved salts with basic seawater of high ionic strength may result in precipitation of manganese and iron minerals (Shoikovitz, 1976).

Sublacustrine spring water containing dissolved calcium carbonate may be cooler than lake water, especially during the summer. Warming of the spring water by mixing may result in precipitation of calcium carbonate, since higher temperature results in loss of CO_2 and decreased solubility of calcium carbonate. The same results are achieved by mixing of relatively warm sublacustrine spring water with cool or very cold lacustrine or ocean water, precipitating either calcite, aragonite, or ikaite.

Pressure reduction and warming of relatively phosphorous-rich seawater attendant to **upwelling** along continental margins leads to the formation of **phosphorites** (Baturin, 1982; Slansky, 1986). Cold water can dissolve more CO_2 and more calcium phosphate (apatite) than warm water. Deep seawater saturated in apatite is slightly alkaline, containing about 0.3 ppm PO_4. Degassing of dissolved CO_2 occurs during upwelling as a consequence of total pressure reduction and temperature increase. This results in a rise of pH, and since apatite is less soluble in alkaline waters, even if they are warmer, there is precipitation of apatite (phosphorite), reducing the concentration of PO_4 to about 0.01 ppm. The zone of phosphate precipitation occurs between about 30 and 500 m. A shallower environment results in consumption of phosphate by phytoplankton during photosynthesis, and a very deep environment with a high content of CO_2 prevents phosphate saturation.

A similar story of upwelling leads to the deposition of **iron formations** (Drever, 1974; Button et al., 1982) and **manganese deposits** (Force and Cannon, 1988; Frakes and Bolton, 1992). Since the sources of these waters are different, the chemistries must be at least slightly different. Differences in chemistry may result in precipitations, but mixing may also result in dissolution of pre-existing rock.

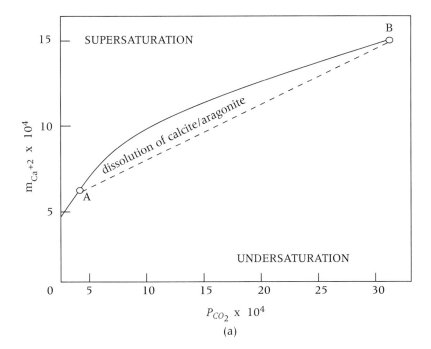

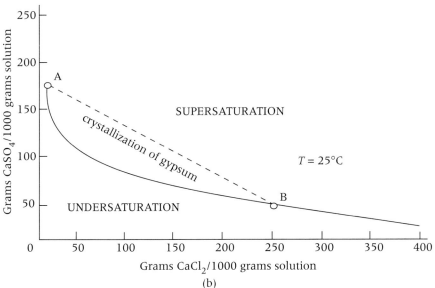

Figure 39.5
Dissolution and precipitation as a consequence of mixing of aqueous solution. Diagrams modified from Runnels (1969).

(a) Solubility of calcite as a non-linear function of partial pressure of CO_2. Mixing of saturated solutions A and B results in a locus of mixed waters along dashed line that are undersaturated in calcite and have the potential to dissolve co-existing calcite. (b) Solubility of gypsum as a function of added calcium chloride containing the common ion calcium. Mixing along line A–B results in supersaturation and potential precipitation of gypsum.

In the iron formation model (Button et al., 1982), ferric hydroxide and silica precipitates in open marine waters. The iron could originate from exhalative sources along active volcanic arcs and rifts (Gross, 1980). The precipitates settle to deep reducing regions of ocean basins. The iron-bearing precipitates are reduced, placing iron back into solution. Upwelling of these waters, enriched in iron, leads to precipitation associated with carbonates (including siderite) on continental shelves.

Continental-derived groundwater interfacing with groundwater of marine origin has led to precipitation of a variety of carbonates including the direct precipitation of dolomite (von der Borch et al. 1975). Mixing of meteoric water with marine water is thought to have resulted in the diagenetic dissolution of halite and precipitation of dolomite and calcite (Hanford and Moore, 1976).

Mixing of Low-Temperature Water with Hydrothermal Water

Mixing of cold meteoric water with hydrothermal solutions can result in precipitations significant in the formation of some economically important rocks

(Chapter 34). Some of these mixing aqueous systems mark the transition from strictly low-temperature rock-forming phenomena to hydrothermal activity. Magmatic water, warm connate water, or meteoric water heated by nearby magmatic systems can mix with cold meteoric water to cause precipitation.

Sublacustrine thermal springs are one example of precipitation induced by mixing, leading to the formation of sulfide deposits (Tiercelin et al., 1989). Precipitation of silica may occur at the sites of sublacustrine silica-bearing hot springs in saline alkaline lakes as a result of mixing. Mixing of the hot spring water with cold lake water results in rapid cooling, a probable drop in pH. Silica precipitates because it is less soluble at lower temperatures and at pH below 9. The silica may initially take the form of a sodium silicate precursor such as magadiite ($NaSi_7O_{13} \cdot 3H_2O$) (Eugster, 1967), or it may occur as a silica gel (Renaut and Owen, 1988).

Exhalative Mn and Fe in hydrothermal **"black smokers"** mixing with cold, oxygen-depleted seawater results in precipitation of pyrite, base metal sulfides, and iron–manganese minerals (Chapter 34). Mixing with oxygenated seawater may cause partial or total oxidation of some of the metals and precipitation of respective oxides, hydroxides, and in some cases silicates. The Algoman type banded iron formations (or **banded ironstone formations**, Young, 1989) may have formed from iron exhalations (Maynard, 1991). The Lake Superior type banded iron formations (Figure 36.8a) may also be dominantly chemical precipitates, especially in view of their varvelike (seasonally controlled) characteristics (James and Trendall, 1982), and these formations may also have an exhalative source for the iron. Evidently Mn can be kept in solution for considerable time and travel some distance from the exhalation source, forming manganese nodules on the ocean floor. Brines originating in the subseafloor at geothermal vents are likely to generate sulfides on the ocean floor, such as the Kuroko deposits of Japan (Ogura, 1972), without mixing with oxygenated seawater.

At Creede, Colorado (Hayba et al., 1985; Plumlee, 1989) heated meteoric waters rich in chlorine, sodium, carbonate, sulfate, and soluble organics probably mixed with cooler dilute meteoric waters. Metal deposition, including gold, occurs as a result of the changes in salinity, pH, and temperature. At Jerrit Canyon, Nevada (Hofstra et al., 1989), gold-bearing **jasperoid** resulted from mixing of two meteoric fluids having contrasting temperatures and water–rock exchange histories. One water was an acid, $Cl–CO_2–H_2S$–rich brine containing gold; the other was an oxidizing, low-salinity, neutral-pH water.

Precipitation Resulting from External Factors Without Mixing

Change in Temperature

For saturated solutions of halite, diurnal or seasonal drops in the temperature of brines in an evaporite environment can result in precipitation because of reduced solubility. Conversely, seawater infiltrating basalt tephra near hot dikes beneath Surtsey volcano (Jakobsson and Moore, 1986) results in precipitation of anhydrite because of reduced solubility at the higher temperatures (Figure 12.7).

If an aqueous solution saturated in calcium carbonate moves into a warmer environment, there is degassing of CO_2 and precipitation of calcite or aragonite.

Change in Pressure

Outgassing of CO_2 from aqueous solutions at or near saturation in calcium carbonate can result from reduction of total pressure. For example, rise of solutions from depth to surface and near-surface regions can result in boiling and precipitation of carbonates in epithermal ore deposit systems and as travertine at hot springs (Steinen et al., 1987). At Mammoth Hot Springs in Yellowstone National Park, aragonite stalagmitic "shrubs" form by such inorganic processes (Penticost, 1990).

The system need not be hydrothermal. Cool groundwater saturated with calcium carbonate entering fractures or caves in carbonate rocks experiences a very slight but significant reduction in pressure. Even the slightest outgassing of CO_2 leads to the precipitation of calcium carbonate and the formation of speleothems (Chapter 40).

Agitation of water saturated in calcium carbonate can result in CO_2 loss and precipitation of calcite. This is essentially a CO_2 degassing resulting from slight change in pressure, triggered by the agitation. Streams issuing from carbonate terrain lose some of their dissolved calcium carbonate as precipitates on rock surfaces where there is turbulent flow.

40

Weathered Rocks

Hematite-bearing zone (darker, above arrows) resulting from oxidation of chalcocite enrichment blanket attendant to lowered water table. New chalcocite zone is beneath the hematite (lighter portion), which extends to protore (chalcopyrite zone). Santa Rita (Chino Mine), New Mexico, 1958. From Anderson (1982.)

Weathering As a Rock-Forming Process

A **weathered rock** may be no more than a basalt having a millimeter-thick limonitic rind, or it may be a smectite-rich quartz-bearing rock containing only vague outlines of feldspar phenocrysts as indicators of the granite porphyry that it used to be.

Chemical weathering is an approach to chemical equilibrium in a low-temperature system involving rocks and minerals, air, water, and, in some cases, organics (Goldich, 1938). The reactive constituents of air are oxygen, water vapor, and carbon dioxide; the reactive components of water are the water itself and its dissociation products, dissolved oxygen, carbon dioxide and its byproducts, chlorine, sulfate, and dissolved metals and semimetals. Rainwater is relatively free of nonvolatile substances, acquiring dissolved inorganic and organic substances as the water becomes part of the meteoric water system interacting with rock materials.

Most weathering reactions are very sluggish, in large part due to kinetic barriers inherent in low-temperature chemical phenomena. Solid weathering products include both open-space precipitates and reactive replacements. The chemical environment in the surface and near-surface region is typically oxidizing because of the presence of free oxygen. Local reducing conditions exist, especially where there is an abundance of organic matter.

Oxidation involves conversion of a wide variety of minerals, including anhydrous silicates, to oxidized and hydrolyzed equivalents. Sulfides are converted to sulfates, and ferrous iron is oxidized to ferric iron in the form of hematite and hydrolyzed minerals such as goethite. There is simple hydration of anhydrite to gypsum, and dehydration of gypsum to anhydrite. Chemical weathering may also involve congruent dissolution, leaving pores and cavities in rocks such as limestones and evaporites that may or may not be filled by later precipitates.

Documentation of weathering in the subaerial environment is much more extensive than beneath the floor of lacustrine and ocean water bodies. Weathering processes in the subaerial **vadose** (water unsaturated) zone involve both gases and water solutions, whereas processes beneath water bodies and below the water table on continental terrain are in the **phreatic** (water saturated) zone and involve water containing dissolved solids and gases.

The temperature and pressure regime of weathering are both low, but there is considerable interfacing with hydrothermal waters (Chapter 34) and low-temperature aqueous precipitating systems (Chapter 39). Weathering affects both bedrock and unconsolidated surface sediments. Weathering processes are transitional to diagenetic processes occurring in buried sediments (Chapters 36–38). For example, accumulations of siliciclastic, calcareous, phosphatic, ferruginous, manganiferous, or gypsiferous sediments may be cemented in the weathering zone, a process that can also be considered very early diagenetic.

Soils are major products of weathering and may be converted into **soilstone** by cementing processes that most commonly generate **ferricrete**, **calcrete**, and **silcrete**. A clue to the soil origin of lithified sediments is the presence of **rhizoliths**, organosedimentary structures, or their molds, produced by mineral replacements of plant roots (Klappa, 1980).

Organic acids commonly form in aerobic soil environments and enhance chemical reactivity by facilitating dissolution. For example, **chelation** is the bonding of metal and semimetal cations to organic molecules excreted by lichens. The resulting H$^+$ ions and chelate are readily soluble in water, participating in weathering reactions.

Weathering includes additional features resulting from precipitation and reaction. Among these are incrustations, nodules, ooids, pisoids, concretions, geodes, veinlets (such as quartz, alunite, and gypsum), and the formation of mineral and fossil casts. Rhythmic precipitations in fluid-saturated rock generates **Liesegang rings**.

Rock–Fluid Reaction in the Weathering Zone

Dry air is not an effective weathering agent, but water containing dissolved oxygen and carbon dioxide, including meteoric water of the atmosphere, fluvial systems, and vadose groundwater zone, all typically being slightly acidic, are quite reactive. Water acidity is increased by the presence of organic acids and sulfuric acid generated by the oxidation of sulfides. The pH–Eh regions of water are shown in Figure 40.1.

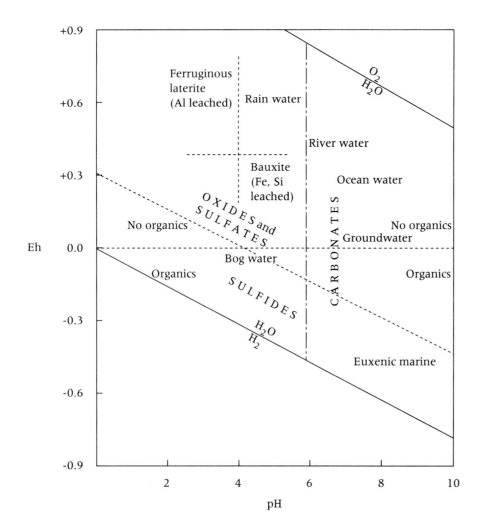

Figure 40.1
Eh–pH diagram showing position of waters, presence and absence of organics, relative stability of carbonate, sulfide, oxide, and sulfate minerals, and the sensitivity of ferruginous and aluminous laterites with respect to these chemical factors. Based on Garrels and Christ (1965); Norton (1973).

Fundamental Reactions

Quartz, plagioclase, alkali feldspar, and carbonate minerals occur in abundance in a variety of rocks. Quartz participates very little in weathering reactions because it is relatively insoluble in the weathering zone. Solution of carbonate minerals is an important weathering process, but reprecipitation of carbonates is also a weathering process.

Feldspars participate in weathering reactions that generate new minerals by way of incongruent dissolution. A few of the more pertinent reactions are given here. K-feldspar or the K-feldspar component of alkali feldspar react with water in an acidic environment to produce sericitic muscovite (or illite, which is lower in K and higher in OH than muscovite):

$$3KAlSi_3O_8 + 12H_2O + 2H^+ \rightarrow$$
K-feldspar
$$KAl_3Si_3O_{10}(OH)_2 + 6H_4SiO_4 + 2K^+$$
muscovite

Muscovite can react with water in an acid environment to form kaolinite:

$$2KAl_3Si_3O_{10}(OH)_2 + 3H_2O + 2H^+ \rightarrow$$
muscovite
$$3Al_2Si_2O_3(OH)_4 + 2K^+$$
kaolinite

The direct reaction from K-feldspar to kaolinite is

$$2KAlSi_3O_8 + 2H^+ + 9H_2O \rightarrow$$
K-feldspar
$$Al_2Si_2O_5(OH)_4 + 4H_4SiO_4 + 2K^+$$
kaolinite

K-feldspar (and albite) can also generate montmorillonites and mixed-layer illite–montmorillonite. Montmorillonite is likely to contain Na, Ca, as well as the K, Fe, and Mg that characterize illites. Kaolinite is devoid of such elements:

Dissolution of plagioclase generates Na and Ca. Montmorillonite commonly is formed at first, but other minerals such as K-smectite, analcite, phillipsite, and calcite may form.

Dissolution of calcite and dolomite can be shown by the reactions

$$CaCO_3 + CO_2 + H_2O \rightarrow$$
calcite
$$Ca^{2+} + 2HCO_3^- CaMg(CO_3)_2 + 2CO_2 + 2H_2O \rightarrow$$
dolomite
$$Ca^{2+} + Mg^{2+} + 4HCO_3^-$$

Any mineral, such as pyrite, biotite, hornblende, and augite, that liberates ferrous iron, typically leads to the formation of iron oxides and iron oxyhydroxides:

$$2Fe^{2+} + 1.5O_2 \rightarrow Fe_2O_3$$
hematite
$$2Fe^{2+} + H_2O + 1.5O_2 \rightarrow 2FeOOH$$
goethite–lepidocrosite

Limonite is mostly a mixture of goethite and clay minerals.

Any sulfide is likely to generate sulfuric acid along with corresponding metal cations in solution or precipitated as oxides:

$$2FeS_2 + 15/2O_2 + 4H_2O \rightarrow Fe_2O_3 + [4SO_4^{2-} + 8H^+]$$
hematite   sulfuric
           acid

Extent of Weathering

A weathered rock may not be much different from its protolith, or it may be so changed that identification of protolith is difficult or impossible. Weathering may (1) modify the attributes of an original rock (montmorillonite pseudomorphs of feldspar phenocrysts), (2) generate a rock from unconsolidated sediments (calcrete from bioclastic remains, silcrete from quartz sands), (3) completely destroy the attributes of a protolith by dissolution and reprecipitation (**dripstone**), or (4) generate nonrock particle accumulations (grus, soils).

Weathering As a Function of Rock Type

Weathering of Basalt

Basalt is the most common igneous rock in ocean basins and is also abundant on continents. Minerals such as calcic plagioclase, hornblende, pyroxene, and olivine have relatively more ionic bonding within their lattices compared to a more covalent component of bonding in siliceous minerals such as quartz, albite, and K-feldspar. As a consequence, weathering reactions proceed rapidly in mafic igneous rocks relative to felsic igneous rocks. Furthermore, basalts may have a glassy matrix and glass is very unstable (soluble) relative to its crystalline equivalent.

Olivine reacts with water as follows:

$$Mg_2SiO_4 + 4H_2O \rightarrow 2Mg^{2+} + 4OH^- + H_4SiO_4$$
forsterite   silicic acid

The magnesium released may generate sepiolite as follows:

$$2Mg^{2+} + 3H_4SiO_4 + 4OH^- \rightarrow Mg_2Si_3O_6(OH)_4 \text{ (sepiolite)} + 6H_2O$$

Magnesium reacting with basaltic glass (**sideromelane**, or **tachylite**) and water may generate a variety of smectite minerals:

$$\text{basaltic glass} + Mg^{2+} + H_2O \rightarrow \text{smectite (montmorillonite)} + Ca^{2+} + K^+ + H_4SiO_4 + H^+ + Fe^{2+} + Mn^{2+}$$

If alteration is submarine, an abundance of Mg ions may result in the formation of chlorite, and potassium from seawater may generate K-rich smectites. Under mildly oxidizing conditions in the ocean, both ferrous and ferric ions may locate in glauconite $[KMgFe(SiO_3)_3 \cdot 3H_2O]$. Near subaerial or submarine hot springs, potassium and iron may locate in jarosite $[KFe_3(OH)_6(SO_4)_2]$. Calcium may end up in minerals such as anhydrite, calcite, Ca–smectite, and phillipsite $[KCa(Al_3Si_5O_{16}) \cdot 6H_2O]$.

Alteration of pillow basalts and basaltic tuffs (Hay and Iijima, 1968) generates an assemblage of zeolites and clay minerals that occurs as a characteristic yellowish brown or reddish brown waxlike or vitreous material known as **palagonite**. **Bentonitic tuff** is a dark olive gray rock containing smectites and amorphous silica as weathering products of glassy volcanic rock.

Weathering of Granitic Rocks

Granites, granodiorites, quartz diorites, and their gneissoid equivalents are abundantly exposed on the continents. Weathering of such rocks occurs where they have been locally exposed in uplifted tectonic blocks and where they have been stripped of their sedimentary rock veneer in cratonic regions.

Weathering of granitic rocks produces rocks ranging from "rotten" granite to well-developed soil profiles (Harriss and Adams, 1966). In areas of significant relief, **grus** may form that is immediately transported away from the site of weathering. Disintegration to grus is closely tied to the reaction of water with biotite, forming hydrobiotite and hydrobiotite–biotite interlayers (Wahrhaftig, 1965; Isherwood and Street, 1976). Expansion of biotite generates cracks in and between adjacent minerals, this physical degradation being accompanied by additional reactions among altered biotites, hornblende, feldspar crystals, and water that infiltrates along these fractures. Minerals such as vermiculite (Figure 40.2), smectite, and sericite form, and an active grus-generating surface is developed (Figure 40.3a). Higher content of biotite in granitic complexes leads to relatively advanced grussification. Additionally, water access along joints localizes grussification and may leave relatively unweathered granitic rock in the interior (**spheroidal weathering**) (Figure 40.3b). Iron staining (especially on quartz crystals) in the form of Liesegang rings may occur within granitic rock that is not yet grussified but has iron mobilized from biotite and hornblende.

Weathering alteration of alkali feldspar is less advanced than plagioclase at the time of grussification. Relatively fresh grains of alkali feldspar are common in gruss along with quartz. This is one of the reasons why arkosic arenites are more common than plagioclase-bearing arenites.

Weathering of Nepheline Syenites

Weathered syenites and nepheline syenites do not have the relicts of quartz crystals that survive weathering of granites. The fundamental reaction equations generating sericite (or illite) at the expense of alkali feldspar apply regardless of the presence or absence of quartz. Silica-undersaturated magmatic rocks containing plagioclase yield smectite clay minerals. Intense leaching in a temperate or tropical environment may convert the illite and montmorillonite to kaolinite. Silica leaching may generate gibbsite, a principal ingredient of bauxites formed from nepheline syenites (Gordon et al., 1958).

Weathering of Silicic Volcanic Rocks

Many silicic pyroclastic flow and lava flow rock contain glass or partially devitrified glass. Glass is particularly unstable in the weathering regime, typically

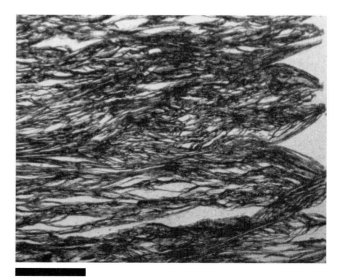

Figure 40.2
Vermiculite with characteristic expanded micaceous texture. PPL. Bar scale = 1 mm.

Figure 40.3
Grussification of granite.
(a) Surface of coarse-grained two-feldspar granite actively supplying grus to regolith. Notch Peak granite, Utah. Scale = U.S. dime (lower right). (b) Contrast between unaltered and grus-generating granodiorite altered laterally from joint surfaces that allowed fluid penetration into the rock. Austin Granodiorite, Nevada. Scale = lens cap (right).

altering to montmorillonites (Papke, 1969). Thorough leaching of the montmorillonite converts it to kaolinite.

Weathering of Siliciclastic Sedimentary Rocks

Siliciclastic sedimentary rocks are characterized by quartz, K-feldspar, and an assortment of micaceous and clay minerals including muscovite, chlorite, and smectites. Plagioclase occurs only in cases of rapid burial where it may be protected from decomposition by weathering and diagenetic processes.

Weathering of quartz arenites is inhibited by the non-reactive quartz, and this is one reason why quartz arenites are cliff-formers. Nevertheless, weathering of quartz sand deposits can result in the formation of silcretes (see page 517) if environmental conditions are appropriate (Thiry and Millot, 1986).

Arkoses, especially those with plagioclase as well as alkali feldspar, are generally more susceptible to weathering reaction than quartz arenites, which involve the generation of clay minerals at the expense of the feldspars. Carbonate cements in quartz sandstones may be removed during weathering, rendering the sandstone friable.

Litharenites and wackes contain a higher proportion of plagioclase and amphiboles than arkoses and are more reactive in the weathering regime. Weathering of mud rocks is limited by the relative nonreactivity of their clay mineral content, being already a product of prior weathering. Nevertheless, remobilization of mud from mud rock can generate some bizarre rock forms (Figure 40.4).

The feldspathic arenites and wackes, as well as shales and siltstones, contain Al-bearing minerals. With the appropriate climatic setting (see later section), weathering may generate bauxitic laterites. The famous Guinean bauxite deposits are developed in shales and slates.

Weathering of Carbonate Sedimentary Rocks

Limestones and dolostones can be cliff formers, just like quartz arenites; yet dissolution of carbonates in the weathering zone is well known. This apparent paradox can be explained as follows. Rainwater on a carbonate rock surface has little or no dissolution power even though the water has a little dissolved atmospheric CO_2 and therefore is very slightly acidic. In a dry climate there is a paucity of water in the first place, and limestones tend to form positive geomorphic forms similar to quartz arenites. In wetter climates, meteoric water that has penetrated carbonate rock along fractures can mix with groundwaters that generally have a higher content of dissolved CO_2. In that case, since an increase in the partial pressure of

Figure 40.4
Redeposition of mud in weathered mud rocks.
(a) Vertically oriented badland-like erosional forms with surficial redeposition of mud in horizontally bedded lacustrine calcareous mud rock. Cathedral Gorge, Panaca, Nevada. Scale = lens cap (bottom). (b) Pisolitic-like surface of bentonitic mudstones resulting from subaerial weathering. Sandy interbeds unaffected. Brushy Basin member of Morrison Formation, near Cortez, Colorado. Scale = U.S. quarter (right).

---

carbon dioxide promotes solution of calcium carbonate (Chapter 12), the formation of **karst terrain** in humid climates is to be expected.

Weathering of carbonate rocks includes all those void and cavernous features resulting from dissolution, as well as all those features resulting from reprecipitation of the dissolved carbonate. Thus, **caliche**, **calcrete**, and **speleothems** (dripstone) are rocks or modification of rocks resulting from weathering.

Because pure carbonate rocks contain very little Al-bearing material, it would seem that bauxites could not form by weathering of carbonate rocks. Surprisingly, there is extensive development of **karst bauxites** (Bardossy, 1982) resulting from weathering in appropriate conditions of climate and drainage (next section), such as the **Terra Rosa** grounds of the Mediterranean. The bauxite is formed from interlayered claystones, volcanic ashes, and argillaceous sediments washed into the depressions of the karst topography.

Weathering of Metamorphic Rocks

The weathering products and resulting weathered rocks derived from metamorphic rock are governed by climate and original rock type. Weathering of serpentinite is not likely to produce a rock anything like the weathered equivalent of a mica schist. On the other hand, the weathering of an amphibolite can be expected to produce soils similar to those derived from basalt, unless the basalt contains glass.

Weathering of Rocks Containing Sulfides

Any rock that contains a sulfide, whether it be a pyrite-bearing black shale or a rock with hypogene hydrothermal sulfide mineralization, is particularly reactive in the weathering regime because of the generation of sulfuric acid. The low resultant pH condition overrides many of the aspects of original rock composition.

Reactions producing sericite, illite, montmorillonite, or kaolinite at the expense of feldspars are driven by the H^+ ion. **Supergene alteration** of hydrothermal deposits depends on this acidic (and oxidizing) condition. Ferruginous and aluminous **laterites** form in a low pH (and high Eh) environment (Figure 40.1), and **gossans** form in the portions of sulfide-bearing ore bodies that are exposed to the same condition (Blanchard, 1968). Gossan rock also forms in pillow basalts as a result of seafloor weathering reactions involving oxidation of prior hydrothermally introduced pyrite (Haymon et al., 1989).

Weathered Rocks

Soilstones

Soils develop by long-term interaction of rocks or rock materials with the atmosphere and water at the Earth's surface. Soils can form on any rock type under the appropriate climate and geomorphic circumstances (Rose et al., 1979; Birkeland, 1984; Hausenbuiller, 1985). Soils are capable of sustaining plant growth, consisting of an organic-dominating **O horizon**, an organic-bearing **A horizon**, an eluvial (leached) **E horizon** (A2 horizon) containing mostly silty quartz, an illuviated (redeposited) **B horizon** containing some humus but mostly clay minerals and iron oxyhydroxides, a **C horizon** that has no clay and no humus but some precipitates such as caliche, and an **R horizon** that is fresh bedrock (Figure 40.5).

There are several variations of this fundamental soil structure, depending on host-rock type, climate, and drainage characteristics (Milnes et al., 1987; Nahon, 1991). For example, **pedalfers** are characterized by an enrichment of kaolinite and iron oxide minerals in the B horizon, and the pedalfer variant known as **podzol** has a white clay-rich A zone and an Fe-rich B zone. Both are common in temperature regions. **Lateritic soil** is characterized by Fe enrichment in a tropical climate. **Pedocal** soils develop in semi-arid climates and have a very thin A horizon (no O horizon), a montmorillonite-rich B horizon, and caliche precipitation in the C horizon.

The significance of these soil profiles with respect to rock-forming processes is that they may become lithified into **soilstone**. Recognition of such soils is important in deciphering modern surface processes as well as **paleosols** that have become buried in the stratigraphic section. **Pedogenic** (soil forming) fabrics, such as **rhizoliths** (**rhizoconcretions**) (concretion-like root casts), are diagnostic of the former presence of soil (Morrison, 1964; Mack, 1992).

Ferricretes (Iron Crusts)

Groundwater carrying ferrous iron may precipitate in an oxidizing environment in permeable regolith (unconsolidated rock debris) forming **ferricretes** (Nahon, 1986, 1991). Parts of the Cohansey sand formation in New Jersey have been cemented by hematite precipitating from groundwater (Figure 40.6a). Ferricretes typically form in lateritic (iron-rich in this case) pedogenic environments of tropical regions, but they may also form in temperate climates. Induration is principally by precipitation of iron oxyhydroxides (Figure 40.6b), although clay minerals are common associates. An idealized ferricrete profile is shown in Figure 40.7, pp. 511–512 in which a progressive destruction of the parent rock (granite, in this case) is traced from initial replacement of feldspars and amphibole by clays and iron oxyhydroxides to dissolution of quartz and finally to the formation of ferruginous nodules and pisoliths characterizing the ferricrete at and near the surface.

Ferrugineous Laterites

Concentration of iron by weathering processes generates lateritic iron-rich soils and major deposits of **ferruginous laterite**. The host rock must be iron bearing, such as basalt or a banded iron formation (Weggen et al., 1990), and the environment must be acidic (pH < 4) and oxidizing (Eh > +0.4), idealized in Figure 40.1 without influence of organic ligands (Norton, 1973).

Preferential leaching of the alkali earth elements, silica, and aluminum must occur leaving iron as an insoluble residue. Preferential solubility depends on

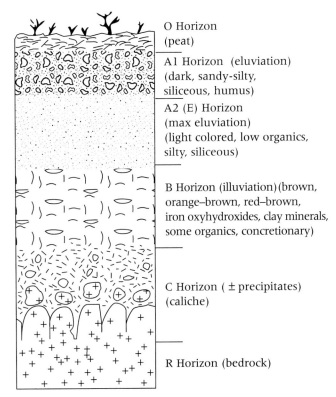

Figure 40.5
Typical soil profile.

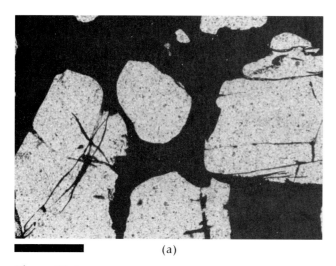 (a)

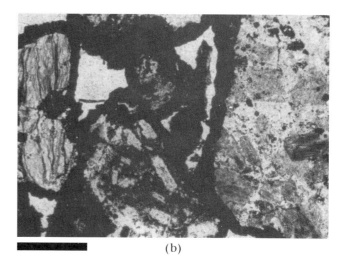 (b)

Figure 40.6
Ferricretes.

(a) Pore and grain-fracture filling hematite in Miocene–Pliocene Cohansey quartz sand formation. This is a ferricrete of meteoric water derivation. PPL. Bar scale = 0.12 mm. Juliustown, New Jersey. (Courtesy D. Davis.) (b) Hematite (black) partially fills porposity in altered porphyritic andesitic regolith breccia. Note argillized relict phenocrysts in fragment center and right. PPL. Bar scale = 0.5 mm. Lower Geiger Grade, Washoe County, Nevada.

---

restricted conditions of pH and Eh and probably also on organic complexing. For example, Al and silica are leached below a pH of 5.5, as indicated for Al by the reaction

$$\underset{\text{gibbsite}}{Al(OH)_3} + 3H^+(aq) \rightarrow Al^{3+}(aq) + 3H_2O \text{ (pH < 5.5)}$$

Manganese laterites do not form since manganese minerals are soluble at low pH.

Lateritic and Karst Bauxites

Any aluminum-bearing rock or unconsolidated rock material is capable of being transformed into bauxite. Most **lateritic bauxites** are formed from sedimentary rocks, particularly those that are arkosic or argillaceous (Bardossy and Aleva, 1990). The **karst bauxites** are a special type that form in carbonates (Bardossy, 1982).

The Arkansas lateritic bauxites (Gordon et al., 1958) are formed from nepheline syenite, although this association accounts for only a few percent of the world's production. Bauxitization is estimated to have taken 1 m.y., and it occurred in the Arkansas region during the Eocene. Bauxite is formed directly from the coarse-grained nepheline syenite and by secondary bauxitization of bauxite materials derived from these primary bauxites and redeposited in stratigraphic succession (Figure 40.8; p. 513). As with any weathering product at the Earth's surface, the weathered materials are susceptible to erosion and redeposition; in the case of the Arkansas bauxite gravels, they were subjected to a second stage of lateritic weathering.

It is well known that climate, hydrogeologic and tectonic position, and rock permeability are key factors in the formation of bauxite. Elevated temperatures are conducive to bacterial breakdown of macroflora (into peat and humus). If there are no organic acids, the pH can be maintained between 4 and 6, and if the Eh is from +0.2 to +0.4, silica and iron go into solution, leaving aluminous minerals such as gibbsite. Formation of bauxites is not necessarily an indication of elevated temperature. Bauxitization of the upper parts of basalt lava flows in New South Wales appears to have occurred under wet, cool to cold climate conditions (Taylor et al., 1992). High rainfall and permeability are conducive to bauxitization because an ongoing chemical environment is maintained in which hydration and leaching play major roles. Bauxite is particularly well developed on topographic highs, well above the water table.

The formations of **pisolites** is characteristic of bauxites (Figure 40.9a,b; p. 514). A sequence of events can be traced in ferruginous laterites (Nahon, 1991) that has its parallel in aluminous laterites. It begins

(Figure 40.10) with dissolution along fissures generating crystalline walls (**septa**) between the original fissure void and minerals being replaced. Anastomosing networks of septa isolate rock material in the form of **alveoles**. Except for quartz, the reactive replacements are from incongruent dissolution. For example, dissolution of alkali feldspar generates gibbsite as potassium goes into solution and is removed. Locally this is the final stage of bauxitization in the Arkansas deposits, and a "granitic" textured bauxite is formed as a result of the pseudomorphic replacement of primary magmatic well-formed alkali feldspar crystals (Gordon, 1958).

Continuation of bauxitization leads to the formation of **nodules** (Figure 40.10; p. 514). Illuviation (deposition) on alvioles, filling fissure voids and alveole voids (void between septa and relict mineral), along with further reactive replacements, generates the nodules. The formation of pisoids (or ooids if 1 mm) is at the expense of the larger nodules by continuing processes of reactive replacement and illuviation. In iron laterites and ferricretes, goethite pisolites typically form from hematite nodules (Nahon, 1991). Bauxite pisoids and ooids are characterized by a **layered cortex** ("banded cortex") structure, resulting from variable crystallinity of gibbsite and variable amounts of iron oxide impurities. **Compound pisolites** contain several smaller pisoids or ooids, co-existing with other pisolites isolated in matrix (cryptocrystalline gibbsite) (Figure 40.9b).

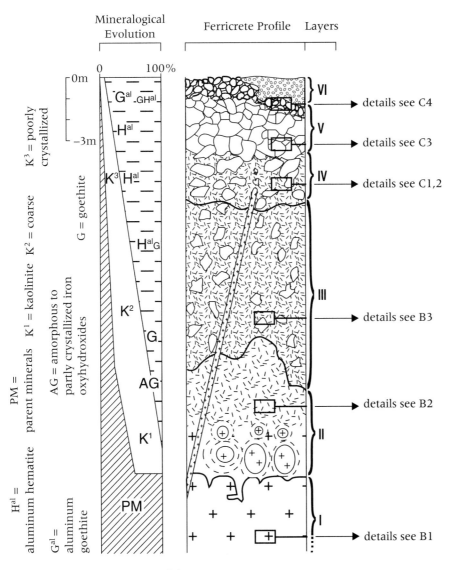

Figure 40.7; text p. 509
Sketches of idealized ferricrete profile and of its mineralogical evolution.

(a) Profile with I = unweathered parent rock, II = weathered rock with preservation of original structure, III = mottled clay layer, IV = transitional layer, and V = ferricrete layer with nodular and pisolitic structure. Mineralogical abundances indicated to the left of column A (PM = parent minerals, K1 = coarse kaolinite, K2 = finer crystalline kaolinite, K3 = poorly crystallized kaolinite, AG = amorphous and poorly crystallized iron oxyhydroxides, G = goethite).
(b) Microscopic textural-mineralogical relations in zones I, II, and III shown in B1, B2, and B3 respectively. (c) Hand specimen scale, structural relations in zone IV shown in C1 and C2, and those in zone V are shown in C3 and C4. (Courtesy Daniel Nahon; Nahon, 1991.)

continued

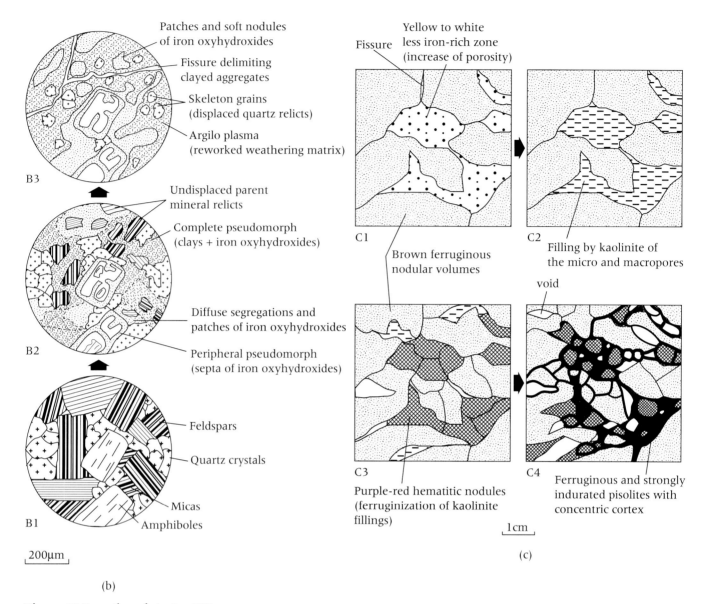

Figure 40.7 continued; text p. 509

Nickel Laterites

In situ weathering of nickeliferous peridotites (0.3–0.4% Ni) or their serpentinized equivalents may generate nickel laterites. Nickel laterites formed over nickel sulfide deposits occur but are rare. As with other laterites, a tropical climate is required in order to leach Ca, Mg, and silica.

Golightly (1981) describes a nickel laterite weathering profile consisting of, from top to bottom: (1) ferricrete cap, (2) limonite zone, (3) nontronite and silica boxwork, and (4) saponite at the base in contact with peridotite. The ferricrete cap consists of colloform goethite in tubes, veinlets, and pisoids with some hematite. The limonite zone contains nickeliferous goethite and amorphous ferric hydroxides and rare gibbsite. The nontronite and silica boxwork zone is localized. The silica is released from silicates and reprecipitated as jaspilitic quartz (layered jasper). **Saprolite** is a clay-rich decomposed rock containing smectites such as nontronite (Fe-rich) and saponite (Mg-rich), with blocky aspect and local preservation of host rock textures and structures. Cracks and voids in the saprolite typically contain **garnierite** $[(Mg,Fe,Ni)_3Si_2O_5(OH)_4]$, pimelite $[Ni_3Si_4O_{10}(OH)_2]$, and chalcedony, all resulting from colloidal precipitation.

Supergene enrichment is greatest in the saprolite zone. Decomposition of tens of meters of host rock, taking place for at least a million years, is required to bring the deposits to ore grade. Karstlike landforms

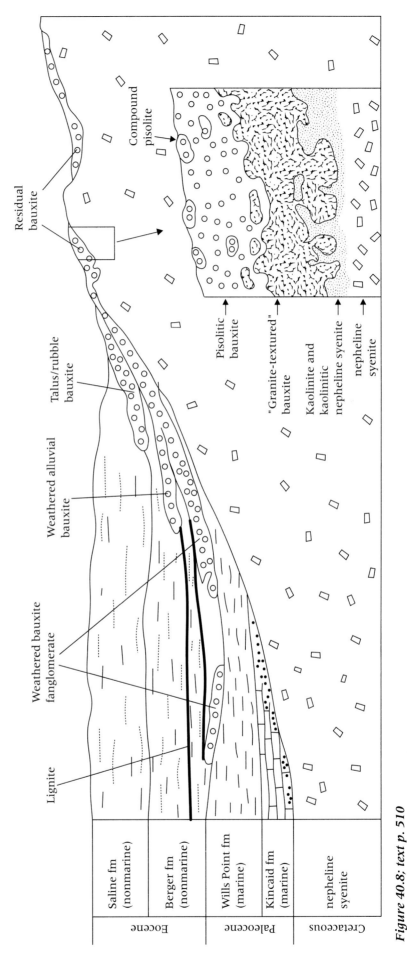

Figure 40.8; text p. 510
Diagrammatic representation of four bauxite occurrences in the Arkansas bauxite region. Modified from Gordon et al. (1958).

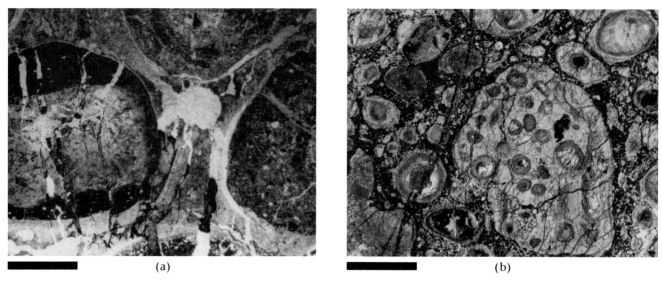

Figure 40.9; text p. 510
Pisolitic bauxite, Little Rock, Arkansas.

(a) Darker portions in pisolites are red-brown hematite-bearing cryptocrystalline gibbsite, which in pisolite left encloses lighter colored cryptocrystalline gibbsite, and in pisolite right occurs as a fragment. White patches and veins are kaolinite. Opaque iron oxide occurs with kaolinite and siderite in some veins (lower center and lower left). PPL. Bar scale = 1 mm. (b) Compound pisolite (center) associated with individual pisoliths in hematite-bearing gibbsite matrix. Bar scale = 1 cm.

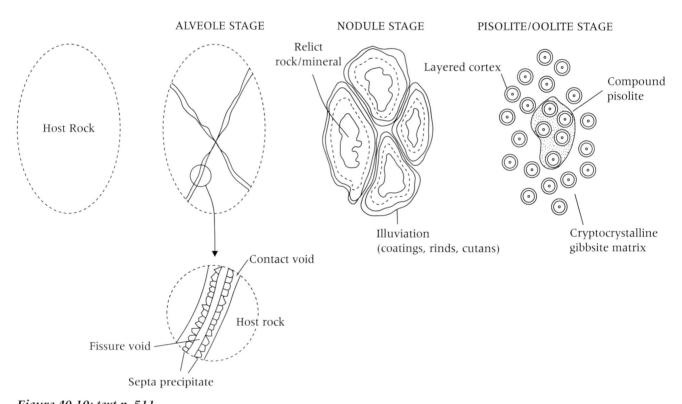

Figure 40.10; text p. 511
Diagrammatic representation of the development of pisoliths in bauxitic laterites. Starting rock is any Al-bearing rock such as a previously formed redeposited bauxite pebble, a compound pisolite, nepheline syenite, arkose, shale, tuff. Fissures initially define alveole that becomes sites of dissolution. Nodules form by illuviation (coatings, caps, cutans) and continued reaction. Pisolites (and olites) form by further reaction and illuviation with nodules.

are developed in part due to enhanced leaching in closely spaced joints in the serpentinized peridotite that occurs on slopes.

Calcretes

Calcium carbonate can be mobile or immobile in the weathering zone, depending on the activity of CO_2-bearing waters. Lithified crusts dominated by calcium carbonate are **calcretes**. They form by (1) in situ reworking of carbonate bedrock, (2) cementing of unconsolidated fragmental or oolitic carbonate rock accumulations, and (3) by cementing of noncarbonated clastic rocks or sediment accumulations. **Paleocalcretes** are indicators of the former presence of a subaerial weathering surface (Warren, 1983; Strong et al., 1992). If they occur in the phreatic groundwater zone, they indicate that submersion has taken place due to rise in the water table. **Penetrative calcretes** result from subsurface calcrete formation associated with downward penetrating root systems (5–6 m) into carbonate rocks forming rhizoliths (Rossinsky et al., 1992).

The formation of calcrete is characterized by precipitation of calcium carbonate either by (1) ascending solutions (by capillary rise, in part), which evaporate and lose CO_2, or (2) by descending solutions in which dilution with water low in CO_2 leads to precipitation. The calcium carbonate precipitates are known as **caliche** and occur as powdering coatings or as laminated or pisolitic **duracrust** of the calcrete variety (Reeves, 1976).

Climate is critical because if it is too wet calcium carbonate is dissolved, whereas if it is too dry dissolution is minor and there is little reprecipitation. Calcrete occurring in deserts is suggestive of a climate change. Climate must be between dry and wet, and there must be a net moisture deficit to prevent long-range leaching of the calcium carbonate precipitates.

Calcretes commonly form in semi-arid climates where temperature is elevated. Cool wet climates are also conducive to the formation of calcrete, especially if there is an abundant source of carbonate rock and a well-drained weathering profile (Strong et al., 1992).

Some of the effects of remobilization of calcium carbonate in limestone are shown in Figure 40.11. Dissolution of calcium carbonate occurs in fractures and along bedding planes where a slightly higher pressure (higher than the ambient atmospheric pressure) favors solution and retention of CO_2. Precipitation of calcium carbonate where solutions exit to the surface is induced by evaporation and loss of CO_2 (Figure 40.11a). Solution and redeposition of limestone, forming a calcrete-like crust is shown in Figure 40.11b.

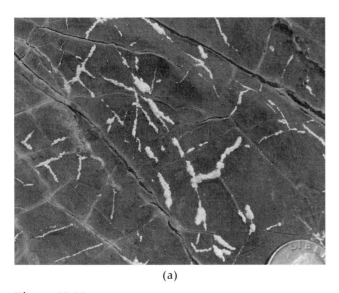
(a)

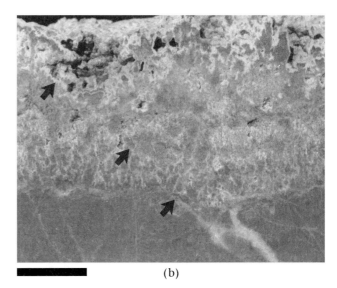
(b)

Figure 40.11
Solution and redeposition in limestones.
(a) Efflorescent calcite along fractures in subaerially exposed limestone, representing precipitation from carbonate-bearing meteoric solutions as a result of evaporation and/or CO_2 degassing. Triassic limestone, West Humboldt Range, Nevada. Scale = U.S. quarter (lower right). (b) Calcrete zone developed residually on limestone as a consequence of limestone (gray) dissolution and calcite (white) reprecipitation. Note patches of limestone and local porosity in calcrete. Prior vein calcite (white) occurs along fracture in limestone (lower center) whose vestige extends through calcrete (arrows). Ely Limestone, Egan Range, Nevada. Bar scale = 1 cm.

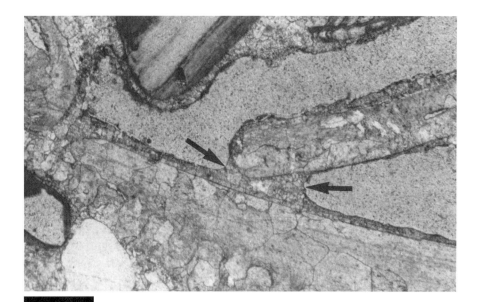

Figure 40.12
Coquina with meniscus sparry calcite cement (between arrows). Bar scale = 0.5 mm. St. Augustine, Florida.

Cementation of beach detritus, such as the shell fragments attached by **meniscus cement** (calcite) (Figure 40.12) can be considered as an early stage of calcrete formation or as syndepositional diagenesis producing a grainstone (coquina). Infiltration of the very porous Ogallala conglomerate with meteoric water containing dissolved calcium carbonate results in the formation of a post-depositional caliche capping (Swineford et al., 1958) (Figure 40.13).

Speleothems

Karst topography is developed in carbonate rocks in which there has been dissolution forming caves that by collapse of their roofs form **sinkholes**. **Karstification** may extend to depth well into the phreatic groundwater zone, wherever the CO_2 content of the water is high enough to cause reaction with the calcium carbonate rock. Caves formed in the phreatic zone may be evacuated of water if there is a lowering of the water table. In that case, caves are particularly susceptible to the formation of **travertine** in the form of **speleothems**. Speleothems include the well-known dripstones known as **stalagmites** and **stalactites**, but also **flowstone** and **cave pearls**. Calcium carbonate cave precipitates resembling grapestones are shown in Figure 40.14. Calcium carbonate precipitates in response to both evaporation and loss of CO_2

Figure 40.13
Caliche originating calcium carbonate (black arrows) precipitated in pores of Ogallala chert pebble conglomerate. Bar scale = 1 cm. New Mexico.

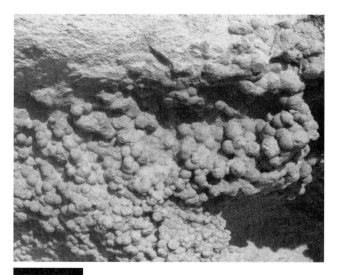

Figure 40.14
Pisolitic reprecipitated calcium carbonate in karstic lacustrine limestone. White Pine Range, Nevada. Bar scale = 2 cm.

Figure 40.15
Calcium carbonate-coated surface and protuburance (lower right) on Navajo Sandstone resulting from precipitation in meteoric water solutions issuing from joints and bedding planes. Zion National Park, Utah.

as calcium carbonate-bearing vadose water issues from fractures and solution avenues into the cave openings. Calcium carbonate precipitates on surfaces of the Navajo Sandstone in Zion National Park (Figure 40.15) result from issuance of calcium carbonate groundwaters from joints and bedding planes.

Dolocrete

Dolomite-cemented horizons occur in fluvial sandstone-mudstone in the Paris Basin (Spötl and Wright, 1992). Interpretation of nodular dolomitic horizons in siltstones and mudstones as vadose-pedogenic is made on the basis of the occurrence of rhizocretions, whereas massive dolomite horizons have no such biogenic association and are interpreted as phreatic dolocretes. Both occurrences of **dolocrete** contain dolomicrospar and some dolomicrite. Glaebules, which are irregular spar-filled cracks, and vuggy porosity are locally abundant in the massive horizons.

Lack of textures indicative of replacement of a calcite precursor point to primary precipitation of (proto)dolomite. Continental origin for the groundwaters is indicated by isotopic data.

Gypcrete

Gyparenite sands may be converted into a duracrust known as **gypcrete**. Jacobson et al. (1988) describe gypcrete crusts that formed in a Holocene playa environment resulting from evaporative concentration of waters discharging from a regional flow system involving solution of Proterozoic evaporites. Gypsum and glauberite [$Na_2Ca(SO_4)_2$] precipitate in the playas, whereas **gypsite** (earthy) and gypcrete (duricrusts) develop on the playa margins and islands. At the playa margins, gypsum crystals occur in **fenestral partings** (small shrinkage cracks) and pore spaces, where they precipitated from both phreatic groundwater and vadose water. The gypsum crystals continued to grow by displacement (no inclusions) of clayey and calcareous sand deposits. These gypsite precipitates are analogous to caliche in carbonate settings. Surface gypcrete forms on the nongypsiferous sediments as a result of several generations of gypsum cementation. Thin crusts of gypcrete also develop on the flanks of gypsum sand dunes.

Silcretes

Siliceous duricrusts formed in pedological environments are **silcretes**. They consist of lamellar, microlamellar, nodular, and pseudobreccia forms rich in opal, opal-CT, chalcedony (fibers length fast), **quartzine** (length slow chalcedony), **lutecite** (chalcedony with inclined extinction), and less commonly limpid (clear) quartz crystals.

Silcretes of the Paris Basin are described as being yellowish with conchoidal fracture (Thiry and Millot, 1986). These silcretes are also characterized by dehydration cracks in opal filled by later stages of opal. Mammillary **microstalactites** of titania (mostly leucoxene) occur in dissolution cavities, which are also sites of what was physical collapse of skeletal (partially dissolved) quartz grains.

The early stage of silicification in regolith is typically laminated or nodular opal-A, probably solidifying from siliceous gels. Subsequent recrystallization to opal-CT, microcrystalline quartz, chalcedony, and clear quartz crystals occurs as the silcrete matures. Silcretes typically form by an ongoing process of dissolution and reprecipitation, typically involving silica dissolved at the top of a profile and reprecipitated at the base.

The source of silica can be from the dissolution of clay minerals in situ, but the presence of skeletal quartz grains in some silcretes indicates direct dissolution of quartz. Secondary overgrowths of quartz (and opal) on skeletal quartz grains of the Fontainbleau quartz sands (Thiry and Millot, 1986) indicate a solution-redeposition process driven by chemical processes alone but which is morphologically similar to those driven by pressure-solution redeposition in deeply buried quartz arenites (Chapters 25, 36). Even though the solubility of quartz is low (6 ppm) in all but very alkaline solutions (Chapter 12), concentration resulting from alternate dry and wet seasonal or climatic periods leads to higher concentrations and eventual

Figure 40.16
SEM view of alunite crystals and bladed silica spherulites (lepispheres). Fossil silcrete interbedded in mid-Tertiary continental noncemented sands and conglomerates of western Portugal. Bar scale = 6 μm. (Courtesy R. Meyer; Meyer and Pena Dos Reis, 1985.)

precipitation, either as quartz directly, or as opal forming from colloids that have flocculated forming hydrated silica gels.

Alunite is reported in silcretes formed on siliciclastic Cenozoic sediments of western Portugal (Meyer and Pena Dos Reis, 1985), resulting from oxidation of sulfide minerals originating in organic-rich sediments. Small, bladed silica spherulites (**lephispheres**) are associated with these well-formed crystals of alunite (Figure 40.16).

Chalcedonic **geodes** (Figure 40.17) commonly occur in the weathering zone in which siliceous gels are formed, but they are not particularly characteristic of silcretes.

Phoscretes

Phoscretes are hardgrounds formed in phosphate-bearing sediments and rocks. In the Georgina Basin, Australia phoscretes form on phosphatic siltstone, phosphatic limestone, and phosphorite (Cook, 1972; Southgate, 1986). The origin of phosphatic sediments generally is agreed to be linked to bacterial degradation of phosphorus- and carbon-bearing sediments and influx of fluorine from seawater (Garrison et al., 1990), typically leading to authigenic carbonate fluorapatite replacements of carbonates (Chapters 37, 39). With subaerial exposure, either by uplift and exhumation of marine phosphatic beds, or by supratidal (storm wave) deposition of organic-bearing sediments (Southgate, 1986), processes best described as vadose diagenesis of these phosphate-bearing sediments can occur. Coatings of crytocrystalline and micritic carbonate fluorapatite on original sediments and the formation of phosphatic pelloids, nodules, and laminae are typical. Laminar crusts formed at the surface are the phoscretes. Subaerial erosion of phoscretes can lead to redeposition of phosphatic particles and other periods of phoscrete formation much as reworked bauxite deposits form (see earlier section).

An interesting occurrence of phoscrete formed by the lateric weathering of an apatite-bearing marble is described by Dahanayake and Subasinghe (1989). Dissolution of calcium carbonate is accompanied by dissolution of apatite and reprecipitation on primary metamorphic apatite crystals. The resulting ooids and pisoids occur in a finely layered phosphatic material that accumulates in sinkholes.

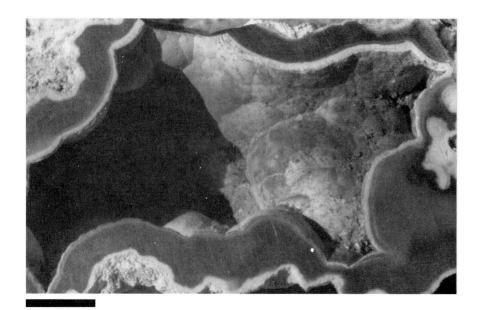

Figure 40.17
Chalcedonic geode resulting from silicification of colonial coral. Bar scale = 1 cm. Tampa Bay, Florida.

Uranium-Enriched Near-Surface Rocks

"Roll front" uranium deposits, formed in sandstones, result from precipitation of carnotite in groundwater aquifers. Uranium, previously introduced by hydrothermal or groundwater flow in the sandstone, is remobilized by throughgoing oxygenated groundwater. Uranium is soluble under these conditions, but as these solutions encounter reducing conditions, such as occur near organic material, uranium and vanadium typically precipitate as the mineral **carnotite**. Concentration of carnotite increases as the solutions continue to pass this redox "front" between oxidizing and reducing environments. Other minerals in the reducing environment include pyrite and other uranium–vanadium minerals. Hematite and goethite, characterize the oxidizing environment.

Supergene Alteration of Sulfide-Bearing Rocks

Weathering of rocks containing sulfide minerals generates sulfuric acid as the metal cations are **hydrolyzed**:

$$2ZnS + 3H_2O + 3.5O_2 \rightarrow 2ZnOH + 2H_2SO_4$$
sphalerite

Sulfuric acid causes relatively rapid and intense alteration of host rock and hydrothermal mineral assemblages. Weathering alteration of sulfide ore deposits has received immense attention because of its economic significance (Titley, 1982; Guilbert and Park, 1986). Since the weathering (a secondary alteration process) is superposed on rocks that have been hydrothermally altered (a primary alteration process), the term **supergene** is used to indicate weathering alteration of mineral deposits.

Supergene alteration, characterized most importantly by oxidation of metal sulfides above the water table, generates an assemblage of metal oxides, sulfates, silicates, and carbonates, (Figure 40.18) that may or may not be ore. More important, from an economic point of view, some of the metal ions are leached from the oxide zone and reprecipitated below the water table in the zone of **supergene enrichment**. Enrichment refers to a higher concentration of metals resulting from addition of metals arriving from the leached zone to rock below the water table that already has a primary metal content resulting from prior hydrothermal activity.

Porphyry copper deposits are the most thoroughly studied supergene alteration of ore deposits. Much current research is on **disseminated gold deposits** of the Carlin type (Kuehn and Rose, 1992). The principles of supergene alteration (oxidation and leaching) apply to any sulfide-bearing rock or ore body, including those in carbonate rock (Scott, 1987) and in massive sulfides exposed to submarine weathering (Herzig et al., 1991). **Leached cappings** typically contain a wide variety of iron oxides and oxyhydroxides. If

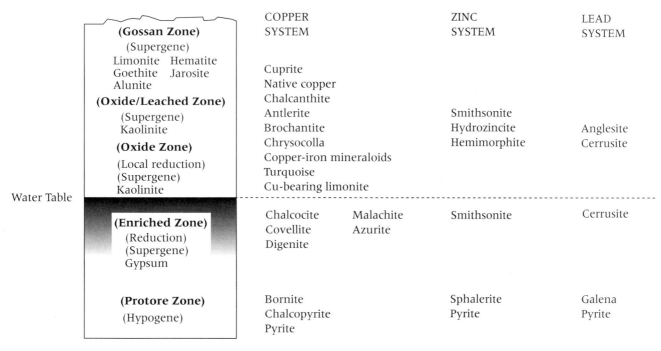

Figure 40.18
Composite sketch of zones produced by supergene alteration of sulfide ore bodies.

leaching of metal cations and other residuals from the oxidation process is extreme, there may be a dominant iron-bearing leached capping known as **gossan**, since metals such as Cu, Pb, and Zn are more soluble than Fe under these conditions. Oxidation and leaching are particularly active in semi-arid temperate climates because of a relatively low water table, allowing for oxidation and meteoric water recharge that flushes the reactive system, and moderately high temperatures that increase reaction rates. High water tables prevent deep oxidation.

Supergene alteration in the porphyry copper systems (Anderson, 1982) generates a brownish or reddish oxide zone (frontispiece photo) containing hematite, and also oxyhydroxide minerals such as goethite and copper-iron mineraloids that are generally referred to as "limonite." Depending on conditions of Eh (or P_{O_2}), pH, and available cations in the oxide zone, there can be an assemblage of metal oxides such as cuprite and tenorite, sulfates such as jarosite, borgstromite-jarosite [$(K,H)Fe_3(SO_4)_2(OH)_6$], antlerite [$Cu_3(OH)_4SO_4$], brochantite [$Cu_4(OH)_6SO_4$], alunite [$KAl_3(OH)_6(SO_4)_2$], carbonates such as malachite and azurite, as well as the silicates chrysocolla (Figure 40.19) and neotocite [$(Cu,Fe,Mn)SiO_2$], and the phosphate turquoise. Some of the pertinent reactions are shown below:

$$2FeS_2 + 7.5O_2 + 4H_2O \rightarrow Fe_2O_3 + 4H_2SO_4 \text{ (aq)}$$
pyrite hematite

$$2CuFeS_2 + 8.5O_2 + 2H_2O \rightarrow$$
chalcopyrite
$$Fe_2O_3 + 2Cu^{2+} + 4SO_4^{2-} + 4H^+$$
hematite

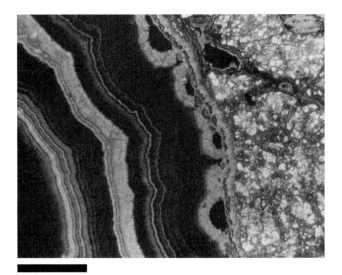

Figure 40.19
Laminated supergene chrysocolla in brecciated quartz diorite porphyry. PPL. Bar scale = 0.5 mm. Yerington, Nevada.

$$Cu^{2+} + H_2O \rightarrow CuO + 2H^+$$
tenorite
$$2Cu^{2+} + CO_3^- + 2(OH)^- \rightarrow Cu_2(OH)_2CO_3$$
malachite
$$2Cu^{2+} + H_2O \rightarrow Cu_2O + 2H^+$$
cuprite
$$Cu^{2+} + H_2O + 0.5O_2 + SiO_2 \rightarrow CuSiO_3 \cdot 2H_2O$$
chrysocolla

Supergene enrichment generates **chalcocite** and a higher grade of ore. Chalcocite occurs locally above the water table in the oxide zone, as does native copper, but it occurs mostly below the water table where the chemical environment is pervasively reducing. Copper ions liberated by reaction of chalcopyrite (or another primary copper mineral) with oxygen in an aqueous environment are fixed into this secondary sulfide in the reducing environment below the water table as shown in the equation:

$$5FeS_2 + 14Cu^{2+} \text{ (from leached zone)} + 14SO_4^{2-}$$
pyrite
$$+ 12H_2O \rightarrow 7Cu_2S + 5Fe^{2+} + 24H^+ + 17SO_4^{2-}$$
chalcocite

Reaction rims of chalcocite on the primary pyrite and chalcopyrite grains just below the water table is characteristic of the supergene enriched zone and diagnostic of the reaction.

Change in the position of the water table leads to reworking of supergene alteration assemblages. A raising of the water table places the oxide zone in a reducing environment, susceptible to accumulation of secondary chalcocite if there is any copper remaining to be leached from the oxide zone still remaining above the water table. More commonly, a lowering of the water table results in second-cycle oxidation characterized by oxidation of supergene chalcocite and generation of hematite (frontispiece photo), as well as oxidization of primary sulfides below the original chalcocite enrichment zone, and even establishment of a new enrichment zone in the deeper regions containing primary sulfides (Anderson, 1982).

The oxidation of both pyrite and chalcopyrite generate sulfuric acid that attacks both hydrothermal and unaltered or previously altered host-rock minerals. For example, porphyry copper mineralization is typically at least partly in a granitic porphyry stock in which the feldspars of both magmatic and hydrothermal derivation are converted to supergene kaolinite and montmorillonite (Figure 40.20). Substitution of H^+ for Na and Ca in feldspars generates alkaline solutions containing NaOH and $Ca(OH)_2$. If the sulfide content of the host rock is low, leaching of metal cations is incomplete and the sulfuric acid is neutralized at an

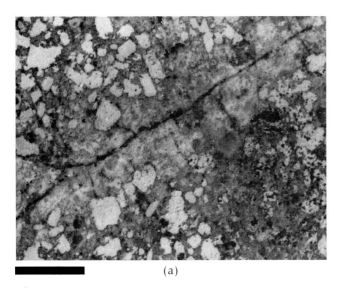

Figure 40.20
Supergene alteration in porphyry copper deposit at Morenci, Arizona.
(a) Sulfide in quartz–K-feldspar vein has been oxidized to iron hydroxides, plagioclase phenocrysts (white) to montmorillonite. K-feldspar along vein is weakly altered to kaolinite. Slab surface. Bar scale = 1 cm. (b) Montmorillonite pseudomorphs of plagioclase of porphyry. Quartz remains unaltered. X-polars. Bar scale = 0.5 mm. (Courtesy Kennecott Copper Corporation.)

early stage by the formation of kaolinite. A higher content of sulfide may lead to complete leaching of metal cations, particularly if there is flushing by recharging meteoric waters (Anderson, 1982).

Oxidation of other metal sulfides parallels that of copper (Figure 40.18). Oxidation of molybdenite typically generates the yellow mineral ferrimolybdite:

$$6MoS_2 + 4Fe^{3+} + 15H_2O + 36O_2 \rightarrow$$
molybdenite
$$2Fe_2(MoO_4)_3 \cdot 15H_2O + 12\ SO_4^{2-} + 36e$$
ferrimolybdite

The simple oxide of galena is soluble in the zone of weathering, but the sulfate and carbonate are stable

Figure 40.21
Hemimorphite (calamine) with other secondary zinc minerals as partial boxwork deriving from sphalerite veins in carbonate gangue. Bar scale = 1 cm.

Figure 40.22
Iron oxyhydroxide–coated siliceous boxwork resulting from dissolution of hydrothermal calcite (note calcite molds). Bar scale = 2 cm. Robinson Mining District, Nevada.

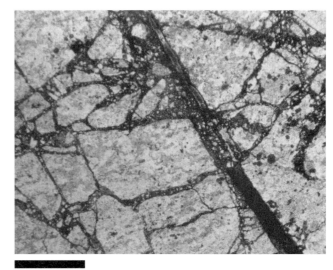

Figure 40.23
Jarosite (black) (actually amber yellow) in brecciated jasperoid resulting from oxidation of sulfide in the presence of K and Fe-bearing meteoric solutions. June-Ellen Mine, Pine Nut Range, Nevada. PPL. Bar scale = 1 mm. (Courtesy R. Robinson.)

and in some cases are ore minerals. The oxidizing reactions are as follows:

$$PbS + 2Fe^{3+} + 3SO_4^{2-} + 1.5O_2 + H_2O \rightarrow$$
galena
$$PbSO_4 + 2Fe^{2+} \; 3SO_4^{2-}$$
anglesite

$$PbS + H_2O + CO_2 + 2O_2 \rightarrow PbCO_3 + SO_4^{2-} + 2H^+$$
galena $\qquad\qquad\qquad\qquad$ cerussite

The simple oxide of zinc is a very high-temperature mineral, not occurring in the weathering zone. At the other extreme, zinc sulfate is soluble in the weathering environment, and the secondary products of

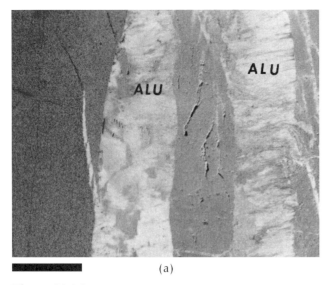
(a)

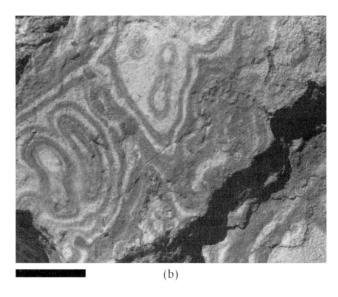
(b)

Figure 40.24
Oxidized sediment-hosted disseminated gold ore.

(a) Supergene alunite veins in oxide ore at Gold Quarry, Nevada, derived from siltstone. Slabbed surface. Bar scale = 1 cm. (Courtesy H. Bonham, Nevada Bureau of Mines and Geology.) (b) Oxidation has generated Leisegang rings in argillaceous silty limestone of the Upper Denay formation. Gold Bar Mine (Atlas), Eureka County, Nevada. Bar scale = 1 cm. (Courtesy J. Tingley, Nevada Bureau of Mines and Geology.)

sphalerite oxidization are represented by carbonates, a sulfate, and a silicate (Figure 40.21), shown here by three unbalanced equations:

$$ZnS \text{ (sphalerite)} + 2Fe^{3+} + 3SO_4^{2-} + H_2O + 1.5O_2 \rightarrow Zn^{2+} + 2Fe^{2+} + 2H^+ + 4SO_4^{2-}$$

$$Zn^{2+} + (CO_2) \rightarrow ZnCO_3 \text{ (smithsonite)} \text{ or } Zn_5(CO_3)_2(OH)_6 \text{ (hydrozincite)} \text{ or } Zn_4Si_2O_7(OH)_2 \text{ (hemimorphite)} \cdot H_2O$$

$$Zn^{2+} + SO_4^{2-} + CaCO_3 + 2H_2O \rightarrow CaSO_4 \text{ (gypsum)} \cdot 2H_2O + ZnCO_3$$

Weathering of quartz–sulfide veins in carbonates may generate a **boxwork** of quartz and limonite with molds of carbonate crystals (Figure 40.22). Oxidation of sulfides in the presence of potassium may generate jarosite (Figure 40.23).

Sediment hosted disseminated gold deposits are characterized by micron-sized gold. The primary hydrothermal ores are dark colored due to the presence of free carbon, but oxidation converts the carbon to CO_2 and the rocks become bleached. Oxidation of primary sulfide minerals associated with the gold generates jarosite, kaolinite, iron oxides–hydroxides, and supergene alunite (Arehart et al., 1992) (Figure 40.24a). Liesegang rings characterize the redistribution of iron oxides in some of the oxidized ores (Figure 40.24b).

Appendix I

Conversion Factors

AREA

1 square inch (in.^{2})   = 6.4516 centimeters squared (cm^2)
1 square centimeter (cm^2) = 10^{-4} m^2

CONCENTRATION

1 microgram (μg)/milliliter (mL)	= 1 part per million (ppm)
	= 0.001 gram (g)/L (liter)
	= 1 milligram (mg)/L
1 weight percent (wt %)	= 0.01 mass fraction = 10^4 ppm = 10^7
1 gram/metric ton	= 1 ppm = 0.001 %
1 part per million (ppm)	= 0.9072 g/short ton = 1 g/metric ton
1 Troy ounce/ton (short)	= 34.286 ppm

DENSITY

density (d) (or rho)   = gms/cc or mg/m^3

LENGTH

1 angstrom (Å)	= 10^{-1} nm = 10^{-4} μm = 10^{-7} mm = 10^{-8} cm
1 nanometer (nm)	= 10 Å = 10^{-9} m
1 micrometer (μm)	= 1 micron (μ) = 1000 nm
	= 10^4 Å = 10^{-3} (0.0001) mm = 10^{-6} m
1 millimeter (mm)	= 1000 μm = 100,000 nm = 0.0394 in.

LENGTH (*continued*)

0.03 millimeter	= 30 microns
1 inch (in.)	= 2.54 cm
1 centimeter (cm)	= 0.39 in.
1 meter (m)	= 3.2808 ft
1 foot (ft)	= 0.3048 m
1 kilometer (km)	= 0.6214 mi (statute) = 3281 ft
1 mile (mi) (statute)	= 1.6093 km = 5280 ft

MASS/WEIGHT

1 gram (g)	= 0.03527 ounce (oz) (Avoirdupois)
	= 0.03215 ounce (Troy)
1 ounce (Avoir)	= 28.3495 g = 1.097 oz (Troy)
1 kilogram (kg)	= 2.20462 lb (Avoir) = 35.27 oz (Avoir)
	= 1.10231×10^{-3}
1 pound (lb) (Avoir)	= 0.4536 kg = 16 oz (avoirdupois)
1 pound (lb) (Troy)	= 12 oz (Troy)
1 short ton	= 2,000 lbs (Avoir)
1 long ton	= 2,240 lbs (Avoir)
1 metric ton	= 2,205 lbs (Avoir)

PRESSURE

1 atmosphere (atm)	= 1.01325 bar = 1.013×10^5 pascals (Pa)
	= 14.6963 psi = 760 mm Hg
1 bar (bar)	= 0.9869 atm = 14.5038 p.s.i. = 10^5 Pa
1 psi	= 0.0689 bar = 0.0680 atm
1 kilogram/square centimeter	= 0.96784 atm = 0.98067 bar = 14.2233 psi
1 kilobar (kb)	= 10 Pa = 986.9 atm
1 pascal (Pa)	= 1 newton/square meter(n/m^2)
1 kilopascal (kPa)	= 10^3 Pa = 10^{-2} bar
1 megapascal (MPa)	= 10^6 Pa = 10 bar
1 gigapascal (GPa)	= 10^9 Pa = 10^4 bar = 10 kbar

TEMPERATURE

1 degree Fahrenheit (1°F)	= 5/9 degree Celsius (°C)
°F	= (9/5) °C + 32°
°C	= (5/9) (°F − 32°)

VISCOSITY

1 poise	= 1 dyne sec/cm^2 = 1 g/cm sec

VOLUME

1 cubic inch (in.^{3})	= 16.4 cm^3
1 liter (L)	= 1000.028 cm^3 = 1.057 quarts (qt)
1 milliliter (mL)	= 1.000028 cm^3 = 1 cubic centimeter (cc)
1 microliter (μL)	= 10^{-3} mL = 10^{-6} L
1 gallon (Imperial)	= 4.546 L
1 gallon (U.S.)	= 3.785 L

Appendix II

Table of Atomic Properties

Element	Symbol	Atomic Number	Atomic Weight[1], $^{12}C=12.000$	Usual Valence	Usual Coordination	Ionic Radius (Å)[2]	First Ionization Potential, eV
Actinium	Ac	89	227.03	3+	—	—	6.90
Aluminum	Al	13	26.98	3+	4	0.47	5.98
				3+	6	0.61	
Antimony	Sb	51	121.75	3+	4	0.85	8.64
Argon	Ar	18	39.95	0	1	—	15.76
Arsenic	As	33	74.92	5+	4	0.42	9.81
				3+	6	0.58	
Barium	Ba	56	137.33	2+	8	1.50	5.21
				2+	12	1.68	
Beryllium	Be	4	9.01	4+	4	0.35	9.32
Bismuth	Bi	83	208.98	3+	6	1.10	7.29
				3+	8	1.19	
Boron	B	5	10.81	3+	3	0.10	8.30
				3+	4	0.20	
Bromine	Br	35	79.90	1−	6	1.88	11.84
Cadmium	Cd	48	112.41	2+	6	1.03	8.99
				2+	8	1.15	
Calcium	Ca	20	40.08	2+	6	1.08	6.11
				2+	8	1.20	
				2+	12	1.43	
Carbon	C	6	12.01	4+	3	0.16	11.26
Cerium	Ce	58	140.12	3+	6	1.09	5.60
				3+	8	1.22	
Cesium	Cs	55	132.91	1+	8	1.82	3.89
				1+	12	1.96	
Chlorine	Cl	17	35.45	1−	6	1.72	13.01
Chromium	Cr	24	51.99	3+	6	0.70	6.76
Cobalt	Co	27	58.93	2+	6	0.73	7.86
Copper	Cu	29	63.55	1+	6	0.96	7.72
				2+	6	0.81	
Dysprosium	Dy	66	162.50	3+	6	0.99	6.80
				3+	8	1.11	
Erbium	Er	68	167.26	3+	6	0.97	6.08
				2+	8	1.08	
Europium	Eu	63	151.97	2+	6	1.25	5.67
				2+	8	1.33	
Fluorine	F	9	18.99	1−	6	1.25	17.42
Gadolinium	Gd	64	157.25	3+	6	1.02	6.16
				3+	8	1.14	
Gallium	Ga	31	69.72	3+	4	0.55	
Germanium	Ge	32	72.61	4+	4	0.48	7.88
				2+	6	0.62	
Gold	Au	79	196.97	3+	4	0.78	9.22
Hafnium	Hf	72	178.49	4+	6	0.79	
				4+	8	0.91	
Helium	He	2	4.00	0	1	—	24.48
Holmium	Ho	67	164.93	3+	6	0.98	

(continued)

Element	Symbol	Atomic Number	Atomic Weight[1], $^{12}C=12.000$	Usual Valence	Usual Coordination	Ionic Radius (Å)[2]	First Ionization Potential, eV
Hydrogen	H	1	1.008	1+	—	—	13.60
					8	1.10	
Indium	In	49	114.82	3+	6	0.88	5.79
				3+	8	1.00	
Iodine	I	53	126.90	1−	6	2.13	10.45
				1−	8	1.97	
Iridium	Ir	77	192.22	4+	6	0.71	
Iron	Fe	26	55.85	2+	6	0.69	7.87
				3+	6	0.63	
Krypton	Kr	36	83.80	0	1	—	14.00
Lanthanum	La	57	138.91	3+	6	1.13	5.61
				3+	8	1.26	
Lead	Pb	82	207.21	2+	6	1.26	7.42
				2+	8	1.37	
				2+	12	1.57	
Lithium	Li	3	6.94	1+	4	0.68	5.39
Lutetium	Lu	71	174.97	3+	6	0.94	—
				3+	8	1.05	
Magnesium	Mg	12	24.31	2+	6	0.80	7.64
				2+	8	0.97	
Manganese	Mn	25	54.94	2+	6	0.75	7.43
				3+	6	0.66	
				4+	6	0.62	
Mercury	Hg	80	200.59	2+	6	1.10	10.43
				2+	8	1.22	
Molybdenum	Mo	42	95.94	4+	6	0.73	7.10
				6+	4	0.50	
				6+	6	0.68	
Neodymium	Nd	60	144.24	3+	6	1.06	5.51
				3+	8	1.20	
Neon	Ne	10	20.18	0	1	—	21.56
Nickel	Ni	28	58.69	2+	6	0.77	7.63
Niobium	Nb	41	92.91	5+	4	0.40	6.88
				5+	6	0.72	
Nitrogen	N	7	14.01	5+	—	0.13	14.53
Osmium	Os	76	190.20	4+	6	0.71	8.50
Oxygen	O	8	15.999	2−	4	1.30	13.61
				2−	6	1.32	
				2−	8	1.34	
Palladium	Pd	46	106.42	4+	4	0.72	8.33
Phosphorus	P	15	30.97	5+	4	0.25	10.48
Platinum	Pt	78	195.08	2+	4	0.68	9.0
				4+	6	0.71	
Polonium	Po	84	209	4+	8	1.16	8.43
Potassium	K	19	39.10	1+	8	1.59	4.34
				1+	12	1.68	
Praesodymium	Pr	59	140.91	3+	6	1.08	5.46
				3+	8	1.22	
Protactinium	Pa	91	231.04	4+	8	1.09	—
Radium	Ra	88	226.03	2+	8	1.56	5.28
				2+	12	1.72	
Radon	Rn	86	222	2+	8	1.56	10.75
				2+	12	1.72	
Rhenium	Re	75	186.21	4+	6	0.71	7.87
Rhodium	Rh	45	102.91	3+	6	0.75	7.46
				4+	6	0.71	
Rubidium	Rb	37	85.47	1+	8	1.68	4.18
				1+	12	1.81	

Element	Symbol	Atomic Number	Atomic Weight[1], $^{12}C=12.000$	Usual Valence	Usual Coordination	Ionic Radius (Å)[2]	First Ionization Potential, eV
Ruthenium	Ru	44	101.07	3+	6	0.76	7.36
				4+	6	0.70	
Samarium	Sm	62	150.36	3+	6	1.04	5.6
				3+	6	1.17	
Scandium	Sc	21	44.96	3+	6	0.83	6.54
				3+	8	0.95	
Selenium	Se	34	78.96	2−	6	1.88	9.75
				2−	8	1.90	
Silicon	Si	14	28.09	4+	4	0.34	8.15
Silver	Ag	47	107.87	1+	4	1.10	7.57
				1+	6	1.23	
				1+	8	1.38	
Sodium	Na	11	22.99	1+	6	1.10	5.14
				1+	8	1.24	
Strontium	Sr	38	87.62	2+	6	1.21	5.69
				2+	8	1.33	
				2+	12	1.48	
Sulfur	S	16	32.07	2−	4	1.56	10.36
				2−	6	1.72	
				2−	8	1.78	
Tantalum	Ta	73	180.95	5+	6	0.72	7.88
				5+	8	0.77	
Technetium	Tc	43	98.91	4+	6	0.72	7.28
Tellurium	Te	52	127.60	2−	—	0.85	9.01
				4+	3	0.60	
Terbium	Tb	65	158.93	3+	6	1.00	5.98
				3+	8	1.12	
Thallium	Tl	81	204.38	1+	6	1.58	6.11
				1+	8	1.68	
				1+	12	1.84	
Thorium	Th	90	232.04	4+	6	1.08	6.95
				4+	8	1.12	
Thulium	Tm	69	168.93	3+	6	0.96	5.81
				3+	8	1.07	
Tin	Sn	50	118.71	2+	8	1.30	7.34
Titanium	Ti	22	47.88	2+	6	0.94	6.82
				3+	6	0.75	
				4+	6	0.69	
Tungsten	W	74	183.85	4+	6	0.73	7.98
				6+	4	0.50	
				6+	6	0.68	
Uranium	U	92	238.03	4+	8	1.08	6.08
				6+	4	0.56	
				6+	6	0.81	
Vanadium	V	23	50.94	5+	4	0.44	6.74
				5+	6	0.62	
Xenon	Xe	54	131.29	0	1	—	12.13
Ytterbium	Yb	70	173.04	3+	6	0.95	6.2
				3+	8	1.06	
Yttrium	Y	39	88.91	3+	6	0.98	6.38
				3+	8	1.10	
Zinc	Zn	30	65.39	2+	4	0.68	9.39
				2+	6	0.83	
Zirconium	Zr	40	91.22	4+	6	0.80	6.84
				4+	8	0.92	

[1] Atomic weights of the elements 1985. *Pure and Appl Chem* 58:1677–1692
[2] Whittaker, E. J. W., and Muntus, R. (1970) Ionic radii for use in geochemistry. *Geochim Cosmochim Acta* 34:945–56.

Appendix III

Geologic Time Scale

ERA	SUB ERA	PERIOD	EPOCH	OROGENY	ROCK FORMATIONS	BIOLOGIC EVOLUTION
CENOZOIC	QUAT	NEOGENE	HOLOCENE	* 0.01	Glacial deposits (N Amer)	Cromagnon Man (0.01 Ma)
			PLEISTOCENE	1.6	Volcanics (w US)	Early Man (0.5 Ma)
	TERTIARY		PLIOCENE	5.3 — Great Basin Rifting	Ogallala cong (c US)	
			MIOCENE	24 — Himalayan	Monterrey sh (CA)	
		PALEOGENE	OLIGOCENE	37 — Alpine	Vicksburg ls (c US)	First rodents
			EOCENE	58 — Laramide	Salt Domes (Gulf Coast)	First horses
			PALEOCENE		Provins ls (Fr)	
					Midway group (s US)	
					Green River sh (UT)	
					Fort Union sh (w US)	
MESOZOIC		CRETACEOUS		Sevier	Pierre sh (w US)	Last ammonoids
					Dover chalk (England)	Last belemnites
					Selma chert (s US)	Last dinosaurs
					Dakota ss (w US)	Last rudists
					Colorado sh (w US)	Expansion of pelecepods
					Mesa Verde ss (w US)	First flowering plants
					Mancos sh (w US)	First placental mammals
					Wealden beds (England)	
					Cordilleran granites	
		JURASSIC		Nevadan	Knoxville fm (CA)	First birds
					Franciscan fm (CA)	First globigerinids
					Portland ls (England)	
					Morrison sh/ss (w US)	
					Oxford clay (England)	
					Navajo/Nugget ss (w US)	
		TRIASSIC		Breakup of Pangaea	Newark red beds (e US)	Last conodonts
					Muschelkalk ls (Euro)	Last fusulinids
					Chugwater ss/sh (w US)	Decline of echinoderms
					Chinle fm (w US)	First coccoliths
					Moenkopi sh (w US)	First dinosaurs
					Luning ls (NV)	First planktic forams
					Thaynes ls (w US)	Early mammals
		PERMIAN		Sonoma — Formation of Pangaea	Zechstein evap (Euro)	Last braciopods
					Castile evaps (TX)	Last trilobites
					Phosphoria (w US)	Abundant forams
					Coconino ss (w US)	
					Dunkard series (e US)	
					Supai sh/ss (w US)	
					Glacial deposits (s hem)	

Time (million annum): 66, 144, 208, 245

(Cordilleran — spanning Cretaceous/Jurassic/Triassic epoch column)

* Quaternary: Holocene and Pleistocene not to scale.

Age (Ma)	Era	Period		Orogeny	Formations	Life
290	PALEOZOIC	CARBONIFEROUS	PENNSYLVANIAN	Alleghenian / Hercynian	Fountain ss (CO) Conemaugh (coal) (e US) Allegheny (coal) (e US) Potsville ss (e US)	Coal swamps Abundant forams First winged insects
320			MISSISSIPPIAN		Madison/Redwall ls (w US) Mauch Chunk sh (e US) Clifton Down ls (England) Pocono ss (e US)	Last graptolites Coal swamps First reptiles
360		DEVONIAN		Antler / Acadian / Caledonian	Catskill redbeds (e US) Chattanooga sh (e US) Old Red ss (Wales) Pilot sh (w US) Onondaga ls (e US) Oriskany ss (e US)	Abundant fish First amphibians First braciopods
408		SILURIAN			Shawangunk cong (e US) Lockport dol (e US) Wenlock sh (Wales) Llandovery fm (Wales)	Decline of trilobites First jawed fish First land plants
438		ORDOVICIAN		Taconian	Martinsburgh sh (e US) Normanskill sh (e US) Queenston redbeds (e US) Beekmantown ls (e US) St. Peter ss (e US) Eureka qtz (NV) Knox dol (e US) Gondwana glacial deposits	Early plants First ammonoids First corals First bryozoa
505		CAMBRIAN			Hales ls (NV) Bonanza King ls (w US) Notch Peak ls (UT) Bonnetere dol (c US) Conasauga ls/sh (e US) Harlech grits (Wales) Burgess sh (Can) Bright Angel sh (w US) Tapeats ss (w US) Prospect Mtn qtzt (w US)	First benthic forams First cephalopods First conodonts First echinoderms First fish First gastropods First graptolites First ostracods First pelecypods First radiolaria First trilobites
570		PROTEROZOIC			Killarney granite (w US) Keweenawan basalt (US) Diamictites Biwabik iron fm (US) Belt group (US/Can) Stromatolites	Calcareous algae Cyanobacteria 1% oxygen in air
2500						
3800		ARCHEAN			Greenstones (Can) Vishnu schist (w US) Laurentian granite (Can)	Anoxic atmosphere First cellular life
4600		Oldest dated rock Formation of Earth				

References

Abed, Abdulkader M., and Mohammed R. Al-Agha. 1989. Petrography, geochemistry and origin of the NW Jordan phosphorites. *Journal of the Geological Society* (London), 146:499–506.

Adams, A. E., W. S. MacKenzie, and C. Guilford. 1984. *Atlas of Sedimentary Rocks Under the Microscope.* New York: Wiley.

Agar, Susan M., David J. Prior, and Jan H. Behrmann. 1989. Back-scatter electron imagery of the tectonic fabrics of some fine-grained sediments: Implications for fabric nomenclature and deformation processes. *Geology,* 17:901–904.

Ague, Jay J., and George H. Brimhall. 1988. Magmatic arc asymmetry and distribution of anomalous plutonic belts in the batholiths of California: Effects of assimilation, crystal thickness, and depth of crystallization. *Geological Society of America Bulletin,* 100:912–927.

Alexopoulos, J. S., R. A. F. Grieve, and P. B. Robertson. 1988. Microscopic lamellar deformation features in quartz: Determinative characteristics of shock-generated varieties. *Journal of Geophysical Research,* 16:796–799.

Allen, J. R. L. 1977. *Physical Processes of Sedimentation.* London: Allen and Unwin.

Allen, Philip A., and John R. Allen. 1990. *Basin Analysis: Principles and Applications.* Cambridge, MA: Blackwell.

Alvarez, W. 1986. Towards a theory of impact crises. *EOS Transactions of the American Geophysical Union,* 67:649–658.

Anderson, A. T. 1983. Oscillatory zoning of plagioclase: Nomarski interference microscopy of etched sections. *American Mineralogist,* 68:125–129.

Anderson, James A. 1982. Characteristics of leached capping and techniques of appraisal. In *Advances in Geology of the Porphyry Copper Deposits: Southwestern North America,* S. R. Titley, ed. Tucson, AZ: University of Arizona Press, pp. 275–295.

Anderson, John J. 1971. Geology of the southwestern High Plateaus of Utah: Bear Valley Formation, an Oligocene-Miocene volcanic arenite. *Geological Society of America Bulletin,* 82:1179–1206.

Anderson, Orson L., and Priscilla C. Grew. 1977. Stress corrosion theory of crack propagation with applications to geophysics. *Reviews of Geophysics and Space Physics,* 15:77–104.

Arehart, Greg B., Stephen E. Kesler, and James R. O'Neil. 1992. Evidence for the supergene origin of alunite in sediment-hosted micro gold deposits, Nevada. *Economic Geology,* 87:263–270.

———, Stephen L. Chryssoulis, and Stephen E. Kesler. 1993. Gold and arsenic in iron sulfides from sediment-hosted disseminated gold deposits: Implications for depositional processes. *Economic Geology,* 88:171–185.

Arthurton, R. S. 1973. Experimentally produced halite compared to Triassic layered halite-rocks from Cheshire, England. *Sedimentology,* 20:145–160.

Ashley, R. P., and M. L. Silberman. 1976. Direct dating of mineralization at Goldfield, Nevada, by potassium-argon and fission-track methods. *Economic Geology,* 71:904–924.

Ashworth, J. R. 1976. Petrogenesis of migmatites in the Huntly-Portsoy area, northeast Scotland. *Mineralogical Magazine,* 40:661–682.

———, ed. 1985. *Migmatites.* Glasgow, Scotland: Blackie.

———, and G. A. Chinner. 1978. Coexisting garnet and cordierite in migmatites from the Scottish Caledonides. *Contributions to Mineralogy and Petrology,* 65:379–394.

Atherton, M. P., and C. D. Gribble. 1983. *Migmatites, Melting and Metamorphism.* Cheshire, England: Shiva.

Augustithus, S. S. 1973. *Atlas of the Textural Patterns of Granites, Gneisses, and Associated Rocks.* New York: Elsevier.

Baars, Donald L. 1983. *The Colorado Plateau, a Geologic History.* Albuquerque, NM: University of New Mexico Press.

Bachman, Steven B. 1978. A Cretaceous and Early Tertiary subduction complex, Mendocino coast, northern California. In *Mesozoic Paleogeography of the Western United States,* David G. Howell and Kristin A. McDougall, eds. Pacific Coast Paleogeography Symposium 2. Los Angeles, CA: Society of Economic Paleotologists and Mineralogists, pp. 419–430.

Bacon, C. R. 1986. Magmatic inclusions in silicic and intermediate volcanic rocks. *Journal of Geophysical Research,* 91:6091–6112.

———, and J. Metz. 1984. Magmatic inclusions in rhyolites, contaminated basalts, and compositional zonation beneath the Coso volcanic field, California. *Contributions to Mineralogy and Petrology,* 85:346–365.

Bagnold, R. A. 1966. An approach to the sediment transport problem from general physics. *U.S. Geological Survey Professional Paper,* 422:37.

Bailey, Edgar H., and Rollin E. Stevens. 1960. Selective staining of K-feldspar and plagioclase on rock slabs and thin sections. *American Mineralogist,* 45:1020–1025.

Ballhaus, Christian G., and Andrew Y. Glikson. 1989. Magma mixing and intraplutonic quenching in the Wingellina Hills Intrusion, Giles Complex, central Australia. *Journal of Petrology,* 30:1443–1469.

Barbarin, Bernard. 1990. Plagioclase xenocrysts and mafic magmatic enclaves in some granitoids of the Sierra Nevada Batholith, California. *Journal of Geophysical Research,* 95:17,747–17,756.

———, and Jean Didier. 1991. Macroscopic features of mafic microgranular enclaves. In *Enclaves and Granite Petrology,* J. Didier and B. Barbarin, eds. Amsterdam: Elsevier, pp. 253–262.

Barbey, Pierre. 1991. Restites in migmatites and autochthonous granites: Their main features and their genesis. In *Enclaves and Granite Petrology,* J. Didier and B. Barbarin, eds. Amsterdam: Elsevier, pp. 479–491.

Bard, J. P. 1986. *Microtextures of Igneous and Metamorphic Rocks.* Dordrecht, Holland: Reidel.

Bardossy, Gyorgy. 1982. *Karst Bauxites.* New York: Elsevier.

———, and G. J. J. Aleva. 1990. *Lateritic Bauxites.* New York: Elsevier.

Barker, A. J. 1990. *Introduction to Metamorphic Textures and Microstructures.* New York: Chapman and Hall.

Barker, Daniel. 1983. *Igneous Rocks.* Englewood Cliffs, NJ: Prentice-Hall, reprinted 1992.

Barnes, Ivan. 1985. Mineral-water reactions in metamorphism and volcanism. *Chemical Geology,* 49:21–29.

Barrow, G. 1893. On an intrusion of muscovite-biotite gneiss in the southeast highlands of Scotland. *Geological Society of London Quarterly Journal,* 49:330–358.

Barth, T. F. W. 1962. A final proposal for calculating the mesonorm of metamorphic rocks. *Journal of Geology,* 70:497–498.

Bateman, P. C. 1988. Stratigraphic classification and nomenclature of igneous and metamorphic rock bodies: Discussion and reply. *Geological Society of America,* 100:995–997. Reply by A. Salvador, Chairman International Subcommission on Stratigraphic Classification.

———, and Bruce W. Chappell. 1979. Crystallization, fractionation, and solidification of the Tuolumne Intrusine Series, Yosemite National Park, California. *Geological Society of America Bulletin,* 90:465–482.

Bathurst, Robin G. C. 1975. *Carbonate Sediments and Their Diagenesis.* New York: Elsevier.

Baturin, G. H. 1982. Phosphorites on the sea floor: Origins, composition and distribution. *Developments in Sedimentology,* no. 33. Amsterdam: Elsevier.

Bayly, M. B. 1960. Modal analysis by point-counter—the choice of sample area. *Journal Geological Society of Australia,* 6:119–130.

Beach, A. 1980. Retrogressive metamorphic processes in shear zones with special reference to the Lewisian complex. *Journal of Structural Geology,* 2:257–263.

Beane, Richard E. 1982. Hydrothermal alteration in silicate rocks. In *Advances in Geology of the Porphyry Copper Deposits: Southwestern North America,* Spencer R. Titley, ed. Tuscon, AZ: University of Arizona Press, pp. 117–137.

Beard, James S. 1990. Partial melting of metabasites in the contact aureoles of gabbroic plutons in the Smartville Complex, Sierra Nevada, California. *Geological Society of America Memoir,* 174:303–313.

Beaty, D. W., C. G. Cunningham, C. W. Naeser, and G. P. Lundis. 1989. Genetic model for the carbonate-hosted Pb-Zn-Cu-Ag-Au manto-chimney deposits of the Gilman Colorado District, based on fluid inclusion, stable isotope, geologic, and fission-track time-temperature studies. In U.S. Geological Survey Research on Mineral Resources, Fifth Annual V. E. McKelvey Forum on Mineral and Energy Resources. *U.S. Geological Survey Circular,* 1035:4.

Bédard, Jean H. J. 1987. The development of compositional and textural layering in Archean komatiites and in Proterozoic komatiitic basalts from Cape Smith, Quebec, Canada. In *Origins of Igneous Layering,* I. Parsons, ed. Dordrecht, Holland: Reidel, pp. 399–418.

Bell, P. M. 1964. High-pressure melting relations for jadeite composition. *Carnegie Institute Washington Yearbook,* 63:171–174.

Bell, T. H., and M. A. Etheridge. 1973. Microstructure of mylonites and their descriptive terminology. *Lithos,* 6:337–348.

———, A. Forde, and N. Hayward. 1992. Do smoothly curving, spiral-shaped inclusions trails signify porphyroblast rotation? *Geology,* 20:59–62.

———, M. J. Rubenach, and P. D. Fleming. 1986. Porphyroblast nucleation, growth and dissolution in regional metamorphic rocks as a function of deformation partitioning during foliation development. *Journal of Metamorphic Petrology,* 4:37–67.

Bentor, Y. K. 1980. Marine phosphorites. *Society of Economic Paleontologists and Mineralogists,* Special Publication, 29.

Bergantz, G. W. 1991. Chemical and physical characterization of plutons. In *Contact Metamorphism,* Derrill M. Kerrick, ed. *Reviews in Mineralogy,* 26.

Best, M. G., Richard Lee Armstrong, William C. Graustein, Glenn F. Embree, and Robert C. Ahlborn. 1974. Mica granites of the Kern Mountains pluton, eastern White Pine County, Nevada: Remobilized basement of the Cordilleran miogeosyncline? *Geological Society of America Bulletin,* 85:1277–1286.

Birkeland, P. W. 1984. *Soils and Geomorphology.* New York: Oxford University Press.

Bischoff, James L., John A. Fitzpatrick, and Robert J. Rosenbauer. 1992. The solubility and stabilization of ikaite ($CaCO_3 \cdot 6H_2O$) from 0° to 25°C: environmental and paleoclimatic implications for thinolite tufa. *Journal of Geology,* 101:21–33.

Blanchard, Roland. 1968. Interpretation of leached outcrops. *Nevada Bureau of Mines Bulletin,* 66.

Blatt, Harvey, Gerard Middleton, and Raymond Murray. 1980. *Origin of Sedimentary Rocks,* 2nd ed. Englewood Cliffs, NJ: Prentice-Hall.

Blenkinsop, T. G., and E. H. Rutter. 1986. Cataclastic deformation of quartzite in the Moine thrust zone. *Journal of Structural Geology,* 8:669–681.

Blichert-Toft, Janne, Charles E. Lesher, and Minik T. Rosing. 1992. Selectively contaminated magmas of the Tertiary East Greenland macrodike complex. *Contribution to Mineralogy and Petrology,* 110:154–172.

Bloss, F. Donald. 1981. *The Spindle Stage Principles and Practice.* Cambridge, MA: Cambridge University Press.

Bodnar, R. J., T. J. Reynolds, and C. A. Kuehn. 1985. Fluid inclusion systematics in epithermal systems. *Reviews in Economic Geology,* 2:73–96.

Boggs, Sam, Jr. 1987. *Principles of Sedimentology and Stratigraphy.* Columbus, OH: Merrill.

———. 1992. *Petrology of Sedimentary Rocks.* New York: Macmillan.

Bonham, Harold F., Jr. 1988. Models for volcanic-hosted epithermal precious metal deposits. In *Bulk Mineable Precious Metal Deposits of the Western United States,* Robert W. Schafer, James J. Cooper, and Peter G. Vikre, eds. Symposium Proceedings April 6–8, 1987. Reno, NV: Geological Society of Nevada, pp. 259–271.

Boone, Gary M. 1962. Potassic feldspar enrichment in magma: Origin of syenite in Deboullie District, northern Maine. *Geological Society of America Bulletin,* 73:1451–1476.

———, David T. Doty, and Matthew T. Heizler. 1989. Hurricane Mountain Formation mélange: Description and tectonic significance of a Penobscottian accretionary complex. *Maine Geological Survey Studies in Maine Geology,* 2:33–83.

———, and E. P. Wheeler II. 1968. Staining for cordierite and feldspars in thin section. *American Mineralogist,* 53:327–331.

Borchert, Hermann, and Richard O. Muir. 1964. *Salt Deposits.* New York: Van Nostrand Reinhold.

Borradaile, Graham J. 1976. A strain study of a granite-granite gneiss transition and accompanying schistosity formation in the Betic orogenic zone, southeast Spain. *Journal of the Geological Society of London,* 132:417–428.

———, M. Brian Bayly, and Chris McA. Powell, eds. 1982. *Atlas of Deformational and Metamorphic Rock Fabrics.* New York: Springer-Verlag.

Bostick, Neely H. 1979. Microscopic measurement of the level of catagenesis of solid organic matter in sedimentary rocks to aid exploration for petroleum and to determine former burial temperatures. *Society of Economic Paleontologists and Mineralogists,* Special Publication, 26: 17–43.

Bottinga, Y., A. Kudo, and D. Weill. 1966. Some observations on oscillatory zoning and crystallization of magmatic plagioclase. *American Mineralogist,* 51:792–806.

Bouchez, Jean Luc, Christian Delas, Gérard Gleizes, Anne Nédélec, and Michael Cuney. 1992. Submagmatic microfractures in granites. *Geology,* 20:35–38.

Boudreau, A. E. 1987. Pattern formation during crystallization and the formation of fine-scale layering. In *Origins of Igneous Layering,* I. Parsons, ed. Dordrecht, Holland: Reidel, pp. 453–471.

Bowen, N. L. 1913. The melting phenomena of the plagioclase feldspars. *American Journal of Science,* 35:577–599.

———. 1914. The system diopside-forsterite-silica. *American Journal of Science,* 38:207–264.

———. 1915. The crystallization of haplobasaltic, haplodioritic, and related magmas. *American Journal of Science,* 40:161–185.

———. 1922. The behavior of inclusions in igneous magmas. *Journal of Geology,* 30:513–570.

———. 1928. *The Evolution of the Igneous Rocks.* Princeton, NJ: Princeton University Press.

———. 1940. Progressive metamorphism of siliceous limestone and dolomite. *Journal of Geology,* 48:225–274.

———. 1956. *The Evolution of the Igneous Rocks.* New York: Dover.

———, and O. F. Tuttle. 1950. The system $NaAlSi_3O_8$–$KAlSi_3O_8$–H_2O. *Geological Society of America Bulletin,* 58:498–511.

Braitsch, O. 1971. *Salt Deposits, Their Origin and Composition.* New York: Springer-Verlag.

Brandeis, Geneviéve, and Claude Jaupart. 1986. On the interaction between convection and crystallization in cooling magma chambers. *Earth and Planetary Science Letters,* 77:345–361.

Brandon, Mark T. 1989. Deformational styles in a sequence of olistromal mélanges, Pacific Rim Complex, western Vancouver Island, Canada. *Geological Society of America Bulletin,* 101:1520–1542.

———, Darrel S. Cowan, and Joseph A. Vance. 1988. The Late Cretaceous San Juan thrust system, San Juan Islands, Washington. *Geological Society of America Special Paper,* 221.

Branney, Michael J. 1991. Eruption and depositional facies of the Whorneyside Tuff Formation, English Lake District: An exceptionally large-magnitude phreatoplinian eruption. *Geological Society of America Bulletin,* 103:886–897.

———, B. Peter Kokelaar, and Brian J. McConnell. 1992. The Bad Step Tuff: A lava-like rheomorphic ignimbrite in a calc-alkaline piecemeal caldera, English Lake District. *Bulletin of Volcanology,* 54:187–199.

Brothers, R. N., and R. H. Grapes. 1989. Clastic lawsonite, glaucophane, and jadeitic pyroxene in Franciscan metagraywackes from the Diable Range, California. *Geological Society of America Bulletin,* 101:14–26.

Brown, G. C., and W. S. Fyfe. 1970. The production of granitic melts during ultrametamorphism. *Contributions to Mineralogy and Petrology,* 28:310–318.

Bruck, P. M. 1974. Granite varieties and structures of the northern and upper Liffey Valley units of the Leinster Batholith. *Geological Survey Ireland Bulletin,* 1:381–393.

Bryan, Wilfred B. 1972. Morphology of quench crystals in submarine basalts. *Journal of Geophysical Research,* 77:5812–5819.

Bryant, Donald G. 1968. Intrusive breccias associated with ore, Warren (Bisbee) Mining District, Arizona. *Economic Geology,* 63:1–12.

Buczynski, Chris, and Henry S. Chafetz. 1991. Habit of bacterially induced precipitates of calcium carbonate and the influence of medium vicosity on mineralogy. *Journal of Sedimentary Petrology,* 61:226–233.

Buddington, A. F. 1948. Origin of granitic rocks of the northwest Adirondacks. In *Origin of Granite,* James Gilluly, Chairman. *Geological Society of America Memoir,* 28:21–43.

———. 1959. Granite emplacement with special reference to North America. *Geological Society of America Bulletin,* 70:671–747.

Bunch, T. E. 1975. Petrography and petrology of basaltic achondrite polymict breccias (howardites). *Proceedings of Sixth Lunar Science Conference,* pp. 469–492.

Bunsen, R. W. 1851. Über die Processe der volkanischen Gesteinsbildungen Islands. *Ann. Phys. Chem.,* 83:197–272.

Burnham, C. W. 1959. Contact metamorphism of magnesian limestones at Crestmore, California. *Geological Society of America Bulletin,* 70:879–920.

———. 1962. Facies classification and types of hydrothermal alteration. *Economic Geology,* 57:768–784.

———. 1979. Magmas and hydrothermal fluids. In *Geochemistry of Hydrothermal Ore Deposits,* H. L. Barnes, ed. New York: Wiley, pp. 37–76.

———, and H. Nekvasil. 1986. Equilibrium properties of granite pegmatite magmas. *American Mineralogist,* 71:239–263.

———, and H. Ohmoto. 1980. Late-stage processes of felsic magmatism. *Mining Geology,* Special Issue, 8:1–11.

Bustillo, M. A., and P. A. Ruiz-Ortiz. 1987. Chert occurrences in carbonate turbidites: Examples from the Upper Jurassic of the Beltic Mountains (southern Spain). *Sedimentology,* 34:611–622.

Button, A., T. D. Brock, P. J. Cook, H. P. Euster, A. M. Goodwin, H. L. James, L. Margules, K. H. Nealson, J. O. Neiagu, A. F. Trendall, and M. F. Walter. 1982. Sedimentary iron deposits, evaporites, and phosphorites. In *Mineral Deposits and Evolution of the Biosphere,* H. D. Holland and M. Schidlowski, eds. New York: Springer-Verlag, pp. 259–273.

Calvache, V., Marta Lucia, and Stanley N. Williams. 1992. Lithic-dominated pyroclastic flows at Galeras volcano, Columbia—an unrecognized volcanic hazard. *Geology,* 20:539–542.

Cameron, E. N., R. H. Jahns, A. H. McNair, and L. R. Page. 1949. Internal structure of granitic pegmatites. *Economic Geology Monograph,* 2.

Candela, Philip A. 1991. Physics of aqueous phase evolution in plutonic environments. *American Mineralogist,* 76:1081–1091.

Carmichael, D. M. 1969. On the mechanism of prograde metamorphic reactions in quartz-bearing pelitic rocks. *Contributions to Mineralogy and Petrology,* 20:244–267.

Carmichael, I. S. E. 1963. The crystallization of feldspar in volcanic acid liquids. *Quarterly Journal of the Geological Society of London,* 119:95–131.

———. 1965. Trachytes and their feldspar phenocrysts. *Mineralogical Magazine,* 34:107–125.

———. 1979. Glass and the glassy rocks. In *The Evolution of the Igneous Rocks*, H. S. Yoder, Jr., ed. Princeton, NJ: Princeton University Press.

———, Francis J. Turner, and John Verhoogan. 1974. *Igneous Petrology*. New York: McGraw-Hill.

Carr, James R., and M. J. Hibbard. 1991. Open-ended mineralogical/textural rock classification. *Computers and Geosciences,* 17:1409–1463.

Carrigan, Charles R., and John C. Eichelberger. 1990. Zoning of magmas by viscosity in volcanic conduits. *Nature,* 343:248–251.

Carten, Richard B., Ennis P. Geraghty, and Bruce M. Walker. 1988. Cyclic development of igneous features and their relationship to high-temperature hydrothermal features in the Henderson porphyry molybdenum deposit, Colorado. *Economic Geology,* 83:266–296.

Carter, J. G., ed. 1991. *Skeletal Biomineralization.* New York: Van Nostrand Reinhold.

Carter, N. L., and G. C. Kennedy. 1964. Origin of diamonds in the Canyon Diablo and Novo Urei meteorites. *Journal of Geophysical Research,* 69:2403–2421.

———, C. B. Officer, C. A. Chesner, and W. I. Rose. 1986. Dynamic deformation of volcanic ejects from the Toba caldera: Possible relevance to Cretaceous/Tertiary boundary phenomena. *Geology,* 14:380–383.

Cartwright, I. 1988. Crystallization of melts, pegmatite intrusion and the Inverian retrogression of the Scourian Complex, northwest Scotland. *Journal of Metamorphic Geology,* 6:77–93.

———, and A. C. Bernicont. 1986. The generation of quartz-normative melts and corundum-bearing restites by crustal anatexis: Petrogenetic modeling based on an example from the Lewisian of northwest Scotland. *Journal of Metamorphic Geology,* 4:79–99.

Cas, R. A. F., and J. V. Wright. 1987. *Volcanic Successions, Modern and Ancient.* London: Allen and Unwin.

Cashman, K. V. 1990. Textural constraints on the kinetics of crystallization of igneous rocks. In *Modern Methods of Igneous Petrology: Understanding Magmatic Processes,* J. Nicholls and J. K. Russell, eds. *Reviews in Mineralogy,* 24:259–314.

———, and George W. Bergantz. 1991. Magmatic processes. *Reviews of Geophysics, Supplement, U.S. National Report to International Union of Geodesy and Geophysics, 1987–1990,* pp. 500–509.

———, and John M. Ferry. 1988. Crystal size distribution (CSD) in rocks and the kinetics and dynamics of crystallization. *Contributions to Mineralogy and Petrology,* 99:401–415.

———, and Richard S. Fiske. 1991. Fallout of pyroclastic debris from submarine volcanic eruptions. *Science,* 253:275–280.

Caskey, S. John, and Richard A. Schweickert. 1992. Mesozoic deformation in the Nevada test site and vicinity: Implications for the structural framework of the Cordilleran fold and thrust belt and Tertiary extension north of Las Vegas Valley. *Tectonics,* 11:1314–1331.

Cerny, P. 1971. Graphic intergrowths of feldspars and quartz in some Czechoslovak pegmatites. *Contributions to Mineralogy and Petrology,* 30:343–355.

Chafetz, Henry S., Patrick F. Rush, and Nancy M. Utech. 1991. Microenvironment controls on mineralogy and habit of $CaCO_3$ precipitates: An example from an active travertine system. *Sedimentology,* 38:107–126.

Chamberlin, T. C. 1897. The method of multiple working hypotheses. *Journal of Geology,* 5:837–848.

Chao, E. C. T., J. J. Fahey, J. Littler, and D. J. Milton. 1962. Stishovite, SiO_2, a very high pressure new mineral from Meteor Crater, Arizona. *Journal of Geophysical Research,* 72:419.

Chapin, Charles E., and Gary R. Lowell. 1979. Primary and secondary flow structures in ash-flow tuffs of the Gribbles Run paleovalley, central Colorado. In *Ash Flow Tuffs,* Charles E. Chapin and Wolfgang E. Elston, eds. *Geological Society of America Special Paper,* 180:137–154.

Chappell, B. W., A. J. R. White, and D. Wyborn. 1987. The importance of residual source material (restite) in granite petrogenesis. *Journal of Petrology,* 28:1111–1138.

Chayes, F. 1952. Relation between composition and indices of refraction in natural plagioclase. *American Journal of Science* (Bowen Volume), 250A:85–106.

———. 1956. *Petrographic Modal Analysis.* New York: Wiley.

Chester, F. M. 1988. The brittle-ductile transition in a deformation-mechanism map for halite. *Tectonophysics,* 154:125–136.

———, M. Friedman, and J. M. Logan. 1985. Foliated cataclasites. *Tectonophysics,* 111:139–146.

Chopra, P. N., and M. S. Paterson. 1984. The role of water in the deformation of dunite. *Journal of Geophysical Research,* 89:7861–7876.

Chowns, T. M., and J. E. Elkins. 1974. The origin of quartz geodes and cauliflower cherts through the silicification of anhydrite nodules. *Journal of Sedimentary Petrology,* 44:885–903.

Clark, Russell G., Jr., and John B. Lyons. 1986. Petrogenesis of the Kinsman intrusive suite: Peraluminous granitiods of western New Hampshire. *Journal of Petrology,* 27:1365–1393.

Clemens, J. D. 1989. The importance of residual source material (restite) in granite petrogenesis: A comment. *Journal of Petrology,* 30:1313–1316.

———, and Victor J. Wall. 1984. Origin and evolutions of a peraluminous silicic ignimbrite suite: The Violet Town volcanics. *Contributions to Mineralogy and Petrology,* 88:354–371.

Cole, Ronald B., and Peter G. DeCelles. 1991. Subaerial to submarine transition in early Miocene pyroclastic flow deposits, southern San Joaquin basin, California. *Geological Society of America Bulletin,* 103:221–235.

Collinson, J. D., and D. B. Thompson. 1989. *Sedimentary Structures,* 2nd ed. London: Unwin Hyman.

Compton, R. R. 1955. Trondjemite batholith near Bidwell Bar, California. *Geological Society of America Bulletin,* 66:9–44.

Congdon, Roger D., and W. P. Nash. 1991. Eruptive pegmatite magma: Rhyolite of the Honeycomb Hills, Utah. *American Mineralogist,* 76:1261–1278.

Coniglio, Mario. 1987. Biogenic chert in the Cow Head Group (Cambro-Ordovician), western Newfoundland. *Sedimentology,* 34:813–823.

Conrad, Mark E., and H. R. Naslund. 1989. Modally-graded rhythmic layering in the Skaergaard intrusion. *Journal of Petrology,* 30:251–269.

Cook, Peter J. 1972. Petrology and geochemistry of the phosphate deposits of Northwest Queensland, Australia. *Economic Geology,* 67:1193–1213.

Cook, Sterling S. 1988. Supergene copper mineralization of the Lakeshore Mine, Pinal County, Arizona. *Economic Geology,* 83:297–309.

Cookro, Theresa M., Miles L. Silberman, and Byron R. Berger. 1988. Gold-tungsten-bearing hydrothermal deposits in the Yellow Pine Mining District. In *Bulk Mineable Precious Metal Deposits of the Western United States,* Robert W. Schafer, James J. Cooper, and Peter G. Vikre, eds. Symposium Proceedings, April 6–8, 1987. Reno, NV: Geological Society of Nevada, pp. 577–613.

Coombs, D. S. 1953. The nature and alteration of some Triassic sediments from Southland, New Zealand. *Royal Society of New Zealand Transactions,* 82.

Cotter, Edward. 1992. Diagenetic alteration of chamositic clay minerals to ferric oxide in oolitic ironstone. *Journal of Sedimentary Petrology,* 62:54–60.

Couturié, J. P. 1969. Sûr l'antériorité du granite porphyroïde de la Margeride par rapport au granite à cordiérite du Velay (Massif Central Francais). *C. R. Academie Science* (Paris), 269D:2298–2300.

Cowan, D. S. 1982. Deformation of partly dewatered and consolidated Franciscan sediments near Piedras Blancas Point, California. In *Trench-Forearc Geology: Sedimentation and Tectonics on Modern and Ancient Active Plate Margins,* J. K. Leggett, ed. *Geological Society of London Special Publication,* 10:439–457.

Cox, S. F., and M. A. Etheridge. 1983. Crack-seal fiber growth mechanisms and their significance in the development of oriented layer silicate microstructures. *Tectonophysics,* 92:147–170.

Creasey, S. C. 1966. Hydrothermal alteration. In *Geology of the Porphyry Copper Deposits,* S. R. Titley and C. L. Hicks, eds. Tuscon, AZ: University of Arizona Press.

Cross, W., J. P. Iddings, L. V. Pirsson, and H. Washington. 1902. The quantitative classification of igneous rocks. *Journal of Geology,* 10:555–690.

———. 1906. The texture of igneous rocks. *Journal of Geology,* 14:692–707.

Curtis, C. D., S. R. Lipshie, G. Ortel, and M. J. Pearson. 1980. Clay orientation in some Upper Carboniferous mudrocks, its relation to quartz content and some inferences about fissility, porosity and compaction history. *Sedimentology,* 27:333–339.

Czamanske, Gerald K., Michael L. Zientek, and Craig E. Manning. 1991. Low-K granophyres of the Stillwater Complex, Montana. *American Mineralogist,* 76:1646–1661.

Dahanayake, Kapila, and S. M. N. D. Subasinghe. 1989. A modern terrestrial phosphorite—an example from Sri Lanka. *Sedimentary Geology,* 61:311–316.

Davis, Richard A., Jr. 1992. *Depositional Systems,* 2nd ed. Englewood Cliffs, NJ: Prentice Hall.

Debat, P., J-C. Soula, L. Kubin, and L-L. Vidal. 1978. Optical studies of natural deformation microstructures in feldspars (gneiss and pegmatites from Occitania, southern France). *Lithos,* 9:133–145.

Decima, Arvedo, J. A. McKenzie, and B. C. Schreiber. 1988. The origin of "evaporative" limestones: An example from the Messinian of Sicily (Italy). *Journal of Sedimentary Petrology,* 58:256–272.

Decker, Robert W., and Robert L. Christiansen. 1984. Explosive eruptions of Kilauea Volcano, Hawaii. In *Explosive Volcanism: Inception, Evolution, and Hazards.* Washington, DC: National Academy of Sciences, pp. 122–132.

Deer, W. A., R. A. Howie, and J. Zussman. 1962a. *Rock-Forming Minerals, Ortho- and Ring Silicates,* vol. 1. London: Longman.

———. 1962b. *Sheet Silicates,* vol. 3. London: Longman.

———. 1963a. *Chain Silicates,* vol. 2. London: Longman.

———. 1963b. *Alkali Feldspars,* vol. 4. New York: Wiley.

———. 1978. *Single Chain Silicates,* vol. 2A. London: Longman.

———. 1982. *Orthosilicates,* vol. 1A. London: Longman

———. 1986. *Disilicate and Ring Silicates,* vol. 1B. London: Longman.

———. 1992. *An Introduction to the Rock-forming Minerals,* 2nd ed. New York: Wiley.

Dell'Angelo, L. N., and J. Tullis. 1988. Experimental deformation of partially melted granitic aggregates. *Journal of Metamorphic Geology,* 6:495–515.

Dence, M. R. 1971. Impact melts. *Journal of Geophysical Research,* 76:5552–5565.

———, and P. B. Robertson. 1989. Shock metamorphism. In *The Encyclopedia of Igneous and Metamorphic Petrology,* D. R. Bowes, ed. New York: Van Nostrand Reinhold, pp. 526–530.

Dethier, David P., David R. Pevear, and David Frank. 1981. Alteration of new volcanic deposits. *U.S. Geological Survey Professional Paper* 1250:649–665.

de Wit, M. J. 1976. Metamorphic textures and deformation: A new mechanism for the development of syntectonic porphyroblasts and its implications for interpreting timing relations in metamorphic rocks. *Geological Journal,* 11:71–102.

Dickson, J. A. D. 1965. A modified staining technique for carbonates in thin section. *Nature,* 205:587.

Dietrich, R. V., and D. M. Sheehan. 1964. Approximate chemical analyses from modal analyses of rocks. *Virginia Polytechnic Institute, Engineering Extension Series,* Circular 1:28.

Dietz, R. S. 1968. Shatter cones in cryptoexplosive structures. In *Shock Metamorphism of Natural Materials,* B. M. French and N. M. Short, eds. Baltimore, MD: Mono Book Corporation, pp. 267–285.

Dimroth, E. 1976. Aspects of sedimentary petrology of cherty iron-formations. In *Handbook of Strata-bound and Stratiform Ore Deposits,* vol. 7, K. H. Wolf, ed. New York: Elsevier.

Dix, Owen R., and M. P. A. Jackson. 1982. Lithology, microstructures, fluid inclusions, and geochemistry of rock salt and of the cap rock contact in Oakwood Dome, East Texas: Significance for nuclear waste storage. *Bureau of Economic Geology, The University of Texas at Austin Report of Investigations,* 120.

Donaldson, C. H. 1982. Origin of some of the Rhum harrisite by segregation of intercumulus liquid. *Mineralogical Magazine,* 45:201–209.

———, and C. M. B. Henderson. 1988. A new interpretation of round embayments in quartz crystals. *Mineralogical Magazine,* 52:27–33.

Donselaar, M. E. 1989. The Cliff House Sandstone, San Juan Basin, New Mexico: Model for the stacking of "transgressive" barrier complexes. *Journal of Sedimentary Petrology,* 59:13–27.

Dorais, Michael J., James A. Whitney, and Michael F. Roden. 1990. Origin of mafic enclaves in the Dinkey Creek pluton, central Sierra Nevada batholith, California. *Journal of Petrology,* 31:853–881.

Dott, R. H. 1964. Wacke, graywacke, and matrix—what approach to immature sandstone classification? *Journal of Sedimentary Petrology,* 34:629–632.

Dougan, T. W. 1979. Compositional and modal relationships and melting reactions in some migmatitic metapelites from New Hampshire and Maine. *American Journal of Science,* 279:897–935.

Dowty, Eric. 1976. Crystal structure and crystal growth II: Sector zoning in minerals. *American Mineralogist,* 61:460–469.

———. 1980. Crystal growth and nucleation theory and the numerical simulation of igneous crystallization. In *Physics of Magmatic Processes,* R. B. Hargraves, ed. Princeton, NJ: Princeton University Press, pp. 419–485.

———, Klaus Keil, and Martin Prinz. 1974. Lunar pyroxene-phyric basalts: Crystallization under supercooled conditions. *Journal of Petrology,* 15:419–453.

Dravis, J. J., and D. A. Yurewicz. 1985. Enhanced carbonate petrography using fluorescence microscopy. *Journal of Sedimentary Petrology,* 55:795–804.

Drever, J. I. 1974. Geochemical model for the origin of Precambrian banded iron formations. *Geological Society of America Bulletin,* 85:1099–1106.

———. 1988. *The Geochemistry of Natural Waters.* Englewood Cliffs, NJ: Prentice Hall.

Druitt, T. H. 1985. Vent evolution and lag breccia formation during the Cape Riva eruption of Santorini, Greece. *Journal of Geology,* 93:439–454.

Duhig, Nathan C., Garry J. Davidson, and Joe Stolz. 1992. Microbial involvement in the formation of Cambrian sea-floor silica-iron oxide deposits, Australia. *Geology,* 20:511–514.

Duke, Edward F., Jack A. Redden, and James J. Papike. 1988. Calamity Peak layered granite-pegmatite complex, Black Hills, South Dakota. I: Structure and emplacement. *Geological Society of America Bulletin,* 100:825–840.

Duke, Michael B. 1965. Metallic iron in basaltic achondrites. *Journal of Geophysical Research,* 70:1523–1527.

———, and Leon T. Silver. 1967. Petrology of eucrites, howardites, and mesosiderites. *Geochimica et Cosmochimica Acta,* 31:1637–1665.

Dunham, R. J. 1962. Classification of carbonate-rocks according to depositional texture. In *Classification of Carbonate Rocks,* W. E. Ham, ed. *American Association of Petroleum Geologists Memoir,* 1:108–121.

Eberz, Gunter W., and Ian A. Nicholls. 1988. Microgranitoid enclaves from the Swifts Creek pluton southeast Australia: Textural and physical constraints on the nature of magma mingling processes in the plutonic environment. *Geologische Rundschau,* 77:713–736.

Eggleton, Richard A., and Peter R. Buseck. 1980. The orthoclase-microcline inversion: A high-resolution transmission electron microscope study and strain analysis. *Contributions to Mineralogy and Petrology,* 74:123–133.

Ehlers, E. G. 1972. *The Interpretation of Geological Phase Diagrams.* San Francisco: Freeman.

———. 1987. *Optical Mineralogy, Mineral Descriptions,* vol. 2. Palo Alto, CA: Blackwell.

Eichelberger, J. C. 1975. Origin of andesite and dacite: Evidence of mixing at Glass Mountain in California and at other circum-Pacific volcanoes. *Geological Society of America Bulletin,* 86:1381–1391.

———. 1980. Vesiculation of mafic magma during replenishment of silicic magma reservoirs. *Nature,* 288:446–450.

Einaudi, Mario T. 1982. Description of skarns associated with porphyry copper plutons. In *Advances in Geology of the Porphyry Copper Deposits, Southwestern North America,* Spencer R. Titley, ed. Tuscon, AZ: University of Arizona Press, pp. 139–183.

Einsele, G. 1992. *Sedimentary Basins.* New York: Springer-Verlag.

Ellis, A. J. 1959. The solubility of calcite in carbon dioxide solutions. *American Journal of Science,* 257:354–365.

———. 1963. The solubility of calcite in sodium chloride at high temperatures. *American Journal of Science,* 261:259–267.

Emeis, K.-C., H.-H. Richnow, and S. Kempe. 1987. Travertine formation in Plitvice National Park, Yugoslavia: Chemical versus biological control. *Sedimentology,* 34:595–610.

Emeleus, C. H. 1987. The Rhum Layered Complex, Inner Hebrides, Scotland. In *Origins of Igneous Layering,* I. Parsons, ed. Dordrecht, Holland: Reidel, pp. 263–286.

Emmons, R. C. 1943. The universal stage. *Geological Society of America Memoir,* 8.

Engelder, James T. 1974. Cataclasts and the generation of fault gouge. *Geological Society of America Bulletin*, 85:1515–1522.

Ernst, W. G. 1960. Diabase-granophyre relations in the Endion sill, Duluth, Minnesota. *Journal of Petrology*, 1:286–303.

———. 1971. Petrologic reconnaissance of Franciscan metagraywackes from the Diablo Range, central California Coast Ranges. *Journal of Petrology*, 34:43–59.

———. 1975. *Subduction Zone Metamorphism.* New York: Halsted.

———. 1976. *Petrologic Phase Equilibria.* San Francisco: Freeman.

Eugster, H. P. 1967. Hydrous sodium silicates from Lake Magadi, Kenya. Precursors of bedded chert. *Science*, 157:1177–1180.

———, and L. A. Hardie. 1975. Sedimentation in an ancient playa lake complex: The Wilkins Peak member of the Green River Formation of Wyoming. *Geological Society of America Bulletin*, 86:319–334.

Evans, Charles C., and Robert N. Ginsburg. 1987. Fabric-selective diagenesis in the Late Pleistocene Miami limestone. *Journal of Sedimentary Petrology*, 57:311–318.

Evans, James P., and Douglas S. Neves. 1992. Footwall deformation along Willard thrust, Sevier orogenic belt: Implications for mechanisms, timing, and kinematics. *Geological Society of America Bulletin*, 104:516–527.

Everden, J. F., and G. T. James. 1964. Potassium-argon dates and the Tertiary floras of North America. *American Journal of Science*, 262:945–974.

Eyles, Carolyn H., and Nicholas Eyles. 1989. The upper Cenozoic White River "tillites" of southern Alaska: Subaerial slope and fan-delta deposits in a strike-slip setting. *Geological Society of America Bulletin*, 101:1091–1102.

Eyles, Nicholas, and Martin B. Lagoe. 1989. Sedimentology of shell-rich deposits (coquinas) in the glaciomarine upper Cenozoic Yakataga Formation, Middleton Island, Alaska. *Geological Society of America Bulletin*, 101:129–142.

Fabriès, Jacques, Oscar Figueroa, and Jean-Pierre Lorand. 1987. Petrology and thermal history of highly deformed mantle xenoliths from the Montferrier basanites, Languedoc, southern France: A comparison with ultramafic complexes from the north Pyrenean zone. *Journal of Petrology*, 28:887–919.

Fang, J. H., and L. Zevin. 1985. Quantitative x-ray diffractometry of carbonate rocks. *Journal of Sedimentary Petrology*, 55:611–612.

Fenn, P. M. 1977. The nucleation and growth of alkali feldspars from hydrous melts. *Contributions to Mineralogy and Petrology*, 15:135–161.

———. 1986. On the origin of graphic granite. *American Mineralogist*, 71:325–330.

Finks, Robert M. 1960. Late Paleozoic sponge faunas of the Texas region. *Bulletin of the American Museum of Natural History, New York*, 120: article 1.

Fisher, A. B., and E. G. Ehlers. 1982. Porphyroblasts and "crystallization force": An experimental analog. *Geology*, 10:394–399.

Fisher, R. V. 1960. Classification of volcanic breccias. *Geological Society of America Bulletin*, 71:973–982.

———. 1961. Proposed classification of volcaniclastic sediments and rocks. *Geological Society of America Bulletin*, 72:1409–1414.

———. 1966. Mechanism of deposition from pyroclastic flows. *American Journal of Science*, 264:350–363.

———. 1979. Models for pyroclastic surges and pyroclastic flows. *Journal of Volcanology and Geothermal Research*, 6:305–318.

———, and H.-U. Schmincke. 1984. *Pyroclastic Rocks.* New York: Springer-Verlag.

Fiske, Richard S., and Tokihiko Matsuda. 1964. Submarine equivalents of ash flows in the Tokiwa Formation, Japan. *American Journal of Science,* 262:76–106.

Fitch, F. J. 1959. Macro point counting. *American Mineralogist,* 44:667–668.

Flinn, Richard A., and Paul K. Trojan. 1990. *Engineering Materials and Their Applications,* 4th ed. Boston: Houghton Mifflin.

Flood, R. H., and S. E. Shaw. 1975. A cordierite-bearing granite suite from the New England batholith, N.S.W. Australia. *Contribution to Mineralogy and Petrology,* 52:157–164.

———, and R. H. Vernon. 1978. The Cooma granodiorite, Australia: An example of *in situ* crustal anatexis? *Geology,* 6:81–84.

Folk, R. L. 1959. Practical petrographic classification of limestone. *American Association of Petroleum Geologists,* 43:1–38.

———. 1962. Spectral subdivisions of limestone types. In Classification of sedimentary rocks, W. E. Ham, ed. *American Association of Petroleum Geologists Memoir,* 1:62–84.

———. 1974. *Petrology of Sedimentary Rocks.* Austin, TX: Hempill.

———, P. B. Andrews, and D. W. Lewis. 1970. Detrital sedimentary rock classification and nomenclature for use in New Zealand. *New Zealand Journal of Geology and Geophysics,* 13:937–968.

Foord, Eugene E., Louis B. Spaulding, Jr., Roger A. Mason, and Robert F. Martin. 1989. Mineralogy and paragenesis of the Little Three Mine Pegmatites, Ramona District, San Diego County, California. *The Mineralogical Record,* 20:101–127.

Force, Reic R., and William F. Cannon. 1988. Depositional model for shallow-marine manganese deposits around black shale basins. *Economic Geology,* 83:93–117.

Fournier, Robert O. 1985. The behavior of silica in hydrothermal solutions. In *Geology and Geochemistry of Epithermal Systems,* B. R. Berger and P. M. Bethke, eds. *Reviews in Economic Geology,* 2:45–59.

———, J. Michael Thompson, Charles G. Cunningham, and Roderick A. Hutchinson. 1991. Conditions leading to a recent small hydrothermal explosion at Yellowstone National Park. *Geological Society of America Bulletin,* 103:1114–1120.

Frakes, Larry, and Barrie Bolton. 1992. Effects of ocean chemistry, sea level, and climate on the formation of primary sedimentary manganese ore deposits. *Economic Geology,* 87:1207–1211.

Francis, W. 1961. *Coal: Its Formation and Composition.* London: Edward Arnold.

Franco, R. R., and J. F. Schairer. 1951. Liquid temperatures in mixtures of the feldspars of soda, potash, and lime. *Journal of Geology,* 59:259–267.

French, B. M. 1968. Shock metamorphism as a geological process. In *Shock Metamorphism of Natural Materials,* B. M. French and N. M. Short, eds. Baltimore, MD: Mono Book Corporation.

———. 1973. Mineral assemblages in diagenetic and low-grade metamorphic iron-formation. *Economic Geology,* 68:1063–1074.

———. 1990. 25 years of the impact-volcanic controversy: Is there anything new under the sun? or inside the Earth? *EOS Transactions, American Geophysical Union,* 71:411–414.

———, and Roger L. Nielsen. 1990. Vredefort bronzite granophyre: Chemical evidence for origin as a meteorite impact melt. *Tectonophysics,* 171:119–138.

———, and N. M. Short, eds. 1968. *Shock Metamorphism of Natural Materials.* Baltimore, MD: Mono Book Corporation.

Friedman, Gerald M. 1959. Identification of carbonate minerals by staining methods. *Journal of Sedimentary Petrology,* 29:87–97.

Frost, T. P., and G. A. Mahood. 1987. Field, chemical and physical contraints on mafic-felsic magma interaction in the Larmark Granodiorite, Sierra Nevada, California. *Geological Society of America Bulletin,* 99:272–291.

Furnes, H., I. B. Fridleifsson, and F. B. Atkins. 1980. Subglacial volcanics—on the formation of acid hyaloclastites. *Journal of Volcanology and Geothermal Research,* 8:95–110.

Fyfe, W. S., N. J. Price, and A. B. Thompson. 1978. *Fluids in the Earth's Crust.* New York: Elsevier.

Galwey, A. K., and K. A. Jones. 1966. Crystal size frequency distribution of garnets in some analyzed metamorphic rocks from Mallaig, Inverness, Scotland. *Geological Magazine,* 103:143–152.

Garrels, R. M., and Charles L. Christ. 1965. *Solutions, Minerals, and Equilibria.* San Francisco: Freeman, Cooper.

Garrison, R. E., M. Kastner, and C. E. Reimers. 1990. Miocene phosphogenesis in California. In *Phosphate Deposits of the World. Vol. 3: Neogene to Modern Phosphorites.* Cambridge, MA: Cambridge University Press.

Gaudemer Y., and P. Tapponnier. 1987. Ductile and brittle deformation in the northern Snake Range, Nevada. *Journal of Structural Geology,* 9:159–180.

Gawthorpe, Robert L. 1987. Burial dolomitization and porosity development in a mixed carbonate-clastic sequence: An example from the Bowland Basin, northern England. *Sedimentology,* 34:533–558.

Gehman, H. M., Jr. 1958. Notch Peak intrusive, Millard County, Utah, geology, petrogenesis, and economic deposits. *Utah Geological and Mineralogical Survey Bulletin,* 62:1–48.

George, R. P., Jr. 1978. Subsolidus deformation mechanisms and hypersolidus creep of peridotite. In *Proceedings of the American Geophysical Union Chapman Conference on Partial Melting in the Earth's Upper Mantle,* H. J. B. Dick, ed. *Oregon Department of Geology and Mineral Industries Bulletin,* 96:233–262.

Gillen, C. 1982. *Metamorphic Geology, an Introduction to Tectonic and Metamorphic Processes.* London: Allen and Unwin.

Gilluly, James. 1946. The Ajo mining district, Arizona. *U.S. Geological Survey Professional Paper,* 209.

Gilmour, I., S. S. Russell, J. W. Arden, M. R. Lee, I. A. Franchi, and C. T. Pillinger. 1992. Terrestrial carbon and nitrogen isotopic ratios from Cretaceous-Tertiary boundary nanodiamonds. *Science,* 258:1624–1626.

Given, R. Kevin, and Bruce H. Wilkinson. 1985. Kinetic control of morphology, composition, and mineralogy of abiotic sedimentary carbonates. *Journal of Sedimentary Petrology,* 55:109–119.

Goldich, S. S. 1938. A study of rock weathering. *Journal of Geology,* 46:17–58.

Goldstein, J. I., D. E. Newbury, P. Echlin, D. C. Joy, C. Fiori, and E. Lifshin. 1981. *Scanning Electron Microscopy and X-ray Microanalysis.* New York: Plenum.

Goldstrand, Patrick M. 1991. Evolution of Late Cretaceous and Early Tertiary basins of southwest Utah based on clastic petrology. *Journal of Sedimentary Petrology,* 62:495–507.

Golightly, J. Paul. 1981. Nickeliferous laterite deposits. *Economic Geology,* 75th Anniversary Volume:710–735.

Gordon, Mackenzie, Jr., Joshua I. Tracey, Jr., and Miller W. Ellis. 1958. Geology of the Arkansas bauxite region. *U.S. Geological Survey Professional Paper,* 299.

Gornitz, V. M., and B. C. Schreiber. 1981. Displacive halite hoppers from the Dead Sea: Some implications for ancient evaporite deposits. *Journal of Sedimentary Petrology,* 51:787–794.

Gow, A. J. 1969. On the rates of growth of grains and crystals in South Polar firn. *Journal of Glaciology,* 8:241–252.

———. 1970. Deep core studies of the crystal structure and fabrics of Antarctic glacier ice. *USA CRREL Research Report,* 282.

———. 1971. Depth-time-temperature relationships of ice crystal growth in polar glaciers. *USA CRREL Research Report,* 300.

———. 1975. Time-temperature dependence of sintering in perennial isothermal snowpacks. Snow mechanics symposium (*Proceedings of the Grindewald Symposium,* April 1974). IAHS-AISH Publications, 114.

———, and T. Williamson. 1976. Rheological implications of the internal structure and crystal fabrics of the West Antarctic ice sheet as revealed by deep core drilling at Byrd Station. *USA CRREL Research Report,* 76–35.

Grant, James A., and B. Ronald Frost. 1990. Contact metamorphism and partial melting of pelitic rocks in the aureole of the Laramie Anorthosite Complex, Morton Pass, Wyoming. *American Journal of Science,* 290:425–472.

Gray, N. H. 1970. Crystal growth and nucleation in two large diabase dikes. *Canadian Journal of Earth Sciences,* 7:366–375.

Green, H. W., II. 1984. "Pressure solution" creep: Some causes and mechanisms. *Journal of Geophysical Research,* 89:4313–4318.

Green, T. A. 1976. Experimental generation of cordierite or garnet-bearing granitic liquids from a pelitic composition. *Geology,* 4:85–88.

Greensmith, J. T. 1989. *Petrology of the Sedimentary Rocks,* 7th ed. London: Unwin Hyman.

Gribble, C. D. 1970. The role of partial fusion in the genesis of certain cordierite-bearing rocks. *Scottish Journal of Geology,* 6:75–82.

———, and Allan J. Hall. 1992. *A Practical Introduction to Optical Mineralogy.* New York: Chapman and Hall.

Grier, Susan. 1983. Tertiary stratigraphy and geologic history of the Sacramento Pass area, Nevada. In *Utah Geological and Mineral Survey Special Studies 59. Guidebook Part I, Geological Society of America Rocky Mountain and Cordilleran Sections Meetings.* Salt Lake City, Utah, pp. 139–160.

Grieve, R. A. F. 1987. Terrestrial impact structures. *Annual Reviews of Earth and Planetary Science,* 15:245–270.

———. 1990. Impact cratering on the Earth. *Scientific American,* 262:66–73.

Griggs, D. T., M. S. Patterson, H. C. Heard, and F. J. Turner. 1960. Annealing recrystallization in calcite crystals and aggregates. *Geological Society of America Memoir,* 79:21–38.

———, F. J. Turner, and H. C. Heard. 1960. Deformation of rocks at 500° to 800°C. *Geological Society of America Memoir,* 79:39–104.

Grogan, Robert M., and James C. Bradbury. 1968. Fluorite-zinc-lead deposits of the Illinois-Kentucky Mining District. In *Ore Deposits of the United States, 1933–1967,* John D. Ridge, ed. New York: American Institute of Mining, Metallurgical, and Petroleum Engineers, Inc., pp. 371–398.

Groshong, Richard H., Jr. 1988. Low-temperature deformation mechanisms and their interpretation. *Geological Society of America Bulletin,* 100:1329–1360.

Gross, Gordon A. 1980. A classification of iron formations based on depositional environments. *Canadian Mineralogist,* 18:215–222.

Grout, Frank F. 1940. *Kemp's Handbook of Rocks.* New York: Van Nostrand Reinhold.

Grutzeck, Michael W. 1986. St. Peter Sandstone: A closer look. *Journal of Sedimentary Petrology,* 56:669–673.

Guilbert, John M., and Charles F. Park. 1985. *The Geology of Ore Deposits.* New York: Freeman.

Gunatilaka, A., A. Al-Zamel, D. J. Shearman, and A. Reda. 1987. A spherulitic fabric in selectively dolomitized siliciclastic crustacean burrows, northern Kuwait. *Journal of Sedimentary Petrology,* 57: 922–927.

Gupta, L. N., and W. Johannes. 1982. Petrogenesis of a stromatic migmatite (Nelaug, southern Norway). *Journal of Petrology,* 23:548–567.

Gutteridge, Peter. 1985. Grain-size measurement from acetate peels. *Journal of Sedimentary Petrology,* 55:595–596.

Hadizadeh, J., and E. H. Rutter. 1983. The low temperature brittle-ductile transition in quartzite and the occurrence of cataclastic flow in nature. *Geologische Rundschau,* 72:493–509.

Haines, M. 1968. Two staining tests for brucite in marble. *Mineralogical Magazine,* 36:886–888.

Hallam, A. 1987. End-Cretaceous mass-extinction event: Argument for terrestrial causation. *Science,* 238:1237–1242.

Hamilton, D. L., and G. M. Anderson. 1967. Effects of water and oxygen pressures on the crystallization of basaltic magmas. In *Basalts: The Poldervaart Treatise on Rocks of Basalt Composition,* H. Hess and A. Poldervaart, eds. New York: Interscience.

———, C. W. Burnham, and E. F. Osborn. 1964. The solubility of water and effects of oxygen fugacity and water content on crystallization in mafic magmas. *Journal of Petrology,* 5:21–39.

———, and W. S. MacKenzie. 1965. Phase-equilibrium studies in the system $NaAlSiO_4$(nepheline)–$KAlSiO_4$(kalsilite)–SiO_2–H_2O. *Mineralogical Magazine,* 34:214–231.

Hanford, C. R. 1981. Coastal sabkha and salt pan deposition of the lower Clear Fork (Permian), Texas. *Journal of Sedimentary Petrology,* 51:761–778.

———, and C. H. Moore, Jr. 1976. Diagenetic implications of calcite pseudomorphs after halite from the Joachim Dolomite (Middle Ordovician), Arkansas. *Journal of Sedimentary Petrology,* 46:387–392.

Hanor, J. S. 1978. Precipitation of beachrock cements: Mixing of marine and meteoric waters vs. CO_2 degassing. *Journal of Sedimentary Petrology,* 48:489–501.

Hardie, Lawrence A. 1984. Evaporites: Marine or non-marine? *American Journal of Science,* 284:193–240.

———. 1987. Dolomitization: A critical view of some current views. *Journal of Sedimentary Petrology,* 57:166–183.

———, and Hans P. Eugster. 1971. The depositional environment of marine evaporites: A case for shallow, clastic accumulation. *Sedimentology,* 16:187–220.

———, Tim K. Lowenstein, and Ronald J. Spencer. 1985. The problem of distinguishing between primary and secondary features in evaporites. In *Sixth International Symposium on Salt, 1983,* vol. 1, B. Charlotte Schreiber and H. Lincoln Harner, eds. Alexandria, VA: The Salt Institute, pp. 11–39.

Harker, Alfred. 1909. *The Natural History of Igneous Rocks.* New York: Macmillan.

Harriss, Robert C., and John A. S. Adams. 1966. Geochemical and mineralogical studies on the weathering of granitic rocks. *American Journal of Science,* 264:146–173.

Hausenbuiller, R. L. 1985. *Soil Science,* 3rd ed. Dubuque, IA: Brown.

Hay, R. L., and A. Iijima. 1968. Nature and origin of palagonite tuffs of the Honolulu Group on Oahu, Hawaii. In *Studies in Volcanology—A Memoir in Honor of Howel Williams. Geological Society of America Memoir,* 116:331–376.

Hayba, Daniel O., Philip M. Bethke, Pamela Heald, and Nora K. Foley. 1985. Geologic, mineralogic, and geochemical characteristics of volcanic-hosted epithermal precious-metal deposits. In *Geology and Geochemistry of Epithermal Sysems,* B. R. Berger and P. M. Bethke, eds. *Reviews in Economic Geology,* 2:129–163.

Haymon, Rachel M., Randolph A. Koski, and Michael J. Abrams. 1989. Hydrothermal discharge zones beneath massive sulfide deposits mapped in the Oman ophiolite. *Geology,* 17:531–535.

Heard, H. C. 1960. Transition from brittle fracture to ductile flow in Solenhofen limestone. In *Rock Deformation,* David Griggs and John Handin, eds. *Geological Society of America Memoir,* 79:193–226.

Hearn, B. Carter, Jr. 1968. Diatremes with kimberlitic affinities in north-central Montana. *Science,* 159:622–625.

Hedenquist, J. W., and R. W. Henley. 1985. Hydrothermal eruptions in the Waitapu geothermal system, New Zealand: Their origin, associated breccias and relation to precious metal mineralization. *Economic Geology,* 80:1640–1668.

Heiken, Grant. 1972. Morphology and petrography of volcanic ashes. *Geological Society of America Bulletin,* 83:1961–1988.

———, et al., eds. 1991. *The Lunar Sourcebook: A User's Guide to the Moon.* Cambridge, MA: Cambridge University Press.

Heimann, R., W. Frank, and A. Willgallis. 1970. Dissolution forms of single crystal spheres of quartz in acid alkali fluoride melts. *Neues Jahrbuch für Mineralogie Monatshefte,* 74–83.

Helgeson, H. C. 1964. *Complexing and Hydrothermal Ore Deposition.* New York: Macmillan.

Helz, Rosalind Tuthill, and Thomas L. Wright. 1992. Differentiation and magma mixing on Kilauea's east rift zone. *Bulletin of Volcanology,* 54:361–384.

Henley, R. W. 1985. The geothermal framework of epithermal deposits. In *Geology and Geochemistry of Epithermal Systems,* B. R. Berger and P. M. Bethke, eds. *Reviews in Economic Geology,* 2:1–24.

Henry, Christopher D. 1990. Case study of an extensive silicic lava: The Bracks Rhyolite, trans-Pecos, Texas. *Journal of Volcanology and Geothermal Research,* 43:113–132.

———, and John A. Wolff. 1992. Distinguishing strongly rheomorphic tuffs from extensive silicic lavas. *Bulletin of Volcanology,* 54:171–186.

Herzig, Peter M., Mark D. Hannington, Steven D. Scott, George Maliotis, Peter A. Rona, and Geoffrey Thompson. 1991. Gold-rich sea-floor gossans in the Troodos Ophiolite and on the Mid-Atlantic Ridge. *Economic Geology,* 86:1747–1755.

Hess, Paul C. 1977. Structure of silicate melts. *Contributions to Mineralogy and Petrology,* 15:162–178.

———. 1989. *Origins of Igneous Rocks.* Cambridge, MA: Harvard University Press.

Hesse, Reinhard. 1987. Selective and reversible carbonate-silica replacements in Lower Cretaceous carbonate-bearing turbidites of the Eastern Alps. *Sedimentology,* 34:1055–1077.

Heydari, Ezat, and Clyde H. Moore. 1989. Burial diagenesis and thermochemical sulfate reduction, Smackover Formation, southeastern Mississippi salt basin. *Geology,* 17:1080–1084.

Hibbard, M. J. 1965. Origin of some alkali feldspar phenocrysts and their bearing on petrogenesis. *American Journal of Science,* 263:245–261.

———. 1971. Evolution of a plutonic complex, Okanogan Range, Washington. *Geological Society of America Bulletin,* 82:3013–3048.

———. 1979. Myrmekite as a marker between preaqueous phase and postaqueous phase saturation in granitic systems. *Geological Society of America Bulletin,* 90:1047–1062.

———. 1980. Indigenous source of late stage dikes and veins in granitic plutons. *Economic Geology,* 75:410–423.

———. 1981. The magma mixing origin of mantled feldspars. *Contributions to Mineralogy and Petrology,* 76:158–170.

———. 1987. Deformation of incompletely crystallized magma systems: Granitic gneisses and their tectonic implications. *Journal of Geology,* 95:543–561.

———. 1991. Textural anatomy of twelve magma-mixed granitoid systems. In *Enclaves and Granite Petrology,* J. Didier and B. Barbarin, eds. Amsterdam: Elsevier, pp. 431–444.

———, and J. J. Sjoberg. 1994. Texture development from incongruent melting of pyroxene in a Vermont limburgite. *Canadian Mineralogist,* in press.

———, and R. Watters. 1985. Fracturing and diking in incompletely crystallized granitic plutons. *Lithos,* 18:1–12.

Hildreth, Wes. 1981. Gradients in silicic magma chambers: Implications for lithospheric magmatism. *Journal of Geophysical Research,* 86:10153–10192.

Hill, Richard Bruce. 1983. Depositional environments of the Upper Cretaceous Ferron sandstone south of Notom, Wayne County, Utah. *Brigham Young University Geology Studies,* 29:59–79.

Hill, Mason L. 1963. Role of classification in geology. In *The Fabric of Geology,* C. C. Albritton, ed. San Francisco: Freeman.

Hiraga, Hiroyuki, and Toshihiko Shimamoto. 1987. Textures of sheared halite and their implications for the seismogenic slip of deep faults. *Tectonophysics,* 144:69–86.

Hobbs, B. E., W. D. Means, and P. F. Williams. 1976. *An Outline of Structural Geology.* New York: Wiley.

Hofstra, A. H., G. P. Landis, R. O. Rye, D. J. Birak, A. R. Dahl, W. E. Daly, and M. B. Jones. 1989. Geology and origin of the Jerritt Canyon sediment-hosted disseminated gold deposits, Nevada. In *U.S. Geological Survey Research on Mineral Resources, Fifth Annual V. E. McKelvey Forum on Mineral and Energy Resources. U.S. Geological Survey Circular,* 1035:30–32.

Hollister, L. S., and M. L. Crawford. 1986. Melt-enhanced deformation: A major tectonic process. *Geology,* 14:558–561.

———, and A. J. Gancarz. 1971. Compositional sector-zoning in clinopyroxene from the Narce area, Italy. *American Mineralogist,* 56:959–979.

Holser, William T. 1966. Bromide geochemistry of salt rocks. In *Second Symposium on Salt. Geology, Geochemistry, and Mining,* vol. 1. Cleveland, OH: North Ohio Geological Society, pp. 248–275.

Hooper, P. R., W. D. Kleck, C. R. Knowles, S. P. Reidel, and R. L. Thiessen. 1984. Imnaha Basalt, Columbia River Basalt Group. *Journal of Petrology,* 25:473–500.

Hopson, R. F., and K. Ramseyer. 1990. Cathodoluminescence microscopy of myrmekite. *Geology,* 18:336–339.

Horowitz, Alan S., and Paul E. Potter. 1971. *Introductory Petrography of Fossils.* New York: Springer-Verlag.

Hörz, F. 1982. Ejecta of the Ries Crater, Germany. *Geological Society of America Special Paper,* 190:39–55.

Houghton, H. F. 1980. Refined techniques for staining plagioclase and alkali feldspar in thin section. *Journal of Sedimentary Petrology,* 50:629–631.

House, William M., and David R. Gray. 1982. Cataclasites along the Saltville thrust, U.S.A., and their implications for thrust-sheet emplacement. *Journal of Structural Geology,* 4:257–269.

Hovorka, S. 1987. Depositional environments of marine-dominated bedded halite, Permian San Andreas Formation, Texas. *Sedimentology,* 34:1029–1054.

———. 1992. Halite pseudomorphs after gypsum in bedded anhydrite—clue to gypsum-anhydrite relationships. *Journal of Sedimentary Petrology,* 62:1098–1111.

Hughes, Andrew D., and David Whitehead. 1987. Glauconitization of detrital silica substrates in the Barton Formation (upper Eocene) of the Hampshire Basin, southern England. *Sedimentology,* 34:825–835.

Hughes, Charles James. 1960. The Southern Mountains Igneous Complex, Isle of Rhum. *Journal of the Geological Society of London,* 116:111–138.

Hughes, J. M., and R. E. Stoiber. 1985. Vanadium sublimates from the fumaroles of Izalco Volcano, El Salvador. *Journal of Volcanologic and Geothermal Research,* 24:283–291.

Humphrey, Fred L. 1960. Geology of the White Pine Mining District, White Pine County, Nevada. *Nevada Bureau of Mines Bulletin,* 57.

Humphries, D. W. 1992. The preparation of thin sections of rocks, minerals, and ceramics. *Royal Microscopical Society Microscopy Handbooks 24.* Oxford: Oxford University Press/Royal Microscopical Society.

Hunt, Charles B., Paul Averitt, and Ralph L. Miller. 1953. Geology and geography of the Henry Mountains Region, Utah. *U.S. Geological Survey Professional Paper,* 228.

Hunter, Robert H. 1987. Textural equilibrium in layered igneous rocks. In *Origins of Igneous Layering,* I. Parsons, ed. Dordrecht, Holland: Reidel, pp. 473–503.

Huppert, Herbert E., and R. Stephen J. Sparks. 1988. The generation of granitic magmas by intrusion of basalt into continental crust. *Journal of Petrology,* 29:599–624.

———, R. S. Sparks, and J. S. Turner. 1984. Some effects of viscosity on the dynamics of replenished magma chambers. *Journal of Geophysical Research,* 89:6857–6877.

Hurlbut, C. S., Jr. 1984. The jeweler's refractometer as a mineralogical tool. *American Mineralogist,* 69:391–398.

Husch, Jonathan M. 1990. Palisades sill: Origin of the olivine zone by separate magmatic injection rather than gravity settling. *Geology* 18:699–702.

Hutchison, Charles S. 1974. *Laboratory Handbook of Petrographic Techniques.* New York: Wiley.

Hutton, Donald H. W. 1988. Granite emplacement mechanisms and tectonic controls: Inferences from deformation studies. *Transactions of the Royal Society of Edinburgh: Earth Sciences,* 79:245–255.

———, and R. J. Reavy. 1992. Strike-slip tectonics and granite petrogenesis. *Tectonics,* 2:960–967.

Ilchik, R. P. 1990. Geology and geochemistry of the Vantage gold deposits, Alligator Ridge-Bald Mountain Mining District, Nevada. *Economic Geology,* 85:50–75.

Irvine, T. N. 1979. Rocks whose composition is determined by crystal accumulation and sorting. In *Evolution of the Igneous Rocks, Fiftieth Anniversary Perspectives,* H. S. Yoder, Jr., ed. Princeton, NJ: Princeton University Press, pp. 245–306.

———. 1982. Terminology for layered intrusions. *Journal of Petrology,* 23:127–162.

Irving, T. H. 1987. Layering and related structures in the Duke Island and Skaergaard intrusions. In *Origins of Igneous Layering,* I. Parsons, ed. Dordrecht, Holland: Reidel.

Isherwood, Dana, and Alayne Street. 1976. Biotite-induced grussification of the Boulder Creek granodiorite, Boulder County, Colorado. *Geological Society of America Bulletin,* 87:366–370.

Izett, G. A. 1990. The Cretaceous/Tertiary boundary interval, Raton Basin, Colorado and New Mexico, and its content of shock-metamorphosed minerals; evidence relevant to the K/T boundary impact-extinction theory. *Geological Society of America Special Paper,* 249.

———, and B. F. Bohor. 1987. Comment and reply on "dynamic deformation of volcanic ejecta from the Toba caldera: Possible relevance to Cretaceous/Tertiary boundary phenomena." *Geology,* 15:90–92.

Jackson, E. D. 1961. Primary textures and mineral associations in the ultramafic zone of the Stillwater Complex, Montana. *U.S. Geological Survey Professional Paper,* 358.

———. 1967. Ultramafic cumulates in the Stillwater, Great Dyke, and Bushveld intrusives. In *Ultramafic and Related Rocks,* P. J. Wyllie, ed. New York: Wiley, pp. 20–38.

———, and D. C. Ross. 1956. A technique for modal analysis of medium- and coarse-grained (3–10mm) rocks. *American Mineralogist,* 41:648–651.

Jackson, K. A., D. R. Uhlmann, and J. D. Hunt. 1967. On the nature of crystal growth from the melt. *Journal of Crystal Growth,* 1:1–36.

Jackson, M. P. A., R. R. Cornelius, C. H. Craig, A. Gansser, J. Stocklin, and C. J. Talbot. 1990. Salt diapirs of the Great Kavir, Central Iran. *Geological Society of America Memoir,* 177.

Jacobson, G., A. V. Arakel, and Chen Yijian. 1988. The central Australian groundwater discharge zone: Evolution of associated calcrete and gypcrete deposits. *Australian Journal of Earth Sciences,* 35:549–565.

Jahns, R. H. 1953. The genesis of pegmatites. I: Occurrence and origin of giant crystals. *American Mineralogist,* 38:563–598.

———. 1955. The study of pegmatites. *Economic Geology,* 50th Anniversary Volume:1025–1130.

———, and C. W. Burnham. 1969. Experimental studies of pegmatite genesis. I: A model for the derivation and crystallization of granitic pegmatites. *Economic Geology,* 64:843–864.

Jakobsson, Sveinn P., and James G. Moore. 1986. Hydrothermal minerals and alteration rates at Surtsey volcano, Iceland. *Geological Society of America Bulletin,* 97:648–659.

James, H. L., and A. F. Trendall. 1982. Banded iron formation: Distribution in time and paleoenvironmental significance. In *Mineral Deposition and the Evolution of the Biosphere,* H. D. Holland and M. Schidlowskii, eds. New York: Springer-Verlag.

James, R. S., and D. L. Hamilton. 1969. Phase relations in the system $NaAlSi_3O_8$-$KAlSi_3O_8$-$CaAl_2Si_2O_8$-SiO_2 at 1 kilobar water vapour pressure. *Contributions to Mineralogy and Petrology,* 21:111–141.

Jaupart, C., and S. Tait. 1990. Dynamics of eruptive phenomena. In *Modern Methods of Mineralogy: Understanding Magmatic Processes,* J. Nicholls and J. K. Russell, eds. *Reviews in Mineralogy,* Paul H. Ribbe, series ed. 24:213–236.

Johannsen, A. 1931, 1937, 1938. *A Descriptive Petrography of the Igneous Rocks,* 4 vols. Chicago: University of Chicago Press.

Johnson, J. G. 1990. Method of multiple working hypotheses: A chimera. *Geology,* 18:44–45.

Johnson, S. E. 1990. Lack of porphyroblast rotation in the Otago schists, New Zealand: Implications for crenulation cleavage development, folding and deformation partitioning. *Journal of Metamorphic Geology,* 8:13–30.

Jones, Noriss W., and F. Donald Bloss. 1980. *Laboratory Manual for Optical Mineralogy.* Minneapolis, MN: Burgess.

Jordan, Peter, and Rolf Nuesch. 1989. Deformational behavior of shale interlayers in evaporite detachment horizons, Jura overthrust, Switzerland. *Journal of Structural Geology,* 11:859–871.

Jurewicz, Stephen R., and E. Bruce Watson. 1984. Distribution of partial melt in a felsic system: The importance of surface energy. *Contributions to Mineralogy and Petrology,* 85:25–29.

Kaczor, Sofia M., Gilbert N. Hanson, and Zell E. Peterman. 1988. Disequilibrium melting of granite at the contact with a basic plug: A geochemical and petrographic study. *Journal of Geology,* 96:61–78.

Kamb, W. B. 1958. Isogyres in interference figures. *American Mineralogist,* 43:1029–1067.

Keith, Terry E. C, Thomas J. Casadevall, and David A. Johnston. 1981. Fumarole encrustations: Occurrence, mineralogy, and chemistry. *U.S. Geological Survey Professional Paper,* 1250:239–250.

Kelsey, C. H. 1965. Calculation of C.I.P.W. norm. *Mineralogical Magazine,* 34:276–282.

Kendall, Christopher G. St. C., and John Warren. 1987. A review of the origin and setting of tepees and their associated fabrics. *Sedimentology,* 34:1007–1027.

Kennedy, W. Q. 1949. Zones of progressive regional metamorphism in the Moine Schists of the western Highlands of Scotland. *Geological Magazine,* 86:43–56.

Kerr, P. F. 1977. *Optical Mineralogy,* 4th ed. New York: McGraw-Hill.

Kidder, D. L., and K. Swett. 1989. Basal Cambrian reworked phosphates from Spitsbergen (Norway) and their implications. *Geological Magazine,* 126:79–88.

Kimberley, Michael M. 1979. Origin of oolitic iron formations. *Journal of Sedimentary Petrology,* 49:111–132.

King, C. 1878. *U.S. Geological Exploration of the Fortieth Parallel,* vol. 1. Washington, DC: U.S. Government Printing Office, pp. 504–529.

Kirkpatrick, R. James. 1975. Crystal growth from the melt: A review. *American Mineralogist,* 60:798–814.

———. 1977. Nucleation and growth of plagioclase, Makaopuhi and Alae lava lakes, Kilauea volcano, Hawaii. *Geological Society of America Bulletin,* 88:78–84.

Klappa, C. F. 1980. Rhizoliths in terrestrial carbonates: Classification, recognition, genesis, and significance. *Sedimentology,* 27:613–629.

Klein, Cornelis, Jr. 1973. Changes in mineral assemblages with metamorphism of some banded Precambrian iron formations. *Economic Geology,* 68:1075–1088.

Knapp, S. T., M. Friedman, and J. M. Logan. 1987. Slip and recrystallization of halite gouge in experimental shear zones. *Tectonophysics,* 135:171–183.

Kodosky, Lawrence, and Mary Keskinen. 1990. Fumarole distribution, morphology, and encrustation mineralogy associated with the 1986 eruptive deposits of Mount St. Augustine, Alaska. *Bulletin Volcanologique,* 52:175–185.

Komar, Paul D. 1972. Flow differentiation in igneous dikes and sills: Profiles of velocity and phenocryst concentration. *Geological Society of America Bulletin,* 83:3443–3448.

Kranz, R. L. 1983. Microcracks in rocks: A review. *Tectonophysics,* 100:449–480.

Krauskopf, K. 1979. *Introduction to Geochemistry,* 2nd ed. New York: McGraw-Hill.

Krumbein, W. C. 1934. Size frequency distributions of sediments. *Journal of Sedimentary Petrology,* 4:65–77.

Kuehn, C. A., and A. W. Rose. 1992. Geology and geochemistry of wall-rock alteration at the Carlin Gold Deposit, Nevada. *Economic Geology,* 87:1697–1721.

Kues, Barry S. 1982. *Fossils of New Mexico.* Albuquerque, NM: University of New Mexico Press.

Kuo, Lung-Chuan, and R. James Kirkpatrick. 1985. Kinetics of crystal dissolution in the system diopside-forsterite-silica. *American Journal of Science,* 285:51–90.

LaBerge, G. L., E. I. Robbins, and T.-M. Hon. 1987. A model for the biological precipitation of Precambrian iron-formations. A: Geological evidence. In *Precambrian Iron-Formations,* P. W. V. Appel and G. L. LaBerge, eds. Athens, Greece: Theophrastus.

Laduron, D. M. 1971. A staining method for distinguishing paragonite from muscovite in thin section. *American Mineralogist,* 56:1117–1119.

Lang, H. M., and G. R. Dunn. 1990. Sequential porphyroblast growth during deformation in a low-pressure metamorphic terrain, Orrs Island–Harpswell Neck, Maine. *Journal of Metamorphic Geology,* 8:199–216.

Larsen, L. H., and A. Poldervaart. 1961. Petrologic study of Bald Rock Batholith, near Bidwell Bar, California. *Geological Society of America Bulletin,* 72:69–92.

Lasemi, Zakaria, Mark R. Boardman, and Philip A. Sandberg. 1988. Cement origin of supratidal dolomite, Andros Island, Bahamas. *Journal of Sedimentary Petrology,* 59:249–257.

———, and Philip A. Sandberg. 1984. Transformation of aragonite-dominated lime muds to microcrystalline limestones. *Geology,* 12:420–423.

Lash, Gary G. 1989. Documentation and significance of progressive microfabric changes in Middle Ordovician trench mudstones. *Geological Society of America Bulletin,* 101:1268–1279.

Laskowski, T. E., D. M. Scotford, and D. E. Laskowski. 1979. Measurement of refractive index in thin sections using dispersion staining and oil immersion techniques. *American Mineralogist,* 64:440–445.

Laznicka, Peter. 1988. *Breccias and Coarse Fragmentites, Petrology, Environments, Associations, Ores.* New York: Elsevier.

Leblanc, M., and P. Billard. 1978. A volcano-sedimentary copper deposit on a continental margin of upper Proterozoic age? Bleida, Anti-Atlas, Morocco. *Economic Geology,* 73:1101–1111.

Lehrman, Norman J. 1986. The McLaughlin Mine Napa and Yolo Counties, California. In *Precious-Metal Mineralization in Hot Springs Systems Nevada-California,* Joseph V. Tingley and Harold F. Bonham, Jr., eds. Nevada Bureau of Mines and Geology, Report 41, pp. 85–126.

LeMaitre, R. W., ed. 1989. *A Classification of Igneous Rocks and Glossary of Terms.* Oxford: Blackwell.

Levin, Jeffrey Ross, and Alan Davis. 1989. The relationship of coal optical fabrics to Alleghenian tectonic deformation in the central Appalachian fold-and-thrust belt, Pennsylvania. *Geological Society of America Bulletin,* 101:1333–1347.

Lewis, R. S., T. Ming, J. F. Wacker, E. Anders, and E. Steel. 1987. Interstellar diamonds in meteorites. *Nature,* 326:160–162.

Lindgren, W. 1933. *Mineral Deposits.* New York: McGraw-Hill.

Lindsley, D. H. 1964. Melting relations of plagioclase at high pressure. *Carnegie Institute Washington Yearbook,* 63:204–205.

Lipman, P. W. 1976. Caldera-collapse breccias in the western San Juan Mountains, Colorado. *Geological Society of America Bulletin,* 87:1397–1410.

———, and D. R. Mullineaux, eds. 1981. The 1980 eruption of Mount St. Helens, Washington. *U.S. Geological Survey Professional Paper,* 1250.

Lipschutz, M. E. 1964. Origin of diamonds in the ureilites. *Science,* 143:1431–1434.

———, and Edward Anders. 1961. On the mechanism of diamond formation. *Science,* 134:2095–2099.

Lister, G. S., and A. W. Snoke. 1984. S-C mylonites. *Journal of Structural Geology,* 6:617–638.

Lockwood, J. P., and J. Moore. 1979. Regional deformation of the Sierra Nevada, California, on conjugate microfault sets. *Journal of Geophysical Research,* 84:6041–6049.

Lofgren, Gary. 1971a. Experimentally produced devitrification textures in naturally rhyolitic glass. *Geological Society of America Bulletin,* 82:111–124.

———. 1971b. Spherulitic textures in glassy and crystalline rocks. *Journal of Geophysical Research,* 76:5635–5648.

———. 1974. An experimental study of plagioclase crystal morphology: Isothermal crystallization. *American Journal of Science,* 274:243–273.

———. 1980. Experimental studies on the dynamic crystallization of silicate melts. In *Physics of Magmatic Processes,* R. B. Hargraves, ed. Princeton, NJ: Princeton University Press, pp. 487–551.

———. 1983. Effect of heterogeneous nucleation on basaltic textures: A dynamic crystallization study. *Journal of Petrology,* 24:229–255.

Logan, B. W. 1987. *The MacLeod Evaporite Basin, Western Australia.* Tulsa, OK: The American Association of Petroleum Geologists.

Long, Philip E., and Bernard J. Wood. 1986. Structures, textures, and cooling histories of Columbia River basalt flows. *Geological Society of America Bulletin,* 97:1144–1155.

Lovering, T. G. 1972. Jasperoid in the U.S.—its characteristics, origins, and economic significance. *U.S. Geological Survey Professional Paper,* 710.

———, and A. V. Heyl. 1974. Jasperoid as a guide to mineralization in the Taylor Mining District and vicinity near Ely, Nevada. *Economic Geology,* 69:46–58.

Lovering, T. S. 1949. Rock alteration as a guide to ore—East Tintic District, Utah. *Economic Geology Monograph,* 1.

———, and E. N. Goddard. 1950. Geology and ore deposits of the Front Range, Colorado. *U.S. Geological Survey Professional Paper,* 223.

Lowell, J. D., and J. M. Guilbert. 1970. Lateral and vertical alteration-mineralization zoning in porphyry ore deposits. *Economic Geology,* 65:373–408.

Lowenstam, Heinz A. 1989. *On Biomineralization.* London: Oxford University Press.

Lowenstein, T. K. 1987. Evaporite depositional fabrics in the deeply buried Jurassic Buckner Formation, Alabama. *Journal of Sedimentary Petrology,* 57:108–116.

———. 1988. Origin of depositional cycles in a Permian "saline giant": The Salado (McNutt zone) evaporites of New Mexico and Texas. *Geological Society of America Bulletin,* 100:592–608.

———. 1989. Origin of ancient potash evaporites: Clues from the modern nonmarine Qaidam Basin of western China. *Science,* 245:1090–1092.

———, and L. A. Hardie. 1985. Criteria for the recognition of salt-pan evaporites: Petrographic and fluid inclusion evidence. *American Journal of Science,* 290:1–42.

———, and Ronald J. Spencer. 1990. Syndepositional origin of potash evaporites: Petrographic and fluid inclusion evidence. *American Journal of Science,* 290:1–42.

Lucas, Stephen E., and J. Casey Moore. 1986. Cataclastic deformation in accretionary wedges: Deep Sea Drilling Project Leg 66, southern Mexico, and on-land examples from Barbados and Kodiak Islands. *Geological Society of America Memoir,* 166:89–103.

Luth, W. C., R. H. Jahns, and O. F. Tuttle. 1964. The granite system at pressures of 4 to 10 kilobars. *Journal of Geophysical Research,* 69:759–773.

Mack, Greg H. 1992. Paleosoils as an indicator of climatic change at the Early–Late Cretaceous boundary, southwestern New Mexico. *Journal of Sedimentary Petrology,* 62:483–494.

MacKenzie, W. S., C. H. Donaldson, and C. Guilford. 1982. *Atlas of Igneous Rocks and Their Textures.* New York: Wiley.

———, and C. Guilford. 1980. *Atlas of Rock-forming Minerals in Thin Section.* New York: Wiley.

Mackin, J. H. 1963. Rational and empirical methods of investigation in geology. In *The Fabric of Geology,* C. C. Albritton, ed. San Francisco: Freeman, pp. 135–163.

———. 1968. Iron ore deposits of the Iron Springs District, southwestern Utah. In *Ore Deposits of the United States 1933/1967,* vol. 2, J. D. Ridge, ed. New York: AIME, pp. 992–1019.

Maddock, R. H. 1983. Melt origin of fault-generated pseudotachylytes demonstrated by textures. *Geology,* 11:105–108.

Madon, Mazlan B. H. J. 1992. Depositional setting and origin of berthierine oolitic ironstones in the Lower Miocene Terengganu Shale, Tenggol Arch, offshore Peninsular Malaysia. *Journal of Sedimentary Petrology,* 62:899–916.

Maher, H. D., Jr. 1987. Kinematic history of mylonitic rocks from the Augusta Fault Zone, South Carolina and Georgia. *American Journal of Science,* 287:795–816.

Maiklem, W. D., D. G. Bebout, and R. P. Glaister. 1969. Classification of anhydrite—A practical approach. *Bulletin of Canadian Petroleum Geology,* 17:194–233.

Mainprice, David, Jean-Luc Bouchez, Philippe Blumenfeld, and José Maria Tubià. 1986. Dominant c slip in naturally deformed quartz: Implications for dramatic plastic softening at high temperature. *Geology,* 14:819–822.

Malavielle, J., and J. F. Ritz. 1989. Mylonitic deformation in décollements: Examples from the Southern Alps, France. *Journal of Structural Geology,* 11:583–590.

Maliva, Robert G., and Raymond Siever. 1988. Mechanism and controls of silicification of fossils in limestones. *Journal of Geology,* 96:387–398.

———. 1989. Nodular chert formation in carbonate rocks. *Journal of Geology,* 97:421–433.

Mandado, J. A., and J. M. Tena. 1986. A peel technique for sulfate (and carbonate) rocks. *Journal of Sedimentary Petrology,* 56:548–549.

Mandelbrot, B. B. 1983. *The Fractal Geometry of Nature*. San Francisco: Freeman.
Maniar, Papu D., and Philip M. Piccoli. 1989. Tectonic discrimination of granitoids. *Geological Society of America Bulletin,* 101:635–643.
Mark, Kathleen. 1987. *Meteorite Craters.* Tuscon: University of Arizona Press.
Marsh, B. D. 1981. On the crystallinity, probability of occurrence, and rheology of lava and magma. *Contributions to Mineralogy and Petrology,* 78:85–98.
———. 1984. Mechanics and energetics of magma formation and ascension. In *Studies in Geophysics, Explosive Volcanism: Inception, Evolution, and Hazards.* F. R. Boyd, Jr., ed. Washington DC: National Academy Press, pp. 67–83.
———. 1987. Magmatic processes. *Reviews in Geophysics,* 25:1043–1053.
———. 1988a. Crystal capture, sorting, and retention in convecting magma. *Geological Society of America Bulletin,* 100:1720–1737.
———. 1988b. Crystal size distributions (CDS) in rocks and the kinetics and dynamics of crystallization. I: Theory. *Contributions to Mineralogy and Petrology,* 99:277–291.
———, and Martin R. Maxey. 1985. On the distribution and separation of crystals in convecting magma. *Journal of Volcanology and Geothermal Research,* 24:95–150.
Marshall, D. J. 1988. *Cathodoluminescence of Geological Materials.* Winchester, MA: Unwin Hyman.
Marshall, L. A., and R. S. J. Sparks. 1984. Origin of some mixed-magma and net-veined ring intrusions. *Journal of the Geological Society of London,* 141:171–182.
Marshall, Royal R. 1961. Devitrification of natural glass. *Geological Society of America Bulletin,* 72:1493–1520.
Martinez, B., and F. Plana. 1987. Quantitative x-ray diffraction of carbonate sediments: Mineralogical analysis through fitting of Lorentzez profiles to diffraction peaks. *Sedimentology,* 34:169–174.
Mason, R. 1978. *Petrology of Metamorphic Rocks.* London: Allen & Unwin.
Maynard, J. B. 1991. Iron: Syngenetic deposition controlled by the evolving ocean-atmosphere system. In *Sedimentary and Diagenetic Mineral Deposits: A Basin Analysis Approach to Exploration,* Eric R. Force, J. James Eidel, and J. Barry Maynard, eds. *Reviews in Economic Geology,* 5:141–145.
Mazzone, Peter, and Stephen E. Haggerty. 1989. Peraluminous xenoliths in kimberlite: Metamorphosed restites produced by partial melting of pelites. *Geochimica et Cosmochimica Acta,* 53:1551–1561.
Mazzullo, Jim, and Stewart Magenheimer. 1986. The original shapes of quartz sand grains. *Journal of Sedimentary Petrology,* 57:479–487.
McBirney, A. R. 1979. Effects of assimilation. In *The Evolution of the Igneous Rocks,* H. S. Yoder, Jr., ed. Princeton, NJ: Princeton University Press.
———. 1984. Rheological properties of magmas. *Annual Reviews of Earth and Planetary Science,* 12:337–357.
———. 1987. Constitutional zone refining of layered intrusions. In *Origins of Igneous Layering,* I. Parsons, ed. Dordrecht, Holland: D. Reidel Publishing Company, pp. 437–451.
———. 1992. *Igneous Petrology,* 2nd ed. Boston, MA: Jones and Bartlett.
———, and R. M. Noyes. 1979. Crystallization and layering of the Skaergaard intrusion. *Journal of Petrology,* 20:487–554.
McBride, Earle F. 1987. Diagenesis of the Maxon Sandstone (Early Cretaceous), Marathon Region, Texas: A diagenetic quartzarenite. *Journal of Sedimentary Petrology,* 57:98–107.

McCallum, I. S. 1987. Petrology of the igneous rocks. *Reviews of Geophysics,* 25:1021–1042.

———, L. D. Raedeke, and E. A. Mathez. 1980. Investigations of the Stillwater Complex. Part I. Stratigraphy and structure of the banded zone. *American Journal of Science,* 280A:59–87.

McCallum, M. E., T. S. Woolsey, and S. A. Schumm. 1976. A fluidization mechanism for subsidence of bedded tuffs in diatremes and related volcanic vents. *Bulletin Volcanologique,* 39:512–527.

McCormick, J. E., L. L. Evans, R. A. Palmer, and F. D. Rasnick. 1971. Environment of the zinc deposits of the Mascot–Jefferson City District, Tennessee. *Economic Geology,* 66:757–762.

McDowell, S. Douglas. 1978. Little Chief granite porphyry: Feldspar crystallization history. *Geological Society of America Bulletin,* 89:33–49.

McKenzie, D. 1984. The generation and compaction of partially molten rock. *Journal of Petrology,* 25:713–765.

McLellan, E. L. 1983. Contrasting textures in metamorphic and anatectic migmatites: An example from the Scottish Caledonides. *Metamorphic Geology,* 1:241–262.

———. 1988. Migmatite structures in the Central Gneiss Complex, Boca de Quadra, Alaska. *Journal of Metamorphic Petrology,* 6:517–542.

McMillan, Kent. 1986. Spatially varied miaroles in the albite porphyry of Cuchillo Mountain, southwestern New Mexico. *American Mineralogist,* 71:625–631.

Mehnert, K. R. 1968. *Migmatites and the Origin of Granitic Rocks.* New York: Elsevier.

———. 1971. *Migmatites and the Origin of Granitic Rocks,* 2nd ed. Amsterdam: Elsevier.

———, W. Busch, and G. Schneider. 1973. Initial melting at grain boundaries of quartz and feldspar in gneisses and granulites. *Neues Jahrbuch für Mineralogie Monatshefte,* 4:165–183.

Melosh, H. J. 1989. *Impact Cratering: A Geologic Process.* New York: Oxford University Press.

Meyer, Robert, and Rui B. Pena Dos Reis. 1985. Paleosols and alunite silcretes in continental Cenozoic of western Portugal. *Journal of Sedimentary Petrology,* 55:76–85.

Miller, Calvin F., Edward F. Stoddard, Larry J. Bradfish, and Wayne A. Dollase. 1981. Composition of plutonic muscovite: Genetic implications. *Canadian Mineralogist,* 19:25–34.

Miller, James D., Jr., and Paul W. Weiblen. 1990. Anorthositic rocks of the Duluth Complex: Examples of rocks formed from plagioclase crystal mush. *Journal of Petrology,* 31:295–339.

Miller, S. L. 1969. Clathrate hydrates of air in Antarctic ice. *Science,* 165:489–490.

Milnes, A. R., R. P. Bourman, and R. W. Fitzpatrick. 1987. Petrology and mineralogy of "laterites" in southern and eastern Australia and southern Africa. *Chemical Geology,* 60:237–250.

Misch, P. 1964. Stable association of wollastonite-anorthite, and other calc-silicate assemblages in amphibolite-facies crystalline schists of Nanga Parbat, northwest Himalayas. *Beiträge zur Mineralogie und Petrographie,* 10:315–356.

———. 1969. Paracrystalline microboudinage of zoned grains and other criteria for synkinematic growth of metamorphic minerals. *American Journal of Science,* 267:43–63.

———. 1971. Porphyroblasts and "crystallization force": Some textural criteria. *Geological Society of America Bulletin,* 82:245–252.

Moffatt, William G., George W. Pearsall, and John Wulff. 1964. *The Structure and Properties of Materials,* vol. 1. New York: Wiley.

Molen, I. van der, and M. S. Paterson. 1979. Experimental deformation of partially-melted granite. *Contributions to Mineralogy and Petrology,* 70:299–318.

Molenaar, N., and A. F. M. de Jong. 1987. Authigenic quartz and albite in Devonian limestones: Origin and significance. *Sedimentology,* 34:623–640.

Montel, J. M., J. Didier, and M. Pichavant. 1991. Origin of surmicaceous enclaves in intrusive granites. In *Enclaves and Granite Petrology,* J. Didier and B. Barbarin, eds. New York: Elsevier, pp. 509–527.

Morad, S. 1988. Albitized microcline grains of post-depositional and probable detrital origins in Brøttum Formation sandstones (Upper Proterozoic), Sparagmite Region of southern Norway. *Geological Magazine,* 125:229–239.

Morrison, R. B. 1964. Lake Lahontan: Geology of southern Carson Desert, Nevada. *U.S. Geological Survey Professional Paper,* 401.

Morse, S. A. 1969. Syenites. *Carnegie Institute Washington Yearbook,* 1967–68:112–120.

———. 1980. *Basalts and Phase Diagrams: An Introduction to the Use of Phase Diagrams in Igneous Petrology.* New York: Springer-Verlag.

Mount, J. 1985. Mixed siliciclastic and carbonate sediments: A proposed first-order textural and compositional classification. *Sedimentology,* 32:435–442.

Muncill, Gregory E., and Antonio C. Lasaga. 1988. Crystal-growth kinetics of plagioclase in igneous systems: Isothermal H_2O-saturated experiments and extension of a growth model to complex silicate melts. *American Mineralogist,* 73:982–992.

Murray, Richard W., David L. Jones, and Marilyn R. Buchholtzten Brink. 1992. Diagenetic formation of bedded chert: Evidence from chemistry of the chert-shale couplet. *Geology,* 20:271–274.

Myers, James D., A. Krishna Sinha, and Bruce D. Marsh. 1984. Assimilation of crustal material by basaltic magma: Strontium isotope and trace element data from the Edgecumbe volcanic fields, southeast Alaska. *Journal of Petrology,* 25:1–26.

Mysen, B. O. 1988. *Structure and Properties of Silicate Melts.* New York: Elsevier.

Nabelek, Peter I., Gilbert N. Hanson, Theodore C. Labotka, and James J. Papike. 1988. Effects of fluids on the interaction of granites with limestones: The Notch Peak stock, Utah. *Contributions to Mineralogy and Petrology,* 99:49–61.

Nahon, Daniel B. 1986. Evolution of iron crusts in tropical landscapes. In *Rates of Chemical Weathering,* Steven M. Colman and David P. Dethier, eds. New York: Academic Press, pp. 169–191.

———. 1991. *Introduction to the Petrology of Soils and Chemical Weathering.* New York: Wiley.

Naney, Michael T., and Samuel E. Swanson. 1980. The effect of Fe and Mg on crystallization in granitic systems. *American Mineralogist,* 65: 639–653.

Naughton, J. J., V. A. Greenberg, and R. Goguel. 1976. Encrustations and fumarolic condensates at Kilauea Volcano, Hawaii: Field, drill-hole and laboratory observations. *Journal of Volcanological and Geothermal Research,* 1:149–165.

Nealson, Kenneth H., and Charles R. Myers. 1990. Iron reduction by bacteria: A potential role in the genesis of banded iron formations. *American Journal of Science,* 290-A:35–45.

Nekvasil, H. 1988. Calculation of equilibrium crystallization paths of compositionally simple hydrous felsic melts. *American Mineralogist,* 73:956–965.

Nelson, C. E., and D. L. Giles. 1985. Hydrothermal eruption mechanisms and hot spring gold deposits. *Economic Geology,* 80:1633–1639.

Nesbitt, H. Wayne. 1980. Genesis of the New Quebec and Adirondack granulites: Evidence for their production by partial melting. *Contributions to Mineralogy and Petrology,* 72:303–310.

Nesse, William D. 1991. *Introduction to Optical Mineralogy,* 2nd ed. New York: Oxford University Press.

Nicholson, Keith. 1989. Early Devonian geothermal systems in northeast Scotland: Exploration targets for epithermal gold. *Geology,* 17:568–571.

Nielsen, Richard L. 1968. Hypogene texture and mineral zoning in a copper-bearing granodiorite porphyry stock, Santa Rita, New Mexico. *Economic Geology,* 63:37–50.

Niggli, P. 1935. Zur mineralogischen Klassifikation der Eruptivgesteine. *Schweizerische Mineralogische und Petrographische Mitteilungen,* 15:295–318.

Nishiyama, Tadao. 1989. Kinetics of hydrofracturing and metamorphic veining. *Geology,* 17:1068–1071.

Nixon, Peter H., ed. 1987. *Mantle Xenoliths.* New York: Wiley.

Norton, Stephen A. 1973. Laterite and bauxite formation. *Economic Geology,* 68:353–361.

O'Brien, Hugh E., Anthony J. Irving, and I. Stewart McCallum. 1988. Complex zoning and resorption of phenocrysts in mixed potassic mafic magmas of the Highwood Mountains, Montana. *American Mineralogist,* 73:1007–1024.

Officer, C. B., and C. C. Drake. 1983. The Cretaceous/Tertiary transition. *Science,* 219:1383–1390.

Ogura, H. 1972. Geology and "kuroko" ore deposits of the Hanaoka-Matsumine mine, northern Japan. *24th International Geological Congress, sec. 4, Mineral Deposits,* pp. 318–325.

Olsen, Sakiko N. 1982. Open- and closed-system migmatites in the Front Range, Colorado. *American Journal of Science,* 282:1596–1622.

Osborne, F. Fritz. 1936. Intrusives of part of the Laurentian complex in Quebec. *American Journal of Science* (series 5), 32:407–434.

Papke, Keith G. 1969. Montmorillonite deposits in Nevada. *Clays and Clay Minerals,* 17:211–222.

Parsons, I., ed. 1987. *Origins of Igneous Layering.* Dordrecht, Holland: Reidel.

———, and A. W. Butterfield. 1981. Sedimentary features of the Nunarssuit and Klokken syenites, south Greenland. *Journal of the Geological Society of London,* 138:289–306.

Pasteris, Jill Dill. 1983. Value of reflected light microscopy in teaching. *Journal of Geological Education,* 31:17–22.

Paterson, S. R. 1988. Cannibal Creek granite: Post-tectonic "ballooning" pluton or pre-tectonic piercement diapir? *Journal of Geology,* 96:730–736.

———, Ron H. Vernon, and Othmar T. Tobisch. 1989. A review of criteria for the identification of magmatic and tectonic foliations in granitoids. *Journal of Structural Geology,* 11:349–363.

Pauly, H. 1963. "Ikaite," a new mineral from Greenland. *Arctic,* 16:263–264.

Pearce, T. H., and A. H. Clark. 1989. Nomarski interference contrast observations of textural details in volcanic rocks. *Geology,* 17:757–759.

Penticost, Allan. 1990. The formation of travertine shrubs: Mammoth Hot Springs, Wyoming. *Geological Magazine,* 127:159–168.

Peterson, Jon S. 1985. Columnar-dendritic feldspars in the lardalite intrusion, Oslo region, Norway: 1. Implications for unilateral solidification of a stagnate boundary layer. *Journal of Petrology,* 26:223–252.

Peterson, Sidney, and Robert T. Pack. 1983. Paleoenvironments of the Upper Jurassic Summerville Formation near Capitol Reef National Park, Utah. *Brigham Young University Geology Studies,* 29:13–26.

Pettijohn, F. J. 1975. *Sedimentary Rocks,* 3rd ed. New York: Harper & Row.

Phillips, E. R. 1964. Myrmekite and albite in some granites of the New England batholith, New South Wales. *Australia Geological Society Journal,* 11:49–60.

———. 1974. Myrmekite—one hundred years later. *Lithos,* 7:181–194.

———, D. M. Ransom, and R. H. Vernon. 1972. Myrmekite and muscovite developed by retrograde metamorphism at Broken Hill, New South Wales. *Mineralogical Magazine,* 38:570–578.

Phillips, G. Neil, Victor J. Wall, and John D. Clemens. 1981. Petrology of the Strathbogie Batholith: A cordierite-bearing granite. *Canadian Mineralogist,* 19:47–63.

Phillips, Wm. Revell. 1971. *Mineral Optics.* New York: Freeman. 249 pp.

———, and Dana T. Griffen. 1981. *Optical Mineralogy, The Nonopaque Minerals.* San Francisco: Freeman.

Philpotts, A. R. 1964. Origin of pseudotachylytes. *American Journal of Science,* 252:577–614.

———, 1976. Silicate liquid immiscibility: Its probable extent and petrogenetic significance. *American Journal of Science,* 276:1147–1177.

———, 1988. *Petrography of Igneous and Metamorphic Rocks.* Englewood Cliffs, NJ: Prentice Hall.

———, 1990. *Principles of Igneous and Metamorphic Petrology.* Englewood Cliffs, NJ: Prentice Hall.

Picard, M. D. 1971. Classification of fine-grained sedimentary rocks. *Journal of Sedimentary Petrology,* 41:179–195.

Pichler, H. 1965. Acid hyaloclastites. *Bulletin of Volcanology,* 28:293–310.

Piper, D. P., and P. J. Rodgers. 1980. Procedure for the assessment of the conglomerate resources of the Sherwood Sandstone group. *Assessment Report Institute Geological Science,* 56:11.

Pirsson, Louis V., and Adolph Knopf. 1947. *Rocks and Rock Minerals.* New York: Wiley.

Pitcher, Wallace S., and Antony R. Berger. 1972. *The Geology of Donegal: A Study of Granite Emplacement and Unroofing.* New York: Wiley-Interscience.

Piwinskii, A. J., and P. J. Wyllie. 1970. Experimental studies of igneous rock series: Felsic body suite from the Needle Point pluton, Wallowa Batholith, Oregon. *Journal of Geology,* 78:52–76.

Plafker, G. 1956. A technique for modal analyses of some fine- and medium-grained (0.1–5mm) rocks. *American Mineralogist,* 41:652–655.

Plumlee, Geoffrey S. 1989. Chemical modeling of ore and gangue deposition in the Creede, Colorado, epithermal system. In *U.S. Geological Survey Research on Mineral Resources, Fifth Annual V. E. McKelvey Forum on Mineral and Energy Resources. U.S. Geological Survey Circular,* 1035:55–57.

Potter, P. E., J. B. Maynard, and W. A. Pryor. 1980. *Sedimentology of Shale.* New York: Springer-Verlag.

Powell, C. McA., and R. H. Vernon. 1979. Growth and rotation history of garnet porphyroblasts with inclusion spirals in a Karakoram schist. *Tectonophysics,* 54:25–43.

Presnall, D. C., and Paul C. Bateman. 1973. Fusion relations in the system $NaAlSi_3O_8$-$CaAl_2Si_2O_8$-$KAlSi_3O_8$-SiO_2 and generation of granitic magmas in the Sierra Nevada Batholith. *Geological Society of America Bulletin,* 84:3181–3202.

Prior, D. J., R. J. Knipe, M. P. Bates, N. T. Grant, R. D. Law, G. E. Lloyd, W. Welbon, S. M. Agar, K. H. Brodie, R. H. Maddock, E. H. Rutter, S. H. White, T. H. Bell, C. C. Ferguson, and J. Wheeler. 1987. Orientation of specimens: Essential data for all field geologists. *Geology,* 15:829–831.

Pyke, R., J. Naldrett, and R. Eckstrand. 1973. Archean ultramafic flows in Munro Township, Ontario. *Geological Society of America Bulletin,* 84:955–978.

Radtke, A. S., R. O. Rye, and F. W. Dickson. 1980. Geology and stable isotope studies of the Carlin gold district, Nevada. *Economic Geology,* 75:641–672.

Railsback, L. Bruce. 1990. Comments and reply on "method of multiple working hypotheses: A chimera." *Geology,* 18:917–918.

Ramberg, Hans. 1962. Intergranular precipitation of albite formed by unmixing of alkali feldspar. *Neues Jahrbuch für Mineralogie Abhandlungen,* 98:14–34.

Ramdohr, P. 1981. *The Ore Minerals and Their Intergrowths,* 2nd ed. London: Pergamon Press.

Ramsay, J. G. 1963. Structure and metamorphism of the Moine and Lewisian rocks of the northwest Caledonides. In *The British Caledonides,* M. R. W. Johnson and F. H. Stewart, eds. London: Oliver and Boyd, pp. 143–175.

———, and Martin I. Huber. 1987. *The Techniques of Modern Structural Geology, Folds and Fractures,* vol. 2. New York: Academic Press.

Read, H. H. 1957. *The Granite Controversy.* New York: Wiley.

Reading, H. G., ed. 1978. *Sedimentary Environments and Facies.* New York: Elsevier.

Reed, Mark H., and Nicolas F. Spycher. 1985. Boiling, cooling, and oxidation in epithermal systems: A numerical modeling approach. In *Geology and Geochemistry of Epithermal Systems,* B. R. Berger and P. M. Bethke, eds. *Reviews in Economic Geology,* 2:249–272.

Reeves, C. C., Jr. 1976. *Caliche: Origin, Classification, Morphology and Uses.* Lubbock, TX: Estacado Books.

Reimold, W. U. 1990. The geochemistry of pseudotachylites in and around the Vredefort "Dome," South Africa. *South Africa Journal of Geology,* 93:350–365.

Reineck, H.-E., and I. B. Singh. 1980. *Depositional Sedimentary Environments, with Reference to Terrigenous Clastics.* New York: Springer-Verlag.

Renaut, Robin W., and R. Bernhart Owen. 1988. Opaline cherts associated with sublacustrine hydrothermal springs at Lake Bogoria, Kenya Rift valley. *Geology,* 16:699–702.

Rigby, J. Keith. 1977. *Field Guide to the Southern Colorado Plateau.* Dubuque, IA: Kendal/Hunt.

Rittmann, A. 1929. Die Zonemethode. *Schweizerische Mineralogische und Petrographische Mitteilungen,* 9:1–46.

Robertson, J. K., and P. J. Wyllie. 1971. Rock-water systems, with special reference to the water-deficient region. *American Journal of Science,* 276:252–277.

Robinson, Gilpin R., and Joseph P. Smoot. 1989. Stratabound copper and zinc mineralization in Triassic lacustrine beds of the Culpepper Basin, Virginia. In *U.S. Geological Survey Research on Mineral Resources, Fifth Annual V. E. McKelvey Forum on Mineral and Energy Resources. U.S. Geological Survey Circular,* 1035:57–59.

Rockhold, J. R., P. I. Nabelek, and M. D. Glascock. 1987. Origin of rhythmic layering in the Calamity Peak satellite pluton of the Harney Peak Granite, South Dakota: The role of boron. *Geochimica et Cosmochimica Acta,* 51:487–496.

Roddy, D. J., R. O. Pepin, and R. B. Merill. 1977. *Impact and Explosion Cratering.* Elmsford, NY: Pergamon Press.

Roedder, Edwin. 1979. Silicate liquid immicibility in magmas. In *The Evolution of the Igneous Rocks,* H. S. Yoder, Jr., ed. Princeton, NJ: Princeton University Press.

Romberger, Samuel B. 1988. Geochemistry of gold in hydrothermal deposits. *U.S. Geological Survey Bulletin,* 1857A:A9–A25.

Rona, Peter A., Kurt Boström, Lucien Laubier, and Kenneth L. Smith, Jr., eds. 1983. *Hydrothermal Process at Seafloor Spreading Centers.* New York: Plenum Press.

Roper, P. J., and D. E. Dunn. 1973. Superposed deformation and polymetamorphism, Brevard Zone, South Carolina. *Geological Society of America Bulletin,* 84:3373–3386.

Rose, Arthur W., Herbert E. Hawkes, and John S. Webb. 1979. *Geochemistry in Mineral Exploration.* New York: Academic Press.

Rosenfeld, J. L. 1970. Rotated garnets in metamorphic rocks. *Geological Society of America Special Paper,* 129.

Ross, Charles A., and June R. P. Ross. 1984. *Geology of Coal.* Benchmark Papers in Geology, vol. 77. Stroudsburg, PA: Hutchinson Ross.

Ross, C. S., and R. L. Smith. 1961. Ash-flow tuffs: Their origin, geologic relations and identification. *U.S. Geological Survey Professional Paper,* 366.

Ross, Martin E. 1986. Flow differentiation, phenocryst alignment, and compositional trends within a dolerite dike at Rockport, Massachusetts. *Geological Society of America Bulletin,* 97:232–240.

Rossinsky, Victor, Jr., Harold R. Wanless, and Peter K. Swart. 1992. Penetrative calcretes and their stratigraphic implications. *Geology,* 20:331–334.

Rouchy, J. M. 1980. The evaporite sequences of the terminal Miocene of Sicily and of southern Spain. In *Evaporite Deposits: Illustrations and Interpretation of Some Environmental Sequences.* Paris: Éditions Technip.

———, A. Laumondals, and E. Groessens. 1987. *The Lower Carboniferous (Visean) Evaporites in Northern France and Belgium: Depositional, Diagenetic and Deformational Guides to Reconstruct a Disrupted Evaporitic Basin,* Lecture Notes in Earth Sciences, vol. 13, T. M. Peryt, ed. *Evaporite Basins.* Berlin and Heidelberg: Springer-Verlag.

Runnells, Donald D. 1969. Diagenesis, chemical sediments, and the mixing of natural waters. *Journal of Sedimentary Petrology,* 39:1118–1201.

Runner, J. J. 1943. Structure and origin of Black Hills Precambrian granite domes. *Journal of Geology,* 51:431–457.

Ruperto, V. L., R. E. Stevens, and M. B. Normans. 1964. Staining of plagioclase and other minerals with F. D. and C. Red No. 2. *U.S. Geological Survey Professional Paper,* 501B:B152–153.

Russell, I. C. 1885. Geological history of Lake Lahontan. *U.S. Geological Survey Monograph,* 11.

Rutter, E. H. 1972. The influence of interstitial water on the rheological behavior of calcite rocks. *Tectonophysics,* 14:13–33.

———. 1986. On the nomenclature and mode of failure in rocks. *Tectonophysics,* 122:381–387.

Ruzyla, K., and D. I. Jezek. 1987. Staining method for recognition of pore space in thin section and polished sections. *Journal of Sedimentary Petrology,* 57:777–778.

Ryan, Michael P., ed. 1990. *Magma Transport and Storage.* New York: Wiley.

Ryder, R. T., T. D. Fouch, and J. H. Ellison. 1976. Early Tertiary sedimentation in the western Unita Basin. *Geological Society of America Bulletin,* 87:496–512.

Sales, R. H., and C. Meyers. 1948. Wall-rock alteration at Butte, Montana. *American Institute of Metallugical and Mining Engineers Transactions,* 178:9–35. Also AIME Technical Publications 2400.

Sandberg, Charles A., and Gilbert Klapper. 1967. Stratigraphy, age, and significance of the Cottonwood Canyon member of the Madison Limestone in Wyoming and Montana. *U.S. Geological Survey Bulletin,* 1251B:1–70.

Sanderson, Ivan D. 1984. Recognition and significance of inherited quartz overgrowths in quartz arenites. *Journal of Sedimentary Petrology,* 54:473–486.

Scambos, J. C., M. C. Loiselle, and D. R. Wones. 1986. The center pond pluton: The restite of the story (phase separation and melt evolution in granitoid genesis). *American Journal of Science,* 286:241–280.

Schairer, J. F. 1957. Melting relations of the common rock-forming silicates. *Journal of the American Ceramic Society,* 40:215–235.

———, and N. L. Bowen. 1947. The system anorthite-leucite-silica. *Society Geological Finlande Bulletin,* 20:67–87.

———, and H. S. Yoder, Jr. 1962. The system diopside-enstatite-silica. *Carnegie Institute Washington Yearbook,* 61:75–82.

Schaller, W. T., and Edward P. Henderson. 1932. Mineralogy of drill cores from the potash field of New Mexico and Texas. *U.S. Geological Survey Bulletin,* 833.

Schieber, Jürgen. 1986. The possible role of benthic microbial mats during the formation of carbonaceous shales in shallow Mid-Proterozoic basins. *Sedimentology,* 33:521–536.

Schmid, R. 1981. Descriptive nomenclature and classification of pyroclastic deposits and fragments: Recommendations of the IUGS subcommission on the systematics of igneous rocks. *Geology,* 9:4–43.

Schmid, S. M. 1982. Microfabric studies as indicators of deformation mechanisms and flow laws operative in mountain building. In *Mountain Building Processes,* K. J. Hsü, ed. New York: Academic Press, pp. 95–110.

———, R. Panozzo, and S. Bauer. 1987. Simple shear experiments on calcite rocks: Rheology and microfabric. *Journal of Structural Geology,* 9:747–778.

Schneiderman, Jill S. 1989. The Ascutbey Mountain breccia: Field and petrologic evidence for an overlapping relationship between Vermont sequence and New Hampshire sequence rocks. *American Journal of Science,* 289:771–811.

Scholl, David W., and William H. Taft. 1964. Algae, contributors to the formation of calcareous tufa, Mono Lake, California. *Journal of Sedimentary Petrology,* 34:309–319.

Scholle, Peter P. A. 1978. A color illustrated guide to carbonate rock constituents, textures, cements and porosities. *American Association of Petroleum Geologists Memoir,* 27.

———, Don G. Bebout, and Clyde H. Moore, eds. 1983. Carbonate depositional environments. *American Association of Petroleum Geologists Memoir,* 33.

———, and Darwin Spearing, eds. 1982. Sandstone depositional environments. *American Association of Petroleum Geologists Memoir,* 31.

Schoneveld, Chr. 1977. A study of some typical inclusion patterns in strongly paracrystalline-rotated garnets. *Tectonophysics,* 38:453–471.

Schreiber, B. Charlotte, and David Walker. 1992. Halite pseudomorphs after gypsum: A suggested mechanism. *Journal of Sedimentary Petrology,* 62:61–70.

Scoffin, Terence P. 1987. *Introduction to Carbonate Sediments and Rocks.* New York: Chapman and Hall.

Scott, K. M. 1987. The mineralogical distribution of pathfinder elements in gossans derived from dolomitic shale-hosted Pb-Zn deposits, northwest Queensland, Australia. *Chemical Geology,* 64:295–306.

Seaman, Sheila J., and Patrick C. Ramsey. 1992. Effects of magma mingling in the granites of Mount Desert Island, Maine. *Journal of Petrology,* 100:395–409.

Sederholm, J. J. 1923. On migmatites and associated Precambrian rock. II. *Bulletin du Commission Géology du Finlande,* 77:1–153.

———. 1967. *Selected Works, Granites and Migmatites.* New York: Wiley.

Sharp, W. E., E. L. Carlson, and I. Kheoruenromne. 1977. A stain test for fluorite. *American Mineralogist,* 62:171–172.

Sharpton, V. L., and B. C. Schuraytz. 1989. On reported occurrence of shock-deformed clasts in the volcanic ejecta from Toba caldera, Sumatra. *Geology,* 17:1040–1043.

———, and P. D. Ward, eds. 1990. Global catastrophes in earth history. *Geological Society of America Special Paper,* 247.

Shaw, H. R. 1974. Diffusion of H_2O in granitic liquids. In *Geochemical Transport and Kinetics,* A. W. Hofmann et al., eds. Publication 634. Washington, DC: Carnegie Institute of Washington, pp. 139–170.

———. 1980. The fracture mechanisms of magma transport from the mantle to the surface. In *Physics of Magmatic Processes,* R. B. Hargraves, ed. Princeton, NJ: Princeton University Press, pp. 201–264.

Shearman, D. J., A. McGugan, C. Stein, and A. J. Smith. 1989. Ikaite, $CaCO_3 \cdot 6H_2O$, precursor of the thinolites in the Quaternary tufas and tufa mounds of the Lahontan and Mono Lake Basins, western United States. *Geological Society of America Bulletin,* 101:913–917.

Shelley, David. 1985. *Optical Mineralogy.* New York: Elsevier.

———. 1992. *Igneous and Metamorphic Rocks Under the Microscope.* New York: Chapman and Hall.

Sheridan, Michael F. 1970. Fumarolic mounds and ridges of the Bishop Tuff, California. *Geological Society of America Bulletin,* 81:851–868.

———. 1979. Emplacement of pyroclastic flows: A review. In *Ash Flow Tuffs,* Charles E. Chapin and Wolfgang E. Elston, eds. *Geological Society of America Special Paper,* 180:125–134.

Shimamoto, Toshihiko. 1989. The origin of S-C mylonites and a new fault-zone model. *Journal of Structural Geology,* 11:51–64.

Shirley, David N. 1987. Differentiation and compaction in the Palisades Sill, New Jersey. *Journal of Petrology,* 28:835–865.

Sholkovitz, E. R. 1976. Flocculation of dissolved organic and inorganic matter during the mixing of river water and seawater. *Geochimica et Cosmochimica Acta,* 40:831–845.

Sibley, D. F., T. A. Vogal, B. M. Walker, and G. Byerly. 1976. The origin of oscillatory zoning in plagioclase: A diffusion and growth controlled model. *American Journal of Science,* 276:273–284.

Sibson, R. H. 1977. Fault rocks and fault mechanisms. *Journal of the Geological Society of London,* 133:191–213.

Silberman, M. L. 1983. Geochronology of hydrothermal alteration and mineralization—Tertiary epithermal precious-metal deposits in the Great Basin. In *Geothermal Resource Council, Special Report,* 13: 287–303.

———, and Byron R. Berger. 1985. Relationship of trace-element patterns to alteration and morphology in epithermal precious-metal deposits. In *Geology and Geochemistry of Epithermal Systems,* B. R. Berger and P. M. Bethke, eds. *Reviews in Economic Geology,* 2:203–232.

Sillitoe, Richard H. 1973. The tops and bottoms of porphyry copper deposits. *Economic Geology,* 68:799–815.

Silver, L. T., and P. H. Schultz, eds. 1982. Geological implications of impacts of large asteroids and comets on the earth. *Geological Society of America Special Paper,* 190.

Simonson, Bruce M. 1987. Early silica cementation and subsequent diagenesis in arenites from four early Proterozoic iron formations of North America. *Journal of Sedimentary Petrology,* 57:494–571.

Simpson, Carol. 1983. Strain and shape-fabric variations associated with ductile shear zones. *Journal of Structural Geology,* 5:61–72.

———. 1985. Deformation of granitic rocks across the brittle-ductile transition. *Journal of Structural Geology,* 7:503–511.

———, and Stefan M. Schmid. 1983. An evaluation of criteria to deduce the sense of motion in sheared rocks. *Geological Society of America Bulletin,* 94:1281–1288.

———, and R. P. Wintsch. 1989. Evidence for deformation-induced K-feldspar replacement by myrmekite. *Journal of Metamorphic Geology,* 7:261–275.

Slansky, Maurice. 1986. *Geology of Sedimentary Phosphates.* New York: Elsevier.

Slemmons, David B. 1962. Determination of volcanic and plutonic plagioclases using a three- or four-axis universal stage. *Geological Society of America Special Paper,* 69.

———. 1963. Plagioclase determination with the aid of the extinction angles in sections normal to (010). A critical comparison of current albite-Carlsbad charts. *American Journal of Science,* 261:157–167.

Smith, G. I., and D. V. Haines. 1964. Character and distribution of nonclastic minerals in the Searles Lake Evaporite Deposit, California. *U.S. Geological Survey Bulletin,* 1881P.

Smith, J. V. 1958. The effect of composition and structural state on the rhombic section and pericline twins of plagioclase feldspars. *Mineralogical Magazine,* 31:914–928.

———, and William L. Brown. 1988. *Feldspar Minerals,* vol. 1, *Crystal Structures, Physical, Chemical and Microtextural Properties,* 2nd rev. extended ed. Berlin: Springer-Verlag.

Smith, M. R., W. R. Wilson, J. A. Benham, C. A. Pescio, and P. Valenti. 1988. The Star Pointer gold deposit Robinson Mining District White Pine County, Nevada. In *Bulk Mineable Precious Metal Deposits of the Western United States,* Robert W. Schafer, James J. Cooper, and Peter G. Vikre, eds. Symposium Proceedings, April 6–8, 1987, Reno, NV: Geological Society of Nevada, pp. 221–231.

Smith, R. L. 1960. Zones and zonal variations in welded ash flows. *U.S. Geological Survey Professional Paper,* 354F.

———. 1979. Ash-flow magmatism. In *Ash Flow Tuffs,* Charles E. Chapin and Wolfgang E. Elston, eds. *Geological Society of America Special Paper,* 180:5–27.

Smithson, S. B. 1963. A point-counter for modal analysis of stained rock slabs. *American Mineralogist,* 48:1164–1166.

Solomon, M. 1963. Counting and sampling errors in modal analysis by point counter. *Journal of Petrology,* 4:367–382.

Sonnenfeld, Peter. 1984. *Brines and Evaporites.* San Diego: Academic Press.

Sorensen, Henning, and Lotte Melchior Larsen. 1987. Layering in the Ilimaussaq alkaline intrusion, south Greenland. In *Origins of Igneous Layering,* I. Parsons, ed. Dordrecht, Holland: Reidel, pp. 1–28.

Soudry, David. 1987. Ultra-fine structures and genesis of the Campanian Negev high-grade phosphorites (southern Israel). *Sedimentology,* 34:641–660.

Southgate, Peter N. 1986. Cambrian phoscrete profiles, coated grains, and microbial processes in phosphogenesis: Georgina Basin, Australia. *Journal of Sedimentary Petrology,* 56:429–441.

Sparks, R. S. J., and Herbert E. Huppert. 1984. Density changes during the fractional crystallization of basaltic magmas: Fluid dynamic implications. *Contributions to Mineralogy and Petrology,* 85:300–309.

———, H. Siguedsson, and L. Wilson. 1977. Magma mixing: A mechanism for triggering acid explosive eruptions. *Nature,* 267:315–318.

———, and J. V. Wright. 1979. Welded air-fall tuffs. In *Ash Flow Tuffs,* Charles E. Chapin and Wolfgang E. Elston, eds. *Geological Society of America Special Paper,* 180:155–166.

Spötl, Christoph. 1989. The Alpine Haselgebirge Formation, northern calcareous Alps (Austria): Permo-Scythian evaporites in an alpine thrust system. *Sedimentary Geology,* 65:113–125.

Sprunt, E. S., and W. F. Brace. 1974. Direct observation of microcavities in crystalline rocks. *International Journal of Rock Mechanics and Mineral Science,* 11:139–150.

Spry, Alan. 1969. *Metamorphic Textures.* London: Pergamon Press.

Steinen, Randolph P., Norman H. Gray, and John Mooney. 1987. A Mesozoic carbonate hot-spring deposit in the Hartford basin of Connecticut. *Journal of Sedimentary Petrology,* 57:319–326.

Steiner, J. C., R. H. Jahns, and W. C. Luth. 1975. Crystallization of alkali feldspar and quartz in the haplogranite system $NaAlSi_3O_8$-$KAlSi_3O_8$-SiO_2-H_2O at 4 kb. *Geological Society of America Bulletin,* 86:83–98.

Stel, H. 1981. Crystal growth in cataclasites: Diagnostic microstructures and implications. *Tectonophysics,* 78:585–600.

Stewart, F. H. 1949. The petrology of the evaporites of the Eskdale no. 2 boring east Yorkshire. I: The lower evaporite bed. *Mineralogical Magazine,* 28:622–675.

Stimac, James A., and David A. Wark. 1992. Plagioclase mantles on sanidine in silicic lavas, Clear Lake, California: Implications for the origin of rapakivi texture. *Geological Society of America Bulletin,* 104:728–744.

Stöffler, D. 1966. Zones of impact metamorphism in the crystalline rocks of the Nördlinger Rise Crater. *Contributions to Mineralogy and Petrology,* 12:15–24.

———. 1972. Deformation and transformation of rock-forming minerals by natural and experimental shock processes. I: Behavior of minerals under shock compression. *Fortschritte Mineralogie,* 49:50–113.

———. 1974. Deformation and transformation of rock-forming minerals by natural and experimental shock processes. II: Physical properties of shocked minerals. *Fortschritte Mineralogie,* 51:256–289.

———, H.-D. Knoll, U. B. Marvin, C. H. Simonds, and P. H. Warren. 1980. Recommended classification and nomenclature of lunar highland rocks—a committee report. In *Proceedings of the Conference on the Lunar Highlands Crust,* J. J. Papike and R. B. Merrill, eds. *Geochimica et Cosmochimica Acta,* Supplement 12:51–70.

Stoiber, R. E., and E. S. Davidson. 1959. Amygdule mineral zoning in the Portage Lava series, Michigan Copper district. *Economic Geology,* 54:1250–1277, 1444–1460.

———, and S. A. Morse. 1981. *Microscopic Identification of Crystals.* Corrected reprint ed. Malabar, FL: Krieger.

———, and William I. Rose, Jr. 1974. Fumarole encrustation at active Central American volcanoes. *Geochimica et Cosmochimica Acta,* 38:495–516.

Straccia, Frances G., Bruce H. Wilkinson, and Gerald R. Smith. 1990. Miocene lacustrine algal reefs—southwestern Snake River Plain, Idaho. *Sedimentary Geology,* 67:7–23.

Streckeisen, A. L. 1967. Classification and nomenclature of igneous rocks. *Neues Jahrbuch für Mineralogie Abhandlungen,* 107:144–240.

———. 1973. Plutonic rocks, classification and nomenclature recommended by the IUGS subcommission on the systematics of igneous rocks. *Geotimes,* 18:26–30.

———. 1976. To each plutonic rock its proper name. *Earth-Science Reviews,* 12:1–33.

———, and R. W. LeMaitre. 1979. A chemical approximation to the model QAPF classification of the igneous rocks. *Neues Jahrbuch für Mineralogie Abhandlungen,* 136:169–206.

Strong, G. E., J. R. A. Giles, and V. P. Wright. 1992. A Holocene calcrete from North Yorkshire, England: Implications for interpreting palaeoclimates using calcretes. *Sedimentology,* 39:333–347.

Sunagawa, Ichiro. 1981. Characteristics of crystal growth in nature as seen from the morphology of mineral grains. *Bulletin Minéralogie,* 104:81–87.

Sutton, John, and Janet Watson. 1951. The pre-Torridonian metamorphic history of the Loch Torridon and Scourie areas in the Northwest Highlands, and its bearing on the chronological classification of the Lewisian. *Geological Society of London Quarterly Journal,* 106:241–307.

———. 1969. Scourian-Laxfordian relationships in the Lewisian of northwest Scotland. In *Age Relations in High-grade Metamorphic Terrains. Geological Association of Canada Special Paper,* 5:119–128.

Swanson, Samuel E. 1977. Relation of nucleation and crystal-growth rate to the development of granitic textures. *American Mineralogist,* 62:966–978.

———, Michael T. Naney, H. R. Westrich, and J. C. Eichelberger. 1989. Crystallization history of Obsidian Dome, Inyo Domes, California. *Bulletin of Volcanology,* 51:161–176.

Swett, Keene, and R. Keith Crowder. 1982. Primary phosphatic oolites from the Lower Cambrian of Spitsbergen. *Journal of Sedimentary Petrology,* 52:587–593.

Swineford, Ada, A. B. Leonard, and J. C. Frye. 1958. Petrology of the Pliocene pisolitic limestones in the Great Plains. *Kansas Geological Survey Bulletin,* 130, part 2:97–116.

Sylvester, A. G., G. Oertel, C. A. Nelson, and J. M. Christie. 1978. Papoose Flat pluton: A granitic blister in the Inyo Mountains, eastern California. *Geological Society of America Bulletin,* 89:1205–1219.

Symonds, Robert. 1993. Scanning electron microscope observations of sublimates from Merapi volcano, Indonesia. *Geochemical Journal (Matsuo Memorial Issue),* 26:337–350.

———, William I. Rose, Mark H. Reed, Frederick E. Lichte, and David L. Finnegan. 1987. Volatization, transport and sublimation of metallic and non-metallic elements in high temperature gases at Merapi Volcano, Indonesia. *Geochimica et Cosmochimica Acta,* 51:2083–2101.

Takenouchi, Sukune, and G. C. Kennedy. 1965. The solubility of carbon dioxide in NaCl solutions at high temperatures and pressures. *American Journal of Science,* 263:445–454.

Talbot, C. J. 1979. Fold trains in a glacier of salt in southern Iran. *Journal of Structural Geology,* 1:5–18.

———. 1981. Sliding and other deformation mechanisms in a glacier of salt, south Iran. In *Thrust and Nappe Tectonics,* K. R. McClay and N. J. Price, eds. International Conference. University of London Department of Geology. *Special Publication Geological Society of London,* 9:173–183.

———, and M. P. A. Jackson. 1987. Internal kinematics of salt diapirs. *American Association of Petroleum Geologists Bulletin,* 71:1068–1093.

Talbot, Christopher, and Victor von Brunn. 1989. Melanges, intrusive and extrusive sediments, and hydraulic arcs. *Geology,* 17:446–448.

Taylor, Graham, R. A. Eggletob, C. C. Holzhauer, L. A. Maconachie, Mark Gordon, M. C. Brown, and K. G. McQueen. 1992. Cool climate lateritic and bauxitic weathering. *Journal of Geology,* 100:669–677.

Taylor, Hugh P., Jr. 1979. Oxygen and hydrogen isotope relationships in hydrothermal mineral deposits. In *Geochemistry of Hydrothermal Ore Deposits,* 2nd ed., H. L. Barnes, ed. New York: Wiley, pp. 236–277.

Thériault, François, and Ian Hutcheon. 1987. Dolomitization and calcitization of the Devonian Grosmont Formation, northern Alberta. *Journal of Sedimentary Petrology,* 57:955–966.

Thiry, Médard, and Georges Millot. 1987. Mineralogical forms of silica and their sequence of formation in silcretes. *Journal of Sedimentary Petrology,* 57:342–352.

Thompson, A. B. 1982. Dehydration melting of pelitic rocks and the generation of H_2O undersaturated granitic liquids. *American Journal of Science,* 282:1567–1595.

Thompson, G. A. 1956. Geology of the Virginia City quadrangle, Nevada. *U.S. Geological Survey Bulletin* 1042C:45–77.

Thompson, J. B., and F. G. Ferris. 1990. Cyanobacterial precipitation of gypsum, calcite, and magnesite from natural alkaline lake water. *Geology,* 18:995–998.

Thy, P., and K. H. Esbensen. 1982. Origin of fine-grained granular rocks in layered intrusions. *Geological Magazine,* 119:405–412.

Tiercelin, Jean-Jacques, Catherine Thouin, Tchibangu Kalala, and Andre Mondeguer. 1989. Discovery of sublacustrine hydrothermal activity and associated massive sulfide and hydrocarbons in the north Tanganyika trough, east Africa Rift. *Geology,* 17:1053–1056.

Tikoff, Basil, and Christian Teyssier. 1992. Crustal-scale, in echelon "P-shear" tensional bridges: A possible solution to the batholithic room problem. *Geology,* 20:927–930.

Tiller, W. A. 1964. Dendrites. *Science,* 146:871–879.

Tilley, C. E. 1924. Contact metamorphism in the Comrie area of the Perthshire highlands. *Quarterly Journal Geological Society of London,* 80:23–66.

Titley, Spencer R. 1982. *Advances in Geology of the Porphyry Copper Deposits, Southwestern North America.* Tuscon, AZ: University of Arizona Press.

———, and Carol L. Hicks. 1966. *Geology of the Porphyry Copper Deposits, Southwestern North America.* Tuscon, AZ: University of Arizona Press.

Tobi, A. C. 1956. A chart for measurement of optical axial angles. *American Mineralogist,* 44:516–519.

———, and H. Kroll. 1975. Optical determination of the An-content of plagioclases twinned by Carlsbad-Law: A revised chart. *American Journal of Science,* 275:731–736.

Tracy, R. J. 1987. Metamorphic geology. *Reviews of Geophysics,* 25:1115–1122.

———, and P. Robinson. 1983. Acadian migmatite types in pelitic rocks of central Massachusetts. In *Migmatites, Melting, and Metamorphism,* M. P. Atherton and C. D. Gribble, eds. Cheshire, England: Shiva, pp. 163–173.

Troger, W. E. 1979. *Optical Determination of Rock-forming Minerals, Part I: Determinative Tables.* English edition of the fourth German edition by H. U. Bambaur, F. Tabotgzky, and H. D. Trochim. Stuttgart: Schwiezerbart'sche Verlagsbuchhandlung.

Tsuchiyama, A. 1985. Dissolution kinetics of plagioclase in the melt of the system diopside-albite-anorthite, and the origin of dusty plagioclase in andesites. *Contributions to Mineralogy and Petrology,* 89:1–16.

———, and E. Takahashi. 1983. Melting kinetics of a plagioclase feldspar. *Contributions to Mineralogy and Petrology,* 84:345–354.

Tucker, Maurice E. 1991. *Sedimentary Petrology: An Introduction to the Origin of Sedimentary Rocks.* Cambridge, MA: Blackwell.

Tullis, J., and R. A. Yund. 1977. Experimental deformation of dry Westerly granite. *Journal of Geophysical Research,* 82:5705–5718.

———. 1989. Hydrolytic weakening of quartz aggregates: The effects of water and pressure on recovery. *Geophysical Research Letters,* 16:1343–1346.

Tunell, G. 1952. The angle between the a-axis and the trace of the rhombic section on the {010}-pinacoid in the plagioclases. *American Journal of Science* (Bowen Volume), 250A:547–552.

Turcotte, D. L., Steven H. Emerman, and D. A. Spence. 1987. Mechanics of dyke injection. In *Mafic Dyke Swarms,* Henry C. Halls and Walter F. Fahrig, eds. *Geological Association of Canada Special Paper,* 34.

Turner, F. J. 1968. *Metamorphic Petrology—Mineralogical and Field Aspects.* New York: McGraw-Hill.

———. 1980. *Metamorphic Petrology—Mineralogical and Field Aspects,* 2nd ed. New York: McGraw-Hill.

———, and L. E. Weiss. 1963. *Structural Analysis of Metamorphic Tectonites.* New York: McGraw-Hill.

Turner, J. S. 1985. Multicomponent convection. *Annual Reviews of Fluid Mechanics,* 17:11–44.

Tuttle, O. F., and N. L. Bowen. 1958. Origin of granite in the light of experimental studies in the system. *Geological Society of America Memoir,* 74.

Vance, J. A. 1962. Zoning in igneous plagioclase: Normal and oscillatory zoning. *American Journal of Science,* 260:746–760.

———. 1965. Zoning in igneous plagioclase: Patchy zoning. *Journal of Geology,* 73:636–651.

———. 1969. On synneusis. *Contributions to Mineralogy and Petrology,* 24:7–29.

van Krevelen, D. W. 1981. *Coal—Typology, Chemistry, Physics, Constitution.* New York: Elsevier.

Vernon, R. H. 1975. *Metamorphic Processes.* New York: Wiley.

———. 1983. *Metamorphic Processes—Reactions and Microtextures.* London: Allen and Unwin.

———. 1991a. Questions about myrmekite in deformed rocks. *Journal of Structural Geology,* 13:979–985.

———. 1991b. Interpretation of microstructures of microgranitoid enclaves. In *Enclaves and Granite Petrology,* J. Didier and B. Barbarin, eds. Amsterdam: Elsevier, pp. 277–291.

———, and W. J. Collins. 1988. Igneous microstructures in migmatites. *Geology,* 16:1126–1129.

Vidal, J-L., L. Kubin, P. Debat, and J-C. Soula. 1980. Deformation and dynamic recrystallization of K-feldspar augen in orthogneiss from Montagne Moire, Occitania, southern France. *Lithos,* 13:247–255.

Vikre, P. C. 1985. Precious metal vein systems in the National district, Humboldt County, Nevada. *Economic Geology,* 80:360–393.

von der Borch, Chris C., David E. Lock, and Douglas Schwebel. 1975. Ground-water formation of dolomite in the Coorong region of South Australia. *Geology,* 3:283–285.

von Platen, H. 1965. Experimental anatexis and genesis of migmatites. In *Controls of Metamorphism,* W. S. Pitcher and G. W. Flinn, eds. New York: Wiley.

Vrolijk, Peter, and Simon M. F. Sheppard. 1991. Syntectonic carbonate veins from the Barbados accretionary prism (ODP Leg 110): Record of palaeohydrology. *Sedimentology,* 38:671–690.

Wager, L. R. 1959. Differing powers of crystal nucleation as a factor for producing diversity in layered igneous intrusions. *Geological Magazine,* 96:75–80.

———, and G. M. Brown. 1968. *Layered Igneous Rocks.* London: Oliver and Boyd.

———, G. M. Brown, and W. J. Wadsworth. 1960. Types of igneous cumulates. *Journal of Petrology,* 1:73–85.

Wahlstrom, Ernest E. 1979. *Optical Crystallography,* 5th ed. New York: Wiley.

Wahrhaftig, C. 1965. Stepped topography of the Sierra Nevada, California. *Geological Society of America Bulletin,* 76:1165–1190.

Walker, D., R. J. Kirkpatrick, J. Longhi, and J. F. Hays. 1976. Crystallization history of lunar picritic basalt sample 12002: Phase-equilibria and cooling-rate studies. *Geological Society of America Bulletin,* 87:646–656.

Walker, G. P. L. 1969. The breaking of magma. *Geological Magazine,* 106:166–173.

Wallace, S. R., N. K. Muncaster, D. C. Jonson, W. B. MacKenzie, A. A. Brookstrom, and V. E. Surface. 1968. Multiple intrusion and mineralization at Climax, Colorado. In *Ore Deposits of the United States, 1939–1967 (Graton-Sales Volume),* J. D. Ridge, ed., New York: American Institute of Mining, Metallurgical, and Petroleum Engineers, pp. 605–640.

Walter, J. V., and H. C. Helgeson. 1977. Calculation of the thermodynamic properties of aqueous silica and the solubility of quartz and its polymorphs at high pressure and temperature. *American Journal of Science,* 277:1315–1351.

Walter, M. R., ed., 1976. *Stromatolites. Developments in Sedimentology,* vol. 20. New York: Elsevier.

Ward, Colin R., ed., 1984. *Coal Geology and Coal Technology.* Boston, MA: Blackwell.

Wardlaw, Norman C. 1968. Carnallite-sylvite relationships in the Middle Devonian Prairie Evaporite Formation, Saskatchewan. *Geological Society of America Bulletin,* 79:1273–1294.

Warren, John K. 1983. Pedogenic calcrete as it occurs in Quaternary calcareous dunes in coastal south Australia. *Journal of Sedimentary Petrology,* 53:787–796.

Waters, A. C., and K. B. Krauskopf. 1941. Protoclastic border of Colville batholith. *Geological Society of America Bulletin,* 52:1355–1418.

Watkinson, D. H., and P. J. Wyllie. 1969. Phase equilibria studies bearing on the limestone-assimilation hypothesis. *Geological Society of America Bulletin,* 80:1565–1576.

Weaver, Charles, E. 1984. *Shale-slate metamorphism in southern Appalachians.* New York: Elsevier.

Weggen, J., Bergisch Gladbach, and I. Valeton. 1990. Polygenetic lateritic iron ores on BIF's in Minas Gerais/Brazil. *Geologische Rundschau,* 79:301–318.

Weijermars, R., and H. E. Rondeel. 1984. Shear band foliation as an indicator of sense of shear: Field observations in central Spain. *Geology,* 12:603–606.

Weiler, Y., E. Sass, and I. Zak. 1974. Halite oolites and ripples in the Dead Sea, Israel. *Sedimentology,* 21:623–632.

Weimer, Robert J., and John H. Hoyt. 1964. Burrows of *Callianassa Major* Say: Geologic indicators of littoral and shallow neritic environments. *Journal of Paleontology,* 38:761–767.

Wenk, H. R. 1978. Are pseudotachylites products of fracture or fusion? *Geology,* 6:507–511.

White, Donald E. 1974. Diverse origins of hydrothermal ore fluids. *Economic Geology,* 69:954–973.

White, Noel C., and Jeffrey W. Hedenquist. 1990. Epithermal environments and styles of mineralization: Variations and their causes, and guidelines for exploration. *Journal of Geochemical Exploration,* 36:445–474.

Whitney, J. A. 1975. The effects of pressure, temperature, and XH_2O on phase assemblages in four synthetic rock compositions. *Journal of Geology,* 83:1–31.

Whitney, P. R., and J. M. McLellend. 1983. Origin of biotite-hornblende-garnet coronas between oxides and plagioclase in olivine metagabbros, Adirondack region, New York. *Contributions to Mineralogy and Petrology,* 82:34–41.

Wiebe, R. A. 1988. Structural and magmatic evolution of a magma chamber: The Newark Island layered intrusion, Nain, Labrador. *Journal of Petrology,* 29:383–411.

———. 1991. Commingling of contrasted magmas and generation of mafic enclaves in granitic rocks. In *Enclaves and Granite Petrology,* J. Didier and B. Barbarin, eds. Amsterdam: Elsevier, pp. 393–402.

———, and Don Snyder. 1993. Slow, dense replenishments of a basic magma chamber: The layered series of the Newark Island layered intrusion, Nain, Labrador. *Contributions to Mineralogy and Petrology,* 113:59–72.

Wilcox, H. G. 1966. Determination of indicatrix orientation and 2V with the spindle stage: A caution and a test. *American Mineralogist,* 51:919–924.

Williams, H., and A. R. McBirney. 1979. *Volcanology.* San Francisco: Freeman.

———, F. J. Turner, and C. M. Gilbert. 1954 (1st ed.), 1982 (2nd ed.). *Petrography,* San Francisco: Freeman.

Williams, L. A., and David A. Crerar. 1985. Silica diagenesis. II: General mechanisms. *Journal of Sedimentary Petrology,* 55:312–321.

———, G. A. Parks, and D. A. Crerar. 1985. Silica diagenesis. I: Solubility controls. *Journal of Sedimentary Petrology,* 55:301–311.

Wilson, C. J. 1975. Preferred orientation in quartz ribbon mylonites. *Geological Society of America Bulletin,* 86:968–974.

———, and I. A. Bell. 1979. Deformation of biotite and muscovite: Optical microstructure. *Tectonophysics,* 58:179–200.

———, and G. P. L. Walker. 1982. Ignimbrite depositional facies: The anatomy of a pyroclastic flow. *Journal of the Geological Society of London,* 139:581–592.

Wilson, Edith Newton, Lawrence A. Hardie, and Owen M. Phillips. 1990. Dolomitization front geometry, fluid flow patterns, and the origin of massive dolomite: The Triassic Latemar Buildup, northern Italy. *American Journal of Science,* 290:741–796.

Winkler, H. G. F. 1970. Ablution of metamorphic facies, introduction of the four divisions of metamorphic stage and of a classification based on isograds in common rocks. *Neues Jahrbuch für Mineralogie Abhandlungen,* 5:189–248.

———. 1979. *Petrogenesis of Metamorphic Rocks,* 5th ed. New York: Springer-Verlag.

———, M. Boese, and T. Marcopoulos. 1975. Low temperature granitic melts. *Neues Jahrbuch für Mineralogie Monatshefte,* 6:245–268.

Wise, D. U., D. E. Dunn, J. T. Engelder, P. A. Geiser, R. D. Hatcher, S. A. Kish, A. L. Odom, and S. Schamel. 1984. Fault-related rocks: Suggestions for terminology. *Geology,* 12:391–394.

Wohletz, K. H., and Robert G. McQueen. 1984. Experimental studies of hydromagmatic volcanism. In *Explosive Volcanism: Inception, Evolution, and Hazards.* Washington, DC: National Academy Press, pp. 158–169.

———, and M. F. Sheridan. 1979. A model of pyroclastic surge. In *Ash Flow Tuffs,* Charles E. Chapin and Wolfgang E. Elston, eds. *Geological Society of America Special Paper,* 180:177–193.

Wojtal, Steven, and Gautam Mitra. 1986. Strain hardening and strain softening in fault zones from foreland thrusts. *Geological Society of America Bulletin,* 97:674–687.

Wolff, J. A., and J. V. Wright. 1981. Rheomorphism of welded tuffs. *Journal of Volcanology and Geothermal Research,* 10:13–34.

Woodland, B. G. 1963. A petrographic of thermally metamorphosed pelitic rocks in the Burke area, northeastern Vermont. *American Journal of Science,* 261:354–375.

Wright, J. V., A. L. Smith, and S. Self. 1980. A working terminology of pyroclastic deposits. *Journal of Volcanological and Geothermal Research,* 8:315–336.

Wu, Schuman, and Richard H. Groshong, Jr. 1991. Low-temperature deformation of sandstone, southern Appalachian fold-thrust belt. *Geological Society of America Bulletin,* 103:861–875.

Wyllie, P. J. 1977. Crustal anatexis: An experimental review. *Tectonophysics,* 43:41–71.

———. 1983. Experimental studies on biotite- and muscovite-granites and some crustal magmatic sources. In *Migmatites, Melting and Metamorphism,* M. P. Atherton and C. D. Gribble, eds. Cheshire, England: Shiva.

———, K. G. Cox, and G. M. Bigger. 1962. The habit of apatite in syenitic systems and igneous rocks. *Journal of Petrology,* 3:238–243.

Yanguas, J. E., and J. J. Dravis. 1985. Blue fluorescent dye technique for recognition of microporosity in sedimentary rocks. *Journal of Sedimentary Petrology,* 55:600–602.

Yardley, B. W. D. 1978. Genesis of the Skagit gneiss migmatites, Washington, and the distinction between possible mechanisms of migmatization. *Geological Society of America Bulletin,* 89:941–951.

———, W. S. MacKenzie, and C. Guilford. 1990. *Atlas of Metamorphic Rocks and Their Textures.* New York: Wiley.

Yoder, H. S., Jr., D. B. Stewart, and J. R. Smith. 1957. Ternary feldspars. *Carnegie Institute Geophysical Laboratory Yearbook,* 56:206–214.

———, and C. E. Tilley. 1962. Origin of basalt magmas: An experimental study of natural and synthetic rock systems. *Journal of Petrology,* 3:342–532.

Young, T. P. 1989. Phanerozoic ironstones: An introduction and review. In *Phanerozoic Ironstones,* T. P. Young and W. E. G. Taylor, eds. *Geological Society of London Special Publication,* 46:9–25.

Zeck, H. P. 1970. An erupted migmatite form Cerro del Hoyazo, southeast Spain. *Contributions to Mineralogy and Petrology,* 26:225–246.

Zen, E-an. 1988. Phase relations of peraluminous granitic rocks and their petrogenetic implications. *Annual Reviews of Earth and Planetary Science,* 16:47.

Zenger, D. H. 1972. Significance of supratidal dolomitization in the geologic records. *Geological Society of America Bulletin,* 83:1–11.

Zies, E. G. 1929. The Valley of Ten Thousand Smokes I, II. *National Geographic Technical Paper, Katmai Series,* 4:1–79.

Zwart, H. J. 1962. On the determination of polymetamorphic mineral associations, and its application to the Bosost area (central Pyrenees). *Geologische Rundschau,* 52:38–65.

———, and T. J. Calon. 1977. Chloritoid crystals from Curaglia: Growth during flattening or pushing aside? *Tectonophysics,* 39:477–486.

Index

Italics indicates a figure or table;
boldface indicates a definition.

a-normal extinction angle method, 77
aa lava, **347**
Abyssal plain, 452
Academy Granite, 226, *229*
Acaustobiolith, *155*
Accessory mineral, 74
Accessory plate, *2*, 22
Accretion wedge, 452
Accretionary lapilli, **375**
Acetate peel, **97**
Achnelith, **375**
Acicular apatite, *247*, **255**, *255*
Acmite, *46*
Actinolite, *43*, *45–46*, 59
Actinolitic hornblende, 43
Activity coefficient, **166**
Acute bisectrix Bxa, **40**
Adcumulate, **240**, *240*
Adiabatic boiling, **398**
Adiabatic decompression, 367
Adiabatic melting, **175**, **357–358**
Adularia, *44*, *46*, **59**, *59*, 420
Aegirine, *45–46*, **59**
Aegirine-augite, *45–46*, **59**, *59*
Agglomerate, *148*
Alaskite, *145*
Albite, *44*, *46*, **59**, *126*
Albite-anorthite system, **177**, *178*
Albite-anorthite-diopside-olivine system, *186*
Albite-anorthite-diopside system, *185*
Albite-silica system, *181*
Algae, 456–458
Algal bioherm, 459
Algal boring, *458*
Algal mat, 459, 477, 479
Algal reef, 459
Alivalite, *147*
Alkali feldspar, *46*, **61**, 69, *128*
Alkali feldspar ovoids, 244, *245*, *249*
Alkali feldspar series, **179–180**, *179*
Allanite, *46*, **74**
Allochem, **153**, **472**
Allogenic, **434**
Allophane, 107, 426
Allotriomorphic granular, **113**
Alluvial fan, **445**
Almandine, *46*, **65**, 282, *306*, *310*, *360*
Alnöite, *146*
Al_2SiO_5 polymorphs, 279, *280*
Alteration
 argillic, 419
 deuteric, **410**, *413*
 K-silicate, 416
 palagonitic, 378
 potassic, 416, *417*
 sericitic, *413*
Alunite, *45*, *46*, **59**, 73, *402*, 421, 518, 522
Alveole, **511**, *514*
Amesite, *46*, *62*
Ammonite, **466**, *467*
Amorphous solids, **73**
Amphibole, *46*
Amphibolite, *149*, **150**
Amygdule, **413**, *414*
Analcime, *44*, *46*, **60**, 74
Analyzing polarizer, **21**
Anatase, *46*, **74**
Anatexis, 357, 363, *363*
Anatexite, **357**
Andalusite, *25*, *44*, *47*, **60**, *60*, 281, *402*, 405
Andesine, *44*, *47*, **60**
Andesite, *146*, 221
Andradite, *47*, **65**
Anglesite, *519*, 522
Anhydrite, 34, *45*, *47*, **60**, *60*
 nodular, **492**, *492*
 recrystallized, *317*
Anisotropic minerals, **12**, 21, 36
Ankerite, *43*
Anorthite, *44*, *47*, **60**
Anorthoclase, *44*, *47*, **61**
Anorthosite, *145*, *147*
Anthophyllite, *45*, **47**, **61**, *61*
Antidune, *441*, *441*
Antigorite, *44*, *47*, **61**
Antiperthite, *128*
Antirapakivi mantling, *245*, **250**
Antlerite, *519*
Apatite, *47*, *73*, **74**
Aphyric, 221
Aplite, **226**, *227*

575

Aplite-pegmatite, 391, *391*
Aplitic granite, *393*
Apparent textural relations, *132, 133, 134, 135, 137*
Aqueous phase, *164, 180, 182, 184*
Aragonite, *45, 47,* **61,** *61*
Arapien Formation, *452*
Arenite, 151, *152,* 154, 381, 461
Arfevdsonite, *47,* **61**
Argentite, *401, 408*
Argillite, *149, 152*
Arkose, **444**
Ash, 373, **374,** 375, 377, 379
Ash cloud, 375
Ash fall, 381–382
Asphaltite, *155*
Assimilation, **261,** *262–270, 350*
  scale of, 271–272
Augen, *303*
Augite, *45, 47,* **61,** *61, 116, 120*
Augitite, *147*
Authigenic, **434,** *473, 475*
Autobreccia, *348*
Autolith, **347,** *383*
Automated image analysis, **96,** 97
Avalanching, 206, 376–377
Azurite, *47*

Back-diking, *244*
Baddeleyite, 329
Bagnold effect, 202
Ball and pillow structure, **438**
Ballistic flight, 329
Banded iron formation, **436,** 502
Banded ironstone-formation, **436**
Bar, 433, **440,** 443, 450
Barite, *44, 47,* 61
Barkevikite, *47,* 228, *231*
Barre Granite, *230*
Barrier bar, **440**
Basalt, *146,* 184, *185–186,* 222–223, *222–223*
  weathering of, **505**
Basanite, *146,* 367
Base surge, *374*
Bauxite, **73,** *504,* **510,** *513*
Beach deposit, 450
Beachrock, *450*
Becke Line, **8,** *8,* **9**
Bedding, 203, 206, 439
  graded, **206,** 440
  imbrication of, **440**
  planar, **439**
Belemnoid, *466*
Bentonite, **73**
Bernoulli Lift Force, **205,** 206
Berthierine, *62,* **438**
Bertrand Lens, **38**
Biaxial minerals. *See* Crystals, biaxial
Binary eutectic, *175*
Bioclastite, 477
Bioherm, **457, 462**
Bioliths, classification of, *155*
Biomicrite, *154, 467–468*

Biomineralization, 473, 475
Biopelites, classification of, *155*
Biosedimentary rocks, **454**
Biostrome, **457, 462**
Biota, 454–455, 475, 477
Biotite, *19, 45, 47,* **62,** *139*
Biotite mantled alkali feldspar/K-feldspar, *246, 249,* **250**
Bioturbation, **462**
Birefraction, *23*
Birefringence, *32*
  determination of, 31
  maximum, **26,** 33
  presenting, **26,** 31, 33
Bishop Tuff, *376*
Bitterns, **482**
Bitumens, 73
Black smoker, *412,* **423**
Blackhawk Formation, 457
Blade-shaped biotite, *247,* **253,** *254*
Blue Gate Shale, *451*
Blue-green algae, *457*
Blueschist, *149*
Boehmite, *48,* 73
Bonitite, *146*
Bonnaza King Formation, *322*
Borax, *48*
Bornite, *519*
Botryodial morphology, 399
Boudinage, 340
Boundstone, *153,* 462
Bowen's reaction series, 226
Bowlingite, *48,* **73**
Boxy cellular plagioclase, **256,** *256*
Brachiopod, **467,** *468*
Breccia
  collapse, 401
  crackle, *385–386,* **401**
  crush, *150,* 319
  fault, *150,* 319, *321*
  glassy, 329, *330*
  gouge, 319
  hydrothermal, **383,** *385*
  lag, 345
  lapilli-tuff, *148*
  magmatic, 345
  micro, *320*
  monolithic, *321*
  mosaic, **401**
  pyroclastic, *148*
  sedimentary, classification of, *153*
  stockwork, *386,* 401
Breccia pipe, **384**
Brine, **481,** *484,* 487
  hydroscopic, *492*
Brittle deformation, **213**
Brittle flow, 213
Brittle fracture, **333**
Brittle-ductile behavior, **213**
Bronzite, *48,* 63
Bronzitite, *147*
Brookite, *48,* 74
Brucite, *44,* 48
Bryozoa, *468,* **469**
Buoyancy, 203, 205, 208, 235, 364
Burial metamorphism, **337**

Bushveld Complex, *288*
Bytownite, *44, 48,* **62**

c surface, *300*
$CaCO_3$-H_2O system, *169*
Calamine, *521*
Calamity Peak pluton, *394*
Calcareous ooids, *473*
Calcareous ooze, *471*
Calcite, 13, *14, 45, 48,* **62,** 117
Calcitization (calcification), **475**
Calcrete, 504, 508, **515**
  paleo, 515
  penetrative, **515**
Caliche, **508,** 515
Callianassa major, 465
Camptonite, *146*
Canaan Peak Formation, *434*
Cancrinite, *48*
Cape Ann Granite, 230, *232*
Capillary action, 202, 515
Carbon, *48,* **73**
Carbonaceous sediments, classification of, *155*
Carbonate compensation depth CCD, **480**
Carbonate sedimentary rocks, classification of, *153–154*
Carbonate-siliciclastic rocks, classification of, *154*
Carbonatite, *145*
Carlin-type ore deposit, *401, 408,* **419**
Carnallite, *48*
Carnotite, 519
$CaSO_4$-H_2O System, **170**
Cassiterite, *48,* **74**
Cataclasite, **151, 319,** *322*
  classification of, *150–151*
Cataclastic flow, **333**
Cathedral Peak Porphyritic Granite, 230, *232*
Cathode luminescence, **96**
Caustobiolith, *155*
Cave pearl, 516
Celadonite, *48,* 73, 74
Celestite, 61
Cellular relation, *142*
Cement, **433,** 471, 473, 475–476, 479
  algae, 457
  meniscus, **479**
  pendulous, **479**
  replacement, *434*
Cement texture, **443**
Cementation, 433
Central illumination, **9**
Cephalopod, **466,** *467*
Cerussite, *48, 519,* 522
Chabazite, *48,* 74
Chalcanthite, *519*
Chalcedony, *48,* **73,** *400*
Chalcocite, 520
Chalcopyrite, *48,* **72,** *519*
Chalk, **461,** *462*
Chamosite, 49, 62, **438**

Channel scar, **441**
Charlevoix Structure, *328*
Charnockite, *145*
Chelation, **504**
Chemical accumulation in magma, *238*
Chemical weathering, **503**
Chert, *49*, **73**, *454*
Chevron halite, **487**, *488*
Chiastolite, **60**, *307*
Chicken-wire structure, **492**, *492*
Chill margins, *228*
Chinle Formation, *448*
Chloraluminate, *427*
Chlorite, *43–44, 49*, **62**, *73*
Chloritoid, *44, 49, 62*, **62**, *307, 310*
Chlorophaeite, *49*, **73**
Chloroplast, *456*
Chondrodite, *49*
Chromite, *49*, **72**
Chrysocolla, *49*, **73**, *520, 519–520*
Chrysotile, *49*
Cinnabar, **72**
Claron Formation, *479*
Clastic dike, **340**
Clastic sedimentary rocks, **431**
    classification of, *151–154*
Clathrate compound, **315**
Clay, *432, 436, 439*, **441**, *444–446, 448, 453*
    flocculation of, **439**
    grain size of, *432, 452*
Clay minerals, *11*, **73**, *431–432*
    orientation of, *8*
Cleavage, *6, 10–11, 18–19, 21, 25–26, 41*
    continuous, *339*
    crack-like disjunctive, *338*
    crenulation, *292, 300, 335, 338*
    differentiated zonal, *338*
    dissolution, *338*
    spaced, *338*
    slaty, *338*
    stylolitic, *339*
Cliff House Sandstone, *134*
Climate as a factor in weathering, *507–510, 512, 515, 520*
Clinoamphibole, *43, 49*
Clinochlore, *49, 62*
Clinoptilolite, *49, 74*
Clinopyroxene, *26, 49, 61*
Clinopyroxenite, *147*
Clinozoisite, *44, 49*, **62**, *63, 282*
Clinton Formation, *437*
CO_2-H_2O system, **168**
Coal, *73, 155*
    anthracite, **477**
    bituminous, *457*, **477**
    bone, *155*
    rank of, **477**
    sub-bituminous, *457*
Coal maceral, **455**
Coal Valley Formation, *456*
Coccolithophore (Coccolith), **461–462**
Cockade structure, **400**, *400*
Coconino Sandstone, *328*
Coesite, **24**, *49, 328, 325, 329*

Cohansey Quartz Sand Formation, *510*
Colemanite, *45, 49*, **62**, *63, 496*
Colloform morphology, *399–400*
Colloid, **165**
Collophane, *49, 73*
Color fringes, **10**
Columbia River Basalt, *223*
Comb structure, *217, 399*
Comb-layering, *129*, **239**, *239*
Comendite, *146*
Comminution, **333**
Common ion effect, **166**, *500*
Compaction, *433*
Complex ion, *397*
Components in phase equilibria, *174–176*
Composition plane, *77, 88*
Compositional gradient, **194**
Comstock Lode, Nevada, *492, 408*, **408**, *418*
Condensing lens, *37*
Conglomerate, *153*
    classification of, *153*
    oligomict, **444**
    ortho, **444**
    para, **444**
    petromict, **444**
Congruent dissolution, **199**
Connate water, *171*, **172**
Conodont, **470**, *470*
Constitutional undercooling, **193**
Continental shelf, *452*
Continental slope, *452*
Convection, **236**
Conway Granite, *230, 232*
Cooling cracks, *340*
Cooling unit, *379*
Coon-tail ore, *421*
Coquina, *467, 477, 516*
Coral, **462**, *463*
Coral reef, *454*, **462**
Cordierite, *44, 49*, **62**, *120, 126, 288, 360, 362, 367*
Coring, **195**, *262*
Cornet halite, **487**, *488*
Corrosion surfaces, **491**
Cortex, *474, 475–476*
Corundum, *44, 50*, **62**, *181*
Cotectic line, **184**, *186, 188–189*
Cotectic surface, *187*
Cottenwood Canyon Member of Madison Limestone, *470*
Covellite, *519*
Cow Head Group, *463*
Crack-seal veins, *340, 401*
Creep, *161, 163, 206*
    aseismic, *320*
    dislocation, **209**
    recrystallization-accommodated dislocation, **209**
Crescummulate, *241*
Cretaceous-Tertiary boundary, *326*
Cristobalite, *50*, **72**, *168, 177–178, 187*
Critical point of water, **164**, *165*
Critical radius, *192*
Cross-bedding, *207*, **439**, *440, 448*
    festooned, *445*

hummocky, *207*
in magmas, *237*
Cross-laminar shell structure, *467*
Cross-lamination, **439**
Cross-stratification, *206–207, 447*
Crossed polars (X-polars), **21**
Crossite, *44, 50*, **62**, *64*
Crush microbreccia, *319*
Crustacean, **465**
Cryptocrystalline aragonite, *459*
Cryptoperthite, *11*
Crystal dispersive pressure, **208**
Crystal dissolution, **198**
Crystal grain, **108**
Crystal growth, **192**, *194*
    continuous, **193**
    layer spreading, **193**
    open space, *403, 404*
    potential space, *403, 404*
    rates of, *193*
Crystal growth morphology, *197*
Crystal integrity. See Grain integrity
Crystal morphology, *114*
    anhedral, **111**, *111, 114*
    astructural, **111**, *114–115*
    boxy cellular, *195, 198, 249, 256, 256*
    cellular, **198**
    dendritic, *196–198, 190*
    euhedral, **111**, *111*
    eustructural, **111**, *112–115*
    horsetail, *196*
    skeletal, *198, 222*
    spongy, *198, 201, 201*
    subhedral, **111**, *111*
    substructural, **111**, *114–115*
Crystal mush, *203*
Crystal nucleation, **190–191**, *192*
Crystal plasticity, **214**
Crystal settling, *207*
Crystal shape. See Crystal morphology
Crystal size distributions, **97**
Crystal twinning. See Twins
Crystal zoning, **113**, *116–117*
    continuous, **115**, *117*
    normal progressive, *116–117, 179*
    normal progressive oscillatory, *117*
    oscillatory, **113**, *116–117*, **179**, *194*
    patchy, **115**, *117*
    plagioclase, *194, 195*
    progressive, *94*, **113**, *117*
    sector, *62*, **115**, *116–117*
    step, **113**, *116–117, 255, 256*
Crystal zoning curve, **94**, *117*
Crystallization force, **297**
Crystals
    ansiotropic, **12**
    biaxial, **14**, *41*
    isotropic, **12**
    unixial, **14**, *26, 41*
Crystallite, *190, 222, 222*
Crystallization, *161*
    force, *297*
    fractional, **235**, *239*
    incongruent, *176*
    latent heat of, *191*
    path of, *176*
    three-stage, *191*

Crystallization (*continued*)
 two-stage, **191**, *224*
Crystallization theory, **192**
Crystallographic orientation analysis, **99**
Crystallographic preferred orientation, **99**, 127, 129, 215, 217
Cummingtonite, *45, 50,* **62**, *64*
Cumulate halite, *487*
Cumulate magmatic rocks, **235**
Cuprite, *50*
Curtis Formation, *320*
Cyanobacteria, **457**

Dacite, *146*
Dalradian, Scotland, *290*
Debris flow, 206–207
Décollement, 318, 340
Dedolomitization, **476**
Deep water basin, 485
Deformation
 brittle, **210**, *210, 212,* **213**, *214,* **333**
 ductile, 210, **213**, *212, 214*
 low-temperature, **332**
 rock, 210
Deformation bands, 115, 214, *214*
Deformation lamellae, 214, *216*
Deformation mechanisms, 332, *333*
Deformation twins, 214
Delta, **450**
Denay Formation, *468,* **522**
Density inversion, 203, 207
Depolymerization, 180
Depositional environment
 abyssal plain, *478*
 bahada, 445
 barrier island, 450
 beach, 441, *478*
 continental shelf, 451, *478*
 continental slope, *478*
 delta, **450**
 eolian, 206, **445**
 fluvial, 440, **445**, 446
 lacustrine, **477**
 lagoonal, 485
 marginal marine, **450**, 477
 marine, **477**
 open marine, **477**
 passive continental margin, 451
 playa, 449, 485
 river, **446**
 salina, **485–486**
 subaerial, 206, **445**
 subaqueous, **206**
 subglacial, **445**
 sublacustrine, 450
 swamp, 454, 477, 479
 tidal flat, **451**
Depositional structure, **439**
Depth hoar, 313
Desiccation cracks, 340
Desilication, 266, *267*
Detachment kinetics, 199
Detrital, 434
Deviatoric stress, 209, **210**

Devitrification, **230**, 234
 textures of, *233, 348*
Diabase, *145,* 226, **228**, *228*
Diagenesis, *171,* 172, **433, 473**
 of calareous rocks, **475**
 of carbonaceous rocks, **477**
 of siliceous rocks, **477**
Diamictite, **444**
Diamond, *325,* 326
Diapir, **203**, 357
 migmatitic, 361
Diaplectic glass, 328
Diaspore, *50,* 73
Diatom, **460**, *461*
Diatomite, **461**, *461*
Diatreme, **203, 384**
Dictyonema flabelliforme graptolite, *470*
Digenite, *519*
Diffusion, 237, 239–240
 chemical, 235, 237
 in crystals, 170, *263*
 double, 236
 in liquids, 179, 192, 237, *263*
Diffusion creep, 214
Diffusion rate
 in fluids, 193
 in silicate melt, 237, 239
 in solids, 194–195
Diffusion-supersaturation, **179, 194**
Diffusional mass transfer, **335**
Dike propagation, 203
Dilational pumping, **202**, 208
Dimensional orientation, **127**, *129,* 276, 290
Dinosaurs, 324
Diopside, *45, 50,* **63**, *64,* 283
Diopside-anorthite system, 175–176, *175*
Diorite, *145*
Direct melting, 262
Disharmonic folding, 338, *338*
Dislocation creep, **209**, 213, *214*
 recrystallization-accommodated, 209
Dislocation glide flow, *214*
Dislocations, 209, 211, *215*
Dispersion, **36**, 38
Displacive growth, **489**, *490*
Disseminated gold deposits, 519
Dissemination, 416
Dissolution, 161, 169, 175, **198**, 489, *491, 501*
 congruent, **199**
 incongruent, **199**
Dissolution cleavage, 338
Dissolution crystal morphology, **199**
Dissolution-crystallization, *138*
Dissolution melting, **199**, 261
Dissolution-reprecipitation, 261
Dissolution texture, 180
Dolerite, *145*
Dolocrete, **517**
Dolomite, *45, 50,* **63**, 170, 283
Dolomitization, 170, **475–476**
Dolostone. *See* Dolomite
Double refraction, **13**, *14,* 37
Drag. *See* Viscous drag
Dravite, 72

Dripstone, **505**
Duchesne River Formation, *447*
Ductile deformation, 210, *212,* 318, *318,* 334
Ductile flow, 213
Ductile strain, **213**
Dumortierite, *50*
Dune, **439**
Dunite, *147*
Duracrust, 515
Dynamomagmatic texture, 353

Earthworm, **462**
Ebb tide, 451
Echinoderm, **469–470**
Eckermanite, 61
Eclogite, 210, *210*
 facies, *286*
Eh, **171**, *172*
Eh-pH diagram, *504*
Elastic limit, 212, *212*
Elastic strain, 212, *213*
Elbait, 72
Electrolyte solutions, 199
Elevated pressure, effect on phase equilibria, **180**
Elutriation, 236
Ely Limestone, *454*
Embryo solid, 192, *192*
Emery Sandstone, *451*
Empirical observation, **131**
Enclaves, **243**, 257–258
Enderbite, *145*
Endicott Diorite, *349*
Endolithic borings, **459**
Enstatite, *24, 44, 50,* **63**, *64*
Enterolithic structure, **492**
Entrada Sandstone, *452*
Environment of deposition. *See* Depositional environment
Enzymes, as catalysts, 167
Eolian, 206
Eolian Sandstone, *436*
Epiclastic volcanic rocks, *380,* **380–381**
Epicontinental seaway, **451**
Epidote, *45, 50,* **63**, *65, 74*
Epithermal system, 409, *412,* **418**, *420–421,* 422
Epithermal vein, 408, 419
Epoxy, *7, 22*
Equal area stereonet, *102*
Equigranular, **109**, *110*
Equilibrium, **278**
 in aqueous solutions, 165
 invariant, 279
 thermodynamic, 282
 univariant, **278**
Equilibrium constant, 165–166
Equilibrium crystallization, 194–195
Equilibrium curve, 167
Erg, **445**
Erosion
 agents of, 445
 fluvial, 206 441, 447
 wind, 445

Erosion channel, 441
Erosion structures, **441**
Erosion surface, 447
Eruption column, **374–375**, *375*
Esker, 445
Essexite, *145*
Etching, 96
Eureka Quartzite, *322*
Eutaxitic structure, **379**
Eutectic, *175*
   binary, 175, *175*
   pseudo ternary, 186
   ternary, 186, 188
Eutectoid, **121**
Euxenic marine, *504*
Evaporation, 313
Evaporite minerals, *483*
Evaporite sedimentary rocks, 155, **481**
Exhalative system, *412,* **420**
Explosion breccia, 203, *383*
Explosive fragmentation, 203–204
Exsolution, *128, 179,* 180
Extinction, *15, 17,* **21***, 28, 30*
   acute, 93–94
   inclined, **25**
   obtuse, 93–94
   negative, **93**
   parallel, **25**
   positive, **93**
   progressive, **115**
   symmetrical, **25**
Extinction angle, **26**
Extinction bands, 29, *29,* 30
Extinction cross, **25**, *25, 39,* 40
Extinction of species, 324
Extinction surfaces, 29–30
Extraterrestrial objects, 324

Fabric, **126–127**, *129,* 144
   linear, 128
   planar, 128
Fairbairn lamellae, 214
Fanglomerate, *446*
Fast wave, **26**
Fault, 339
   aseismic, **213**
   brittle, *151,* 214
   ductile, *151,* 214
   hydroplastic, 335
   seismic, 211, 320
Fault breccia, 203, *321*
Fault gouge, *150*
Fayalite, *50*
Fecal pellets, 455, 462, 465, 471
Feldspar, *50*
Feldspathoid, 189
Fenestrate, **469**
Ferricrete, 504, **509**, *510–511*
Ferrimolybdite, 521
Ferroactinolite, **43**
Ferron Sandstone, *451*
Ferrosilite, **66**
Ferruginous laterite, *504,* **509**
Ferruginization, **476–477**
Fiamme, *376,* **379**

Fibrolite, *50*
Filter pressing, 203, *208,* 236
Firn, **314**
Fish, **471**, *471*
Fish bone, *471*
Fish scale, *471*
Fissility, **439**
Flagella, **456**, 462
Flagstaff Limestone, 466, *466*
Flame structure, **438**
Flattening, 295, 297–298
Flint, **73**
Flocculation, 439
Flood plain, 443–448
Flood tide, 451
Flow
   ash, 206
   cataclastic, **333**
   channel, 446
   current, **207**, 431
   debris, **206–207**, 445
   differentiation, 208
   directional, 439–440
   ductile, 213
   ebb, **207**
   flood, **207**
   fluid, **204**
   fluvial, **207**
   grain, 206, 440
   gravity, **206**
   homogenous, 341
   laminar, 204
   oscillatory, **204**
   pyroclastic, 377–379, 381
   sand, 206
   streamline, 204
   tidal, 207
   turbidity, **207**, 431, 441
   turbulent, **204**
Flow alignment, *222*
Flow regime, **440**, *441*
Flow velocity, 205, 440, 448, 450
Flowstone, 516
Fluid
   Newtonian, 205, 236
   non-Newtonian, 205
Fluid inclusion, 397
Fluid relocation textures, **353–354**, *354–356*
Fluidization, 204
Fluorapatite, *44, 50,* **63**
Fluorite, *11, 44, 50,* **65**, *421, 427*
Flutecasts, **441**
Fluvial conglomerate, *447*
Fluvial process, 431, 441, 446–447
Fluvial Sandstone, *447–448*, 444
Fluvial siltstone, *448*
Fluvial system, 433, 449–450
Foam structure, 217
Foidite, *146*
Foidolilite, *145*
Foidolite, *145*
Fold-thrust belt, 332
Folds, 337
   disharmonic, 338
   flexural slip, 337
   isoclinal, *317,* 318

   kink, **338**
   recumbent, *317*
Foliation, 298
   shear band, 296, *300*
Foraminifera, **460**, *460*
Force, 204
Forceful injection, **370**
Foreland basin, 451
Forsterite, *50*
Fossil flora, *456*
Fossilized wood, *456*
Fountain Formation, *443*
Foyaite, 189
Fractional crystallization, **207**, **235**
Fracture, 339
   extension, 339
   propagation, 203
Fragmentation, brittle, **345**
Fragments
   juvenile, 374
   lithic, 374, 444
   vitric, 374
Framestone, 462
Franciscan Formation, *332*
Francolite, **455**, 470, 477
Free energy
   surface, *192*
   volum, *192*
Friable, **433**
Frictional heat, **211**
Frictional strength, 205
Frosted grains, **445**
Frustule, **460**, 461
Fumarole, **424**
   rootless, **424**
   solfataric, **425**
Fumarole exhalation, *424*
Fumarole incrustation, 425
Fumarole mound, **424**
Fumarole sublimate, *427*
Fumarolic vent, *425*
Fumarolite, **424**, *426*
Fundamental constituents or materials of rocks, 1, **107**, *157*
Fundamental rock-forming processes, 1, 159
   classification of, **161**
Furrow, *441*
Fusinite, **455**, *457*, 477
Fusulinid foraminifera, **460**, *460*

Gabbro, *145*, 184
Gabbronorite, 147
Galena, *50,* **73**
Garnet, *44, 51,* **65**, *120*
Garnet amphibolite, *365*
Garnet granulite, *210*
Garnierite, *51,* **73**
Gaseous exhalites, 424
Gastropod, *465, 466*
Gedrite, 61
Gel, **165**
General rock-forming processes, 1
Geode, 518, *518*
Geothermal gradient, **211**

Geothermal system, *412*
Geothermal water, *171,* 173, 396
Geyserite, **420**
Gibbsite, **51**, 73
Gile Mountain Formation, 275, 306
Gillarch pharyngeal, *471*
Glacial ice, 217, 312
Glacier, 217
Glaciofluvial system, 445
Glass, *51*
   devitrification of, **230**
   diaplectic, 328
   non-porphyritic, 221
   porphyritic, 222, *222*
   refractive index of, 7, *10,* 73
   shock, 73
   thetomorphic, 328
   volcanic, 73
Glass breccia, 329, *330*
Glass shards, 374, *377*
Glass transformation temperature, **230**
Glauconite, *51,* **73**
Glaucophane, *45, 51,* **65**, *65*
Glide, intracrystalline, *291*
Gneiss, *150*
   dynamomagmatic, **351**, *351–352*
   granitic, *149*
   homogenous, 351, *352*
   lit-par-lit, *368*, *368*
   pencil, *303*
   quartzo-feldspathic, *149*
Goethite, *51,* 74
Gold, invisible, 387
Goniatites, 467
Gornergletscher, *312*
Goshvicthys Parvis knightia, 471
Gossan, **508**, *519*
Graded bedding, 206, 440, *440,*
   in magma, 237
Grain
   as constituents of rocks, **108**
   crystal, **108**
   discontinuities in, **115**, *118*
   lithic, 431
   mineral, **143**
   non-mineral, **143**
   recycled, 433
   rounded, 434
Grain boundary sliding, 214
Grain cracking, 319, 333
Grain flow, 206
Grain integrity, **111**, **112**, *112,* 198
Grain morphology, **111**, *111–114.* See
   also Crystal morphology
Grain recycling, 433
Grain rotation, 334
Grain rounding, 434–435
Grain settling, 440
Grain shape. *See* Grain morphology
Grain size, **108–109**, *110,* 144
Grain size sorting, 117, **118–119**, *121,*
   126
Grain size variation, 221
Grainstone, *153*, 465–466, *466*
Granite, *145,* 182, 185, 187
Granite de la Margeride, *237*
Granite du Velay, *367*

Granite of Lac Du Bonnet, *354*
Granite porphyry, 191
Granite system, *181,* 187
Granodiorite, *145,* 187
Granofels, *149,* **150**
Granophyre, **121**, 234
Granular
   allotriomorphic, 113
   equigranular, **109**, *110*
   hypidiomorphic, 113
   inequigranular, **109–110**
   panidiomorphic, 113
   serriate inequigranular, **109**, *110*
Granular texture. *See* Texture, granular
Granulation, **333**
Granule ironstone-formation, **436**
Granulite, 301
Graphic texture, *392*
Graphic-vermicular intergrowth, *392*
Graphicgrain, *124*
Graphite, *51,* **73**
Graptolite, **470**, *470*
Gravel, *442*
Gravitational force, 204
Gravitational instability, **207**
Gravity flow, 206
Gravity force, 204
Gravity settling, 205, 235
Graywacke, 357, **444**
Great Salt Lake, *482*
Green River Formation, *449,* 456
Greenalite, 437–438
Greenschist, *149*
Greenstone, *149*
Grossularite, *51,* **65**, 289
Ground surge deposit, **378**
Groundwater, 171, *171*
Grunerite, 63, 438
Grus, **506**, *507*
Gypcrete, **517**
Gypsrudite, 497, *497*
Gypsum, *44, 51,* **65**, *66,* 170, *519*
   solubility in H_2O, *170*

H_2O, in the geologic environment, *171*
Half Dome Granodiorite, 230, *232*
Halimeda, 457, *458*
Halite, *44, 51,* **66**, 316, 427, *427*
   hopper, 487
   solubility of, 166, *167*
Halite-H_2O system, 166, *167*
Halite raft, **487**, *488*
Harrisite, 241
Harrisitic layering, *147,* 239
Harzburgite, *147*
Hastingsite, *51,* 66
Haughton Structure, *327*
Hauynite (hauyne), *44, 51,* **66**
Hawaiite, *146*
Hedenbergite, *45, 51,* **66**
Hematite, *51*
   earthy, 74
   specular, *43*
Hemimorphite, *51,* 521
Herringbone pattern, 487, *488*

Heteradcumulate, **240**
Heulandite, *51,* 74
Homogeneous magmatic rocks, 221
Hopper crystal, **193**
   bismuth, *193*
   halite, 487
Hornblende, *26, 45, 51,* **66**, *66,* 120
Hornblende-biotite zones in alkali
   feldspar, *251,* 246
Hornblende-mantled biotite, **254**, *247,*
   254
Hornblende-mantled quartz, **251**, *246,*
   252
Hornblendite, 147
Hornfels, *149,* **150**, 305
Horsetail crystal morphology, *196*
Hot spring, 396
Howardite, 329, **331**, *331*
Humulith, *155*
Humus, 509–510
Hunter Creek Formation, *456*
Hyaloclasite, **378**
Hybrid magmatic rocks, 261
Hybrogenic biotite, **254**
Hydraulic fracturing, **202**, 334
Hydraulic weakening, 334
Hydrobiotite, **74**
Hydrocarbon, 73
Hydrogrossular, *52,* **65**
Hydrohalite, **167**
Hydrolysis, **199**
Hydrolytic weaking, **213**, *214*
Hydromuscovite, **68**
Hydroscopic brine, **492**
Hydrothermal alteration, 401, 408
Hydrothermal breccia, **383**, *385*
Hydrothermal explosion, *384*
Hydrothermal precipitation, *399*
Hydrothermal solution, 396–398
Hydrothermal system, **409**, *409–411,*
   boiling, *419*
Hydrothermal water, *171,* 173
Hydrozincite, *519*
Hypersolvus Granite, 230, *232*
Hypersthene, *44, 52,* **66**, *67*
Hypidiomorphic granular, **113**
Hypothermal, 409

Ice, *52*
   bubbly, 313, *314–315*
   glacial, 312
   polygonal, 315, *315–316*
   sieve, 313, *314*
   spongy cellular, 313, *313*
Ichnofossil, **462**
Idaite, *126*
Iddingsite, *52, 68,* **74**
Idocrase, *44, 52*
Igneous rocks, *171,* 173
   classification of, *145–148*
   layered, *235,* **240**
   non-porphyritic, 221, *223, 226, 230*
   porphyritic, 221–222, *228*
Ignimbrite, **376**
Ijolite, *145*

Ikaite, **498**
Illite, *52*, 73
Illumination, *15*, *17*, 21, 28
   maximum, *24–25*, 30
   partial, *30*
Ilmenite, *52*, **73**, 74
Image analysis, **96**
Image analyzer, *2*
Impact crater, *324*, 326
Impact melt, 324
Impactite, **324**
   extraterrestrial, 329
Including texture, **119**, *124–125*
Inclusions, *345–346*, 349
   cognate, **347**, *349*
   micro, 258, *258*
   mini, 258, *258*
   xenolithic, **261**, *347*, 350, *350*
Incongruent crystallization freezing, **177**
Incongruent direct melting, 201
Incongruent dissolution, **199**
Incongruent dissolution Rapakivi mantling, **249**, *245*, *249*, 250
Incorporating growth, **489**, *490*
Interface attachment kinetics, 193
Interference colors, *31–32*, 35
   anomalous, **36**
   orders of, *31–32*
   presenting, **31**
Interference figures, **37**, 38, *37–39*, 40
Intertidal zone, 450, 497
Intraclast, 153, *154*, 471
Intraclastite, 477
Intramicrite, *154*
Intrasparite, *154*
Intratelluric crystals, **375**
Invertebrate exoskeletons, 455
Ionic strength, **166**
Iron crust, 509
Iron formation, *437*
Ironstone, 155, *437*
   oolitic, **438**
Isogyre, **37**
Isotropic minerals, **12**
IUGS igneous rock classification, 144, 147, *145–147*

Jadeite, *45*, *52*, **66**, *67*
Jarosite, *45*, *52*, **74**
Jasper, **73**
Jasperoid, *421*
Joint, 339
Jotunite, *145*
Juvenile water, *171*, 173

K-feldspar, *52*, 177
Kaersutite, *52*, 69
Kamacite, **331**, *331*
Kame, 445
Kaolinite, *52*, *73*, 405, 407, 427
Karst bauxite, **508**
Karst terrain, **508**

Karst topography, 508
Karstification, 516
Karsting, **401**
Kayenta Formation, *443*, *448–449*
KCl/HCl ratio, *405*, 407
Kenyte, 222, *222*
Kernite, *52*
Kerogen, 73, **477**
Key Largo Limestone, *458*
Keyhole Canyon Granite, 227, *230*
Kimberlite, *146*, 383
Kinetic sieving, **440**
Kingdom Animalia, **462**
Kingdom Fungi
Kingdom Monera, **456**
Kingdom Planta
Kingdom Protista, **460**
Kinkbands, **115**, 214, *216*
Komatiite, *146*, *196*, 241
Kyanite, *11*, *44*, *52*, **66**, *67*, 210–211, *210*, *216*, 278–279, *280*

Labradorite, *44*, *52*, **66**
Lagoon, 485
Lahar, **382**
Lamellar relation, *125*
Lamprophyre, *146*
Lane Monzonite, *383*
Lapilli, *148*, **374–375**
Lapillistone, *148*
Larnite, *52*
Larvikite, *145*
Late magmatic, **389**
Latent heat of crystallization, 191
Lateral secretion, 419
Lateral-vertical secretion deposits, **418**
Laterite
   aluminous, **508**, 510
   ferrugineous, **509**, 510–511
   nickel, **512**
Latite, *146*
Laumontite, *52*, 74
Lava
   aa, *347*
   phoehoe, 222, *222*
Lava flow, 223, 226
Lawsonite, *45*, *53*, *63*, **67**, *68*, *289*, 289
Layering
   comb, 239, *239*
   cumulate, 237
   orbicular, 239, *239*
   symmetrical, **399**
Leached capping, **519**
Lechatelierite, *325*, 326, 329
Lee side, **206**, 440
Leinster batholith, 230
Length fast, **34**
Length slow, **34**
Lens
   Bertrand, *2*
   condensing, *38*
   objective, *2*, *9*, *22*, *37*, *38*
   ocular, *2*, *9*, *22*
Lepidochrosite, *53*, 74
Lepidolite, *53*

Lepisphere, *518*
Leucite, *44*, *53*, **67**, *177*, *179*, *187*, 189
Leucite-silica system, **176**, *177*
Leucocratic, 226
Leucosome, **361**
Leucoxene, *53*, 73
Lherzolite, *147*
Liesegang rings, **504**, *522*
Light
   monochromatic, *27–29*
   sodium, **29**, *29*, 30
   white, *30*, *31*
Lignite, *457*
Lignitic mud rock, 455
Limburgite, *146*
Limestone, *454*, *455*, *463*, *467*, *473*, *476*
   classification of, 73, *153*
   corraline, 462
   fossiliferous, 454
   mylonitic, *340*
   oolitic, 477, 479
   organic, **477**
   shelf, 480
   siliceous, 475
Limonite, *53*
Liquefaction, **206**, 207
Liquid immiscibility, 235
Liquidus, *175*
Lisagang rings, *129*
Litharenite, *151*
Lithic fragment, 374
Lithification, 431, 433, 445
Lithons, **321**
Lithophysae, **233–234**, *233*, 410
Lithosphere, 357
Lizardite, 61
Loadcast, **438**
Loess, **445**
Low-temperature deformation regime, **341**

Maceral, **455**
Mafic igneous rocks. *See* Igneous rocks
Magma
   crystallization of, 230, 233
   felsic, 227
   mafic, 226
   peraluminous, 359
   replenishment of, 208
   solubility of H_2O in, 180, 182, *182–183*
Magma bombs, **374**
Magma chamber, 207
Magma droplets, **375**
Magma fracturing, 345
Magma lumps, **375**
Magma mingling, **242–243**, *243*
Magma mixing, **242**
Magmatic
   early, **389**
   late, **389**
   post, **389**
Magmatic breccia, 345, *346–347*
Magmatic convection, 208
Magmatic crystallization, 174

Magmatic layering, *129*
Magmatic rocks. *See* Igneous rocks
Magmatic stoping, **203**
Magmatic water, 173
Magnesite, *45, 53,* **67**
Magnetite, *53,* **73–74**
Malachite, *53*
Malignite, *145*
Mancos Shale Formation, *452*
Manganese deposits, 500
Manganite, *53,* 74
Mangerite, *145*
Manicouagan Structure, *329*
Mantle texture, **121, 126,** *125*
Marble, *149,* **150**
Margarite, *45, 53,* **67**
Marialite, **71**
Marl, *436, 444, 450, 461, 477*
Maskelynite, *329*
Matrix, **95**
  of igneous rocks, 222–226, *228*
  of sedimentary rocks, 444
McLaughlin gold mine, *422*
McNutt Potash Zone, Salado Formation, *490*
Mechanical accumulation of crystals, 235–237, *238*
Mechanical energy, 210
Mechanical fragmentation, 153
Mechanical interaction, 163
Mechanical stage, 95
Meionite, **71**
Mélange, 318, *335, 452*
Melanosome, **361,** *363*
Melilite, *44, 53,* **67**
Melilitolite, *147*
Melt extraction, *363,* 364–365
Melt solutions, **161**
Melting, 163
  adiabatic, **357**
  direct, 198, **200**
  friction, 357
  impact, 357
  incongruent, 176, 201
  incongruent direct, 201
Menefee Formation, *457*
Meniscus cement, *516*
Merapi Volcano, *427*
Merced Formation, *467*
Merwinite, *53*
Mesothermal, *409,* **418**
Metalliferous brines, 423
Metamorphic differentiation, **276**
Metamorphic facies, **283, 285,** *286*
Metamorphic facies series, *285–286*
Metamorphic grade, 280, **283, 285**
Metamorphic history, **298,** *305–310*
Metamorphic mineral assemblage, 282
Metamorphic mineral reactions, **278,** 280, *284–285*
Metamorphic regime, **275**
Metamorphic rocks, 275
  classification of, *149*
Metamorphic tectonites, 298
Metamorphic water, *171,* 173
Metamorphism
  Barovian, *286*
  burial, *308*
  contact, *286*
  dynamothermal, 275, *293–294,* 304–307, *309*
  field of, *276*
  isochemical, 276
  lower limit of, **275**
  metasomatic, **276,** *298*
  monocyclic, **304**
  polycyclic, 303, *304*
  progressive, 298
  protoliths of, **276,** *277*
  retrogressive, **303**
  shock, **324**
  static thermal, 275, *289,* 304–307
  thermal, *289–290, 294*
  upper limit of, **275**
Metarhyolite, *301*
Metastability, 170, 278
Metatuff, *308*
Meteor (Barringer) Crater, *324,* 328
Meteor impact, 204, 326
Meteorite, *324,* **331**
Method of multiple working hypotheses, 131
Miarolitic cavity, 227, 230, *394,* 410
Mica beards, **339**
Mica fish, *300*
Mica glimmer plate, **36**
Michel-Lévy chart, *32*
Micrite, *154,* **471**
Microaplite, **354**
Microbreccia, *150, 320, 349, 354*
Microcataclasite, *321, 323*
Microcline, *44, 53, 59,* **68,** *126, 231*
Microcline cryptoperthite, *53*
Microcline perthite, *393*
Microconcretions, 438
Microcracking, **333**
Microfault, 115
Microfaulting, 334
Microfracturing, 334
  transgranular, 334
Microfractures, 11
Microphenocryst, 191, *222,* 225
Microvein, *119, 285, 287, 353, 354, 391*
  indigenous, **390**
Migma, 361
Migmatite, 149
  agmatitic, **347,** *369*
  anatectic, *357, 364*
  arteritic, 368
  dynamic, *363*
  injection, 203, **368,** *368*
  lit-par-lit injection, *369*
Mililitolite, *147*
Mineral phases, 175–176, 180, 184
Mineralogical maturity, **433**
Mineralogical norm, 143
Minerals
  anisotropic, **12**
  birefringence of, **26, 31,** *33, 36*
  gangue, 397
  isotropic, **12**
  natural color of, **18,** *36*
  opaque, **11**
  rock-forming, 6, 16, *18, 33, 43*
  staining of, 63, 96
Mineralogical mode, **95**
Minette, *146*
Minimum melt, 186, 359
Minnesotaite, *437,* 438
Mississagi Quartzite, *327*
Mississippi Valley type ore deposit, 419
Mode. *See* Mineralogical mode
Moenkopi Formation, *439, 448, 452*
Monazite, *53,* **74,** *145*
Monomineralic rock, 240
Monticellite, *53*
Montmorillonite, *53,* 73, *402, 405*
Monzodiorite, *145*
Monzogabbro, *145*
Monzonite, *145*
Moraine, 445
Morrison Formation, *448*
Mosaic breccia, *209, 349,* 350
Mowrey Shale, *471*
Mud rocks, classification of, *152–153*
Mudcrack, **438,** *438*
Mugearite, *146*
Muscovite, *45, 53,* **68,** *133, 139, 139*
Mylonite, *149,* **150,** *214,* 302
Myrmekite, **119,** *127,* 130, **136–138,** *226, 227, 228, 355, 390, 392*

NaCl-H$_2$O system, **166**
Nannoplankton, 462, 480
Native copper, *519*
Natrolite, *54,* 74
Natural levee, *448*
Nautiloid, *466,* 467
Navajo Sandstone, *517*
Neoblasts, **217,** *218*
Neosome, **361**
Neotocite, **520**
Nepheline, *44, 54,* **68,** *116,* 189, 228
Nepheline syenite, *228, 231*
  weathering of, 506, *513*
Net-veined complex, *243, 243*
Newtonian fluids, 205
Nodule, **511,** *511–512,* 514
Nomarski interference contrast microscopy, **96**
Non-Newtonian fluids, 205
Nontronite, *54*
Nördlinger Ries Crater, 329
Nordmarkite, *145*
Norite, *147*
Normative classification. *See* Rock classification
North Horn Formation, *447*
Northern Snake Range décollement, *300, 302, 335, 340*
Nosean, *54, 66, 116*
Nucleation, 168, 170, 176
  free energy of activation for, **190**
  heterogenous, **190**
  homogenous, **190**
Nucleation density, **191**
Nucleation energy, 170
Nucleation rate, 168, 184, 188, 191–192

Nucleation theory, **190**
Nucleus
　critical size of, 192, *192*
　stability of, *192*

Oblique illumination, 9, **10**, *10*
Obsidian, **221**, **222**, *222*
Obstacle scour, *438*, **441**
Obtuse bisectrix, **40**
Oceanite, 146
Ocellar hornblende-mantled quartz, 246, *252*, **253**
Ogallala Conglomerate, *516*
Oikocryst (also Oikograin), *124*, 184, **199**, 226, 227, 241, *241*
Oligoclase, 44, 54, **68**
Olistostrome, **335**
Olivine, 45, 54, **68**, *68*
Omphacite, 45, 54, **68**
Onaping Formation, *330*
Oncoid, **459**, *459*
Ooid, *154*, 474
Ooidal ironstone, 438
Oolite, 153, *474*
Oomicrite, *154*
Oosparite, *154*
Ooze, 455, 460–462, 471, 473, 477
Opal, 54, 74, **74**, *167*, 168, *215*
Opalite, 397
Opdalite, *154*
Open marine environment, **451**
Open space precipitation, 403
Ophiolite, 237, *237*
Ophiomorpha, 465, *465*
Ophitic texture, 184, 227
Optic angle, 38, *39*, **40**, *40*
Optic axis, **14**, *23*
Optic sign, **16**, 38, **41**, *41*, 42
Optic train, *22*
Optical mineralogy
　conoscopic, 37
　orthoscopic, 6
Oracle Granite, *323*
Orbicular layering, 239
Orbicular structure, *129*, 239
Orcas Chert, *480*
Ore deposits, 396, 409
Ore minerals, 397, 401, 405, 408–409, 419
Orientation analysis, 97
Orogenesis, 303–304
Orthite, *54*
Orthoamphibole, 54
Orthoclase, 44, 54, 59, **69**, *69*, 188–189
Orthocumulate, **239**
Orthopyroxene, 54
Orthopyroxenite, *147*
Orthoquartzite, **6**
Oscillatory crystallization, 239, *239*
Ostia, **462**, *463*
Ostracod, **465**, *465*
Osumi granodiorite, 228, *231*
Outwash deposits, 445
Overprinting, **407–408**

Oxygen isotopic ratios, 396–397
Oxyhornblende, 45, 54, **69**

Packstone, *153*, 479
Pahoehoe, 222, *222*
Palagonite, 54, **74**, **506**
Paleosol, 509
Paleosome, **361**
Palynomorph, **456**
Panidiomorphic granular, **113**
Pantellerite, 146
Paragenetic sequence, **405**
Paragonite, 55
Pargasite, 55
Passive continental margin, 451
Patchy extinction, *214*
Patchy zoning, *89*, **115**, *117*
Path of changing melt composition, 176
Path of crystallization, **176**
Path difference, **27**, *32*
Path of melting, **176**
Peat, **445**
Pebble, **441**
Pedalfer, **509**
Pedocal, **509**
Pedogenic, **509**, 517
Pegmatite, *357*, 389
　magmagenic, **389**
Pegmatite zoning, 393, *394*
Pelagic biota, 454
Pelagic clay, 473
Pelagic organism, 460
Pelagic sediment, 477, 480
Pele's hair, 375
Pele's tears, 375
Pelecypod, **467**, *467*
Pelloid, 153, *154*
Pencil gneiss, *303*
Penninite, 55, *62*
Periclase, 55
Peridotite, *147*
Peritectic, **177**, *177*
Perlite, *383*
Permeability, 203
Perovskite, 55, **75**
Perthite, 121, *128*, 180
　crypto, 185
Petersburg meteorite, 329, *331*
Petrofabric analysis, **98**
Petrographic microscope, *2*, 6
Petromict conglomerate, **444**
pH *172*
Phacolith, **370**
Phasal (phase) difference, **28**, 31
Phase analysis, **95–96**
Phase diagram
　binary, *175*
　ternary, *184*
Phase equilibria, **175**
Phase state, **175**
Phengite, 55, **69**
Phenocryst, **191**, 221
Phlogopite, 45, 55, **69**
Phonolite, *146*
Phoscrete, 472, **518**

Phosphatic nodule, 472
Phosphatic ooids, 473
Phosphatic ooze, **471**
Phosphatization, **477**
Phosphopeloid, 472
Phosphorite, **73**, 155, *472*, 500
Photosynthesis, 456–457, 462
Phreatic zone, **503**
Phreatomagmatic, **377**
Phreatoplinian, **378**
Phyllite, 149, 150, *152*, 292
Phyllonite, *150*
Phylum Annelida, 462
Phylum Arthropoda, 463, 465
Phylum Brachiopoda, 467
Phylum Bryozoa, 469
Phylum Cnidaria, 462
Phylum Conodonta, 470
Phylum Echinodermata, 469
Phylum Hemichordata, 470
Phylum Mollusca, 465–466
Phylum Porifera, 462
Phyteral, **455**
Phytoclast, **456**
Phytoplankton, 477
Picrite, *146*
Piedmontite, 55, **74**
Pigeonite, 45, 55, **69**, *69*
Pillow structure, 243
Pimelite, **512**
Pine Hollow Formation, *435–436*
Pisoid, **459**, *459*
Pisolite, *512–514*
Plagioclase, 55, *126*
　birefringence of, *86*
　composition of, 76, *86*, *91*
　optical orientation of, *85*
Plagioclase-clinopyroxene-olivine system, **184**
Plagioclase-clinopyroxene system, 184, *185*
Plagioclase series, 76, **177**
Plagioclase system, *178*, *181*
Plagioclase zoning, *178*, 248, *256*
Plagioclase zoning curve, **94**
Plagiogranite, 145
Planar deformation features, 328
Planar fabric, 128
Plankton, 455, 460, 462, 471, 480
Playa, 485
Pleochroism, **18**
　formula of, *21*
Plinian air fall, *374*
Pluton, 203, 208, 221, 227
Plutonic igneous rocks, 221
　classification of, *145*, 147
Pocket zone, **395**, 410
Poikigrain, **119**, *124*
Poikilitic, **119**, *224*, *225*, *255*
Point bar, **440**
Point counting, **95**
Point defects, 209
Polarized light, **12**, 13, 16, 18–19, 21, 29, 37, 96
Polarizer
　analyzing, *2*, **21**, *22*, 29
　substage, *21–22*

Polarizing microscope, *2*
 optic train of, *22*
Polygonal mosaic, *195*, **217**, *218, 240, 315–316*
Polygonalization, *214*
Polyhalite, *55*, *494*
Polymerization. *See* Silicate polymerization
Porcellanite, *155*
Pore water pressure, 334
Porkchop Geyser, *384*
Porosity, 207, 433, 435
Porphyrograin, **109**, **111**, 257, *257*
Porphyry, 224, *224, 226*
Porphyry copper deposit, 519–520
Porphyry copper hydrothermal system, *409*
Porphyry hydrothermal systems, **415**
Porphyry molybdenum deposit, 416, *417*
Porphyry-pluton system, *410*, 415
Potential space, 403
Powellite, 55
Preble Formation, *336*
Preferred dimensional orientation, **97**, 290, *291–292*
Preferred grain location, *292*
Preferred lattice orientation, 97, **99**, *100–102*, 129
Preferred location, **127**, *129*, 290
Prehnite, *45*, 55, **69**
"Present is the key to the past", 131
Pressure quench, **184**
Pressure shadow, 335, 355, *358*
Pressure solution, **211**, *214, 291*, 335, *336*
Prochlorite, 55, *62*
Progressive extinction, **214**, *216*
Protodolomite, **476**
Protolith, *276*, 278
Protore, *519*
Provenance, **431**, *432*
Pseudoleucite, 55, **67**, **74**
Pseudomorphic replacement, 285, *402*
Pseudonodule, **438**
Pseudosubgrains, *214*
Pseudotachylite, **73**, *150*, 323, 327, *327*
Psilomelane, 74
Pumice, **221**, *222*, 223, 410
 fragments, 374, *376*
Pumpellyite, *44*, 55, **69**
Pyrite, 55, **73**
Pyroclastic eruption, 204, *374*
Pyroclastic fall
 Hawaiian, 377
 Plinian, 377
 Strombolian, 377
 Surtseyan, 378
Pyroclastic flow, **378–379**
Pyroclastic rocks, **373**
 classification of, *148*
Pyroclastic surge, **378**
Pyroclasts, 374
Pyrolusite, 55, 74
Pyrope, *56*, **65**
Pyrophyllite, *56*, **74**, *405*

Pyroxene, *56*
Pyroxenite, *147*

Quartz, *15, 24, 44, 56*, **69**, *70*
 authigenic, 473, 475, *475*
 high-T (beta), *178*
 low-T (alpha), *178*
 micropoikilitic, 234
 polymorphic forms of, *178*
 resorption of, *200*
 solubility of, **167**, *168*
Quartz-alkali feldspar-plagioclase system, **187**, *188*
Quartz-alkali feldspar system, **185**
Quartz arenite, *151*
Quartz overgrowth, **433**, 434–435, *435–436*
Quartz ribbons, 214, *216*
Quartz subgrains, *216*
Quartzine, **517**
Quartzite, *149, 215, 218*
Quaternary cotectic line, **187**
Quaternary two-feldspar "granite" system, **187–188**, *189*

Radiolaria, **460**, *460*
Rainpit, **438**
Ralstonite, 427
Rapakivi mantling, **244**, *125, 245, 249*
Rapakivi-antirapakivi double mantle, *125, 245*, **250**, *251*
Rational thought, **131**
Reaction
 dehydration, **281**
 dissociation, 281
 hydration, **281**
 sense of, 285, *287–288*
Reactive replacement, 161, 401–403, *403, 492*
Recovery, **209**, *214*, **217**
Recrystallization, 161, **209**, *214, 218*, 289
 annealing, **209**, 289
 dynamic, **209**
 dynamothermal, **209**, 290
 mimetic, *291*
 by nucleation, **217**
 by subgrain rotation, **217**
 syndeformational, **290**, **295**
Red algae, **456**
Red-1 retardation plate, 33, *34*
Redox potential, **171**
Reflectance, **477**
Reflectance fabrics, **332**
Refractive index, **7**, *18*
 relative, *9–10*
Regolith, **509**, 517
Relief, **11**, *11*
Relief scan, **11**
Reniform structure, 399
Replacement deposit, 418
Replacive growth, **489**, *490*
Resistor minerals, 366

Restite, 361
Restitite, **361**, **365–366**, *365*
 xenocrystic, 366
Retardation, **27**, *31, 34*
Rheomorphism, **379**
 of tuffs, 376
Rhizoconcretion, 509
Rhizolith, **504**, 509
Rhyodacite, *146*, 187
Rhyolite, *146*, 185, 223
Rib Hill Sandstone, *383*
Ribbon chert, 337
Ribbon structure, *216*, **241**
Richterite, 56
Rico Formation, *432*
Riebeckite, *44*, 56, **70**, 223
Ries Crater, 329
Rill, **441**
Ringing Rocks Magmatic Complex, *196*
Ripple cross-lamination, *433*, **439**
Ripple marks, 206, **439**
River, 446
Rock
 classifiable attributes, **143–144**
 tensile strength of, 202
Rock classification
 biolith, *155*
 biopelite, *155*
 clastic sedimentary, *151–153*
 igneous, 144, *145–147*
 ironstone, *154*
 limestone, *154*
 lunar igneous, 149
 mafic igneous, *147*
 metamorphic, *149*
 phosphorite, *155*
 plutonic, *145*
 pyroclastic, *148*
 traditional, 141, *143–144*
 ultramafic, *147*
 volcanic, *146*
Rock salt, 337
Rock slide, **206**
Rock-forming processes, 1, 161, *162*, 163
Rock-forming systems, 1
Roll front uranium deposit, **519**
Rutile, *56*, **74**, **75**

s plane, *300*
Sabkha, 451, 485, *486*
St. Peter Sandstone, *435*
Salado evaporite, *490*
Salina, 485, *486*
Saline lake, 485
Salite, *56*, *63*
Salt Wash Member Morrison Formation, *448*
Salt diapir, 316–318
Salt dome, 317
Salt glacier, *317*
Salt lake, 485
Saltation, **206**
Sand bar, 433, **440**, 446, 450

Sand dune, 206
Sand flow, 206
Sand volcano, **438**
Sand wave, **439**
Sanding, **401**
Sandstone, classification of, *151–152*
Sanidine, *44, 56,* **70**, *70, 187,* 224
Sanidine-albite system, **179–180**, *179*
Santiaguito Dome, *424*
Saponite, *56*
Sapphirine, *56*
Saprolite, **512**
Saprolith, *155*
Scalarituba, *277*
Scaly fabric, 335
Scanning electron microscope, **96**
Scapolite, *45, 56,* **70**
Scheelite, *56*
Schist, *149*
  hornfelsic, *307, 350*
Schistosity, 298, *301*
Schlieren, 236, *237,* 257
Schorl, *56,* 72
Scopulite, *190*
Scoria, 223
Scour-and-fill, in magma, 237
Scour marks, **441**
Secondary election images, 97
Secondary K-feldspar, *417*
Secondary twinning, 214, *216*
Sediment. *See* Grain
Sedimentary environment. *See* Depositional environment
Sedimentary facies, 433, 445
Sedimentary rocks, classification of, *151–155*
Scolithos, 462, *464*
Sedimentary regime, 433
Sedimentary structures, 432, 438–439
Sedimentary volcano, 335
Sedimentation, 431
Sedimentology, 431
Selenite, *492*
Semi-automatic point counter, **95**
Semigraphic, *127*
Semipolygonal arcs, **296**, *297*
Sepiolite, 505
Septa, **511**, *514*
Sequential crystallization, 139
Seriate porphyrograined, **111**
Sericite, *56,* **68**, **74**
Serpentine, *56,* 61
Serpentinite, *149*
Shale, classification of, *155*
Shape, of grains. (*See* Grain morphology)
Shape analysis, **97**
Shatter cone, 327, *327*
Shear, sense of, **296**, *299–300*
Shear band foliation, **296**, *300*
Shear stress, **204**
Shinarump Conglomerate, *447–448*
Shrimp burrows, 134–135, **465**, *465*
Shock metamorphism, **324**, *325,* 326
Shock waves, **324**, 326
Shonkinite, *145*
Shoshonite, *146*

Siderite, *45, 57,* **71**
Sideromelane, **73**
Silcrete, 504, **517**
Silica
  amorphous, 167–168
  biogenic, 167
  colloidal, 168
  solubility of, *168*
Silica-aqueous phase system, **167**
Silica colloid, 168
Silica polymorphs, 167, *178*
Silicate depolymerization, **180**, 182, 186
Silicate melt, 161, **174**, 180, 199
Silicate polymerization, 168, 199
Siliceous ooze, 460, 462, 471, 473, 477
Siliceous sinter, *384,* 396, **420**, *422*
Silicic acid, 167
Siliciclastic depositional environment, **444**
Siliciclastic sedimentary rocks, 438, 443
Siliciclastic sediments, 431, 434, 439
Siliciclastic texture, **441**
Silicification, *474,* **475**
Sillimanite, *45, 57,* **71**, *71,* 139, *139,* 210, 278–279, *280*
Siltstone, *152*
Silver Creek Granite, *354*
Simultaneous crystallization, 139
Sinkhole, **516**
Sintering, 313
SiO$_2$-H$_2$O System, 167, *168*
Siphonodella sandbergi Zone, *470*
Siphonodella sp., *470*
Size analysis, 97
Skaergaard magmatic body, 239
Skarn, 414, *415–416*
Slate, *149,* **150**, *152, 292*
Slickenite, **320**
Slickensides, 319, *319–321*
Slip plane, 214
Slump scar, **441**
Smackover Formation, *495*
Small-circle girdles, 316
Smartville Ophiolite, *237*
Smectite, 426, 503, 505–506, 512
Smithsonite, *57,* 519
Snow crystals, 312
Sodalite, *57,* 66
Soda-niter, *57*
Sodium light, 27, 29, *29,* 31
Soft sediment deformation, 207, 335, **438**
Soil, **509**
Soil horizons, **509**, *509*
Soilstone, **504**, **509**
Sole marks, **441**
Solfatara, **425**
Solid solution, 176
Solidus curve, *175*
Solubility
  of anhydrite in water, *170*
  of calcite in H$_2$O, *166, 169*
  of calcite in NaCl solutions, *166*
  of CO$_2$ in pure H$_2$O, *168*
  of gypsum in water, *170*

of H$_2$O in magma, *182*
of silica polymorphs, *168*
Solubility product, 165
Solutions
  aqueous, **165**
  congruent, **165**
  incongruent, **165**
Solvus, **180**, 182, 186
Sorting, 206
Sparite, 454, *463, 468,* 469, 473, 476
Specular hematite, **73**
Speleothem, 172, **516**
Spessartite, *57,* **65**
Sphalerite, *57*
Sphene, *45, 57,* **71**, **75**, *325*
Sphene-centered ocellar texture, **253**, *246, 253*
Sphericity, 435
Spherulite, **232**, *233*
  lithophysal, *233,* 234
Spherulitic crystal growth, *196*
Spiculite, *463*
Spinel, *44, 57,* **71**, *186*
Spinifex texture, **241**
Spodumene, *57*
Sponge, **462**, *463*
Spongin, **462**
Spongy cellular morphology, 199–200, *201,* 222, 256, 257, 313, *313*
Spongy cellular plagioclase, **248**, **257**
Springer stock, *413*
Spurrite, 57
Stalactite, 516
Stalagmite, 516
Staurolite, *44, 57,* **71**, *71,* 283, *301,* 308, *310*
Steamboat Springs, *422*
Stilbite, *57,* 74
Stillwater Complex, *241*
Stilpnomelane, *45, 57,* **71**, 438
Stishovite, *57,* 325, 328–329
Stockwork, 401, 416
Stoke's Law, **205**
Stone Mountain Granite, 228
Stoss side of dune, **206**
Strain, of rocks, 212, **213**
Strain energy, 210
Strain rate, 211, 213, 215
Strain shadow, 335, 338
Stratabound deposit, 419
Stress, of rocks, 209, 210, **210**
Stress corrosion cracking, **202**
Stromatolite, **459**
Strombolian eruption, **377**
Stylolite, 338, *339*
Subarkose, *151–152,* **444**
Subduction complex, 453
Subgrain, 214, *216,* 218
Subgrain rotation, 217
Subgrain rotation recrystallization, 217
Sublimation, 161, 163, 424, **425**
Sublitharente, *151–152*
Submarine trench, 451, 453
Submicroscopic material, **107**
Substage condenser, 2
Sudbury, Ontario *227*
Suevite, **329**

Sulfate reduction, 482
Sulfur, 57, **74**
Summerville Formation, *452*
Supercritical aqueous phase, **164–165**, *165*
Supergene alteration, **508**, *519*, 520, *521*
Supergene enrichment, **519**
Superheating of magma, 243, 250, 255
Supraglacial deposits, **449**
Surface energy, 210
Surmicaceous enclaves, 366, *367*
Surtseyan eruption, **378**
Sutro Tuff, *380*, *440*
Sutured contacts, 335
Snowball texture, 295, *295*
Syenite, *145*
Sylvite, *57*
Symplectic, **121**
Syncrystalline microboudinage, **295**, *295–296*
Synneusis, 225

Tachylite, 222, *222*
Taconite, *437*
Talc, *45*, *57*, **71**, 282
Talus, 445
Tapeats Sandstone, *451*
Tectonic fragmentation, **203**
Tectonic overpressure, 211
Tectonics, *214*, 298
  transpressional, 272
Tectonite, 298
TEM bright field micrograph, *215*
Tenoumer, Mauritania crater, *330*
Tepee structure, *481*, **489**
Tephra, **374**
Tephra chronology, 382
Tephrite, *146*
Ternary cotectic line, **184**, 187
Ternary eutectic, **186**, 188
Ternary "granite" system, 186, *187*, 189
Ternary phase diagram, **184**
Terra Rosa ground, **508**
Texture
  antirapakivi, **250**, *245*
  aphyric, 223
  axiolitic, *233*
  cataclastic, 320
  cellular, 121, *248*, *250*, 256, *256*, *257*
  dynamomagmatic, **353**
  eutaxitic, **379**
  granophyric, 224, *229*, 234
  granular, **108–109**, *110*, **118–119**, *123*
  graphic, 224, *224*
  including, **119**, *124*
  intergranular, 119, *123*
  lamellar, 121, *125*
  lensoid, 121, *125*
  mantle, 121, *125*, *229*, 287, *288*
  micrographic, 188
  myrmekitic, 119, *124*, *127*
  ocellar, *246*, *252*, **253**
  oikocryst, 184
  ophitic, **119**, 184, *331*
  phyric, 223
  poikilitic, 119, 234, 255
  poikiloblastic, **288**, *288*
  polygonal mosaic, **289**
  porphyritic, 222, *222*
  porphyroblastic, **288**, *294*, 297–298, *303*
  rapakivi, 244, *245*, **244–249**, *250*
  rod, 121, *125*
  semigraphic, *127*
  side-by-side granular, 224, **289**
  siliciclastic, 441
  snowball, **295**, *295*
  superposed, **121**
  sutured, 335
  symplectic, *121*
  transecting, **119**, *123*
  vermicular, 119, *127*, 224
  vermicular-graphic, *224*, 226, *229*
Textural interpretation, 130
Textural maturity, **433**
Thenardite, *57*
Theralite, *145*
Thermal inertia, 203
Thermal undercooling, **192**
Thermodynamic calculation, 278–279, 282–283
Thin section, 7
Thinardite, 427
Thinolite tufa, **498**, *499*
Tholeiite, *146*
Thompsonite, *57*, **74**
Three-stage crystallization, **191**, *224*
Tidal flat, **451**
Tidal flow, 207
Tillite, **445**
Titanite, *58*, 61
Titus Canyon Formation, *321*
Tonalite, *145*
Tool mark, **441**
Topaz, *58*
Tourmaline, *20*, *45*, *58*, **71**, *403*
Trace fossils, 462, *464–465*
Trachyandesite, *146*
Trachybasalt, *146*
Trachyte, *146*
Traction carpet, 206
Transecting texture, **119**, *123*
Travertine, 396, **422**, *423*, 516
Tremolite, *45*, *58*, 283
Trichites, **222**, *222*
Tridymite, *44*, *58*, **72**, *177–178*, *187*, *215*, *380*
Trilobite, **463**, *464*
Troctolite, *147*
Trona, *58*
Trondhjemite, *145*
Tufa, 423, *498–500*
Tuff, *148*
  bentonitic, **506**
Tuff cones, 378
Tuff ring, 378
Turbidite, **207**, 453, 480
Turbidity flow, **207**

Twin disappearance zone, *92*, 94
Twin gliding, 214
Twin plane, 21, 25
Twins, 77, 78, *78–84*, 84, *87–90*, *93*, 115, 117, *120*, *122*, 216

Udden-Wentworth grain size scale, **441**, *442*
Ulexite, *58*
Ultrabasic igneous rocks, classification of, *147*
Ultracataclasite, *150*
Ultrametamorphism, **357**
Ultramylonite, *150*
Undercooling of melt, 191
Uniformitarianism, principle of, **131**
Units (as rock-forming materials), **107–108**
Universal stage, 6, 77, **98–99**
Upwelling, 500
Utricles, **457**, *458*, 459
Uvarovite, *58*, **65**

Vadose zone, **503**
Valmey Formation, *339*
Vapor phase cavities, **410**
Vapor phase crystallization, **379**, 410
Vein deposit, 418
Velay Granite, *367*
Vermicular relation, *58*, **65**
Vermiculargrain, **119**, *124*
Vermiculite, *58*, **73**, *506*
Vertebrate endoskeleton, 455
Vesicle, *222–223*, **222–223**, *233*, **410**
Vesiculation, 233, **373**
Vesuvianite, **66**
Vibration directions, *15*, **16**, *17*, *34–35*
Vigeland sculptures, *228*, *231*
Viscosity of fluids, 205–206, **205–206**
Viscous drag, 205, 208
Vitric fragments, 374
Vitrinite, **455**, *457*, **477**
Vitrophyre, **191**, *379*
Viviparus, *466*
Volatilization, **198**
Volcanic ash, 203, *376*, *381*
Volcanic breccia, 203
Volcanic glass, 73
Volcanic igneous rocks, 221
Volcanic intramagmatic, 409, **410**
Volume free energy, **190**
Vortices, 204
Vorticifex Carinifex newberryi, *466*
Vredefort Structure, *327*
Vug, *399*, *399–401*
Vulcanian eruption, **377**

Wacke, *152*, **444**
Wackestone, *153*, 479
Wairakite, *58*
Wake, 204

Water
  boiling point of, **164**
  critical point of, 164, *165*
  in the geologic environment, 171, *171*, 173
  phase relations of, *165*
Water table, 503, 510, 515–516, 520
Wave (optical), **26–27**
Wave action, 207
Wave interference, 29, *30*
Wavebase, 204
Weathered rocks, **503**
Weathering, chemical, **503**
Websterite, *147*

Wehrlite, *147*
Welded tuff, *376*, **379**
Westerly Granite, 226, *227*
Wheeler Shale, *464*
White light, **30**, 36, 40
White mica, *58*, 73
Wollastonite, *45*, *58*, 72, *72*, *120*, 281–282
Work hardening, **217**

X-ray diffraction phase analysis, 96
Xenocryst, **261**

Xenolith, **261**, **347**, **350**, *350*, 367
Xenotime, *58*, **75**

Zeolite, 73, **74**
Zircon, *58*, **75**, *196*
Zoisite, *44*, *58*, **72**, *72*
Zoning
  crystal. *See* Crystal zoning
  mineral, **405**
  spatial, **405**
Zooecia, *468*, **469**